Origin and Evolution of Precambrian High-Grade Gneiss Terranes, with Special Emphasis on the Limpopo Complex of Southern Africa

edited by

Dirk D. van Reenen
Department of Geology
University of Johannesburg
Auckland Park
South Africa

Jan D. Kramers
Department of Geology
University of Johannesburg
and School of Geosciences
University of the Witwatersrand, Johannesburg
South Africa

Stephen McCourt
School of Geological Sciences
University of KwaZulu Natal
Durban
South Africa

Leonid L. Perchuk
Department of Petrology
Moscow State University
and the Institute of Experimental Mineralogy
Russian Academy of Sciences
Chernogolovka
Moscow District
Russia

THE
GEOLOGICAL
SOCIETY
OF AMERICA®

Memoir 207

3300 Penrose Place, P.O. Box 9140 ▪ Boulder, Colorado 80301-9140, USA

2011

Copyright © 2011, The Geological Society of America (GSA), Inc. All rights reserved. GSA grants permission to individual scientists to make unlimited photocopies of one or more items from this volume for noncommercial purposes advancing science or education, including classroom use. For permission to make photocopies of any item in this volume for other noncommercial, nonprofit purposes, contact The Geological Society of America. Written permission is required from GSA for all other forms of capture or reproduction of any item in the volume including, but not limited to, all types of electronic or digital scanning or other digital or manual transformation of articles or any portion thereof, such as abstracts, into computer-readable and/or transmittable form for personal or corporate use, either noncommercial or commercial, for-profit or otherwise. Send permission requests to GSA Copyright Permissions, 3300 Penrose Place, P.O. Box 9140, Boulder, Colorado 80301-9140, USA. GSA provides this and other forums for the presentation of diverse opinions and positions by scientists worldwide, regardless of their race, citizenship, gender, religion, or political viewpoint. Opinions presented in this publication do not reflect official positions of the Society.

Copyright is not claimed on any material prepared wholly by government employees within the scope of their employment.

Published by The Geological Society of America, Inc.
3300 Penrose Place, P.O. Box 9140, Boulder, Colorado 80301-9140, USA
www.geosociety.org

Printed in U.S.A.

GSA Books Science Editors: Marion E. Bickford and Donald I. Siegel

Library of Congress Cataloging-in-Publication Data

Origin and evolution of Precambrian high-grade gneiss terranes, with special emphasis on the Limpopo complex of Southern Africa / edited by Dirk D. van Reenen ... [et al.].
    p. cm. — (Memoir ; 207)
  Includes bibliographical references.
  ISBN 978-0-8137-1207-9 (cloth)
  1. Gneiss—South Africa—Limpopo Region. 2. Gneiss—Limpopo Belt (South Africa and Zimbabwe) 3. Geology, Stratigraphic—Precambrian. 4. Geology—South Africa—Limpopo Region. 5. Geology—Limpopo Belt (South Africa and Zimbabwe) I. Van Reenen, D. D.

QE475.G55O75 2011
552′.4—dc22

2010046615

Cover: Outcrop of Sand River Gneiss demonstrating the principle of linking distinct fabric-forming and magmatic events with distinct segments of a multistage pressure-temperature (P-T) path in the polymetamorphic Central Zone, Limpopo Complex, South Africa (courtesy Jan Kramers).

Contents

Dedication to Leonid L. Perchuk v
    D.D. van Reenen, J.D. Kramers, and S. McCourt

Introduction vii
    D.D. van Reenen, J.D. Kramers, and S. McCourt

  1. Microstructures of melt-bearing regional metamorphic rocks 1
      R.H. Vernon

  2. Petrological and experimental application of REE- and actinide-bearing accessory minerals to the study of Precambrian high-grade gneiss terranes 13
      D. Harlov

  3. Fluids in granulites 25
      J.L.R. Touret and J.-M. Huizenga

  4. Fluid-absent melting versus CO_2 streaming during the formation of pelitic granulites: A review of insights from the cordierite fluid monitor 39
      M.J. Rigby and G.T.R. Droop

  5. Local mineral equilibria and P-T paths: Fundamental principles and applications to high-grade metamorphic terranes 61
      L.L. Perchuk

  6. The geochronology of the Limpopo Complex: A controversy solved 85
      J.D. Kramers and H. Mouri

  7. High-pressure and ultrahigh-temperature metamorphism of Precambrian high-grade terranes: Case study of the Limpopo Complex 107
      T. Tsunogae and D.D. van Reenen

  8. Granite emplacement and the retrograde P-T-fluid evolution of Neoarchean granulites from the Central Zone of the Limpopo Complex 125
      J.M. Huizenga, L.L. Perchuk, D.D. van Reenen, Y. Flattery, D.A. Varlamov, C.A. Smit, and T.V. Gerya

  9. Intracrustal radioactivity as an important heat source for Neoarchean metamorphism in the Central Zone of the Limpopo Complex 143
      M.A.G. Andreoli, G. Brandl, H. Coetzee, J.D. Kramers, and H. Mouri

  10. A review of Sm-Nd and Lu-Hf isotope studies in the Limpopo Complex and adjoining cratonic areas, and their bearing on models of crustal evolution and tectonism 163
      J.D. Kramers and A. Zeh

11. *Thrust exhumation of the Neoarchean ultrahigh-temperature Southern Marginal Zone,
    Limpopo Complex: Convergence of decompression-cooling paths in the hanging wall and
    prograde P-T paths in the footwall*...189
    D.D. van Reenen, C.A. Smit, L.L. Perchuk, C. Roering, and R. Boshoff

12. *Neoarchean to Paleoproterozoic evolution of the polymetamorphic Central Zone of
    the Limpopo Complex* 213
    C.A. Smit, D.D. van Reenen, C. Roering, R. Boshoff, and L.L. Perchuk

13. *Archean magmatic granulites, diapirism, and Proterozoic reworking in
    the Northern Marginal Zone of the Limpopo Belt* 245
    T.G. Blenkinsop

14. *Heterogeneous strain and polymetamorphism in high-grade terranes: Insight into
    crustal processes from the Athabasca Granulite Terrane, western Canada, and
    the Limpopo Complex, southern Africa* 269
    K.H. Mahan, C.A. Smit, M.L. Williams, G. Dumond, and D.D. van Reenen

15. *Formation and evolution of Precambrian granulite terranes: A gravitational
    redistribution model* 289
    L.L. Perchuk and T.V. Gerya

16. *Tectonic models proposed for the Limpopo Complex: Mutual compatibilities
    and constraints* 311
    J.D. Kramers, S. McCourt, C. Roering, C.A. Smit, and D.D. van Reenen

Dedication to Leonid L. Perchuk

Dirk D. van Reenen
Department of Geology, University of Johannesburg, Auckland Park, South Africa

Jan D. Kramers
*Department of Geology, University of Johannesburg, and
School of Geosciences, University of the Witwatersrand, South Africa*

Stephen McCourt
School of Geological Sciences, University of KwaZulu Natal, Durban, South Africa

This Geological Society of America (GSA) Memoir, *Origin and Evolution of Precambrian High-Grade Gneiss Terranes, with Special Emphasis on the Limpopo Complex of Southern Africa*, is dedicated to the memory of Leonid L. Perchuk, who was born 20 November 1933 in Odessa, Ukraine, and who passed away 19 June 2009, in Johannesburg, South Africa. Leonid obtained his master's degree in 1956 from Odessa State University, Ukraine, his Ph.D. in 1963 from the USSR Academy of Sciences (supervisor Prof. D.S. Korzhinskiy), and the doctor of science degree in 1968 from the USSR Academy of Sciences (now Russian Academy of Sciences) with the topic of "Equilibria of rock-forming minerals." His career as an active scientist spanned 45 yr, and until his untimely death as the result of a heart attack at the age of 75, he still had the dual appointment as Head of Petrology at Moscow State University and Head Scientist at the Institute of Experimental Mineralogy, Russian Academy of Sciences, Chernogolovka, Moscow District.

Within geology, the subdiscipline of granulite facies metamorphism and the evolution of Precambrian granulite terranes is well established owing to scientific contributions of many outstanding scientists in diverse fields such as experimental petrology, mineralogy, geochemistry, geochronology, rock deformation, and deep geophysics. Leonid was one of these outstanding scientists, and his major contributions to three diverse branches of modern geoscience—petrology, geochemistry, and geodynamics—were made possible by his great scientific curiosity and the fact that he dedicated his entire life to understanding the geodynamic evolution of the Earth's crust and mantle. He created the first semiempirical system of internally consistent mineralogical thermometers and barometers, and formulated the main principle of redistribution of components between coexisting minerals with temperature and pressure (Perchuk, 1969, 1971, 1991). This formed the basis for the derivation of the first P-T paths calculated for eclogites more than 30 yr ago (Perchuk, 1976, 1977, 1986, 1989). P-T paths are now essential tools for understanding the complex geodynamic processes that create high-grade gneiss terranes–orogenic belts (Perchuk, 2005; Perchuk et al., 2008; Perchuk and van Reenen, 2008). Leonid also made major contributions to the formulation of the concept of gravitational redistribution of material within the Earth's continental crust as a viable mechanism for the origin and evolution of Precambrian high-grade gneiss terranes located between granite-greenstone cratons (Perchuk, 1989, 1991; Perchuk et al., 2001; Gerya et al., 2000, 2004).

van Reenen, D.D., Kramers, J.D., and McCourt, S., 2011, Dedication to Leonid L. Perchuk, *in* van Reenen, D.D., Kramers, J.D., McCourt, S., and Perchuk, L.L., eds., Origin and Evolution of Precambrian High-Grade Gneiss Terranes, with Special Emphasis on the Limpopo Complex of Southern Africa: Geological Society of America Memoir 207, p. v–vi, doi: 10.1130/2010.1207(000). For permission to copy, contact editing@geosociety.org. © 2011 The Geological Society of America. All rights reserved.

Leonid Perchuk visited the Department of Geology, Rand Afrikaans University (now the University of Johannesburg) annually for 14 yr since 1995 to collaborate on the Limpopo project. He was still working on this GSA Memoir until a few hours before his heart attack in the early hours of 13 June 2009. At the time of his death Leonid had published 18 books and more than 400 papers in international journals. He was a regular invited speaker at international conferences and continued to coconvene special sessions at such conferences. Leonid received many awards during his long scientific career, including the following:

- Elected to the International Higher Education Academy of Sciences (1993);
- Member of the Russian Academy of Natural Sciences (1995) and Soros Special Professor (1995);
- Gold Medal and Diploma as "The Most Recognized Scientist of Russia" from the President of the Russian Federation (1999);
- The Lomonosov Prize from Moscow State University (2000);
- The Korzhinskiy Prize from the Russian Academy of Sciences (2001);
- Diploma and Medal from the Russian Ministry of Natural Resources (2008);
- Special Professor in the Department of Geology, University of Johannesburg, South Africa.

Last, but not least, Leonid was an outstanding teacher and was always surrounded by enthusiastic young scientists with whom he generously shared his ideas, knowledge, and time. His scientific school included many successful scientists that are now distributed all over the world.

References Cited

Gerya, T.V., Perchuk, L.L., van Reenen, D.D., and Smit, C.A., 2000, Two-dimensional numerical modeling of pressure-temperature-time paths for the exhumation of some granulite facies terrains in the Precambrian: Journal of Geodynamics, v. 30, p. 17–35, doi:10.1016/S0264-3707(99)00025-3.

Gerya, T.V., Perchuk, L.L., Maresch, W.V., and Willner, A.P., 2004, Inherent gravitational instability of hot continental crust: Implications for doming and diapirism in granulite facies terrains, *in* Whitney, D.L., Teyssier, C., and Siddoway, C.S., eds., Gneiss Domes in Orogeny: Geological Society of America Special Paper 380, p. 97–115.

Perchuk, L.L., 1969, The effect of temperature and pressure on the equilibria of natural Fe-Mg minerals: International Geology Review, v. 11, p. 875–901, doi:10.1080/00206816909475127.

Perchuk, L.L., 1971, Crystallochemical problems in the theory of phase correspondence: Geokhimia, v. 1, p. 23–38.

Perchuk, L.L., 1976, Gas-mineral equilibria and possible geochemical model of the Earth's Interior: Physics of the Earth and Planetary Interiors, v. 13, p. 232–239, doi:10.1016/0031-9201(76)90097-2.

Perchuk, L.L., 1977, Thermodynamic control of metamorphic processes, *in* Saxena, S.K., and Bhattacharji, S., eds., Energetics of Geological Processes: New York, Springer-Verlag, p. 285–352.

Perchuk, L.L., 1986, The course of metamorphism: International Geology Review, v. 28, p. 1377–1400, doi:10.1080/00206818609466374.

Perchuk, L.L., 1989, P-T-fluid regimes of metamorphism and related magmatism with specific reference to the granulite-facies Sharyzhalgay complex of Lake Baikal, *in* Daly, J.S., Cliff, R.A., and Yardley, B.W.D., eds., Evolution of Metamorphic Belts: Geological Society [London] Special Publication 43, p. 275–291.

Perchuk, L.L., 1991, Derivation of thermodynamically consistent system of geothermometers and geobarometers for metamorphic and magmatic rocks, *in* Perchuk, L.L., ed., Progress in Metamorphic and Magmatic Petrology: Cambridge, UK, Cambridge University Press, p. 93–112.

Perchuk, L.L., 2005, Configuration of P-T trends as a record of high-temperature polymetamorphism: Doklady Russian Academy of Sciences, Earth Sciences, v. 401, p. 311–314.

Perchuk, L.L., and van Reenen, D.D., 2008, Comments on "P-T record of two high-grade metamorphic events in the Central Zone of the Limpopo Complex, South Africa": Reply: Lithos, v. 106, p. 403–410, doi:10.1016/j.lithos.2008.07.011.

Perchuk, L.L., Gerya, T.V., van Reenen, D.D., and Smit, C.A., 2001, Formation and dynamics of granulite complexes within cratons: Gondwana Research, v. 4, p. 729–732, doi:10.1016/S1342-937X(05)70524-4.

Perchuk, L.L., van Reenen, D.D., Varlamov, D.A., van Kal, S.M., Tabatabaeimanesh, and Boshoff, R., 2008, P-T record of two high-grade metamorphic events in the Central Zone of the Limpopo Complex, South Africa: Lithos, v. 103, p. 70–105.

Introduction

Dirk D. van Reenen
Department of Geology, University of Johannesburg, Auckland Park, South Africa

Jan D. Kramers
Department of Geology, University of Johannesburg, and School of Geosciences, University of the Witwatersrand, Johannesburg, South Africa

Stephen McCourt
School of Geological Sciences, University of KwaZulu Natal, Durban, South Africa

EVOLUTION OF PRECAMBRIAN HIGH-GRADE GNEISS TERRANES

Granite-greenstone cratons and associated high-grade gneiss terranes have for many years been the focus of basic geological research aimed at understanding the early stages of the Earth's geological history and at finding answers to questions such as whether plate tectonics operated in the Neoarchean. However, cratons have for a long time been studied mainly independently from the adjacent high-grade terranes (e.g., Vielzeuf and Vidal, 1990; De Wit and Ashwal, 1997) with the result that their relationship remained largely unknown. Some researchers (e.g., Mason, 1973) believed that low-grade granite-greenstone terranes demonstrate continuous metamorphic transitions with adjacent high-grade terranes, thus supporting a crust-on-edge relationship. Other researchers preferred a situation in which granulites were formed from low-grade granite-greenstone precursors as the result of the low-pressure transformation of amphibolite facies rocks into garnet-free charnockites, a process that resulted from the influx of CO_2 at constant pressure and temperature (Janardhan et al., 1982; Newton, 1986, 1995; Stahle et al., 1987; Raith et al., 1989).

However, research that focused on systematic field and structural studies supported by input from geochemical, geophysical, geochronological, and petrological data (physicochemical analyses, fluid inclusion studies, construction of pressure-temperature [P-T] paths), as well as on numerical modeling of the exhumation of granulite facies complexes, showed that high-grade terranes are invariably separated from adjacent low-grade granite-greenstone cratons by crustal-scale ductile shear zones, and that the crustal units on either side of these tectonic boundaries reflect very different evolutionary histories (Krill, 1985; Perchuk, 1989; Spear, 1993; Touret, 2001; Barbey and Raith, 1990; McCourt and van Reenen, 1992; Spear and Florence, 1992; van Reenen and Smit, 1996; Perchuk and Krotov, 1998; van Reenen et al., 1987, 1988, 1990; Aftalion et al., 1991; De Wit et al., 1992; Roering et al., 1992a, 1992b; Smit et al., 1992; Treloar et al., 1992; Bibikova et al., 1993; Mkweli et al., 1995; Perchuk et al., 1996, 2000a, 2000b; Percival et al., 1997; Gerya et al., 2000; Kreissig et al., 2000, 2001; Nguuri et al., 2001). More importantly, the results obtained from comparative studies of various Precambrian high-grade terranes between granite-greenstone cratons (e.g., the Limpopo Complex of southern Africa, the Kanskiy Complex of eastern Siberia, the Lapland Complex of Fennoscandia, and the Athabasca Granulite Terrane of western Canada) showed that these distinct terranes share fundamental geologic features that strongly point to similar evolutionary histories (van Reenen et al., 1987, 2008; Roering et al., 1992a; Perchuk et al., 1996, 2000a, 2000b, 2001, 2002, 2003; Gerya et al., 2000; Smit et al., 2000, 2001; Mahan et al., 2008).

THE LIMPOPO COMPLEX OF SOUTHERN AFRICA

The Limpopo Complex of southern Africa is an excellent natural field laboratory for studying deep crustal processes involved in the formation and evolution of complexly deformed and metamorphosed Precambrian high-grade gneiss terranes between granite-greenstone cratons. It exhibits (1) a continuous cross section from the low-grade granite-greenstone terrane of the Kaapvaal Craton in the south through the high-grade Limpopo Complex and into the low-grade granite-greenstone terrane of the Zimbabwe Craton in the north, (2) well-exposed tectonic boundaries with low-grade granite-greenstone rocks from the adjacent cratons, (3) a wide range of thermodynamic parameters that include ultrahigh temperature (UHT) and high pressure (HP) signatures, (4) monometamorphic (Southern and Northern Marginal Zones) and polymetamorphic (Central Zone) subzones, and (5) well-exposed outcrops. In addition, researchers have access to a large database that includes geological maps published at different scales as well as data pertaining to petrological, geochemical, geochronological, and geophysical studies plus numerical modeling of P-T paths (van Reenen et al., 1992, and Kramers et al., 2006, and references therein; James et al., 2001; Zeh et al., 2004; Perchuk et al., 2008; Perchuk and van Reenen, 2008; van Reenen et al., 2008).

Despite the large database available, researchers still disagree on many aspects of the geology of the Limpopo Complex (e.g., Kramers et al., 2006, and references therein; van Reenen et al., 2008). This continuing debate in the literature involves the timing, style, and number of high-grade deformation and metamorphic events that affected the different subzones that make up the Limpopo Complex as well as their relationship to one another, the significance of major Neoarchean anatectic events that affected the high-grade rocks of the Central Zone of the Limpopo Complex, the status of the ca. 2.61 Ga Bulai pluton as an important tectono-metamorphic time marker in the Central Zone, and the basis for the construction of P-T diagrams for studying the geodynamic evolution of the Limpopo Complex (Perchuk and van Reenen, 2008; Zeh and Klemd, 2008; Rigby, 2009). This lack of consensus is best demonstrated by the variety of models that have been proposed for the evolution of the Limpopo Complex. These include Himalayan-type models (Light, 1982; Roering et al., 1992b; Treloar et al., 1992), a gravitational redistribution model (Gerya et al., 2000), a simple terrane accretion model (Rollinson, 1993), a highly complex Turkic-type accretion model (Barton et al., 2006), and a transpressive model (Holzer et al., 1998; Schaller et al., 1999).

The collaborative and interdisciplinary project aimed at studying the tectonic evolution of the Limpopo Complex culminated in a field workshop (van Reenen and Boshoff, 2008) that focused on the following: First, to assess the advances made toward a better understanding of the complex geological issues involved in the study of Precambrian polymetamorphic high-grade gneiss terranes since the previous Limpopo field workshop (van Reenen et al., 1992). Second, to highlight the critical role of heterogeneous deformation in the formation and preservation of multiple granulite-facies events on scales that range from the regional to the thin section. Third, to demonstrate that conventional thermobarometry coupled with the principle of local equilibrium allows the successful construction of multistage P-T paths for polymetamorphic granulites characterized by complex reaction textures. The major goal, however, was to emphasize the critical importance of integrated geological studies as the basis for the construction of deformational (D)–pressure (P)–temperature (T)–time (t) (D-P-T-t) diagrams as essential tools to unravel the evolution of complexly deformed high-grade polymetamorphic terranes, with the Limpopo Complex as the classic example.

MEMOIR 207

The Memoir is divided into four sections. The first section (five chapters) deals with topics that are generic to the study of high-grade gneiss terranes. The chapters in sections 2 and 3 are directly related to the Limpopo Complex. Section 2 (five chapters) deals with general aspects of the Limpopo Complex, whereas section 3 (four chapters) is a detailed discussion of the evolution of the three subzones (Southern Marginal Zone, Central Zone, Northern Marginal Zone) that constitute the Limpopo Complex. This section also includes a discussion of the critical role of heterogeneous deformation related to the evolution of the Athabasca Granulite Terrane of western Canada and the Limpopo Complex. The final section (two chapters) deals with models proposed for the evolution of the Limpopo Complex and the way forward.

The first section opens with a contribution by *Vernon* on three-mineral aggregates (quartz, K-feldspar, and sodic plagioclase) in veinlets as the most reliable criteria for the former presence of felsic melt in regional migmatites. Several other criteria that may be reliable are also discussed, but the author emphasizes that all criteria must be handled with care. *Harlov* reviews the current role of rare earth element (REE)– and actinide-bearing accessory minerals as both geochronological markers and recorders of geochemical processes in the study of Precambrian high-grade gneiss terranes. The papers by *Touret and Huizenga,* and *Rigby and Droop,* discuss fluid inclusion evidence and cordierite volatile content, respectively, for and against the involvement of fluids in granulite formation. Touret and Huizenga point out that fluid inclusions and many metasomatic features observed in granulite terranes can only be explained by large-scale movement of high-salinity aqueous fluids and, to a lesser extent, CO_2, implying that lower-crustal granulites are not as dry as previously assumed. They furthermore emphasize that owing to the complexity of the tectono-metamorphic histories of granulite terrains, fluid inclusion data alone are not sufficient criteria but need to be integrated with geochemical and mineralogical studies done on the same rock samples. By contrast, Rigby and Droop show that in most published case studies the cordierite volatile contents, calculated activities, and melt-H_2O contents are entirely consistent with a mode of origin of granulites that is dominated by fluid-absent melting processes. Their data furthermore suggest that most fluid-saturated granulites formed initially under fluid-absent conditions but subsequently became fluid saturated along a retrograde path. The contribution by *Perchuk* is devoted to the construction of P-T paths for metapelitic granulites based on classical geothermometry and geobarometry coupled with the concept of local mineral equilibria. The fundamental philosophical approach of treating zoned crystals such as garnet as preserving a series of equilibrium states has been, and still is, questioned by different researchers. Perchuk discusses the theoretical thermodynamic basis for this approach and demonstrates that the resulting methodology is the only rigorous means for unraveling the P-T history of high-grade polymetamorphic rocks with reaction textures.

The second set of papers opens with a review by *Kramers and Mouri* of 50 yr of geochronological studies in the Limpopo Complex. A controversy existed chiefly for the Central Zone: Zircon U-Pb dates on (predominantly anatectic) granitoids mainly defined the Neoarchean tectonometamorphic event, whereas the latter 2.0 Ga episode was dated chiefly by work on metamorphic minerals. This can be resolved by considering the various chronometers in the context of reheating, deformation, and resultant fluid access, as well as new mineral growth, whereby both episodes involved tectonism as well as metamorphism. *Tsunogae and van Reenen* discuss evidence in support of a Neoarchean high-pressure (P >14 kbar, T ~900 °C) and ultrahigh-temperature (T >1000 °C, P ~10 kbar) signature for the Central Zone of the Limpopo Complex. *Huizenga et al.* use geothermobarometric and fluid inclusion data to study the P-T evolution of metapelitic gneisses that occur as enclaves in and the host rock to the ca. 2.61 Ga Bulai granitic pluton. Their data show that the enclaves and country rock preserve evidence for similar decompression cooling P-T paths, thus providing supporting evidence that this important time marker intruded already deformed and metamorphosed high-grade Central Zone rocks. The enclaves alone also preserve evidence of a small but significant sub-isobaric (P ~5.5 kbar) heating (~50 °C) event that reflects the emplacement of the Bulai pluton. *Andreoli et al.* discuss the proposal that intracrustal radioactivity could have been the main heat source for the Neoarchean high-grade event that affected the Central Zone. Furthermore, they argue that lower crustal partial melting could have led to diapirism of the lower and middle crust through a density inversion and a strong lowering of the lower crustal viscosity. Steeply plunging structures in the Central Zone mapped as aligned sheath folds could be a consequence of diapirism. *Kramers and Zeh* review published whole rock Sm-Nd and zircon Lu-Hf data that, viewed in combination, provide insight into the crustal evolution and tectonic processes involved in the formation and evolution of the Limpopo Complex. An important conclusion from this study is that the data indicate the Northern Marginal Zone to be an accretionary margin of the Zimbabwe Craton prior to a hypothetical Neoarchean continent collision. Thus, in a collision model, the Zimbabwe Craton would be the overriding plate. The data also do not require any of the three zones of the Limpopo Complex to be separate terranes, or the discrimination of terranes within the Central Zone. Strong indications for a mantle-derived component in the igneous rocks of the Mahalapye Complex during the ca. 2.02 Ga Paleoproterozoic reworking of the Limpopo Complex suggest a connection between the magmatism and the ca. 2.05 Ga Bushveld event.

The third set of papers focuses on the evolution of the three subzones that compose the Limpopo Complex. *Van Reenen et al.* discuss the thrust-controlled exhumation of the Southern Marginal Zone as part of a single Neoarchean high-grade event. This event commenced with crustal thickening at ca. 2.72 Ga and ended at ca. 2.62 Ga with the convergence of decompression-cooling P-T paths in the hanging wall and prograde P-T paths in the footwall of the Hout River Shear Zone, the thrust sense structure that bounds the high-grade terrane in the south. The contribution by *Smit et al.* focuses on the polymetamorphic evolution of the geologically much more complex Central Zone of the Limpopo Complex. They highlight the critical role that heterogeneous strain has played in the development and preservation of multiple granulite facies events at different scales including the thin section scale. Linking distinct fabric forming events and related mineral assemblages with distinct P-T-t paths and deformation events (D-P-T-t diagram) provides an essential tool for studying the protracted polymetamorphic Neoarchean => Paleoproterozoic evolution of the Central Zone in the interval 2.68–2.0 Ga. The Neoarchean evolution can be linked to the collision of the Kaapvaal Craton with the proto-Zimbabwe Craton, whereas the Paleoproterozoic reworking of the Central Zone can in part be linked to magmatic underplating as the result of the emplacement of the Bushveld Igneous Complex at ca. 2.05 Ga. *Blenkinsop* discusses the evolution of the Northern Marginal Zone, in which magmatic enderbites and charnockites occur as multiple diapiric intrusions at a scale of several kilometers, interspersed with vertically aligned septa of supracrustal rocks. This zone appears to represent a deeper level of the Zimbabwe Craton margin, thickened and uplifted by horizontal shortening and steep extrusion. In the final paper of this section *Mahan et al.* present examples of multiscale heterogeneous deformation from two classic deep-crustal granulite terranes, the Athabasca Granulite Terrane of western Canada and the Limpopo Complex in southern Africa, which can form the basis for refining important natural tools (D-P-T-t diagrams) for unraveling complex tectonic histories in polyphase metamorphic terranes.

Memoir 207 concludes with two contributions devoted to models for the evolution of the Limpopo Complex. *Perchuk and Gerya* offer a fresh view of the gravitational redistribution model as applied to the evolution of Precambrian high-grade terranes within granite-greenstone cratons. They point out that crustal thickening and gravitational processes need not be mutually exclusive mechanisms because the early crustal thickening stage of orogeny mainly involves crustal thickening mechanisms, whereas gravitational mechanisms (doming and diapirism) played the dominant role during uplift and exhumation. In the final paper of the volume *Kramers et al.* discuss geological constraints on the different models that have been proposed for the evolution of the Limpopo Complex. These include Himalayan-type crustal thickening models, a gravitational redistribution model, a

Paleoproterozoic transpressive model, and both simple and highly complex (Turkic-type) terrane accretion models. Many, or parts, of these are mutually compatible. For instance, Paleoproterozoic transpression can be superimposed on a Neoarchean collision model. Some models, such as the multi-terrane accretion models, appear inconsistent with the constraints. No single proposed model on its own, however, can account for all observations. This contribution also includes suggestions for the way forward.

ACKNOWLEDGMENTS

DDvR acknowledges the financial support of the Limpopo project by the National Research Foundation (NRF) from 1983 (Gun 2049254), and the support from the Faculty of Science of the University of Johannesburg (previously the Rand Afrikaans University). LLP acknowledges support through the Program of the RF President entitled "Leading Research Schools of Russia" (grant no. 1949.2008.15.5). The guest editors appreciate all authors for their contributions to this GSA Memoir, and also acknowledge GSA Books Science Editor Pat Bickford for his support of this project. All papers within this volume were vetted by at least two independent reviewers. We appreciate their support of this Memoir. A project of this nature could not have reached completion without the support of several colleagues. André Smit, René Boshoff, Chris Roering, and Jan-Marten Huizenga were not only involved as authors and co-authors of different papers but also made significant contributions to the preparation of the Memoir. Michael Chakuparira assisted with the drafting and redrafting of figures for different papers. We are grateful to Taras Gerya for his help in producing final manuscripts for the two papers with Leonid Perchuk as first author. René Boshoff reprocessed all papers in order to ensure a final uniform volume. A special word of thanks is due to Johan Wolfaardt (Sand River Safaris, Musina), Dawie Groenewald (Out of Africa Safaris, Musina), and Neels van Wyk (Farm Kilimanjaro, Alldays district) for their hospitality during years of fieldwork in the Central Zone.

REFERENCES CITED

Aftalion, M., Bibikova, E.V., Bowes, D.R., Hopgood, A.M., and Perchuk, L.L., 1991, Timing of early Proterozoic collisional and extentional events in the granulite-gneiss-charnockite-granite complex, Lake Baikal, SSR: U-Pb, Rb-Sr, and Sm-Nd study: Journal of Geology, v. 99, p. 851–862, doi:10.1086/629556.

Barbey, P., and Raith, M., 1990, The granulite belt of Lapland, in Vielzeuf, D., and Vidal, Ph., eds., Granulites and Crustal Evolution: Dordrecht, Kluwer, NATO ASI Ser. C, v. 311, p. 111–132.

Barton, J.M., Jr., Klemd, R., and Zeh, A., 2006, The Limpopo Belt: A result of Archean to Proterozoic, Turkic-type orogenesis?, in Reimold, W.U., and Gibson, R.L., eds., Processes on the Early Earth: Geological Society of America Special Paper 405, p. 315–332.

Bibikova, E.V., Melnikov, V.F., and Avakyan, K.Kh., 1993, Lapland granulites: Petrology, geochemistry and isotopic age: Petrology, v. 1, p. 215–234.

De Wit, M.J., and Ashwal, L.D., eds., 1997, Greenstone Belts: Oxford, UK, Clarendon Press, 809 p.

De Wit, M.J., van Reenen, D.D., and Roering, C., 1992, Geologic observations across a tectono-metamorphic boundary in the Babangu area, Giyani (Sutherland) Greenstone Belt, South Africa: Precambrian Research, v. 55, p. 111–122, doi:10.1016/0301-9268(92)90018-J.

Gerya, T.V., Perchuk, L.L., van Reenen, D.D., and Smit, C.A., 2000, Two-dimensional numerical modeling of pressure-temperature-time paths for the exhumation of some granulite facies terrains in the Precambrian: Journal of Geodynamics, v. 30, p. 17–35, doi:10.1016/S0264-3707(99)00025-3.

Gerya, T.V., Perchuk, L.L., Maresch, W.V., and Willner, A.P., 2004, Inherent gravitational instability of hot continental crust: Implications for doming and diapirism in granulite facies terrains, in Whitney, D.L., Teyssier, C., and Siddoway, C.S., eds., Gneiss Domes in Orogeny: Geological Society of America Special Paper 380, p. 97–115.

Holzer, L., Frey, R., Barton, J.M., Jr., and Kramers, J.D., 1998, Unraveling the record of successive high-grade events in the Central Zone of the Limpopo belt using Pb single phase dating of metamorphic minerals: Precambrian Research, v. 87, p. 87–115, doi:10.1016/S0301-9268(97)00058-2.

James, D.E., Fouch, M.J., Van Decar, J.C., van der Lee, S., and Kaapvaal Seismic Group, 2001, Tectospheric structure beneath southern Africa: Geophysical Research Letters, v. 28, p. 2485–2488.

Janardhan, A.S., Newton, R.C., and Hansen, E.C., 1982, The transformation of amphibolite facies gneiss to charnockite in southern Karnataka and northern Tamil Nadu, India: Contributions to Mineralogy and Petrology, v. 79, p. 130–149, doi:10.1007/BF01132883.

Kramers, J.D., McCourt, S., and van Reenen, D.D., 2006, The Limpopo Belt, in Anhaeusser, C.R., and Thomas, R.J., eds., Geology of South Africa: Johannesburg, Geological Society of South Africa, and Pretoria, Council for Geoscience, p. 209–236.

Kreissig, K., Nögler, T.F., Kramers, J.D., van Reenen, D.D., and Smit, A.S., 2000, An isotopic and geochemical study of the northern Kaapvaal craton and the Southern Marginal Zone of the Limpopo Belt: Are they juxtaposed terranes?: Lithos, v. 50, p. 1–25, doi:10.1016/S0024-4937(99)00037-7.

Kreissig, K., Holzer, L., Frei, R., Villa, I.M., Kramers, J.D., Kröner, A., Smit, A.S., and van Reenen, D.D., 2001, Geochronology of the Hout River shear zone and metamorphism in the Southern Marginal Zone of the Limpopo Belt, South Africa: Precambrian Research, v. 109, p. 145–173, doi:10.1016/S0301-9268(01)00147-4.

Krill, A.G., 1985, Svecokarelian thrusting with thermal inversion in the Karasjok-Levajok area of the northern Baltic Shield: Norges Geologiske Undersøkelse, v. 403, p. 89–101.

Light, M.P.R., 1982, The Limpopo Mobile Belt—A result of continental collision: Tectonics, v. 1, p. 325–342, doi:10.1029/TC001i004p00325.

Mahan, K.H., Goncalves, P., Flowers, R.M., Williams, M.L., and Hoffman-Setka, D., 2008, The role of heterogeneous strain in the development and preservation of a polymetamorphic record in high-P granulites, western Canadian shield: Journal of Metamorphic Geology, v. 26, p. 669–695, doi: 10.1111/j.1525-1314.2008.00783.x.

Mason, R., 1973, The Limpopo Mobile Belt, Southern Africa: Philosophical Transactions of the Royal Society of London, v. A273, p. 463–485.

McCourt, S., and van Reenen, D.D., 1992, Structural geology and tectonic setting of the Sutherland greenstone-belt, Kaapvaal Craton, South Africa: Precambrian Research, v. 55, p. 93–110, doi:10.1016/0301-9268(92)90017-I.

Mkweli, S., Kamber, B., and Berger, M., 1995, Westward continuation of the craton–Limpopo Belt tectonic break in Zimbabwe and new age constraints on the timing of the thrusting: Journal of the Geological Society [London], v. 152, p. 77–83, doi:10.1144/gsjgs.152.1.0077.

Newton, R.C., 1986, Fluids in granulite facies metamorphism: Advances in Physical Geochemistry, v. 5, p. 36–59.

Newton, R.C., 1995, Simple-system mineral reactions and high-grade metamorphic fluids: European Journal of Mineralogy, v. 7, p. 861–881.

Nguuri, T., Gore, J., James, D.E., Webb, S.J., Wright, C., Zengeni, T.G., Gwavava, O., Snoke, J.A., and Kaapvaal Seismic Group, 2001, Crustal structure beneath southern Africa and its application for the formation and evolution of the Kaapvaal and Zimbabwe cratons: Geophysical Research Letters, v. 28, p. 2501–2504, doi:10.1029/2000GL012587.

Perchuk, L.L., 1969, The effect of temperature and pressure on the equilibria of natural Fe-Mg minerals: International Geology Review, v. 11, p. 875–901, doi:10.1080/00206816909475127.

Perchuk, L.L., 1971, Crystallochemical problems in the theory of phase correspondence: Geokhimia, v. 1, p. 23–38.

Perchuk, L.L., 1976, Gas-mineral equilibria and possible geochemical model of the Earth's Interior: Physics of the Earth and Planetary Interiors, v. 13, p. 232–239, doi:10.1016/0031-9201(76)90097-2.

Perchuk, L.L., 1977, Thermodynamic control of metamorphic processes, in Saxena, S.K., and Bhattacharji, S., eds., Energetics of Geological Processes: New York, Springer-Verlag, p. 285–352.

Perchuk, L.L., 1986, The course of metamorphism: International Geology Review, v. 28, p. 1377–1400, doi:10.1080/00206818609466374.

Perchuk, L.L., 1989, P-T-fluid regimes of metamorphism and related magmatism with specific reference to the granulite-facies Sharyzhalgay complex of Lake Baikal, in Daly, J.S., Cliff, R.A., and Yardley, B.W.D., eds., Evolution of Metamorphic Belts: Geological Society [London] Special Publication 43, p. 275–291.

Perchuk, L.L., 1991, Derivation of thermodynamically consistent system of geothermometers and geobarometers for metamorphic and magmatic rocks, in Perchuk, L.L., ed., Progress in Metamorphic and Magmatic Petrology: Cambridge, UK, Cambridge University Press, p. 93–112.

Perchuk, L.L., 2005, Configuration of P-T trends as a record of high-temperature polymetamorphism: Doklady Russian Academy of Sciences, Earth Sciences, v. 401, p. 311–314.

Perchuk, L.L., and Krotov, A.V., 1998, Petrology of the mica schists of the Tanaelv belt in the southern tectonic framing of the Lapland granulite complex: Petrology, v. 6, p. 149–179.

Perchuk, L.L., and van Reenen, D.D., 2008, Comments on "P–T record of two high-grade metamorphic events in the Central Zone of the Limpopo Complex, South Africa": Reply: Lithos, v. 106, p. 403–410, doi:10.1016/j.lithos.2008.07.011.

Perchuk, L.L., Gerya, T.V., van Reenen, D.D., Safonov, O.G., and Smit, C.A., 1996, The Limpopo metamorphic complex, South Africa: 2. Decompression/cooling regimes of granulites and adjusted rocks of the Kaapvaal craton: Petrology, v. 4, p. 571–599.

Perchuk, L.L., Gerya, T.V., van Reenen, D.D., Krotov, A.V., Safonov, O.G., Smit, C.A., and Shur, M.Yu., 2000a, Comparative petrology and metamorphic evolution of the Limpopo (South Africa) and Lapland (Fennoscandia) high-grade terrains: Mineralogy and Petrology, v. 69, p. 69–107.

Perchuk, L.L., Gerya, T.V., van Reenen, D.D., Smit, C.A., and Krotov, A.V., 2000b, P-T paths and tectonic evolution of shear zones separating high-grade terrains from cratons: Examples from Kola Peninsula (Russia) and Limpopo Region (South Africa): Mineralogy and Petrology, v. 69, p. 109–142, doi:10.1007/s007100050020.

Perchuk, L.L., Gerya, T.V., van Reenen, D.D., and Smit, C.A., 2001, Formation and dynamics of granulite complexes within cratons: Gondwana Research, v. 4, p. 729–732, doi:10.1016/S1342-937X(05)70524-4.

Perchuk, L.L., Tokarev, D.A., van Reenen, D.D., Varlamov, D.A., Gerya, T.V., Sazonova, L.V., Fan, V.I., Smit, C.A., Brink, M.C., and Bisschoff, A.A., 2002, The dynamic and thermal history of the explosion structure Vredefort in the Kaapvaal Craton, South Africa: Petrology, v. 10, p. 395–432.

Perchuk, L.L., Sazonova, L.V., van Reenen, D.D., and Gerya, T.V., 2003, Ultramylonites and their significance for the understanding of the history of the Vredefort impact structure, South Africa: Petrology, v. 11, no. 2, p. 114–129.

Perchuk, L.L., van Reenen, D.D., Varlamov, D.A., van Kal, S.M., Tabatabaeimanesh, and Boshoff, R., 2008, P-T record of two high-grade metamorphic events in the Central Zone of the Limpopo Complex, South Africa: Lithos, v. 103, p. 70–105.

Percival, J.A., Roering, C., van Reenen, D.D., and Smit, C.A., 1997, Tectonic evolution of associated greenstone belts and high-grade terrains, in De Wit, M.J., and Ashwal, L.D., eds., Greenstone Belts: Oxford, UK, Oxford University Press, p. 398–421.

Raith, M., Hoernes, S., Stähle, H.J., and Klatt, 1989, Contrasting mechanisms of charnockite formation in the amphibolite to granulite transition zones of southern India, in Bridgewater, D., ed., Fluid Movements, Element Transport and the Composition of the Deep Crust: Dordrecht, Kluwer, NATO ICI Ser. C281, p. 29–39.

Rigby, M.J., 2009, Conflicting P-T paths within the Central Zone of the Limpopo Belt: A consequence of different thermobarometric methods: Journal of African Earth Sciences, v. 54, p. 111–126, doi:10.1016/j.jafrearsci.2009.03.005.

Roering, C., van Reenen, D.D., Smit, C.A., Barton, J.M., de Beer, J.H., De Wit, M.J., Stettler, E.H., van Schalkwyk, J.F., Stevens, G., and Pretorius, S., 1992a, Tectonic model for the evolution of the Limpopo Belt: Precambrian Research, v. 55, p. 539–552.

Roering, C., van Reenen, D.D., de Wit, M.J., Smit, C.A., de Beer, J.H., and Van Schalkwyk, J.F., 1992b, Structural geological and metamorphic significance of the Kaapvaal Craton–Limpopo Belt contact: Precambrian Research, v. 55, p. 69–80, doi:10.1016/0301-9268(92)90015-G.

Rollinson, H.R., 1993, A terrane interpretation of the Archaean Limpopo Belt: Geological Magazine, v. 130, p. 755–765, doi:10.1017/S001675680002313X.

Schaller, R.M., Steiner, O., and Studer, I., 1999, Exhumation of Limpopo Central Zone granulites and dextral continent-scale transcurrent movement at 2.0 Ga along the Palala Shear Zone, Northern Province, South Africa: Precambrian Research, v. 96, p. 263–298, doi:10.1016/S0301-9268(99)00015-7.

Smit, C.A., Roering, C., and van Reenen, D.D., 1992, The structural framework of the southern margin of the Limpopo Belt, South Africa, in van Reenen, D.D., Roering, C., Ashwal, L.D., and De Wit, M.J., eds., The Archaean Limpopo Granulite Belt: Tectonics and Deep Crustal Processes: Precambrian Research, v. 55, p. 51–67.

Smit, C.A., van Reenen, D.D., Gerya, T.V., Varlamov, D.A., and Fed'kin, A.V., 2000, Structural-metamorphic evolution of the Southern Yenisey Range of Eastern Siberia: Implications for the emplacement of the Kanskiy granulite complex: Mineralogy and Petrology, v. 69, p. 35–67, doi:10.1007/s007100050018.

Smit, C.A., van Reenen, D.D., Gerya, T.V., and Perchuk, L.L., 2001, P-T conditions of decompression of the Limpopo high grade terrain: Record from shear zones: Journal of Metamorphic Geology, v. 19, p. 249–268, doi:10.1046/j.0263-4929.2000.00310.x.

Spear, F.S., 1993, Metamorphic phase equilibria and pressure–temperature–time paths: Washington, D.C., Mineralogical Society of America Monograph, v. 22, 799 p.

Spear, F.S., and Florence, F.P., 1992, Thermobarometry in granulites: Pitfalls and new approaches: Precambrian Research, v. 55, p. 209–241, doi:10.1016/0301-9268(92)90025-J.

Stahle, H.J., Raith, M., Hoernes, S., and Delfs, A., 1987, Element mobility during incipient granulite formation at Kabbaldurga, southern India: Journal of Petrology, v. 28, p. 803–834.

Touret, J.L.R., 2001, Fluids in metamorphic rocks: Lithos, v. 55, p. 1–25, doi:10.1016/S0024-4937(00)00036-0.

Treloar, P.J., Coward, M.J., and Harris, B.W., 1992, Himalayan-Tibetan analogies for the evolution of the Zimbabwe craton and Limpopo belt: Precambrian Research, v. 55, p. 571–587, doi:10.1016/0301-9268(92)90046-Q.

van Reenen, D.D., and Boshoff, R., 2008, Limpopo International Field Workshop, July 13–17, 2008: Department of Geology, University of Johannesburg, unpublished field guide, 130 p.

van Reenen, D.D., and Smit, C.A., 1996, The Limpopo Metamorphic Belt, South Africa: 1. Geological setting and relationship of the Granulite Complex with the Kaapvaal and Zimbabwe Cratons: Petrology, v. 4, p. 610–618.

van Reenen, D.D., Barton, J.M., Jr., Roering, C., Smit, C.A., and van Schalkwyk, J.F., 1987, Deep-crustal response to continental collision: The Limpopo Belt of southern Africa: Geology, v. 15, p. 11–14, doi:10.1130/0091-7613(1987)15<11:DCRTCC>2.0.CO;2.

van Reenen, D.D., Roering, C., Smit, C.A., Van Schalkwyk, J.F., and Barton, J.M., Jr., 1988, Evolution of the northern high-grade margin of the Kaapvaal Craton, South Africa: Journal of Geology, v. 96, p. 549–560, doi:10.1086/629251.

van Reenen, D.D., Roering, C., Brandl, G., Smit, C.A., and Barton, J.M., Jr., 1990, The granulite facies rocks of the Limpopo belt, southern Africa, in Vielzeuf, D., and Vidal, Ph., eds., Granulites and Crustal Evolution: Dordrecht, Kluwer, NATO ASI Ser. C, v. 311, p. 257–289.

van Reenen, D.D., Roering, C., Ashwal, L.D., and De Wit, M.J., eds., 1992, The Archean Limpopo granulite belt: Tectonics and deep crustal processes: Precambrian Research, v. 55, Special Issue, p. 1–4, doi:10.1016/0301-9268(92)90009-D.

van Reenen, D.D., Boshoff, R., Smit, C.A., Perchuk, L.L., Kramers, J.D., McCourt, S.M., and Armstrong, R.A., 2008, Geochronological problems in the Limpopo complex, South Africa: Gondwana Research, v. 14, p. 644–662, doi:10.1016/j.gr.2008.01.013.

Vielzeuf, D., and Vidal, Ph., 1990, Granulites and Crustal Evolution: Dordrecht, Kluwer, NATO ICI Ser. C, v. 311, 585 p.

Zeh, A., and Klemd, R., 2008, Comments on "P-T record of two high-grade metamorphic events in the Central Zone of the Limpopo Complex, South Africa": Comment: Lithos, v. 106, p. 399–402, doi:10.1016/j.lithos.2008.07.008.

Zeh, A., Klemd, R., Buhlmann, S., and Barton, J.M., 2004, Pro- and retrograde P–T evolution of granulites of the Beit Bridge Complex (Limpopo Belt, South Africa): Constraints from quantitative phase diagrams and geotectonic implications: Journal of Metamorphic Geology, v. 22, p. 79–95, doi:10.1111/j.1525-1314.2004.00501.x.

MANUSCRIPT ACCEPTED BY THE SOCIETY 24 MAY 2010   Printed in the USA

Microstructures of melt-bearing regional metamorphic rocks

R.H. Vernon*

Department of Earth & Planetary Sciences and National Key Centre for Geochemical Evolution and Metallogeny of Continents, Macquarie University, Sydney, 2109, Australia

ABSTRACT

The most reliable microstructural criterion for the former presence of felsic melt in regional migmatites is a three-mineral (quartz, K-feldspar, and sodic plagioclase) aggregate in veinlets. Several other criteria are potentially reliable, namely: (1) euhedral crystals of feldspar (precipitated from liquid) or peritectic minerals (e.g., garnet, cordierite, orthopyroxene, K-feldspar) lining felsic "protoleucosomes"; (2) inclusion-free euhedral overgrowths of feldspar (precipitated from liquid) or peritectic minerals (e.g., garnet, cordierite, orthopyroxene, K-feldspar) on residual grains of the same minerals with abundant inclusions in the mesosome; (3) aligned, euhedral feldspar crystals; (4) simple twinning in K-feldspar; (5) dihedral angles of ≤ 60° subtended where a grain of feldspar and/or quartz (inferred to have pseudomorphed former melt) meets two grains of other minerals; (6) cuspate volumes of quartz, K-feldspar or sodic plagioclase, especially where surrounded by grains inferred to have been residual during melting; (7) veinlets of inferred former melt (now mineral pseudomorphs consisting of one of quartz, K-feldspar or sodic plagioclase, preferably, though less commonly, involving two or three of these minerals) along grain boundaries or along inferred former intragranular fractures; (8) biotite pseudomorphed by feldspar; (9) veinlets of plagioclase that is more sodic than plagioclase grains in the adjacent rock; (10) plagioclase with oscillatory zoning; (11) microgranophyric intergrowths of quartz and alkali feldspar in patches or veinlets between primary grains; (12) symplectic replacement aggregates that can be explained by reactions between peritectic grains and cooling melt; and (13) melanosome patches and layers, from which leucosome has been extracted. However, all these criteria must be interpreted with care. Some other proposed criteria are questionable, for example: (1) random mineral distributions; (2) grain-size increase; (3) interstitial grains; (4) corroded relics of inferred reactant mineral grains surrounded by areas of quartz, K-feldspar, or sodic plagioclase; (5) projections into a mineral grain; (6) lobes of myrmekite; and (7) plagioclase rims with a constant sodic composition occurring on plagioclase cores that are more calcic and/or of variable composition.

*rvernon@els.mq.edu.au

INTRODUCTION

This review evaluates microstructural criteria commonly used for inferring the former presence of small amounts of hydrous silicate liquid (melt) in high-grade regional metamorphic rocks. Many contact metamorphic rocks preserve microstructural evidence of former melt, but the evidence is commonly removed from regional metamorphic rocks, owing to post-melting deformation and recrystallization-neocrystallization. In such rocks, leucosome veins, peritectic minerals, and depleted whole-rock chemical compositions may be the only evidence of former melt. Nevertheless, some regional metamorphic rocks (especially in low-pressure–high-temperature terranes) retain microstructural features that are consistent with the former presence of melt, even where obvious leucosome layers are scarce or absent.

Microstructural evidence is especially useful for identifying former melt in rocks that have melted only slightly, residual melt in migmatites from which melt has been largely extracted, migmatites into which melt has drained back from leucosome veins or networks (Sawyer, 2001, p. 292), and migmatites penetrated by small amounts of melt from an external source. Careful interpretation is important, because microstructural criteria are often used, not only to identify partly melted rocks but also to infer flow of melt through crustal terranes.

Some microstructural criteria appear to be generally reliable, but others are dubious, and even the more reliable ones need care in their application. Many papers that apply microstructural criteria have been published (e.g., Ashworth, 1976; Dougan, 1979, 1983; Platten, 1982, 1983; Kenah and Hollister, 1983; McLellan, 1983, 1988; Vernon et al., 1990; Grant and Frost, 1990; Laporte, 1994; Laporte et al., 1997; Pattison and Harte, 1998; Vernon and Collins, 1988; Sawyer, 1999, 2000, 2001; Holness and Clemens, 1999; Dallain et al., 1999; Vernon, 1999; Rosenberg and Riller, 2000; Clemens and Holness, 2000; Vernon and Johnson, 2000; Rosenberg, 2001; Guernina and Sawyer, 2003; Vernon et al., 2003; Harris et al., 2003; Holness, and Sawyer, 2008; Hasalová et al., 2008a, 2008b); and this review provides a brief but comprehensive summary and critique of the criteria proposed.

MICROSTRUCTURAL CRITERIA FOR THE FORMER PRESENCE OF MELT IN MIGMATITES

The following microstructural criteria for the presence of small amounts of former melt in metamorphic rocks have been suggested. Some criteria are also applicable to clearly segregated, outcrop-scale leucosomes, though in this paper the emphasis will be on rocks without obvious leucosome layers. Such rocks can be called *mesosome—residuum* in the terminology favored by Sawyer (2008)—if leucosome is also present in the rock, which is common. They cannot be called *palaeosome* unless a transition to an unmodified original composition can be observed. The felsic patches in the mesosomes may be small and irregular to diffuse, and these were called *protoleucosomes* by Sawyer (2001).

The microstructural criteria are observed most clearly in rocks that have not undergone post-melting, solid-state deformation, which can obliterate microstructural evidence if sufficiently intense. Many of the criteria involve single-mineral volumes of quartz, K-feldspar, or sodic plagioclase that are inferred to represent former melt. However, this inference depends on the assumption that, as only one mineral is typically present (e.g., K-feldspar), the other two co-precipitating minerals (e.g., quartz and sodic plagioclase) must have crystallized on existing residual grains of the same minerals. In places, plagioclase can be reasonably inferred to have crystallized on existing plagioclase grains, because of more sodic rims on the primary metamorphic plagioclase (Sawyer, 2001). If all three minerals are in contact, the inference of melt is much stronger, because they crystallize simultaneously near and at the solidus. Such evidence has been observed in some granites deformed in the presence of small amounts of residual melt (Bouchez et al., 1992; Vernon et al., 2004) and in some partly melted contact metamorphic rocks (Holness and Clemens, 1999; Clemens and Holness, 2000), but has not been reported in regional migmatites, as far as I am aware.

Reliable Criteria

Apart from the granitic (three-mineral) veins mentioned previously, the following microstructural criteria appear to be potentially the most reliable. Nevertheless, care needs to be taken with the application of several of them, as explained below.

1. In some contact metamorphic migmatites and uncommon regional metamorphic migmatites, small felsic patches (sometimes poetically referred to as "pools") or "protoleucosomes" of inferred former melt (now pseudomorphs consisting of one of quartz, K-feldspar, or sodic plagioclase), edged by grains with euhedral outlines (Figs. 1, 2), have been inferred to be due to crystal growth from melt (e.g., Pattison and Harte, 1988, p. 481; McLellan, 1988, p. 525; Sawyer, 2001, p. 294; Brown, 2001; Johnson et al., 2004, fig. 2f.). The euhedral boundaries belong to K-feldspar or plagioclase (precipitated from the felsic melt) or peritectic minerals, such as plagioclase, K-feldspar, cordierite, garnet, or orthopyroxene (Kenah and Hollister, 1983; Vernon and Collins, 1988; Vernon et al., 1990, figs. 11.11, 11.18, 11.19, 11.20; Sawyer, 2001, fig. 1; Rosenberg and Berger, 2001; Vernon and Johnson, 2000; Vernon et al., 2003, fig. 15; Marchildon and Brown, 2003, fig. 5; Droop et al., 2003, figs. 3b, 4d; Vernon, 2004, figs. 4.80, 4.81; Vernon et al., 2008, fig. 10).

2. Euhedral, inclusion-free overgrowths of peritectic K-feldspar, cordierite, or garnet on inclusion-rich residual grains of the same minerals in the mesosome at the edges of inferred former melt patches or veinlets (Fig. 2) have been interpreted as precipitates from melt (Vernon, 1999; Vernon and Johnson, 2000; Vernon, 2004, fig. 4.81; Vernon et al., 2008, fig. 10).

3. Aggregates of aligned, euhedral, undeformed plagioclase crystals are consistent with magmatic flow (e.g., Paterson et al., 1989; Vernon, 2000, 2004) and therefore have been used as indicators of former melt in migmatites (Collins et al., 1989; Sawyer,

2001, p. 294, fig. 3c). Magmatic flow can occur in diffuse zones of melt accumulation (protoleucosomes) in migmatites, as discussed and illustrated by Sawyer (2001).

4. Simple twinning of K-feldspar in leucosome patches reflects a magmatic origin (Vernon, 1986), as this feature is typical of igneous K-feldspar (owing to difficult nucleation being assisted by twinned nuclei) but is rare or absent in metamorphic K-feldspar (Vernon, 1999, 2004). However, the absence of simple twinning does not necessarily indicate a nonmagmatic origin, as the K-feldspar can nucleate on existing untwinned K-feldspar grains (Vernon, 1999, 2004).

5. Dihedral angles of ≤ 60°, formed where one grain of quartz, K-feldspar, or sodic plagioclase (inferred to have been precipitated from melt) meets two grains of other solid minerals, are commonly used as indicators of the former presence of partial melt (e.g., Rosenberg and Riller, 2000; Holness and Clemens, 1999; Clemens and Holness, 2000; Harris et al., 2003). These angles are distinctly smaller than equivalent angles (103° to 115°) that involve the same minerals in melt-free high-grade metamorphic rocks (Vernon, 1968). As explained by Smith (1953), if the dihedral angle made by melt against two solid grains is 60° or less it spreads along grain edges (though not covering grain boundaries), forming a continuous flow-channel network through which melt may move, if sufficiently low in viscosity, though the rock behaves as a solid.

Measured dihedral angles of ≤ 60° made by quartz, K-feldspar, or sodic plagioclase are consistent with the former presence of felsic melt, provided it can be justifiably inferred that the other two minerals that necessarily co-precipitated from the melt crystallized somewhere else in the rock, as discussed previously. Small dihedral angles can lead to cuspate shapes of inferred melt pseudomorphs (see next criterion).

Photographic evidence of melt "wetting" grain boundaries has been obtained from the products of partial melting experiments on natural felsic rocks (Mehnert et al., 1973; Büsch et al., 1974; Jurewicz and Watson, 1984, 1985; Laporte, 1994; Laporte and Watson, 1995) and in the products of melting experiments on organic analogue materials (Means, 1989, fig. 7b; Walte et al., 2003, 2005). For example, Laporte (1994) measured melt versus quartz-quartz dihedral angles of 12° to 18° (varying with the experimental temperature and the water content).

Rosenberg and Riller (2000) measured dihedral angles of feldspar versus quartz-quartz in statically recrystallized, partly melted granite, obtaining an angle (corrected statistically) of 27°. Holness and Clemens (1999) and Clemens and Holness (2000) found that the feldspar versus quartz-quartz dihedral angle (θ) in a partly melted feldspathic metaquartzite in the Ballachulish contact metamorphic aureole, Scotland, is bimodal, with peaks of ~110° and ~40–60°. The large angles were inferred to represent solid-state grain-boundary adjustment (e.g., Vernon, 1968), and the smaller angles were inferred to represent grain-boundary adjustment in the presence of melt that solidified rapidly enough to preserve the small angles. In the Ballachulish aureole rocks, small angles also occur in cuspate extensions of original clastic feldspar grains, which were interpreted as representing initiation of melting (Holness and Clemens, 1999; Clemens and Holness, 2000). These extensions are more elongate along quartz-quartz grain boundaries with a larger proportion of melt; small feldspar grains (presumably representing cross sections through the projections) occur at quartz triple junctions. The projections probably represent the feldspar component of a former felsic melt,

Figure 1. K-feldspar with crystal faces against quartz in a leucosome patch in a regional metamorphic migmatite, Snowy Mountains, southeastern New South Wales, Australia. Crossed polars; base of photo, 1.7 mm.

Figure 2. Euhedral K-feldspar (Kfs) and cordierite (Crd) against quartz (Qtz) in a patch of leucosome in metapelite, Cooma complex, southeastern New South Wales, Australia. Also shown in the leucosome is an inclusion-free overgrowth of the K-feldspar (center) on optically continuous mesosome K-feldspar with many inclusions (top left). The quartz has been recrystallized in response to minor late deformation. Crossed polars; base of photo, 7 mm.

the quartz component having crystallized on clastic quartz grains. Solid-state grain boundary adjustment after the feldspar crystallized was inferred to have caused breakup of the feldspar projections into isolated but crystallographically continuous grains.

Small dihedral angles in migmatites are sometimes inferred by inspection (e.g., Hasalová et al., 2008a, p. 38) or estimated from grain shapes (Dallain et al., 1999, p. 66) without making measurements. However, because thin and polished sections are two-dimensional slices through three-dimensional grain aggregates, the interfacial angles seen in these sections are apparent angles. In sections that are not perpendicular to grain boundaries, interfacial angles can appear to be smaller than they actually are, as shown by, for example, angles between cleavage planes in pyroxene and amphibole, as well as angles between intersecting twins in plagioclase. Therefore, simple inspection of interfacial angles in a thin section can be misleading. Unless all three interfaces involved are fortuitously perpendicular to the section, an image of the true dihedral angle can only be obtained with a universal microscope stage, which takes advantage of the thickness (typically 0.03 mm) of "thin sections" of non-opaque minerals. Grain boundaries oblique to the section appear relatively broad and diffuse but appear as sharp, thin lines when rotated to the vertical on a universal stage, so that their orientation can be measured accurately (Kretz, 1966; Vernon, 1968, 1970, 1997). To obtain the true value, interfacial angles must be measured right at the triple junction, which can be difficult for curving interfaces, and especially difficult for small grains.

Flat-stage measurements give an average dihedral angle close to the true mean value (Riegger and van Vlack, 1960; Harris et al., 2003) but give no reliable indication of the influence of structural anisotropy on the spread of angles about the mean (Vernon, 1997). The use of a universal stage avoids the necessity of statistical corrections of flat-stage measurements (e.g., Hunter, 1987; Elliott et al., 1997), as pointed out by Vernon (1997). Holness (2008) used a universal stage to determine the true spread of angles for several natural melt-solid systems, including quartz (mean angle 18°, with a standard deviation of 9°) and plagioclase (mean angle 25°, with a standard deviation of 11°).

An additional problem is that relatively small dihedral angles, even if observed and measured correctly, are not necessarily those achieved when melt was present, owing to possible solid-state grain boundary adjustment during slow cooling, which would tend to change the angles toward those observed in unmelted quartzofeldspathic rocks (e.g., Vernon, 1968; Harris et al., 2003; Holness et al., 2005; Holness, 2006, 2008; Holness and Sawyer, 2008). Such adjustment would be expected in migmatites in slowly cooled terranes, but much less so in contact metamorphic environments, for which small dihedral angles have been measured, as noted previously. Harris et al. (2003) interpreted average angles of 87° to 95° as resulting from the pseudomorphing of former melt by feldspar; however, presumably some later solid-state adjustment had occurred, because, though these angles are smaller than the typical solid-state quartz-feldspar angles, they are much larger than the 60° angles expected for melt pseudomorphs. Similarly, Holness and Sawyer (2008) found that true plagioclase versus quartz-quartz dihedral angles (measured with a universal stage) averaged 68° (with a standard deviation of 17°) in a partly melted contact metamorphic rock, and averaged 112° (with a standard deviation of 16°) in an unmelted amphibolite facies regional metamorphic rock in which grain boundaries closely approach minimum-energy configurations. The 68° average represents an incomplete attempt to change small low-energy melt angles to much larger solid-solid angles.

Though all published examples of dihedral measurements for melt pseudomorphs in quartzofeldspathic rocks have been made for contact metamorphic rocks, Guernina and Sawyer (2003) and Hasalová et al. (2008a) observed inferred melt pseudomorphs with apparently small dihedral angles in regional metamorphic rocks.

6. Cuspate areas (concave outward) of quartz, K-feldspar, or sodic plagioclase (Fig. 3) commonly are inferred to represent former melt that has partly penetrated along grain boundaries, especially where surrounded by "embayed" grains inferred to have been partly melted (e.g., Harte et al., 1991; Sawyer, 1999, p. 274, 2001, p. 294; Holness and Clemens, 1999; Clemens and Holness, 2000; Marchildon and Brown, 2001, p. 227–228; Holness and Watt, 2002; Holness, 2008; Holness and Sawyer, 2008). For

Figure 3. Fingers of inferred former melt (now K-feldspar), enclosing rounded grains of plagioclase, extending along plagioclase-plagioclase and biotite-plagioclase boundaries, with finely tapered ends, suggesting consistently small dihedral angles, in a granulite-facies migmatite from the Ashuanipi Subprovince, Canada. The rounded plagioclase shapes could be due to corrosion during partial melting but could also represent low-energy solid-state inclusion shapes or even low-energy solid-liquid crystal shapes, according to Holness (2006). Also present are several examples of grains projecting into other grains, which could be taken to indicate partial melting, but the projections could equally well be normal solid-state partial inclusions. For example, the biotite grains projecting into quartz (Qtz), below the center, cannot be products of partial melting, as biotite is a reactant in the melt-producing reaction (Sawyer, 2001, p. 295). Crossed polars; base of photo, 1 mm. Photo by Ed Sawyer.

example, using CL-BSE imaging, Rosenberg and Riller (2000) observed interstitial pockets of K-feldspar and subordinate plagioclase at quartz triple junctions in statically recrystallized, partly melted granite, the feldspar shapes having "negative" (concave outward) curvatures. Similar patches with pointed extensions along quartz and quartz-feldspar grain boundaries (owing to small dihedral angles of the cuspate mineral against the two grains of the other minerals involved) were described in partly melted contact metamorphic rocks by Holness and Clemens (1999) and Clemens and Holness (2000). The dihedral angles at the points of the extensions can be typical melt versus solid-solid angles in contact metamorphic rocks, but they become larger in more slowly cooled rocks (e.g., Holness, 2008, fig. 4.10c) and tend to approach low-energy solid angles in regional metamorphic migmatites, as noted previously.

Cuspate melt patches tend to be pervasively distributed in grain boundaries and in small elongate patches mostly parallel to the foliation (Sawyer, 2001). Melt patches with similar shapes have been produced by elimination of grains during grain growth in analogue melting experiments (Walte et al., 2003). Feldspar typically forms the pseudomorphs in quartz-dominated volumes and vice versa (Harte et al., 1991; Rosenberg and Riller, 2000; Holness and Sawyer, 2008).

Holness and Sawyer (2008) observed that melt pockets bound by fewer than four original grains tend to be pseudomorphed by single grains, whereas larger melt pockets crystallize as polymineralic aggregates. From this observation, they inferred that melt pseudomorphs are formed preferentially in the smallest pores in the slowest cooled rocks, owing to an increasing barrier to crystal nucleation for smaller pore sizes.

However, cuspate and lobate shapes can also result from partial inclusions of one mineral in another, formed during growth of metamorphic minerals in the absence of melt. For example, the indented quartz and feldspar boundaries in the aplite of Figure 4 are not due to later melting but mainly to mutual interference of simultaneously growing primary grains, possibly aided by some later solid-state deformation. Similarly, the indented quartz and feldspar boundaries in the granofels of Figure 5 are due to simultaneous solid-state metamorphic growth, not melting. Therefore, care needs to be taken when using this criterion. It is most reliable where the boundaries of the cuspate areas taper into very small, true dihedral angles (e.g., Holness and Clemens, 1999, fig. 4; Sawyer, 2001), as shown in Figure 3.

7. Films or veinlets of inferred former melt (now pseudomorphs consisting of one of quartz, K-feldspar, or sodic plagioclase, less commonly with two or preferably three of these minerals) along grain boundaries or along inferred former intragranular fractures (Mehnert et al., 1973; Rosenberg and Riller, 2000; Sawyer, 1999, p. 274; 2001, p. 294, fig. 1; Rosenberg and Berger, 2001; Holness and Watt, 2002; Guernina and Sawyer, 2003, fig. 6; Hasalová et al., 2008a; Holness and Sawyer, fig. 8c), as shown in Figure 6, appear to be reliable indicators of former melt movement. Melt films spreading along the edges of quartz-feldspar grains and in cracks were produced in partial melting experiments by Mehnert et al. (1973), Büsch et al. (1974), and Laporte (1994). Plagioclase films along K-feldspar grain boundaries (Hasalová et al., 2008a, p. 33, fig. 5a) and seams of K-feldspar and sodic plagioclase (Rosenberg and Riller, 2000),

Figure 4. Aplite, Tioga Pass, Sierra Nevada, California, USA, showing strongly indented quartz-feldspar grain boundaries, resulting partly from simultaneous growth of quartz, K-feldspar, and sodic plagioclase from felsic magma at conditions of relatively fast cooling, possibly aided by later solid-state deformation, evidence of which is provided by microscopically visible microcline twinning (e.g., Eggleton and Buseck, 1980; Fitz Gerald and McLaren, 1982; Bell and Johnson, 1989; Vernon, 2004). The rock has not undergone any post-crystallization partial melting. Crossed polars; base of photo, 4 mm.

Figure 5. Aggregate of quartz, plagioclase, and biotite in an unmelted amphibolite facies granofels from the Broken Hill area, western New South Wales, Australia. The quartz-plagioclase boundaries vary from smoothly curved to broadly indented, with the result that quartz locally forms inclusions and partial inclusions in the plagioclase. Crossed polars; base of photo, 4 mm.

aligned approximately normal to the foliation, are consistent with space being made for melt by deformation.

8. Single-mineral feldspar pseudomorphs of biotite (Fig. 6) also appear to be indicators of former melt in appropriately high-grade rocks (Sawyer, 2001).

9. Veinlets of plagioclase with a more sodic composition than that of plagioclase in the surrounding rock (Rosenberg and Riller, 2000; Haslová et al., 2008a, p. 38) are consistent with crystallization from a melt, provided that crystallization of quartz and K-feldspar elsewhere, from the same melt, can be inferred from reliable evidence discussed in this review.

10. Oscillatory zoning typically, if not invariably, reflects crystallization in liquid (e.g., Vernon, 1976, p. 91; 2004, p. 262–268), and so plagioclase with oscillatory zoning in regional metamorphic leucosomes (e.g., Dougan, 1979, p. 903) is consistent with crystallization from a melt.

11. Microgranophyric intergrowths of quartz and alkali feldspar, in patches or veinlets between primary grains, are consistent with cotectic crystallization of felsic melt (e.g., Pattison and Harte, 1988, p. 481; Grant and Frost, 1990, p. 467; Harte et al., 1991; Holness and Clemens, 1999; Clemens and Holness, 2000; Holness and Watt, 2002; Holness, 2008; Holness and Sawyer, 2008). For example, advanced melting has been inferred to have produced "microgranitic" patches and veinlets containing 40 vol% quartz and 60 vol% feldspar (both K-feldspar and albite) in a partly melted feldspathic metaquartzite in the Ballachulish contact metamorphic aureole, Scotland (Holness and Clemens, 1999; Clemens and Holness, 2000).

However, granophyric intergrowths appear to be confined to shallow to mid-crustal rocks (Holness and Sawyer, 2008), presumably because grain growth in regional metamorphic terranes tends to convert micrographic aggregates with large grain-boundary areas to lower-energy polygonal aggregates (e.g., Vernon, 2004, fig. 4.65).

12. Symplectic replacement aggregates and coronas that can be explained by reactions between peritectic grains and cooling melt are also consistent with the former presence of melt (e.g., Ashworth, 1976, p. 677; Sawyer, 1999, p. 274; Jones and Brown, 1990, fig. 3f), as summarized by Holness (2008, p. 57). The reactions responsible for these microstructures are often referred to as "back-reactions," though probably they rarely follow the same pressure-temperature paths as their supposedly equivalent prograde melting reactions (Brown, 2002; Vernon and Clarke, 2008, p. 192–193), partly because of some melt loss prior to cooling (e.g., White, 2008, p. 90–92). Such reactions may obliterate former melt microstructures (Sawyer, 2001; Holness, 2008).

13. Patches and layers rich in peritectic minerals (melanosomes) in migmatites (Kenah and Hollister, 1983; Tracy and Robinson, 1983; Sawyer et al., 1999, p. 227; White et al., 2004; Vernon and Clarke, 2008, p. 177–178) represent sites of partial melting, from which leucosome melt has been extracted (Sawyer, 1999, p. 272).

Unreliable or Doubtful Criteria

The following microstructural criteria appear to be less reliable or even unreliable.

1. Ashworth (1976, p. 664–666), McLellan (1983, 1988), and Ashworth and McLellan (1985) suggested that *random* spatial distributions of minerals are characteristic of melt crystallization in leucosomes, in contrast to *dispersed* (regular) distributions formed by solid-state neocrystallization and grain growth, and *aggregate* distributions that can form during solid-state deformation and metamorphic differentiation. However, Hasalová et al. (2008a) observed a dispersed distribution in migmatites from the Bohemian Massif, which they inferred to be due to the nucleation of new minerals along boundaries between like mineral grains, though they acknowledged the possible effect of late solid-state grain growth. If their interpretation is correct, a random grain distribution should not be regarded as a reliable criterion of the existence of former melt. However, in view of the evidence of some solid-state deformation in these rocks (see next section), the use of random distributions as a criterion of former melt cannot be dismissed at present.

2. A grain-size increase has been suggested as being characteristic of incipient melting (e.g., Ashworth, 1976, p. 664; Dougan, 1979, p. 903; McLellan, 1983, 1988; Ashworth and McLellan, 1985; Grant and Frost, 1990, p. 467; Sawyer, 2008), which is supported by the common observation that leucosomes tend to be coarser grained than mesosomes. On the other hand, Hasalová et al. (2008a) inferred a grain-size decrease associated with inferred increasing amounts of melt, which implies that grain

Figure 6. Granulite-facies migmatite from the Ashuanipi Subprovince, Canada, consisting mainly of polygonal grains of quartz and feldspar, with dispersed grains of orthopyroxene (high relief) and biotite (elongate crystals). Veinlets and elongate patches of inferred former melt occur along some of the grain boundaries and in trans-granular fracture fillings. Partial melting of biotite (preserving the elongate grain shape) is also evident (center). Photo by Ed Sawyer. Crossed polars, with one-wave quartz plate; base of photo, 3 mm. From Vernon (2004, fig. 4.82) and Vernon and Clarke (2008, fig. 4.10).

size should not be regarded as a reliable criterion for distinguishing the presence of former melt. However, at least some of the grain-size reduction reported by Hasalová et al. (2008a) could be due to post-melting mylonitic deformation, evidence of which is provided by abundant unmelted myrmekite aggregates replacing K-feldspar (Hasalová et al., 2008a, figs. 5C, 5D). Although myrmekite has been reported in apparently undeformed granites, most occurrences are in rocks that show at least microstructural evidence of deformation, such as microscopically visible microcline twinning, feldspar fracturing, quartz recovery and recrystallization, grain-size reduction, mica recrystallization-neocrystallization, and solid-state foliation development (e.g., Hanmer, 1982; LaTour, 1987; Vernon et al., 1983; Simpson, 1985; Simpson and Wintsch, 1989; Vernon, 1991, 2004; Tsurumi et al., 2003; Ree et al., 2005; Menegon et al., 2006). Deformation assists the access of fluids that appear to be necessary for transport of chemical components, as myrmekite formation is a locally metasomatic process (e.g., Ashworth, 1972; Phillips, 1980; Vernon et al., 1983; Simpson, 1985; La Tour, 1987; Simpson and Wintsch, 1989; Yuguchi and Nishiyama, 2008).

3. "Interstitial" grains in some migmatitic gneisses have been regarded as having crystallized from melt (e.g., Pattison and Harte, 1988, p. 481; Hasalová et al., 2008a, figs. 4a, 4b). An extension of this criterion is the "oikocrystic" K-feldspar (less commonly plagioclase or quartz) criterion suggested by Grant and Frost (1990, p. 467). Crystallization of melt may appear to be a reasonable interpretation of these microstructures, because the terms *interstitial*, *poikilitic*, or *oikocrystic*, as used for igneous rocks, generally carry the implication that the included or interstitial minerals finish crystallizing before the other minerals. However, this is not necessarily so (e.g., Vernon, 2004, p. 102–108), especially in metamorphic rocks (Vernon et al., 2008), for which the appearance of a mineral as "interstitial" (that is, as small grains at triple junctions or along grain boundaries of larger grains) can be due to a variety of factors, such as nucleation rates, growth rates, and mineral proportions, which are independent of melting. Therefore, this criterion is not reliable on its own.

However, if the "interstitial" mineral contains inclusions with crystal faces, such as euhedral cordierite, orthopyroxene, plagioclase, or biotite, as described for contact metamorphic migmatites by Grant and Frost (1990, p. 445), crystallization in the presence of melt is much more likely, especially for minerals that are not normally euhedral in metamorphic rocks. This is effectively the same as "reliable" criterion 1.

4. Rounded, apparently corroded relics of inferred reactant mineral grains surrounded by areas of quartz, K-feldspar, or sodic plagioclase have been interpreted as resulting from partial melting (e.g., Grant and Frost, 1990, p. 467; Sawyer, 2001, p. 294). Good examples are irregularly shaped relics of quartz and feldspar in glass formed in the melting experiments of Mehnert et al. (1973, fig. 6) and elongate pseudomorphs of biotite illustrated by Sawyer (2001, fig. 1d), as shown in Figure 6. On the other hand, Holness et al. (2005), Holness (2006, 2008), and Holness and Sawyer (2008) contended that the equilibrium shapes of many minerals in melt tend to be rounded or to have some rounded boundary parts (though planar boundaries are generally formed in natural magmas, presumably for kinetic reasons), which could also support the partial melting interpretation, though for a different reason.

However, many enclosed mineral grains in unmelted metamorphic rocks have also been described as being "rounded and corroded." Moreover, rounded shapes are characteristic of inclusions of many minerals in unmelted high-grade metamorphic rocks, which have been inferred to result from solid-state grain-boundary adjustment (Kretz, 1966; Vernon, 1968). Therefore, care has to be taken to distinguish truly corroded shapes from the rounded shapes of normal mineral inclusions or partial inclusions that are compatible with the main minerals in the assemblage. The problem of distinguishing between corroded relics, true inclusions, and later replacing minerals (pseudo-inclusions) applies to metamorphic rocks generally and can be difficult to solve, as discussed by Vernon et al. (2008). Consequently, inclusion shapes on their own constitute an unreliable criterion for the former presence of melt.

5. Lobate projections or indentations of quartz, K-feldspar, or sodic plagioclase into another mineral have been used as indicators of replacement by partial melt (Hasalová et al., 2008a, figs. 4c, 4d, 5e; 2008b, figs. 3e, 3f), but identical microstructures are formed by lobate boundaries and partial inclusions in some unmelted metamorphic rocks (Fig. 5). For example, the biotite grains projecting into quartz in Figure 3 cannot be products of partial melting, as biotite is a reactant in the melt-producing reaction (Sawyer, 2001, p. 295); therefore, they are better interpreted as partial inclusions.

Moreover, quartz and feldspar grain boundaries are commonly deeply indented in aplites (Fig. 4), for which no suggestion of partial melting is entertained; as stated previously, the indentations are due to competition for space between simultaneously crystallizing minerals with boundaries that have almost equally high interfacial free energies in a situation in which insufficient time is available for average interfacial energies to be minimized. A similar situation applies to some unmelted metamorphic rocks in which minimization of interfacial free energies has not been attained. For example, the lobate grain boundaries in the rock shown in Figure 5 could not have formed by partial melting because the metamorphic grade was too low.

6. An especially doubtful variant of the lobate projection criterion is the projection of myrmekite lobes into K-feldspar, as proposed by Hasalová et al. (2008a, p. 38), who stated that this microstructure is "similar to microstructures described as typical of minerals reacting with melt." Vanderhaeghe (2001, p. 216) also suggested myrmekite as a criterion for melt in migmatites. On the contrary, a great deal of evidence indicates that myrmekite is formed by melt-absent, solid-state reactions, probably generally assisted by the presence of hydrous fluid (e.g., Becke, 1908; Hubbard, 1966; Phillips and Ransom, 1968; Phillips et al., 1972; Ashworth, 1972; Phillips, 1974, 1980; Haapala, 1977; Hanmer, 1982; Vernon et al., 1983; Nold, 1984; Simpson, 1985;

La Tour, 1987; Simpson and Wintsch, 1989; Stel and Breedveld, 1990; Vernon et al., 1990; Hopson and Ramseyer, 1990; Vernon, 1991, 2004; Tsurumi et al., 2003; Ree et al., 2005; Yuguchi and Nishiyama, 2008). The growth process is broadly similar to the formation of pearlite (a symplectic intergrowth of iron carbide and alpha iron) in steel (e.g., Chalmers, 1959) and to the subsolidus formation of many other symplectites in metamorphic rocks (e.g., Griffin, 1971; Vernon, 1978, 2004, p. 242–254; Vernon and Pooley, 1981; Mongkoltip and Ashworth, 1983; Droop and Bucher-Nurminen, 1984; Carswell et al., 1989; Harley et al., 1990; Raith et al., 1997; Pitra and de Waal, 2001; Daczko et al., 2002; Johnson et al., 2004; Harley, 2008).

Glass was produced along the boundaries between the quartz rods and sodic plagioclase in preexisting myrmekite in rocks partly melted in the experiments of Mehnert et al. (1973, p. 174, fig. 4d). Therefore, myrmekite lobes without mineral pseudomorphs of such glass films in migmatites indicate postmelting growth. This commonly occurs during solid-state, mylonitic deformation (e.g., Vernon et al., 1983; Simpson, 1985; La Tour, 1987; Simpson and Wintsch, 1989; Vernon, 1991, 2004; Tsurumi et al., 2003; Ree et al., 2005; Menegon et al., 2006), as noted previously.

7. Plagioclase rims with a constant sodic composition, occurring on plagioclase cores that are more calcic and/or of variable composition (McLellan, 1983, p. 257; Sawyer, 1998; Marchildon and Brown, 2001; Hasalová et al., 2008a, p. 38), are also consistent with crystallization from a pervasive melt. In general, plagioclase with normal zoning is consistent with crystallization from a melt (e.g., Ashworth, 1976, p. 675; Dougan, 1979, p. 903; McLellan, 1988, p. 525), whereas prograde metamorphic plagioclase can show reverse zoning or be unzoned. However, this criterion does not exclude different origins and so should not be used on its own.

CONCLUSIONS

The most reliable microstructural criterion of the former presence of felsic melt is a three-mineral (quartz, K-feldspar, and sodic plagioclase) aggregate in veinlets (Bouchez et al., 1992; Vernon et al., 2004). This is because in a felsic silicate melt, quartz and both feldspars crystallize together, at least at a late stage, to the solidus. In the few described examples, quartz and K-feldspar commonly occur together in the veinlets, and sodic plagioclase is in contact with both these minerals but precipitates on plagioclase grains in the vein walls (Bouchez et al., 1992; Vernon et al., 2004).

In the absence of a three-mineral aggregate, the most reliable criteria are (1) euhedral crystal faces of feldspar (precipitated from liquid) or peritectic minerals (e.g., garnet, cordierite, orthopyroxene, K-feldspar) against inferred former melt; (2) clear (inclusion-free) euhedral overgrowths of feldspar (precipitated from liquid) or peritectic minerals (e.g., garnet, cordierite, orthopyroxene, K-feldspar) on relics of the same minerals with abundant inclusions in the mesosome; (3) aligned, euhedral feldspar crystals; (4) simple twinning in K-feldspar; (5) dihedral angles of ≤ 60° subtended where a grain of feldspar or quartz (inferred to have pseudomorphed former melt) meets two grains of other minerals; (6) cuspate volumes of quartz, K-feldspar, or sodic plagioclase, especially where surrounded by grains inferred to have been partly melted; (7) veinlets of inferred former melt (now pseudomorphs consisting of one of quartz, K-feldspar, or sodic plagioclase, preferably, though less commonly, involving two or three of these minerals) along grain boundaries or along inferred former intragranular fractures; (8) biotite pseudomorphed by feldspar; (9) veinlets of plagioclase that is more sodic than plagioclase grains in the adjacent rock; (10) plagioclase with oscillatory zoning; (11) microgranophyric intergrowths of quartz and alkali feldspar in patches or veinlets between primary grains; (12) symplectic replacement aggregates that can be explained by reactions between peritectic grains and cooling melt; and (13) melanosome patches and layers from which leucosome has been extracted. However, all these criteria should be interpreted with care, and it would be preferable to have several of them fulfilled.

Unreliable or doubtful criteria include (1) random mineral distributions; (2) grain-size increase; (3) interstitial quartz and feldspar grains; (4) corroded relics of reactant mineral grains surrounded by areas of quartz, K-feldspar, or sodic plagioclase; (5) lobate grain projections; (6) lobes of myrmekite; and (7) plagioclase rims with a constant sodic composition occurring on plagioclase cores that are more calcic and/or of variable composition. Though some of these criteria can add support to more reliable criteria, they are too questionable to be used on their own.

REFERENCES CITED

Ashworth, J.A., 1972, Myrmekites of exsolution and replacement origins: Geological Magazine, v. 109, p. 45–62, doi:10.1017/S0016756800042266.

Ashworth, J.R., 1976, Petrogenesis of migmatites in the Huntly–Portsoy area, north-east Scotland: Mineralogical Magazine, v. 40, p. 661–682, doi:10.1180/minmag.1976.040.315.01.

Ashworth, J.R., and McLellan, E.L., 1985, Textures, in Ashworth, J.R., ed., Migmatites: Glasgow, Blackie, p. 80–203.

Becke, F., 1908, Über Myrmekit: Mineralogische und Petrographische Mitteilungen, v. 27, p. 377–390.

Bell, T.H., and Johnson, S.E., 1989, The role of deformation partitioning in the deformation and recrystallization of plagioclase and K-feldspar in the Woodroffe Thrust mylonite zone, central Australia: Journal of Metamorphic Geology, v. 7, p. 151–168, doi:10.1111/j.1525-1314.1989.tb00582.x.

Bouchez, J.-L., Delas, C., Gleizes, G., Nedelec, A., and Cuney, M., 1992, Submagmatic microfractures in granites: Geology, v. 20, p. 35–38, doi:10.1130/0091-7613(1992)020<0035:SMIG>2.3.CO;2.

Brown, M., 2001, Orogeny, migmatites and leucogranites: A review: Proceedings of the Indiana Academy of Sciences, v. 110, p. 313–336.

Brown, M., 2002, Retrograde processes in migmatites and granulites revisited: Journal of Metamorphic Geology, v. 20, p. 25–40, doi:10.1046/j.0263-4929.2001.00362.x.

Büsch, W., Schneider, G., and Mehnert, K.R., 1974, Initial melting at grain boundaries. Part II: Melting of rocks of granodioritic, quartzdioritic and tonalitic composition: Neues Jahrbuch für Mineralogie-Monatshefte, v. 1974, p. 345–370.

Carswell, D.A., Möller, C., and O'Brien, P.J., 1989, Origin of sapphirine-plagioclase symplectites in metabasites from Mitterbachgraben, Dunkelsteinerwald granulite complex, Lower Austria: European Journal of Mineralogy, v. 1, p. 455–466.

Chalmers, B., 1959, Physical Metallurgy: New York, Wiley & Sons, 468 p.

Clemens, J.D., and Holness, M.B., 2000, Textural evolution and partial melting of arkose in a contact aureole: A case study and implications: Visual Geosciences, v. 5, p. 1–14, doi:10.1007/s10069-000-0004-1.

Collins, W.J., Flood, R.H., Vernon, R.H., and Shaw, S.E., 1989, The Wuluma granite, Arunta block, central Australia: An example of in situ, near-isochemical granite formation in a granulite-facies terrane: Lithos, v. 23, p. 63–83, doi:10.1016/0024-4937(89)90023-6.

Daczko, N.R., Clarke, G.L., and Klepeis, K.A., 2002, Kyanite-paragonite–bearing assemblages, northern Fiordland, New Zealand: Rapid cooling of the lower crustal root to a Cretaceous magmatic arc: Journal of Metamorphic Geology, v. 20, p. 887–902, doi:10.1046/j.1525-1314.2002.00421.x.

Dallain, C., Schulmann, K., and Ledru, P., 1999, Textural evolution in the transition from subsolidus annealing to melting process, Velay Dome, French Massif Central: Journal of Metamorphic Geology, v. 17, p. 61–74, doi:10.1046/j.1525-1314.1999.00176.x.

Dougan, T.W., 1979, Compositional and modal relationships and melting reactions in some migmatitic pelites from New Hampshire and Maine: American Journal of Science, v. 279, p. 897–935.

Dougan, T.W., 1983, Textural relations in melanosomes of selected specimens of migmatitic pelitic schists: Implications for leucosome-generating process: Contributions to Mineralogy and Petrology, v. 83, p. 82–98, doi:10.1007/BF00373082.

Droop, G.T.R., and Bucher-Nurminen, K., 1984, Reaction textures and metamorphic evolution of sapphirine-bearing granulites from the Gruf Complex, Italian Central Alps: Journal of Petrology, v. 25, p. 766–803.

Droop, G.T.R., Clemens, J.D., and Dalrymple, D.J., 2003, Processes and conditions during contact anatexis, melt escape and restite formation: The Huntly Gabbro Complex, NE Scotland: Journal of Petrology, v. 44, p. 995–1029, doi:10.1093/petrology/44.6.995.

Eggleton, R.A., and Buseck, P.R., 1980, The orthoclase-microcline inversion: A high-resolution TEM study and strain analysis: Contributions to Mineralogy and Petrology, v. 74, p. 123–133, doi:10.1007/BF01131998.

Elliott, M.T., Cheadle, M.J., and Jerram, D.A., 1997, On the identification of textural disequilibrium in rocks using dihedral angle measurements: Geology, v. 25, p. 355–358, doi:10.1130/0091-7613(1997)025<0355:OTIOTE>2.3.CO;2.

Fitz Gerald, J.G., and McLaren, A.C., 1982, The microstructures of microcline from some granitic rocks and pegmatites: Contributions to Mineralogy and Petrology, v. 80, p. 219–229, doi:10.1007/BF00371351.

Grant, J.A., and Frost, B.R., 1990, Contact metamorphism and partial melting of pelitic rocks in the aureole of the Laramie anorthosite complex, Morton Pass, Wyoming: American Journal of Science, v. 290, p. 425–472.

Griffin, W.L., 1971, Genesis of coronas in anorthosites of the upper Jotun Nappe, Indre Sogn, Norway: Journal of Petrology, v. 12, p. 219–243.

Guernina, S., and Sawyer, E.W., 2003, Large-scale melt-depletion in granulite terranes: An example from the Archean Ashuanipi Subprovince of Quebec: Journal of Metamorphic Geology, v. 21, p. 181–201, doi:10.1046/j.1525-1314.2003.00436.x.

Haapala, I., 1977, Petrography and Geochemistry of the Eurajoki Stock, a Rapakivi-Granite Complex with Greisen-Type Mineralization in Southwestern Finland: Bulletin of the Geological Survey of Finland, no. 286, 128 p.

Hanmer, S., 1982, Microstructure and geochemistry of plagioclase and microcline in naturally deformed granite: Journal of Structural Geology, v. 4, p. 197–214, doi:10.1016/0191-8141(82)90027-X.

Harley, S.L., 2008, Refining the P-T records of UHT crustal metamorphism: Journal of Metamorphic Geology, v. 26, p. 125–154, doi:10.1111/j.1525-1314.2008.00765.x.

Harley, S.L., Hensen, B.J., and Sheraton, J.W., 1990, Two-stage decompression in orthopyroxene-sillimanite granulites from Forefinger Point, Enderby Land, Antarctica: Implications for the evolution of the Archaean Napier Complex: Journal of Metamorphic Geology, v. 8, p. 591–613, doi:10.1111/j.1525-1314.1990.tb00490.x.

Harris, N., McMillan, A., Holness, M., Uken, R., Watkeys, M., Rogers, N., and Fallick, A., 2003, Melt generation and fluid flow in the thermal aureole of the Bushveld Complex: Journal of Petrology, v. 44, p. 1031–1054, doi:10.1093/petrology/44.6.1031.

Harte, B., Pattison, D.R.M., and Linklater, C.M., 1991, Field relations and petrography of partially melted pelitic and semi-pelitic rocks, in Voll, G., Töpel, J., Pattison, D.R.M., and Seifert, F., eds., Equilibrium and Kinetics in Contact Metamorphism: The Ballachulish Igneous Complex and Its Aureole: Berlin, Springer-Verlag, p. 181–219.

Hasalová, P., Schulmann, K., Lexa, O., Stípská, P., Hrouda, F., Ulrich, S., Haloda, J., and Tycová, P., 2008a, Origin of migmatites by deformation-enhanced melt infiltration: A new model based on quantitative microstructural analysis: Journal of Metamorphic Geology, v. 26, p. 29–53.

Hasalová, P., Stípská, P., Powell, R., Schulmann, K., Janousek, V., and Lexa, O., 2008b, Transforming mylonitic metagranite by open-system interactions during melt flow: Journal of Metamorphic Geology, v. 26, p. 55–80.

Holness, M.B., 2006, Melt–solid dihedral angles of common minerals in natural rocks: Journal of Petrology, v. 47, p. 791–800, doi:10.1093/petrology/egi094.

Holness, M.B., 2008, Decoding migmatite microstructures, in Sawyer, E.W., and Brown, M., eds., Working with Migmatites: Mineralogical Association of Canada Short Course Series, v. 38, p. 57–76.

Holness, M.B., and Clemens, J.D., 1999, Partial melting of the Appin Quartzite driven by fracture-controlled H_2O infiltration in the aureole of the Ballachulish Igneous Complex, Scottish Highlands: Contributions to Mineralogy and Petrology, v. 136, p. 154–168, doi:10.1007/s004100050529.

Holness, M.B., and Sawyer, E.W., 2008, On the pseudomorphing of melt-filled pores during the crystallization of migmatites: Journal of Petrology, v. 49, p. 1343–1363, doi:10.1093/petrology/egn028.

Holness, M.B., and Watt, G.R., 2002, The aureole of the Triagh Bhan na Sgurra sill, Isle of Mull: Reaction-driven micro-cracking during pyrometamorphism: Journal of Petrology, v. 43, p. 511–534, doi:10.1093/petrology/43.3.511.

Holness, M.B., Cheadle, M.J., and McKenzie, D., 2005, On the use of changes in dihedral angle to decode late-stage textural evolution in cumulates: Journal of Petrology, v. 46, p. 1565–1583, doi:10.1093/petrology/egi026.

Hopson, R.F., and Ramseyer, K., 1990, Cathodoluminesence microscopy of myrmekite: Geology, v. 18, p. 336–339, doi:10.1130/0091-7613(1990)018<0336:CMOM>2.3.CO;2.

Hubbard, F.H., 1966, Myrmekite in charnockite from south-west Nigeria: American Mineralogist, v. 51, p. 762–773.

Hunter, R.H., 1987, Textural equilibrium in layered igneous rocks, in Parsons, I., ed., Origins of Igneous Layering: Dordrecht, Netherlands, Reidel, p. 473–503.

Johnson, T., Brown, M., Gibson, R., and Wing, B., 2004, Spinel-cordierite symplectites replacing andalusite: Evidence for melt-assisted diapirism in the Bushveld Complex, South Africa: Journal of Metamorphic Geology, v. 22, p. 529–545, doi:10.1111/j.1525-1314.2004.00531.x.

Jones, K.A., and Brown, M., 1990, High-temperature 'clockwise' P–T paths and melting in the development of regional migmatites: An example from southern Brittany, France: Journal of Metamorphic Geology, v. 8, p. 551–578, doi:10.1111/j.1525-1314.1990.tb00486.x.

Jurewicz, S.R., and Watson, E.B., 1984, Distribution of partial melt in a felsic system: the importance of surface energy: Contributions to Mineralogy and Petrology, v. 85, p. 25–29.

Jurewicz, S.R., and Watson, E.B., 1985, The distribution of partial melt in a granite system: The application of liquid phase sintering theory: Geochimica et Cosmochimica Acta, v. 49, p. 1109–1121.

Kenah, P., and Hollister, L.S., 1983, Anatexis in the Central Gneiss Complex, British Columbia, in Atherton, M.P., and Gribble, C.D., eds., Migmatites, Melting and Metamorphism: Nantwich, Cheshire, UK, Shiva Publishing, p. 142–162.

Kretz, R., 1966, Interpretation of the shape of mineral grains in metamorphic rocks: Journal of Petrology, v. 7, p. 68–94.

Laporte, D., 1994, Wetting behaviour of partial melts during crustal anatexis: The distribution of hydrous silicic melts in polycrystalline aggregates of quartz: Contributions to Mineralogy and Petrology, v. 116, p. 486–499, doi:10.1007/BF00310914.

Laporte, D., and Watson, E.B., 1995, Experimental and theoretical constraints on melt distribution in crustal sources: The effect of crystalline anisotropy on melt interconnectivity: Chemical Geology, v. 124, p. 161–184, doi:10.1016/0009-2541(95)00052-N.

Laporte, D., Rapaille, C., and Provost, A., 1997, Wetting angles, equilibrium melt geometry, and the permeability threshold of partially molten crustal protoliths, in Bouchez, J.-L., Stephens, W.E., and Hutton, D.E., eds., Granite: From Melt Segregation to Emplacement Fabrics: Dordrecht, Netherlands, Kluwer, p. 31–54.

La Tour, T.E., 1987, Geochemical model for the symplectic formation of myrmekite during amphibolite-grade progressive mylonitization of granite: Geological Society of America Abstracts with Programs, v. 19, p. 741.

Marchildon, N., and Brown, M., 2001, Melt segregation in late syn-tectonic anatectic migmatites: An example from the Onawa contact aureole,

Maine, USA: Physics and Chemistry of the Earth, v. 26, p. 225–229, doi:10.1016/S1464-1895(01)00049-7.

Marchildon, N., and Brown, M., 2003, Spatial distribution of melt-bearing structures in anatectic rocks from Southern Brittany, France: Implications for melt transfer at grain- to orogen-scale: Tectonophysics, v. 364, p. 215–235, doi:10.1016/S0040-1951(03)00061-1.

McLellan, E.L., 1983, Contrasting textures in metamorphic and anatectic migmatites—An example from the Scottish Caledonides: Journal of Metamorphic Geology, v. 1, p. 241–262, doi:10.1111/j.1525-1314.1983.tb00274.x.

McLellan, E.L., 1988, Migmatite structures in the Central Gneiss Complex, Boca de Quadra, Alaska: Journal of Metamorphic Geology, v. 6, p. 517–542, doi:10.1111/j.1525-1314.1988.tb00437.x.

Means, W.D., 1989, Synkinematic microscopy of transparent polycrystals: Journal of Structural Geology, v. 11, p. 163–174, doi:10.1016/0191-8141(89)90041-2.

Mehnert, K.R., Büsch, W., and Schneider, G., 1973, Initial melting at grain boundaries of quartz and feldspar in gneisses and granulites: Neues Jahrbuch für Mineralogie-Monatshefte, v. 1973, p. 165–182.

Menegon, L., Pennachionni, G., and Stünitz, H., 2006, Nucleation and growth of myrmekite during ductile shear deformation in metagranites: Journal of Metamorphic Geology, v. 24, p. 553–568.

Mongkoltip, P., and Ashworth, J.R., 1983, Quantitative estimation of an open-system symplectite-forming reaction: Restricted diffusion of Al and Si in coronas around olivine: Journal of Petrology, v. 24, p. 635–661.

Nold, J.L., 1984, Myrmekite in Belt Supergroup metasedimentary rocks—Northeast border zone of the Idaho batholith: American Mineralogist, v. 69, p. 1050–1052.

Paterson, S.R., Vernon, R.H., and Tobisch, O.T., 1989, A review of criteria for the identification of magmatic and tectonic foliations in granitoids: Journal of Structural Geology, v. 11, p. 349–363, doi:10.1016/0191-8141(89)90074-6.

Pattison, D.R.M., and Harte, B., 1988, Evolution of structurally contrasting anatectic migmatites in the 3-kbar Ballachulish aureole, Scotland: Journal of Metamorphic Geology, v. 6, p. 475–494, doi:10.1111/j.1525-1314.1988.tb00435.x.

Phillips, E.R., 1974, Myrmekite—One hundred years later: Lithos, v. 7, p. 181–194, doi:10.1016/0024-4937(74)90029-2.

Phillips, E.R., 1980, On polygenetic myrmekite: Geological Magazine, v. 117, p. 29–36, doi:10.1017/S0016756800033070.

Phillips, E.R., and Ransom, D.M., 1968, The proportionality of quartz in myrmekite: American Mineralogist, v. 53, p. 1411–1413.

Phillips, E.R., Ransom, D.M., and Vernon, R.H., 1972, Myrmekite and muscovite developed by retrograde metamorphism at Broken Hill, New South Wales: Mineralogical Magazine, v. 38, p. 570–578, doi:10.1180/minmag.1972.038.297.05.

Pitra, P., and de Waal, S.A., 2001, High-temperature, low-pressure metamorphism and development of prograde symplectites, Marble Hall Fragment, Bushveld Complex (South Africa): Journal of Metamorphic Geology, v. 19, p. 311–325.

Platten, I.M., 1982, Partial melting of feldspathic quartzite around late Caledonian minor intrusions in Appin, Scotland: Geological Magazine, v. 119, p. 413–419, doi:10.1017/S0016756800026327.

Platten, I.M., 1983, Partial melting of semipelite and the development of marginal breccias around late Caledonian minor intrusions in the Grampian Highlands of Scotland: Geological Magazine, v. 120, p. 37–49, doi:10.1017/S0016756800025012.

Raith, M., Karmakar, S., and Brown, M., 1997, Ultra-high-temperature metamorphism and multistage decompressional evolution of sapphirine granulites from the Palni Hill Ranges, southern India: Journal of Metamorphic Geology, v. 15, p. 379–399, doi:10.1111/j.1525-1314.1997.00027.x.

Ree, J.-H., Kim, H.S., Han, R., and Jung, H., 2005, Grain-size reduction of feldspars by fracturing and neocrystallization in a low-grade granitic mylonite and its rheological effect: Tectonophysics, v. 407, p. 227–237, doi:10.1016/j.tecto.2005.07.010.

Riegger, O.K., and van Vlack, L.H., 1960, Dihedral angle measurements: Metallurgical Society Transactions, AIME, v. 21, p. 933–935.

Rosenberg, C.L., 2001, Deformation of partially molten granite: A review and comparison of experimental and natural case studies: International Journal of Earth Sciences, v. 90, p. 60–76, doi:10.1007/s005310000164.

Rosenberg, C.L., and Berger, A., 2001, Syntectonic melt pathways in granitic gneisses, and melt-induced transitions in deformation mechanisms: Physics and Chemistry of the Earth, v. 26, p. 287–293, doi:10.1016/S1464-1895(01)00058-8.

Rosenberg, C.L., and Riller, U., 2000, Partial melt topology in statically and dynamically recrystallized granite: Geology, v. 28, p. 7–10, doi:10.1130/0091-7613(2000)28<7:PTISAD>2.0.CO;2.

Sawyer, E.W., 1998, Formation and evolution of granite magmas during crustal reworking: The significance of diatexites: Journal of Petrology, v. 39, p. 1147–1167.

Sawyer, E.W., 1999, Criteria for the recognition of partial melting: Physics and Chemistry of the Earth, v. 24, p. 269–279, doi:10.1016/S1464-1895(99)00029-0.

Sawyer, E.W., 2000, Grain-scale and outcrop-scale distribution and movement of melt in a crystallising granite: Royal Society of Edinburgh Transactions: Earth Sciences, v. 91, p. 73–85.

Sawyer, E.W., 2001, Melt segregation in the continental crust: Distribution and movement of melt in anatectic rocks: Journal of Metamorphic Geology, v. 19, p. 291–309, doi:10.1046/j.0263-4929.2000.00312.x.

Sawyer, E.W., 2008, Working with migmatites; nomenclature for the constituent parts, in Sawyer, E.W., and Brown, M., eds., Working with Migmatites: Mineralogical Association of Canada Short Course Series, v. 38, p. 1–28.

Sawyer, E.W., Dombrowski, C., and Collins, W.J., 1999, Movement of melt during synchronous regional deformation and granulite facies anatexis, an example from the Wuluma Hills, central Australia, in Castro, A., Fernandez, C., and Vigneresse, J.-L., eds., Understanding Granites: Integrating New and Classical Techniques: Geological Society [London] Special Publication 158, p. 221–237.

Simpson, C., 1985, Deformation of granitic rocks across the brittle-ductile transition: Journal of Structural Geology, v. 7, p. 503–511, doi:10.1016/0191-8141(85)90023-9.

Simpson, C., and Wintsch, R.P., 1989, Evidence for deformation-induced K-feldspar replacement by myrmekite: Journal of Metamorphic Geology, v. 7, p. 261–275, doi:10.1111/j.1525-1314.1989.tb00588.x.

Smith, C.S., 1953, Microstructure: Transactions of the American Society for Metals, v. 45, p. 533–575.

Stel, H., and Breedveld, M., 1990, Crystallographic orientation patterns of myrmekitic quartz: A fabric memory in annealed ribbon gneisses: Journal of Structural Geology, v. 12, p. 19–28, doi:10.1016/0191-8141(90)90045-Z.

Tracy, R.J., and Robinson, P., 1983, Acadian migmatite types in pelitic rocks of central Massachusetts, in Atherton, M.P., and Gribble, C.D., eds., Migmatites, Melting and Metamorphism: Nantwich, Cheshire, UK, Shiva Publishing, p. 163–173.

Tsurumi, J., Hosonuma, H., and Kanagawa, K., 2003, Strain localization due to a positive feedback of deformation and myrmekite-forming reaction in granite and aplite mylonites along the Hatagawa Shear Zone of NE Japan: Journal of Structural Geology, v. 25, p. 557–574, doi:10.1016/S0191-8141(02)00048-2.

Vanderhaeghe, O., 2001, Melt segregation, pervasive melt migration and magma mobility in the continental crust: The structural record from pores to orogens: Physics and Chemistry of the Earth, v. 26, p. 213–223, doi:10.1016/S1464-1895(01)00048-5.

Vernon, R.H., 1968, Microstructures of high-grade metamorphic rocks at Broken Hill, Australia: Journal of Petrology, v. 9, p. 1–22.

Vernon, R.H., 1970, Comparative grain-boundary studies in some basic and ultrabasic granulites, nodules and cumulates: Scottish Journal of Geology, v. 6, p. 337–351.

Vernon, R.H., 1976, Metamorphic Processes: London, Allen and Unwin, 247 p.

Vernon, R.H., 1978, Pseudomorphous replacement of cordierite by symplectic intergrowths of andalusite, biotite and quartz: Lithos, v. 11, p. 283–289, doi:10.1016/0024-4937(78)90035-X.

Vernon, R.H., 1986, K-feldspar megacrysts in granites—Phenocrysts, not porphyroblasts: Earth-Science Reviews, v. 23, p. 1–63.

Vernon, R.H., 1991, Questions about myrmekite in deformed rocks: Journal of Structural Geology, v. 13, p. 979–985, doi:10.1016/0191-8141(91)90050-S.

Vernon, R.H., 1997, On the identification of textural disequilibrium in rocks using dihedral angle measurements: Geology, v. 25, p. 1055, doi:10.1130/0091-7613(1997)025<1055:OTIOTD>2.3.CO;2.

Vernon, R.H., 1999, Quartz and feldspar microstructures in metamorphic rocks: Canadian Mineralogist, v. 37, p. 513–524.

Vernon, R.H., 2000, Review of microstructural evidence of magmatic and solid-state flow: Electronic Geosciences, v. 5, p. 2.

Vernon, R.H., 2004, A Practical Guide to Rock Microstructure: Cambridge, UK, Cambridge University Press, 594 p.

Vernon, R.H., 2007, Problems in identifying restite in S-type granites of southeastern Australia, with speculations on sources of magma and enclaves: Canadian Mineralogist, v. 45, p. 147–178, doi:10.2113/gscanmin.45.1.147.

Vernon, R.H., and Clarke, G.L., 2008, Principles of Metamorphic Petrology: Cambridge, UK, Cambridge University Press, 446 p.

Vernon, R.H., and Collins, W.J., 1988, Igneous microstructures in migmatites: Geology, v. 16, p. 1126–1129, doi:10.1130/0091-7613(1988)016<1126:IMIM>2.3.CO;2.

Vernon, R.H., and Johnson, S.E., 2000, Transition from gneiss to migmatite and the relationship of leucosome to peraluminous granite in the Cooma Complex, SE Australia, in Jessell, M.W., and Urai, J.L., eds., Stress, Strain and Structure: A Volume in Honour of W.D. Means: Journal of the Virtual Explorer, v. 2, p. 26 (CD with printed abstract).

Vernon, R.H., and Pooley, G.D., 1981, SEM/microprobe study of some symplectic intergrowths replacing cordierite: Lithos, v. 14, p. 75–82, doi:10.1016/0024-4937(81)90038-4.

Vernon, R.H., Williams, V.A., and D'Arcy, W.F., 1983, Grainsize reduction and foliation development in a deformed granitoid batholith: Tectonophysics, v. 92, p. 123–145, doi:10.1016/0040-1951(83)90087-2.

Vernon, R.H., Clarke, G.L., and Collins, W.J., 1990, Local mid-crustal granulite facies metamorphism and melting: An example in the Mount Stafford area, central Australia, in Ashworth, J.R., and Brown, M., eds., High-Temperature Metamorphism and Crustal Anatexis: London, Unwin Hyman, p. 272–319.

Vernon, R.H., Collins, W.J., and Richards, S.W., 2003, Contrasting magmas in metapelitic and metapsammitic migmatites in the Cooma Complex, Australia: Visual Geosciences, v. 8, p. 45–54, doi:10.1007/s10069-003-0010-1.

Vernon, R.H., Johnson, S.E., and Melis, E.A., 2004, Emplacement-related microstructures in the margin of a deformed tonalite pluton: The San José pluton, Baja California, México: Journal of Structural Geology, v. 26, p. 1867–1884, doi:10.1016/j.jsg.2004.02.007.

Vernon, R.H., White, R.W., and Clarke, G.L., 2008, False metamorphic events inferred from misinterpretation of microstructural evidence and *P-T* data: Journal of Metamorphic Geology, v. 26, p. 437–449, doi:10.1111/j.1525-1314.2008.00762.x.

Walte, N.P., Bons, P.D., Passchier, C.W., and Koehn, D., 2003, Disequilibrium melt distribution during static recrystallization: Geology, v. 31, p. 1009–1012, doi:10.1130/G19815.1.

Walte, N.P., Bons, P.D., and Passchier, C.W., 2005, Deformation of melt-bearing systems—Insight from *in situ* grain-scale analogue experiments: Journal of Structural Geology, v. 27, p. 1666–1679, doi:10.1016/j.jsg.2005.05.006.

White, R.W., 2008, Insights gained from the petrological modelling of migmatites: Particular reference to mineral assemblages and common replacement textures: Mineralogical Association of Canada Short Course Notes, no. 38, p. 77–93.

White, R.W., Powell, R., and Halpin, J.A., 2004, Spatially-focussed melt formation in aluminous metapelites from Broken Hill, Australia: Journal of Metamorphic Geology, v. 22, p. 825–845, doi:10.1111/j.1525-1314.2004.00553.x.

Yuguchi, T., and Nishiyama, T., 2008, The mechanism of myrmekite formation deduced from steady-diffusion modelling based on petrography: Case study of the Okueyama granitic body, Kyushu, Japan: Lithos, v. 106, p. 237–260, doi:10.1016/j.lithos.2008.07.017.

MANUSCRIPT ACCEPTED BY THE SOCIETY 24 MAY 2010

Petrological and experimental application of REE- and actinide-bearing accessory minerals to the study of Precambrian high-grade gneiss terranes

Daniel Harlov
Deutsches GeoForschungsZentrum, Telegrafenberg, D-14473 Potsdam FR, Germany

ABSTRACT

The current role of rare-earth-element- (REE-) and actinide-bearing accessory minerals as both geochronological markers and recorders of geochemical processes is reviewed in this paper. The minerals covered include the most common REE- and actinide-bearing accessory minerals found in high-grade rocks—i.e., monazite, xenotime, apatite, huttonite, thorite, zircon, allanite, and titanite. The goal of this review is to describe the most recent research developments regarding these minerals and their interaction with each other as well as their role as geochronometers. These results are then applied to two cross sections of lower crust from Rogaland–Vest Agder, SW Norway, and Tamil Nadu, S India (granulite and amphibolite facies) with regard to seeing how REE-accessory minerals relate to each other as a function of metamorphic grade. In either traverse the same relationships are seen between monazite, fluorapatite, allanite, and titanite. Namely, in the amphibolite-facies zone, the REE are hosted by titanite and allanite, whereas in the clinopyroxene-in transition zone between the amphibolite-facies and granulite-facies zones, monazite is the stable REE-bearing phase in the region of the orthopyroxene-in isograd. In the granulite-facies zone the REE are hosted by fluorapatite as opposed to monazite. These similarities suggest that mineral hosting of REE follows certain general trends, which are a function of metamorphic grade, whole rock chemistry, and intergranular fluid chemistry.

INTRODUCTION

Though REE- and actinide-bearing accessory minerals generally make up <5% of the minerals in high-grade metamorphic rocks, they can give significant information about a rock during its initial formation and subsequent geological evolution. This information takes the form of both geochronology (e.g., crystallization age and/or the age of various metamorphic events) and geochemistry (P-T-X constraints, fluid monitors, etc.).

The extensively studied common actinide- and REE-bearing accessory orthophosphate minerals, such as fluorapatite, monazite, and xenotime, can be utilized to date multiple metamorphic fluid-rock interaction events (e.g., Williams et al., 2007) through the redistribution, addition, or substitution of Th, U, Si, and Ca via coupled dissolution-reprecipitation processes (Harlov and Hetherington, 2010), allow for temperature estimation via monazite-xenotime thermometry (Gratz and Heinrich, 1997, 1998; Andrehs and Heinrich, 1998; Heinrich et al., 1997), monazite-garnet thermometry (Pyle et al., 2001), and xenotime-garnet thermometry (Pyle and Spear, 2000), as well as give information on the chemistry and nature of the metasomatizing fluids (e.g., Harlov et al., 2005; Harlov and Hetherington, 2010). Rapid transport of Th and Si in monazite or xenotime would not be possible by simple solid-state diffusion, because it has been shown that Th, U, and Pb diffusion rates in monazite (Cherniak et al., 2004; Cherniak and Pyle, 2008) and xenotime (Cherniak, 2006) are essentially negligible over geological time scales at temperatures beyond 800–1000 °C, even during hydrothermal experiments at low pressures (<500 MPa). This also holds true for equally negligible high-temperature solid-state diffusion of REE in fluorapatite (Cherniak, 2000, 2005) where redistribution of Th, REE, Si, and Na is again best accomplished through fluid-aided coupled dissolution-reprecipitation (Harlov et al., 2005).

Accessory actinide- and (Y + REE)-bearing silicate minerals such as zirconolite (e.g., Gieré et al., 1998), titanite (e.g., Harlov et al., 2006a), allanite (e.g., Finger et al., 1998; Wing et al., 2003), and zircon (Hoskin and Schaltegger, 2003) allow for constraints to be placed on various actinide and (Y + REE) activities as well as oxygen and H_2O fugacities during episodes of fluid-rock interaction. They also give information on the chemical nature of the fluids themselves as well as allow for fluid-rock interaction events to be dated. Understanding the chemical and physical relationship between these minerals as a function of metamorphic grade can add further insights into REE and actinide transport as a function of fluid-melt composition. With this in mind, the goal of this paper is to make a general review regarding our current understanding of REE- and actinide-bearing accessory mineral equilibria in high-grade rocks and their subsequent phase relationships with each other.

Fluorapatite-Monazite-Xenotime

The occurrence of monazite [(Ce, LREE) (PO_4)] and/or xenotime [(Y, HREE) (PO_4)] as inclusions in fluorapatite [$Ca_5(PO_4)_3$(F,Cl,OH)] has been widely documented in high-grade metamorphic rocks. The inclusions typically occur in granulite-facies rocks as opposed to amphibolite-facies rocks where such inclusions are distinctly absent (Pan et al., 1993; Harlov and Förster, 2002b, 2003; Harlov et al., 2007a; Hansen and Harlov, 2007) (Fig. 1). Monazite and xenotime inclusions can also form in fluorapatite from igneous rocks directly after crystallization during cooling (Harlov et al., 2002a) as well as in chlorapatite from gabbros and layered inclusions (Harlov et al., 2002b). Inclusion formation has been directly linked to metasomatism by either external or internal H_2O-bearing fluids, though there is at least one exception to this general observation. Here monazite, zircon, and minor xenotime were overgrown by fluorapatite to form inclusions (cf. Harlov et al., 2008). This general hypothesis is supported by experiments, which indicate that formation and growth of monazite and xenotime inclusions in apatite are the result of coupled dissolution-reprecipitation processes during metasomatic alteration (Harlov et al., 2002b; Harlov and Förster, 2003; Harlov et al., 2005).

During metasomatic alteration, Si and Na are removed from the apatite without the concurrent removal of (Y + REE). This results in a charge imbalance, because (Y + REE) are stabilized in the apatite structure via the coupled substitutions Si^{4+} + (Y + REE)$^{3+}$ = P^{5+} + Ca^{2+} and Na^+ + (Y + REE)$^{3+}$ = 2 Ca^{2+} (Pan and Fleet, 2002). Removal of Na and Si, but not (Y + REE), allows for the fluid-aided nucleation and growth of monazite and/or xenotime either as inclusions in the apatite or as rim grains on the surface of the apatite. Formation of rim grains is apparently due to the diffusion of a part of the (Y + REE) toward the apatite grain surface, where it collects in sufficient concentrations so as to promote the nucleation and growth of monazite or xenotime. As a consequence, the composition of the monazite and/or xenotime normally reflects the (Y + REE + Th + U) abundances in the host apatite from which it is derived. This is seen in their generally low to negligible Th abundances (e.g., Harlov and Förster, 2002b; Harlov et al., 2002a; Harlov et al., 2005; Hansen and Harlov, 2007), although there can be exceptions to this observation if the fluorapatite contains unusually high concentrations of Th, such as in the case of the Durango fluorapatite (cf. Harlov and Förster, 2003). Whereas commonly semi-euhedral to irregular in shape, a substantial minority of monazite and xenotime inclusions can also grow elongated with respect to the c-axis of the apatite (Figs. 1A, 1B). Semi-euhedral to irregular shaped monazite inclusions have been demonstrated to grow in fluid-filled voids as small clumps of euhedral crystals (Fig. 2B; see discussion in Harlov et al., 2005).

Fluids found to induce formation of monazite and/or xenotime inclusions in apatite or as grains along apatite rims include H_2O, H_2O/CO_2 fluids, and KCl + H_2O (Harlov et al., 2002b; Harlov and Förster, 2003) as well as 1N H_2SO_4, and 1N HCl (Harlov et al., 2005). Other fluids, such as $CaCl_2$ + H_2O or NaCl + H_2O, inhibit the growth of these inclusions and rim grains owing to the ability of Ca and Na to enter the apatite structure and charge balance or replace the (Y + REE). Monazite and xenotime inclusions and rim grains can also form in fluorapatite from metapelites, which have undergone partial melting during granulite-facies metamorphism (Harlov et al., 2007a). In such cases, the fluorapatite can also undergo partial dissolution in the H_2O-alkali-SiO_2-rich melt.

Regions of the apatite affected by coupled dissolution-reprecipitation are characterized by a pervasive, interconnected micro- and nano-porosity, which allows for fluids to infiltrate (Harlov et al., 2005) (Fig. 2). The presence of an interconnected fluid medium greatly speeds up mass transfer allowing for the rapid (hours–days) growth of monazite and/or xenotime inclusions by utilizing the available P, Y, and REE, while the micro- and nano-pores provide random nucleation sites (Fig. 2B). Monazite and xenotime inclusions, as well as rim grains, can form over a wide pressure-temperature (P-T) range, i.e., 100–1000 MPa and 100–900 °C (see review in Harlov et al., 2002b; Harlov et al., 2005). Consequently, whether or not nucleation will occur is highly dependent on the level of reactivity between the fluid and the apatite (cf. Harlov et al., 2002b; Harlov and Förster, 2003; Harlov et al., 2005).

In corroboration with what has already been concluded from nature (e.g., Harlov and Förster, 2002b; Harlov et al., 2002a, 2002b; Hansen and Harlov, 2007) and experimentally (Harlov et al. 2002b; Harlov and Förster, 2003; Harlov et al., 2005), monazite and xenotime associated with fluorapatite can serve as a valuable "fingerprint" for recording metasomatic events in high-grade metamorphic rocks. The presence of coexisting monazite and xenotime inclusions in fluorapatite, as well as rim grains, can give some indication of the temperature of the metasomatic overprint via monazite-xenotime geothermometry (cf. Gratz and Heinrich, 1997, 1998; Heinrich et al., 1997). In such cases, coexisting monazite and xenotime inclusions tend to give higher temperatures as opposed to rim grains (e.g., Harlov and Förster, 2002b). More importantly, the association of monazite and/or xenotime with fluorapatite or chlorapatite helps to place constraints on the chemistry of the infiltrating fluids responsible for the metasomatism of the apatite as well as the rock as a whole. In this respect, fluorapatite from high-grade rocks that contain monazite and/or xenotime inclusions/rim grains potentially represents a record of metasomatic events over a wide range of temperatures and specific

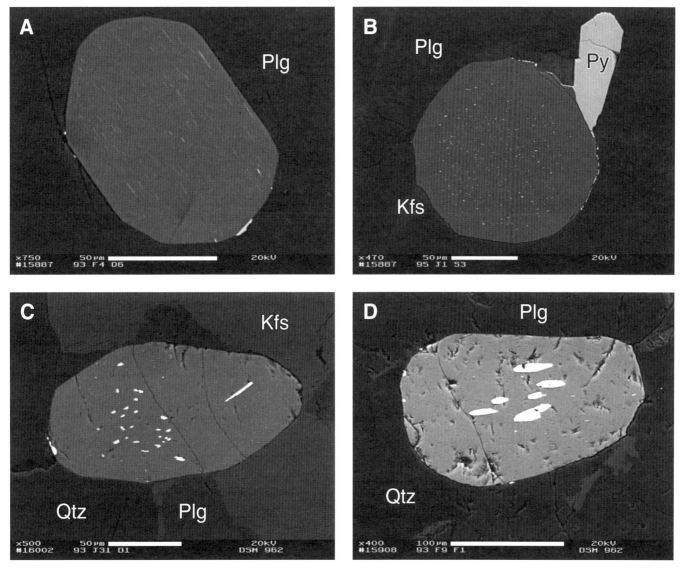

Figure 1. High contrast BSE pictures of monazite inclusions in fluorapatite from Archean-age, granulite-facies tonalitic rocks along a traverse of lower crust, Tamil Nadu, south India (Hansen and Harlov, 2007). These inclusions can take the form of numerous small monazite inclusions elongated with respect to the c-axis (Figs. 1A, 1B). In Figure 1A the fluorapatite grain is cut approximately parallel to the c-axis, whereas in Figure 1B the grain is cut perpendicular to the c-axis. Monazite inclusions in Figures 1C and 1D show increased coarsening and consolidation. Kfs—K-feldspar, Plg—plagioclase, Py—pyrite, Qtz—quartz.

fluid compositions in addition to serving as a major host for light rare earth elements (LREE).

Monazite-Xenotime-ThSiO$_4$

In metamorphic rocks, monazite and xenotime may grow either as the net product of a series of mineral reactions or be inherited from igneous and/or sedimentary precursors. In igneous rocks, typically peraluminous granites and granitoids, monazite and xenotime tend to be among the first minerals to crystallize out of the melt (Bea, 1996). In either case, any Th in the system will be strongly partitioned into the monazite as the monoclinic huttonite (ThSiO$_4$) and/or cheralite [CaTh(PO$_4$)$_2$] component via the coupled substitution reactions Th^{4+} + Si^{4+} = REE^{3+} + P^{5+} and Th^{4+} + Ca^{2+} = 2 REE^{3+}, respectively (Förster, 1998; Zhu and O'Nions, 1999; Seydoux-Guillaume et al., 2002a). To a lesser extent, Th can also be partitioned into xenotime as the tetragonal thorite (ThSiO$_4$) component via the coupled substitution Th^{4+} + Si^{4+} = REE^{3+} + P^{5+} (Hetherington and Harlov, 2008). In contrast to monazite, xenotime commonly takes more U than Th (Seydoux-Guillaume et al., 2002b; Kositcin et al., 2003; Krenn et al., 2008). Experimentally, it has been demonstrated that monazite may be partially overgrown or partially replaced by pure monoclinic huttonite via coupled dissolution-reprecipitation, suggesting that ThSiO$_4$ grains associated with partially altered monazite in nature are actually huttonite as opposed to lower grade tetragonal thorite (Harlov et al., 2007b). Fluid-aided alteration of Th-bearing xenotime in pegmatites causes thorite and uraninite (UO$_2$) inclusions to form in a manner similar to that seen for the formation of monazite and xenotime inclusions in apatite (Hetherington and Harlov, 2008). The mechanism behind the formation of these inclusions is similar, invoking a fluid-filled, pervasive porosity resulting from coupled dissolution-reprecipitation and the subsequent nucleation of thorite and uraninite inclusions in a random sampling of these pores. The current role of monazite and xenotime in dating from one to multiple events during

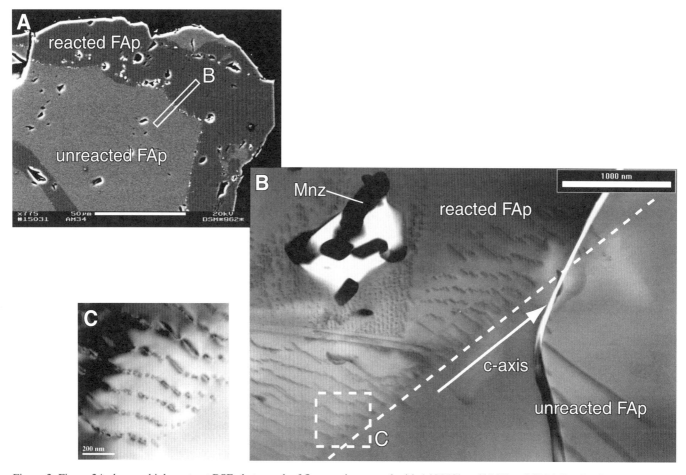

Figure 2. Figure 2A shows a high contrast BSE photograph of fluorapatite reacted with 1 N HCl at 600 °C and 500 MPa (Harlov et al., 2005). Dark regions have reacted with the HCl solution and are depleted in (Y + REE) + Si + Na + S + Cl, i.e., have been metasomatized. Small bright grains outlining the reaction front are monazite. Figure 2B shows part of a transmission electron microscope (TEM) foil cut at the boundary between the reacted and unreacted fluorapatite in the region shown in Figure 2A. The TEM foil shows a series of subparallel nano-channels and a cluster of euhedral monazite crystals in a cavity filled with an amorphous quenched material. A close-up of the nano-channels is shown in Figure 2C. FAp—fluorapatite, Mnz—monazite.

high-grade metamorphism and deformation, either singly (e.g., Pyle and Spear, 2003; Mahan et al., 2006a, 2006b; Williams et al., 2007; Dumond et al., 2008; Hetherington et al., 2008) or in conjunction with allanite (Janots et al., 2009) or fluorapatite (Finger and Krenn, 2007), has increased the number of geochronological tools currently available. This, coupled with the concurrent utilization of both minerals in geothermometry (Gratz and Heinrich, 1997, 1998; Heinrich et al., 1997; Pyle and Spear, 2000; Pyle et al., 2001), as well as tracking their metasomatic derivation from apatite as inclusions and rim grains (Harlov et al., 2005), has greatly increased the amount of information that can be extracted from these two common accessory minerals.

Once formed, metamorphic monazite can be partially or totally altered with respect to the Th and (Y + REE) distribution and content. In contrast, such alteration with respect to Th is commonly not observed in metamorphic xenotime, as it incorporates considerably less Th than coexisting monazite in the same rock (e.g., Franz et al., 1996). During alteration, Th is gained or lost by monazite in variable amounts such that domains consisting of curvilinear intergrowths with sharp compositional boundaries, either depleted or enriched in $ThSiO_4$-$CaTh(PO_4)_2$, are developed in the body of the original monazite grain (Fig. 3). The conclusion reached by most studies of these textures hypothesizes that this alteration is a metasomatically induced process owing to the presence of locally mobile Th, Si, Ca, P, and (Y + REE) in a grain boundary–pore fluid that is reactive with respect to the monazite (e.g., Bingen and van Breemen, 1998). However, speculations on the exact nature of this fluid

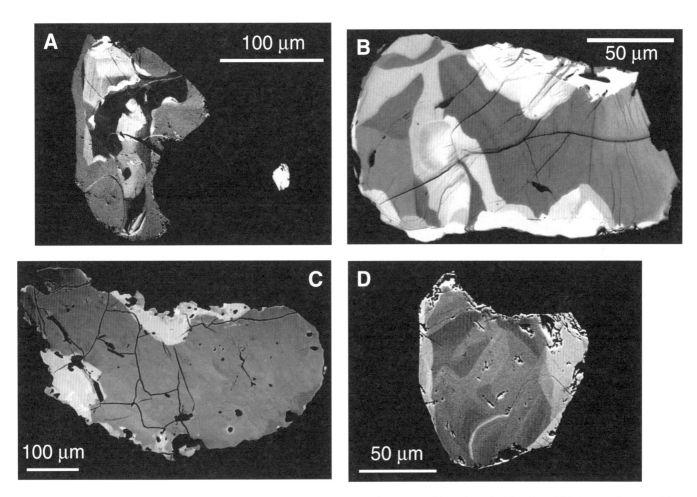

Figure 3. Examples from nature of partially metasomatized monazite in both metamorphic and granitoid rocks, which have undergone either Th gain or loss. Figure 3A shows partially metasomatized monazite, with Th-enriched embayments, in granulite-facies metabasites; Val Strona high-grade traverse, Ivrea-Verbano Zone, Northern Italy (Förster and Harlov, 1999). Some Th-enriched regions contain up to 30% of the huttonite component. The metabasites have been metasomatized by fluids both from intercalated metapelite layers as well as from emplacement of the mafic formation at the base of the traverse. Figure 3B shows partially metasomatized monazite in clinopyroxene-bearing amphibolite facies and orthopyroxene-bearing granulite-facies orthogneisses; Rogaland–West Agder terrane, SW Norway (Bingen and van Breemen, 1998). The surrounding pore fluids are ascribed to being responsible for the formation of both Th-enriched and Th-depleted embayments along the monazite grain rim. Figure 3C shows a metasomatized monazite from a pelitic schist, Cavendish Formation (Star Hill, Vermont), Chester Dome, northeastern flank (Pyle et al., 2005). Figure 3D shows a metasomatized monazite grain from an amphibolite-facies tonalite that forms part of a traverse of Archean lower crust, Tamil Nadu, south India (Hansen and Harlov, 2007).

are more vague. In general these intergrowths tend to give different electron microprobe ages than areas of the monazite grain unaffected by metasomatic alteration (Poitrasson et al., 1996; Seydoux-Guillaume et al., 2003; Goncalves et al., 2005; Williams et al., 2006, 2007; Dumond et al., 2008).

In order to determine whether or not metasomatic alteration could be responsible for the textures shown in Figure 3, a series of experiments was performed at granulite-facies temperatures and pressures (900 °C and 1000 MPa) utilizing a natural monazite from a Brazil beach sand, which is relatively homogeneous with respect to Th. The experiments utilized a variety of alkali-bearing fluids, i.e., 2N NaOH, 2N KOH, and $Na_2Si_2O_5$ + H_2O (Harlov and Hetherington, 2010; Harlov et al., 2011). The source of Si in these experiments was quartz or $Na_2Si_2O_5$, whereas the source of Th was the monazite itself. These fluid compositions were chosen as ones that might possibly exist along grain boundaries in a feldspar-rich rock infiltrated by H_2O-bearing fluids under a variety of metamorphic conditions ranging from low to high grade.

In each experiment the monazite reacted with the fluid such that limited metasomatized regions of variable enrichment in the huttonite component occurred, resulting in textures similar to those seen in nature (compare Figs. 3 and 4; Harlov and Hetherington, 2010). The metasomatized regions are separated from the non-metasomatized regions by sharp, curvilinear, or straight compositional boundaries. Some metasomatized regions also show peculiar enrichments in the huttonite component that appear to be crystallographically controlled. The most likely explanation is that the total dissolution of small monazite grain fragments along with some limited dissolution of the larger monazite grains release Th into the solution during the first stages of the experiment. There is no evidence for the leaching of Th from the monazite, because unmetasomatized areas of the monazite have the same composition as the original monazite. In a manner similar to that seen in the apatite metasomatism experiments, the alkali-bearing fluid appears to have attacked the monazite grains in a coupled process through the partial dissolution of the

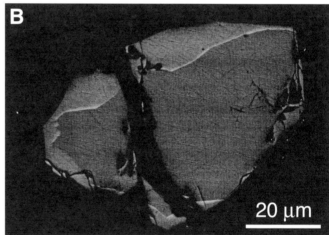

Figure 4. Examples of experimentally metasomatized monazite grains (Harlov and Hetherington, 2010). Figure 4A shows a monazite grain metasomatized in a $Na_2Si_2O_5$ + H_2O solution at 900 °C and 1000 MPa for eight days. Figure 4B shows a monazite grain metasomatized in 2N NaOH solution at 900 °C and 1000 MPa for 25 days. Figure 4C shows a monazite grain metasomatized in 2N KOH solution at 900 °C and 1000 MPa for 25 days. Bright areas are enriched in Th + Si relative to the darker unaltered monazite.

preexisting monazite, followed concurrently by the reprecipitation of new monazite enriched in the $ThSiO_4$ component. This resulted in the pseudomorphic, partial replacement of the original monazite (Harlov and Hetherington, 2010). It also presumes that Th and Si quickly reached saturation or possible super-saturation in the fluid, followed by strong partitioning into the reprecipitated monazite during this coupled dissolution-reprecipitation process (see discussion in Harlov and Hetherington, 2010; Harlov et al., 2011; Putnis, 2002, 2009).

In the $Na_2Si_2O_5+H_2O$ experiments, the Pb was removed from the altered $ThSiO_4$-enriched areas to below EMP detection limits whereas unaltered areas in the monazite retained their original Pb content (Harlov and Hetherington, 2010). This has far-reaching implications with respect to dating. Presuming that Th-enriched (or -depleted) textures in natural monazite are due to fluid-induced alteration, the total removal of Pb during alteration would reset the Th-U-Pb clock so that these textures could be used to date metasomatic events over the geologic history of the monazite grain and potentially the host rock itself. When combined with in situ analytical techniques, the complex compositional zoning typically observed in natural monazite could then be used to constrain the ages for a sequence of geologic or tectonic events (Williams et al., 2007; Dumond et al., 2008).

Zircon, Titanite, and Allanite

The role of zircon in high-grade metamorphic rocks is primarily that of a geochronometer (see review by Davis et al., 2003), though it does contain significant trace amounts of Ti, HREE, and Y in addition to Th, U, and Hf. These trace elements can show relative enrichment or depletion in zircon depending on the geochemical environment of the rock (e.g., Cherniak and Watson, 2003; Schulz et al., 2006; Harley and Kelly, 2007; Kebede et al., 2007; Wu et al., 2008), or, as in the case of Ti, act as a geothermometer in the presence of coexisting rutile in both metamorphic and igneous rocks (Watson et al., 2006; Ferry and Watson, 2007; Cherniak and Watson, 2007; Bin et al., 2008; Hiess et al., 2008; Ferriss et al., 2008). Zircon can also be metasomatically altered via coupled dissolution-reprecipitation processes with respect to trace element compositions (Geisler et al., 2007; Rubatto et al., 2008). Utilizing zircon as a monitor of Y + HREE, Th, and U mobility, both as a function of changes in P-T as well as whole rock and fluid chemistry along a traverse of metamorphic rocks ranging from granulite to amphibolite facies, has not yet been done, although preliminary attempts have been made in this direction (Harlov and Hansen, 2007; Harlov et al., 2010).

Titanite takes in only trace amounts of high field strength elements (HFSE), but it can accommodate considerably greater amounts of REE (Tiepolo et al., 2002) than zircon, ranging from 0.5 up to 3 or more oxide weight percentages (Harlov et al., 2006a; Aleinikoff et al., 2002). Subsequently, the lack of a principal REE host in high-grade rocks, such as monazite, xenotime, or allanite, allows titanite to become relatively enriched in REE (see discussion in Hansen and Harlov, 2007). This ability to take in REE, as well as Th and U, allows for titanite to be used to date geological processes under high-grade conditions as well as to document these processes with regard to REE mobility and the probable role of fluids (e.g., Buick et al., 2007; Storey et al., 2007). Titanite can participate in mineral-mineral, mineral-melt, and mineral-fluid equilibria that can be quite sensitive to intensive parameters such as pressure, temperature, oxygen fugacity (fO_2), and water fugacity (fH_2O) (e.g., Wones, 1989; Xirouchakis et al., 2001a, 2001b; Tropper et al., 2002; Troitzsch and Ellis, 2002; Harlov et al., 2006a). For mafic rocks relatively rich in FeO, reactions among titanite, magnetite, ilmenite, clinopyroxene, olivine, and quartz can constrain P-T-fO_2 conditions, at which these rocks form (Wones, 1989; Xirouchakis and Lindsley, 1998). For example, titanite-bearing, oxide-silicate equilibria have been shown to indicate relative oxidizing conditions (Carmichael and Nicholls, 1967), with the assemblage titanite-magnetite-quartz acting as a boundary between relative reducing and oxidizing conditions (Wones, 1989). More recently, the abundance of Zr in titanite (in the presence of zircon) was experimentally calibrated as a geothermobarometer (Hayden et al., 2008), further adding another dimension to the multiple uses of this important mineral.

In contrast to titanite or zircon, allanite and REE-enriched epidote (cf. Gieré and Sorensen, 2004) are two of the few silicate minerals that can serve as a principal host for REE as well as Th and U in high-grade rocks in the absence of monazite or xenotime (e.g., Harlov et al., 2006b). Similar to monazite or xenotime, the presence of Th and U also allows allanite to be used for dating (e.g., Gregory et al., 2007), although dating this mineral can be problematic (cf. Gieré and Sorensen, 2004). Recently, several groups of workers have begun to work out the phase relations between allanite and monazite as a function of P-T, both from natural observation and from experiments (Wing et al., 2003; Janots et al., 2006, 2007, 2008; Krenn and Finger, 2007). These studies have essentially plotted out a stability field for allanite relative to monazite in P-T space such that the conversion of allanite to monazite or vice versa may be used to help to chart out either prograde or retrograde metamorphic P-T paths. Such studies might also help to explain the presence of common, very thin allanite reaction rims on fluorapatite from rocks along a cross section of lower Archean crust, which Hansen and Harlov (2007) attributed to secondary alteration during retrograde metamorphic decompression and cooling.

DISCUSSION

The physical-chemical relationships between REE- and actinide-bearing minerals as a function of metamorphic grade can be seen in two independent investigations of high-grade felsic orthogneisses from Rogaland–Vest Agder, southwestern Norway (Bingen and van Breemen, 1998) and Tamil Nadu, south India (Hansen and Harlov, 2007). Both studies involve well-sampled traverses across a continuous transition from orthopyroxene-bearing granulite grade to amphibolite grade punctuated by a clinopyroxene-in isograd. Each traverse has been interpreted to represent a cross section of lower crust.

Rogaland–Vest Agder, SW Norway

The Rogaland–Vest Agder gneiss complex is the westernmost terrane of the Sveconorwegian Province and lies along the southwest coast of Norway (Bingen and van Breemen, 1998). It consists of elongate units of banded, migmatitic, granitic, and augen gneisses, intercalated with minor units of amphibolite, quartzite, calc-silicate, and marble. The Rogaland–Vest Agder gneiss complex is characterized from east to west by three metamorphic zones, which are separated by two isograds. Zone 1 is amphibolite facies and is separated from Zone 2 by a clinopyroxene-in isograd. Zone 2 is characterized by the prograde crystallization of clinopyroxene from amphibole and biotite. An orthopyroxene-in isograd marks the beginning of granulite-facies Zone 3. It is characterized by the formation of orthopyroxene at the expense of biotite and, to a lesser extent, hornblende.

REE-accessory-mineral behavior follows certain patterns (Bingen and van Breemen, 1998). Amphibolite-facies Zone 1 is characterized by abundant coarse-grained titanite in the form of isolated grains in the matrix. Titanite modal abundance shows a precipitous drop approximately in the location of the clinopyroxene-in isograd. This is accompanied by a sudden increase in (Fe, Ti) oxides, which may have resulted from the release of Ti during breakdown of titanite. In Zones 2 and 3, titanite is found only as small inclusions inside of other minerals, and its modal abundance decreases as a function of increasing pyroxene relative to hornblende. Allanite is also present in Zone 1. Like titanite, its abundance suddenly drops at the clinopyroxene-in isograd. While present in small amounts in Zone 2, it completely disappears at the orthopyroxene-in isograd. Although fluorapatite abundance is independent of metamorphic grade, its middle-heavy (M-H) REE content increases steeply with increasing metamorphic grade, punctuated by an increase in LREE and Th at the orthopyroxene-in isograd. Monazite and thorite are detected in small amounts close to the clinopyroxene-in isograd. Both minerals show a maximum abundance in Zone 2 between the clinopyroxene-in and orthopyroxene-in isograds (Bingen and van Breemen, 1998).

The appearance of monazite and thorite correlates with the breakdown and gradual disappearance of both titanite and allanite. There is also a tight correlation between the fluorapatite M-H REE content and the modal decrease of titanite, allanite, and hornblende in Zones 1 and 2. This suggests that the growth of monazite is linked to REE substitution in apatite, and subsequently also associated with the breakdown of titanite, allanite, and hornblende via a reaction such as:

Bingen and van Breemen (1998) propose that this reaction probably occurs via a metamorphic fluid, as pervasive mass transport must occur throughout the volume of rock in which the titanite, allanite, and fluorapatite are randomly scattered. One possible source for this fluid is hornblende breakdown, although the influence of external fluids flowing across metamorphic isograds cannot be discounted. At the orthopyroxene-in isograd, the increase in the LREE and Th content of the fluorapatite results either from the breakdown of a significant proportion of previously formed monazite and/or from the breakdown of allanite that survived prograde upper amphibolite-facies metamorphism.

Tamil Nadu, South India

The Tamil Nadu traverse, south India, consists of a 95-km-long traverse made up predominantly of intermediate to felsic orthogneisses along with subordinate amounts of mafic gneisses and minor amounts of metasediments (Hansen and Harlov, 2007). The traverse appears to represent a continuous, ~12-km-thick cross section of Archean lower crust. This section is obliquely tilted on its side at a shallow angle and is cut off at the base by a major shear zone of probable Proterozoic age. Metamorphic grade ranges from amphibolite facies in the north to granulite facies in the south. The traverse was divided into three zones by Hansen and Harlov (2007): the Northern Amphibolite Facies Zone (NAF), the Central Granulite Facies Zone (CGF), and the Southern Granulite Facies Zone (SGF). The quartzo-feldspathic rocks of the NAF contain biotite ± amphibole. Epidote is present in the northern half, with clinopyroxene appearing approximately in the center of the NAF and increasing in abundance until the first appearance of an orthopyroxene-in isograd marks the beginning of the CGF several kilometers south of the clinopyroxene-in isograd. In the first few kilometers of the CGF, clinopyroxene is the dominant pyroxene, with orthopyroxene becoming more abundant and clinopyroxene less abundant as metamorphic grade increases farther south. Concurrently, the abundance of biotite and amphibole decrease, indicating their probable conversion to ortho- and clinopyroxene with no real evidence of partial melting having occurred during the process. The start of the CGF is marked by a large modal increase in garnet coupled with very minor remnants of biotite and amphibole. Orthopyroxene is the dominant pyroxene, with minor to absent clinopyroxene. Oxide and sulfide minerals show definitive trends with coexisting hemo-ilmenite, magnetite, pyrrhotite, and pyrite common in the SGF (Harlov et al., 1997; Harlov and Hansen, 2005). This assemblage is replaced by coexisting magnetite, ilmenite, and pyrite at the center of the CGF. With decreasing metamorphic grade, magnetite-ilmenite-pyrite abundances also decrease such that in

$$3\ (MREE\text{-}HREE)_2O_3 + 3\ (LREE)_2O_3 + 2\ Ca_5(PO_4)_3(F,OH) + 6\ SiO_2$$
in hornblende and titanite   in allanite   in fluorapatite   in quartz

$$= 6\ (LREE)PO_4 + 2\ Ca_2(MREE\text{-}HREE)_3(SiO_4)_3(F,OH) + 6\ CaO$$
monazite   in fluorapatite   in plagioclase.

(1)

the NAF they are relatively rare. In addition, biotite and amphibole are enriched in Fe and Mn, whereas the Ti has presumably gone into titanite. Whole rock compositional trends indicate that relative to the NAF, most of the CGF and all of the SGF are depleted in Rb, Cs, Th, and U, whereas the REE remain relatively constant in abundance across the entire traverse.

Similar to what is seen in the Rogaland–Vest Agder traverse, the REE-bearing accessory minerals follow a certain pattern as a function of metamorphic grade (Hansen and Harlov, 2007). Titanite predominates in the NAF, typically rimming ilmenite. It is enriched in LREE to several oxide weight percentages. Allanite also occurs both as discrete grains and as monazite rims. At the orthopyroxene-in isograd, both titanite and allanite have mostly disappeared. Discrete grains of monazite, both zoned and unzoned, appear at the center of the NAF. Going up grade they disappear by the time the center of the CGF is reached. In general, discrete monazite grains are found mostly in the clinopyroxene-rich zone, which also contains the orthopyroxene isograd separating the NAF from the CGF. Monazite inclusions in fluorapatite and/or rim grains associated with fluorapatite first appear in the NAF in the immediate vicinity of the orthopyroxene-in isograd and are a regular feature in samples from the CGF and SGF. Experimental evidence (e.g., Harlov et al., 2005) indicates that these monazite inclusions and rim grains were metasomatically induced to form from the fluorapatite. This, in turn, implies that the original fluorapatite was substantially more enriched in REE than the inclusion-absent fluorapatite from the NAF. This is especially obvious when monazite inclusions and rim grains are compositionally reintegrated back into the fluorapatite host in order to obtain the original REE abundances (cf. Hansen and Harlov, 2007).

Whereas whole rock analysis indicates relatively constant REE abundances across the traverse, metamorphic grade dictates where and how, i.e., in which minerals, these REE are housed. In the granulite facies of the SGF and CGF, the REE are hosted in fluorapatite. At the clinopyroxene-rich transition between the CGF and NAF, monazite becomes the new stable host for REE, followed at lower grade by titanite as well as allanite in the lowest grade parts of the NAF. This REE accessory mineral trend, as a function of metamorphic grade, is similar to what is seen in the Rogaland–Vest Agder traverse and suggests a reaction sequence similar to that expressed by reaction (1). Again, the role of fluids is important. Hansen and Harlov (2007) speculated that mineral and whole-rock trace element trends seen along this traverse of lower Archean crust were the product of both externally (KCl-NaCl brines and CO_2) and internally (biotite and hornblende breakdown) derived fluids, possibly accompanied by partial melting.

REE Accessory Minerals, REE Mobility, and Metamorphic Grade

In both the Rogaland–Vest Agder and Tamil Nadu traverses, the same physical-chemical relationships are observed between monazite, fluorapatite, allanite, and titanite. In the amphibolite-facies section of the traverse the REE are hosted by titanite and allanite. In the clinopyroxene-rich transition zone, where the orthopyroxene-in isograd marks the boundary between the amphibolite-facies and granulite-facies zones, monazite is the stable REE-bearing phase. In the granulite-facies zone the REE are hosted by fluorapatite. This suggests that hosting of REE by rocks of broadly similar whole rock composition follows the same mineralogical trends as a function of metamorphic grade.

Similar observations can be made for other traverses of lower crust. For example, Harlov and Förster (2002a, 2002b), in a study of the Carboniferous Val Strona traverse, Ivrea-Verbano Zone, northern Italy, noted the presence of abundant monazite inclusions in fluorapatite in the granulite-facies portion of the traverse, independent grains of monazite in a clinopyroxene-rich zone, which contains the orthopyroxene-in isograd, and abundant titanite, commonly seen rimming ilmenite, in the amphibolite-facies portion of the traverse. Fluorapatite in the amphibolite-facies rocks is devoid of monazite inclusions and contains considerably lower REE abundances than fluorapatite in the granulite-facies part of the traverse. Titanite contains moderate amounts of REE and, from its shear abundance in the amphibolite-facies rocks can easily serve as a storehouse for the REE in general (Harlov et al., 2006a).

Other recent studies of high-pressure and ultrahigh-pressure metapelitic rocks from the Bohemian Massif have demonstrated that, with increasing pressure, fluorapatite will take increasing amounts of REE coupled with the breakdown of monazite (Finger and Krenn, 2007; Krenn et al., 2009). Such a scenario fits in with studies of the relative stability of monazite and allanite, which suggest that the stability field of allanite relative to monazite is approximately low to middle amphibolite facies (Janots et al., 2007). Monazite appears to be the stable REE accessory mineral phase, beginning in upper amphibolite facies and continuing into lower granulite facies; then fluorapatite becomes the principal LREE-bearing phase at higher pressure upper granulite-facies and eclogite-facies conditions with garnet probably acting as the principal host for the HREE. Both studies confirm the basic observations from the Rogaland–Vest Agder, Tamil Nadu, and Val Strona traverses and further reinforce the idea that REE storage in high-grade rocks is first and foremost a function of P-T followed by local rock chemistry. In that regard, in addition to their role as well studied geochronometers, monazite, fluorapatite, allanite, and titanite can also provide detailed geochemical information regarding REE mobility and storage as a function of P-T and, in particular, provide important information regarding the composition and nature of high-grade intergranular fluids. Although this later role is still not fully understood, it is in this direction that studies of these minerals in high-grade rocks are currently making rapid progress.

ACKNOWLEDGMENTS

This review of REE accessory minerals in high-grade metamorphic rocks has benefited from critical reviews of the manuscript by Erwin Krenn, Edward Hansen, Greg Dumond, Michael Williams, and Emilie Janots.

REFERENCES CITED

Aleinikoff, J.N., Wintsch, R.P., Fanning, C.M., and Dorais, M.J., 2002, U-Pb geochronology of zircon and polygenetic titanite from the Glastonbury Complex, Connecticut, USA; an integrated SEM, EMPA, TIMS, and SHRIMP study: Chemical Geology, v. 188, p. 125–147, doi:10.1016/S0009-2541(02)00076-1.

Andrehs, G., and Heinrich, W., 1998, Experimental determination of REE distributions between monazite and xenotime: Potential for temperature-calibrated geochronology: Chemical Geology, v. 149, p. 83–96, doi:10.1016/S0009-2541(98)00039-4.

Bea, F., 1996, Residence of REE, Y, Th and U in granites and crustal protoliths; implications for the chemistry of crustal melts: Journal of Petrology, v. 37, p. 521–552, doi:10.1093/petrology/37.3.521.

Bin, F., Page, F.Z., Cavosie, A.J., Fournelle, J., Kita, N.T., Lackey, J.S., Wilde, S.A., and Valley, J.W., 2008, Ti-in-zircon thermometry: Applications and limitations: Contributions to Mineralogy and Petrology, v. 156, p. 197–215, doi:10.1007/s00410-008-0281-5.

Bingen, B., and van Breemen, O., 1998, U-Pb monazite ages in amphibolite- to granulite-facies orthogneiss reflect hydrous mineral breakdown reactions: Sveconorwegian Province of SW Norway: Contributions to Mineralogy and Petrology, v. 132, p. 336–353, doi:10.1007/s004100050428.

Buick, I.S., Hermann, J., Maas, R., and Gibson, R.L., 2007, The timing of sub-solidus hydrothermal alteration in the Central Zone, Limpopo Belt (South Africa); constraints from titanite U-Pb geochronology and REE partitioning: Lithos, v. 98, p. 97–117, doi:10.1016/j.lithos.2007.02.002.

Carmichael, I.S.E., and Nicholls, J., 1967, Iron-titanium oxides and oxygen fugacities in volcanic rocks: Journal of Geophysical Research, v. 72, p. 4665–4687, doi:10.1029/JZ072i018p04665.

Cherniak, D.J., 2000, Rare earth element diffusion in apatite: Geochimica et Cosmochimica Acta, v. 64, p. 3871–3885, doi:10.1016/S0016-7037(00)00467-1.

Cherniak, D.J., 2005, Uranium and manganese diffusion in apatite: Chemical Geology, v. 219, p. 297–308, doi:10.1016/j.chemgeo.2005.02.014.

Cherniak, D.J., 2006, Pb and rare earth element diffusion in xenotime: Lithos, v. 88, p. 1–14, doi:10.1016/j.lithos.2005.08.002.

Cherniak, D.J., and Pyle, J.M., 2008, Th diffusion in monazite: Chemical Geology, v. 256, p. 52–61, doi:10.1016/j.chemgeo.2008.07.024.

Cherniak, D.J., and Watson, E.B., 2003, Diffusion in zircon, in Hanchar, J.M., and Hoskin, J.M., eds., Zircon: Reviews in Mineralogy and Geochemistry, v. 53, p. 113–143.

Cherniak, D.J., and Watson, E.B., 2007, Ti diffusion in zircon: Chemical Geology, v. 242, p. 470–483, doi:10.1016/j.chemgeo.2007.05.005.

Cherniak, D.J., Watson, E.B., Grove, M., and Harrison, T.M., 2004, Pb diffusion in monazite: A combined RBS/SIMS study: Geochimica et Cosmochimica Acta, v. 68, p. 829–840, doi:10.1016/j.gca.2003.07.012.

Davis, D., Williams, I.S., and Krogh, T., 2003, Historical development of zircon geochronology, in Hanchar, J.M., and Hoskin, J.M., eds., Zircon: Reviews in Mineralogy and Geochemistry, v. 53, p. 145–181.

Dumond, G., McLean, N., Williams, M.L., Jercinovic, M.J., and Bowring, S.A., 2008, High-resolution dating of granite petrogenesis and deformation in a lower crustal shear zone: Athabasca granulite terrane, western Canadian Shield: Chemical Geology, v. 254, p. 175–196, doi:10.1016/j.chemgeo.2008.04.014.

Ferriss, E.D.A., Essene, E.J., and Becker, U., 2008, Computational study of the effect of pressure on the Ti-in-zircon geothermometer: European Journal of Mineralogy, v. 20, p. 745–755, doi:10.1127/0935-1221/2008/0020-1860.

Ferry, J.M., and Watson, E.B., 2007, New thermodynamic models and revised calibrations for the Ti-in-zircon and Zr-in-rutile thermometers: Contributions to Mineralogy and Petrology, v. 154, p. 429–437, doi:10.1007/s00410-007-0201-0.

Finger, F., and Krenn, E., 2007, Three metamorphic monazite generations in a high-pressure rock from the Bohemian Massif and the potentially important role of apatite in stimulating polyphase monazite growth along a PT loop: Lithos, v. 95, p. 103–115, doi:10.1016/j.lithos.2006.06.003.

Finger, F., Broska, I., Robers, M.P., and Schermailer, A.S., 1998, Replacement of primary monazite by apatite-allanite-epidote coronas in an amphibolite-facies granite gneiss from the eastern Alps: American Mineralogist, v. 83, p. 248–258.

Förster, H.-J., 1998, The chemical composition of REE-Y-Th-U-rich accessory minerals from peraluminous granites of the Erzgebirge-Fichtelgebirge region, Germany. Part I: The monazite-(Ce)–brabantite solid solution series: American Mineralogist, v. 83, p. 259–272.

Förster, H.-J., and Harlov, D.E., 1999, Monazite-(Ce)–huttonite solid solutions in granulite-facies metabasites from the Ivrea-Verbano Zone, Italy: Mineralogical Magazine, v. 63, p. 587–594.

Franz, G., Andrehs, G., and Rhede, D., 1996, Crystal chemistry of monazite and xenotime from Saxothuringian-Moldanubian metapelites, NE Bavaria, Germany: European Journal of Mineralogy, v. 8, p. 1097–1118.

Geisler, T., Schaltegger, U., and Tomaschek, F., 2007, Re-equilibration of zircon in aqueous fluids and melts: Elements, v. 3, p. 43–50, doi:10.2113/gselements.3.1.43.

Gieré, R., and Sorensen, S.S., 2004, Allanite and other REE-rich epidote-group minerals, in Liebscher, A., and Franz, G., eds., Epidotes: Washington, D.C., Reviews in Mineralogy and Geochemistry, Mineralogical Society of America, v. 56, p. 431–493.

Gieré, R., Williams, C.T., and Lumpkin, G.R., 1998, Chemical characteristics of natural zirconolite: Schweizerische Mineralogische und Petrographische Mitteilungen, v. 78, p. 433–459.

Goncalves, P., Williams, M.L., and Jercinovic, M.J., 2005, Electron-microprobe age mapping of monazites: American Mineralogist, v. 90, p. 578–585, doi:10.2138/am.2005.1399.

Gratz, R., and Heinrich, W., 1997, Monazite-xenotime thermobarometry: Experimental calibration of the miscibility gap in the binary system $CePO_4$-YPO_4: American Mineralogist, v. 82, p. 772–780.

Gratz, R., and Heinrich, W., 1998, Monazite-xenotime thermometry. III. Experimental calibration of the partitioning of gadolinium between monazite and xenotime: European Journal of Mineralogy, v. 10, p. 579–588.

Gregory, C.J., Rubatto, D., Allen, C.M., Williams, I.S., Hermann, J., and Ireland, T., 2007, Allanite micro-geochronology; a LA-ICP-MS and SHRIMP U/Th/Pb study: Chemical Geology, v. 245, p. 162–182, doi:10.1016/j.chemgeo.2007.07.029.

Hansen, E.C., and Harlov, D.E., 2007, Phosphate, silicate and fluid composition across an amphibolite- to granulite-facies transition, Tamil Nadu, India: Journal of Petrology, v. 48, p. 1641–1680, doi:10.1093/petrology/egm031.

Harley, S.L., and Kelly, N.M., 2007, The impact of zircon-garnet REE distribution data on the interpretation of zircon U/Pb ages in complex high-grade terrains; an example from the Rauer Islands, East Antarctica: Chemical Geology, v. 241, p. 62–87, doi:10.1016/j.chemgeo.2007.02.011.

Harlov, D.E., and Förster, H.-J., 2002a, High-grade fluid metasomatism on both a local and regional scale: The Seward Peninsula, Alaska, and the Val Strona di Omegna, Ivrea-Verbano Zone, northern Italy. Part I: Petrography and silicate mineral chemistry: Journal of Petrology, v. 43, p. 769–799, doi:10.1093/petrology/43.5.769.

Harlov, D.E., and Förster, H.-J., 2002b, High-grade fluid metasomatism on both local and a regional scale: The Seward Peninsula, Alaska, and the Val Strona di Omegna, Ivrea-Verbano zone, northern Italy. Part II: Phosphate mineral chemistry: Journal of Petrology, v. 43, p. 801–824, doi:10.1093/petrology/43.5.801.

Harlov, D.E., and Förster, H.J., 2003, Fluid-induced nucleation of (Y+REE)-phosphate minerals within apatite: Nature and experiment. Part II. Fluorapatite: American Mineralogist, v. 88, p. 1209–1229.

Harlov, D.E., and Hansen, E.C., 2005, Oxide and sulphide isograds along a late Archean, deep-crustal profile in Tamil Nadu, south India: Journal of Metamorphic Geology, v. 23, p. 241–259, doi:10.1111/j.1525-1314.2005.00574.x.

Harlov, D.E., and Hansen, E.C., 2007, Trends in phosphate and silicate mineral chemistry across a section of Archean crust, Tamil Nadu, south India: The role of fluids in regional granulite-facies metamorphism: Geophysical Research Abstracts, v. 9, abs. EGU2007-A-06248.

Harlov, D.E., and Hetherington, C.J., 2010, Partial high-grade alteration of monazite using alkali-bearing fluids: Experiment and nature: American Mineralogist, v. 95, p. 1105–1108.

Harlov, D.E., Newton, R.C., Hansen, E.C., and Jarnardhan, A.S., 1997, Oxide and sulphide minerals in highly oxidized Rb-depleted Archean granulites of the Shevaroy Hills Massif, South India: Oxidation states and the role of metamorphic fluids: Journal of Metamorphic Geology, v. 15, p. 701–717, doi:10.1111/j.1525-1314.1997.00046.x.

Harlov, D.E., Andersson, U.B., Förster, H.-J., Nyström, J.O., Dulski, P., and Broman, C., 2002a, Apatite-monazite relations in the Kiirunavaara magnetite-apatite ore, northern Sweden: Chemical Geology, v. 191, p. 47–72, doi:10.1016/S0009-2541(02)00148-1.

Harlov, D.E., Förster, H.-J., and Nijland, T.G., 2002b, Fluid-induced nucleation of REE-phosphate minerals in apatite: Nature and experiment. Part I. Chlorapatite: American Mineralogist, v. 87, p. 245–261.

Harlov, D.E., Wirth, R., and Förster, H.-J., 2005, An experimental study of dissolution-reprecipitation in fluorapatite: Fluid infiltration and the formation of monazite: Contributions to Mineralogy and Petrology, v. 150, p. 268–286, doi:10.1007/s00410-005-0017-8.

Harlov, D., Tropper, P., Seifert, W., Nijland, T., and Förster, H.-J., 2006a, Formation of Al-rich titanite (CaTiSiO$_4$O–CaAlSiO$_4$OH) reaction rims on ilmenite in metamorphic rocks as a function of fH$_2$O and fO$_2$: Lithos, v. 88, p. 72–84, doi:10.1016/j.lithos.2005.08.005.

Harlov, D.E., Johansson, L., van den Kerkhof, A., and Förster, H.-J., 2006b, Transformation of a granitic gneiss to charnockite, Söndrum stenhuggeriet, Halmstad, SW Sweden: The role of advective fluid flow and diffusion during localized solid state dehydration: Journal of Petrology, v. 47, p. 3–33, doi:10.1093/petrology/egi062.

Harlov, D.E., Marshall, H., and Hanel, R., 2007a, Fluorapatite-monazite relationships in granulite-facies metapelites, Schwarzwald, southwest Germany: Mineralogical Magazine, v. 71, p. 223–234, doi:10.1180/minmag.2007.071.2.223.

Harlov, D.E., Wirth, R., and Hetherington, C.J., 2007b, The relative stability of monazite and huttonite at 300–900 °C and 200–1000 MPa: Metasomatism and the propagation of metastable mineral phases: American Mineralogist, v. 92, p. 1652–1664, doi:10.2138/am.2007.2459.

Harlov, D.E., Prochazka, V., Förster, H.-J., and Dobroslav, M., 2008, Origin of monazite-xenotime-zircon-fluorapatite assemblages in the peraluminous Melechov granite massif, Czech Republic: Mineralogy and Petrology, v. 94, p. 9–26, doi:10.1007/s00710-008-0003-8.

Harlov, D., Dunkley, D., Hansen, E., and Hokada, T., 2010, Zircon mineral trace element chemistry as a function of metamorphic grade along a traverse of lower Archean crust, Tamil Nadu, south India: Geophysical Research Abstracts, v. 12, EGU2010-5941.

Harlov, D.E., Wirth, R., and Hetherington, C.J., 2011, Fluid-mediated partial alteration of monazite: The role of coupled dissolution-reprecipitation in element redistribution and mass transfer: Contributions to Mineralogy and Petrology (in press).

Hayden, L.A., Watson, E.B., and Wark, D.A., 2008, A thermobarometer for sphene (titanite): Contributions to Mineralogy and Petrology, v. 155, p. 529–540, doi:10.1007/s00410-007-0256-y.

Heinrich, W., Andrehs, G., and Franz, G., 1997, Monazite-xenotime miscibility gap thermometry. I. An empirical calibration: Journal of Metamorphic Geology, v. 15, p. 3–16, doi:10.1111/j.1525-1314.1997.t01-1-00052.x.

Hetherington, C.J., and Harlov, D.E., 2008, Metasomatic thorite and uraninite inclusions in xenotime and monazite from granitic pegmatites; Hidra anorthosite massif, southwestern Norway: Mechanics and fluid chemistry: American Mineralogist, v. 93, p. 806–820, doi:10.2138/am.2008.2635.

Hetherington, C.J., Jercinovic, M.J., Williams, M.L., and Mahan, K., 2008, Understanding geologic processes with xenotime: Composition, chronology, and a protocol for electron probe microanalysis: Chemical Geology, v. 254, p. 133–147, doi:10.1016/j.chemgeo.2008.05.020.

Hiess, J., Nutman, A.P., Bennett, V.C., and Holden, P., 2008, Ti-in-zircon thermometry applied to contrasting Archean metamorphic and igneous systems: Chemical Geology, v. 247, p. 323–338, doi:10.1016/j.chemgeo.2007.10.012.

Hoskin, P.W.O., and Schaltegger, U., 2003, The composition of zircon and igneous and metamorphic petrogenesis, in Hanchar, J.M., and Hoskin, P.W.O., eds., Zircon: Reviews in Mineralogy and Geochemistry, v. 53, p. 27–62.

Janots, E., Negro, F., Brunet, F., Goffé, B., Engi, M., and Bouybaouène, M.L., 2006, Evolution of the REE mineralogy and monazite stability in HP-LT metapelites of the Sebtide complex (Rif, Morocco): Lithos, v. 87, p. 214–234, doi:10.1016/j.lithos.2005.06.008.

Janots, E., Brunet, F., Goffé, B., Poinssot, C., Burchard, M., and Cemic, L., 2007, Thermochemistry of monazite-(La) and dissakisite-(La): Implications for monazite and allanite stability in metapelites: Contributions to Mineralogy and Petrology, v. 154, p. 1–14, doi:10.1007/s00410-006-0176-2.

Janots, E., Engi, M., Berger, A., Allaz, J., Schwarz, J.O., and Spandler, C., 2008, Prograde metamorphic sequence of REE minerals in pelitic rocks of the Central Alps; implications for allanite-monazite-xenotime phase relations from 250 to 610 °C: Journal of Metamorphic Geology, v. 26, p. 509–526, doi:10.1111/j.1525-1314.2008.00774.x.

Janots, E., Engi, M., Rubatto, D., Berger, A., Gregory, C., and Rahn, M., 2009, Metamorphic rates in collisional orogeny from in situ allanite and monazite dating: Geology, v. 37, p. 11–14, doi:10.1130/G25192A.1.

Kebede, T., Horie, K., Hidaka, H., and Terada, K., 2007, Zircon microvein in peralkaline granitic gneiss, western Ethiopia; origin, SHRIMP U/Pb geochronology and trace element investigations: Chemical Geology, v. 242, p. 76–102, doi:10.1016/j.chemgeo.2007.03.014.

Kositcin, N., McNaughton, N.J., Griffin, B.J., Fletcher, I.R., Groves, D.I., and Rasmussen, B., 2003, Textural and geochemical discrimination between xenotime of different origin in the Archaean Witwatersrand Basin, South Africa: Geochimica et Cosmochimica Acta, v. 67, p. 709–731, doi:10.1016/S0016-7037(02)01169-9.

Krenn, E., and Finger, F., 2007, Formation of monazite and rhabdophane at the expense of allanite during Alpine low temperature retrogression of metapelitic basement rocks from Crete, Greece; microprobe data and geochronological implications: Lithos, v. 95, p. 130–147, doi:10.1016/j.lithos.2006.07.007.

Krenn, E., Ustaszewski, K., and Finger, F., 2008, Detrital and newly formed metamorphic monazite in amphibolite-facies metapelites from the Motajica Massif, Bosnia: Chemical Geology, v. 254, p. 164–174, doi:10.1016/j.chemgeo.2008.03.012.

Krenn, E., Janák, M., Fritz, F., Boska, I., and Konecny, P., 2009, Two types of metamorphic monazite with contrasting La/Nd, Th and Y signature in an ultra high-pressure metapelite from the Pohorje Mountains, Slovenia: Indications for a pressure-depended REE exchange between apatite and monazite?: American Mineralogist, v. 94, p. 801–815, doi:10.2138/am.2009.2981.

Mahan, K.H., Williams, M.L., Flowers, R.M., Jercinovic, M.J., Baldwin, J.A., and Bowring, S.A., 2006a, Geochronological constraints on the Legs Lake shear zone with implications for regional exhumation of lower continental crust, western Churchill Province, Canadian Shield: Contributions to Mineralogy and Petrology, v. 152, p. 223–242, doi:10.1007/s00410-006-0106-3.

Mahan, K.H., Goncalves, P., Williams, M.L., and Jercinovic, M.J., 2006b, Dating metamorphic reactions and fluid flow: Application to exhumation of high-P granulites in a crustal-scale shear zone, western Canadian Shield: Journal of Metamorphic Geology, v. 24, p. 193–217, doi:10.1111/j.1525-1314.2006.00633.x.

Pan, Y., and Fleet, M.E., 2002, Composition of the apatite-group minerals: Substitution mechanisms and controlling factors, in Kohn, M.J., Rakovan, J., and Hughes, J.M., eds., Phosphates: Geochemical, Geobiological and Materials Importance: Washington, D.C., Reviews in Mineralogy, Mineralogical Society of America, v. 48, p. 13–49.

Pan, Y., Fleet, M.E., and Macrae, N.D., 1993, Oriented monazite inclusions in apatite porphyroblasts from the Hemlo gold deposit, Ontario, Canada: Mineralogical Magazine, v. 57, p. 697–707, doi:10.1180/minmag.1993.057.389.14.

Poitrasson, F., Chenery, S., and Bland, D.J., 1996, Contrasted monazite hydrothermal alteration mechanisms and their geochemical implications: Earth and Planetary Science Letters, v. 145, p. 79–96, doi:10.1016/S0012-821X(96)00193-8.

Putnis, A., 2002, Mineral replacement reactions: From macroscopic observations to microscopic mechanisms: Mineralogical Magazine, v. 66, p. 689–708.

Putnis, A., 2009, Mineral replacement reactions: Thermodynamics and kinetics of water-rock interaction: Reviews in Mineralogy and Geochemistry (Oelkers, E.H., and Schott, J., eds.), v. 70, p. 87–124.

Pyle, J.M., and Spear, F.S., 2000, An empirical garnet (YAG)–xenotime thermometer: Contributions to Mineralogy and Petrology, v. 138, p. 51–58, doi:10.1007/PL00007662.

Pyle, J.M., and Spear, F.S., 2003, Four generations of accessory phase growth in low-pressure migmatites from SW New Hampshire, USA: American Mineralogist, v. 88, p. 338–351.

Pyle, J.M., Spear, F.S., Rudnick, R.L., and McDonough, W.F., 2001, Monazite-xenotime-garnet equilibrium in metapelites and a new monazite-garnet thermometer: Journal of Petrology, v. 42, p. 2083–2107, doi:10.1093/petrology/42.11.2083.

Pyle, J.M., Spear, F.S., Wark, D.A., Daniel, C.G., and Storm, L.C., 2005, Contributions to precision and accuracy of monazite microprobe ages: American Mineralogist, v. 90, p. 547–577, doi:10.2138/am.2005.1340.

Rubatto, D., Muentener, O., Barnhoorn, A., and Gregory, C., 2008, Dissolution-reprecipitation of zircon at low-temperature, high-pressure conditions (Lanzo Massif, Italy): American Mineralogist, v. 93, p. 1519–1529, doi:10.2138/am.2008.2874.

Schulz, B., Klemd, R., and Braetz, H., 2006, Host rock compositional controls on zircon trace element signatures in metabasites from the Austroalpine basement: Geochimica et Cosmochimica Acta, v. 70, p. 697–710, doi:10.1016/j.gca.2005.10.001.

Seydoux-Guillaume, A.M., Paquette, J.L., Wiedenbeck, M., Montel, J.M., and Heinrich, W., 2002a, Experimental resetting of the U-Th-Pb systems in monazite: Chemical Geology, v. 191, p. 165–181, doi:10.1016/S0009-2541(02)00155-9.

Seydoux-Guillaume, A.M., Wirth, R., Heinrich, W., and Montel, J.M., 2002b, Experimental determination of thorium partitioning between monazite and xenotime using analytical electron microscopy and X-ray diffraction Rietveld analysis: European Journal of Mineralogy, v. 14, p. 869–878, doi:10.1127/0935-1221/2002/0014-0869.

Seydoux-Guillaume, A.M., Goncalves, P., Wirth, R., and Deutsch, A., 2003, Transmission electron microscope study of polyphase and discordant monazites: Site-specific specimen preparation using the focused ion beam technique: Geology, v. 31, p. 973–976, doi:10.1130/G19582.1.

Storey, C.D., Smith, M.P., and Jeffries, T.E., 2007, In situ LA-ICP-MS U/Pb dating of metavolcanics of Norrbotten, Sweden; records of extended geological histories in complex titanite grains: Chemical Geology, v. 240, p. 163–181, doi:10.1016/j.chemgeo.2007.02.004.

Tiepolo, M., Oberti, R., and Vannucci, R., 2002, Trace-element incorporation in titanite; constraints from experimentally determined solid/liquid partition coefficients: Chemical Geology, v. 191, p. 105–119, doi:10.1016/S0009-2541(02)00151-1.

Troitzsch, U., and Ellis, D.J., 2002, Thermodynamic properties and stability of AlF-bearing titanite $CaTiOSiO_4$-$CaAlFSiO_4$: Contributions to Mineralogy and Petrology, v. 142, p. 543–563.

Tropper, P., Manning, C.E., and Essene, E.J., 2002, The substitution of Al and F in titanite at high pressure and temperature: Experimental constraints on phase relations and solid solution properties: Journal of Petrology, v. 43, p. 1787–1814, doi:10.1093/petrology/43.10.1787.

Watson, E.B., Wark, D.A., and Thomas, J.B., 2006, Crystallization thermometers for zircon and rutile: Contributions to Mineralogy and Petrology, v. 151, p. 413–433, doi:10.1007/s00410-006-0068-5.

Williams, M.L., Jercinovic, M.J., Gonclaves, P., and Mahan, K., 2006, Format and philosophy for collecting, compiling, and reporting microprobe ages: Chemical Geology, v. 225, p. 1–15, doi:10.1016/j.chemgeo.2005.07.024.

Williams, M.L., Jercinovic, M.J., and Hetherington, C.J., 2007, Microprobe monazite geochronology: Understanding geologic processes by integrating composition and chronology: Annual Review of Earth and Planetary Sciences, v. 35, p. 137–175, doi:10.1146/annurev.earth.35.031306.140228.

Wing, B., Ferry, J.M., and Harrison, T.M., 2003, Prograde destruction and formation of monazite and allanite during contact and regional metamorphism of pelites: Petrology and geochronology: Contributions to Mineralogy and Petrology, v. 145, p. 228–250.

Wones, D.R., 1989, Significance of the assemblage titanite + magnetite + quartz in granitic rocks: American Mineralogist, v. 74, p. 744–749.

Wu, Y., Zheng, Y., Gao, S., Jiao, W., and Liu, Y., 2008, Zircon U-Pb age and trace element evidence for Paleoproterozoic granulite-facies metamorphism and Archean crustal rocks in the Dabie Orogen: Lithos, v. 101, p. 308–322, doi:10.1016/j.lithos.2007.07.008.

Xirouchakis, D., and Lindsley, D.H., 1998, Equilibria among titanite, hedenbergite, fayalite, quartz, ilmenite and magnetite: Experiments and internally consistent thermodynamic data for titanite: American Mineralogist, v. 83, p. 712–725.

Xirouchakis, D., Lindsley, D.H., and Andersen, D.J., 2001a, Assemblages with titanite ($CaTiOSiO_4$), Ca-Mg-Fe olivine and pyroxenes, Fe-Mg-Ti oxides, and quartz: Part I. Theory: American Mineralogist, v. 86, p. 247–253.

Xirouchakis, D., Lindsley, D.H., and Frost, R.B., 2001b, Assemblages with titanite ($CaTiOSiO_4$, Ca-Mg-Fe olivine and pyroxenes, Fe-Mg-Ti oxides, and quartz: Part II. Application: American Mineralogist, v. 86, p. 254–264.

Zhu, X.K., and O'Nions, R.K., 1999, Zonation of monazite in metamorphic rocks and its implications for high temperature thermochronology: A case study from the Lewisian terrain: Earth and Planetary Science Letters, v. 171, p. 209–220, doi:10.1016/S0012-821X(99)00146-6.

MANUSCRIPT ACCEPTED BY THE SOCIETY 24 MAY 2010

Fluids in granulites

Jacques L.R. Touret*
Musée de Minéralogie, Ecole des Mines, 60, Blvd Saint-Michel, 75006 Paris, France

Jan-Marten Huizenga*
Department of Geology, School of Environmental Sciences and Development, North-West University, Potchefstroom 2520, South Africa

ABSTRACT

Since the discovery of CO_2 fluid inclusions in granulites, the role of fluids in the formation of these rocks has been widely studied. Owing to the complexity of the tectono-metamorphic history of granulite terrains, fluid inclusion data alone are not sufficient. They need to be integrated with geochemical and mineralogical studies done on the same rock samples. A clear understanding of the tectono-metamorphic history of granulite terranes is also indispensable. The widespread occurrence of CO_2 and the later discovered high-salinity aqueous fluid inclusions support the idea that the lower crust underwent fluid flow and that both carbonic and brine fluids played a role in its formation. Both low-H_2O-activity fluids play a similar role in destabilizing hydrous mineral phases. Furthermore, experimental studies have shown that brine fluids have a much larger geochemical effect on granulites than initially expected. These fluids are far more mobile in the lower crust compared with CO_2 and also have the capability for dissolving numerous minerals. As in the example of the Limpopo Complex, fluid inclusions and many metasomatic features observed in granulite terranes can thus be explained only by large-scale movement of high-salinity aqueous fluids and, to a lesser extent, CO_2, implying that lower-crustal granulites are not as dry as previously assumed. Similar brines and CO_2-rich fluids are also found in mantle material, most likely derived from deeply subducted supracrustal protoliths.

INTRODUCTION

The role of fluids in the formation of granulites has been a matter of discussion for many years. On the one hand, some investigators argue for a dry lower crust. This interpretation is largely based on experimental petrological work and the physical properties of the lower crust (e.g., Thompson, 1983; Clemens and Vielzeuf, 1987; Stevens and Clemens, 1993; Yardley and Valley, 1997). In this view, granulite metamorphism is the result of vapor or fluid absent from the melting processes. Others, on the other hand, favor granulite metamorphism in which low-H_2O-activity fluids do play an essential role (e.g., Touret, 1981, 1986; Newton et al., 1998; Harlov and Wirth, 2000; Perchuk et al., 2000a). The purpose of this chapter is to give an overview of the constraints on this debate about the role of fluids in granulite metamorphism that can be offered by fluid inclusion research, with special reference to the Limpopo high-grade terrane.

---

*E-mails: jacques.touret@ensmp.fr; jan.huizenga@nwu.ac.za.

Touret, J.L.R., and Huizenga, J.-M., 2011, Fluids in granulites, in van Reenen, D.D., Kramers, J.D., McCourt, S., and Perchuk, L.L., eds., Origin and Evolution of Precambrian High-Grade Gneiss Terranes, with Special Emphasis on the Limpopo Complex of Southern Africa: Geological Society of America Memoir 207, p. 25–37, doi:10.1130/2011.1207(03). For permission to copy, contact editing@geosociety.org. © 2011 The Geological Society of America. All rights reserved.

Fluid Inclusion Studies: Some General Considerations

Fluid inclusions, which occur in most rock-forming minerals as minute cavities containing liquid and/or vapor or solid phases, can provide valuable information on all steps of the rock's evolution. These inclusions were among the first objects discovered under a polarizing microscope (Sorby, 1858), but for a number of reasons, both instrumental and theoretical, they remained (and, to some extent, still remain) largely ignored in most petrological studies. The situation, however, changed markedly in the 1970s, when a generation of new instruments (heating-freezing microscopic stages, later completed by Raman and infrared micro-spectroscopy) was introduced. In addition, a better understanding of low-temperature fluid-phase behavior, in particular for carbonic fluids (e.g., Van den Kerkhof, 1990; Thiéry et al., 1994) and fluid-rock interaction for a wide range of pressure-temperature (P-T) conditions improved the interpretation of fluid inclusion data significantly. These developments allowed the integration of fluid inclusion data into a general scheme of petrological and geochemical studies.

Fluid inclusion studies are considered by many researchers as being highly specialized and needing to be done by experts. Fluid inclusion studies require adequate sampling and a time-consuming detailed study of individual samples. As a result, fluid inclusion studies are not always done, although they may add essential data. In many cases the fluid inclusion study is done independently by a specialist who is not involved with other aspects of the study, which may result in interpretation problems. This situation is obviously not satisfactory; inclusions are part of the rock, and, as such, they should be studied as systematically and in detail as in any other aspect. Moreover, as has been advocated in a number of publications (e.g., Touret, 2001, and references therein), fluid inclusion data can only be interpreted appropriately in combination with other geological data obtained from the same samples. In other words, the best chance of success requires that mineralogical, geochemical, and fluid inclusion studies are done by the same person or by a team of researchers working closely together.

Fluid inclusion studies have demonstrated the almost systematic occurrence of unexpected (i.e., not inferred from the mineral assemblage) fluids in the lower crustal or mantle rocks: CO_2 in granulites (Touret, 1971) and charnockites (Santosh, 1986), N_2 in eclogites (Andersen et al., 1989), brines in diamonds (Izraeli et al., 2001, 2004), etc. These findings have raised a marked interest within the geological community as well as some doubts for numerous researchers on the issue of whether these inclusions are really representative remnants of fluids trapped at depth. These reservations include the fact that at deep crustal levels, open spaces will collapse, preventing the existence of a free fluid phase, or that they can only exist in small, chemically inert, isolated pockets (e.g., Yardley and Valley, 1997). Another objection against the reliability of fluid inclusions representing equilibrium fluids trapped at high pressure and temperature is that they cannot survive uplift to the Earth's surface without any modification of their composition or density.

It is true that fluid inclusions in minerals are not simple, perfectly closed containers that give a direct indication of which fluids percolate through the rock system at the time of their formation. Factors complicating the interpretation of fluid inclusion data may include selective fluid trapping, selective water leakage (Bakker and Jansen, 1990), an increase of the fluid salinity from quartz recovery (Van den Kerkhof et al., 2004), fluid reactions with the mineral host or, above all, "transposition," i.e., the formation of successive fluid-inclusion generations at changing pressure and temperature conditions (e.g., Hollister and Crawford, 1981; Roedder, 1984; Shepherd et al., 1985; Andersen et al., 2001).

Fluid inclusions can be, as a first approximation, considered constant volume and constant composition (i.e., constant molar volume or density) systems. This principle requires the following assumptions: (1) The volume of the cavity does not change after the fluid inclusion has been formed, (2) the fluid inclusion does not leak, and (3) the chemical composition of the fluid in the inclusion does not change. Constant density fluids show an approximately linear relationship between pressure and temperature: the isochore (Fig. 1). The isochore principle has some important implications for the interpretation of fluid inclusion data. First, fluid inclusions of a particular molar volume can be trapped at different pressure and temperature conditions along the corresponding isochore; i.e., high-density fluid inclusions do not necessarily indicate a high pressure and temperature of trapping (e.g., Morrison and Valley, 1988). Second, high-density fluid inclusions that are trapped at high pressure and temperature can be preserved during exhumation if the retrograde P-T path of the rock is approximately parallel to that of the isochore. Third, the distribution of fluid inclusion densities can be an important

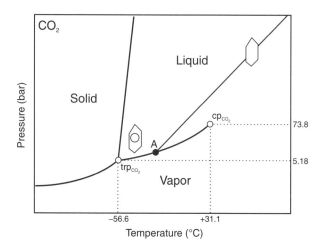

Figure 1. Low-temperature–pressure part of the CO_2 phase diagram. The isochore represents the univariant part of the homogeneous fluid inclusion; A—pressure and temperature of homogenization into the liquid phase; cp_{CO_2}—critical point of CO_2; trp_{CO_2}—triple point of CO_2.

tool in reconstructing the retrograde P-T path of a rock (Fig. 2) (e.g., Vityk and Bodnar, 1995).

FLUID INCLUSION STUDIES IN HIGH-GRADE METAMORPHIC ROCKS

Fluid inclusion studies in high-grade metamorphic rocks are more complicated than those in low-grade rocks. During high-grade metamorphism, rocks are typically subjected to prograde grain-size scale deformation and net-transfer reactions, making the preservation of fluid inclusions highly unlikely. During retrograde metamorphism, high-temperature grain-size-scale deformation may still occur, depending on the exhumation mechanism. Furthermore, retrograde hydration reactions may also occur, depending on the presence of aqueous fluids. In addition, the structural-metamorphic history of high-grade metamorphic rocks is generally complex. The rocks may have undergone more than one deformation event (possibly associated with infiltration of fluids) either in the deep crust under highly ductile conditions or during exhumation under brittle-ductile or brittle conditions. During exhumation, the rocks may also have been subjected to other geological processes such as, for example, contact metamorphic events related to magma emplacement and meteoric fluid infiltration (Huizenga et al., this volume). The identification of different fluid inclusion generations, and their association with specific tectono-metamorphic events, is, therefore, the most important and also the most difficult aspect of any fluid inclusion study in high-grade metamorphic rocks. We will use some examples from the Limpopo high-grade metamorphic terrane to illustrate the above mentioned issues.

Fluids in the Limpopo High-Grade Terrane

The Limpopo high-grade terrane is subdivided into three subzones (Fig. 3, inset): the Southern Marginal Zone (SMZ), the Central Zone (CZ), and the Northern Marginal Zone (NMZ) (e.g., van Reenen et al., 1990). The Southern and Northern Marginal Zones are the high-grade metamorphic equivalents of the greenstone-gneiss terranes of the adjacent Kaapvaal and Zimbabwe Cratons, respectively, whereas the Central Zone comprises metasedimentary (e.g., marble, calc-silicate gneisses, garnet-biotite gneisses, banded iron formation) and meta-igneous rocks (e.g., van Reenen et al., 1990). Numerous fluid inclusion studies were carried out in the Southern and Central Zones, of which the results are summarized in Table 1.

Southern Marginal Zone

The Southern Marginal Zone (SMZ) (Fig. 3) is characterized by a northern high-grade granulite (800–900 °C) and a southern lower grade, hydrated granulite zone (~600 °C), and separated from each other by a retrograde orthoamphibole isograd (van Reenen, 1986). The SMZ is separated from the Kaapvaal Craton by the Hout River Shear Zone, which developed during the exhumation of the granulite terrane (Fig. 3). In the SMZ smaller thrust and strike-slip shear zones are present that are also associated with the exhumation of the granulites (Fig. 3) (Smit and van Reenen, 1997). The granulites of the SMZ (Fig. 3) are typically

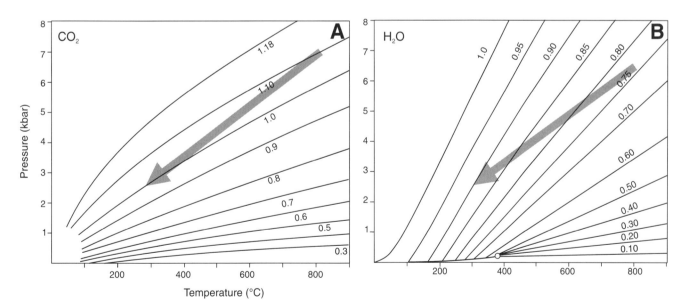

Figure 2. Cooling-decompression retrograde pressure-temperature (P-T) path superimposed on CO_2 (A) and H_2O isochores (B). During cooling and decompression, the CO_2 fluid trapped in the inclusion will become overpressured, whereas H_2O will become underpressured, which may lead to a volume change in the inclusion and thus a readjustment of the fluid inclusion density. Note that for the same decompression-cooling path (gray arrow), the density change for the H_2O (20% increase) is larger and opposite from that for CO_2 (9% decrease). In extreme cases, over- and underpressure may result in explosion and implosion of the inclusion, respectively.

characterized by high-grade granite-greenstone lithologies. Peak metamorphic conditions were followed by, depending on the structural setting, either decompressional cooling or a combination of early decompressional cooling, followed by near-isobaric cooling (Perchuk et al., 2000b).

Cooling and decompression took place during uplift and thrusting of the granulites along shear zones onto the relatively cold Kaapvaal Craton. This resulted in simultaneous cooling and heating of the hanging wall (granulites) and footwall (granite-greenstone terrane of the Kaapvaal Craton), respectively. The SMZ was affected by only one event of high-grade metamorphism, which makes it different from the Central Zone (CZ) of the Limpopo high-grade terrane that has been affected by (at least) two high-grade tectono-metamorphic events.

Evidence for peak-metamorphic fluids in the Southern Marginal Zone. Fluid inclusions and reaction textures. Fluid inclusion studies have provided evidence for the presence of brines and carbonic CO_2-rich fluids, existing under conditions of immiscibility at the peak of granulite facies metamorphism (Table 1). Figure 4A shows pseudo-secondary fluid inclusions in orthopyroxene that constitute numerous unidentified solid phases, together with a pure low-dense CO_2 fluid phase. Similar fluid inclusions were also found in matrix quartz (Fig. 4B). These inclusions clearly indicate the coexistence of a highly saline fluid with almost pure CO_2 (<5 mol% CH_4) at the peak of metamorphism. There is no petrographic evidence for fluid-fluid immiscibility, but considering the high salinity of the brines, this is almost unavoidable (Johnson, 1991).

Other evidence of the presence of high-salinity aqueous fluids is the common observation of high-temperature reaction textures in granulites in the SMZ (Fig. 5). Both perthitic and antiperthitic feldspars occur at the contact of the matrix quartz with garnet and as rims around quartz inclusions within garnet porphyroblasts. Similar K-feldspar reaction textures have been

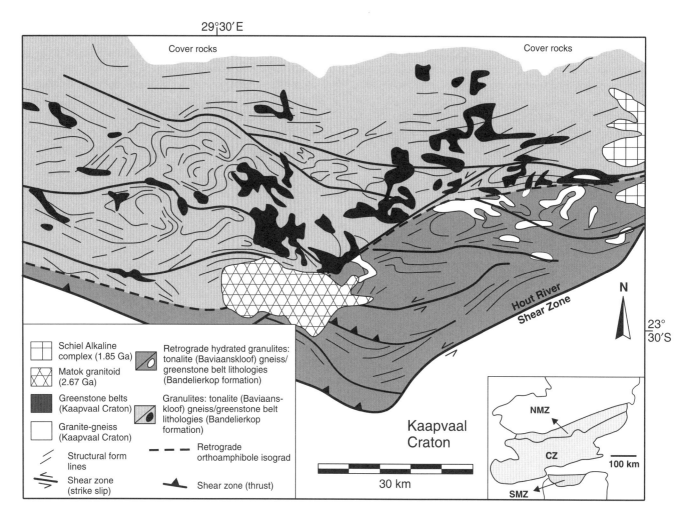

Figure 3. Geological map of the Southern Marginal Zone (SMZ) of the Limpopo Complex (modified from Smit and van Reenen, 1997), showing the granulite and retrograde hydrated granulite terranes, separated from each other by the retrograde orthoamphibole isograd. CZ—Central Zone; NMZ—Northern Marginal Zone.

The mineral assemblages found in the hydrated part of the SMZ were used to calculate H_2O activities in both mafic rocks and metapelites (e.g., Van den Berg and Huizenga, 2001). These calculations show that the retrograde hydration event is related to two compositionally different fluids present at the same time, one with an H_2O activity of 0.1–0.2, and one with an H_2O activity of 0.8–0.9, suggesting that at least two different fluid sources were available during retrograde metamorphism. The low H_2O activity fluid corresponds to the CO_2-rich fluids and the highly saline fluids. The high H_2O activity fluid, on the other hand, corresponds to the low-salinity aqueous fluids.

The most likely crustal fluid sources include devolatilization reactions in the footwall of the Hout River Shear Zone (i.e., the greenstones of the Kaapvaal Craton), crystallizing granitic melts in the middle crust releasing water-rich fluids (Stevens, 1997), and meteoric water infiltration along shear zones (e.g., Yardley et al., 2000). The emplacement of hot granulites onto the Kaapvaal Craton during exhumation (van Reenen and Hollister, 1988) initiates devolatilization reactions in the underlying greenstone belt lithologies, producing H_2O-CO_2 fluids. These fluids migrate upward along active shear zones into the granulites that are uplifted, and may become progressively enriched in CO_2 as water is removed by retrograde hydration reactions. In addition to this process, water-rich fluids may be introduced into the middle crust either from the surface through shear zones (Yardley et al., 2000) or from crystallizing granitic melts in the middle crust (Stevens, 1997). These fluids may either remain water rich or become more enriched in carbonic fluid species as a result of water-graphite interaction (Stevens, 1997; Huizenga et al., this volume). The exact fluid composition will then depend on the availability of graphite and the prevailing oxygen fugacity. Irrespective of whether a water-rich fluid or an aqueous-carbonic fluid is formed, both fluids will additionally be affected by retrograde hydration reactions. The water-rich fluid will then evolve into a high-salinity aqueous fluid (e.g., Bennett and Barker, 1992; Markl and Bucher, 1998; Markl et al., 1998), whereas the aqueous-carbonic fluid becomes more carbonic. From the above it is clear that different fluid sources and fluid-rock interaction have resulted in compositionally different fluid inclusions. Furthermore, fluid infiltration through shear zones is variable on a small scale (e.g., Pili et al., 1997), which will contribute even further to the complexity of the interpretation of the fluid inclusion results.

Central Zone

The Central Zone (CZ) is situated between the Northern and Southern Marginal Zones of the Limpopo Complex terrane (Fig. 3). The CZ has three distinct structural domains (Smit et al., this volume). The first domain comprises large-scale isoclinal folds and sheath folds, whereas the second domain comprises the major SW-NE–trending Tshipise Straightening Zone (TsSZ), which bounds the first domain in the south. These structures developed before ca. 2.6 Ga (van Reenen et al., 2008). Superimposed onto these early structural features is a system of discrete high-grade shear zones that reflect evidence for a superimposed tectono-metamorphic event dated at ca. 2.0 Ga (van Reenen et al., 2008). High-grade gneisses associated with ca. 2.6 Ga structures show relatively high pressure during decompression cooling, whereas sheared gneisses that developed within ca. 2.0 Ga shear zones indicate a relatively low pressure during decompression cooling (Boshoff et al., 2006; van Reenen et al., 2008).

The high-pressure P-T path is linked to the low-pressure P-T path by an isobaric (5.5 kbar) heating path that occurred at ca. 2.0 Ga, resulting in the widespread formation of polymetamorphic granulites in the CZ (Boshoff et al., 2006; van Reenen et al., 2008). In addition, certain areas have also been subjected to mid-crustal local contact metamorphism such as the area around the ca. 2.61 Ga Bulai intrusive (e.g., Huizenga et al., this volume). The above shows that the structural-metamorphic complexity of the CZ makes the interpretation of fluid inclusion data far more complicated compared with those of the SMZ.

Fluids in the Central Zone rocks. Numerous fluid inclusion studies have been done on rocks from the CZ (Table 1). Only two of those studies (Hisada et al., 2005; Huizenga et al., this volume) were done on rocks for which their timing in the structural-metamorphic history is known. The other studies were done on rocks that cannot be placed in the structural-metamorphic context, and their results are therefore difficult, if not impossible, to interpret.

The ca. 2.6 Ga metapelitic rocks that were studied by Huizenga et al. (this volume) show similar results compared with those of the SMZ. Peak metamorphic fluids that were found in high-temperature Mg-rich garnet include CO_2. Brine inclusions in quartz blebs in garnet, and the presence of metasomatic K-feldspar veining around quartz (Fig. 5), which is identical to the ones that occur in the SMZ, indicate the coexistence of this fluid with CO_2. The studied metapelites were subjected to contact metamorphism during the emplacement of the Bulai granitoid intrusive. This has resulted in a second generation of garnet that has lower Mg concentrations compared with the first garnet generation. This garnet is unfortunately fluid-inclusion free, so one cannot determine in which fluid regime the Bulai emplacement took place.

The retrograde fluid evolution cannot be established with certainty. Although the studied samples do not show any evidence for a ca. 2.0 Ga deformation or metamorphism, it cannot be excluded that existing fluid inclusions were affected by this event or that a fluid phase infiltrated these rocks during the ca. 2.0 Ga event. Therefore, the interpretation of fluid inclusions in matrix quartz becomes almost impossible.

Hisada et al. (2005) studied ca. 2.0 Ga migmatites and metapelites, which are reworked 2.6 Ga granulites. Most of the fluid inclusions in these rocks were observed in quartz, which makes it impossible to put them in a metamorphic context. Metamorphic zoning textures in the rocks, however, are proof that a water-rich fluid was present during migmatization and retrograde metamorphism. CO_2 was also present, but the exact relationship between CO_2 and H_2O could not be established.

THE ROLE OF FLUIDS IN GRANULITES

Regional granulite terranes comprise mainly high-grade metamorphosed supracrustal rocks (such as those in the Limpopo Complex), whereas lower crustal xenoliths in volcanic material include mafic igneous rocks derived from the crust-mantle boundary (Bohlen and Mezger, 1989). This suggests that the continental lower crust is granulitic and interacts with the underlying continental upper mantle. These findings imply a division of the lower continental crust in two entities, an upper (U) lower crust, and a lower (L) lower crust, respectively (Fig. 6A). The possible existence of remnants of a free fluid phase in granulites thus relates to a larger scale issue: the role of fluids in the deepest part of the continental crust. The conventional view, best illustrated by the classical model of Etheridge et al. (1983), is that a free fluid phase, mainly aqueous, occurs only at peak conditions in the upper part of the continental crust (Fig. 6B). In this model, however, a free H_2O-rich fluid phase does not exist in mid-crustal migmatites, the source of most granite. H_2O can occur in mica and amphibole or, when these minerals are broken down, by partial melting reactions immediately dissolved in the granitic melts. This marks the onset of the "vapor (or fluid)-absent" regime, supposed to extend into the lower crust and underlying mantle (H_2O barrier in Fig. 6A).

Fluid-inclusion studies, combined with petrologic observations, however, require substantial modifications to this model. These fluids must have a low-water activity in order for them to coexist with granitic melts and anhydrous mineral phases: Any H_2O-rich fluid will immediately induce partial melting of the surrounding rock, generating a granitic melt that will dissolve the interstitial fluid (e.g., Stevens and Clemens, 1993).

Pure CO_2 and high-salinity brines meet these conditions, and although it has taken some time for this to be widely accepted, there is little doubt that both were present during

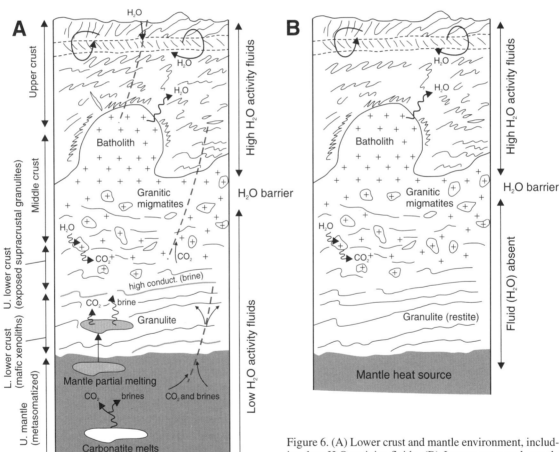

Figure 6. (A) Lower crust and mantle environment, including low-H_2O-activity fluids. (B) Lower crust and mantle environment, excluding an H_2O fluid (i.e., fluid absent model). H_2O barrier: crustal environment dominated by partial melting in which a free high-H_2O-activity fluid cannot exist. Dashed lines indicate shear-fault zones. Note that CO_2 derived from subducted carbonates is not indicated in this diagram. See text for further explanation.

most of the metamorphic evolution, notably at peak conditions. Virtually all regionally exposed granulite terranes that have been studied so far contain pure CO_2 inclusions of variable density, reaching extreme values (>1.1 g/cm^3) in the case of the superdense inclusions (e.g., Tsunogae et al., 2002; Santosh and Tsunogae, 2003). It has been argued (Lamb et al., 1987) that these inclusions were late, trapped during the final stage of retrograde evolution. Although this may be true for secondary CO_2-rich fluid inclusions in quartz, this cannot be the case for primary CO_2-rich fluid inclusions in peak metamorphic minerals, of which the pressure-temperature data obtained from fluid inclusions show a perfect match with those obtained from mineral geothermobarometry (e.g., Touret and Hartel, 1990; Santosh et al., 2008). In these cases the existence of peak metamorphic CO_2 fluids at high and ultrahigh temperatures cannot be questioned.

Brine inclusions occur in much smaller quantities and are much smaller and therefore more difficult to detect. Some contain only a small amount of aqueous liquid at room temperature, in many cases without a vapor bubble. Very often the cavity is squeezed around a number of solid minerals, commonly carbonates and/or salts (Touret and Huizenga, 1999; Van den Berg and Huizenga, 2001). In some cases high-salinity fluid inclusions may be metastable. In these initially halite-free inclusions a halite crystal will only grow upon heating during microthermometric heating experiments (e.g., see figs. 2b and 2c in Huizenga et al., 2005).

There is, however, compelling evidence that, unlike CO_2, brine remnants found in inclusions represent a major fraction of the fluids, which existed at peak or close to peak conditions. As in Limpopo, the widespread occurrence of microtextures like myrmekites or K-feldspar veining indicates extensive metasomatic phenomena, which can only have been caused by a brinelike fluid having circulated at intergranular boundaries. Summarizing, we consider that the existence of both granulite fluids is proven. They are either directly observed as remnants still present in protected areas (e.g., CO_2 primary inclusions in the core of garnet) or, for the brines indirectly, by the traces that they have left in the rock mineral assemblage.

It must be mentioned that there is a notable difference among the relative abundance of both (CO_2 and brine) inclusion types, depending on granulite P-T conditions. Well preserved large (30–50 μm in size) CO_2 inclusions dominate in high- to ultrahigh temperature granulites, typically emplaced at 900 to >1000 °C and at a depth <30 km (5–8 kbar). The P-T data obtained from these inclusions show a good match with mineral data. They occur in quartz and in other minerals, notably pyroxene, feldspar, and cordierite. They can be abundant, representing up to 5 vol% of the mineral host, provided that no recrystallization has occurred. In high-pressure granulites, on the other hand, brine remnants are more abundant and, moreover, they predate CO_2 inclusions, which occur only during the final stage of the P-T path, corresponding to a significant temperature increase, i.e., granulitized eclogites. Generally speaking, it can be concluded that CO_2 inclusions occur mainly in domains that have undergone partial melting (e.g., Touret and Dietvorst, 1983; Olsen, 1987; Whitney, 1992), whereas brine inclusions are much more widespread and more transposed.

Fluid Quantity

Fluid inclusions alone are unable to give information about the absolute fluid amount present at the time of their formation. For CO_2, one can obtain a lower constraint on the amount of fluid that was present from the fluid preserved in the inclusions, which in a few cases is surprisingly high. In some high-temperature granulites (e.g., Indian charnockites and/or enderbites) the presence of primary inclusions in garnet cores amounts to a few weight percentages, high enough to suggest that the fluid amount at peak conditions was quite large. Evidence provided by isotope signatures of carbon (Hoefs and Touret, 1975) and helium (e.g., Dunai and Touret, 1993) shows that CO_2 has mainly a mantle origin, brought into the lower crust by magmatic melts, mainly gabbroic or intermediate in composition, which are also responsible for the high temperature of granulite metamorphism (e.g., Touret and Huizenga, 1999). Therefore, it is not surprising that large amounts of CO_2 fluid inclusions are always found in intrusive rocks, which escaped postmagmatic recrystallization. In the case of the Southern Marginal Zone of the Limpopo Complex there is certainly a clear indication that internally derived fluids were present as well.

The case of brines is far more complicated. For example, Bennett and Barker (1992), Markl and Bucher (1998), and Markl et al. (1998) showed that brines could be formed from a water-rich fluid, which becomes enriched in salts through progressive retrograde hydration reactions. Van den Kerkhof et al. (2004) also demonstrated that high-salinity fluids may be formed during quartz recovery in granulites. Furthermore, it is also true that many brine inclusions appear to be re-equilibrated during post-metamorphic evolution. However, in some cases, brine inclusions are directly linked to their protolith (e.g., former evaporites) so that they must be remnants of the pre-metamorphic sedimentary fluids (e.g., Touret and Dietvorst, 1983). Then, when they followed a metamorphic P-T path parallel to the fluid isochore, minute salinity differences could be preserved, even in rocks that went through extreme metamorphic conditions. The best example to illustrate the preservation of sedimentary fluids through the metamorphic cycle include high- to ultrahigh-pressure rocks of the Dabie Shan in China, where the stable isotope oxygen signature points to pre-metamorphic interaction with fresh (meteoric) waters and also seawater (Fu et al., 2002).

Overall, the fluids preserved in inclusions are quite small, and their interpretation is not straightforward. The extent of mineral reactions at intergranular boundaries suggests that the amount that was present cannot be neglected. The saline aqueous fluids circulate easily along grain boundaries (Watson and Brenan, 1987) and induce mineralogical changes, which remain quite apparent when the deep rocks have been exhumed to the Earth's surface. Feldspar micro-veining is an example of these metasomatic mineralogical changes, which have been described in detail for many regions such as the Ivrea Zone (Harlov and Wirth, 2000), in Sri Lanka (Perchuk et al., 2000a), and the Limpopo Complex (Van den Berg and Huizenga, 2001), but it is our experience that they occur virtually in any granulite terrane. Moreover, a number of other features point to large-scale activity of saline aqueous fluids during peak and early retrograde granulite conditions. These include incipient charnockites (i.e., fluid-assisted dehydration zones) (e.g., Perchuk and Gerya, 1993), regional oxidation of granulite terranes (Harlov et al., 1997), and carbonated megashear zones (e.g., Newton, 1989; Newton and Manning, 2002). All these features, surprisingly underestimated in most recent metamorphic petrology textbooks, have been discussed in detail elsewhere in the literature (e.g., Newton et al., 1998).

Another important observation that suggests the possible existence of abundant lower crustal brines is obtained from seismic reflectivity and/or electrical conductivity. The presence of horizontal reflectors and large, regional-size conductive layers in tectonically active areas and extensional basins (Wannamaker et al., 1997, 2004) are interpreted to have been caused by the presence of large volumes of saline fluids.

Summarizing, besides CO_2, large amounts of saline fluids did exist at peak granulite conditions. Except for a few remnants preserved in inclusions, they were expelled from the rock system during post-metamorphic uplift, leaving only the traces that indicate that large fluid quantities must have percolated through these "dry" rocks. A large amount of fluid may have remained in the lower crust for a long period of time. This is supported by the abundance of CO_2 inclusions found in late shear zones and, above all, the existence of high-conductivity layers in Paleozoic granulite terranes after thermal re-equilibration (e.g., Touret and Marquis, 1994).

Now that we have showed the evidence for the systematic occurrence of two granulite fluid types, namely dense, pure CO_2 and high-salinity aqueous brines, it must be emphasized that their respective roles were very different owing to their contrasting possibilities for dissolving minerals or elements at high pressure and temperature. CO_2 has a low solubility for most mineral phases and elements and acts almost as an inert component, only reducing the H_2O activity to stabilize anhydrous mineral assemblages. Mineral (e.g., calcite, anhydrite, corundum, quartz) and element solubilities (including Al), on the other hand, are high in brines at granulite facies P-T conditions. This fluid is thus able to induce large-scale metasomatic effects, for which the ones described in the literature so far are probably only the tip of the iceberg. A notable amount of experimental data has been produced by Newton and Manning (2002, 2005, 2006, 2008), and Tropper and Manning (2007) at the University of California. We can foresee that these results will change significantly our perception of lower crustal processes in the near future.

Lower Crust–Mantle Connection

Fluid inclusions identical to those found in granulites have also been found in upper mantle rocks that occur as xenoliths in basalts. Pure CO_2 inclusions were first discovered by E. Roedder (1965) in mantle xenoliths from Hawaii basalts. His findings were later confirmed when similar inclusions were found in xenoliths within many other alkali basalts. The density of these inclusions corresponds to a depth of formation of ~30 km. However, this result does not relate to the formation conditions but rather reflects the maximum internal pressure (~10 kb) that an inclusion can resist while embedded in hot lava during eruption. This implies, as discussed in detail elsewhere (Touret, 2010; Touret et al., 2010), that the level at which free CO_2 occurs may be much deeper, most likely at a pressure of ~20 kbar at which the mineral phases have equilibrated. In fact, the only fluid inclusions formed at mantle conditions that can survive uplift are those that occur in diamonds. Minute fluid inclusions are abundant in some (cloudy) diamonds, but their small size and physical properties make their study difficult. Nevertheless, O. Navon and co-workers at the Hebrew University of Jerusalem found an impressive list of high-density fluids that show a continuous compositional range between carbonatitic and saline end members (e.g., Izraeli et al., 2001, 2004). Furthermore, alkali chlorides have been found in kimberlites, either as mineral phases or in melt inclusions (Kamenetsky et al., 2004, 2009; Maas et al., 2005). A close relationship also exists between brines and carbonatites, which is indicated by the common occurrence of immiscible brines in primary carbonatite minerals such as apatite (e.g., Morogan and Lindblom, 1995). Although carbonatites and kimberlites are relatively rare at the Earth's surface, the widespread evidence of mantle metasomatism, assumed by most workers to be caused by carbonatite melts (e.g., Coltorti and Grégoire, 2008), indicates that they are probably common in the mantle. In support of this, experiments have shown that carbonatitic melts do have a great potential to metasomatize the mantle owing to their high mobility (Hammouda and Laporte, 2000). The breakdown of carbonate mineral phases, liberating the CO_2 found in inclusions, depends on the pressure, temperature, and oxygen fugacity and may thus vary locally within the mantle (e.g., Dalton and Wood, 1995).

CONCLUSIONS

We believe that these mantle fluids, CO_2 and brines, are the major source of granulite fluids. As we have mentioned previously, the situation is more complicated for brines, for which not only the density but also the composition changes continuously during rock evolution. Together with the findings of alkali chlorides in kimberlites, carbonatites, and diamonds it is now clear that both fluids play a very important role in the lower crust and mantle. Clearly, finding adequate chemical tracers (e.g., Cl stable isotopes) is a major challenge for present-day geochemistry.

Both CO_2 and chlorine were probably introduced into the lower crust and mantle through plate tectonic processes since the

Archean (e.g., Santosh and Omori, 2008). With respect to CO_2, this option was discussed in detail and shown to be viable by Santosh and Omori (2008). A complete discussion would be beyond the scope of this chapter, but generally it can be argued that our findings agree with a scenario first envisaged by Menzies et al. (1985), more recently by Santosh and Omori (2008), and further expressed to us by R.C. Newton (June 2009, personal commun.). The subducted carbonate-rich sediments can either release CO_2 into the lower crust from decarbonation reactions during subduction and/or react with mantle peridotite in the shelf region of the carbonated solidus. The resulting carbonatite magmas, which are enriched in H_2O, alkalis, and halogens, metasomatize the overlying mantle wedge, creating enriched mantle. This material is intrinsically unstable, being rich in hyperfusibles and, probably, radioactivity. It will eventually melt to form alkaline basalts, outgassing early, being rich in CO_2 and Cl, and stall out in the lower crust. There, these mantle fluids may mix with other internally derived fluids, which are produced by prograde metamorphic reactions and modified by fluid-rock interaction, or (for CO_2) expelled from saturated granitic melts. The "dry" lower crust is without doubt a place where extensive fluid-dominated processes have taken place.

ACKNOWLEDGMENTS

We would like to thank the editors for giving us the opportunity to write this chapter. We would like to thank R.C. Newton and D.D. van Reenen for discussion and for their comments on an earlier version of this chapter. Critical comments and suggestions by P. Barbey, C. Manning, and M. Santosh helped us to improve the chapter significantly.

REFERENCES CITED

Andersen, T., Burke, E.A.J., and Austrheim, H., 1989, Nitrogen-bearing, aqueous fluid inclusions in some eclogites from the Western Gneiss Region of the Norwegian Caledonides: Contributions to Mineralogy and Petrology, v. 103, p. 153–165, doi:10.1007/BF00378501.

Andersen, T., Frezzotti, M.-L., and Burke, E.A.J., eds., 2001, Fluid Inclusions: Phase Relationships–Methods–Applications: Lithos, v. 55, 320 p.

Bakker, R.J., and Jansen, J.B.H., 1990, Preferential water leakage from fluid inclusions by means of mobile dislocations: Nature, v. 345, p. 58–60, doi:10.1038/345058a0.

Bennett, D.G., and Barker, A.J., 1992, High salinity fluids: The result of retrograde metamorphism in thrust zones: Geochimica et Cosmochimica Acta, v. 56, p. 81–95, doi:10.1016/0016-7037(92)90118-3.

Bohlen, S., and Mezger, K., 1989, Origin of granulite terranes and the formation of the lowermost continental crust: Science, v. 244, p. 326–329, doi:10.1126/science.244.4902.326.

Boshoff, R., van Reenen, D.D., Smit, C.A., Perchuk, L.L., Kramers, J.D., and Armstrong, R.A., 2006, Geological history of the Central Zone of the Limpopo Complex: The west Alldays area: Journal of Geology, v. 114, p. 699–716, doi:10.1086/507615.

Clemens, J.D., and Vielzeuf, D., 1987, Constraints on melting and magma production in the crust: Earth and Planetary Science Letters, v. 86, p. 287–306, doi:10.1016/0012-821X(87)90227-5.

Coltorti, M., and Grégoire, M., 2008, Metasomatism in oceanic and continental lithospheric mantle: Introduction, in Coltorti, M., and Grégoire, M., eds., Metasomatism in Oceanic and Continental Lithospheric Mantle: Geological Society [London] Special Publication 293, p. 1–9.

Dalton, J.A., and Wood, B.J., 1995, The stability of carbonate under upper-mantle conditions as a function of temperature and oxygen fugacity: European Journal of Mineralogy, v. 7, p. 883–891.

Dunai, T.J., and Touret, J.L.R., 1993, A noble gas study of a granulite sample from the Nilgiri Hills, southern India—Implications for granulite formation: Earth and Planetary Science Letters, v. 119, p. 271–281, doi:10.1016/0012-821X(93)90138-Y.

Etheridge, M.A., Wall, V.J., and Vernon, R.H., 1983, The role of the fluid phase during regional metamorphism and deformation: Journal of Metamorphic Geology, v. 1, p. 205–226, doi:10.1111/j.1525-1314.1983.tb00272.x.

Franz, L., and Harlov, D.E., 1998, High-grade K-feldspar veining in granulites from the Ivrea-Verbano Zone, Northern Italy: Fluid flow in the lower crust and implications for granulite facies genesis: Journal of Geology, v. 106, p. 455–472.

Fu, B., Zheng, Y.-F., and Touret, J.L.R., 2002, Petrological, isotopic and fluid inclusion studies of eclogites from Sujiahe, NW Dabie Shan (China): Chemical Geology, v. 187, p. 107–128, doi:10.1016/S0009-2541(02)00014-1.

Fyfe, W.S., 1973, The granulite facies, partial melting and the Archean crust: Philosophical Transactions of the Royal Society of London, ser. A, v. 273, p. 457–462.

Hammouda, T., and Laporte, D., 2000, Ultrafast mantle impregnation by carbonatite melts: Geology, v. 28, p. 283–285, doi:10.1130/0091-7613(2000)28<283:UMIBCM>2.0.CO;2.

Harlov, D.E., and Förster, H.-J., 2002, High-grade fluid metasomatism on both a local and regional scale: The Seward Peninsula, Alaska, and the Val Strona di Omegna, Ivrea-Verbano Zone, northern Italy. Part I: Petrography and silicate mineral chemistry: Journal of Petrology, v. 43, p. 769–799, doi:10.1093/petrology/43.5.769.

Harlov, D.E., and Wirth, R., 2000, K-feldspar-quartz and K-feldspar-plagioclase phase boundary interactions in garnet-orthopyroxene gneisses from the Val Strona di Omegna, Ivrea-Verbano Zone, northern Italy: Contributions to Mineralogy and Petrology, v. 140, p. 148–162, doi:10.1007/s004100000185.

Harlov, D.E., Newton, R.C., Hansen, E.C., and Janardhan, A.S., 1997, Oxide and sulphide minerals in highly oxidized, Rb-depleted, Archaean granulites of the Shevaroy Hills Massif, South India: Oxidation states and the role of metamorphic fluids: Journal of Metamorphic Geology, v. 15, p. 701–717, doi:10.1111/j.1525-1314.1997.00046.x.

Harlov, D.E., Hansen, E.C., and Bigler, C., 1998, Petrologic evidence for K-feldspar metasomatism in granulite facies rocks: Chemical Geology, v. 151, p. 373–386, doi:10.1016/S0009-2541(98)00090-4.

Hisada, K., and Miyano, T., 1996, Petrology and microthermometry of aluminous rocks in the Botswana Limpopo Central Zone: Evidence for isothermal decompression and isobaric cooling: Journal of Metamorphic Geology, v. 14, p. 183–197, doi:10.1046/j.1525-1314.1996.05857.x.

Hisada, K., Miyano, T., and van Reenen, D.D., 1994, Fluid inclusion study on gedrite formation in metapelites near Maratele in the Limpopo Central Zone, eastern Botswana: Science Reports from the Institute of Geoscience: University of Tsukuba, v. 15, p. 1–7.

Hisada, K., Perchuk, L.L., Gerya, T.V., van Reenen, D.D., and Paya, B.K., 2005, P-T-fluid evolution in the Mahalapye Complex, Limpopo high-grade terrain, eastern Botswana: Journal of Metamorphic Geology, v. 23, p. 313–334, doi:10.1111/j.1525-1314.2005.00579.x.

Hoefs, J., and Touret, J.L.R., 1975, Fluid inclusion, and carbon isotope study from Bamble granulites (South Norway): Contributions to Mineralogy and Petrology, v. 52, p. 165–174, doi:10.1007/BF00457292.

Hoerness, S., Lichtenstein, U., van Reenen, D.D., and Mokgatlha, K., 1995, Whole-rock/mineral O-isotope fractionations as a tool to model fluid-rock interaction in deep seated shear zones of the Southern Marginal Zone of the Limpopo Belt, South Africa: South African Journal of Geology, v. 98, p. 488–497.

Hollister, L., and Crawford, M.L., 1981, eds., Short Course in Fluid Inclusions: Applications to Petrology: Calgary, Mineralogical Association of Canada, 304 p.

Huizenga, J.M., Gutzmer, J., Banks, D.A., and Greylin, L.N., 2005, The Paleoproterozoic carbonate-hosted Pering Zn-Pb deposit, South Africa. II: Fluid inclusion, fluid chemistry and stable isotope constraints: Mineralium Deposita, v. 40, p. 686–706, doi:10.1007/s00126-005-0015-9.

Huizenga, J.-M., Perchuk, L.L., van Reenen, D.D., Flattery, Y., Varlamov, D.A., Smit, C.A., and Gerya, T.V., 2011, this volume, Granite emplacement and the retrograde P-T-fluid evolution of Neo-Archean granulites from the

Central Zone of the Limpopo Complex (South Africa), *in* van Reenen, D.D., Kramers, J.D., McCourt, S., and Perchuk, L.L., eds., Origin and Evolution of Precambrian High-Grade Gneiss Terranes, with Special Emphasis on the Limpopo Complex of Southern Africa: Geological Society of America Memoir 207, doi:10.1130/2011.1207(08).

Izraeli, E.S., Harris, J.W., and Navon, O., 2001, Brine inclusions in diamonds: A new upper mantle fluid: Earth and Planetary Science Letters, v. 187, p. 323–332, doi:10.1016/S0012-821X(01)00291-6.

Izraeli, E.S., Harris, J.W., and Navon, O., 2004, Fluid and mineral inclusions in cloudy diamonds from Koffiefontein, South Africa: Geochimica et Cosmochimica Acta, v. 68, p. 2561–2575, doi:10.1016/j.gca.2003.09.005.

Johnson, E.L., 1991, Experimentally determined limits for H_2O-CO_2-NaCl immiscibility in granulites: Geology, v. 19, p. 925–928, doi:10.1130/0091-7613(1991)019<0925:EDLFHO>2.3.CO;2.

Kamenetsky, M.B., Sobolev, A.V., Kamenetski, V.S., Maas, R., Danyushevsky, L.V., Thomas, R., Pokhilenko, N.P., and Sobolev, N.V., 2004, Kimberlite melts rich in alkali chlorides and carbonates: A potent metasomatic agent in the mantle: Geology, v. 32, p. 845–848, doi:10.1130/G20821.1.

Kamenetsky, V.S., Roland, R., Kamenetsky, M.B., Paton, C., Phillips, D., Golovin, A.V., and Gornova, M.A., 2009, Chlorine from the mantle: Magmatic halides in the Udachnaya-East kimberlite, Siberia: Earth and Planetary Science Letters, v. 285, p. 96–104, doi:10.1016/j.epsl.2009.06.001.

Lamb, V.M., Brown, P.E., and Valley, J., 1987, Post-metamorphic CO_2-rich inclusions in granulites: Contributions to Mineralogy and Petrology, v. 96, p. 485–495, doi:10.1007/BF01166693.

Maas, R., Kamenetsky, M.B., Sobolev, A.V., Kamenetsky, V.S., and Sobolev, N.V., 2005, Sr, Nd, and Pb isotope evidence for a mantle origin of alkali chlorides and carbonates in the Udachnaya kimberlite, Siberia: Geology, v. 33, p. 549–552, doi:10.1130/G21257.1.

Markl, G., and Bucher, K., 1998, Composition of fluids in the lower crust inferred from metamorphic salt in lower crustal rocks: Nature, v. 391, p. 781–783, doi:10.1038/35836.

Markl, G., Ferry, J., and Bucher, K., 1998, Formation of saline brines and salt in the lower crust by hydration reactions in partially retrogressed granulites from the Lofoten Islands, Norway: American Journal of Science, v. 298, p. 705–757.

Menzies, M.A., Kempton, P., and Duncan, M., 1985, Interaction of continental lithosphere and asthenospheric melts below the Geronimo Volcanic Field, Arizona, USA: Journal of Petrology, v. 26, p. 663–693.

Morogan, V., and Lindblom, S., 1995, Volatiles associated with the alkaline-carbonatite magmatism at Alnö, Sweden: A study of fluid and solid inclusions in minerals from the Laångarsholment ring complex: Contributions to Mineralogy and Petrology, v. 122, p. 262–274, doi:10.1007/s004100050126.

Morrison, J., and Valley, J.W., 1988, Post-granulite facies fluid infiltration in the Adirondack Mountains: Geology, v. 16, p. 513–516, doi:10.1130/0091-7613(1988)016<0513:PGFFII>2.3.CO;2.

Newton, R.C., 1989, Fluids in metamorphism: Annual Review of Earth and Planetary Sciences, v. 17, p. 385–410, doi:10.1146/annurev.ea.17.050189.002125.

Newton, R.C., and Manning, C.E., 2002, Experimental determination of calcite solubility in NaCl-H_2O solutions at deep crust/upper mantle pressures and temperatures: Implications for metasomatic processes in shear zones: American Mineralogist, v. 87, p. 1401–1409.

Newton, R.C., and Manning, C.E., 2005, Solubility of anhydrite, $CaSO_4$, in NaCl-H_2O solutions at high pressure and temperatures: Applications to fluid-rock interaction: Journal of Petrology, v. 46, p. 701–716, doi:10.1093/petrology/egh094.

Newton, R.C., and Manning, C.E., 2006, Solubilities of corundum, wollastonite and quartz in H_2O-NaCl solutions at 800°C and 10 kbar: Interaction of simple minerals with brines at high pressure and temperature: Geochimica et Cosmochimica Acta, v. 70, p. 5571–5582, doi:10.1016/j.gca.2006.08.012.

Newton, R.C., and Manning, C.E., 2008, Solubility of corundum in the system Al_2O_3-SiO_2-H_2O-NaCl at 800 °C and 10 kbar: Chemical Geology, v., 249, p. 250–261.

Newton, R.C., Aranovich, L.Ya., Hansen, E.C., and Vandenheuvel, B.A., 1998, Hyper-saline fluids in Precambrian deep-crustal metamorphism: Precambrian Research, v. 91, p. 41–63, doi:10.1016/S0301-9268(98)00038-2.

Ohmoto, H., and Kerrick, D., 1977, Devolatilisation equilibria in graphitic systems: American Journal of Science, v. 277, p. 1013–1044.

Olsen, S.N., 1987, The composition and role of fluids in migmatites: A fluid inclusion study of the Front Range rocks: Contributions to Mineralogy and Petrology, v. 96, p. 104–120, doi:10.1007/BF00375531.

Perchuk, L.L., and Gerya, T.V., 1993, Fluid control of charnockitisation: Chemical Geology, v. 108, p. 175–186, doi:10.1016/0009-2541(93)90323-B.

Perchuk, L.L., Safonov, O.G., Gerya, T.V., Fu, B., and Harlov, D.E., 2000a, Mobility of components in metasomatic transformation and partial melting of gneisses: An example from Sri Lanka: Contributions to Mineralogy and Petrology, v. 140, p. 212–232, doi:10.1007/s004100000178.

Perchuk, L.L., Gerya, T.V., van Reenen, D.D., Krotov, A.V., Safonov, O.G., Smit, C.A., and Shur My., 2000b, Comparable petrology and metamorphic evolution of the Limpopo (South Africa) and Lapland (Fennoscandia) high-grade terrains: Mineralogy and Petrology, v. 69, p. 69–107, doi:10.1007/s007100050019.

Pili, E., Sheppard, S.M.F., Lardeaux, J.-M., Martelat, J.E., and Nicollet, C., 1997, Fluid flow vs. scale of shear zones in the lower continental crust and the granulite paradox: Geology, v. 25, p. 15–18, doi:10.1130/0091-7613 (1997)025<0015:FFVSOS>2.3.CO;2.

Powell, R., Will, T.M., and Phillips, G.N., 1991, Metamorphism in Archaean greenstone belts: Calculated fluid compositions and implications for gold mineralization: Journal of Metamorphic Geology, v. 9, p. 141–150, doi:10.1111/j.1525-1314.1991.tb00510.x.

Roedder, E., 1965, Liquid CO_2 inclusions in olivine-bearing nodules and phenocrysts from basalts: American Mineralogist, v. 50, p. 1746–1782.

Roedder, E., 1984, Fluid Inclusions: Reviews in Mineralogy: Washington, D.C., Mineralogical Society of America, v. 12, 644 p.

Santosh, M., 1986, Carbonic metamorphism of charnockites in the southwestern Indian shield: A fluid inclusion study: Lithos, v. 19, p. 1–10, doi:10.1016/0024-4937(86)90011-3.

Santosh, M., and Omori, S., 2008, CO_2 flushing: A plate tectonic perspective: Gondwana Research, v. 13, p. 86–102, doi:10.1016/j.gr.2007.07.003.

Santosh, M., and Tsunogae, T.J., 2003, Extremely high density pure CO_2 fluid inclusions in a garnet granulite from southern India: Journal of Geology, v. 111, p. 1–16.

Santosh, M., Tsunogae, T., Ohyama, H., Sato, K., Li, J.H., and Liu, S.J., 2008, Carbonic metamorphism at ultrahigh-temperatures: Evidence from the North China Craton: Earth and Planetary Science Letters, v. 266, p. 149–165, doi:10.1016/j.epsl.2007.10.058.

Shepherd, T.J., and Rankin, A.H., and Alderton, D.H.M., 1985, A Practical Guide to Fluid Inclusion Studies: Glasgow, Blackie and Son, 239 p.

Smit, C.A., and van Reenen, D.D., 1997, Deep crustal shear zones, high-grade tectonites, and associated metasomatic alteration in the Limpopo Belt, South Africa: Implications for deep crustal processes: Journal of Geology, v. 105, p. 37–57, doi:10.1086/606146.

Sorby, H.C., 1858, On the microscopic structures of crystals, indicating the origin of minerals and rocks: Geological Society of London Quarterly Journal, v. 14, p. 453–500, doi:10.1144/GSL.JGS.1858.014.01-02.44.

Stevens, G., 1997, Melting, carbonic fluids and water recycling in the deep crust: An example from the Limpopo Belt, South Africa: Journal of Metamorphic Geology, v. 15, p. 141–154, doi:10.1111/j.1525-1314.1997.00010.x.

Stevens, G., and Clemens, J.D., 1993, Fluid-absent melting and the role of fluids in the lithosphere: A slanted summary?: Chemical Geology, v. 108, p. 1–17, doi:10.1016/0009-2541(93)90314-9.

Thiéry, R., Van den Kerkhof, A.M., and Dubessy, J., 1994, vX properties of CH_4-CO_2 and CO_2-N_2 fluid inclusions: Modelling for $T < 31°C$ and $P < 400$ bars: European Journal of Mineralogy, v. 6, p. 753–772.

Thompson, A.B., 1983, Fluid-absent metamorphism: Journal of the Geological Society [London], v. 140, p. 533–547, doi:10.1144/gsjgs.140.4.0533.

Touret, J.L.R., 1971, Le faciès granulite en Norvège Méridionale. II: Les inclusions fluides: Lithos, v. 4, p. 423–436, doi:10.1016/0024-4937 (71)90125-3.

Touret, J.L.R., 1981, Fluid inclusions in high grade metamorphic rocks, *in* Hollister, L., and Crawford, M.L., eds., Short Course in Fluid Inclusions: Applications to Petrology: Mineralogical Association of Canada, v. 6, p. 13–36.

Touret, J., 1986, Fluid inclusions in rocks from the lower continental crust: Geological Society [London] Special Publication 24, p. 161–172, doi:10.1144/GSL.SP.1986.024.01.15.

Touret, J.L.R., 2001, Fluids in metamorphic rocks: Lithos, v. 55, p. 1–25, doi:10.1016/S0024-4937(00)00036-0.

Touret, J.L.R., 2010, Mantle to lower-crust fluid/melt transfer through granulite metamorphism: Russian Geology and Geophysics (in press).

Touret, J.L.R., and Dietvorst, P., 1983, Fluid inclusions in high grade anatectic metamorphites: Journal of the Geological Society [London], v. 140, p. 635–649, doi:10.1144/gsjgs.140.4.0635.

Touret, J.L.R., and Hartel, T., 1990, Synmetamorphic fluid inclusions in granulites, *in* Vielzeuf, D., and Vidal, Ph., eds., Granulites and Crustal Evolution: NATO ASI Ser. C, v. 311, p. 397–417.

Touret, J.L.R., and Huizenga, J.M., 1999, Intraplate magmatism at depth: High-temperature lower crustal granulites: Journal of African Earth Sciences, v. 27, p. 376–382.

Touret, J.L.R., and Marquis, G., 1994, Fluides profonds et conductivité électrique de la croûte continentale inférieure: Comptes Rendus de l'Académie des sciences, v. 318, p. 1469–1482.

Touret, J.L.R., Grégoire, M., and Teitchou, M., 2010, Was the lethal eruption of lake Nyos related to CO_2/H_2O density inversion?: Comptes Rendus de l'Académie des sciences (in press).

Tropper, P., and Manning, C.E., 2007, The solubility of corundum in H_2O at high pressure and temperature and its implications for Al mobility in the deep crust and upper mantle: Chemical Geology, v. 240, p. 54–60, doi:10.1016/j.chemgeo.2007.01.012.

Tsunogae, T., and van Reenen, D.D., 2007, Carbonic fluid inclusions in sapphirine+quartz bearing garnet granulite from the Limpopo Belt, southern Africa: Journal of Mineralogical and Petrological Sciences, v. 102, p. 57–60.

Tsunogae, T., Santosh, M., Osanai, Y., Owada, M., Toyoshima, T., and Hokada, T., 2002, Very high-density carbonic fluid inclusions in sapphirine-bearing granulites from Tonagh Island in the Archean Napier Complex, East Antarctica: Implications for CO_2 infiltration during ultrahigh-temperature (T>1100°C) metamorphism: Contributions to Mineralogy and Petrology, v. 143, p. 279–299.

Van den Berg, R., and Huizenga, J.M., 2001, Fluids in granulites of the Southern Marginal Zone of the Limpopo Belt, South Africa: Contributions to Mineralogy and Petrology, v. 141, p. 529–545.

Van den Kerkhof, A.M., 1990, Isochoric phase diagrams in the systems CO_2-CH_4 and CO_2-N_2: Applications to fluid inclusions: Geochimica et Cosmochimica Acta, v. 54, p. 621–629, doi:10.1016/0016-7037(90)90358-R.

Van den Kerkhof, A.M., Kronz, A., Simon, K., and Scherer, T., 2004, Fluid-controlled quartz recovery in granulite as revealed by cathodoluminescence and trace element analysis (Bamble sector, Norway): Contributions to Mineralogy and Petrology, v. 146, p. 637–652, doi:10.1007/s00410-003-0523-5.

van Reenen, D.D., 1986, Hydration of cordierite and hypersthene and a description of the retrograde orthoamphibole isograd in the Limpopo Belt, South Africa: American Mineralogist, v. 71, p. 900–915.

van Reenen, D.D., and Hollister, L.S., 1988, Fluid inclusions in hydrated granulite facies rocks, Southern Marginal Zone of the Limpopo Belt, South Africa: Geochimica et Cosmochimica Acta, v. 52, p. 1057–1064, doi:10.1016/0016-7037(88)90260-8.

van Reenen, D.D., Roering, C., Brandl, G., Smit, C.A., and Barton, J.M., 1990, The granulite facies rocks of the Limpopo Belt, Southern Africa, *in* Vielzeuf, D., and Vidal, P., eds., Granulites and Crustal Evolution: Dordrecht, Netherlands, Kluwer, NATO-ASI Series, v. C311, p. 257–289.

van Reenen, D.D., Pretorius, A.I., and Roering, C., 1994, Characterization of fluids associated with gold mineralisation and with regional high-temperature retrogression of granulites in the Limpopo Belt, South Africa: Geochimica et Cosmochimica Acta, v. 58, p. 1147–1159, doi:10.1016/0016-7037(94)90578-9.

van Reenen, D.D., Boshoff, R., Smit, C.A., Perchuk, L.L., Kramers, J.D., McCourt, S., and Armstrong, R.A., 2008, Geochronological problems related to polymetamorphism in the Limpopo Complex, South Africa: Gondwana Research, v. 14, p. 644–662, doi:10.1016/j.gr.2008.01.013.

Vityk, M.O., and Bodnar, R.J., 1995, Do fluid inclusions in high-grade metamorphic terranes preserve peak metamorphic density during retrograde decompression?: American Mineralogist, v. 80, p. 641–644.

Wannamaker, P.E., Doerner, W.M., Stodt, J.A., and Johnston, J.M., 1997, Subdued state of tectonism of the Great Basin interior relative to its eastern margin based on deep resistivity structure: Earth and Planetary Science Letters, v. 150, p. 41–53, doi:10.1016/S0012-821X(97)00076-9.

Wannamaker, P.E., Caldwell, T.G., Doerner, W.M., and Jiracek, G.R., 2004, Fault zone fluids and seismicity in compressional and extensional environments inferred from electrical conductivity: The New Zealand Southern Alps and U.S. Great Basin: Earth, Planets, and Space, v. 56, p. 1171–1176.

Watson, E.B., and Brenan, J.M., 1987, Fluids in the lithosphere 1. Experimentally determined wetting characteristics of CO_2-H_2O fluids and their implication for fluid transport, host-rock physical properties and fluid inclusion formation: Earth and Planetary Science Letters, v. 85, p. 497–515, doi:10.1016/0012-821X(87)90144-0.

Whitney, D.L., 1992, Origin of CO_2-rich fluid inclusions in leucosomes from the Skagit migmatites, North Cascades, Washington, USA: Journal of Metamorphic Geology, v. 10, p. 715–725, doi:10.1111/j.1525-1314.1992.tb00118.x.

Yardley, B.W.D., and Valley, J.W., 1997, The petrologic case for a dry lower crust: Journal of Geophysical Research, v. 102, p. 12,173–12,185, doi:10.1029/97JB00508.

Yardley, B., Gleeson, S., Bruce, S., and Banks, D., 2000, Origin of retrograde fluids in metamorphic rocks: Journal of Geochemical Exploration, v. 69–70, p. 281–285, doi:10.1016/S0375-6742(00)00132-1.

MANUSCRIPT ACCEPTED BY THE SOCIETY 24 MAY 2010

Fluid-absent melting versus CO_2 streaming during the formation of pelitic granulites: A review of insights from the cordierite fluid monitor

M.J. Rigby
Department of Geology, University of Pretoria, Lynwood Road, Pretoria, 0002, South Africa, and Runshaw College, Langdale Road, Leyland, Lancashire, PR25 3DQ, UK

G.T.R. Droop
School of Earth, Atmospheric and Environmental Science, University of Manchester, Oxford Road, Manchester, M13 9PL, UK

ABSTRACT

An extensive database of published cordierite volatile contents from twenty-eight different high-grade terranes is used to investigate whether CO_2 streaming or fluid-absent melting processes prevail during the formation and evolution of granulites. In most case studies the cordierite volatile contents, calculated activities, and melt-H_2O contents are entirely consistent with a mode of origin dominated by fluid-absent melting processes. In accordance with published experimental and theoretical evidence these data suggest that CO_2 is not a prerequisite for granulite formation. Even in cases in which cordierite preserves high CO_2 contents this does not necessarily imply that fluid-saturated conditions prevailed. A cordierite may be CO_2-rich but can still be fluid undersaturated and preserve H_2O contents that equilibrated with melts formed from fluid-absent melting. In fluid-saturated case studies the mechanism and relative timing of saturation should be evaluated. It is evident from the data that most fluid-saturated granulites formed initially under fluid-absent conditions but subsequently became fluid saturated along a retrograde path.

INTRODUCTION

The role of CO_2 during partial melting and granulite formation has long been the subject of intense controversy. The concept of a CO_2-rich free-fluid phase accompanying high-grade metamorphism was initially introduced by Touret (1971), who found that fluid inclusions across an amphibolite- to granulite-facies transition recorded a change from an H_2O-dominated environment to a CO_2-rich fluid regime. The ubiquity of CO_2-rich fluid inclusions within granulite minerals, worldwide, led to the hypothesis of carbonic metamorphism or CO_2 streaming, which stipulates that the lower crust is pervasively flushed with hot, possibly mantle-derived, CO_2 (Newton et al., 1980). In addition to evidence from fluid inclusion studies, the controversial experimental data of Peterson and Newton (1989, 1990) further supported the idea of carbonic metamorphism by suggesting that partial melting of the assemblage quartz plus phlogopite mica was actually enhanced by CO_2. The principal alternative proposal for granulite formation suggests that the low aH$_2$O values recorded in many high-grade terranes are not a consequence of effective dilution through CO$_2$-streaming but are an intrinsic result of the partitioning of H$_2$O into melt, especially melt generated in situ by anatexis (e.g., Fyfe, 1973; Clemens, 1990), though the passage of hot, dry, externally derived magma has also been proposed as a mechanism for reducing aH$_2$O (e.g., Morrison and Valley, 1988). In the case of anatexis (arguably the more important of the two melt-related processes), rocks heated to modest temperatures undergo devolatilization and initial partial melting in the presence of a hydrous fluid (e.g., Thompson,1983; Stevens and Clemens, 1993; Clemens and Droop, 1998). However, the consumption of H$_2$O by "wet"-melting reactions could potentially exhaust the local fluid supply, leading to melting at higher temperatures via incongruent, fluid-absent melting reactions (e.g., Powell, 1983; Clemens and Vielzeuf, 1987; Stevens and Clemens, 1993). There is an extensive body of experimental data that suggests that, at temperatures above the wet solidus, hydrous silicates such as biotite break down via incongruent melting reactions to produce anhydrous phases and a strongly H$_2$O-undersaturated granitic melt (e.g., Clemens and Wall, 1981; Thompson, 1982; Grant, 1985; Clemens and Vielzeuf, 1987; Le Breton and Thompson, 1988; Clemens, 1989, 1992, 1995; Carrington and Harley, 1995; Clemens et al., 1997; Stevens et al., 1997). Furthermore, the experiments of Clemens (1993), which were designed to replicate and reproduce the results of Peterson and Newton (1989, 1990), demonstrated that CO$_2$ did *not* flux during melting. Furthermore, Clemens and Watkins (2001) provide an additional perspective, based on temperature-water (T-H$_2$O) variation in granitic magmas, which demonstrates the only way to obtain high-T magmas with low H$_2$O content is if they formed by fluid-absent melting. Despite the wealth of convincing theoretical and experimental evidence *against* widespread CO$_2$ streaming during granulite formation, the concept of carbonic metamorphism still prevails in the recent literature (e.g., Tsunogae et al., 2002; Fonarev et al. 2003; Török et al., 2005; Cuney et al., 2007; Santosh et al., 2008; Ohyama et al., 2008; Santosh and Omori, 2008). The aim of this paper is to assess whether evidence from the cordierite fluid monitor can provide new constraints on an old debate and maybe to finally lay the controversy to rest.

ASSESSING THE PRESENCE OR ABSENCE OF A FLUID PHASE DURING METAMORPHISM AND PARTIAL MELTING

Primary evidence for the existence of a fluid phase at the thermal peak of metamorphism is rare. Fluid inclusions (see Huizenga and Touret, this volume) may preserve evidence of a fluid phase during the peak of metamorphism and partial melting. However, some fluid inclusions in metamorphic rocks may have been trapped late in the rocks' history and therefore only record evidence of a late-stage retrograde fluid that was not in equilibrium with the granulite-facies assemblage (e.g., Kreulen, 1987; Morrison and Valley, 1988). Furthermore, Morgan et al. (1993) demonstrated that the compositions of fluid inclusions in quartz and olivine can re-equilibrate owing to hydrogen diffusion in or out of the inclusion. In the absence of reliable primary evidence, the surest indicators of fluid presence are the magnitudes of the activities *(a)* of the volatile species. A peak metamorphic aH$_2$O value close to unity is good evidence that the mineral assemblage in the rock crystallized in the presence of an aqueous fluid (e.g., Holland, 1979; Moazzen et al., 2001; Rigby et al., 2008a). Similarly, low values (<<1) for both aH$_2$O and aCO$_2$ would imply fluid absence (e.g., Lamb and Valley, 1988; Stevens and Clemens, 1993; Clemens and Droop, 1998; Droop et al., 2003). The metamorphic fluid is commonly more complex than a simple binary H$_2$O-CO$_2$ fluid (e.g., Aranovich and Newton, 1998; Shmulovich and Graham, 2004; Gerya et al., 2005), but provided that it can be demonstrated or justifiably assumed that the activities of minor volatile components such as NaCl, CH$_4$, etc., were also small (e.g., Moazzen et al., 2001) then calculated aH$_2$O and aCO$_2$ values can provide useful insights into the fluid regime during metamorphism and anatexis (e.g., Powell, 1983; Lamb and Valley, 1988; Vry et al., 1990; Clemens and Droop, 1998; Harley et al., 2002; Rigby et al., 2008a).

Traditionally, aH$_2$O and aCO$_2$ have been estimated by applying thermodynamic methods to mineral-fluid equilibria. The first step is to estimate peak metamorphic pressure-temperature (P-T) conditions by applying standard thermobarometric methods (e.g., Essene, 1989; Droop, 1989; Rigby et al., 2008b; Powell and Holland, 2008; Rigby, 2009). Once P and T are known, dehydration equilibria can be used to estimate aH$_2$O and decarbonation equilibria to estimate aCO$_2$. This is a powerful methodology and has been used, for example, to demonstrate that aH$_2$O decreased across the transition from amphibolite-facies to granulite-facies rocks at Broken Hill in the Willyama terrane, Australia (Phillips, 1981). The problem with the traditional method, when attempting to assess the presence or absence of a fluid, is that it is difficult to estimate both aH$_2$O and aCO$_2$ simultaneously in a single rock sample. This is because few rocks contain assemblages that permit both pure dehydration and pure decarbonation equilibria to be written among end members of the minerals present. Mixed-volatile equilibria (e.g., Kerrick, 1974) can be written for many marbles and calc-silicate rocks but are unsatisfactory in this respect because they can only provide upper limits on the aH$_2$O and aCO$_2$, corresponding to fluid-saturated conditions. The mineral cordierite, however, provides a solution to the problem.

THE CORDIERITE FLUID MONITOR

Introduction

Cordierite, (Mg,Fe)$_2$Al$_4$Si$_5$O$_{18}$.nH$_2$O.mCO, contains a large channel site at the center of its six-membered ring structure that can accommodate molecular H$_2$O and CO$_2$ (e.g., Wood and Nassau, 1967; Goldman et al., 1977; Aines and Rossman, 1984). In fluid-present environments the total volatile content of cordierite is variable but intrinsically dependent on P, T, and XCO$_2$

in the coexisting fluid (Mirwald and Schreyer, 1977; Kurepin, 1985; Johannes and Schreyer, 1981; Carey, 1995; Harley and Carrington, 2001; Thompson et al., 2001; Harley et al., 2002). In fluid-absent, melt-present environments the cordierite volatile content depends on the partitioning of the volatile species between cordierite and the coexisting melt (Stevens et al., 1995; Carrington and Harley, 1996; Harley and Carrington, 2001; Thompson et al., 2001; Harley et al., 2002). Providing that the channel constituents are faithfully retained during equilibration and any subsequent P-T evolution (Harley et al., 2002; Rigby and Droop, 2008), cordierite is a mineral that should provide useful insight into the fluid regime developed during metamorphism and anatexis (Stevens et al., 1995; Carrington and Harley, 1996; Kalt, 2000; Harley and Carrington, 2001; Thompson et al., 2001; Harley et al., 2002; Harley, 2004; Harley and Thompson, 2004; Harley, 2008; Rigby et al., 2008a; Rigby and Droop, 2008).

Cordierite-Volatile-Melt Equilibria: A Thermodynamic Basis for Determining Fluid Conditions from Cordierite Volatile Contents

Cordierite-H_2O Equilibria

The dehydration of Mg-cordierite can be represented by the reaction:

$$Mg_2Al_4Si_5O_{18}.H_2O = Mg_2Al_4Si_5O_{18} + H_2O. \quad (1)$$

For any balanced dehydration reaction, the following is approximately true at equilibrium:

$$0 = \Delta H°_{1,298} - \Delta S°_{1,298} + (P-1)\Delta V°_s + RT\ln K + nRT\ln(fH_2O_{P,T}/fH_2O_{1,T}), \quad (2)$$

where P is pressure, T is absolute temperature, $\Delta H°_{1,298}$ is the standard-state enthalpy change of reaction, $\Delta S°_{1,298}$ is the standard-state entropy change of reaction, $\Delta V°_s$ is the volume change of reaction, R is the gas constant, K is the equilibrium constant, n is the number of molecules of H_2O evolved per mole of reactant, $fH_2O_{P,T}$ is the fugacity of H_2O at the P and T of interest, and $fH_2O_{1,T}$ is the fugacity of H_2O at 1 bar and the T of interest. The experimental studies of Carey (1995), Skippen and Gunter (1996), and Harley and Carrington (2001) support the assertion that there is no volume change in the solids of reaction (1).

The equilibrium constant for reaction (1) is given by:

$$K = \left(\frac{a_{Mgcrd} \cdot a_{H_2O}}{a_{Mgcrd.H_2O}}\right), \quad (3)$$

where aMgcrd is the activity of anhydrous cordierite, aMgcrd.H_2O is the activity of hydrous cordierite, and aH_2O is the activity of H_2O. Any substitution of Fe for Mg is assumed to reduce the aMgcrd and aMgcrd.H_2O by the same amount. The experimental studies of Carey and Navrotsky (1992), Carey (1995), Skippen and Gunter (1996), and Harley and Carrington (2001) support an ideal, one-site mixing model, defining n as the number of molecules of H_2O per formula unit of 18-oxygen volatile-free cordierite. Thus: aMgcrd.H_2O = XH_2O^{crd} and aMgcrd = 1 − XH_2Ocrd and

$$K = \left[\frac{(1-X^{crd}_{H_2O}) \cdot a_{H_2O}}{X^{crd}_{H_2O}}\right] = \left[\frac{(1-n) \cdot a_{H_2O}}{n}\right] \quad (4)$$

and

$$0 = \Delta H° - T\Delta S° + RT\ln\{(1-n).aH_2O.f_{P,T}/n.f_{1,T}\}. \quad (5)$$

Recognizing that $f_{1,T}$ is ~1 bar (Burnham et al., 1969) and rearranging yields:

$$aH_2O = [n/(1-n)]\{\exp[(-\Delta H°/RT)+(\Delta S°/R)]/fH_2O\}. \quad (6)$$

Consequently aH_2O is proportional to $n/(1–n)$ at a specified P-T condition. This formulation is essentially equivalent to those derived by Kurepin (1985), Carey (1995), Skippen and Gunter (1996), and Harley and Carrington (2001).

Harley and Carrington (2001) modeled the maximum volatile content of cordierite coexisting with pure H_2O over the P-T range 2–8 kbar and 500–1000 °C. From the regression of their own experimental data and those of Mirwald et al. (1979), Boberski and Schreyer (1990), Carey (1995), and Skippen and Gunter (1996), they recovered best-fit $\Delta H°$ and $\Delta S°$ values of 34.95 (±2.66) Jmole^{-1} and 97.69 (±2.74) JK^{-1} mole^{-1}, respectively.

Providing that independent P-T estimates are available, equation (6), coupled with the thermodynamic data of Harley and Carrington (2001), can be used to calculate aH_2O in natural cordierites whose H_2O contents (n) have been determined. Furthermore, the maximum H_2O content (n saturation) at any given P and T can be modeled by setting aH_2O = 1 (corresponding to H_2O-saturated conditions) and rearranging the equation to solve for n.

Cordierite-CO_2 Equilibria

For any decarbonation reaction, the following is approximately true at equilibrium:

$$0 = \Delta H°_{1,298} - \Delta S°_{1,298} + (P-1)\Delta V°_s + RT\ln K + mRT\ln(fCO_{2P,T}/fCO_{2 1,T}), \quad (7)$$

where P is pressure, T is absolute temperature, $\Delta H°_{1,298}$ is the standard state enthalpy change of reaction, $\Delta S°_{1,298}$ is the standard state entropy change of reaction, ΔV_s is the volume change of reaction for solids, R is the universal gas constant, K is the equilibrium constant, m is the number of molecules of CO_2 evolved per mole of reactant, $fCO_{2P,T}$ is the fugacity of CO_2 at the P and T of interest, and $fCO_{21,T}$ is the fugacity of CO_2 at 1 bar and the T of interest. The assumptions inherent in this equation are similar to those made for the aH_2O equation. Consider the following reaction:

$$Mg_2Al_4Si_5O_{18} \cdot CO_2 = Mg_2Al_4Si_5O_{18} + CO_2. \quad (8)$$

cordierite　　　　cordierite　　　fluid

Assuming that the $Mg_2Al_4Si_5O_{18}$ part of the cordierite molecule has the same molar heat capacity in both CO_2-bearing and non-CO_2-bearing cordierite, and that free CO_2 has the same molar heat capacity as the CO_2 in the channel site of CO_2-bearing cordierite, $\Delta H°$ and $\Delta S°$ may be regarded as independent of T (Thompson et al., 2001).

The equilibrium constant for reaction (8) is given by:

$$K = \left(\frac{a_{Mgcrd} \cdot a_{CO_2}}{a_{Mgcrd.CO_2}}\right), \quad (9)$$

where aMgcrd is the activity of non-CO_2-bearing cordierite, aMgcrd.CO_2 is the activity of CO_2-bearing cordierite, and aCO$_2$ is the activity of CO_2. As the substitution of a given amount of Fe for Mg can be assumed to reduce the aMgcrd and aMgcrd.CO_2 by the same amount, then cordierite activity can be modeled simply as a function of the occupancy of the channel site. Using a one molecule, ideal one-site model, consistent with the experimental data of Thompson et al. (2001), and defining m as the number of molecules of CO_2 per formula unit of 18-oxygen volatile-free cordierite, then: aMgcrd.CO_2 = XCO_2^{crd} and aMgcrd = $1 - XCO_2$crd.

Thus:

$$K = \left[\frac{(1 - X_{CO_2}^{crd}) \cdot a_{CO_2}}{X_{CO_2}^{crd}}\right] = \left[\frac{(1 - m) \cdot a_{CO_2}}{m}\right]. \quad (10)$$

Harley et al. (2002) include a pressure term ascribed to a solid volume change in reaction 8, where $\Delta V° = 2.513$ Jbar^{-1}mole^{-1} and derived the following expression:

$$aCO_2 = [m/(1-m)]\{\exp[((-\Delta H° + P\Delta V°)/RT) + (\Delta S°/R)]/fCO_2. \quad (11)$$

Providing that independent P-T estimates are available, equation (11), coupled with the thermodynamic data of Thompson et al. (2001), can be used to calculate aCO_2 in natural cordierites whose CO_2 contents have been analyzed. Moreover, the maximum CO_2 contents (m saturation) at any given P and T can be modeled by setting the $aCO_2 = 1$ (corresponding to CO_2-saturated conditions) and rearranging the equation to solve for m. This procedure, as outlined by Harley et al. (2002), yields the CO_2-saturation isopleths. Harley et al. (2002) explain that the CO_2 model is not as well constrained toward lower temperatures as is the H_2O model, because it is derived only from experiments at T >800 °C. However, for present purposes this is inconsequential, as the model is applied to granulite facies rocks, whose temperature range falls within that of the experimental data.

Cordierite H_2O-CO_2 Saturation Surfaces

Harley et al. (2002) outlined a method for modeling H_2O-CO_2 cordierite saturation surfaces. The results are consistent with the experimental data of Carey (1995), Skippen and Gunter (1996), Carrington and Harley (1996), Harley and Carrington (2001), and perhaps most importantly, Thompson et al. (2001). The first step is to use the sub-regular solution model for nonideal mixing in a binary H_2O-CO_2 fluid (Holland and Powell, 1990) to calculate compatible aH_2O-aCO_2 pairs at various P-T-XCO_2 conditions relevant to the field area under investigation. Then, assuming that ideal H_2O-CO_2 mixing occurs in the channels of the cordierites, equations (6) and (11) can be used to determine the n and m contents of the cordierite at the appropriate, *compatible* activity values, as defined by the solution model. For a cordierite coexisting with an H_2O-CO_2 fluid at a specified P-T condition, this enables construction of a cordierite channel volatile composition diagram that will depict the numbers of molecules of H_2O (n), CO_2 (m), and $H_2O + CO_2$ ($n + m$) per formula unit for a given channel XCO_2 (crd) [= $m/(m + n)$] value. An example of this modeling is illustrated in Figure 1. Measured volatile contents of natural cordierites from a variety of metamorphic terranes can be compared with modeled saturation surfaces at the P-T conditions of interest to discriminate between fluid-absent and fluid-present conditions (Harley et al., 2002). Fluid-present conditions can be inferred if the *mean* total volatile content (incorporating error in the analysis and P-T estimates) intersects the total volatile saturation curve at the P and T of interest. If the mean total volatile content plots above the saturation curve of interest, and does not intersect the curve when the error is taken into account, the sample would be deemed oversaturated (fluid

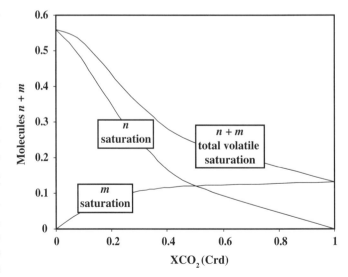

Figure 1. Cordierite volatile saturation surfaces at 2 kbar and 800 °C, based on the modeling outlined by Harley et al. (2002).

present). Fluid-absent conditions would be inferred if the *mean* total volatile content, which incorporates the error in the analysis and P-T conditions, fails to intersect the saturation curve of interest (Rigby and Droop, 2008).

Cordierite-Melt Relationships

For cordierite and granitic melt in equilibrium at any given P and T, the $a\text{H}_2\text{O}$ defined by cordierite in equation (6) must equal the $a\text{H}_2\text{O}$ defined by the melt H_2O contents. Using data obtained for the distribution of H_2O between cordierite and melt (D_w experiments) at various P-T conditions, Harley and Carrington (2001) refined the Burnham (1994) model of H_2O solubility in granitic melts:

$$a\text{H}_2\text{O} = k_w(X_w)^2 \ [\text{at } X_w < 0.5] \quad (12)$$

and

$$a\text{H}_2\text{O} = 0.25k_w\{\exp[(6.52-(2667/T))X_w]\} \ [\text{at } X_w > 0.5], \quad (13)$$

where X_w is the mole fraction of water in an 8-oxygen formula unit melt, and K_w is a coefficient defined empirically at the P-T conditions of interest from H_2O saturation experiments. Equations (12) and (13) can be used in melt-present areas to estimate the H_2O contents of granitic melts coexisting with cordierite. Providing that hydrous phases such as biotite have not been exhausted, a melt phase in a fluid-undersaturated volume of rock will have an H_2O content that approaches the minimum possible value for the specified P-T conditions (Stevens and Clemens, 1993; Harley and Carrington, 2001). Assuming this to be true, Harley and Carrington (2001) used their D_w relationships to calculate the H_2O content of a cordierite that could coexist at a given P-T with a melt that has its minimum possible value at the same P-T. This rationale has been used to produce a contoured P-T diagram (Fig. 2) depicting cordierite-melt-H_2O relationships over a P-T range relevant to granulite-facies metamorphism. Figure 2 utilizes the minimum wt% H_2O melt isopleths of Johannes and Holtz (1996) and can be used to assess whether the measured volatile data are consistent with independent P-T conditions and melt-H_2O contents. The measured cordierite H_2O contents and independent P-T conditions will only be deemed compatible if the measured *mean* H_2O value falls within the error of the independent P-T estimates. The assumption inherent in this approach is that, at a given P and T in fluid-undersaturated areas, the melt H_2O content always has a minimum or close-to-minimum possible value.

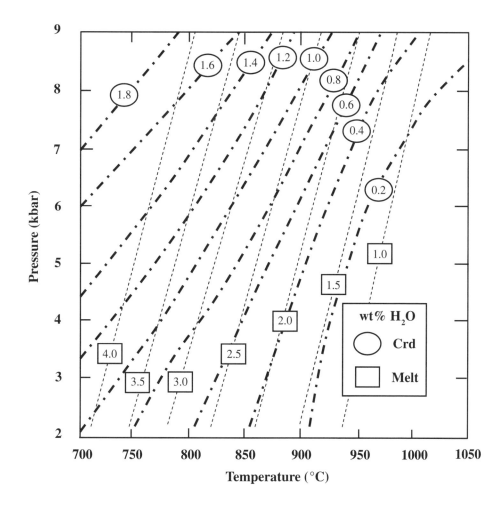

Figure 2. Contoured pressure-temperature (P-T) diagram depicting cordierite-melt H_2O relationships for conditions in which the melt has the minimum possible H_2O content at a given P-T. The minimum weight percentages of H_2O melt isopleths (dashed lines) are after Johannes and Holtz (1996). The cordierite (Crd) H_2O isopleths (dot-dashed lines) were calculated using the D_w relationships of Harley and Carrington (2001).

CORDIERITE VOLATILE CONTENTS AND THE FLUID CONDITIONS RECORDED IN HIGH-GRADE TERRANES

Figure 3 represents a comprehensive database of published cordierite volatile contents analyzed either by Secondary Ion Mass Spectrometry (SIMS) (e.g., Harley and Carrington, 2001; Harley et al., 2002; Harley and Thompson, 2004; Harley, 2004, 2008) or Fourier-Transform Infrared spectroscopy (FTIR) (e.g., Kalt, 2000; Rigby et al., 2008a; Rigby and Droop, 2008). The database contains 1111 individual analyses from 113 samples collected from 28 different high-grade or melt-bearing terranes. Inspection of Figure 3 shows that the cordierite volatile contents are highly variable, with CO_2 ranging from 0 to 1.97 wt% and H_2O contents from 0.16 to 2.69 wt%. Despite this spread, the vast majority of samples preserve volatile contents that cluster between 0.5 and 1.5 wt% H_2O, with CO_2 contents that are generally <0.3 wt%. Furthermore, there appears to be a broad antithetic relationship; cordierites that are rich in CO_2 tend to be relatively H_2O depleted and vice versa. In order to assess the significance of the preserved volatile contents, the aH_2O, aCO_2, and wt% H_2O (melt) have been calculated for each sample, using the methods outlined previously. The data are reported in Tables 1–3. Furthermore, measured values of n plus m are compared with the volatile

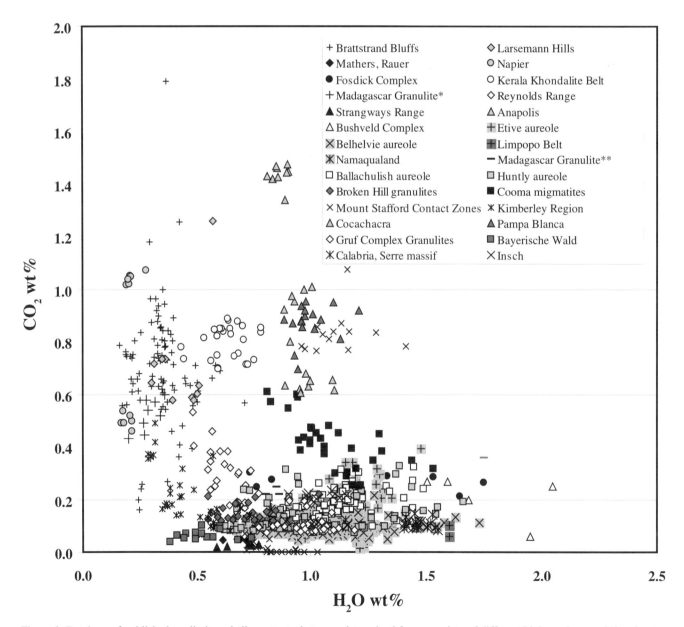

Figure 3. Database of published cordierite volatile contents that were determined from a variety of different high-grade or melt-bearing terranes. Refer to the references in Tables 1–3 for the relevant background information and source of the original data.

saturation surfaces at the P and T of interest (generally the estimated P-T conditions at the thermal peak of metamorphism) to discriminate between fluid-present and fluid-absent conditions (Figs. 4–6). In fluid-absent cases, the rationale behind Figure 2 is used to assess whether the cordierite volatile data are consistent with fluid-absent melting at the independently determined P-T of interest. Owing to the large number of samples from different study areas, the database has been split into three sections, mainly for descriptive purposes and clarity. The data will be presented and discussed in terms of the following: (1) low-pressure (<4.5 kbar) granulites and migmatites, (2) medium-pressure (>5 kbar) granulites, and (3) ultrahigh-temperature (UHT) granulites (T >900 °C). The relevant background information, mineral assemblages, and P-T conditions for each area are too voluminous to report here. However, the references to this information are provided in Tables 1–3.

Low-Pressure Granulites and Migmatites

Cordierite Volatile Data

The cordierites from Huntly, Belhelvie, Kimberley, zone 4 of the Mount Stafford contact aureole, Broken Hill, and Ballachulish (Table 1) are characterized by low to moderate H_2O contents (0.301–1.488 wt%) and low CO_2 contents (0.001–0.387 wt%). For the estimated P-T conditions of formation, for each area respectively, this yields aH_2O values ranging from 0.094 to 0.625 and aCO_2 values of 0.045 to 0.386. A comparison of the mean total volatile contents to modeled H_2O-CO_2 saturation surfaces at the P-T of interest shows that all these samples are fluid undersaturated (Fig. 4).

In the Etive aureole (Droop and Treloar, 1981) the H_2O contents gradually decrease up-grade from 1.23 to 0.836 wt%. Immediately adjacent to the igneous contact, H_2O contents rise again to values of ~1.263 wt%. CO_2 contents are typically low, with values ranging from 0.072 to 0.120 wt%. However, immediately adjacent to the igneous contact, CO_2 contents increase to 0.249 wt%. For estimated P-T conditions of formation the volatile contents record an up-grade decrease in aH_2O from 0.810 to 0.537. The samples immediately adjacent to the igneous contact record aH_2O values that are in excess of unity. The aCO_2 is fairly constant, with values ranging from 0.100 to 0.180. Once again, the exceptions are the samples immediately adjacent to the igneous contact, where aCO_2 values range from 0.400 to 0.641. A comparison of the mean total volatile content to modeled

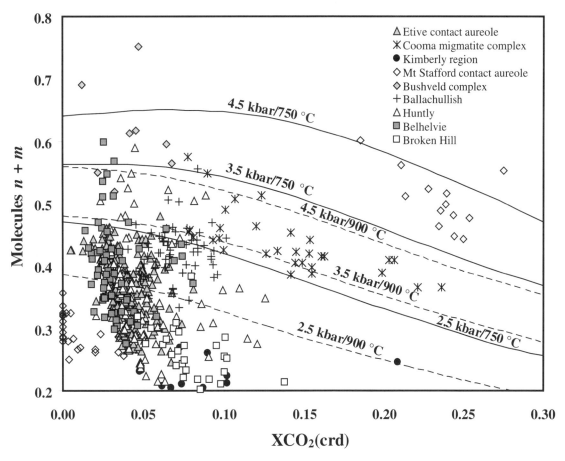

Figure 4. Comparison of the total volatile content (molecules $n + m$ p.f.u.) of cordierite (crd) from low-pressure terranes versus modeled volatile saturation surfaces at the P-T of interest; p.f.u.—per formula unit.

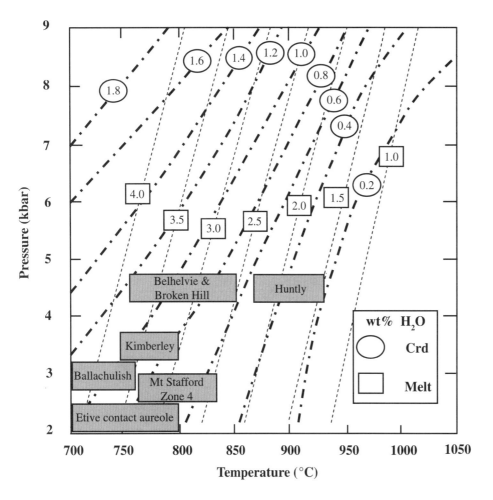

Figure 5. Contoured P-T diagram depicting cordierite-melt H_2O relationships for conditions in which the melt has the minimum possible H_2O content at a given P-T. The minimum weight percentages of H_2O melt isopleths (dashed lines) are after Johannes and Holtz (1996). The cordierite (Crd) H_2O isopleths (dot-dashed lines) were calculated using the D_w relationships of Harley and Carrington (2001). The gray boxes represent the independently estimated P-T conditions for each of the field areas, respectively (the extent of the box in the P-T space indicates the uncertainty).

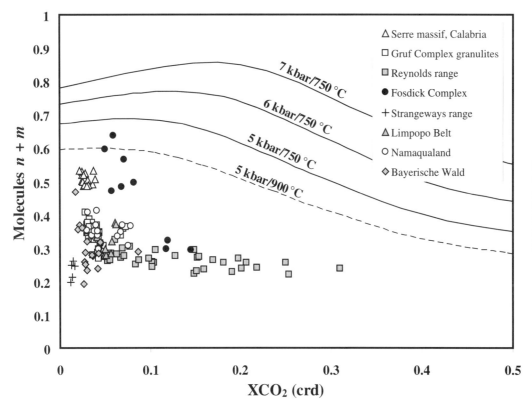

Figure 6. Comparison of the total volatile content (molecules $n + m$ p.f.u.) of cordierite (crd) from medium-pressure terranes versus modeled volatile saturation surfaces at the P-T of interest.

TABLE 1. CORDIERITE VOLATILE DATA FOR LOW-PRESSURE (<4.5 kbar) GRANULITES AND MIGMATITES

Sample	H$_2$O (wt%)	CO$_2$ (wt%)	n (pfu)	m (pfu)	m+n (pfu)	XCO$_2$	P (kbar)	T (°C)	aH$_2$O	aCO$_2$	H$_2$O (wt%)	D$_w$
					Cordierite						Melt	

Huntly, Newer Gabbro Complex, NE Scotland (Droop et al., 2003[PT]; Rigby and Droop, 2008[CV])

BQ38 (12)	0.974	0.099	0.333	0.014	0.347	0.040	4.5	900	0.397	0.091	4.03	4.14
	0.101	0.011	0.010	0.041	0.042	–	–	–	0.060	0.011	0.47	–
BQ41 (8)	0.914	0.169	0.313	0.024	0.336	0.070	4.5	900	0.360	0.158	3.74	4.09
	0.060	0.080	0.015	0.030	0.034	–	–	–	0.034	0.077	0.26	–
1003 (12)	0.856	0.109	0.293	0.015	0.308	0.050	4.5	900	0.327	0.100	3.49	4.07
	0.058	0.029	0.011	0.030	0.031	–	–	–	0.031	0.027	0.24	–
CASB5 (12)	0.983	0.101	0.332	0.014	0.346	0.040	4.5	900	0.392	0.092	4.00	4.07
	0.064	0.017	0.010	0.031	0.032	–	–	–	0.038	0.016	0.30	–
FOW1 (8)	1.032	0.082	0.350	0.011	0.361	0.032	4.5	900	0.427	0.075	4.27	4.13
	0.100	0.010	0.010	0.040	0.042	–	–	–	0.063	0.009	0.50	–
PIR1 (12)	0.755	0.160	0.258	0.022	0.281	0.080	4.5	900	0.275	0.149	3.07	4.07
	0.130	0.070	0.014	0.050	0.052	–	–	–	0.066	0.066	0.52	–
PIR101 (8)	0.706	0.094	0.242	0.013	0.255	0.052	4.5	900	0.251	0.086	2.88	4.09
	0.066	0.015	0.010	0.031	0.033	–	–	–	0.031	0.014	0.25	–
BQH1 (15)	0.860	0.118	0.294	0.017	0.311	0.053	4.5	900	0.330	0.109	3.50	4.07
	0.100	0.010	0.010	0.040	0.042	–	–	–	0.038	0.039	0.30	–

Belhelvie, Newer Gabbro Complex, NE Scotland (Droop and Charnley, 1985[PT]; Rigby and Droop, 2008[CV])

BELHM1 (7)	1.095	0.185	0.369	0.026	0.395	0.065	4.5	800	0.368	0.150	3.78	3.45
	0.134	0.058	0.045	0.008	0.046	–	–	–	0.071	0.017	0.73	–
BELSQ2 (12)	1.039	0.057	0.350	0.008	0.358	0.022	4.5	800	0.340	0.045	3.56	3.42
	0.124	0.013	0.042	0.002	0.042	–	–	–	0.065	0.004	0.68	–
BELHM2c (12)	1.056	0.090	0.356	0.012	0.368	0.034	4.5	800	0.348	0.072	3.62	3.43
	0.136	0.020	0.046	0.003	0.046	–	–	–	0.070	0.006	0.73	–
BELHM3a (12)	1.436	0.102	0.484	0.014	0.498	0.028	4.5	800	0.590	0.082	5.54	3.86
	0.162	0.022	0.055	0.003	0.055	–	–	–	0.101	0.007	1.12	–
BELSP9a (18)	0.988	0.082	0.333	0.011	0.344	0.033	4.5	800	0.315	0.066	3.36	3.40
	0.110	0.022	0.037	0.003	0.037	–	–	–	0.058	0.035	0.68	–

Etive thermal aureole, Scotland (Droop and Moazzen, 2007[PT]; Rigby et al., 2008a[CV])

MM135A (11)	1.230	0.115	0.425	0.016	0.441	0.037	2.2	685	0.810	0.144	5.09	4.14
	0.093	0.031	0.039	0.004	0.039	–	–	–	0.105	0.055	1.32	–
MM188C (12)	1.188	0.080	0.409	0.011	0.420	0.027	2.2	685	0.758	0.100	4.81	4.05
	0.094	0.022	0.039	0.003	0.039	–	–	–	0.103	0.038	1.08	–
MM174B (12)	1.154	0.120	0.397	0.017	0.414	0.041	2.2	685	0.722	0.150	4.61	4.00
	0.053	0.034	0.029	0.005	0.029	–	–	–	0.074	0.061	0.90	–
MD12 (8)	1.203	0.114	0.414	0.016	0.430	0.037	2.2	690	0.784	0.144	4.92	4.09
	0.088	0.044	0.037	0.006	0.037	–	–	–	0.100	0.078	1.14	–
MM171 (18)	1.160	0.072	0.405	0.010	0.415	0.025	2.2	700	0.772	0.095	4.83	4.17
	0.072	0.031	0.033	0.004	0.033	–	–	–	0.090	0.057	1.07	–
MM144 (11)	1.040	0.113	0.359	0.016	0.375	0.043	2.2	705	0.643	0.150	4.15	3.99
	0.051	0.033	0.028	0.005	0.028	–	–	–	0.071	0.062	0.61	–
MM195A (12)	1.092	0.104	0.377	0.015	0.392	0.037	2.2	710	0.704	0.139	4.47	4.10
	0.067	0.022	0.032	0.003	0.032	–	–	–	0.084	0.041	0.83	–
MM169 (11)	1.020	0.117	0.355	0.017	0.372	0.045	2.2	715	0.648	0.160	4.19	4.10
	0.073	0.036	0.034	0.005	0.034	–	–	–	0.087	0.071	0.69	–
MM166A (10)	0.845	0.092	0.292	0.013	0.305	0.042	2.2	730	0.503	0.129	3.44	4.08
	0.104	0.016	0.042	0.002	0.042	–	–	–	0.103	0.032	0.36	–
MM164 (10)	1.056	0.107	0.365	0.015	0.380	0.040	2.2	730	0.699	0.152	4.42	4.19
	0.131	0.030	0.050	0.004	0.050	–	–	–	0.136	0.060	0.92	–
MM143 (12)	0.906	0.111	0.311	0.016	0.327	0.048	2.2	735	0.556	0.158	3.69	4.08
	0.093	0.028	0.039	0.004	0.039	–	–	–	0.098	0.056	0.47	–
MM194B (10)	1.018	0.107	0.352	0.015	0.367	0.041	2.2	740	0.676	0.155	4.30	4.22
	0.072	0.027	0.033	0.004	0.033	–	–	–	0.090	0.056	0.79	–
MM166E (28)	0.876	0.095	0.303	0.013	0.316	0.042	2.2	745	0.548	0.139	3.65	4.16
	0.095	0.027	0.039	0.004	0.039	–	–	–	0.101	0.056	0.46	–
MM166F (9)	0.933	0.113	0.322	0.016	0.338	0.047	2.2	745	0.598	0.166	3.90	4.18
	0.065	0.020	0.031	0.003	0.031	–	–	–	0.083	0.042	0.56	–
MM166C (8)	0.836	0.090	0.289	0.013	0.302	0.042	2.2	745	0.512	0.133	3.47	4.15
	0.111	0.019	0.044	0.003	0.044	–	–	–	0.111	0.039	0.40	–

(Continued)

TABLE 1. CORDIERITE VOLATILE DATA FOR LOW-PRESSURE (<4.5 kbar) GRANULITES AND MIGMATITES (Continued)

Sample	Cordierite											Melt	
	H_2O (wt%)	CO_2 (wt%)	n (pfu)	m (pfu)	$m+n$ (pfu)	XCO_2	P (kbar)	T (°C)	aH_2O	aCO_2		H_2O (wt%)	D_w

Etive thermal aureole, Scotland (Droop and Moazzen, 2007[PT]**; Rigby et al., 2008a**[CV]**) (Continued)**

MM166D (11)	0.874	0.100	0.304	0.014	0.318	0.045	2.2	745	0.549	0.148		3.63	4.15
	0.054	0.017	0.029	0.002	0.029	—*	—	—	0.074	0.035		0.43	—
MD19 (12)	0.856	0.091	0.296	0.013	0.309	0.042	2.2	750	0.537	0.135		3.58	4.18
	0.090	0.012	0.038	0.002	0.038	—	—	—	0.098	0.025		0.44	—
MD18 (12)	1.088	0.118	0.377	0.017	0.393	0.043	2.2	765	0.797	0.183		4.85	4.46
	0.094	0.036	0.039	0.005	0.039	—	—	—	0.117	0.079		1.30	—
MM198c (10)	0.971	0.088	0.332	0.012	0.344	0.036	2.2	780	0.680	0.139		4.27	4.39
	0.063	0.013	0.031	0.002	0.031	—	—	—	0.089	0.030		0.82	—
MM197A (14)	0.984	0.103	0.336	0.014	0.351	0.041	2.2	780	0.693	0.163		4.32	4.39
	0.089	0.026	0.038	0.004	0.038	—	—	—	0.110	0.057		0.88	—
MM193Y (12)	1.190	0.233	0.412	0.033	0.445	0.074	2.2	800	1.005	0.400		5.85	4.91
	0.100	0.057	0.041	0.008	0.042	—	—	—	0.142	0.138		2.64	—
MD9A (9)	1.263	0.249	0.437	0.035	0.472	0.075	2.2	1000	1.783	0.641		10.51	8.32
	0.110	0.081	0.044	0.011	0.045	—	—	—	0.253	0.294		12.61	—
MD9B (9)	1.014	0.188	0.351	0.027	0.377	0.071	2.2	1000	1.242	0.480		7.09	6.99
	0.117	0.078	0.046	0.011	0.047	—	—	—	0.231	0.280		4.46	—

Broken Hill, Willyama Complex, NSW (Phillips, 1981[PT]**; Harley et al., 2002**[CV]**)**

BH-ea3 (7)	0.644	0.162	0.217	0.023	0.24	0.094	4.5	800	0.175	0.123		2.50	3.90
	0.064	0.035	0.022	0.005	0.022	—	—	—	0.034	0.04		0.30	—
BH-ea6 (6)	0.586	0.132	0.198	0.018	0.217	0.085	4.5	800	0.156	0.096		2.30	4.00
	0.049	0.013	0.017	0.002	0.017	—	—	—	0.026	0.017		0.30	—
BH-baz (6)	0.744	0.16	0.252	0.022	0.274	0.08	4.5	800	0.212	0.119		2.80	3.80
	0.059	0.032	0.02	0.004	0.02	—	—	—	0.035	0.036		0.30	—
BH-alm (8)	0.841	0.138	0.286	0.019	0.305	0.062	4.5	800	0.252	0.102		3.20	3.80
	0.106	0.038	0.036	0.005	0.036	—	—	—	0.062	0.042		0.60	—

Cooma Migmatites, southern NSW (Flood and Vernon, 1978[PT]**; Ellis and Obata, 1992**[PT]**; Harley et al., 2002**[CV]**)**

C-IBA (5)	0.885	0.582	0.305	0.082	0.388	0.213	3.5	725	0.321	0.653		3.00	3.40
	0.086	0.037	0.03	0.005	0.005	—	—	—	0.064	0.094		0.40	—
C-6r (4)	1.338	0.305	0.463	0.043	0.507	0.086	3.5	725	0.631	0.272		5.10	3.80
	0.197	0.04	0.068	0.006	0.068	—	—	—	0.184	0.06		1.00	—
C-8a (8)	1.03	0.504	0.359	0.069	0.428	0.161	3.5	725	0.41	0.445		3.60	3.50
	0.102	0.29	0.035	0.04	0.053	—	—	—	0.087	0.375		0.60	—
C-8ba (7)	1.04	0.455	0.36	0.065	0.425	0.152	3.5	725	0.412	0.414		3.60	3.50
	0.081	0.029	0.028	0.004	0.028	—	—	—	0.073	0.06		0.50	—
C-8b (8)	1.217	0.324	0.422	0.046	0.467	0.097	3.5	725	0.534	0.286		4.40	3.60
	0.106	0.068	0.037	0.01	0.038	—	—	—	0.107	0.094		0.60	—

Mount Stafford Contact Zones, Anmatjira Range, Arunta Complex (Greenfield et al., 1998[PT]**; Harley et al., 2002**[CV]**)**

ISB26 (6) zone 5	0.964	0.749	0.333	0.106	0.44	0.241	3.5	680	0.331	0.634		4.30	4.50
	0.084	0.045	0.029	0.006	0.03	—	—	—	0.062	0.09		0.60	—
1210x (8) zone 5	1.109	0.812	0.383	0.115	0.499	0.232	3.5	680	0.412	0.694		5.20	4.70
	0.137	0.097	0.047	0.014	0.049	—	—	—	0.107	0.134		1.00	—
ISB15 (5) zone 4	0.909	0.001	0.314	0.001	0.314	0.000	2.5	800	0.579	0.001		4.00	4.40
	0.077	0.001	0.027	0.001	0.027	—	—	—	0.105	0.001		0.50	—
1190f1 (9) zone 4	0.873	0.006	0.302	0.001	0.303	0.003	2.5	800	0.546	0.01		3.90	4.40
	0.079	0.017	0.027	0.003	0.027	—	—	—	0.103	0.033		0.50	—
1190f2 (7) zone 4	0.798	0.043	0.276	0.006	0.282	0.022	2.5	800	0.481	0.06		3.50	4.40
	0.059	0.022	0.021	0.003	0.021	—	—	—	0.078	0.043		0.40	—
isb27 (4) zone 4	0.75	0.005	0.259	0.001	0.26	0.003	2.5	800	0.442	0.009		3.30	4.40
	0.067	0.004	0.023	0.001	0.023	—	—	—	0.081	0.011		0.40	—

(Continued)

TABLE 1. CORDIERITE VOLATILE DATA FOR LOW-PRESSURE (<4.5 kbar) GRANULITES AND MIGMATITES (Continued)

Sample	Cordierite									Melt	
	H_2O (wt%)	CO_2 (wt%)	n (pfu)	m (pfu)	m+n (pfu)	XCO_2	P (kbar)	T (°C)	aH_2O	aCO_2	H_2O (wt%)	D_w

Kimberley Region, Springvale pluton, Halls Creek Mobile Belt (Bodorkos et al., 2000[PT]; Harley et al., 2002[CV])

GW-A1 (8)	0.58	0.132	0.199	0.018	0.217	0.083	3.5	775	0.203	0.124	2.20	3.90
enclave 1	*0.059*	*0.099*	*0.02*	*0.013*	*0.024*	–	–	–	*0.039*	*0.105*	*0.30*	–
GW-B (5)	0.606	0.144	0.208	0.02	0.228	0.089	3.5	775	0.215	0.141	2.20	3.60
enclave 1	*0.105*	*0.018*	*0.036*	*0.002*	*0.036*	–	–	–	*0.065*	*0.029*	*0.50*	–
GW-C (8)	0.66	0.124	0.226	0.017	0.243	0.072	3.5	775	0.24	0.117	2.50	3.80
enclave 1	*0.077*	*0.023*	*0.026*	*0.003*	*0.027*	–	–	–	*0.053*	*0.033*	*0.40*	–
GW-E (5)	0.363	0.189	0.124	0.026	0.151	0.175	3.5	775	0.116	0.181	1.50	4.10
enclave 2	*0.027*	*0.033*	*0.009*	*0.005*	*0.01*	–	–	–	*0.017*	*0.05*	*0.20*	–
GW-F (9)	0.422	0.213	0.145	0.03	0.174	0.169	3.5	775	0.139	0.211	1.80	4.20
enclave 2	*0.041*	*0.057*	*0.014*	*0.008*	*0.016*	–	–	–	*0.025*	*0.084*	*0.30*	–
GW-G (8)	0.301	0.387	0.103	0.054	0.157	0.344	3.5	775	0.094	0.386	1.40	4.60
enclave 2	*0.023*	*0.049*	*0.008*	*0.007*	*0.01*	–	–	–	*0.014*	*0.084*	*0.20*	–

Ballachulish thermal aureole (Pattison and Harte, 1988[PT]; Rigby and Droop, 2008[CV])

MRB6 (5)	1.488	0.241	0.373	0.038	0.411	0.092	3	690	0.480	0.257	3.80	2.55
	0.063	*0.029*	*0.028*	*0.004*	*0.028*	–	–	–	*0.121*	*0.036*	*0.46*	–
MRB7 (6)	1.054	0.214	0.370	0.031	0.401	0.077	3	690	0.475	0.208	3.77	3.58
	0.108	*0.038*	*0.041*	*0.006*	*0.042*	–	–	–	*0.181*	*0.046*	*0.68*	–
MRB16G (4)	1.108	0.194	0.390	0.028	0.417	0.067	3	700	0.526	0.193	4.08	3.68
	0.079	*0.024*	*0.033*	*0.004*	*0.033*	–	–	–	*0.114*	*0.026*	*0.49*	–
MRB5 (6)	0.939	0.169	0.330	0.024	0.354	0.069	3	700	0.406	0.168	3.36	3.58
	0.040	*0.003*	*0.022*	*0.001*	*0.022*	–	–	–	*0.105*	*0.007*	*0.35*	–
MRB 2c (7)	0.968	0.108	0.340	0.016	0.356	0.044	3	700	0.425	0.106	3.47	3.59
	0.099	*0.002*	*0.039*	*0.001*	*0.039*	–	–	–	*0.180*	*0.044*	*0.63*	–
MRB16F (4)	1.202	0.221	0.423	0.032	0.454	0.070	3	715	0.625	0.230	4.67	3.89
	0.035	*0.018*	*0.021*	*0.003*	*0.021*	–	–	–	*0.070*	*0.021*	*0.33*	–
MRB16D (5)	1.073	0.191	0.377	0.038	0.415	0.092	3	715	0.517	0.279	4.02	3.75
	0.106	*0.071*	*0.041*	*0.010*	*0.042*	–	–	–	*0.153*	*0.078*	*0.61*	–
MRB16E (8)	1.157	0.159	0.407	0.023	0.430	0.053	3	720	0.592	0.166	4.47	3.86
	0.087	*0.020*	*0.035*	*0.003*	*0.035*	–	–	–	*0.121*	*0.022*	*0.54*	–
MRB27 (7)	1.182	0.267	0.415	0.038	0.454	0.085	3	720	0.613	0.284	4.60	3.90
	0.143	*0.037*	*0.053*	*0.005*	*0.053*	–	–	–	*0.172*	*0.039*	*0.79*	–
MRB3a (8)	1.059	0.176	0.372	0.025	0.398	0.063	3	725	0.518	0.186	4.03	3.80
	0.027	*0.018*	*0.019*	*0.003*	*0.020*	–	–	–	*0.078*	*0.023*	*0.31*	–
MRB4 (8)	1.036	0.133	0.364	0.019	0.383	0.050	3	750	0.529	0.148	4.10	3.95
	0.066	*0.011*	*0.029*	*0.002*	*0.029*	–	–	–	*0.112*	*0.014*	*0.46*	–
MRB16A (10)	1.164	0.200	0.409	0.029	0.438	0.066	3	750	0.640	0.225	4.77	4.10
	0.115	*0.019*	*0.044*	*0.003*	*0.044*	–	–	–	*0.151*	*0.023*	*0.72*	–

Bushveld Complex, high-grade contact zone (Waters and Lovegrove, 2002[PT])

Bush (7)	1.679	0.178	0.587	0.025	0.613	0.042	3	750	1.314	0.243	9.36	5.57
	0.233	*0.089*	*0.081*	*0.013*	*0.082*	–	–	–	*0.257*	*0.106*	*1.83*	–
Bush (7)	1.679	0.178	0.587	0.025	0.613	0.042	3.5	750	1.101	0.198	8.60	5.12
	0.233	*0.089*	*0.081*	*0.013*	*0.082*	–	–	–	*0.216*	*0.086*	*1.69*	–

Insch, Newer Gabbro Complex

ICLA-1 (7)	1.006	0.164	0.339	0.023	0.362	0.062	5	800	0.324	0.133	3.44	3.41
	0.061	*0.053*	*0.021*	*0.007*	*0.022*	–	–	–	*0.020*	*0.043*	*0.21*	–
RH3A (7)	1.084	0.185	0.365	0.026	0.391	0.065	5	800	0.365	0.151	3.75	3.46
	0.101	*0.058*	*0.034*	*0.008*	*0.035*	–	–	–	*0.034*	*0.047*	*0.35*	–

Note: Italicized numbers refer to the 1σ uncertainty. [PT] refers to the reference for the background information and independent P-T conditions. [CV] refers to the reference from which the measured cordierite volatile data were taken.
*_ = not determined; n—number of molecules of H_2O evolved per mole of reactant; m—number of molecules of CO_2 per formula unit (pfu) of 18-oxygen volatile-free cordierite.

saturation surfaces at the P-T of interest shows that the lowest grade samples approach fluid saturation, whereas up-grade, at temperatures >700 °C, the regime becomes progressively fluid undersaturated. Conversely, samples immediately adjacent to the igneous contact are markedly oversaturated with respect to their P-T conditions of formation. Similarly, samples from the Bushveld high-grade contact zone preserve high H_2O contents (1.435–2.047 wt%) that yield unrealistic aH_2O values up to 1.314. The CO_2 contents are typically low (0.059–0.268), yielding a mean aCO_2 value of 0.167. The mean total volatile contents plot above the saturation curve of interest, which implies that the samples are markedly oversaturated with respect to their P-T conditions of formation.

The Cooma migmatites and diatexites from zone 5 of the Mount Stafford contact aureole are not strictly granulites but record moderate H_2O contents (0.885–1.338 wt%) and moderately high CO_2 contents (0.312–0.812 wt%). For the estimated P-T conditions of formation this yields aH_2O values ranging from 0.321 to 0.631 and aCO_2 values from 0.272 to 0.694. Incorporating the errors in the analyses and P-T estimates, the total volatile contents of the Cooma migmatites approach fluid saturation, whereas the zone 5 samples from Mount Stafford are unequivocally fluid saturated.

Fluid Regime

The low aH_2O values and total volatile contents reported for Huntly, Belhelvie, Kimberley, zone 4 of the Mount Stafford contact aureole, Broken Hill, and Ballachulish imply that fluid-undersaturated conditions prevailed. The theoretical constraints outlined in Figure 5 provide an independent check on the validity of this interpretation. In all cases, apart from Huntly and Kimberley, the measured cordierite volatile contents yield calculated melt H_2O contents that are entirely compatible with the independent P-T constraints and experimentally determined minimum melt H_2O contents (Johannes and Holtz, 1996). This good agreement suggests that in each of the aforementioned melt-bearing terranes, fluid-absent melting processes operated during the granulite-grade metamorphism.

In the Huntly case, the estimated H_2O contents of granitic melts coexisting with cordierite range from 2.88 to 4.27 wt%. Figure 5 predicts that at 900 °C and 4.5 kbar (Droop et al., 2003), cordierite should contain ~0.35 wt% H_2O, and the coexisting melt should have ~1.7 wt% H_2O. This implies that either the peak temperature estimates reported by Droop et al. (2003) are too high or that cordierite has re-equilibrated during cooling. Melts in communication with cordierite during cooling or uplift will continuously exchange H_2O in order to maintain the D_w equilibrium (Harley and Carrington, 2001). The P-T conditions for Huntly are particularly well constrained, and therefore it seems unlikely that the lower temperatures implied by the cordierite volatile data reflect peak metamorphic conditions. The apparent overestimation of both the cordierite H_2O content and the calculated melt H_2O content could be due to re-equilibration as the rocks cooled (Rigby and Droop, 2008). Conversely, the Kimberley rocks preserve cordierite H_2O contents that yield melt H_2O contents that are far too low for the estimated P-T conditions of formation. In cases such as this, the only viable but poorly understood explanation is that the cordierite must have lost some H_2O (Harley et al., 2002). Although this may be disconcerting, the good agreement, in other areas, between the calculated melt H_2O contents and the independently measured melt-H_2O contents suggests that the cordierite fluid monitor is reliable in most cases (Harley et al., 2002; Rigby and Droop, 2008).

In demonstrably fluid-absent cases, petrological evidence for fluid-absent melting must also be considered. Petrological data from each of the aforementioned fluid-absent case studies (references from Table 1) does suggest the operation of fluid-absent melting reactions. In the case of Ballachulish the good agreement between estimated melt-H_2O and experimentally determined minimum melt-H_2O contents would generally support the contention that cordierite H_2O contents were representative of peak metamorphic values. Unfortunately this good agreement is illusory. Pattison and Harte (1988) proposed that fluid-present melting in quartz-rich layers was fluxed by H_2O from nearby dehydrating quartz-absent rocks. Similarly, Holness and Clemens (1999) demonstrate that melting of the Appin quartzite was driven by fracture-controlled H_2O infiltration, which is incompatible with the observation that cordierites in melt-present rocks are fluid-undersaturated (Rigby and Droop, 2008). Conversely, the cordierite volatile contents from successive contact metamorphic zones of the Etive aureole provide evidence for fluid-present conditions at 685 °C that were followed, up-grade, by predominantly fluid-absent conditions, which is a similar evolution to that reported from Ballachulish. The measured cordierite volatile contents and estimated melt H_2O contents from fluid-undersaturated samples are entirely consistent with the theoretical constraints imposed by Figure 5. However, the data are also consistent with the operation of independently determined fluid-present and fluid-absent melting reactions thought to operate in the aureole (Droop and Moazzen, 2007). Whereas this holds true for most Etive samples, the high aH_2O values and total volatile contents of those immediately adjacent to the igneous contact (cf. Holness and Clemens, 1999) imply oversaturation with respect to the P-T conditions of formation. The oversaturation recorded in these samples may be satisfactorily explained by the fact that the cordierite did not acquire its fluid content at the assumed peak of metamorphism. As the igneous body cooled and crystallized it would have expelled free H_2O, which may have locally re-saturated the cordierites adjacent to the contact (Moazzen et al., 2001; Rigby et al., 2008a). If one ignores the possibility of cooling and re-equilibration with an externally derived fluid at lower temperatures, then this will cause overestimation of the aH_2O and melt H_2O content for the assumed peak metamorphic conditions (Harley et al., 2002). The high total volatile contents and aH_2O values from the Bushveld contact zone also imply that oversaturated conditions were attained, which suggests that fluid infiltration from either the adjacent Rustenburg Layered

Suite or the dehydrating country rocks may have locally re-saturated these rocks during cooling.

The Cooma migmatites and diatexites from zone 5 of the Mount Stafford contact aureole also preserve evidence of fluid-saturated conditions. However, unlike the high-grade Bushveld and highest-grade Etive samples that have been locally re-saturated with a predominantly H_2O-rich fluid, these samples preserve a record of a moderately CO_2-rich fluid. The cordierite H_2O contents of the Cooma migmatites imply that coexisting granitic melts contained between 2.86 and 5.10 wt% H_2O, values that are broadly consistent with the constraints imposed by Figure 5. However, Ellis and Obata (1992) suggest that the Cooma migmatites formed as a result of fluid-present melting under high (0.9) aH_2O conditions. Considering the preserved CO_2 contents and low aH_2O values inferred from cordierite compositions, such a mechanism is implausible (Harley et al., 2002). The zone 5 diatexites from Mount Stafford preserved similarly high CO_2 contents, with aCO_2 reaching 0.694. The aH_2O is typically low (~0.331), with coexisting melts preserving between 4.3 and 5.20 wt% H_2O. Greenfield et al. (1998) attribute the extensive melting observed in zone 5 to open-system behavior, which is similar to Ballachulish but with a carbonic fluid. If this assertion is correct, it may also explain the high CO_2 contents recorded in the cordierites.

Medium-Pressure Granulites

Cordierite Volatile Contents

The cordierites from the Gruf Complex, Calabria, Fosdick Complex, Reynolds Range, Strangways Range, Bayerische Wald, Namaqualand, and the Limpopo Belt (Table 2) preserve H_2O contents of 0.425–1.6 wt%, typically with low CO_2 contents (0.023–0.256 wt%). For the estimated P-T conditions in each of the areas, these data yield aH_2O values ranging from 0.128 to 0.620, and aCO_2 values between 0.013 and 0.398. A comparison of the mean total volatile content of each sample to modeled H_2O-CO_2 saturation surfaces at the P-T conditions of interest shows that all samples are markedly fluid undersaturated (Fig. 6). Samples from the Larsemann Hills and Brattstrand Bluffs demonstrate variability, indicating fluid saturation in some cases and fluid absence in others.

Fluid Regime

The low aH_2O values and total volatile contents of cordierite from the Fosdick Complex, Strangways Range, Bayerische Wald, and Namaqualand imply that fluid-undersaturated conditions were attained. Inspection of Figure 7 reveals that the measured H_2O contents are entirely consistent with the independently determined melt H_2O contents at the P-T conditions of interest. This supports the conclusion that they formed through fluid-absent melting processes. The cordierite volatile contents preserved in the Reynolds Range and Prydz Bay (Brattstrand Bluffs and Larsemann Hills) samples are too low for the specified P-T conditions, and therefore volatile leakage is probable

(Harley et al., 2002). The volatile contents of the Gruf Complex and Calabrian granulites also appear to be inconsistent with the independent P-T constraints. However, in these cases the P-T path for each area (Droop and Bucher-Nurminen, 1984, and Schenk, 1984, respectively) must be taken into account. Application of the initial P-T conditions (corresponding, in the Gruf case, to initial growth of coarse-grained cordierite: superscript 1 in Fig. 7) would yield results inconsistent with the constraints imposed by Figure 7. However, compelling textural and mineralogical evidence indicates that cordierite growth continued as the rocks underwent near-isothermal decompression (Droop and Bucher-Nurminen, 1984; Schenk, 1984). Providing that cordierite was in communication with melt during decompression, then re-equilibration should have taken place at the "new" P-T conditions as both the melt and cordierite H_2O content would have continuously changed in order to maintain the D_w equilibrium (Harley and Carrington, 2001; Harley et al., 2002). By ignoring the later decompression and subsequent exchange of H_2O, the application of the initial P-T conditions can yield large underestimates in both aH_2O and the melt H_2O content (Rigby and Droop, 2008). Taking this into account, the post-decompression P-T conditions (superscript 2 in Fig. 7) of Calabria and the Gruf Complex are consistent with the preserved volatile contents in each case.

The Limpopo Belt Al, Mg-granulite samples preserve cordierite H_2O contents of 1.6 wt%. At the inferred P-T conditions of initial cordierite formation (7 kbar and 800 °C; Droop, 1989), cordierite should contain ~1.4 wt% H_2O (Fig. 6). This implies that the cordierite volatile contents have either re-equilibrated during cooling or that the P-T conditions determined for initial cordierite formation are incorrect. Droop's (1989) data do not preclude initial cordierite formation at pressures as high as 8 kbar, which would be compatible with the observed H_2O contents. However, Droop (1989) suggests that the Limpopo granulites underwent isothermal decompression. If cordierite was in communication with melt during decompression, then re-equilibration would be likely to have preserved lower H_2O contents. If the first decompression-cooling path of Perchuk et al. (2008) is invoked, terminating at 5–6 kbar and 680–700 °C, the volatile contents would be compatible with the independent retrograde P-T constraints. Harley (2008) demonstrated that cordierite volatile contents can be used to refine P-T paths. However, this is difficult to implement in polymetamorphic terranes such as the Limpopo Belt, especially when different authors espouse different P-T paths (Rigby et al., 2008d, 2010).

UltraHigh-Temperature Granulites

Cordierite Volatile Contents

The cordierites from the ultrahigh-temperature (UHT) terranes generally preserve the highest CO_2 contents, with values up to 1.47 wt% (Table 3). The exceptions are the Madagascan* granulites reported in Rigby and Droop (2008) and the sample from the Mather Peninsula from Harley et al. (2002). H_2O contents are typically low to moderate, with values ranging from

TABLE 2. CORDIERITE VOLATILE DATA FOR MEDIUM-PRESSURE (>5 kbar) GRANULITES

Sample	H_2O (wt%)	CO_2 (wt%)	Cordierite n (pfu)	m (pfu)	$m+n$ (pfu)	XCO_2	P (kbar)	T (°C)	aH_2O	aCO_2	Melt H_2O (wt%)	D_w
Gruf Complex, Central Italian Alps (Droop and Bucher-Nurminen, 1984[PT]; Rigby and Droop, 2008[CV])												
AL18 (8)	1.058	0.084	0.355	0.012	0.366	0.032	5	750	0.300	0.066	3.43	3.24
	0.073	0.004	0.033	0.010	0.034	–*	–	–	0.031	0.059	0.24	–
AL20 (10)	0.989	0.091	0.331	0.012	0.344	0.036	5	750	0.332	0.061	3.46	3.50
	0.080	0.030	0.035	0.011	0.036	–	–	–	0.030	0.055	0.28	–
14271 (20)	0.970	0.115	0.325	0.016	0.341	0.046	5	750	0.291	0.084	3.42	3.52
	0.068	0.036	0.032	0.011	0.034	–	–	–	0.029	0.057	0.24	–
Calabria, Serre Massif (Schenk, 1984[PT]; Rigby and Droop, 2008[CV])												
Cal29C (8)	1.462	0.098	0.491	0.014	0.504	0.027	4	730	0.606	0.083	5.29	3.62
	0.050	0.013	0.017	0.002	0.017	–	–	–	0.048	0.046	0.42	–
Cal32 (8)	1.479	0.114	0.497	0.016	0.512	0.031	4	730	0.620	0.097	5.39	3.65
	0.056	0.021	0.019	0.003	0.019	–	–	–	0.051	0.063	0.44	–
Fosdick Complex, West Antarctica (Smith, 1996[PT]; Harley et al., 2002[CV])												
Leucosome (8)	1.244	0.256	0.423	0.039	0.462	0.085	5	750	0.358	0.168	3.90	3.10
	0.392	0.039	0.133	0.006	0.134	0.033	–	–	0.234	0.042	1.80	–
SE Reynolds Range, Arunta Complex (Buick et al., 1998[PT]; Harley et al., 2002[CV])												
9017 (39)	0.685	0.229	0.232	0.033	0.264	0.124	5	775	0.157	0.149	2.20	3.20
Leucosome in shears	0.109	0.115	0.037	0.016	0.04	0.068	–	–	0.045	0.095	0.50	–
Strangways Range, Arunta Complex (Goscombe, 1992[PT]; Harley et al., 2002[CV])												
9052 (5)	0.69	0.023	0.229	0.003	0.232	0.014	6	875	0.147	0.013	2.30	3.30
Grt-Opx-Crd patch	0.092	0.005	0.031	0.001	0.031	0.004	–	–	0.03	0.004	0.30	–
Bayerische Wald (Kalt, 2000[PT,CV])												
BW-06	0.435	0.045	0.220	0.006	0.226	0.027	6	850	0.131	0.024	2.11	4.85
	0.120	0.004	0.061	0.001	0.061	–	–	–	0.026	0.006	0.42	–
BW-13	0.470	0.072	0.227	0.053	0.279	0.170	6	850	0.136	0.223	2.16	4.61
	0.060	0.019	0.029	0.014	0.032	–	–	–	0.027	0.056	0.43	–
BW-18	0.600	0.093	0.288	0.013	0.301	0.043	6	850	0.187	0.053	2.68	4.46
	0.060	0.035	0.029	0.005	0.029	–	–	–	0.037	0.013	0.54	–
BW28	0.520	0.085	0.266	0.018	0.283	0.061	6	850	0.168	0.072	2.49	4.78
	0.040	0.034	0.020	0.007	0.022	–	–	–	0.034	0.018	0.50	–
BW-36	0.570	0.054	0.288	0.080	0.368	0.217	6	850	0.188	0.350	2.68	4.70
	0.080	0.009	0.040	0.013	0.043	–	–	–	0.038	0.087	0.54	–
BW-44	0.710	0.048	0.369	0.044	0.413	0.095	6	850	0.272	0.183	3.48	4.90
	0.120	0.007	0.062	0.006	0.063	–	–	–	0.054	0.046	0.70	–
BW-45	0.540	0.055	0.271	0.090	0.361	0.249	6	850	0.173	0.398	2.53	4.69
	0.040	0.002	0.020	0.003	0.020	–	–	–	0.035	0.099	0.51	–
BM-01	0.425	0.057	0.217	0.044	0.260	0.155	6	850	0.128	0.183	2.09	4.91
	0.030	0.014	0.015	0.011	0.019	–	–	–	0.026	0.046	0.42	–
E-26	0.490	0.050	0.243	0.007	0.250	0.028	6	850	0.149	0.028	2.30	4.69
	0.040	0.006	0.020	0.001	0.020	–	–	–	0.030	0.007	0.46	–
E-22	0.520	0.070	0.270	0.010	0.280	0.036	6	850	0.172	0.041	2.52	4.85
	0.050	0.013	0.026	0.002	0.026	–	–	–	0.034	0.010	0.50	–
Limpopo Belt granulites, Beitbridge (Droop, 1989[PT]; Rigby et al., 2008c[CV])												
D1016	1.600	0.100	0.536	0.014	0.550	0.025	6	800	0.478	0.052	5.41	3.38
	0.050	0.009	0.017	0.001	0.017	–	–	–	0.021	0.017	0.24	–
BB14b	1.600	0.060	0.536	0.008	0.544	0.015	6	800	0.478	0.031	5.41	3.38
	0.180	0.002	0.060	0.001	0.060	–	–	–	0.076	0.004	0.86	–
Namaqualand												
DM-9 (10)	0.871	0.106	0.298	0.015	0.313	0.043	–	–	–	–	–	–
	0.094	0.038	0.032	0.005	0.032	–	–	–	–	–	–	–
WW-8 (18)	0.917	0.118	0.314	0.017	0.330	0.050	–	–	–	–	–	–
	0.072	0.041	0.025	0.006	0.025	–	–	–	–	–	–	–

Note: [PT] refers to the reference for the background information and independent P-T conditions. [CV] refers to the reference from which the measured cordierite volatile data were taken. Italicized numbers refer to the 1σ uncertainty.
*– = not determined; n—number of molecules of H_2O evolved per mole of reactant; m—number of molecules of CO_2 per formula unit (pfu) of 18-oxygen volatile-free cordierite.

0.194 to 1.005 wt%. For the estimated P-T conditions in each of the areas, these data yield extremely low aH_2O values that range from 0.025 to 0.253 and a wide range of aCO_2 values (0.050–0.800). A comparison of the mean total volatile content with the saturation surface of interest shows that the Madagascan, Napier Complex, Mather, Kerala Khondalite Belt, Cocachacra, and Pampa Blanca granulites are fluid undersaturated (Fig. 8). In the case of Anapolis, fluid-saturated conditions were unequivocally attained, as shown by the high CO_2 contents of the cordierites.

Fluid Regime

The cordierites from Madagascar preserve H_2O contents that are incompatible with the independent melt H_2O contents at the P-T of interest. Therefore it is likely that these samples have lost some H_2O (Harley et al., 2002; Rigby and Droop, 2008). In spite of this negative outcome, a comparison of the data from Rigby and Droop (2008; i.e., Madagascar*) to that obtained by Harley et al. (2002; i.e., Madagascar**) does highlight a critical issue. Samples from the same region, containing similar mineral assemblages that have evolved under the same or comparable P-T conditions, yield markedly different cordierite volatile contents. Even considering that H_2O may have been lost from some cordierites, the difference in measured CO_2 contents emphasizes the localized and variable composition of the fluid phase. In field areas where the cordierite volatile data are consistent with independent melt H_2O contents and P-T conditions, one must proceed with caution in concluding that a particular fluid regime prevailed over a wide area, especially when only a few samples have been investigated (Rigby and Droop, 2008).

Cordierite from the Napier Complex, Mather, Kerala Khondalite Belt, Cocachacra, and Pampa Blanca all preserve H_2O contents that are compatible with the independently determined melt H_2O contents (Fig. 9; Table 3), suggesting that fluid-absent melting processes were responsible for their formation. Harley and Thompson (2004) and Harley (2004) raise an important issue with regard to preserved volatile contents of the Napier Complex granulites; a cordierite can be CO_2-rich but remain strictly fluid undersaturated. Thus the preservation of relatively CO_2-rich cordierites does not necessarily imply fluid saturation or that CO_2-streaming processes operated. It is thus crucial to assess whether the preserved volatile contents reflect fluid-absent or fluid-present conditions at the P-T of interest. Despite the preservation of high CO_2 contents, the preserved H_2O contents are compatible with the independent melt H_2O contents. This

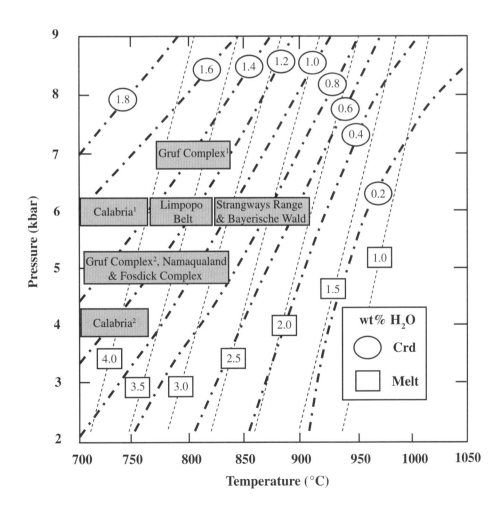

Figure 7. Contoured P-T diagram depicting cordierite-melt H_2O relationships for conditions in which the melt has the minimum possible H_2O content at a given P-T. The minimum weight percentages of H_2O melt isopleths (dashed lines) are after Johannes and Holtz (1996). The cordierite H_2O isopleths (dot-dashed lines) were calculated using the D_w relationships of Harley and Carrington (2001). The gray boxes represent the independently estimated P-T conditions for each of the field areas, respectively (the extent of the box in the P-T space indicates the uncertainty). Superscripts 1 and 2 refer to pre- and post-decompression P-T conditions, respectively.

TABLE 3. CORDIERITE VOLATILE DATA FOR ULTRA-HIGH-TEMPERATURE GRANULITES

Sample	Cordierite										Melt	
	H_2O (wt%)	CO_2 (wt%)	n (pfu)	m (pfu)	$m+n$ (pfu)	XCO_2	P (kbar)	T (°C)	aH_2O	aCO_2	H_2O (wt%)	D_w
SE Madagascar* (Markl et al., 2000[PT]; Rigby and Droop, 2008[CV])												
15834B (8)	0.890	0.080	0.304	0.011	0.316	0.036	6	920	0.238	0.050	3.19	3.58
	0.056	0.014	0.019	0.002	0.019	–*	–	–	0.023	0.008	0.31	–
15834A (8)	0.927	0.091	0.317	0.013	0.330	0.039	6	920	0.253	0.057	3.32	3.59
	0.066	0.017	0.023	0.002	0.023	–	–	–	0.028	0.008	0.37	–
15832 (10)	0.805	0.144	0.275	0.020	0.295	0.067	6	920	0.207	0.091	2.89	3.58
	0.063	0.076	0.022	0.011	0.024	–	–	–	0.025	0.019	0.35	–
Prydz Bay migmatitic granulites, Antarctica (Fitzsimons, 1996[PT]; Harley et al., 2002[CV]; Harley, 2004[CV])												
BB2 (26) Brattstrand Bluffs	0.305	0.781	0.104	0.109	0.213	0.513	6	860	0.055	0.499	1.20	4.00
	0.068	0.156	0.023	0.022	0.034	0.075	–	–	0.018	0.156	0.30	–
BB4 (17) Brattstrand Bluffs	0.494	0.583	0.168	0.081	0.25	0.325	6	860	0.096	0.361	1.70	3.50
	0.09	0.118	0.031	0.016	0.038	0.045	–	–	–	0.122	0.40	–
15 samples (62) Brattstrand Bluffs	0.398	0.707	0.136	0.099	0.234	0.423	6	860	0.075	0.447	1.50	3.70
	0.081	0.131	0.028	0.018	0.026	0.076	–	–	0.028	0.12	0.40	–
LarsBZ (6)	0.487	0.706	0.168	0.096	0.264	0.364	5.5	825	0.09	0.416	1.70	3.50
	0.065	0.273	0.022	0.037	0.053	0.063	–	–	0.021	0.249	0.30	–
L909(4) Larsemann Hills	0.33	0.707	0.112	0.099	0.211	0.468	5.5	825	0.056	0.43	1.30	3.90
	0.035	0.055	0.012	0.008	0.015	0.03	–	–	0.011	0.069	0.20	–
Mather Peninsula, Rauer Islands (Harley, 1998[PT]; Harley et al., 2002[CV]; Harley, 2004[CV])												
SH/87/218 (8) Grt-Opx-Crd-Sil	0.698	0.055	0.233	0.008	0.241	0.031	8.5	920	0.091	0.021	1.80	2.60
	0.067	0.021	0.022	0.003	0.024	0.008	–	–	0.009	0.011	0.30	–
Napier Complex, Enderby Land (Harley, 1998[PT]; Harley, 2004[CV]; Harley, 2008[CV])												
49752 Mt Sones (8) Grt-Spr-Sil-Crd	0.194	0.498	0.064	0.068	0.132	0.513	8.5	1000	0.025	0.256	0.85	4.40
	0.021	0.026	0.007	0.004	0.01	0.018	–	–	0.008	0.024	0.10	–
49354 Mt Hardy (6) Spr-Qz-Crd	0.215	1.04	0.072	0.142	0.214	0.664	8.5	1000	0.028	0.576	0.96	4.30
	0.031	0.021	0.01	0.003	0.013	0.026	–	–	0.011	0.025	0.10	–
Kerala Khondalite Belt, Southern India (Nandakumar and Harley, 2000[PT]; Harley, 2004[CV])												
SHED10 (10) Grt-Crd leucosome	0.636	0.854	0.219	0.12	0.34	0.355	6.5	900	0.129	0.557	2.00	3.10
	0.036	0.022	0.012	0.006	0.01	0.011	–	–	0.013	0.019	0.30	–
Chittikara-SLH (12) Grt-Crd leucosome	0.633	0.762	0.218	0.108	0.326	0.33	6.5	900	0.128	0.49	2.00	3.10
	0.112	0.053	0.039	0.012	0.02	0.043	–	–	0.03	0.043	0.60	–
SE Madagascar** (Markl et al., 2000[PT]; Harley et al., 2002[CV])												
patch migmatite (8) Crd leucosome	0.287	0.527	0.095	0.072	0.167	0.431	6	920	0.057	0.344	1.30	4.50
	0.05	0.065	0.017	0.009	0.022	0.029	–	–	0.016	0.074	0.30	–
Cocachacra Area (Martignole and Martelat, 2003[PT]; Harley, 2008[CV])												
mar93 (8)	0.995	0.636	0.331	0.086	0.417	0.207	8	900	0.158	0.258	2.69	2.71
	0.072	0.023	0.082	0.049	0.096	–	–	–	0.011	0.022	0.19	–
mar206 (8)	0.933	0.956	0.310	0.130	0.440	0.295	8	900	0.144	0.408	2.52	2.71
	0.044	0.048	0.057	0.081	0.099	–	–	–	0.009	0.041	0.16	–
Pampa Blanca Area (Martignole and Martelat, 2003[PT]; Harley, 2008[CV])												
mar102 (8)	0.969	0.835	0.323	0.114	0.436	0.261	8	900	0.152	0.350	2.63	2.71
	0.075	0.077	0.084	0.096	0.127	–	–	–	0.012	0.023	0.21	–
mar103 (8)	1.005	0.907	0.335	0.124	0.458	0.270	8	900	0.161	0.385	2.73	2.71
	0.051	0.041	0.064	0.074	0.098	–	–	–	0.016	0.026	0.27	–
Anapolis (Baldwin et al., 2005[PT]; Harley, 2008[CV])												
ML67 (8)	0.900	1.470	0.299	0.200	0.499	0.401	8	1000	0.176	0.800	0.58	0.64
	0.034	0.021	0.011	0.003	0.012	–	–	–	0.031	0.022	0.10	–

Note: [PT] refers to the reference for the background information and independent P-T conditions; [CV] refers to the reference from which the measured cordierite volatile data were taken. Italicized numbers refer to the 1σ uncertainty.
*– = not determined; n—number of molecules of H_2O evolved per mole of reactant; m—number of molecules of CO_2 per formula unit (pfu) of 18-oxygen volatile-free cordierite.

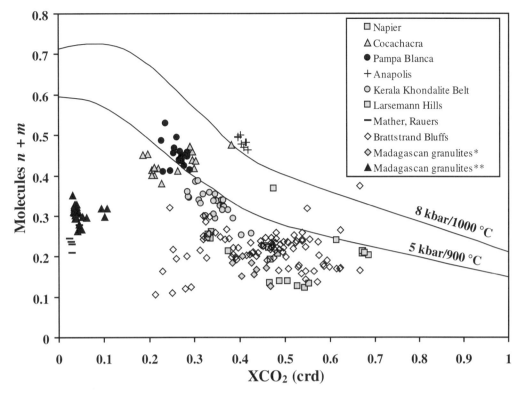

Figure 8. Comparison of the total volatile content (molecules $n + m$ p.f.u.) of cordierite from ultrahigh-temperature (UHT) terranes versus modeled volatile saturation surfaces at the P-T of interest.

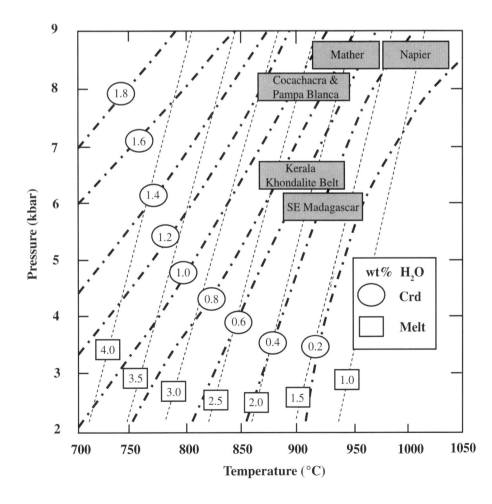

Figure 9. Contoured P-T diagram depicting cordierite-melt H_2O relationships for conditions in which the melt has the minimum possible H_2O content at a given P-T. The minimum weight percentages of H_2O melt isopleths (dashed lines) are after Johannes and Holtz (1996). The cordierite (Crd) H_2O isopleths (dot-dashed lines) were calculated using the D_w relationships of Harley and Carrington (2001). The gray boxes represent the independently estimated P-T conditions for each of the field areas, respectively (the extent of the box in the P-T space indicates the uncertainty).

further supports the notion that fluid-absent melting processes prevailed during their formation. The same conclusion applies to the Kerala Khondalite Belt, Cocachacra, and Pampa Blanca granulites. This conclusion is, however, contradictory to what is commonly inferred from fluid-inclusion studies. For example, Tsunogae et al. (2002) interpret high-density, CO_2-rich fluid inclusions in gneisses from the Napier Complex to reflect UHT metamorphism and granulite formation in the presence of a free carbonic fluid phase. As demonstrated previously, a cordierite can be CO_2 rich but fluid undersaturated. Applying similar logic, one must ask whether CO_2-rich fluid inclusions necessarily imply attainment of fluid-saturated conditions. Harley (2004) suggests that the cordierite and fluid-inclusion evidence from the Napier Complex can be reconciled if the CO_2-rich inclusions are interpreted to have been produced from post-UHT fluids released during the crystallization of nearly dry (<1 wt% H_2O) but CO_2-bearing melts during near-isobaric cooling. Moreover, Harley (2004) states that fluid inclusions formed at this stage in the P-T history would necessarily have been high density and appear "primary" if trapped in quartz as the remaining melt crystallized or migrated along grain boundaries. However, Cuney et al. (2007) would argue against such a mechanism because fluid inclusions commonly have densities that are compatible with peak rather than retrograde metamorphic conditions. There is, however, an alternative and potentially more viable explanation. Stevens and Clemens (1993) argue that even when CO_2-rich fluid inclusions are demonstrably primary in origin, this is not necessarily robust evidence for fluid-present conditions. In anatectic rocks, H_2O in a binary H_2O-CO_2 fluid is preferentially incorporated into the melt (e.g., Clemens, 1984; Clemens, 1993; Johannes and Holtz, 1996; Harley and Carrington, 2001), leaving a CO_2-enriched fluid residuum at grain boundaries that possess high-dihedral angles (Watson and Brenan, 1987). These CO_2-rich volatile dregs will appear primary in origin and have densities that are compatible with peak metamorphic conditions. These observations are commonly interpreted to reflect the presence of a pervasive CO_2-rich fluid (e.g., Tsunogae et al., 2002; Török et al., 2005; Satish-Kumar, 2005; Cuney et al., 2007; Santosh et al., 2008; Ohyama et al., 2008; Santosh and Omori, 2008). However, these interpretations are flawed because they do not consider the role of partial melting and its effect on H_2O activity and fluid composition. To justifiably conclude that fluid-saturated conditions prevailed, the measured total volatile content of cordierite must be compared with the saturation surfaces at the P and T of interest. Where fluid-saturated conditions are implied, for example in the Anapolis granulites, which preserve the highest CO_2 contents, the only viable explanation is that fluid infiltration must have occurred. This conclusion is supported by petrographic observations of symplectites and corona textures (Baldwin et al., 2005) and P-$a$$H_2O$ pseudosection modeling (Baldwin et al., 2005; Harley, 2008), which indicate that the cordierite-bearing mineral assemblages in the Anapolis granulites could only have developed through fluid infiltration at ~9.4 kbar (Harley, 2008). This raises arguably the most important question: Have demonstrably fluid-saturated samples been fluid saturated throughout their evolution? As highlighted above, in the case of Anapolis, the answer is that the rocks were not always fluid saturated. However, the cordierite volatile contents *alone* will not provide this insight. P-T modeling, in particular the use of P/T – $a$$H_2O$ or P/T – H_2O_{bulk} pseudosections, should be constructed in order to constrain the evolution of the observed mineral assemblages with regard to potential variations in either bulk H_2O or $a$$H_2O$ (e.g., Baldwin et al. 2005). Once this has been done, the cordierite volatile contents can be used to refine the P-T path and to support or amend the conclusions drawn from the pseudosection modeling (Harley, 2008).

FLUID-ABSENT MELTING OR CO_2 STREAMING DURING GRANULITE FORMATION?

A summary of the fluid conditions attained in each of the study areas is presented in Table 4. The cordierite fluid monitor does not always yield useful results, as some of the study areas yield volatile contents that are too low for their estimated P-T conditions of formation, and therefore the only viable explanation is that they have lost some H_2O. Harley et al. (2002) and Rigby and Droop (2008) stipulate that it is crucial to determine whether inconsistencies in the cordierite volatile data reflect post-peak-metamorphic leakage or alternative problems such as the application of erroneous peak P-T conditions or cordierite volatile re-equilibration with melt as conditions evolved along a P-T path. The various case studies presented here highlight a number of these critical issues. The cordierite volatile contents of samples from the Gruf Complex, Calabria, Huntly, and Limpopo would all be inconsistent with the independently determined peak metamorphic conditions if the possibility of subsequent re-equilibration during cooling or decompression were ignored.

Despite the potential problems related to leakage and re-equilibration, in the vast majority of case studies the calculated $a$$H_2O$ values and measured total volatile contents imply that fluid-absent conditions prevailed. The internally consistent cordierite volatile and melt-H_2O data support the contention that the independent P-T estimates applied to each of the field areas were valid. Moreover, the measured cordierite volatile contents are representative of peak metamorphic values attained through cordierite having equilibrated with an H_2O-undersaturated granitic melt that formed by fluid-absent, incongruent melting. In concert with the low $a$$CO_2$ values reported from many terranes, these data collectively argue against any significant influx of CO_2 from either deep, possibly mantle-derived sources or local magmatic bodies. The only conclusion that can be drawn from this body of evidence is that the presence of CO_2 is *not* a prerequisite for granulite formation. The cordierite volatile data support the extensive body of experimental and theoretical evidence that unequivocally demonstrates that granulites form principally through fluid-absent melting processes (e.g., Fyfe, 1973; Clemens and Wall, 1981; Thompson, 1982; Grant, 1985; Le Breton and Thompson, 1988; Clemens and Vielzeuf, 1987; Clemens, 1990, 1993, 1995;

TABLE 4. SUMMARY OF FLUID CONDITIONS DEVELOPED IN EACH OF THE STUDY AREAS

Fluid undersaturated (H_2O dominated)
(1) Belhelvie, Newer Gabbro Complex, NE Scotland
(2) Etive thermal aureole, NW Scotland
(3) Broken Hill, Willyama Complex, NSW, Australia
(4) Mount Stafford, contact zone 4, Arunta Complex, Australia
(5) Bayerische Wald, Germany
(6) Insch, Newer Gabbro Complex, NE Scotland
(7) Fosdick Complex, Antarctica
(8) Strangways Range, Arunta Complex, Australia
(9) Namaqualand, South Africa

Fluid undersaturated (CO_2 dominated)
(10) Pampa Blanca, Peru
(11) Cocachacra, Peru
(12) Mount Sones, Napier Complex, Enderby Land, Antarctica
(13) Mount Hardy, Napier Complex, Enderby Land, Antarctica
(14) Kerala Khondalite Belt, Southern India

Fluid undersaturated (modified P-T)
(15) Huntly, Newer Gabbro Complex, NE Scotland
(16) Calabria, Serre Massif, Italy
(17) Limpopo Belt granulites, Zimbabwe
(18) Gruf Complex, Central Italian Alps

Fluid present (H_2O rich; re-saturated)
(1) Bushveld Complex, high-grade contact zone, South Africa
(2) Etive thermal aureole (samples within 15 m of the contact), NW Scotland

Fluid present (CO_2 rich: re-saturated)
(1) Cooma Migmatites, southern NSW, Australia
(2) Mount Stafford Contact Zone 5, Arunta Complex, Australia
(3) Anapolis, Brazil
(4) Etive thermal aureole (samples within 3 m of the igneous contact), NW Scotland

Leakage
(1) Kimberley Region, Springvale pluton, Halls Creek Mobile Belt, Australia
(2) Ballachulish thermal aureole, NW Scotland
(3) SE Madagascar
(4) Brattstrand Bluffs, Prydz Bay migmatitic granulites, Antarctica
(5) Larsemann Hills, Prydz Bay migmatitic granulites, Antarctica
(6) SE Reynolds Range, Arunta Complex, Australia

Carrington and Harley, 1995; Clemens et al., 1997; Stevens et al., 1997; Clemens and Droop, 1998; Clemens and Watkins, 2001; Nair and Chacko, 2002; Johnson et al., 2001; Droop et al., 2003; White et al., 2005; Diener et al., 2008). It is particularly illuminating that some cordierites preserve high CO_2 contents but still remain strictly fluid undersaturated. The low H_2O wt% contents preserved in CO_2-rich cordierites are still compatible with the independently determined melt-H_2O contents, thus supporting a mechanism of formation involving fluid-absent melting. Low aH_2O values and preferentially CO_2-enriched cordierite can all be explained within a framework dominated by fluid-absent melting processes.

In fluid-present environments it is crucial to identify any potential mechanisms for generating such conditions. The localized increase in CO_2 and/or total volatile contents of cordierite in samples immediately adjacent to igneous bodies (e.g., Etive, zone 5 of Mount Stafford, and the Bushveld contact zone) highlights such an issue. In the Etive case, there is no difference in bulk-rock chemistry that would adequately explain the localized increase in CO_2, e.g., from decarbonation of graphite-rich pelites. Therefore, considering the proximity of the samples to the igneous body, the high CO_2 contents and unrealistic aH_2O values may be attributed to the input of magmatically derived fluids (Rigby et al., 2008a). A similar conclusion may be drawn for the fluid-saturated samples from the Bushveld, zone 5 of Mount Stafford, and Cooma. However, for fluid-saturated samples that are distant from magmatic bodies (e.g., Anapolis), a plausible explanation is less apparent. The cordierite volatile data and independent petrological data suggest that fluid infiltration is the only viable mechanism for generating the preserved cordierite volatile contents and mineral assemblages developed in the Anapolis granulites (Baldwin et al., 2005; Harley, 2008). However, prior to fluid infiltration the Anapolis granulites had evolved under fluid-undersaturated conditions (Baldwin et al., 2005; Harley, 2008). Only at a later stage, after initial granulite formation, did CO_2-rich fluids play a role in stabilizing cordierite in reaction coronas around garnet and sillimanite. Thus the outstanding issues are not whether CO_2 streaming *causes* granulite formation, but where does the CO_2 come from, and how does it gain ingress to the rock? The former point is still a matter of debate (see the discussion in Cesare et al., 2005), whereas the latter is partly addressed by the experiments of Watson and Brenan (1987), which indicate that CO_2 cannot wet grain boundaries of quartz, so that ingress is likely to be fracture controlled.

CONCLUDING REMARKS

The cordierite fluid monitor can be applied to a wide range of metamorphic environments, yielding results that are generally compatible with independent lines of evidence. The cordierite volatile data, in most cases, suggest that granulites formed thorough fluid-absent melting processes and that CO_2 streaming was not involved. In fluid-present environments it is essential to evaluate whether the rock was fluid-saturated throughout the metamorphism or whether the granulite formed initially through fluid-absent melting and was then re-saturated at a later stage, owing to fluid infiltration along a retrograde path. Neither fluid-inclusion studies nor cordierite volatile data alone will solve this problem. In fluid-saturated environments the cordierite volatile data must be used in conjunction with P/T-aH$_2$O/aCO$_2$ modeling of the mineral assemblages in order to accurately constrain the P-T-X$_{fluid}$ history.

ACKNOWLEDGMENTS

M.J. Rigby would like to acknowledge a University of Pretoria research development grant, which enabled this research to be conducted. Simon Harley is thanked for providing a large portion of the cordierite volatile data. John Clemens, Dan Harlov, and Konstantin Podlesskii are thanked for their reviews, which improved the clarity of the manuscript.

REFERENCES CITED

Aines, R.D., and Rossman, G.R., 1984, The high-temperature behaviour of water and carbon dioxide in cordierite and beryl: American Mineralogist, v. 69, p. 319–327.

Aranovich, L.Y., and Newton, R.C., 1998, Reversed determination of the reaction: Phlogopite + quartz = enstatite + potassium feldspar + H$_2$O in the ranges 750–875°C and 2–12 kbar at low H$_2$O activity with concentrated KCl solutions: American Mineralogist, v. 83, p. 193–204.

Baldwin, J.A., Powell, R., Brown, M., Moraes, R., and Fuck, R.A., 2005, Modelling of mineral equilibria in ultrahigh temperature metamorphic rocks from the Anapolis-Itaucu Complex, central Brazil: Journal of Metamorphic Geology, v. 23, p. 511–531, doi:10.1111/j.1525-1314.2005.00591.x.

Boberski, C., and Schreyer, W., 1990, Synthesis and water contents of Fe^{2+}-bearing cordierites: European Journal of Mineralogy, v. 2, p. 565–584.

Bodorkos, S., Cawood, P.A., Oliver, N.H.S., and Nemchin, A.A., 2000, Rapidity of orogenesis in the Palaeoproterozoic Halls Creek Orogen, northern Australia: Evidence from SHRIMP zircon data, CL zircon images and mixture modelling studies: American Journal of Science, v. 300, p. 60–82.

Buick, I.S., Cartwright, I., and Harley, S.L., 1998, The retrograde P-T-t path for low-pressure granulites from the Reynolds Range, central Australia, petrological constraints and implications for low-P/high-T metamorphism: Journal of Metamorphic Geology, v. 16, p. 511–529, doi:10.1111/j.1525-1314.1998.00152.x.

Burnham, C.W., 1994, Development of the Burnham model for prediction of H$_2$O solubility in magmas, in Carroll, M.R., and Holloway, J.R., eds., Volatiles in Magmas: Washington, D.C., Mineralogical Society of America, Reviews in Mineralogy, v. 30, p. 123–129.

Burnham, C.W., Holloway, J.R., and Davis, N.F., 1969, Thermodynamic Properties of Water to 1000°C and 10,000 Bars: Geological Society of America Special Paper 132, 96 p.

Carey, J.W., 1995, A thermodynamic formulation of hydrous cordierite: Contributions to Mineralogy and Petrology, v. 119, p. 155–165, doi:10.1007/BF00307278.

Carey, J.W., and Navrotsky, A., 1992, The molar enthalpy of dehydration of cordierite: American Mineralogist, v. 77, p. 930–936.

Carrington, D.P., and Harley, S.L., 1995, Partial melting and phase relations in high-grade metapelites—An experimental petrogenetic grid in the KFMASH system: Contributions to Mineralogy and Petrology, v. 120, p. 270–291, doi:10.1007/BF00306508.

Carrington, D.P., and Harley, S.L., 1996, Cordierite as a monitor of fluid and melt water contents in the lower crust: An experimental calibration: Geology, v. 24, p. 647–650, doi:10.1130/0091-7613(1996)024<0647:CAAMOF>2.3.CO;2.

Cesare, B., Meli, S., Nodari, L., and Russo, U., 2005, Fe^{3+} reduction during biotite melting in graphitic metapelites: Another origin of CO$_2$ in granulites: Contributions to Mineralogy and Petrology, v. 149, p. 129–140, doi:10.1007/s00410-004-0646-3.

Clemens, J.D., 1984, Water contents of silicic to intermediate magmas: Lithos, v. 17, p. 273–287, doi:10.1016/0024-4937(84)90025-2.

Clemens, J.D., 1989, The importance of residual source material (restite) in granite petrogenesis: A comment: Journal of Petrology, v. 30, p. 1313–1316.

Clemens, J.D., 1990, The granulite-granite connection, in Vielzeuf, D., and Vidal, P., eds., Granulites and Crustal Evolution: Dordrecht, Netherlands, Kluwer, NATO ASI Ser., p. 25–36.

Clemens, J.D., 1992, Partial melting and granulite genesis: A partisan overview: Precambrian Research, v. 55, p. 297–301, doi:10.1016/0301-9268(92)90029-N.

Clemens, J.D., 1993, Experimental evidence against CO$_2$-promoted deep crustal melting: Nature, v. 363, p. 336–338, doi:10.1038/363336a0.

Clemens, J.D., 1995, Phlogopite stability in the silica-saturated portion of the system KAlO$_2$-MgO-SiO$_2$-H$_2$O: New data and a reappraisal to 1.5 Gpa: American Mineralogist, v. 80, p. 982–997.

Clemens, J.D., and Droop, G.T.R., 1998, Fluid, P-T paths and the fates of anatectic melts in the Earth's crust: Lithos, v. 44, p. 21–36, doi:10.1016/S0024-4937(98)00020-6.

Clemens, J.D., and Vielzeuf, D., 1987, Constraints on melting and magma production in the crust: Earth and Planetary Science Letters, v. 86, p. 287–306, doi:10.1016/0012-821X(87)90227-5.

Clemens, J.D., and Wall, V.D., 1981, Origin and crystallization of some peraluminous (S-type) granitic magmas: Canadian Mineralogist, v. 19, p. 111–131.

Clemens, J.D., and Watkins, J.M., 2001, The fluid regime of high-temperature metamorphism during granitoid magma genesis: Contributions to Mineralogy and Petrology, v. 140, p. 600–606.

Clemens, J.D., Droop, G.T.R., and Stevens, G., 1997, High-grade metamorphism, dehydration and crustal melting: A reinvestigation based on new experiments in the silica-saturated portion of the system KAlO$_2$-MgO-SiO$_2$-H$_2$O-CO$_2$ at P < 1.5 Gpa: Contributions to Mineralogy and Petrology, v. 129, p. 308–325, doi:10.1007/s004100050339.

Cuney, M., Coulibaly, Y., and Boiron, M.-C., 2007, High-density early CO$_2$ fluids in the ultrahigh-temperature granulites of Ihouhaouene (In Ouzzal, Algeria): Lithos, v. 96, p. 402–414, doi:10.1016/j.lithos.2006.11.009.

Diener, J.F.A., White, R.W., and Powell, R., 2008, Granulite facies metamorphism and subsolidus fluid-absent reworking, Strangways Range, Arunta Block, central Australia: Journal of Metamorphic Geology, v. 26, p. 603–622, doi:10.1111/j.1525-1314.2008.00782.x.

Droop, G.T.R., 1989, Reaction history of garnet-sapphirine granulites and conditions of Archaean high-pressure granulite-facies metamorphism in the Central Limpopo Mobile Belt, Zimbabwe: Journal of Metamorphic Geology, v. 7, p. 383–403, doi:10.1111/j.1525-1314.1989.tb00604.x.

Droop, G.T.R., and Bucher-Nurminen, K., 1984, Reaction textures and metamorphic evolution of sapphirine-bearing granulites from the Gruf Complex, Italian Central Alps: Journal of Petrology, v. 25, p. 766–803.

Droop, G.T.R., and Charnley, N.R., 1985, Comparative geobarometry of pelitic hornfelses associated with the Newer Gabbros: A preliminary study: Journal of the Geological Society [London], v. 142, p. 53–62, doi:10.1144/gsjgs.142.1.0053.

Droop, G.T.R., and Moazzen, M., 2007, Contact metamorphism and partial melting of Dalradian pelites and semipelites in the southern sector of the Etive aureole: Scottish Journal of Geology, v. 43, p. 155–179, doi:10.1144/sjg43020155.

Droop, G.T.R., and Treloar, P.J., 1981, Pressures of metamorphism in the thermal aureole of the granite Etive complex: Scottish Journal of Geology, v. 17, p. 85–102, doi:10.1144/sjg17020085.

Droop, G.T.R., Clemens, J.D., and Dalrymple, D.J., 2003, Processes and conditions during contact anatexis, melt escape and restite formation: The Huntly Gabbro Complex, NE Scotland: Journal of Petrology, v. 60, p. 1–35.

Ellis, D.J., and Obata, M., 1992, Migmatite and melt segregation at Cooma, New South Wales: Transactions of the Royal Society of Edinburgh, Earth Sciences, v. 83, p. 95–106.

Essene, E.L., 1989, The current status of thermobarometry in metamorphic rocks, in Daly, J.S., Cliff, R.A., and Yardley, B.W.D., eds., Evolution of Metamorphic Belts: Geological Society [London] Special Publication 43, p. 1–44.

Fitzsimons, I.C.W., 1996, Metapelitic migmatites from Brattstrand Bluffs, east Antarctica—Metamorphism, melting and exhumation of the mid-crust: Journal of Petrology, v. 37, p. 395–414, doi:10.1093/petrology/37.2.395.

Flood, R.H., and Vernon, R.H., 1978, The Cooma granodiorite, Australia: An example of in situ crustal anatexis?: Geology, v. 6, p. 81–84, doi:10.1130/0091-7613(1978)6<81:TCGAAE>2.0.CO;2.

Fonarev, V.I., Santosh, M., Vasiukova, O.V., and Filimonov, M.B., 2003, Fluid evolution and exhumation path of the Trivandrum Granulite Block, southern India: Contributions to Mineralogy and Petrology, v. 145, p. 339–354, doi:10.1007/s00410-003-0456-z.

Fyfe, W.S., 1973, The granulite facies, partial melting and Archean crust: Philosophical Transactions of the Royal Society of London, ser. A, v. 273, p. 457–461.

Gerya, T.V., Maresch, W.V., Burchard, M., Zakhartchouk, V., Doltsinis, N.L., and Fockenberg, T., 2005, Thermodynamic modeling of solubility and speciation of silica in H_2O-SiO_2 fluid up to 1300°C and 20 kbar based on the chain reaction formalism: European Journal of Mineralogy, v. 17, p. 269–283, doi:10.1127/0935-1221/2005/0017-0269.

Goldman, D.S., Rossman, G.R., and Dollase, W.A., 1977, Channel constituents in cordierite: American Mineralogist, v. 62, p. 1144–1157.

Goscombe, B., 1992, Silica-undersaturated sapphirine, spinel and kornerupine granulite facies rocks, NE Strangways Range, Central Australia: Journal of Metamorphic Geology, v. 10, p. 181–201.

Grant, J.A., 1985, Phase equilibria in low-pressure partial melting of pelitic rocks: American Journal of Science, v. 285, p. 409–435.

Greenfield, J.E., Clarke, G.L., and White, R.W., 1998, A sequence of partial melting reactions at Mt Stafford, central Australia: Journal of Metamorphic Geology, v. 16, p. 363–378, doi:10.1111/j.1525-1314.1998.00141.x.

Harley, S.L., 1998, Ultrahigh temperature granulite metamorphism (1050°C, 12 kbar) and decompression in garnet (Mg-70)-orthopyroxene-sillimanite gneisses from the Rauer Group: East Antarctica: Journal of Metamorphic Geology, v. 16, p. 541–562, doi:10.1111/j.1525-1314.1998.00155.x.

Harley, S.L., 2004, Extending our understanding of ultrahigh temperature crustal metamorphism: Journal of Mineralogical and Petrological Sciences, v. 99, p. 140–158, doi:10.2465/jmps.99.140.

Harley, S.L., 2008, Refining the P-T records of UHT crustal metamorphism: Journal of Metamorphic Geology, v. 26, p. 125–154, doi:10.1111/j.1525-1314.2008.00765.x.

Harley, S.L., and Carrington, D.P., 2001, The distribution of H_2O between cordierite and granitic melt: H_2O incorporation in cordierite and its application to high-grade metamorphism and crustal anatexis: Journal of Petrology, v. 42, p. 1595–1620, doi:10.1093/petrology/42.9.1595.

Harley, S.L., and Thompson, P., 2004, The influence of cordierite on melting and mineral-melt equilibria in UHT metamorphism: Philosophical Transactions of the Royal Society of Edinburgh, Earth Sciences, v. 95, p. 87–99, doi:10.1017/S0263593304000100.

Harley, S.L., Thompson, P., Henson, B.J., and Buick, I.S., 2002, Cordierite as a sensor of fluid conditions in high-grade metamorphism and crustal anatexis: Journal of Metamorphic Geology, v. 20, p. 71–86, doi:10.1046/j.0263-4929.2001.00344.x.

Holland, T.J.B., 1979, High water activities in the generation of high pressure kyanite eclogites of Tauern Window, Austria: Journal of Geology, v. 87, p. 1–27, doi:10.1086/628388.

Holland, T.J.B., and Powell, R., 1990, An enlarged and updated internally consistent thermodynamic dataset: Journal of Metamorphic Geology, v. 8, p. 89–124, doi:10.1111/j.1525-1314.1990.tb00458.x.

Holness, M.B., and Clemens, J.D., 1999, Partial melting of the Appin Quartzite driven by fracture-controlled H_2O infiltration in the aureole of the Ballachulish Igneous Complex, Scottish Highlands: Contributions to Mineralogy and Petrology, v. 136, p. 154–168, doi:10.1007/s004100050529.

Johannes, W., and Holtz, F., 1996, Petrogenesis and Experimental Petrology of Granitic Rocks: Berlin, Springer-Verlag, 335 p.

Johannes, W., and Schreyer, W., 1981, Experimental introduction of CO_2 and H_2O into Mg-cordierite: American Journal of Science, v. 281, p. 299–317.

Johnson, T.E., Hudson, N.F.C., and Droop, G.T.R., 2001, Partial melting in the Inzie Head gneisses: The role of water and a petrogenetic grid in KFMASH applicable to anatectic pelitic migmatites: Journal of Metamorphic Geology, v. 19, p. 99–118, doi:10.1046/j.0263-4929.2000.00292.x.

Kalt, A., 2000, Cordierite channel volatiles as evidence for dehydration melting: An example from high-temperature metapelites of the Bayerische Wald (Variscan belt, Germany): European Journal of Mineralogy, v. 12, p. 987–998.

Kerrick, D.M., 1974, Review of mixed-volatile (H_2O-CO_2) equilibria: American Mineralogist, v. 59, p. 729–762.

Kreulen, R., 1987, Thermodynamical calculations of the C-O-H system applied to fluid inclusions: Are fluid inclusions unbiased samples of ancient fluids?: Chemical Geology, v. 61, p. 59–64, doi:10.1016/0009-2541(87)90027-1.

Kurepin, V.A., 1985, H_2O and CO_2 contents of cordierite as an indicator of thermodynamic conditions of formation: Geochemistry International, v. 22, p. 148–156.

Lamb, W.M., and Valley, J.W., 1988, Granulite-facies amphibole and biotite equilibria and calculated peak metamorphic water activities: Contributions to Mineralogy and Petrology, v. 100, p. 349–360, doi:10.1007/BF00379744.

Le Breton, N., and Thompson, A.B., 1988, Fluid-absent (dehydration) melting of biotite in metapelites in the early stages of crustal anatexis: Contributions to Mineralogy and Petrology, v. 99, p. 226–237, doi:10.1007/BF00371463.

Markl, G., Bäuerle, J., and Grujic, D., 2000, Metamorphic evolution of Pan-African granulite facies metapelites from southern Madagascar: Precambrian Research, v. 102, p. 47–68, doi:10.1016/S0301-9268(99)00099-6.

Martignole, J., and Martelat, J.E., 2003, Regional-scale Grenvillian-age UHT metamorphism in the Mollendo-Camana block (basement of the Peruvian Andes): Journal of Metamorphic Geology, v. 21, p. 99–120, doi:10.1046/j.1525-1314.2003.00417.x.

Mirwald, P.W., and Schreyer, W., 1977, Die stabile and metastabile Abbaureaktion von Mg-cordierit in Talk, Disthen und Quartz und ihre Abhangigkeit von Gleichgewichtswassergehalt des Cordierites: Fortschritte der Mineralogie, v. 55, p. 97–99.

Mirwald, P.W., Maresch, W.V., and Schreyer, W., 1979, Der Wassergehalt von Mg-cordierite zwischen 500 und 800 °C sowie 0.5 und 11 kbar: Fortschritte der Mineralogie, v. 57, p. 101–102.

Moazzen, M., Droop, G.T.R., and Harte, B., 2001, Abrupt transition in H_2O activity in the melt-present zone of a thermal aureole. Evidence from H_2O contents of cordierites: Geology, v. 29, p. 311–314, doi:10.1130/0091-7613(2001)029<0311:ATIHOA>2.0.CO;2.

Morgan, G.B., VI, I-Ming Chou, Pasteris, J.D., and Olsen, S.N., 1993, Re-equilibration of CO_2 fluid inclusions at controlled hydrogen fugacities: Journal of Metamorphic Geology, v. 11, p. 155–164, doi:10.1111/j.1525-1314.1993.tb00137.x.

Morrison, J., and Valley, J.W., 1988, Post-granulite facies fluid infiltration in the Adirondack Mountains: Geology, v. 16, p. 513–516, doi:10.1130/0091-7613(1988)016<0513:PGFFII>2.3.CO;2.

Nair, R., and Chacko, T., 2002, Fluid-absent melting of high-grade semi-pelites: P-T constraints on orthopyroxene and implication for granulite formation: Journal of Petrology, v. 43, p. 2121–2142, doi:10.1093/petrology/43.11.2121.

Nandakumar, V., and Harley, S.L., 2000, A reappraisal of the pressure-temperature path of granulites from the Kerala Khondalite Belt, southern India: Journal of Geology, v. 108, p. 687–703, doi:10.1086/317947.

Newton, R.C., Smith, J.V., and Windley, B.F., 1980, Carbonic metamorphism, granulites and crustal growth: Nature, v. 288, p. 45–50, doi:10.1038/288045a0.

Ohyama, H., Tsunogae, T., and Santosh, M., 2008, CO_2-rich fluid inclusions in staurolite and associated minerals in a high-pressure ultrahigh-temperature granulite from the Gondwana suture in southern India: Lithos, v. 101, p. 177–190, doi:10.1016/j.lithos.2007.07.004.

Pattison, D.R.M., and Harte, B., 1988, Evolution of structurally contrasting anatectic migmatites in the 3kbar Ballachulish aureole, Scotland: Journal of Metamorphic Geology, v. 6, p. 475–494, doi:10.1111/j.1525-1314.1988.tb00435.x.

Perchuk, L.L., van Reenen, D.D., Varlamov, D.A., van Kal, S.M., Tabatabaeimanesh, and Boshoff, R., 2008, P-T record of two high-grade metamorphic events in the Central Zone of the Limpopo Complex, South Africa: Lithos, v. 103, p. 70–105, doi:10.1016/j.lithos.2007.09.011.

Peterson, J.W., and Newton, R.C., 1989, CO_2-enhanced melting of biotite-bearing rocks at deep-crustal pressure-temperature conditions: Nature, v. 340, p. 378–380, doi:10.1038/340378a0.

Peterson, J.W., and Newton, R.C., 1990, Experimental biotite-quartz melting in the KFMASH-CO_2 system and the role of CO_2 in the petrogenesis of granites and related rocks: American Mineralogist, v. 75, p. 1029–1042.

Phillips, G.N., 1981, Water activity changes across an amphibolite-granulite facies transition, Broken Hill, Australia: Contributions to Mineralogy and Petrology, v. 75, p. 377–386, doi:10.1007/BF00374721.

Powell, R., 1983, Fluids and melting in upper amphibolite facies conditions: Journal of the Geological Society [London], v. 140, p. 629–634, doi:10.1144/gsjgs.140.4.0629.

Powell, R., and Holland, T.J.B., 2008, On thermobarometry: Journal of Metamorphic Geology, v. 26, p. 155–179, doi:10.1111/j.1525-1314.2007.00756.x.

Rigby, M.J., 2009, Conflicting P-T paths in the Central Zone of the Limpopo Belt: A consequence of different thermobarometric methods?: Journal of African Earth Sciences, v. 54, no. 5, p. 111–126, doi:10.1016/j.jafrearsci.2009.03.005.

Rigby, M.J., and Droop, G.T.R., 2008, The cordierite fluid monitor: Case studies for and against its potential application: European Journal of Mineralogy, v. 20, p. 693–712, doi:10.1127/0935-1221/2008/0020-1867.

Rigby, M.J., Droop, G.T.R., and Bromiley, G., 2008a, Variations in fluid activity across the Etive thermal aureole, Scotland: Evidence from cordierite volatile contents: Journal of Metamorphic Geology, v. 26, p. 331–346, doi:10.1111/j.1525-1314.2007.00752.x.

Rigby, M.J., Mouri, H., and Brandl, G., 2008b, P-T conditions and the origin of quartzo-feldspathic veins in metasyenites from the Central Zone of the Limpopo Belt, South Africa: South African Journal of Geology, v. 111, p. 313–332, doi:10.2113/gssajg.111.2-3.313.

Rigby, M.J., Droop, G.T.R., Plant, D., and Gräser, P., 2008c, Electron probe micro-analysis of oxygen in cordierite: Potential implications for analysis of volatiles in minerals: South African Journal of Geology, v. 111, p. 239–250, doi:10.2113/gssajg.111.2-3.239.

Rigby, M.J., Mouri, H., and Brandl, G., 2008d, A review of the P-T-t evolution of the Limpopo Belt: Constraints for a tectonic model: Journal of African Earth Sciences, v. 50, p. 120–132, doi:10.1016/j.jafrearsci.2007.09.010.

Rigby, M.J., Eriksson, P.G., and Mavimbela, P.K., 2010, Comments on "Structural and compositional constraints on the emplacement of the Bushveld Complex" by B. Clarke, R. Uken and J. Reinhardt: Lithos, v. 115, p. 272–275.

Santosh, M., and Omori, S., 2008, CO_2 windows from mantle to atmosphere: Models on ultrahigh-temperature metamorphism and speculations on the link with melting of snowball Earth: Gondwana Research, v. 14, p. 82–96, doi:10.1016/j.gr.2007.11.001.

Santosh, M., Tsunogae, T., Ohyama, H., Sato, K., Li, J.H., and Liu, S.J., 2008, Carbonic metamorphism at ultrahigh-temperatures: Evidence from North China Craton: Earth and Planetary Science Letters, v. 266, p. 149–165, doi:10.1016/j.epsl.2007.10.058.

Satish-Kumar, M., 2005, Graphite-bearing CO_2-fluid inclusions in granulites: Insights on graphite precipitation and carbon isotope evolution: Geochimica et Cosmochimica Acta, v. 69, p. 3841–3856, doi:10.1016/j.gca.2005.02.007.

Schenk, V., 1984, Petrology of felsic granulites, metapelites, metabasics, ultramafics, and metacarbonates from Southern Calabria (Italy): Prograde metamorphism, uplift and cooling of a former lower crust: Journal of Petrology, v. 25, p. 225–298.

Shmulovich, K.I., and Graham, C.M., 2004, An experimental study of phase equilibria in the systems H_2O-CO_2-$CaCl_2$ and H_2O-CO_2-NaCl at high pressures and temperatures (500–800 °C, 0.5–0.9 GPa): Geological and geophysical applications: Contributions to Mineralogy and Petrology, v. 146, p. 450–462, doi:10.1007/s00410-003-0507-5.

Skippen, G.B., and Gunter, A.E., 1996, The thermodynamic properties of H_2O in magnesian and iron cordierite: Contributions to Mineralogy and Petrology, v. 124, p. 82–89, doi:10.1007/s004100050175.

Smith, C.H., 1996, H_2O-CO_2 contents of cordierite in migmatites of the Fosdick Mountains: Marie Byrd Land: Terra Antartica, v. 3, p. 11–22.

Stevens, G., and Clemens, J.D., 1993, Fluid-absent melting and the roles of fluids in the lithosphere: A slanted summary?: Chemical Geology, v. 108, p. 1–17, doi:10.1016/0009-2541(93)90314-9.

Stevens, G., Clemens, J.D., and Droop, G.T.R., 1995, Hydrous cordierite in granulite and crustal magma production: Geology, v. 23, p. 925–928, doi:10.1130/0091-7613(1995)023<0925:HCIGAC>2.3.CO;2.

Stevens, G., Clemens, J.D., and Droop, G.T.R., 1997, Melt production during granulite-facies anatexis: Experimental data from primitive metasedimentary protoliths: Contributions to Mineralogy and Petrology, v. 128, p. 352–370, doi:10.1007/s004100050314.

Thompson, A.B., 1982, Dehydration melting of pelitic rocks and the generation of H_2O-undersaturated granitic liquids: American Journal of Science, v. 124, p. 1567–1595.

Thompson, A.B., 1983, Fluid-absent metamorphism: Journal of the Geological Society [London], v. 140, p. 533–547, doi:10.1144/gsjgs.140.4.0533.

Thompson, P., Harley, S.L., and Carrington, D.P., 2001, The distribution of H_2O-CO_2 between cordierite and granitic melt under fluid-saturated conditions at 5 kbar and 900°C: Contributions to Mineralogy and Petrology, v. 142, p. 107–118.

Török, K., Dégi, J., Szép, A., and Marosi, G., 2005, Reduced carbonic fluids in mafic granulite xenoliths from the Bakony-Balaton Highland Volcanic Field, W-Hungary: Chemical Geology, v. 223, p. 93–108, doi:10.1016/j.chemgeo.2005.05.010.

Touret, J., 1971, Le facies granulite en Norvege meridionale II: Les inclusions fluides: Lithos, v. 4, p. 423–436, doi:10.1016/0024-4937(71)90125-3.

Tsunogae, T., Santosh, M., Osanai, Y., Owada, M., Toyoshima, T., and Hokada, T., 2002, Very high-density carbonic fluid inclusions in sapphirine-bearing granulites from Tonagh Island in the Archean Napier Complex, East Antarctica: Implications for CO_2 infiltration during ultrahigh-temperature (T>1,100 °C) metamorphism: Contributions to Mineralogy and Petrology, v. 143, no. 3, p. 279–299.

Vry, J.K., Brown, P.E., and Valley, J.W., 1990, Cordierite volatile content and the role of CO_2 in high-grade metamorphism: American Mineralogist, v. 75, p. 71–88.

Waters, D.J., and Lovegrove, D.P., 2002, Assessing the extent of disequilibrium and overstepping of prograde metamorphic reactions in metapelites from the Bushveld Complex aureole, South Africa: Journal of Metamorphic Geology, v. 20, p. 135–149, doi:10.1046/j.0263-4929.2001.00350.x.

Watson, E.B., and Brenan, J.M., 1987, Fluids in the lithosphere, 1. Experimentally determined wetting characteristics of CO_2–H_2O fluids and their implications for fluid transport, host-rock physical properties and fluid inclusion formation: Earth and Planetary Science Letters, v. 85, p. 497–515, doi:10.1016/0012-821X(87)90144-0.

White, R.W., Pomroy, N.E., and Powell, R., 2005, An in-situ metatexite-diatexite transition in upper amphibolite facies rocks from Broken Hill, Australia: Journal of Metamorphic Geology, v. 23, p. 579–602, doi:10.1111/j.1525-1314.2005.00597.x.

Wood, D.L., and Nassau, K., 1967, Infrared spectra of foreign molecules in beryl: Journal of Chemical Physics, v. 47, p. 2220–2228, doi:10.1063/1.1703295.

Manuscript Accepted by the Society 24 May 2010

Local mineral equilibria and P-T paths: Fundamental principles and applications to high-grade metamorphic terranes

Leonid L. Perchuk*

Department of Petrology, Geological Faculty, Moscow State University, Leninskie Gory, Moscow, 119192, Russia;
Institute of Experimental Mineralogy, Russian Academy of Sciences, Chernogolovka, Moscow District, 142432, Russia; and
Department of Geology, University of Johannesburg, P.O. Box 524, Auckland Park, 2006, Johannesburg, South Africa

ABSTRACT

This paper is a review of fundamental principles and rules for reconstructing pressure-temperature (P-T) paths followed by crystalline rocks. The fundamental principles are (1) the Prigogine-Korzhinskii principle of local equilibrium, and (2) the phase correspondence principle for coexisting minerals (Perchuk, 1971, 1977). These principles have consequences (rules) that are particularly useful for the complex assemblages that occur in Precambrian high-grade terranes. The resulting methodology is the only rigorous means of unraveling the history of polymetamorphic rocks. To assure the accuracy of any thermobarometric methodology it is essential to consider geological structures, microstructural analysis, calculations of mineral modes in terms of temperature and pressure, fluid inclusion data, and numerical modeling of P-T paths. The sequence in which these different approaches are considered has important consequences for the accurate reconstruction of P-T paths for polymetamorphic high-grade rocks. Several examples of P-T path reconstructions for mono- and polymetamorphic granulite facies complexes are considered in detail.

INTRODUCTION

There are two basic methods of reconstructing physico-chemical conditions of metamorphic environments. The classical method reviewed here involves careful observation to discriminate the minerals that were in equilibrium at particular points in the metamorphic history of a rock. Calibrated thermobarometers are then used to establish physical conditions of these reference points, and thermodynamic modeling can, in some cases, be used to obtain detailed information on the continuous pressure-temperature-time (P-T-τ) path of the rock. The difficulties with this method are that it is laborious and requires considerable petrologic expertise, but it is theoretically rigorous. The alternative is to use thermodynamic methods to predict mineralogy. Typically, users of such methods assume bulk equilibrium. Because these methods are easily automated by computer, they are seductive in seeming to offer the possibility of mechanizing the labor of traditional metamorphic petrology. Several dangers lurk in this attraction; the most serious is the bulk equilibrium assumption. Metamorphic rocks are by definition disequilibrium systems; thus it is disingenuous to expect that metamorphic systems fossilize a bulk equilibrium state. This is especially true of metamorphic rocks formed at the high and low temperature extremes of the metamorphic spectrum. Unfortunately the ease with which

*posthumous

automated phase equilibrium methods can be applied has led to a situation in which petrologists have begun to neglect the importance of distinguishing local and bulk equilibrium, which is arguably the most important task in petrologic analysis.

Recent discussions (e.g., Zeh and Klemd, 2008; Perchuk and van Reenen, 2008) have shown that during the last decades the education of petrologists in fundamental sciences has been less than satisfactory. This failing has resulted from deficiency in textbooks on thermodynamics of geological processes, most of which were published in the last century (e.g., Ramberg, 1952; Korzhinskii, 1959, 1970; Saxena, 1973; Perchuk and Ryabchikov, 1976; Wood and Fraser, 1977; Yardley, 1989; Spear, 1993, etc.). On the other hand, some extraordinary advances have been made by several outstanding scientists who produced powerful tools for understanding the thermo- and geodynamic history of metamorphic and metasomatic rocks (e.g., Karpov et al., 1971; Berman, 1991; Aranovich and Podlesskii, 1989; Holland and Powell, 1998; Gerya et al., 2000, 2004; Connolly, 2005). A number of thermodynamic data sets have been created a basis for computing internally consistent thermodynamic properties of rock-forming minerals, gases, and melts (partly). However, not all the above tools allow *direct* calculations of accurate P-T-(a_i^{fl}) paths on the basis of true mineral equilibria occurring in metamorphic and magmatic rocks. The fundamental laws must be used because most of the minerals in the rocks are inhomogeneous in their compositions. In this case one is able to compute thermodynamic parameters for the origin and evolution of a distinct magmatic or metamorphic system. Unfortunately, path modeling is not always accompanied by detailed petrologic studies that include the application of the fundamental laws and rules. At the time when the first internally consistent system of mineral thermometers and barometers was published, I wrote the following alert: ". . . the reader must be admonished against indiscriminate and needless use of the instruments" (Perchuk, 1969, p. 898). Unfortunately this warning has not been heeded by everybody in our geological community, which in some cases has resulted in the incorrect interpretations of geological and petrological data. This situation arose 40 yr ago and remains the same today. Because of this, my colleague and the co-editor of this volume, Dirk D. van Reenen, urged me for many years to write a paper on the fundamental principles, rules, and approaches on which the present thermodynamic calculations for petrological purposes are based.

In many aspects this chapter resulted from my lectures to participants (mainly students) and visitors to the Limpopo field workshop (14–19 July 2008), which was accompanied by long daily discussions.

The Appendix to this chapter consists of three tables: Table A1 lists the mineral abbreviations that are shown in italic type throughout the chapter. Table A2 defines the thermodynamic symbols used throughout. Table A3 displays an internally consistent thermodynamic data set for reactions used for thermobarometry of granulites and for calculations of isopleths on P-T diagrams.

MAJOR PRINCIPLES AND RULES OF CLASSIC THERMODYNAMICS RELATED TO GEOLOGICAL PROCESSES

Principle 1: The Gibbs Formulation

At the macroscopic scale any equilibrium preserves no memory of how it was reached; i.e., in each small volume there is no change in the amount of a particular reactant or product with time (τ):

This principle is based on an idea that equilibrium is not a function of kinetics, and therefore the following Gibbs-Korzhinskii free energy (K) equation for a system is independent of time:

$$dK = VdP - SdT + \sum_{n=1}^{k}\mu_n dm_n - \sum_{i=1}^{j} m_i d\mu_i, \quad (1)$$

where k is the number of inert components, and j is the number of perfectly mobile components in the system. For this equation the phase rule should be written as follows:

$$F = c - j + 2 - \Phi, \quad (2)$$

where F is the number of freedom degrees, c is the total number of components ($c = k + j$), and Φ is the number of phases in the system. It should be noted that many of the programs for calculating the external parameters for natural systems are not explicitly based on equations (1) and (2) and therefore cannot be applied for systems with perfectly mobile components, i.e., mainly for metasomatic systems. Otherwise, one can get, for example, significant erroneous differences in pressures for metasomatic zones and host metamorphic rocks, which were formed simultaneously at the same P-T conditions.

The question arises: How can Principle 1 be applied for the derivation of a P-T-time path if no time coordinate (τ) is present in equations (1) and (2)? The second thermodynamic principle related to metamorphic and magmatic systems addresses this question.

Principle 2: Local Equilibria versus the Non-Equilibrated System

At the macroscopic scale, but in a limited volume of a system, the gradients of external parameters and chemical potentials of chemical elements might be considered negligible, and therefore we can apply the principle of local (Prigogine's term) or mosaic (Korzhinskii's term) equilibrium. The general thermodynamic theory of the evolution of a system is based on the Prigogine criterion:

$$\frac{d_X S}{d\tau} \leq 0 \quad (3)$$

$$\frac{d_X S}{d\tau} = \int_V \sum_\alpha \left(J_\alpha \frac{dX_\alpha}{d\tau} \right) dV, \quad (4)$$

where J_α and X_α are the fluxes, and their associated forces (according to Onsager's terminology) $\frac{d_X S}{d\tau}$ is the entropy production in the system related to changes in the forces X_α, and V is the volume of the system. Thus, the rate of change of entropy of the system from one state to the other is described by the following equation (Ivanov et al., 1997):

$$\frac{dS}{d\tau} = \frac{d_X S}{d\tau} + \frac{d_J S}{d\tau}, \qquad (5)$$

$$\frac{d_J S}{d\tau} = \int_V \sum_\alpha \left(X_\alpha \frac{dJ_\alpha}{d\tau} \right) dV, \qquad (6)$$

where $\frac{d_J S}{d\tau}$ is the entropy production in the system related to changes in the fluxes J_α. In terms of the linear Onsager thermodynamics of irreversible processes (Glensdorf and Prigogine, 1973), the relation between the forces and fluxes is the following:

$$J_\alpha = L_{\alpha\beta} X_\beta, \qquad (7)$$

where summation is performed over the repeated subscript β, $L_{\alpha\beta} = L_{\beta\alpha}$. Now by the use of equations (4)–(7) we can get the Prigogine theorem as the following inequality (Ivanov et al., 1997):

$$\frac{d_X S}{d\tau} = \frac{d_J S}{d\tau} = \frac{1}{2}\frac{dS}{d\tau} \leq 0, \qquad (8)$$

which follows from the general evolution criterion (3) for linear Onsager systems. We consider the evolution of a small volume (a point) of a mineral system that is in *local equilibrium* with surrounding small volumes. We also assume that microscopic parts of the system reach *local equilibrium* much earlier than when *global equilibrium* is established among the parts. We can therefore calculate temperature, chemical potential, and other thermodynamic parameters of the point. Such an approach can be used for any system in order to involve the time flow via the Prigogine theorem (we refer here to true mosaic equilibrium systems in which entropy may be generated by irreversible transfer of entropy, mass, or volume between local domains). For example, Korzhinskii (1970) and Thompson (1959) successfully used this approach to describe metasomatic zonation, i.e., for systems with perfectly mobile components (see equations 1 and 2). This approach can also be used to describe the propagation of microcracks via the evolution of a *fracture* point in the minerals, rocks, and even in composite materials (Ivanov et al., 1997).

Equation (8) allows the derivation of a P-T-τ path via local mineral equilibria within both equilibrium and nonequilibrium systems—a typical case for a (poly)metamorphic rock (Perchuk, 1973, 2005). In addition, Principle 2 has been well demonstrated experimentally at high P-T parameters for the *Crd = Grt + Qtz + Sil* net-transfer reaction (Aranovich and Podlesskii, 1983), and for exchange equilibria in both *Bt-Grt-Crd* (Perchuk and Lavrent'eva, 1983) and *Opx-Grt-Hbl* (Perchuk and Lavrent'eva, 1990) systems. (See Table A1 in the Appendix for full names of abbreviated mineral terms in italic type throughout this chapter. See also Table A2, which defines the thermodynamic symbols used.)

MAJOR RULES OF CLASSICAL THERMODYNAMICS RELATED TO EQUILIBRIA OF ROCK-FORMING MINERALS

Two rules given below reflect the equilibrium principle for systems with solid solutions. Such systems will always tend to partition elements between different phases so as to minimize the free energy of the system.

Sobolev-Ramberg Rule

This rule was formulated independently by two great petrologists, Vladimir S. Sobolev (1948) and Hans Ramberg (1952), on the basis of regularities in the compositions of coexisting rock-forming minerals: *At moderate P-T parameters exchange equilibria are displaced to the side of the formation of the (alumino) silicate solid solution composed of a radical of a strong acid* (A) *with the metal of a low electronegativity* (Me_1), *plus the (alumino)silicate composed of a weak acid radical* (B) *with a metal of higher electronegativity* (Me_2), *i.e., the reaction*

$$Me_2A + Me_1B => Me_2B + Me_1A \qquad (9)$$

must be displaced to the right side. There are a few physico-chemically sound exceptions to this rule that have been discussed in classical papers and textbooks (e.g., Ramberg, 1952; Perchuk and Ryabchikov, 1976). At the time of its discovery, the rule explained systematic relationships between mineral compositions observed by many petrologists. For instance, in a majority of cases the following inequality

$$X_{Mg}^{Crd} > X_{Mg}^{Chl} > X_{Mg}^{Bt} > X_{Mg}^{Hbl} > X_{Mg}^{Cpx} > X_{Mg}^{Opx} > X_{Mg}^{Grt} \qquad (10)$$

was well known but not understood. The rule clearly demonstrated that we are dealing with *exchange equilibria* (9) of Fe-Mg minerals.

Phase Correspondence Rule

In general, with increasing temperature, chemical bonds between a strong cation and hypothetical (alumino) silicic acids become stronger (and those of a less electropositive isomorphic cation respectively weaker) in the following sequence:

framework (alimino)silicates → sheet (alimino)dimetasilicates → band (alimino)metasilicate → chain (alumino)metasilicate → diortho(alumino)silicate → orthosilicate

(separately for hydrous and anhydrous minerals). With temperature variations the strength of chemical bonds in the coexisting minerals will change in proportion to the relative difference in strength of chemical bonds between the silicates in this series. Consequently, ΔS and ΔH values for the exchange reactions and, thus, the effects of redistribution of cations should also be proportional to this difference (Perchuk, 1971, 1977). On the other hand, each group of (alumino)silicates is composed of minerals of different crystal families. For example, nepheline and K-feldspar have completely different structures but belong to the framework aluminosilicate group. The question arises: To what side is the exchange reaction

$$NaAlSiO_4 (Ne) + KAlSi_3O_8(Or/San) = KAlSiO_4(Ks) + NaAlSi_3O_8(Ab) \quad (11)$$

displaced (in terms of relative abundances of K- and Na- end members in coexisting *Ne-Ks* and *Or/San-Ab* solid solutions)? This question can also be posed for the following exchange reaction between pyroxenes:

$$CaMgSi_2O_6(Di) + Fe_2Si_2O_6 (Fs) = CaFeSi_2O_6(Hed) + Mg_2Si_2O_6(En). \quad (12)$$

According to the phase correspondence rule the direction of the displacement must be predictable. The tables that account for crystallochemical and structural properties of rock-forming minerals, coexisting in crystalline rocks (separately for water-bearing and water-free minerals), were first compiled in 1977 (Perchuk, 1977). These have been modernized, and the simplified table composed for Fe-Mg minerals is discussed in this chapter. In Table 1 each mineral has a unique number and occupies one cell according to water content and crystallochemical and structural properties. Consequently different silicate properties depend systematically on the position within the table. For example, Figure 1A demonstrates correlations between numbers of silicates in Table 1 and the V_{Mg}–V_{Fe} values for their end members. On the basis of Table 1 we can also qualitatively predict thermodynamic effects of any Fe-Mg exchange equilibrium (Perchuk, 1977): *the more significant the difference of the numbers (ΔNo) in Table 1, the larger both the ΔS and ΔH values for the exchange reactions must be*. Figure 1B demonstrates linear correlations of ΔNo with the entropy (ΔS) and enthalpy (ΔH) for several *Grt*-bearing exchange equilibria chosen from Table 1. ΔS and ΔH effects for exchange reactions within the groups of water-bearing and water-free minerals from Table 1 show good correlation (Perchuk, 1989, 1991):

H_2O-Bearing Minerals

$$\Delta H = 359.16\Delta No + 0.890.5; r^2 = 0.987$$

$$\Delta S = 0.15\Delta No + 0.945; r^2 = 0.917$$

H_2O-Free Minerals

$$\Delta H = 1477.2\Delta No - 365.18; r^2 = 0.997$$

$$\Delta S = 1.304\Delta No - 1.769; r^2 = 0.966$$

Some of the ΔH and ΔS values show reasonable accuracy. However, these equations *cannot* serve as the basis for mineral thermometry: They just demonstrate the efficiency of the phase correspondence rule for exchange equilibria between complex mineral solid solutions.

A special problem is the position of cordierite in the considered system of exchange equilibria. Theoretically it is a water-free framework silicate. However, any natural cordierite contains nonstructural water. This is why cordierite *(Crd)* occupies cell no. 1 in Table 1. On the other hand, thermodynamic properties of the Crd_{Fe-Mg} solid solution depend on the volatile (H_2O, CO_2, etc.) content in its lattice channels (Aranovich and Podlesskii, 1989). Because volatiles are not structural constituents of cordierite, this mineral occupies an intermediate position between water-bearing and water-free silicates. This leads to the rather nonsystematic deviation of ΔH and ΔS effects from their general correlations with ΔNo, and therefore *Crd* is excluded from Figure 1B as well as from the above correlations.

Thus, the redistribution of isomorphic elements among coexisting minerals is governed by three basic factors: (1) relative strength of silicic radicals, (2) structures, and (3) water contents. The greater the difference in the relative values (or ΔNo) of these factors, the more pronounced the redistribution effect would be. In general the degree of redistribution of two isomorphic components between two silicates with temperature will be higher with the greater difference in the strengths of silicic radicals and in the

TABLE 1. SYSTEMATICS OF Fe-Mg ROCK-FORMING SILICATES AND ALUMINOSILICATES BASED ON THE PHASE CORRESPONDENCE RULE (PERCHUK, 1977)

No.	Mineral
1	Cordierite
2	Talc
3	Serpentine
4	Chlorite
5	Biotite
6	Hornblende
7	Na-amphibole
8	Cummingtonite
9	Orthoamphibole
10	Tourmaline
11	Chloritoid
12	Staurolite
13	Kornerupine
14	Vesuvian
15	Ossumilite
16	Clinopyroxene
17	Orthopyroxene
18	Sapphirine
19	Olivine
20	Garnet

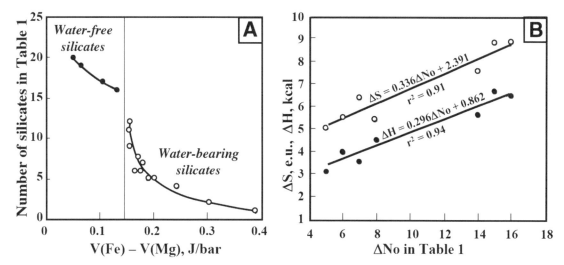

Figure 1. Correlation between numbers of silicates in Table 1 and the $V_{Mg}-V_{Fe}$ values for their end members (A), and the dependence of differences in silicate numbers (ΔNo) in Table 1 on both the entropy (ΔS) and the enthalpy (ΔH) for selected Fe-Mg exchange equilibria involving garnet (B). These silicates occur in gneisses and schists from metamorphic complexes of different grades. Values for $V_{Mg}-V_{Fe}$ are calculated using the database of Holland and Powell (1998); ΔS and ΔH are taken from Perchuk (1989).

symmetrical categories of the minerals. The greatest effects were found for the equilibria between hydrous and anhydrous minerals (Perchuk, 1969, 1977). According to this rule, changes in temperature lead to the redistribution of *all* isomorphic elements according to their relative electro-negativities, thus increasing the redistribution of Fe and Mg from the Korzhinskii effect of the interaction of acid-basic components. Consequently, we can evaluate overall the entropy and enthalpy effects for a number of exchange reactions (9) between pairs of coexisting Fe-Mg and K-Na silicates-aluminosilicates (Perchuk, 1977, 1989, 1991).

Rule no. 2 has laid the foundation for the first internally consistent system of mineral thermometers (Perchuk, 1969, 1977, 1991), many of which are still valid, even in terms of their accuracy. This system was, at the time, based on (1) existing experimental data, and (2) *wet chemical analyses of coexisting* minerals (e.g., Perchuk, 1976; Perchuk and Ryabchikov, 1968, and references therein). In a later part of this chapter I shall illustrate how the above rules work for high-grade metamorphic terranes (HGT).

CLASSICAL METHOD OF P-T-τ PATHS DERIVATION

Theoretical

The first P-T paths were calculated for eclogites more than 35 yr ago (for example, Perchuk, 1973, 1976, 1977, 1986, 1989), and then for high-grade rocks (e.g., Perchuk et al., 1983, 1985, 1989; Spear and Selverstone, 1983); many of these P-T paths are still valid. However, discussions on the techniques for their derivation continue even today (see, e.g., Frost and Chacko, 1989; Spear and Florence, 1992; Spear, 1993; Powell and Holland, 2008; Zeh and Klemd, 2008; Perchuk and van Reenen, 2008). Two major approaches for the derivation of a P-T path exist. The first is based on thermobarometry of local mineral equilibria (e.g., Perchuk and van Reenen, 2008, and references therein), and the second is derived from the Escola mineral facies principle, i.e., finding a stable mineral assemblage at a given P-T, and the bulk rock composition (e.g., Holland and Powell, 1998; Connolly, 2005). As P-T trajectories reflect a polymetamorphic process (Perchuk, 2005; Perchuk et al., 2006), it is pertinent to review our method for the correct derivation of P-T-τ paths.

Since 1983 (e.g., Perchuk et al., 1983) we have typically used metapelites for deriving P-T paths, because pelites are common in high-grade terranes and commonly contain similar or identical mineral assemblages. The key divariant mineral equilibria were studied experimentally (Aranovich and Podlesskii, 1989; Perchuk and Lavrent'eva, 1983, 1990) and incorporated into thermodynamically consistent systems of mineral thermometers and barometers (e.g., Aranovich and Podlesskii, 1989; Perchuk, 1986, 1989, 1991; Perchuk et al., 1983, 1985, 1996, 2000a, 2006, 2008; Smit et al., 2001; van Reenen et al., 2004), which are based on the following exchange reactions–mineral thermometers:

$$\tfrac{1}{3}Prp + \tfrac{1}{3}Ann = \tfrac{1}{3}Alm + \tfrac{1}{3}Phl \qquad (r1)$$

$$\tfrac{1}{3}Prp + \tfrac{1}{2}Crd_{Fe} = \tfrac{1}{3}Alm + \tfrac{1}{2}Crd_{Mg} \qquad (r2)$$

$$\tfrac{1}{3}Prp + Fs = \tfrac{1}{3}Alm + En \qquad (r3)$$

$$\tfrac{1}{2}Crd_{Mg} + \tfrac{1}{3}Ann = \tfrac{1}{2}Crd_{Fe} + \tfrac{1}{3}Phl \qquad (r4)$$

$$\tfrac{1}{2}Crd_{Mg} + Fs = \tfrac{1}{2}Crd + En, \qquad (r5)$$

whereas the net-transfer reactions (involving the Mg and Al end members of solid solutions):

$$\tfrac{1}{3}Prp + \tfrac{2}{3}Sil + \tfrac{5}{6}Qtz = \tfrac{1}{2}Crd_{Mg} \tag{r6}$$

$$Prp + Kfs + H_2O = Phl + \tfrac{1}{2}Sil + \tfrac{5}{2}Qtz \tag{r7}$$

$$\tfrac{1}{2}Crd_{Mg} = En + \tfrac{2}{3}Qtz + OK \tag{r8}$$

$$Prp + \tfrac{3}{2}Qtz = 2En + \tfrac{1}{2}Crd_{Mg} \tag{r9}$$

$$Prp = 3En + OK \tag{r10}$$

are effective for mineral barometry. Table A3 in the Appendix contains a thermodynamic data set for (r1)–(r10) end-member reactions and the mixing properties for their participants. These data can be used for the derivation of isopleths practically for each net-transfer reaction from the set (r1)–(r10). The intersection point on the P-T plane of two corresponding isopleths, calculated thermodynamically on the basis of reactions (r1)–(r5) and (r6)–(r10), defines the P-T parameters for a *local equilibrium* in the rock studied petrologically in detail.

Thus, our approach allows calculating the pressures along a P-T path using temperature estimates based on any complementary mineral thermometer (r1–r5 in Table A3), or by arbitrarily prescribing the temperature values within the T_{max} and T_{min} by deriving them from microprobe profiles across mineral porphyroblasts according to the above phase correspondence rule. In fact, we can confidently estimate the parameters at the beginning (equilibria of porphyroblasts) and the end (mostly reaction textures) of a specific P-T path, whereas the uncertainties may be hidden for the rest of the path. However, using data from Table A3, we can calculate both the pressure and the compositions of coexisting Fe-Mg minerals in the presence of garnet for any given local equilibrium at T and X_{Mg}^{Grt}. For example, many high-grade aluminous gneisses belonging to the system SiO_2-Al_2O_3-MgO-FeO-K_2O-H_2O are characterized by zoned garnet coexisting with biotite and cordierite. Biotite commonly occurs in many generations. This creates difficulties for identification of the local equilibrium composition of biotite, e.g., the uncertainty of correctly identifying the X_{Mg}^{Bt} for a specific biotite generation in equilibrium with known X_{Mg}^{Grt}.

In order to solve this problem, principles of classical equilibrium thermodynamics (e.g., Korzhinskii, 1959; Perchuk and Zverovich, 1962) for systems with solid solutions can be applied. The calculations are based on the thermodynamic equilibrium condition that can be formulated for any end-member reaction as

$$\Delta G_r^{P,T} = \Delta H_r^{P_o,T_o} - T\Delta S_r^{P_o,T_o} + \int_{P_o}^{P} \Delta V_r^{P,T} dP + \int_{T_o}^{T} \Delta Cp_r^{P_o,T} dT - T\int_{T_o}^{T} \frac{\Delta Cp_r^{P_o,T}}{T} dT + \sum_{i=1}^{M} c_i RT \ln a_i = 0, \tag{13}$$

$$\Delta H_r^{P_o,T_o} = \sum_{i=1}^{M} c_i H_i^{P_o,T_o}, \tag{14}$$

$$\Delta S_r^{P_o,T_o} = \sum_{i=1}^{M} c_i S_i^{P_o,T_o}, \tag{15}$$

$$\Delta V_r^{P,T} = \sum_{i=1}^{M} c_i V_i^{P,T}, \tag{16}$$

$$\Delta Cp_r^{P_o,T} = \sum_{i=1}^{M} c_i Cp_r^{P_o,T}, \tag{17}$$

where $\Delta G_r^{P,T}$ is the Gibbs free energy change of the reaction at the given pressure (P) and temperature (T); M is the number of chemical species (end members) participating in the reaction; c_i is the stoichiometric coefficient for the i-th end member of the reaction (coefficients on the left-hand side of the reaction are taken with minus); a_i is the activity of the i-th end member; $\Delta H_r^{P_o,T_o}$ and $\Delta S_r^{P_o,T_o}$ are enthalpy and entropy changes, respectively, of the reaction at standard pressure (P_o) and temperature (T_o); $\Delta V_r^{P,T}$ is the volume change of the reaction at given P and T; $\Delta Cp_r^{P_o,T}$ is the isobaric heat capacity change of the reaction at P_o and T; $H_i^{P_o,T_o}$, $S_i^{P_o,T_o}$, $V_i^{P,T}$ and $Cp_i^{P_o,T}$ are, respectively, the enthalpy of formation, entropy, volume, and isobaric heat capacity of the i-th end member (e.g., Holland and Powell, 1998). Where $\Delta V_r^{P,T}$ and $Cp_i^{P_o,T}$ remain constant (e.g., Table A3), equation (13) simplifies to

$$\Delta G_r^{P,T} = \Delta H_r^{P_o,T_o} - T\Delta S_r^{P_o,T_o} + \Delta V_r^{P_o,T_o}(P - P_o) + \Delta Cp_r^{P_o,T_o}\left(T - T_o - T\ln\frac{T}{T_o}\right) + \sum_{i=1}^{M} c_i RT \ln a_i = 0, \tag{18}$$

where $\Delta V_r^{P_o,T_o}$ and $\Delta Cp_r^{P_o,T_o}$ are, respectively, volume and isobaric heat capacity changes of the reaction at P_o and T_o. Data for $\Delta H_r^{P_o,T_o}$, $\Delta S_r^{P_o,T_o}$, $\Delta V_r^{P_o,T_o}$ and $\Delta Cp_r^{P_o,T_o}$ for reactions (r1)–(r10) and formulas for computing activities of different end members ($RT \ln a_i$) are given in Table A3.

For the given local garnet-bearing assemblage, the thermodynamic equilibrium condition (18) should be formulated for all *linearly independent* end-member reactions among (r1)–(r10). In the case of coexisting Grt, Crd, and Bt, the number of linearly independent exchange reactions is *two* (r1 and r2, or r1 and r4, or r2 and r4) and at a given pressure (these exchange reactions are almost independent of pressure because of very small volumetric effects, cf. $\Delta V_r^{P_o,T_o}$ in Table A3) one can compute *two* unknown parameters, for example, temperature and X_{Mg}^{Bt} based on compositions of coexisting Grt and Crd. Such computation can be based on the simultaneous iterative solving of two thermodynamic equilibrium equations (18) written for two exchange reactions. When using internally consistent thermodynamic databases for mineral reactions (such as, e.g., Table A3), the result of the calculation will not depend on the choice of any specific pairs of reactions. When *Sil* and *Qtz* are also present in the local equilibrium assemblage, the linearly independent net-transfer reaction r6 is also possible between minerals (especially when respective reactive relationship between *Grt, Crd, Sil,* and *Qtz* are documented locally, e.g., Perchuk et al., 1989, 1996). In this case the number of independent thermodynamic equilibrium equations increases to *three*, and thus three unknown parameters can be calculated (e.g., *P*, *T*, and X_{Mg}^{Bt}) based on the compositions of coexisting *Grt* and *Crd*. Finally, the presence of *Kfs* in the same local equilibrium assemblage (*Grt+Sil+Crd+Bt+Qtz+Kfs*) adds one more independent reaction (r7) that determines water activity ($a_{H_2O}^{fl}$) in the metamorphic fluid that coexisted with this assemblage.

Similarly, when the composition of equilibrated *Crd* is uncertain, using any garnet composition along a profile across the porphyroblast, together with the composition of locally equilibrated biotite (e.g., the core of a relatively large grain included in the *Grt*; see discussion in Spear, 1993), we can calculate the temperature using the *Bt-Grt* thermometer (r1 in Table A3). This temperature value allows the calculation of the X_{Mg}^{Crd}, *P*, and $a_{H_2O}^{fl}$ (based on reactions r2, r6, and r7, respectively) at which all minerals from the *Bt-Kfs-Grt-Crd-Sil-Qtz* assemblage were in local equilibrium.

In the case of the *Grt + Opx + Crd + Bt + Qtz ± Kfs* paragenesis, if chemical profiles across all *contacting* Fe-Mg minerals are known, we can easily calculate T_{max} and T_{min} from reactions r5, r2, and/or r1. With reaction r9 we can calculate T_{max}–P_{max} parameters, suggesting that the cores of *contacting* orthopyroxene and cordierite porphyroblasts reflect local equilibria with X_{Mg}^{Grt} = maximum value. Using chemical profiles across *contacting* garnet, cordierite, and biotite porphyroblasts, we can also calculate minimal P-T parameters, which are related to the latest stage of metamorphic retrogression. Any intermediate pressure between T_{max} and T_{min} along the P-T path can also be easily calculated, if we are able to identify corresponding compositions of coexisting minerals from equilibrium (r9) at a given temperature along the chemical profiles, and independently test these data using the *Bt-Grt* exchange equilibrium. In addition, using compositions of contacting cordierite and orthopyroxene from symplectic textures of reaction r9, we can also calculate P-T parameters for each local equilibrium (r9) and make a decision concerning either isobaric heating or cooling during the subsequent evolution of the rock (e.g., Perchuk et al., 2006).

Our method is similar to other analytical methods in which both P-T parameters and some mineral compositions are calculated simultaneously on the basis of a system of independent end-member reactions. For example, the only difference between our approach and Spear's (1993) method is that Spear extracted independent *S*, *V*, and *H* data for end-member species (equations 14–16) instead of constraining data for reactions. It is indeed essential to apply such classical local equilibrium approaches, in particular, to high-grade rocks, which preserve a range of local equilibrium assemblages and mineral compositions (e.g., Spear, 1993; Gerya and Maresch, 2004; Hisada et al., 2005), commonly reflecting polymetamorphic evolution (Perchuk, 2005; Perchuk and van Reenen, 2008; Perchuk et al., 2006).

Practical

What should not be lacking from modern petrological studies is a thorough understanding of the reaction history of a rock and thus how to interpret the mineralogy and chemistry to know what calculations to make. Our method is based on the comprehensive study of a rock chosen as a recorder of the thermodynamic and geodynamic history of a high-grade terrane. The following four major steps should form the basis for the derivation of a P-T path (e.g., Perchuk et al., 1996, 2000a, 2006; Smit et al., 2001; Gerya and Maresch, 2004; van Reenen et al., 2004).

The sequence of events related to the formation and deformation of the reaction textures preserved in the rock must be established.

Metamorphism is always coupled with rock deformation, and each metamorphic stage corresponds to a distinct deformational stage. This is why many geochronologic studies of metamorphic complexes are always linked to a field structural geological examination. Unfortunately this obvious rule is often ignored in many petrological papers, or—in contrast—driven to meaningless complexity. So, workers sometimes manage to distinguish numerous metamorphic and deformational stages, cycles, etc. (sometimes, dozens of them; see, e.g., Aftalion et al., 1991), whereas others (e.g., Davidson, 1984; Smit and van Reenen, 1997; van Reenen et al., 2004) use only *major* deformational events, which recorded the general but systematic orientation of structural elements (lineation, large fold axes, general schistosity, etc.). The most complex problem is related to polymetamorphic HGTs that preserve many geological indicators of both the first high-temperature (HT) event and the second HT overprint. For example, in the Sand River granoblastic gneisses (e.g., Fig. 2A) from the Central Zone (CZ) of the Limpopo Complex, sheath folds were formed during the D2 regional deformational event (ca. 2.65 Ga). Then, ~650 m.y. later the folds were transformed into boudins (e.g., Fig. 2B) during the second HT metamorphic shear deformational event (D3). The boudin (Fig. 2B) is composed of a typical "straight gneiss" (Davidson, 1984) with an HT "blastomylonitic" texture. It is impossible to visually determine whether these events were synchronous or not. More information can be obtained only by geological mapping with the application of conventional techniques for revealing the spatial distribution of planar and linear features of crystalline rocks (field measurements and their stereographic interpretations). Thus the key metapelite samples for deriving a P-T-τ path must be collected from an outcrop that has been thoroughly examined using structural geological methods, allowing the sample to be linked to a specific deformational event. This should be supported by microstructural analysis to reveal the crystallization succession of minerals (in thin sections), starting from the earliest stages in terms of oriented mineral inclusions in garnet and/or orthopyroxene

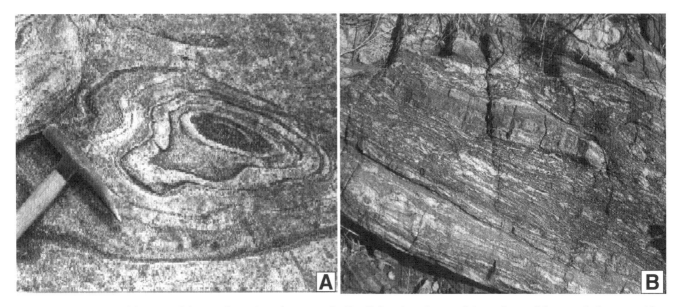

Figure 2. Macro-textural features of the transformation of common *Bt-Grt-Crd* gneisses into straight gneisses of the same bulk composition under granulite-facies conditions, Central Zone of the Limpopo Complex, southern Africa (Perchuk et al., 2006). See Table A1 (Appendix) for mineral abbreviations. (A) Small D2/M2 sheath fold developed in Sand River gneisses. (B) Boudin of an original sheath fold (ca. 2.65 Ga, D2/M2), resembling the fold in panel 2A, which was deformed in the course of the second high-temperature (ca. 2 Ga, D3/M3) event.

(e.g., Smit et al., 2001), followed by the development of different reaction textures during the late stages of a deformation event (e.g., Harley, 1989; Perchuk, 1989; Perchuk et al., 1989, 1996; Vernon et al., 2008).

The chemical evolution of rock-forming minerals should be inferred, based on detailed microprobe profiling to discriminate between exchange and net transfer reaction mechanisms.

Microprobe analyses should be conducted to identify trends in the compositional evolution of minerals participating in both the exchange and net-transfer fluid-mineral reactions. The most efficient technique is detailed microprobe profiling of the largest *contacting* grains of these minerals (these are often porphyroblasts) of variable composition in order to (1) reproduce chemical zoning and accompanied fluid-mineral inclusions, and (2) determine the number of generations of the same mineral, which could be produced during younger metamorphic stages. In addition, thorough petrographic studies, X-ray mapping, and cathodoluminescence (CL) imaging can be used to distinguish different generations of minerals. The best way is to locate, in a thin section, clear reaction textures (e.g., r6, r7, r11 in Table A3) surrounding garnet porphyroblasts, which may at least indicate the latest stage of the monometamorphic P-T evolution of a rock (e.g., Vernon et al., 2008). In the samples collected from the periphery of granulite complexes, reversible reactions (e.g., *Crd* ⇔ *Grt+Sil+Qtz*) that indicate the change from a decompression cooling (DC) regime to an isobaric cooling (IC) regime often can be identified (Perchuk et al., 1989, 1996, and references therein).

Local P-T parameters should be calculated for the different stages of the formation of mineral assemblages using the "core + core versus rim + rim" procedure for each given generation of the contacting minerals.

Mineral compositions of cores and rims should be selected using the microprobe profiling procedure. Equilibrium compositions of cores with the X_{Mg}^{Grt} = maximum flat chemical profile with a central plateau should be used for *contacting* porphyroblasts. The rim exchange and net transfer equilibria compositions (reaction textures) should be unaffected by late exchange diffusion in order to avoid the influence of this factor on thermobarometry (commonly it is a ~20–30-µm-wide zone). The calculation of local P-T parameters demands an extensive microprobe database (and possibly also X-ray maps) of both the porphyroblasts and the small grains of different generations participating in the inferred reactions within the same compositional band of the rock. It should be reiterated that calculation of P-T points is not a simple formal procedure and should be based on the deciphered chemical evolution of rock-forming minerals (see step 2). The key to success is to determine how the composition of the minerals under observation evolved and to use that knowledge to select the points for P-T calculation.

Derivation of the P-T path on the basis of calculated P-T parameters of local mineral equilibria.

The consistency of the P-T trajectory should be tested against both the mineral zoning profiles and isopleths calculated for the pertinent divariant assemblages and/or reaction textures. Again, porphyroblasts commonly preserve evidence for the earlier stages of the P-T evolution, whereas fine-grained mineral aggregates, formed as a result of net transfer reactions, could provide information on the latest stages of the P-T evolution. Because of the spatial localization of mineral equilibria with decreasing temperature, different parts of the aggregates may record information for different stages of the P-T history. Therefore, in order to use the compositions of small grains from reaction textures, the local presence of all phases participating in the corresponding "barometric" reactions is required.

In addition, as mentioned above, because of the mass-balance principle, the derived P-T path must be consistent with calculated isolines of mineral modes in the relevent divariant assemblages, because the net-transfer reactions modify the qualitative mineral proportions (modes). For example, the amount of garnet in divariant equilibria diminishes during DC or isobaric heating (IH) regimes, and increases during IC of metapelitic gneiss. One can plot the derived P-T trajectory on a diagram with equal polymineralic modes in the rock and check whether this trajectory matches such relations (e.g., Gerya, 1991; Perchuk et al., 1989, 1996, 2000a; Gerya and Maresch, 2004; Smit et al., 2001). Of course for this purpose the researcher can also use the bulk rock composition and apply the "pseudosections" approach from the THERMOCALC tool (Holland and Powell, 1998; Carson et al., 1999; Wei et al., 2003) or the PERPLE_X program (e.g., Connolly, 2005), both containing the mass-balance calculations. Applicability of mass-balance calculations should indeed be taken with caution, because not all of the rock reacts at the same time. The isolines of mineral modes are a function of the *local reactive bulk composition,* which is difficult to know and changes as the reactions proceed. Systematic testing of local bulk compositions based on X-ray mapping of reaction textures may also be considered.

Thus, the major approach relates to the derivation of P-T-τ paths based on divariant assemblages in metapelites of similar mineralogy collected from different geological units, and applying the calculation of P-T parameters for equilibria applicable to the *local* volume of the rock. Rule 2 suggests that decreasing temperature leads to redistribution of Mg from water-free minerals (B) into water-bearing minerals (A), whereas Fe moves in the opposite direction; i.e., the integrated reaction (9) must be displaced to the right side. Therefore, exchange reactions (r1) and (r2) must also be displaced to the right side during the retrograde stage of metamorphism. This behavior is in contrast to simultaneously operating net-transfer reactions (r6), (r7), and (r9), minerals of which must become more Fe-rich with the consumption of garnet because of the mass-balance principle (Fig. 3A). Thus, ΔG for the exchange equilibrium is "competing" with ΔG of the net-transfer reaction within a local volume via the mineral growth mechanisms. If the temperature effect on reactions (r1) and (r2) becomes more prominent than the effect of both P and T on reactions (r6) and (r7), the summarized X_{Mg} change of coexisting water-bearing minerals must be negligible or higher, whereas no diffusion effects pertain to this local equilibrium process. The replacements of the *Grt* rims by *Crd* and/or *Bt* thus commonly proceeds via their growth in the course of net-transfer reactions (r6), (r7), and (r9).

In most studies based on the described approach the final P-T configuration is close to a straight line that is oriented almost parallel to the isopleths of water-bearing minerals from reactions (r6) and/or (r7). This result was criticized on the basis of kinetic reasons, i.e., owing to the higher rate of Mg and Fe diffusion in cordierite in comparison with coexisting garnet (e.g., Frost and Chacko, 1989). Rule 2, however, requires a linear correlation between P and T that reflects decreasing temperature for the exchange equilibrium (r2) that operates simultaneously with a net-transfer reaction (r6). Although decreasing temperature leads to an increase of X_{Mg}^{Crd} and a decrease of X_{Mg}^{Grt}, the net-transfer reaction simulates a decrease of X_{Mg} of both the minerals (Fig. 3E) with decreasing pressure. As the result of simultaneously operating reactions (r2) and (r6), X_{Mg}^{Grt} systematically decreases (Fig. 3E, left thick arrow), whereas X_{Mg}^{Crd} shows little change (right vertical arrow in Fig. 3F). Projections of diagrams E and F onto the P-T plane thus result in a *straight* decompression-cooling P-T path (Fig. 4). This straight path is based on thermodynamics of simultaneously and locally operating exchange and net-transfer reactions, but is unrelated to kinetics.

Tests

Fluid Inclusions (FLINC)

To check the configuration and accuracy of the derived P-T path (Figs. 4–7), detailed fluid inclusion (FLINC) studies and numerical modeling of the P-T path are highly recommended. Using FLINC data, the researcher can obtain information not only on the composition of the locally equilibrated fluid, but also on the P-T parameters at which the fluid was captured (e.g., Hisada et al., 2005; Huizenga et al., this volume). If minerals contain *primary* fluid inclusions, the P-T trend can be obtained for the rock without utilizing mineral barometry. For this purpose, data on the densities of these inclusions should be used in combination with pressure-independent mineralogical thermometers (Perchuk, 1986, 1989). The theoretical basis of this approach is obvious enough: Metamorphic reactions proceed not only simultaneously with deformation events but also in the presence of a fluid. Hence, a change of mineral compositions in local equilibria with a fluid must be in correlation with its density during each stage of a metamorphic event, i.e., along the P-T path derived on the basis of mineral thermobarometry. Inasmuch as minerals recrystallize and capture fluids during exhumation, this fluid is always in local equilibrium with the host mineral grain. There is a possibility of determining the composition and other characteristics of this fluid using (1) primary fluid inclusions in minerals (e.g., Hisada et al., 2005, and references therein), and (2) by direct thermodynamic calculations of $a_{H_2O}^{fl}$ via the equilibrium $Grt + Kfs + H_2O = Bt + Qtz + Al_2SiO_5$ (r12 in Table A3) using, for example, the GEOPATH computer package (Gerya and Perchuk, 1990). The use of FLINC data for thermobarometry has some limitations related to the preservation

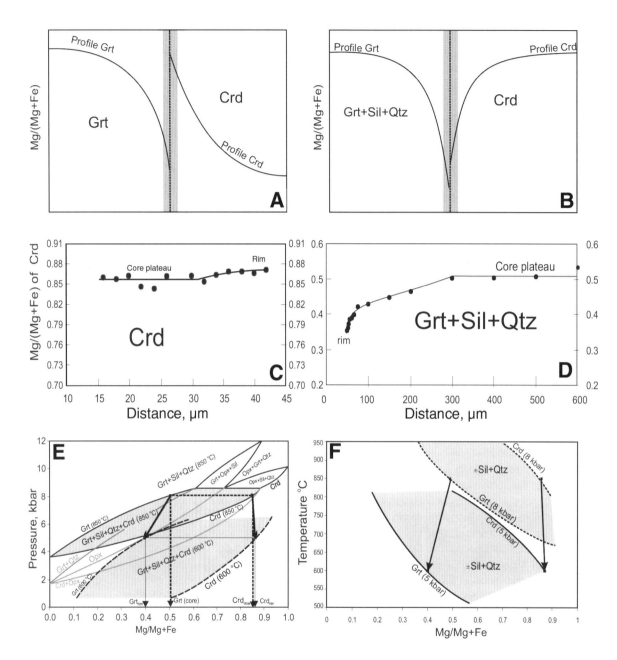

Figure 3. Major systematic chemical zoning of coexisting garnet and cordierite porphyroblasts in a metapelite (A–D), and corresponding P-X_{Mg} (E) and T-X_{Mg} (F) diagrams. See Table A1 (Appendix) for mineral abbreviations. (A) Illustration of the phase correspondence rule in terms of chemical zoning of contacting Grt and Crd porphyroblasts: A decrease of X_{Mg}^{Grt} and an increase of X_{Mg}^{Crd} resulted from decreasing temperature (P = constant). (B) Decrease of both X_{Mg}^{Grt} and X_{Mg}^{Crd} with decreasing pressure (T = constant) in assemblages with Qtz and Sil; shaded area is a possible zone of late post-metamorphic diffusion. Diagrams C and D demonstrate true mineral zoning in metapelite TOV13 from the Central Zone of the Limpopo Belt, southern Africa; diagram C shows a slight chemical zoning of the Crd grain in contact with a large garnet porphyroblast with a significant X_{Mg} decrease; both show flat plateaus and no diffusion profiles. Diagrams E and F demonstrate a cooperative influence (arrows) of P and T on the compositions of contacting Grt and Crd. For further discussion, see text.

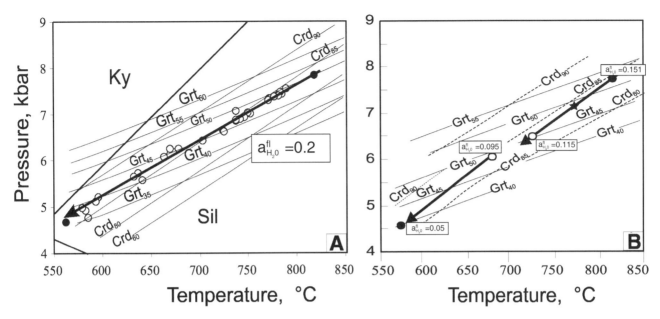

Figure 4. Evolutionary P-T paths for metapelitic gneiss (sample TOV13) from the Central Zone of the Limpopo high-grade metamorphic terrane (HGT), southern Africa, derived at $a_{H_2O}^{fl} = 0.2$ (diagram A), and at $a_{H_2O}^{fl} = 0.0004T(°C) - 0.1759$ (diagram B). The P-T paths are calculated on the basis of reaction (r7) (Perchuk et al., 2008). Thin dashed lines in both diagrams are N_{Mg}^{Grt} isopleths, whereas the solid lines are the N_{Mg}^{Crd} isopleths. Black circles correspond to minimal and maximal P-T parameters in diagrams E and F from Figure 3, whereas coordinates of all other data points (open circles) are calculated on the basis of intermediate compositions of the minerals along profiles in Figures 3C and 3D. The black star on the upper path in diagram B is related to the example discussed in the text.

of primary inclusions (such might decrepitate during decompression of the host rocks).

Figure 8 demonstrates the results of combining FLINC data with Bt-Grt-Crd mineral thermobarometry (Table A3) for deriving a P-T path. This figure also shows the subsequent comparison of the FLINC induced P-T path with the P-T path derived on the basis of the thermobarometric method under discussion. This procedure is based on the suggestion that the highest density of the CO_2 portion of an inhomogeneous fluid corresponds to the highest P-T parameters for each given local fluid-mineral equilibrium. Thus peaks A, B, C, and D (Fig. 8A) must correspond to the same peaks in diagrams B and C, and, therefore, to coordinates of the same data points in diagram D. If the suggestion is correct, the independent calculations of P-T parameters for the local *mineral* equilibria must plot on the FLINC induced P-T path. The diagram in Figure 8D demonstrates the proper consistency of independently obtained results.

Numerical

Whenever possible, calculated P-T paths should be tested by means of numerical geodynamic simulations (see, e.g., Gerya et al., 2000; Gerya and Maresch, 2004). A coincidence between the P-T paths derived from natural mineral assemblages of rocks and from numerical geodynamic simulations (based on the densities and rheological characteristics of these rocks) provides additional support for the accuracy of the paths. If all of the aforementioned conditions are fulfilled, the researcher has a convincing argument that the derived P-T trajectory is accurate and requires only geological and geodynamic interpretation. Consistency of results on the derivation of a P-T path linking the method of local equilibria to the results of numerical experiments for a granulite facies terrane is well exemplified for the Limpopo Complex (Gerya et al., 2000; Perchuk and Gerya, this volume).

Geochronological

In many cases the combination of structural data with U-Pb isotopic ages of zircon and aluminosilicate minerals (Frei et al., 1997) allows the identification of the sequence of deformational and metamorphic stages and events in one HG complex, and thus the age of a P-T path (e.g., Kreissig et al., 2000, 2001; van Reenen et al., 2004, 2008; Boshoff et al., 2006). In high-grade *polymetamorphic* complexes, however, we have to measure the isotopic age of a P-T path using growth zoning of both zircon and the associated garnet on the basis of approaches and procedures developed by Rubatto (2002) and Rubatto and Hermann (2007). This allows additional information on high-grade polymetamorphism reflected by a single rock sample (e.g., Hermann and Rubatto, 2003).

Thus, to make sure that the derivation of a P-T path is correct, the specific consideration of geological structures, microstructural analysis, calculations of mineral modes in terms of T and P, FLINC data, and numerical modeling of the P-T path must be done. All these studies are particularly crucial for HG polymetamorphism (Perchuk, 2005).

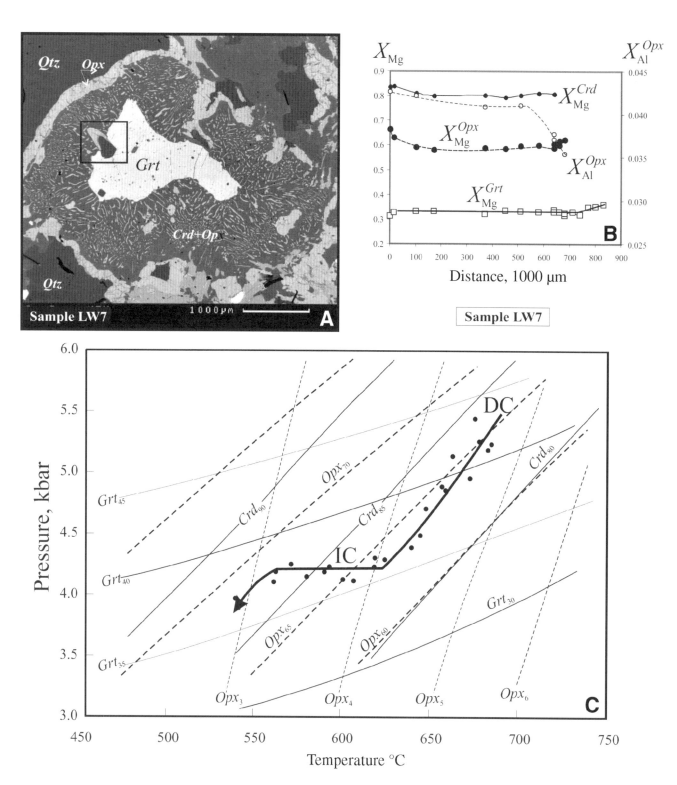

Figure 5. The perfect reaction texture (A), chemical zoning of coexisting *Fe-Mg* minerals (B), and a P-T path (C) for metapelite (sample LW7) from the South Marginal Zone (SMZ) of the Limpopo HGT (Perchuk et al., 1996). (A) Reaction texture *Grt + Qtz → Opx + Crd* (r9); inset in black frame is evidence for the formation of *Opx* as the first zone (corona) of the metasomatic reaction (r9). (B) Change in compositions of garnet, cordierite, and orthopyroxene in the reaction texture (r9); subscripts at *Crd, Grt,* and *Opx* isopleths that are participants of the reaction (r9) denote Mg numbers of the minerals, whereas the subscript at the *Opx* steep isolines denotes $N_{OK}^{Opx} = 50Al/(0.5Al+Mg+Fe)$ in *Opx* (Aranovich and Podlesskii, 1989).

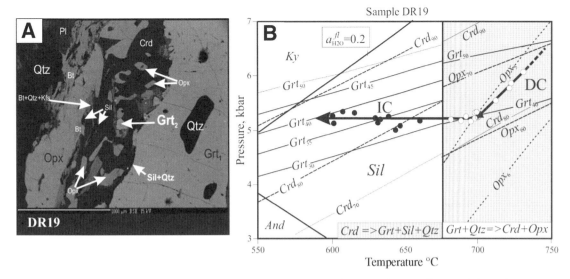

Figure 6. Unique reaction texture (A); compositional and P-T evolution (B) of metapelite DR19 from the Southern Marginal Zone of the Limpopo Complex (Perchuk et al., 1996). The position of data points in diagram B corresponds to calculated P-T parameters for each given local equilibrium according to the chemical profiles. The gray block in the P-T diagram defines the stability field of the assemblage $Crd + Opx + Crd + Grt$; the assemblage $Grt + Sil + Crd + Qtz$ is stable in the white portion of the diagram. Note that the isobaric cooling (IC) part of the P-T path correctly follows the decompression cooling (DC) portion at ~5 kbar.

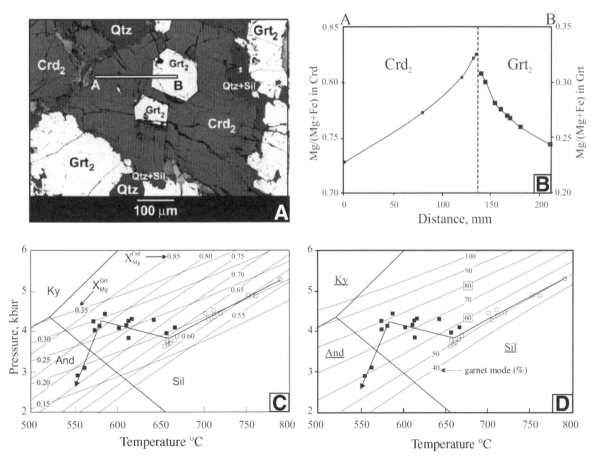

Figure 7. Metamorphic evolution of metapelitic granulite (sample A-275) from the Yenisey Range, eastern Siberia (Gerya and Maresch, 2004). (A) Back-scattered electron images. (B) Chemical zoning of newly formed idioblastic garnet (Grt_2) and Crd_2. (C) Kinked P-T path that is plotted at $a_{H_2O}^{fl}$ onto the P-T plane with the X_{Mg} isopleths for garnet and cordierite (in presence of $Sil+Qtz$). (D) Relationships of the P-T path with the Grt isomodes (atomic percentages).

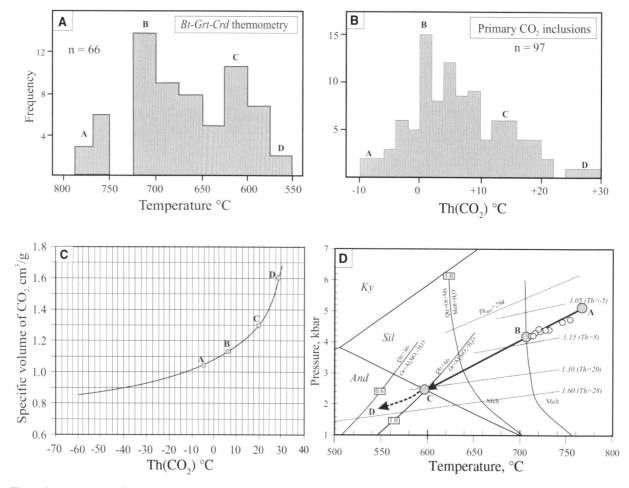

Figure 8. An example of the derivation of a P-T path using a combination of mineral thermo(baro)metry and fluid inclusion data for metapelites and migmatites from the Mahalapye block, the Central Zone (CZ) of the Limpopo Complex (modified from Hisada et al., 2005). (A) Histogram for mineral thermometric data from metapelites and associated migmatites. (B) Histogram based on measurements of primary fluid inclusions in Qtz and And. (C) Empirical correlation between V_{CO2} and Th (CO_2) (homogenization temperature), after Tomilenko and Chupin (1983). (D) P-T path derived on the basis of Bt-Grt-Crd thermometry (diagram A) and Th (CO_2) data for primary fluid inclusions; open circles reflect P–T parameters calculated using data for the reaction $Grt + Qtz + Al_2SiO_5 = Crd$ (see Table A3) at distinct water activities (Hisada et al., 2005); solid and dashed arrowed lines are calculated for Sil and And, respectively. Small open circles represent thermobarometric P-T calculations for cores of coexisting Grt and Crd. Coordinates of the gray circles A, B, C, and D are obtained on the basis of corresponding data from diagrams A–C with an accuracy of ±15 °C and ±1 kbar. Lines of dehydration reaction and melting at indicated water activity, and also phase transitions for Al_2SiO_5 are calculated from the database of Holland and Powell (1998). For further explanation, see text.

EXAMPLES FROM MONOMETAMORPHIC HIGH-GRADE TERRANES

Decompression-Cooling (DC) P-T Paths

Perchuk et al. (1989, 1996) demonstrated that the rule "one HG complex = one PT path" (e.g., Harley, 1989; Perchuk, 1989) is not correct for granulite complexes worldwide and that both DC and IC paths may occur in one terrane, or even in one sample collected close to the contact with the adjacent craton. Moreover, a model for the formation of Precambrian HGTs (e.g., Perchuk, 1989; Perchuk et al., 2001) suggests not only movement of granulites toward the cool country rocks (IC P-T path) but also their partial sinking during subsequent exhumation. In this case the P-T path must have a kinked shape. Indeed, all three kinds of P-T trajectories occur worldwide in HGTs. I shall next demonstrate the presence of reaction textures (r6) and (r9) and of chemical zoning of the associated Fe-Mg minerals in the course of their formation.

Consider the following P-T paths and their possible geothermodynamic explanations: (1) DC paths that reflect the exhumation and cooling of an HG terrane; (2) an early DC1 path reflecting the early exhumation and cooling stage of an HG terrane followed by an IC path that culminated with the second stage of exhumation (DC2); (3) a DC1 path with a kinked compression cooling (CC) stage followed by a DC2 stage, this P-T path reflects the early exhumation stage (DC1) followed by sinking-burial and cooling (CC), which culminated with the latest exhumation stage (DC2) of an HG terrane during a single tectono-metamorphic event.

Only the detailed approach used for the derivation of the most common DC path (case 1) will be described, because the approaches for the composite paths are similar, if not identical.

Aluminous gneisses are good examples for demonstrating the efficiency of our method for calculating a P-T path. Let us consider a rock belonging to the system Si-Al-Mg-Fe-K-H-O and characterized by zoned garnet that coexists with biotite and cordierite in the presence of sillimanite, quartz, and K-feldspar, i.e., the assemblage

$$Bt + Sil + Qtz + Grt + Crd + Or + H_2O. \quad (19)$$

However, the minerals from assemblage (19) are not necessarily in local equilibrium in the rock in which reaction textures corresponding to end-member reactions (r5) and (r6) are common. We can thus consider rock assemblage (19) as two independent local equilibrium assemblages:

$$Sil + Qtz + Grt + Crd \quad (20)$$

and

$$Bt + Sil + Qtz + Grt + Or + H_2O. \quad (21)$$

These assemblages can be described thermodynamically (equation 18) by reactions (r2), (r6) and (r1), (r7), respectively.

We also can confidently estimate the parameters at the beginning and the end of a P-T path, whereas the uncertainties may be unclear for the rest of the path. Systematic probe profiling across contacting porphyroblasts allows identifying the most Mg-rich core of garnet porphyroblasts that defines the peak metamorphic conditions. If at maximal X_{Mg}^{Grt} and at certain compositions of contacting cordierite and biotite (or Crd-Bt inclusions in the Grt) equilibrium conditions (18) for end-member exchange reactions (r1) and (r2) give the same temperature estimates, the compositions of Grt, Crd, and Bt can be considered to be in equilibrium, possibly reflecting maximal P and T.

One of the best examples is metapelite TOV13 from the Musina area in the Central Zone of the Limpopo Complex, southern Africa (Perchuk et al., 2008). This sample contains all minerals from the assemblage (19). Garnet coexists with Sil, Qtz, and Crd (X_{Mg} = 0.841) and shows maximal X_{Mg}^{Grt} = 0.513 (a short flat profile) at P = 7.74 kbar and T = 817 °C, estimated by solving simultaneously two thermodynamic equilibrium equations (equation 18) composed for reactions (r2) and (r6) (Table A3). Therefore in local equilibrium X_{Mg}^{Bt} must be 0.691, as follows from the thermodynamic equilibrium condition for exchange reaction (r1). Indeed, we can easily find these minerals in direct contact with the garnet (Perchuk et al., 2008). Similarly, using data obtained from probe profiling of coexisting garnet, cordierite, and biotite we can find *minimal* P-T parameters of 4.6 kbar and ~560 °C as well as the corresponding equilibrium compositions of coexisting minerals (Perchuk et al., 2008). Using data from Table A3 (Appendix) for end-member reactions (r1), (r2), (r6), and (r7), we can furthermore calculate pressure $a_{H_2O}^{fl}$ and the compositions of coexisting cordierite and biotite for *any local equilibrium* within the temperature range 817–560 °C, because garnet shows systematic zoning from the core to the rim (Fig. 3D). For this calculation we initially assume that the garnet composition changes linearly with temperature. This assumption should be further tested by a systematic study of coexisting Grt and Crd compositions in Grt + Crd + Sil + Qtz reaction textures corresponding to simultaneously operating end-member reactions (r6) and (r2). Because of negligible volume changes for exchange reactions r2 in Table A3 (Appendix), we can calculate equilibrium compositions of cordierite at a given temperature independent of pressure. So, the Grt porphyroblast with X_{Mg}^{Grt} = 0.495 contains inclusions of Sil and Qtz and is in contact with a large (0.6 × 0.2 mm) cordierite grain. If the grain shows systematic compositional change along the microprobe profile, X_{Mg}^{Crd} = 0.848 is the only equilibrium composition. In contrast to cordierite, biotite commonly occurs in many generations that may create difficulties in identification of the Bt composition in its local equilibrium with $Grt_{0.495}$ and $Crd_{0.848}$. Theoretical X_{Mg}^{Bt} at a given temperature must be 0.715. Indeed, biotite of X_{Mg}^{Bt} = 0.714 (probe analysis #vK785) *coexists* with Crd (#vK104) and Grt (#Dp.36) in sample TOV13 as one of the earliest Bt generations that shows no notable chemical zoning (Perchuk et al., 2008). This allows the simultaneous calculation of both the water activity and pressure via thermodynamic equilibrium conditions (18) for end-member reactions (r6) and (r7) at T = 770 °C and other given compositional parameters. Equation (19) involved in these calculations can, in particular, be solved with the GEOPATH program (Gerya and Perchuk, 1990): calculated P = 7.17 kbar and $a_{H_2O}^{fl}$ = 0.132 at the given T = 770 °C. Based on this approach we can calculate numerous *theoretical* thermodynamic and compositional parameters and compare them with *true* mineral compositions (obtained with microprobe) and their local positions (e.g., along zoning profiles) in coexisting minerals.

Thus, if minimal and maximal parameters for a P-T trajectory are known, other data points for the path can be calculated theoretically for any given composition of garnet (see also Spear and Selverstone, 1983; Spear, 1993). If these compositions of Fe-Mg minerals really exist in the rock, we can use theoretical compositions for the correct derivation of the P-T path. While doing so, we should remember that the configuration of a P-T path depends greatly on $a_{H_2O}^{fl}$, partial melting of the gneiss during exhumation, and other thermogeodynamic parameters. Diagrams A and B in Figure 4 demonstrate the influence of $a_{H_2O}^{fl}$ on the P-T path configuration. For example, at $a_{H_2O}^{fl}$ = 0.2, all calculated pressures and temperatures for local mineral equilibria from end-member reactions (r1), (r2), (r6), and (r7) lie strongly along the P-T path (Fig. 4A). However, maximal ($a_{H_2O}^{fl}$ = 0.151) and minimal ($a_{H_2O}^{fl}$ = 0.05) P-T parameters (Perchuk et al., 2008), calculated on the basis of reaction (r7) within a wide range of temperature, allow much more information to be derived for the evolution of a TOV13 rock type. For example, a sharp change of $a_{H_2O}^{fl}$ within the temperature range 680–730 °C suggests that the break in the TOV13 P-T path is related to partial melting of Neoarchean gneisses in the Central Zone of the Limpopo Complex (Perchuk et al., 2008, their figure 11). This is another example that demonstrates the strong relationship between detailed petrographic observations of different generations of reacting minerals in the assemblage (19) and the derivation of a P-T path. Similar relations were shown for different Precambrian complexes in many published papers (e.g., Perchuk et al., 1989, 1996, 2000a, 2000b, 2008; Smit et al., 2001; van Reenen et al., 2004; Gerya and Maresch, 2004).

All the above theoretical and practical discussions are applicable if the rock is homogeneous in terms of mineral distribution. If the metamorphic rock is thinly banded, the calculations must be applied to each given band with a defined bulk composition. This rule, though, is also crucial for all other approaches (Berman, 1991; Spear, 1993; Holland and Powell, 1998; Connolly, 2005).

Decompression-Cooling P-T Path Complicated by (Sub)Isobaric-Cooling (IC) Trajectory

The isobaric-cooling (IC) type of P-T paths (Sandiford and Powell, 1986; Ellis, 1987; Harley, 1989) suggests a decrease in temperature of a metamorphic complex at a given depth owing either to some cooling agents or to a specific regime in subduction-collision zones, at which the vertical movement of the complex was halted to allow heat being lost slowly. Perchuk et al. (1989, 1996, 2000a) showed that the IC path reflects a later stage of the DC P-T path that resulted from gravitational mechanisms of redistribution of material in the Earth's crust. The DC → IC path is the reflection of movement of metamorphic blocks within an uprising granulite complex toward the contact with the sinking cool greenstone cratonic wall rocks. The IC portion of the DC path is well recorded by both reaction textures and the chemical zoning of minerals from both the granulites and the wall rocks. Two perfect examples are shown in Figures 5 and 6.

Figure 5A demonstrates a reaction texture (r9) that occurs in many granulite facies terranes worldwide (e.g., Harley, 1989; Perchuk, 1989; Perchuk et al., 2006, and references therein). The textures consist of two parts: coronas (*Opx±Pl*) and symplectites (*Opx+Crd*). A systematic change in compositions of the *Grt* rim and adjacent *Crd-Opx* symplectites allows the reconstruction of P-T parameters at the latest stage of exhumation (e.g., Perchuk, 1989; Vernon et al., 2008). Figure 5B shows a flat X_{Mg}^{Grt} profile across the relict garnet grain. Using these data and thermodynamic data for the reaction (r9) from Table A3 (Appendix), the DC → IC evolutionary P-T path for the rock (arrow in Fig. 5C)

has been calculated. Both parts of the trajectory reflect one stage of the Limpopo SMZ exhumation at ca. 2.67 Ga (van Reenen et al., 1992; van Reenen and Smit, 1996). The geodynamic explanation of the P-T path inflection during a single stage exhumation has been formulated on the basis of petrologic studies (Perchuk et al., 1996) and two-dimensional (2D) numerical modeling (Gerya et al., 2000).

Figure 6 demonstrates another example of the transition from DC → IC trajectories. The reaction texture $Qtz + Grt_1 \rightarrow Crd_2 + Opx_2$ (diagram A) is subsequently followed by the reaction texture $Crd_2 \rightarrow Qtz + Sil + Grt_2$. Apart from chemical zoning of the Fe-Mg minerals, the DC → IC transition in diagram B properly explains the formation of the second generation of garnet (Grt_2) after Crd_2 with decreasing temperature. Thus, two linked P-T diagrams in Figure 6B reflect subsequently operating reactions (r9) and (r6) during the IC stage exhumation that is linked to the final subhorizontal movement of the granulite complex (Gerya et al., 2000; van Reenen et al., this volume).

Finally, consider the derivation of a kinked P-T path that suggests a change from the DC regime to the CC regime that is in turn followed by the second DC granulite exhumation, but at lower temperatures. Figure 7 demonstrates a unique reaction texture (A), chemical zoning of the contacting minerals (B), and the corresponding kinked P-T path (C and D) for granulites from the Yenisey HGT (Perchuk et al., 1989; Smit et al., 2000; Gerya and Maresch, 2004). Figure 7A clearly shows the newly formed perfectly idioblastic garnet (Grt_2) in association with *Sil* (2) and *Qtz* (2), which developed after cordierite. The most exiting *Crd-Grt* relationships are recorded in their direct contact: X_{Mg} of both the minerals *increases,* reflecting an increase of pressure at decreasing temperatures that shifted reaction (r6) to the left side. This is a good example of the total absence of late stage Fe-Mg diffusion between *Crd* and *Grt*. Thus, the mechanism that controlled both the exchange and net-transfer reactions involving the Fe-Mg minerals in this granulite is only *crystal growth*. The DC2 portion of the P-T path in Figures 7C and 7D relates to a very late stage of exhumation, when andalusite appeared in the rocks (Gerya and Maresch, 2004). On the other hand, both the reaction texture and the P-T path of Figure 7 support the gravitational redistribution model for the exhumation of Precambrian HG terranes (Perchuk, 1989). This is the subject of a separate chapter in this volume (Perchuk and Gerya, this volume).

EXAMPLES FROM POLYMETAMORPHIC HIGH-GRADE TERRANES

The method of derivation of P-T paths for polymetamorphic HGTs is the same as discussed above. The major problem is to identify two or more HG tectono-metamorphic events. If these events can be distinguished, there is no problem in deducing the paths for each event separately. The best example of this procedure is provided by granulites from the core of the Vredefort impact crater in South Africa. These HG rocks were first formed in the course of regional metamorphism at ca. 3.1 Ga and underwent the second high-grade event behind the shock wave more than 1000 m.y. later (Gibson and Reimold, 2001). Detailed petrologic studies of the evolution of the Vredefort HG rocks showed that each event is recorded in both the textures and the crater core mineral assemblages, allowing the derivation of related P-T paths (Perchuk et al., 2002, 2003). Figure 9 clearly demonstrates (see explanations in the caption) typical reaction textures and the two-stage P-T evolution of the rocks. The major feature of the P-T path in Figure 9 is the isobaric-heating–generated shock wave that passed through the rocks. The composite P-T path in Figure 9 should be considered to reflect the result of a strong petrologic field experiment for HG polymetamorphic events.

It is well known that in order to preserve any distinct isotopic age the "clock-mineral" must be totally recrystallized at the same P-T parameters at which other local equilibria have been reached. One of the best sources of the age data is zircon. Unfortunately, zircon weakly reacts with other minerals in the course of regional metamorphism in the presence of an alkali-poor H_2O–CO_2 fluid. However, zircon quickly recrystallizes in the presence of alkali-rich fluids or granitic melts (e.g., during partial melting). In contrast, garnet easily grows (in the field of its stability) during regional metamorphism, trapping liquid CO_2 inclusions (e.g., Tomilenko and Chupin, 1983; Hisada et al., 2005). Despite the fact that geochronology is based on single crystal approaches, it should also be considered within the context of classical principles and rules as formulated in this chapter. For example, garnet of the sheared metapelite (sample T73) from the Central Zone (CZ) of the Limpopo Complex yielded the Pb-Pb age of 2023 ± 11 Ma (van Reenen et al., 2008), whereas the cores of zircons from the same sample showed highly discordant $^{207}Pb/^{206}Pb$ ages varying within 3.49 and 2.59 Ga; and only extremely narrow rims of the zircons yielded an age of ca. 2 Ga (Boshoff et al., 2006). Thus, in a single sample *zircon* records at least three tectono-metamorphic events that are well known for the Limpopo Complex (Barton et al., 1994; Kröner et al., 1999; van Reenen et al., 2008), whereas *garnet* formed (recrystallized?) only during the third event. In many other examples, however, the isotopic age of garnet, depending on the degree of overprint, may vary between ca. 2.6 and ca. 2 Ga (e.g., Boshoff et al., 2006; van Reenen et al., 2008). Because garnet is the permanent participant of most of the exchange and net-transfer reactions, the question arises as in which stage of metamorphic evolution a "*Grt*-bearing reaction" took place? The best answer to this question was provided by Rubatto (2002): A distinct partition coefficient $K_{REE}^{Zrn/Grt}$ provides information on the synchronous reactions with *Zrn* and *Grt* involved, and we can obtain an accurate isotopic age for the rock from a polymetamorphic HGT (e.g., Hermann and Rubatto, 2003; Buick et al., 2006; Rubatto and Hermann, 2007).

Structural-petrologic studies are also powerful tools for discriminating between different HG events in one HGT (Perchuk and van Reenen, 2008). As a rule, *structural* indicators (lineation, fold axes, schistosity, etc.) of different HG events preserve different unrepeated orientations that can be properly seen on a common stereographic projection (e.g., Boshoff et al., 2006). For example, a typical orientation of lineations (L2) and schistosity (S2) for the Neoarchean gneisses in the CZ of the Limpopo Complex, southern Africa, varies within narrow limits; L2/S2 = ~40°–45°/~200°–220°. This D2 structural pattern was completely established near peak metamorphic conditions. About 650 m.y. later this structural pattern was overprinted by a new L3/S3 pattern characterized by a flat (NNW 12°–15°) lineation (van Reenen et al., 2004; Boshoff et al., 2006). This conclusion is supported by independent structural data produced for a metapelitic gneiss (sample 98 Ma-55) with the description ". . . a mineral lineation defined by the preferred orientation of sillimanite and (locally) orthoamphibole plunges 15°/335°" and ". . . trace element zoning patterns in garnet and the age of accessory phases are most consistent with a single tectonometamorphic event at ~2.03 Ga" (Buick et al., 2006, p. 152 and 150, respectively). These data are consistent with both structural and isotopic age data for the T73 gneiss sample (see previous paragraph). For this sample the rare earth element (REE) distribution among coexisting minerals has never been studied, whereas isobaric heating during the D3/M3 event is well recognized (Boshoff et al., 2006). In addition, using the data set from Table A3 (Appendix) and Buick's data on mineral chemistry for sample 98 Ma-55, we have calculated P-T parameters of equilibria (r2), (r3), and (r6) and plotted them onto a diagram with the P-T path for the Baklykraal structure (van Reenen et al., 2004). Figure 10A demonstrates the position of the data point (closed circle) on the isobaric heating portion of the composite D3/M3 P-T path for this shear structure (e.g., Perchuk et al., 2008).

Now let us consider another well-documented example from the easternmost part of the CZ of the Limpopo Complex (Boshoff et al., 2006; Perchuk et al., 2006, 2008). Figures 3 and 4 demonstrate the opposite influence of T and P on simultaneously operating reactions (r2) and (r6) in sample TOV13, containing large amounts of garnet and small amounts of cordierite (Perchuk et al., 2008). Sample TOV13 was collected from the Ha-Tshanzi sheath fold, whose structural pattern is similar to L2 and S2 indicators of the D2/M2 event (see stereographic projection in Fig. 10A). Near the Ha-Tshanzi structure is the Campbell crossfold. This fold underwent the D3/M3 overprint (van Reenen et al., 2008) that resulted in (1) strong shearing of the D2/M2 metapelitic gneisses, and (2) growth of cordierite by replacement of *Grt*, *Sil*, and *Qtz* (Perchuk et al., 2008). The bulk compositions of samples collected from both structures are similar (Fig. 10B). In Figure 10B the minimal P-T parameters for sample TOV13 from the D2/M2 Ha-Tshanzi sheath fold correspond approximately to the maximal P-T parameters for gneiss from the D3/M3 Campbell shear structure. This suggests intensive isobaric heating of the Campbell shear structure (sample 06-19) during the D3/M3 high-grade event. Thus, the configurations of both *integrated* P-T paths in Figures 10A and 10B are similar to the trajectory in Figure 9, which was chosen above as the standard case for the polymetamorphic history of high-grade rocks.

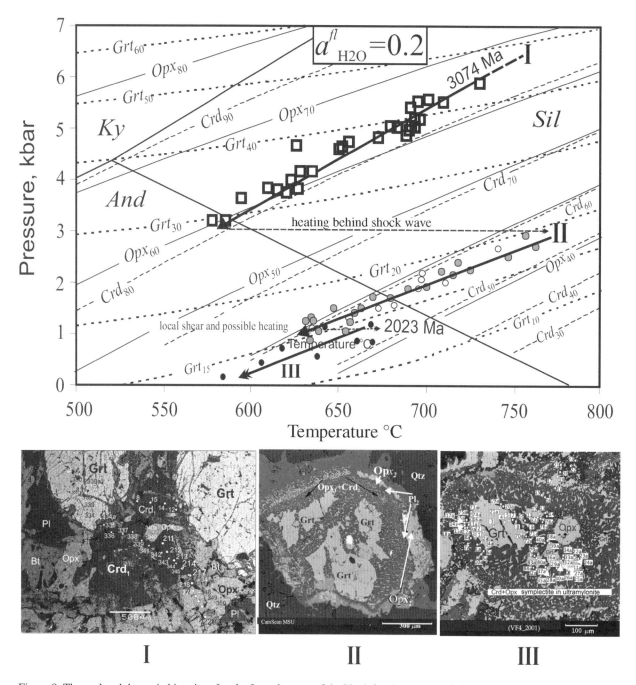

Figure 9. Thermal and dynamic histories of rocks from the core of the Vredefort Dome recorded in textures and mineral compositions of the high-grade metapelites during the pre-impact P-T path (more than 3074 Ma), and the post-impact P-T paths (2023 Ma) (Perchuk et al., 2002, 2003). The N_{Mg} isopleths for Fe-Mg minerals in the reaction $Grt + Qtz \rightarrow Crd + Opx$ are calculated, assuming that $a^{fl}_{H_2O} = 0.2$. Note that the minimal P-T parameters for the 3074 Ma path (stage I, granoblastic texture) approximately correspond to the maximal P-T parameters recorded in the 2023 Ma post-impact symplectites (stage II), which suggests isobaric heating. The P-T path for stage II intersects the isopleths on the right side of the reaction. Open rectangles represent P-T parameters for the local equilibrium $Grt_1 + Qtz = Crd_1 + Opx_1$ during the exhumation of the granulite at ca. 3.1 Ga (stage I). Gray circles show the same assemblage of symplectitic textures formed at 2023 Ma in the same metapelite behind the shock wave. The black circles (stage III) reflect the formation of the ultramylonitic texture at the end of stage II owing to shear heating. Age data are taken from Gibson and Reimold (2001).

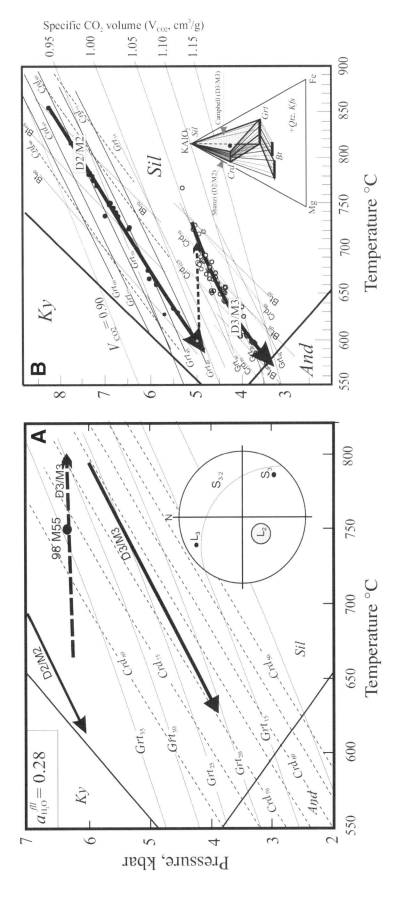

Figure 10. Integrated P-T paths for polymetamorphic rocks from central (A) and eastern (B) portions of the Central Zone (CZ) of the Limpopo Complex, southern Africa (modified from Perchuk et al., 2008). (A) Schematic position of the P-T path for the D2/M2 granoblastic gneiss (sample JC1) and closely associated, strongly sheared metapelitic (straight) gneiss T18 (van Reenen et al., 2004) from the Baklykraal structure. Stereographic inset in the diagram reflects the orientation of lineation (L) and schistosity for two events. Closed circles in diagram B are characterized by P-T parameters estimated with the GEOPATH program (Gerya and Perchuk, 1990). (B) P-T paths for the rocks from the D2/M2 (ca. 2.6 Ga) Shanzi sheath fold (see Fig. 4) and the closely associated D3/M3 (ca. 2 Ga) Campbell shear structure. Bulk compositions of the metapelitic rocks are similar, as is well seen in the paragenetic triangle inset (black circle). The minimal P-T parameters for the D2/M2 Ha-Tshanzi sheath fold approximately correspond to the maximal P-T parameters of the D3/M3 Campbell shear structure, suggesting isobaric heating during the D3/M3 high-grade event. Configurations of both integrated P-T paths in diagrams A and B are similar in shape to the united trajectory in Figure 9.

Most metapelites from the CZ of the Limpopo Complex are affected by the D3/M3 metamorphic fluids to lesser or greater degrees. In many cases the degree of influence of the latest HG fluid activity correlates with structural parameters in general and with linear indicators in particular. This regularity was well illustrated by van Reenen et al. (2004) and Boshoff et al. (2006). Even in these cases, rims of zircons reflect an isotopic age of ca. 2 Ga, whereas garnets (Pb-Pb leaching method) show imprecise ages between 2.6 and 2.0 Ga. Errors in determination of the age may approach ~200 m.y. (Boshoff et al., 2006). These errors could be potentially related to different generations of garnet that participated in the reaction $Crd = Grt + Qtz + Sil$ and which reflected a systematic change of local P-T parameters (Fig. 10) with time.

It must be remembered that isobaric heating is only one among a range of possible geodynamic phenomena. No doubt other tectonic models, such as isothermal compression-decompression or compression-heating, may lead to a different P-T record. But in any case, P-T paths are the sole recorders of the polymetamorphic nature of a crystalline complex.

CONCLUSIONS

1. The principle of local mineral equilibria, in conjunction with thermodynamic data, is the most effective tool for estimating the correct $P\text{-}T\text{-}a_i^{fl}$ parameters and the derivation of P-T-τ paths for *mono-* and *polymetamorphic rocks* on the basis of classical thermobarometry. In the case of using local mineral equilibria, no principal differences exist between the end-member reaction approach discussed here and other thermodynamic methods aimed to reconstruct the thermo-geodynamic history of a crystalline rock.

2. To obtain an accurate P-T-τ path, a sequence of operations is recommended. They combine field and microstructural studies, fluid inclusion data, calculations of mineral modes in terms of T and P, thermodynamic calculations of $P\text{-}T\text{-}a_i^{fl}$ parameters from thermodynamic data, numerical geodynamic modeling tests of the derived P-T path, etc.

3. Correctly derived P-T paths are adequate to reveal the essential features of high-grade *polymetamorphic* events in granulite facies terranes. Structural and isotopic age data are useful additional sources of information for deducing the sequence of tectono-metamorphic events in a given HG complex.

ACKNOWLEDGMENTS

This study was supported by RFBR (grant nos. 08-05-00354 and 09-05-00991), the Program of the RF President "Leading Research Schools of Russia" (grant no. 1949.2008.5), and by NRF grant no. 68288 to Dirk van Reenen. Owing to Leonid Perchuk's untimely death on 19 June 2009 the volume editors called upon Taras Gerya to prepare a final version of this chapter based on detailed comments and corrections suggested by F.S. Spear and J. Connolly. J. Connolly kindly agreed to approve the final version of the chapter. The volume editors would also like to thank Ron Vernon for valuable advice concerning the handling of the chapter, and Jan Martin Huizenga for his advice after having read through the final version.

APPENDIX.

TABLE A1. MINERAL ABBREVIATIONS USED

Symbol	Explanation
Ab	Albite
Alm	Almandine
An	Anorthite
And	Andalusite
Ann	Annite
Bt	Biotite
Chl	Chlorite
Cpx	Clinopyroxene
Crd	Cordierite
Di	Diopside
En	Enstatite
fl	Fluid
Fs	Ferrosilite
Grs	Grossular
Grt	Garnet
Hbl	Hornblende
Hed	Hedenbergite
Kfs	K-feldspar
Ks	Kalsilite
Ky	Kyanite
Ne	Nepheline
OK	"Orthocorundum" (as an Al_2O_3 end member of Opx)
Opx	Orthopyroxene
Or	Orthoclase
Phl	Phlogopite
Pl	Plagioclase
Prp	Pyrope
Qtz	Quartz
San	Sanidine
Sil	Sillimanite
Zrn	Zircon

Symbol	Definition
TABLE A2. THERMODYNAMIC SYMBOLS USED	
a_i	Activity of i-th component
a_{Crd}	Activity of dry cordierite in the system $Crd_{dry} - Crd_{H_2O-CO_2}$
$a_{H_2O}^{fl}$	Activity of water in a fluid (fl)
Cp	Isobaric heat capacity
G	Gibbs free energy (Gibbs potential)
H	Enthalpy
m_i	Mass of i-th component
N_i^{ϕ}	$100 \cdot X_i^{\phi}$
N_{Mg}^{ϕ}	$100 \cdot X_{Mg}^{\phi}$ (Mg number)
N_{OK}^{Opx}	$100 \cdot X_{OK}^{Opx}$
P	Pressure
R	Gas constant
S	Entropy
T	Temperature
Th	Homogenization temperature
V	Volume
X_i^{ϕ}	Mole fraction of i-th component in the phase
$X_{H_2O}^{fl}$	Mole fraction of water in a fluid (fl)
X_{OK}^{Opx}	0.5Al/(0.5Al+Fe+Mg) of Opx
μ_i	Chemical potential of i-th component

TABLE A3. INTERNALLY CONSISTENT THERMODYNAMIC DATA SET FOR REACTIONS USED FOR THERMOBAROMETRY OF GRANULITES AND FOR CALCULATIONS OF ISOPLETHS ON P-T DIAGRAMS (PERCHUK ET AL., 2008, AND REFERENCES THEREIN)

No.	Reactions*	$\Delta H_r^{P_o,T_o}$** (Cal)	$\Delta S_r^{P_o,T_o}$** (Cal/K)	$\Delta V_r^{P_o,T_o}$** (Cal/bar)	$\Delta Cp_r^{P_o,T_o}$** (Cal/K)
r1	$\frac{1}{3}$ Prp + $\frac{1}{3}$ Ann = $\frac{1}{3}$ Alm + $\frac{1}{3}$ Phl	−7843	−5.699	0.02500	0
r2	$\frac{1}{3}$ Prp + $\frac{1}{2}$ Crd$_{Fe}$ = $\frac{1}{3}$ Alm + $\frac{1}{2}$ Crd$_{Mg}$	−6134	−2.668	−0.03535	0
r3	$\frac{1}{3}$ Prp + Fs = $\frac{1}{3}$ Alm + En	−4766	−2.654	−0.02342	0
r4	$\frac{1}{2}$ Crd$_{Mg}$ + $\frac{1}{3}$ Ann = $\frac{1}{2}$ Crd$_{Fe}$ + $\frac{1}{3}$ Phl	−1.7461	−2.8546	0.058235	0
r5	$\frac{1}{2}$ Crd$_{Mg}$ + Fs = $\frac{1}{2}$ Crd + En	1368	0.014	0.01193	0
r6	$\frac{1}{3}$ Prp + $\frac{2}{3}$ Sil + $\frac{5}{6}$ Qtz = $\frac{1}{2}$ Crd$_{Mg}$	51	4.620	0.63827	0
r7	Prp + Kfs + H$_2$O = Phl + $\frac{1}{2}$ Sil + $\frac{5}{2}$ Qtz	−23,595	−31.214	0.547	0
r8	$\frac{1}{2}$ Crd$_{Mg}$ = En + $\frac{2}{3}$ Qtz + OK	6096	−4.897	−0.42628	0
r9	Prp + $\frac{3}{2}$ Qtz = 2En + $\frac{1}{2}$ Crd$_{Mg}$	−3311	3.028	0.76483	0
r10	Prp = 3En + OK	2785	−1.869	0.32855	5.88
r11†	$\frac{1}{3}$ Grs + $\frac{2}{3}$ Sil + $\frac{1}{3}$ Qtz = An	2722	10.266	0.435	0
r12†	$\frac{1}{2}$ Grs + Prp + Kfs + H$_2$O = Phl + $\frac{3}{2}$ An + $\frac{3}{2}$ Qtz	−19,512	−15.815	1.1995	0

*Activities of end members of solid solutions:

Garnet: $RT\ln a_{Prp}^{Grt} = 3RT\ln X_{Mg}^{Grt}$, $RT\ln a_{Alm}^{Grt} = 3RT\ln X_{Fe}^{Grt}$, $RT\ln a_{Crs}^{Grt} = 3RT\ln X_{Ca}^{Grt} + (2.484+0.26T)(X_{Ca}^{Grt})^2(2- X_{Ca}^{Grt})$ + $5784 − 1.24T$ X_{Mg}^{Grt} $(1 − X_{Ca}^{Grt}) − (234+0.75T) + [(0.075–4.566)(0.2 − X_{Ca}^{Grt})^2]P$, where X_{Mg}^{Grt} = Mg/(Fe+Mg+Ca), X_{Fe}^{Grt} = Fe/(Fe+Mg+Ca), and X_{Ca}^{Grt} = Ca/(Fe+Mg+Ca).

Cordierite: $RT\ln a_{Crd} = 2RT\ln X_{Mg}^{Crd} − 1333 + 0.617T − 0.336P + 1026(1− X_{H_2O}^{fl}) + 472(1− X_{H_2O}^{fl})^2$, $RT\ln a_{Crd} = 2RT\ln X_{Fe}^{Crd} − 1333 + 0.617T − 0.336P + 1026(1− X_{H_2O}^{fl}) + 472(1− X_{H_2O}^{fl})^2$, where X_{Mg}^{Crd} = Mg/(Fe+Mg); X_{Fe}^{Crd} = Fe/(Fe+Mg); $X_{H_2O}^{fl}$ = H$_2$O/(H$_2$O+CO$_2$), which is the mole fraction of water in the metamorphic fluid coexisting with Crd, $X_{H_2O}^{fl} \approx a_{H_2O}^{fl}$ is assumed as a first approximation.

Orthopyroxene: $RT\ln a_{En} = RT\ln X_{En} + (X_{Fs})^2(1.86T−2533) + X_{Fs}X_{OK} \times (2671+1.86T) − (1237+0.1425P)(X_{OK})^2$; $RT\ln a_{Fs} = RT\ln X_{Fs} + (X_{En})^2(1.86T−2533) + X_{En}X_{OK}(1.86T−7937) − (6441+0.1425P)(X_{OK})^2$; $RT\ln a_{OK} = RT\ln X_{OK} − 6441(X_{Fs})^2 − X_{En}X_{Fs}(6145+1.86T) − 1237(X_{En})^2 − 0.1425P(1− X_{OK})^2$, where X_{En} = Mg/(Fe+Mg+Al/2); X_{Fs} = Fe/(Fe+Mg+Al/2); X_{OK} = Al/(2Fe+2Mg+Al).

Biotite: $RT\ln a_{Phl}^{Bt} = 3RT\ln X_{Mg}^{Bt}$, $RT\ln a_{Ann}^{Bt} = 3RT\ln X_{Fe}^{Bt}$, where X_{Mg}^{Bt} = Mg/(Fe+Mg), X_{Fe}^{Bt} = Fe/(Fe+Mg).

K-feldspar: A sub-regular model for the high albite-sanidine solid solution was used (see also Zyrianov et al., 1978) for calculations of $a_{H_2O}^{fl}$ via reaction (r7) with chemical compositions of Kfs in direct contact with the Bt + Sil + Qtz assemblage developed around garnet.

Plagioclase: $RT\ln a_{An}^{Pl} = RT\ln X_{An}^{Pl} + (X_{Ab}^{Pl})^2[(1980−1.52T)+2 X_{An}^{Pl}(4680−2.348T)]$, where X_{Ab}^{Pl} = Na/(Na+Ca), X_{An}^{Pl} = Ca/(Na+Ca).

**P_o = 1 bar; T_o = 970 K.

†Reactions (r11) and (r12) have been used for calculating pressure at $X_{Ca}^{Grt} > 0.05$.

REFERENCES CITED

Aftalion, M., Bibikova, E.V., Bowes, D.R., Hopgood, A.M., and Perchuk, L.L., 1991, Timing of early Proterozoic collisional and extensional events in the granulite-gneiss-charnockite-granite complex, Lake Baikal, USSR: A U-Pb, Rb-Sr, and Sm-Nd study: Journal of Geology, v. 99, p. 851–861, doi:10.1086/629556.

Aranovich, L.Y., and Podlesskii, K.K., 1983, The cordierite-garnet-sillimanite-quartz equilibrium: Experiments and calculations, in Saxena, S.K., ed., Kinetics and Equilibrium in Mineral Reactions: Berlin, Springer, Advances in Physical Geochemistry, v. 3, p. 179–198.

Aranovich, L.Y., and Podlesskii, K.K., 1989, Geothermobarometry of high-grade metapelites: Simultaneously operating reactions, in Daly, J.S., Cliff, R.A., and Yardley, B.W.D., eds., Evolution of Metamorphic Belts: Geological Society [London] Special Publication 42, p. 41–65.

Barton, J.M., Jr., Holzer, L., Kamber, B., Doig, R., Kramers, J.D., and Nyfeler, D., 1994, Discrete metamorphic events in the Limpopo belt, southern Africa: Implications for the application of P-T paths in complex metamorphic terrains: Geology, v. 22, p. 1035–1038, doi:10.1130/0091 -7613(1994)022<1035:DMEITL>2.3.CO;2.

Berman, R.G., 1991, Thermobarometry using multiequilibrium calculations: A new technique with petrologic applications: Canadian Mineralogist, v. 29, p. 833–855.

Boshoff, R., van Reenen, D.D., Smit, C.A., Perchuk, L.L., Kramers, J., and Armstrong, R., 2006, Geologic history of the Central Zone of the Limpopo complex: (1) The West Alldays area: Journal of Geology, v. 114, p. 699–716, doi:10.1086/507615.

Buick, I.S., Hermann, J., Williams, I.S., Gibson, R.L., and Rubatto, D., 2006, A SHRIMP U–Pb and LA-ICP-MS trace element study of the petrogenesis of garnet–cordierite–orthoamphibole gneisses from the Central Zone of the Limpopo Belt, South Africa: Lithos, v. 88, p. 150–172, doi:10.1016/j.lithos.2005.09.001.

Carson, C.J., Powell, R., and Clark, G.L., 1999, Calculated mineral equilibriafor eclogites in the $CaO–Na_2O–FeO–MgO–Al_2O_3–SiO_2–H_2O$ system: Application to the Pouébo Terrane, Pam Penninsula, New Caledonia: Journal of Metamorphic Geology, v. 17, p. 9–24, doi:10.1046/j.1525 -1314.1999.00177.x.

Connolly, J.A.D., 2005, Computation of phase equilibria by linear programming: A tool for geodynamic modeling and its application to subduction zone decarbonation: Earth and Planetary Science Letters, v. 236, p. 524–541, doi:10.1016/j.epsl.2005.04.033.

Davidson, A., 1984, Tectonic boundaries within the Grenville Province of the Canadian Shield: Journal of Geodynamics, v. 1, p. 433–444, doi:10.1016/0264-3707(84)90018-8.

Ellis, D.J., 1987, Origin and evolution of granulites in normal and thickened crust: Geology, v. 15, p. 167–170, doi:10.1130/0091-7613(1987)15<167: OAEOGI>2.0.CO;2.

Frei, R., Villa, I.M., Nägler, Th.F., Kramers, J.D., Przybylowicz, W.J., Prozesky, V.M., Hofmann, B.A., and Kamber, B.S., 1997, Single mineral dating by the Pb-Pb step-leaching method: Assessing the mechanisms: Geochimica et Cosmochimica Acta, v. 61, p. 393–414, doi:10.1016/S0016-7037(96)00343-2.

Frost, B.R., and Chacko, T., 1989, The granulite uncertainty principle: Limitations on thermobarometry in granulites: Journal of Geology, v. 97, p. 435–450, doi:10.1086/629321.

Gerya, T.V., 1991, Mass-balance based method of physicochemical simulation of metamorphic reactions: Contributions to Physicochemical Petrology, v. 16, p. 112–127.

Gerya, T.V., and Maresch, W.V., 2004, Metapelites of the Kanskiy granulite complex (eastern Siberia): Kinked P-T paths and geodynamic model: Journal of Petrology, v. 45, p. 1393–1412, doi:10.1093/petrology/egh017.

Gerya, T.V., and Perchuk, L.L., 1990, GEOPATH: A new computer program for geothermobarometry and related calculations with the IBM PC computer: Beijing, IMA, General Meeting, 15th, Abstracts, v. 2, p. 1010.

Gerya, T.V., Perchuk, L.L., van Reenen, D.D., and Smit, C.A., 2000, Two-dimensional numerical modelling of pressure-temperature-time paths for the exhumation of some granulite facies terrains in the Precambrian: Journal of Geodynamics, v. 30, p. 17–35, doi:10.1016/S0264-3707(99)00025-3.

Gerya, T.V., Perchuk, L.L., Maresch, W.V., and Willner, A.P., 2004, Inherent gravitational instability of hot continental crust: Implications for doming and diapirism in granulite facies terrains, in Whitney, D.L., Teyssier, C., and Siddoway, C.S., eds., Gneiss Domes in Orogeny: Geological Society of America Special Paper 380, p. 97–115.

Gibson, R.L., and Reimold, W.U., 2001, The Vredefort Impact Structure, South Africa: The Scientific Evidence and Two-Day Excursion Guide: Council of Geosciences of South Africa Memoir 92, 111 p.

Glensdorf, P., and Prigogine, I., 1973, Thermodynamic Theory of Structure, Stability, and Fluctuations: Moscow, Mir, 279 p. [in Russian].

Harley, S.L., 1989, The origin of granulites: A metamorphic perspective: Geological Magazine, v. 126, p. 215–231, doi:10.1017/S0016756800022330.

Hermann, J., and Rubatto, D., 2003, Relating zircon and monazite domains to garnet growth zones: Age and duration of granulite facies metamorphism in the Val Malenco lower crust: Journal of Metamorphic Geology, v. 21, p. 833–852.

Hisada, K., Perchuk, L.L., Gerya, T.V., van Reenen, D.D., and Paya, B.K., 2005, P–T–fluid evolution in the Mahalapye Complex, Limpopo high-grade terrane, eastern Botswana: Journal of Metamorphic Geology, v. 23, p. 313–334, doi:10.1111/j.1525-1314.2005.00579.x.

Holland, T.J.B., and Powell, R., 1998, An internally consistent thermodynamic data set for phases of petrological interest: Journal of Metamorphic Geology, v. 16, p. 309–343, doi:10.1111/j.1525-1314.1998.00140.x.

Huizenga, J.-M., Perchuk, L.L., van Reenen, D.D., Flattery, Y., Varlamov, D.A., Smit, C.A., and Gerya, T.V., 2011, this volume, Granite emplacement and the retrograde P-T-fluid evolution of Neoarchean granulites from the Central Zone of the Limpopo Complex, in van Reenen, D.D., Kramers, J.D., McCourt, S., and Perchuk, L.L., eds., Origin and Evolution of Precambrian High-Grade Gneiss Terranes, with Special Emphasis on the Limpopo Complex of Southern Africa: Geological Society of America Memoir 207, doi:10.1130/2011.1207(08).

Ivanov, V.V., Klimov, V.I., and Chernikova, T.M., 1997, The Hurst statistics of the time flow of structural damage of composites as a measure of the evolution of a fracture spot: Journal of Applied Mechanics and Technical Physics, v. 38, p. 136–139, doi:10.1007/BF02468282.

Karpov, I.K., Kiselev, A.I., and Letnikov, F.A., 1971, Chemical Thermodynamics in Petrology and Geochemistry: Irkutsk, Irkutsk Publication, 256 p. [in Russian].

Korzhinskii, D.S., 1959, Physicochemical Basis of the Analysis of the Paragenesis of Minerals: New York, Consulting Bureau, 142 p.

Korzhinskii, D.S., 1970, Theory of Metasomatic Zoning: Oxford, UK, Clarendon Press, 162 p.

Kreissig, K., Nögler, T.F., Kramers, J.D., van Reenen, D.D., and Smit, C.A., 2000, An isotopic and geochemical study of the northern Kaapvaal craton and the Southern Marginal Zone of the Limpopo Belt: Are they juxtaposed terranes?: Lithos, v. 50, p. 1–25, doi:10.1016/S0024 -4937(99)00037-7.

Kreissig, K., Holzer, L., Frei, R., Villa, I.M., Kramers, J.D., Kröner, A., Smit, C.A., and van Reenen, D.D., 2001, Geochronology of the Hout River shear zone and metamorphism in the Southern Marginal Zone of the Limpopo Belt, South Africa: Precambrian Research, v. 109, p. 145–173, doi:10.1016/S0301-9268(01)00147-4.

Kröner, A., Jaeckel, P., Brandl, G., Nemchin, A.A., and Pidgeon, R.T., 1999, Single zircon ages for granitoid gneisses in the Central Zone of the Limpopo Belt, southern Africa, and geodynamic significance: Precambrian Research, v. 93, p. 299–337, doi:10.1016/S0301-9268(98)00102-8.

Perchuk, L.L., 1969, The effect of temperature and pressure on the equilibria of natural Fe-Mg minerals: International Geology Review, v. 11, p. 875–901, doi:10.1080/00206816909475127.

Perchuk, L.L., 1971, Crystallochemical problems in the theory of phase correspondence: Geokhimiya, v. 1, p. 23–38.

Perchuk, L.L., 1973, Thermodynamic Regime of Petrogenesis: Moscow, Nauka Press, 316 p.

Perchuk, L.L., 1976, Gas-mineral equilibria and possible geochemical model of the Earth's interior: Physics of the Earth and Planetary Interiors, v. 13, p. 232–239, doi:10.1016/0031-9201(76)90097-2.

Perchuk, L.L., 1977, Thermodynamic control of metamorphic processes, in Saxena, S.K., and Bhattacharji, S., eds., Energetic Geological Processes: New York, Springer-Verlag, p. 285–352.

Perchuk, L.L., 1986, The course of metamorphism: International Geology Review, v. 28, p. 1377–1400, doi:10.1080/00206818609466374.

Perchuk, L.L., 1989, P-T-fluid regimes of metamorphism and related magmatism with specific reference to the Baikal Lake granulites, in Daly, S., Yardley, D.W.D., and Cliff, B., eds., Evolution of Metamorphic Belts: Geological Society [London] Special Publication 2, p. 275–291.

Perchuk, L.L., 1991, Derivation of thermodynamically consistent system of geothermometers and geobarometers for metamorphic and magmatic rocks, in Perchuk, L.L., ed., Progress in Metamorphic and Magmatic Petrology: Cambridge, UK, Cambridge University Press, p. 93–112.

Perchuk, L.L., 2005, Configuration of P-T trends as a record of high-temperature polymetamorphism: Doklady Russian Academy of Sciences, Earth Science, v. 401, p. 311–314.

Perchuk, L.L., and Gerya, T.V., 2011, this volume, Formation and evolution of Precambrian granulite terranes: A gravitational redistribution model, in van Reenen, D.D., Kramers, J.D., McCourt, S., and Perchuk, L.L., eds., Origin and Evolution of Precambrian High-Grade Gneiss Terranes, with Special Emphasis on the Limpopo Complex of Southern Africa: Geological Society of America Memoir 207, doi:10.1130/2011.1207(15).

Perchuk, L.L., and Lavrent'eva, I.V., 1983, Experimental investigation of exchange equilibria in the system cordierite-garnet-biotite, in Saxena, S.K., ed., Kinetics and Equilibrium in Mineral Reactions: Berlin, Springer, Advances in Physical Geochemistry, v. 3, p. 199–239.

Perchuk, L.L., and Lavrent'eva, I.V., 1990, Experimental study of mineral equilibria in the system garnet-orthopyroxene-amphibole: International Geology Review, v. 32, p. 486–507, doi:10.1080/00206819009465793.

Perchuk, L.L., and Ryabchikov, I.D., 1968, Mineral equilibria in the system nepheline-alkali feldspar-plagioclase and their petrological significance: Journal of Petrology, v. 9, p. 123–167.

Perchuk, L.L., and Ryabchikov, I.D., 1976, Phase Correspondence in the Mineral Systems: Moscow, Nedra Press, 302 p.

Perchuk, L.L., and van Reenen, D.D., 2008, P-T record of two high-grade metamorphic events in the Central Zone of the Limpopo Complex, South Africa: Lithos, v. 106, p. 403–410, doi:10.1016/j.lithos.2008.07.011.

Perchuk, L.L., and Zverovich, E.I., 1962, The derivation of formulae for calculation of univariant reactions for multicomponent systems using simple computers: Geokhimia, v. 1, p. 82–89.

Perchuk, L.L., Lavrent'eva, I.V., Aranovich, L.Y., and Podlesskii, K.K., 1983, Biotite-Garnet-Cordierite Equilibria and Evolution of Metamorphism: Moscow, Nedra Press, 200 p.

Perchuk, L.L., Aranovich, L.Ya., Podlesskii, K.K., Lavrent'eva, I.V., Gerasinov, B.Yu., Fed'kin, V.V., Kitsal, V.I., Karsalrov, L.P., and Berdnikov, N.V., 1985, Precambrian granulites of the Aldan shield, eastern Siberia, USSR: Journal of Metamorphic Geology, v. 3, p. 265–310, doi:10.1111/j.1525-1314.1985.tb00321.x.

Perchuk, L.L., Gerya, T.V., and Nozhkin, A.D., 1989, Petrology and retrogression in granulites of the Kanskiy Formation, Yenisey Range, Eastern Siberia: Journal of Metamorphic Geology, v. 7, p. 599–617, doi:10.1111/j.1525-1314.1989.tb00621.x.

Perchuk, L.L., Gerya, T.V., van Reenen, D.D., Safonov, O.G., and Smit, C.A., 1996, The Limpopo metamorphic complex, South Africa: 2. Decompression/cooling regimes of granulites and adjusted rocks of the Kaapvaal craton: Petrology, v. 4, p. 571–599.

Perchuk, L.L., Gerya, T.V., van Reenen, D.D., and Krotov, A.V., Safonov, O.G., Smit, C.A., and Shur, M. Yu., 2000a, Comparative petrology and metamorphic evolution of the Limpopo (South Africa) and Lapland (Fennoscandia) high-grade terrains: Mineralogy and Petrology, v. 69, p. 69–107.

Perchuk, L.L., Gerya, T.V., van Reenen, D.D., Smit, C.A., and Krotov, A.V., 2000b, P-T paths and tectonic evolution of shear zones separating high-grade terrains from cratons: Examples from Kola Peninsula (Russia) and Limpopo Region (South Africa): Mineralogy and Petrology, v. 69, p. 109–142, doi:10.1007/s007100050020.

Perchuk, L.L., Gerya, T.V., van Reenen, D.D., and Smit, C.A., 2001, Formation and dynamics of granulite complexes within cratons: Gondwana Research, v. 4, p. 729–732, doi:10.1016/S1342-937X(05)70524-4.

Perchuk, L.L., Tokarev, D.A., van Reenen, D.D., Varlamov, D.A., Gerya, T.V., Sazonova, L.V., Fan, V.I., Smit, C.A., Brink, M.C., and Bisschoff, A.A., 2002, The dynamic and thermal history of the explosion structure Vredefort in the Kaapvaal Craton, South Africa: Petrology, v. 10, p. 395–432.

Perchuk, L.L., Sazonova, L.V., van Reenen, D.D., and Gerya, T.V., 2003, Ultramylonites and their significance for the understanding of the history of the Vredefort impact structure, South Africa: Petrology, v. 11, p. 114–129.

Perchuk, L.L., Gerya, T.V., van Reenen, D.D., and Smit, C.A., 2006, P-T paths and problems of high-temperature polymetamorphism: Petrology, v. 14, p. 117–153, doi:10.1134/S0869591106020019.

Perchuk, L.L., van Reenen, D.D., Varlamov, D.A., van Kal, S.M., Boshoff, R., and Tabatabaeimanesh, S.M., 2008, P-T record of two high-grade metamorphic events in the Central Zone of the Limpopo Complex, South Africa: Lithos, v. 103, p. 70–105, doi:10.1016/j.lithos.2007.09.011.

Powell, R., and Holland, T.J.B., 2008, On thermobarometry: Journal of Metamorphic Geology, v. 26, p. 155–179, doi:10.1111/j.1525-1314.2007.00756.x.

Ramberg, H., 1952, Chemical bonds and distribution of cations in silicates: Journal of Geology, v. 60, p. 48–65.

Rubatto, D., 2002, Zircon trace element geochemistry: Partitioning with garnet and the link between U–Pb ages and metamorphism: Chemical Geology, v. 184, p. 123–138, doi:10.1016/S0009-2541(01)00355-2.

Rubatto, D., and Hermann, J., 2007, Experimental zircon/melt and zircon/garnet trace element partitioning and implications for the geochronology of crustal rocks: Chemical Geology, v. 241, p. 38–61, doi:10.1016/j.chemgeo.2007.01.027.

Sandiford, M., and Powell, R., 1986, Deep crustal metamorphism during continental extension: Modern and ancient examples: Earth and Planetary Science Letters, v. 79, p. 151–158, doi:10.1016/0012-821X(86)90048-8.

Saxena, S.K., 1973, Crystalline Solutions: Berlin, Springer-Verlag, 188 p.

Smit, C.A., and van Reenen, D.D., 1997, Deep crustal shear zones, high-grade tectonites and associated alteration in the Limpopo belt, South Africa: Implication for deep crustal processes: Journal of Geology, v. 105, p. 37–57, doi:10.1086/606146.

Smit, C.A., van Reenen, D.D., Gerya, T.V., Varlamov, D.A., and Fed'kin, A.V., 2000, Structural–metamorphic evolution of the Southern Yenisey Range of Eastern Siberia: Implications for the emplacement of the Kanskiy granulite complex: Mineralogy and Petrology, v. 69, p. 35–67, doi:10.1007/s007100050018.

Smit, C.A., van Reenen, D.D., Gerya, T.V., and Perchuk, L.L., 2001, P-T conditions of decompression of the Limpopo high grade terrane: Record from shear zones: Journal of Metamorphic Geology, v. 19, p. 249–268, doi:10.1046/j.0263-4929.2000.00310.x.

Sobolev, V.S., 1948, Energy of crystalline lattice and iron distribution laws in mineralogy: Mineralogicheskiy Sbornik, v. 2, p. 25–42.

Spear, F.S., 1993, Metamorphic phase equilibria and pressure–temperature–time paths: Washington, D.C., Mineralogical Society of America, 799 p.

Spear, F.S., and Florence, F.P., 1992, Thermobarometry in granulites: Pitfalls and new approaches: Precambrian Research, v. 55, p. 209–241, doi:10.1016/0301-9268(92)90025-J.

Spear, F.S., and Selverstone, J., 1983, Quantitative P-T paths from zoned minerals: Theory and tectonic applications: Contributions to Mineralogy and Petrology, v. 83, p. 348–357, doi:10.1007/BF00371203.

Thompson, J.B., Jr., 1959, Local equilibrium in metasomatic processes, *in* Abelson, P.H., ed., Researches in Geochemistry: New York, John Wiley, p. 427–457.

Tomilenko, A.A., and Chupin, V.P., 1983, Thermometry, Barometry and Fluid Chemistry of Metamorphic Complexes: Novosibirsk, Nauka Press, 200 p.

van Reenen, D.D., and Smit, C.A., 1996, The Limpopo Metamorphic Belt, South Africa: 1. Geological setting and relationship of the Granulite Complex with the Kaapvaal and Zimbabwe Cratons: Petrology, v. 4, p. 619–618.

van Reenen, D.D., Roering, C., Ashwal, L.D., and de Wit, M.J., eds., 1992, The Archean Limpopo granulite belt: Tectonics and deep crustal processes: Precambrian Research, Special Issue, v. 55, p. 209–241.

van Reenen, D.D., Perchuk, L.L., Smit, C.A., Boshoff, R., Varlamov, D.A., Huizenga, J.M., and Gerya, T.V., 2004, Structural and P-T evolution of a major cross fold in the Central Zone of the Limpopo high-grade terrain, South Africa: Journal of Petrology, v. 45, p. 1413–1439, doi:10.1093/petrology/egh028.

van Reenen, D.D., Boshoff, R., Smit, C.A., Perchuk, L.L., Kramers, J.D., McCourt, S.M., and Armstrong, R.A., 2008, Geochronological problems related to polymetamorphism in the Limpopo complex, South Africa: Gondwana Research, v. 14, p. 644–662, doi:10.1016/j.gr.2008.01.013.

van Reenen, D.D., Smit, C.A., Perchuk, L.L., Roering, C., and Boshoff, R., 2011, this volume, Thrust exhumation of the Neoarchean ultrahigh-temperature Southern Marginal Zone, Limpopo Complex: Convergence of decompression-cooling paths in the hanging wall and prograde P-T paths in the footwall, *in* van Reenen, D.D., Kramers, J.D., McCourt, S., and Perchuk, L.L., eds., Origin and Evolution of Precambrian High-Grade Gneiss Terranes, with Special Emphasis on the Limpopo Complex of Southern Africa: Geological Society of America Memoir 207, doi:10.1130/2011.1207(11).

Vernon, R.H., White, R.W., and Clarke, G.L., 2008, False metamorphic events inferred from misinterpretation of microstructural evidence and P–T data: Journal of Metamorphic Geology, v. 26, p. 437–449, doi:10.1111/j.1525-1314.2008.00762.x.

Wei, C.J.J., Powell, R., and Zhang, L.F., 2003, Eclogites from the South Tianshan, NW China: Petrological characteristics and calculated mineral equilibria in the $Na_2O-CaO-FeO-MgO-Al_2O_3-SiO_2-H_2O$ system: Journal of Metamorphic Geology, v. 21, p. 163–179, doi:10.1046/j.1525-1314.2003.00435.x.

Wood, B.J., and Fraser, D.G., 1977, Elementary Thermodynamics for Geologists: Oxford, UK, Oxford University Press, 303 p.

Yardley, B.W.D., 1989, An Introduction to Metamorphic Petrology: Essex, UK, Longman Science and Technical, 264 p.

Zeh, A., and Klemd, R., 2008, P-T record of two high-grade metamorphic events in the Central Zone of the Limpopo Complex, South Africa: Comments: Lithos, v. 106, p. 399–402, doi:10.1016/j.lithos.2008.07.008.

Zyrianov, V.N., Perchuk, L.L., and Podlesskii, K.K., 1978, Nepheline-alkali feldspar equilibria: 1. Experimental data and thermodynamic calculations: Journal of Petrology, v. 19, p. 1–44.

The geochronology of the Limpopo Complex: A controversy solved

Jan D. Kramers*
Geology Department, University of Johannesburg, P.O. Box 524, Auckland Park, 2006, Johannesburg, South Africa, and School of Geosciences, University of the Witwatersrand, Private Bag 3, Wits 2050, South Africa

Hassina Mouri
Geology Department, University of Johannesburg, P.O. Box 524, Auckland Park, 2006, Johannesburg, South Africa

ABSTRACT

The results of 50 years of geochronological work in the Limpopo Complex are reviewed. The data define three main age clusters. The oldest, at ca. 3.3 Ga, exists in the Central and Southern Marginal Zones and is defined by magmatic zircon dates. The second, with a genuine spread between 2.7 and 2.55 Ga, occurs in all three zones. It was a period of high-grade regional metamorphism with intense deformation and widespread anatexis, dated also mainly (but not exclusively) by zircon U-Pb. The third cluster is well constrained at 2.02 ± 0.02 Ga in the Central Zone by zircon overgrowths, sparse magmatic zircons, monazite, apatite, Sm-Nd and Lu-Hf garnet dating, Pb/Pb discrete phase and stepwise leaching dating of garnet and titanite, and hornblende Ar/Ar dating. The Paleoproterozoic dates from metamorphic minerals are particularly associated with zones of intense transcurrent shearing at high-grade metamorphism. In the Northern Marginal Zone this event is more protracted, from 2.08 to 1.94 Ga, and defined in medium- to low-grade shear zones. In the Southern Marginal Zone it is absent. The evidence for both Neoarchean and Paleoproterozoic mineral ages, both defining high-grade tectono-metamorphic events, is in part paradoxical and has led to controversies as to the age of a proposed collisional orogeny. Studying the mineral dates in their tectonic context leads to the conclusion that fluid access in deformation, rather than mere reheating, mainly caused their partial resetting in the Paleoproterozoic event. This allows the controversy to be resolved.

INTRODUCTION

From many papers and reviews of the geochronology of the Limpopo Complex it is apparent that in the Central Zone (CZ) at least three clusters of dates emerge, one at ca. 3.3–3.2 Ga, one from 2.7 to 2.55 Ga, and one at 2.0 Ga (e.g., Eglington and Armstrong, 2004). In the Northern Marginal Zone (NMZ) the last two are recorded, and in the Southern Marginal Zone (SMZ), only the middle one. Some studies have resolved further events, at 3.1 Ga, 2.5 Ga (Mouri et al., 2009), and 2.2 Ga (Kamber et al., 1995a). The 3.3–3.2 Ga period is exclusively, and the 2.7–2.55 Ga period chiefly, defined by zircon U-Pb and Rb-Sr whole-rock dates. Many of the dated rocks of the second period are, however, anatectic granitoids with deformed xenoliths, and this period also includes some metamorphic mineral dates. The 2.0 episode was

---
*jkramers@uj.ac.za

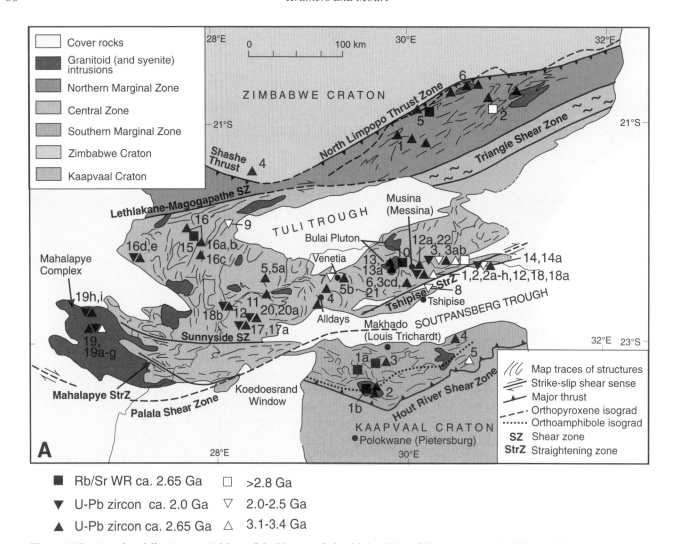

Figure 1 (*Continued on following page*). Maps of the Limpopo Belt with localities of dated samples. (A) Zircon U-Pb and whole-rock Rb-Sr dates. (B) Dates for metamorphic minerals. Numbers next to symbols refer to Figures 2 and 3 and their captions.

mainly detected by various dating methods for metamorphic and fabric-forming minerals, with a minor contribution from zircon U-Pb dates on anatectic melt patches. The 2.7–2.55 and 2.0 Ga age clusters therefore both indicate syntectonic metamorphism, and this has led to much controversy on tectonic models for the Limpopo Belt and the timing of the modeled events.

In this paper we review the geochronological work in the Limpopo Complex to date, and the interpretations as they have, along with more general ideas on the meaning of age data, evolved over time. By studying and comparing the results obtained with different dating techniques, or the same techniques in different tectonic settings in this polymetamorphic province, we can interpret them in the context of magmatism, metamorphic reheating-cooling, deformation, fluid access, and recrystallization. It can be demonstrated that the apparently conflicting aspects of the geochronological data in the Limpopo Complex can be fully reconciled in the framework of two major tectono-metamorphic events at ca. 2.65 Ga and at 2.0 Ga.

Geochronological Research in the Limpopo Complex

The age determination results obtained in the Limpopo Complex to date are summarized in Figures 1–3. In Figure 1 a broad overview of sample localities is given, with symbols indicating methods used and age ranges. Figure 2 shows zircon U-Pb and Rb-Sr whole-rock dates, which are normally (but not exclusively) interpreted as magmatic intrusion ages, whereas Figure 3 shows dates obtained from metamorphic and fabric-forming minerals, and a few thin-slice Rb-Sr dates. A table of the data shown in these figures can be obtained from the GSA Data Repository[1].

---

[1]GSA Data Repository Item 2011042, Table DR1: Geochronological data from the Limpopo Complex, is available at www.geosociety.org/pubs/ft2011.htm, or on request from editing@geosociety.org, Documents Secretary, GSA, P.O. Box 9140, Boulder, CO 80301-9140, USA.

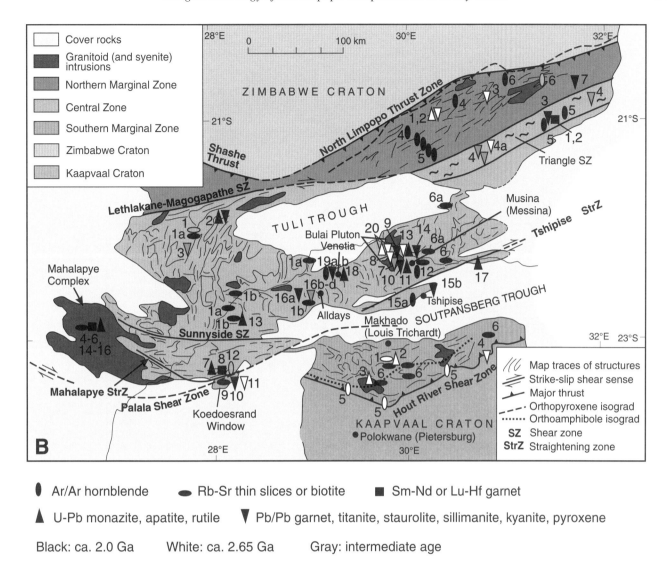

● Ar/Ar hornblende ● Rb-Sr thin slices or biotite ■ Sm-Nd or Lu-Hf garnet
▲ U-Pb monazite, apatite, rutile ▼ Pb/Pb garnet, titanite, staurolite, sillimanite, kyanite, pyroxene

Black: ca. 2.0 Ga   White: ca. 2.65 Ga   Gray: intermediate age

Figure 1 (*Continued*).

Early Work

The first radiometric age from the Limpopo Complex was a ca. 2.0 Ga U-Th-Pb date from a pegmatitic titanite in the Beitbridge area (Holmes and Cahen, 1957). Further dates close to this age were obtained by Van Breemen et al. (1966) by the same method, as well as Rb-Sr from biotites. Van Breemen and Dodson (1972) confirmed the 2.0 Ga date for a larger number of biotite–whole-rock and biotite-feldspar pairs (1a: 4 in Fig. 3B, 6 in Fig. 3C) but also generated whole-rock Rb-Sr dates of 2.69 ± 0.06 Ga from a combination of samples from the Bulai and Singelele Gneisses near Musina (10 in Fig. 2B). Following the field evidence of Bahnemann (1972), who had shown that the Singelele Gneisses were emplaced syntectonically, they proposed that the older age must date a high-grade tectono-metamorphic event, the "Limpopo Orogeny" as defined by MacGregor (1953). Further Rb-Sr whole-rock studies affirmed the Neoarchean tectono-metamorphic activity and its widespread nature. Barton et al. (1979b) found the same date for the Bulai granitoid gneiss on its own (10a in Fig. 2B). In the NMZ an enderbite body at Bangala Dam yielded an age of 2.88 ± 0.05 Ga (Hickman, 1978; 2 in Fig. 2A), and in the SMZ the syntectonic Matok granitoid gave 2.62 ± 0.08 Ga (1b in Fig. 2C), whereas a thin-slice study of metapelites in the Bandelierkop Quarry (1 in Fig. 3D) yielded 2.59 ± 0.07 Ga (Barton et al., 1983).

For the mineral ages clustering about 2.0 Ga, Van Breemen and Dodson (1972) concluded, on the basis of their sharp isochronism over a large area, that these dates indicated a separate thermal event rather than late uplift and cooling after the Neoarchean, as could have been suggested by the thermochronological approach (Armstrong et al., 1966; Jäger, 1967; Harper, 1967). In a regional study following this work, Barton and van Reenen (1992b) reported biotite–whole-rock Rb-Sr dates from the CZ (1b in Fig. 3B, 6a in Fig. 3C), the SMZ (6 in Fig. 3D), and the northern Kaapvaal Craton. In the CZ, these dates range from 1.95 to 2.02 Ga, and in the

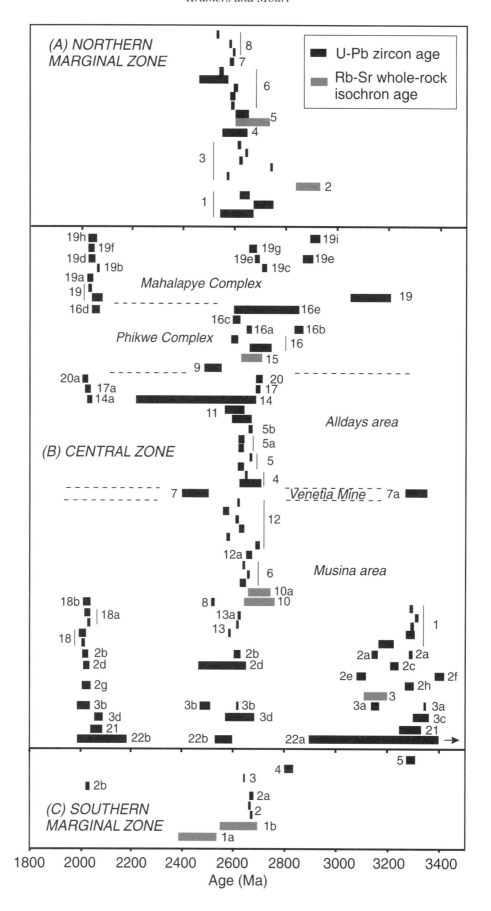

Figure 2. Compilation of dates relating to rock formation for samples from the Limpopo Belt. (A) *Northern Marginal Zone:* (1) Enderbites and charnockites, Mwenezi River section, Berger et al. (1995), multigrain ID-TIMS discordia ages. (2) Bangala Dam, Rb-Sr whole-rock (WR) age, Hickman (1978). (3) Single zircon evaporation ages, Renco Mine area, Blenkinsop et al. (2004). (4) Shashe Granite, W. Zimbabwe, multigrain ID-TIMS, Holzer et al. (1999). (5) Rb-Sr in suite of charnockites and Razi Granite, S. Buhwa, combined, and multigrain ID-TIMS zircon U-Pb on microgranite crosscutting North Limpopo thrust, Mkweli et al. (1995). (6) Razi Granite, multigrain ID-TIMS, Frei et al. (1999). (7) Samba Granite, multigrain ID-TIMS, Kamber et al. (1996). (8) Two syntectonic granites and a post-tectonic pegmatite near Renco Mine, SIMS, Blenkinsop et al. (2004). (B) *Central Zone:* (1) Sand River Gneiss, zircon SIMS and evaporation dates, Kröner et al. (1999). (2a) Sand River Gneiss, zircon cores. (2b) Sand River Gneiss, zircon rims, LA-ICP-MS, Zeh et al. (2007). (2c) Sand River Gneiss, 120 grains from 4 samples, Causeway locality near Musina. (2d) Ranges of overgrowths in same sample. (2e) Granitic leucosome, Causeway locality. (2f) Xenocrysts in above. (2g) Tonalitic leucosome, Causeway locality. (2h) Xenocrysts in above. (2c–2h) Zeh et al. (2010), LA-ICP-MS. (3) Messina Layered Intrusion, WR Rb-Sr, Barton et al. (1979a). (3a) Messina Layered Intrusion, anorthosite, zircon cores. (3b) Zircon rims, Mouri et al. (2009), SIMS. (3c) Messina Layered Intrusion, anorthosite, 11 grains from 2 samples, intrusion age. (3d) Ranges of overgrowths in same sample, Zeh et al. (2010), LA-ICP-MS. (4) Alldays Gneiss. (5) Zanzibar Gneiss, Kröner et al. (1999). (5a) Zanzibar Gneiss. (5b) Regina Gneiss, Zeh et al. (2007), LA-ICP-MS. (6) Verbaard Gneiss, Kröner et al. (1999) and Jaeckel et al. (1997), ID-TIMS, evaporation and SIMS. (7) Granite at Venetia Mine, SIMS, Barton et al. (2003). (7a) Amphibolite in Krone Metamorphic Terrane, near Venetia, SIMS, Chudy et al. (2008). (8) Tshipise Gneiss, SIMS, Kröner et al. (1999). (9) Swejane Granite, Phikwe Complex, Botswana, multigrain ID-TIMS, Holzer et al. (1999). (10) Rb-Sr WR on suite of Singelele (Malala Drift) and Bulai gneisses combined, Van Breemen and Dodson (1972): (10a) Rb-Sr WR, Bulai granite, Barton et al. (1979b). (11) Two sets of SIMS dates, Malala Drift quartzofeldspathic gneisses, Avoca Structure, Boshoff et al. (2006). (12) Six dates, Malala Drift quartz-feldspar (q-f) gneisses, SIMS, Jaeckel et al. (1997) and Kröner et al. (1999): (12a) Singelele (Malala Drift q-f) Gneiss, type locality near Musina, LA-ICP-MS, Zeh et al. (2007). (13) Enderbitic (older) and granitic (younger) Bulai Gneiss, single grain ID-TIMS, Barton et al. (1994): (13a) Granitic Bulai Gneiss, LA-ICP-MS, Zeh et al. (2007). (14) Range for detrital zircons in Metapsammite, Gumbu Group: (14a) Overgrowths, SIMS, Buick et al. (2003). (15, 16) Phikwe Complex, Botswana: (15) Anorthositic and granitoid gneisses, Hickman and Wakefield (1975). (16) Granites, Phikwe Complex, McCourt and Armstrong (1998), SIMS: (16a) Granites, Phikwe Town. (16b) Xenocryst in same. (16c) Granite, 10 km S of Phikwe. (16d) Granite, western Phikwe Complex. (16e) Range of xenocrysts in same. (16a–16e) Zeh et al. (2009), LA-ICP-MS. (17) Zircon cores, garnet-cordierite-orthoamphibole gneiss, Gumbu Group, 190 km W of Musina, SIMS, Buick et al. (2003): (17a) Zircon overgrowths, same sample. (18) Leucocratic melt patches: (18a) Metapelites, Causeway locality near Musina, Jaeckel et al. (1997). (18b) Anatectic granite, evaporation + SIMS, Kröner et al. (1999). (19) Mahalapye Complex, Botswana, granite (with xenocryst zircon cores) and migmatite, SIMS, McCourt and Armstrong (1998): (19a) Mahalapye Complex, Botswana, Mokgware Granite. (19b) Mahalapye Complex, Botswana, biotite gneiss, Lose Quarry. (19c) Mahalapye Complex, Botswana, leucosome, Lose Quarry, Zeh et al. (2007), LA-ICP-MS. (19d) Lose Quarry granodiorite, 2 concordant grains, Millonig et al. (2010). (19e) Xenocrysts in this rock. (19f) Mahalapye Granite (4 concordant grains). (19g) Xenocrysts in this rock. (19d–19g) Millonig et al. (2010), LA-ICP-MS. (19h) Mokgware Granite. (19i) Xenocryst in same, Zeh et al. (2009), LA-ICP-MS. (20, 20a) Magmatic cores and metamorphic rims in mafic granulite, near Swartwater, SIMS, Mouri et al. (2008). (21) Quartzite 6 km S of Causeway locality near Musina; upper and lower concordia intercepts of 5 single zircon ID-TIMS analyses, Barton and Sergeev (1997). (22a, 22b) Quartzites 2 km N and 6 km S of Causeway locality, LA-ICP-MS dates, Zeh et al. (2008, 2010): (22a) Range of detrital zircon dates (>300 grains, extends to ca. 3.9 Ga; see also Armstrong et al., 1988). (22b) Zircon rim dates. (C) *Southern Marginal Zone:* (1) Granitoids and gneisses, WR Rb-Sr, Barton et al. (1983): (1a) Palmietfontein Granite (post-tectonic). (1b) Mafic and porphyritic facies of Matok intrusion (syntectonic). (2) Matok Complex, porphyritic granite; charnoenderbite (older) and granite (younger) of Matok Pluton, ID-TIMS, Barton et al. (1992): (2a) Matok Pluton, granite. (2b) Entabeni Granite, postorogenic, Zeh et al. (2009), LA-ICP-MS. (3) Zircon evaporation (5 grains), Bandelierkop Quarry; leucosome or intrusive vein, Kreissig et al. (2001). (4) Granodioritic gneiss, granulite zone. (5) Tonalitic gneiss, orthoamphibolite zone; both N. Giyani Greenstone Belt; zircon ID-TIMS and evaporation, Kröner et al. (2000). ID-TIMS—isotope dilution thermal ionization mass spectrometry; LA-ICP-MS—laser-ablation inductively coupled plasma-mass spectrometry; SIMS—secondary ionization mass spectrometry.

SMZ from 1.99 to 2.06 Ga. In the Kaapvaal Craton this "blanket" ca. 2.0 Ga age for biotites persists as far south as the Murchison greenstone belt. In contrast to the preferred interpretation of Van Breemen and Dodson (1972), Barton and van Reenen (1992b) regard their results as reflecting a late uplift at ca. 2.0 Ga over a large area, rather than as a separate event.

In the westernmost part of the CZ at Phikwe (Botswana), Hickman and Wakefield (1975) found that widely spaced gneiss samples yielded an Rb-Sr whole-rock age of 2.66 ± 0.06 Ga (15 in Fig. 2B), whereas a suite of mm–spaced slices of a sheared leucocratic gneiss sample yielded an Rb-Sr date of 2.16 ± 0.14 Ga (95% confidence, recalculated using "Isoplot," Ludwig, 2000; 1 in Fig. 3B) and documenting incomplete resetting of Rb-Sr systematics at ca. 2.0 Ga at the centimeter scale. Unlike Van Breemen and Dodson (1972), these authors found evidence of deformation (open folding and shearing) at ca. 2.0 Ga, and could thus show that the younger event was not a static reheating episode. In the first Sm-Nd isotope study of garnets from metamorphic terranes, Van Breemen and Hawkesworth (1980) determined two garnet whole-rock two-point ages of 1.97 and 1.99 Ga (1 in Fig. 3C), identical within error, from rotated garnet porphyroclasts in the Triangle Shear Zone, between the CZ and the NMZ in the eastern part of the Limpopo Complex. They interpreted this result as dating the synmetamorphic shearing, thus confirming that the 2.0 Ga episode in the Limpopo Complex was a true tectono-metamorphic event.

Rb-Sr Whole-Rock and Zircon Dates

Following the early work, much emphasis was placed on the dating of intrusive magmatic rocks that apparently postdated a high-grade tectono-metamorphic event, or appeared syntectonic, and thus could serve as time markers for that event. For the NMZ (Fig. 2) these are shown in Figure 2A and include further Rb-Sr whole-rock (Mkweli et al., 1995, 5) and U-Pb TIMS multigrain zircon dates (Berger et al., 1995, 1; Frei et al.,

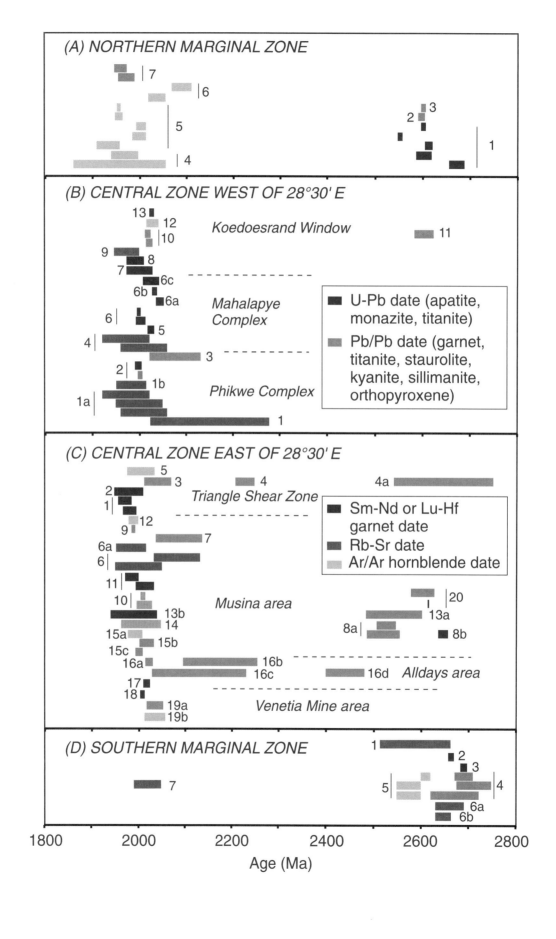

Figure 3. Compilation of dates relating to metamorphism in the Limpopo Belt. (A) *Northern Marginal Zone:* (1, 2) U-Pb (monazite), SIMS and Pb/Pb stepwise leaching (SL); orthopyroxene dates, granodiorite, Mundi River, Kamber et al. (1998). (3) Pb/Pb SL of garnet, near Renco Mine, Blenkinsop and Frei (1996). (4) N of Transition Zone, Bubi and Mwenezi Rivers, 2 laser fusion Ar/Ar isochrons, Kamber et al. (1996). (5) Transition Zone, 5 Ar/Ar stepheating dates, fabric-defining hornblende in amphibolite, Kamber et al. (1995b). (6) Transition Zone, 2 Ar/Ar stepheating dates, fabric-defining hornblende in amphibolite, Kamber et al. (1995a). (7) Pb/Pb SL titanite, retrograde shear zones N of Transition Zone, Kamber et al. (1996). (B) *Central Zone, West of 28°30′E:* (1) Phikwe Complex, Botswana, thin-slice date of Hickman and Wakefield (1975). (1a) Biotite dates, Phikwe and N Swartwater, Van Breemen and Dodson (1972). (1b) Range of 5 Rb-Sr biotite whole-rock (WR) dates from various lithologies and localities, Barton and van Reenen (1992b). (2) Phikwe Complex, Pb/Pb SL titanite and U-Pb apatite in Swejane granite, Holzer et al. (1999). (3) Pb/Pb SL garnet, Selebi orebody, Holzer et al. (1999). (4–6, 6a–6c) Mahalapye Complex, Botswana, migmatites: (4) Rb-Sr WR-feldspar dates, Van Breemen and Dodson (1972). (5) Sm-Nd garnet-leucosome-WR, Chavagnac et al. (2001). (6) U-Pb monazite, Chavagnac et al. (2001) and apatite, Holzer et al. (1999): (6a) Monazites from biotite-garnet gneiss and leucosome, LA-ICP-MS. (6b) Monazite and xenotime, metapelite, in situ LA-ICP-MS. (6c) Garnet Lu-Hf isochron from biotite-garnet gneiss. (6a–6c) Millonig et al. (2010). (7–12) Koedoesrand Window, near Lephalale (Ellisras): (7) Quartzofeldspathic gneiss, N Palala Shear Zone, U-Pb rutile, Holzer et al. (1998). (8) Migmatitic metapelite, Sm-Nd garnet-sill-biotite–WR, Chavagnac et al. (2001). (9) Palala Shear Zone, syntectonic biotite, Schaller et al. (1999). (10) Calc-silicate rocks, Pb/Pb SL titanite, Holzer et al. (1999), Schaller et al. (1999). (11) Boudin in Palala Shear Zone, Pb/Pb SL garnet (Schaller et al., 1997). (12) Two hornblendes, identical age, retrogressed charnockite and two pyroxene-amphibole gabbro at N edge of Palala Shear Zone and adjacent Central Zone (Belluso et al., 2000). (13) Monazite, garnet-cordierite-orthoamphibole gneiss, 190 km W of Musina, Buick et al. (2006). (C) *Central Zone, East of 28°30′E:* (1–5): Mylonites, Triangle Shear Zone, Zimbabwe: (1) Sm-Nd garnet-WR, Van Breemen and Hawkesworth (1980). (2–5), Kamber et al. (1995a): (2) Sm-Nd mineral isochron. (3) Pb/Pb garnet-feldspar–WR (3 samples). (4) Pb/Pb garnet-feldspar–WR (3 samples). (4a) Pb/Pb garnet-feldspar–WR (1 sample). (5) Ar/Ar in fabric-forming hornblende (2 samples). (6) Biotite-WR, Artonvilla Mine, Musina, Van Breemen and Dodson (1972); (6a) Range of 7 biotite-WR dates from various localities and lithologies, Barton and van Reenen (1992b). (7) Semipelitic enclave in Bulai Gneiss, Pb/Pb SL enstatite, Holzer (1998). (8a, 8b) Metapelitic xenolith in Bulai Gneiss: (8a) Pb/Pb SL on garnet and sillimanite, Holzer et al. (1998). (8b) U-Pb LA-ICP-MS in monazite, Millonig et al. (2008). (9) Pb/Pb SL titanite, retrograde reaction, Holzer (1998). (10, 11) Calc-silicate in migmatite in Ha-Tshanzi structure, Musina area, Holzer et al. (1998): (10) Pb/Pb SL on garnet and titanite (smaller error). (11) U-Pb, monazite and apatite (younger). (12) Sand River Gneiss, 2 identical Ar/Ar hornblende step-heating results, Holzer (1998). (13a) Metapelite garnet + sillimanite + cordierite, 5 km NW of Beitbridge, S. bank Limpopo, Pb/Pb SL garnet, Holzer (1998). (13b) Same sample, U-Pb monazite in matrix, Holzer et al. (1996). (14) Pb/Pb garnet-WR, Beit Bridge, Kamber et al. (1995a). (15a–15c) Tshipise Straightening Zone, Holzer (1998): (15a) Sheared orthogneiss, Christmas Farm 8 km W of Tshipise, Ar/Ar stepheating, hornblende, 2 samples identical age. (15b) Same locality, Pb/Pb SL titanite in synkinematic leucosome. (15c) 5 km NE of Tshipise, titanite, coarse-grained relicts in shear zone. (16a–16d) Baklykraal structure, western limb, Pb/Pb SL on garnets, Boshoff et al. (2006): (16a) Shear zone, monometamorphic, probably Gumbu Group. (16b, 16c) Metapelites with complex interference folding, polymetamorphic. (16d) Singelele-type gneiss. (17) Gumbu Group, calc-silicate rock U-Pb titanite, Buick et al. (2003). (18) Monazite, metapelite, Krone Metamorphic Terrane, Venetia area, LA-ICP-MS, Chudy et al. (2008). (19a) Two garnet Pb/Pb SL dates, metapelite, Venetia Mine, Rigby et al. (2011). (19b) Range of Ar/Ar dates on fabric-forming hornblende (3 samples) and muscovite (1 sample), Venetia Mine, Barton et al. (2003). (20) Monazite (SIMS) and garnet Pb/Pb SL dates, metapelite xenolith in Singelele Gneiss, N Bulai Pluton, van Reenen et al. (2008). (D) *Southern Marginal Zone:* (1) Granulitic metapelite, Bandelierkop Quarry, Rb-Sr on thin slices, Barton et al. (1983b). (2) Metapelite, Bandelierkop Quarry, U-Pb monazite. (3) Matok intrusion, metapelitic enclave, monazite. (4) Pb/Pb SL garnet, staurolite, and kyanite just south of Hout River Shear Zone, Khavagari Hills, Giyani Greenstone Belt. (5) Amphibolites in Hout River Shear Zone, Ar/Ar stepheating dates. (2–5) Kreissig et al. (2001). (6a, 6b) Muscovite Rb-Sr ages from pegmatites intruding shear zones in Hout River system, Barton and van Reenen (1992a): (6a) Range of 3 samples on N edge of Giyani Greenstone Belt. (6b) Sample from N edge of Renosterkoppies Greenstone Belt, N of Polokwane. (7) Range of 7 Rb-Sr biotite-WR dates from various lithologies and localities, Barton and van Reenen (1992b).

1995, 6; Holzer et al., 1999, 4; Mkweli et al., 1995, 5) on magmatic enderbites and charnockites. The apparent ages produced show a spread well outside analytical error, from ca. 2.52 to 2.72 Ga. Single zircon studies (Kamber et al., 1996, 7; Blenkinsop et al., 2004, 8) confirmed the main range. In the SMZ (Fig. 2C), magmatic emplacement ages ca. 2.65 Ga include zircon U-Pb dates for the syntectonic Matok intrusion (Barton et al., 1992; Zeh et al., 2009, 2, 2a) and a zircon Pb/Pb evaporation date on a leucocratic vein in the Bandelierkop Quarry (Kreissig et al., 2001, 3 in Fig. 3C). The 2.0 Ga Entabeni granite (Zeh et al., 2009, 2b) is wholly post-tectonic and appears to be related to the Schiel alkaline complex. Two sets of zircon U-Pb results from gneisses in the granulite and the orthoamphibole zone in the eastern SMZ gave 2.816 ± 0.016 and 3.28 ± 0.02 Ga ages, respectively (Kröner et al., 2000, 4, 5), representing so far the only reliable dates for the gneisses themselves.

In the CZ, many studies have generated a large amount of data (Fig. 2B), in which the few Rb-Sr whole-rock results again fall in the same main age range as the many zircon U-Pb dates. Here the cluster about 2.65 Ga is again dominant, with a spread from ca. 2.55 to ca. 2.72 Ga, but in addition there are data clusters about ca. 3.3, ca. 3.15, ca. 2.48, and ca. 2.0 Ga. The oldest two clusters from intracrustal rocks represent the Sand River Gneiss (Kröner et al., 1999, 1; Zeh et al., 2007, 2a; Zeh et al., 2010, 2c, 2e, 2f, 2h) the Messina Layered Suite (Barton et al., 1979a, 3; Barton, 1996; Mouri et al., 2009, 3a; Zeh et al., 2010, 3c), and an amphibolite in the Krone Metamorphic Terrane at the Venetia Mine (Chudy et al., 2008, 7a): These are all interpreted as dating magmatic emplacement. The old dates for the Messina Layered Suite represent to date the only minimum age for the metasediments of the Mount Dowe and Malala Drift Groups, which it intrudes (Bahnemann, 1972; Barton et al., 1979a). Note that in one case (3a), zircons 3.34 Ga and 3.15 Ga in age are present in the same anorthosite, whereby the younger group has been interpreted as having resulted from remelting or fluid activity in a high-grade metamorphic event (Mouri et al., 2009). Further zircon dates belonging to the older clusters are from xenocryst cores in the Mahalapye Complex, Botswana (McCourt and Armstrong, 1998, 19) and detrital zircon grains in quartzites of the Beit Bridge Complex (Barton and Sergeev, 1997, 21; Zeh et al., 2008, 2010; Armstrong et al., 1988, 22a). The two latter data sets include dominant populations at 3.15–3.3 and 3.4–3.6 Ga as well as several grains older than 3.6 Ga and back to 3.88 Ga. There is an unresolved contradiction between the 3.15–3.3 Ga detrital data set in the quartzites, and the oldest zircon core ages of 3.34 Ga (with U contents of ~50 ppm or lower, typical of mafic rocks, and therefore unlikely to be xenocrysts) in the Messina Layered Suite (see above), as this suite is considered to be intrusive in the quartzites.

The cluster about 2.65 Ga contains data from all over the CZ. It includes a whole rock Rb-Sr date and zircons from the Bulai Pluton (Barton et al., 1979b, 10a, 1994, 13; Zeh et al., 2007, 13a) and zircons from Singelele-type quartzofeldspathic gneisses (Jaeckel et al., 1997; Kröner et al., 1999, 12; Zeh et al., 2007, 12a; Boshoff et al., 2006, 11), as well as the "grey" Alldays, Zanzibar, Verbaard, and Regina Gneisses (Jaeckel et al., 1997; Kröner et al., 1999, 4, 5, 6; Zeh et al., 2007, 5a, 5b) and mafic granulites (Buick et al., 2006, 17; Mouri et al., 2008, 20). Zircon rims on older cores from the Sand River Gneisses (Zeh et al., 2007, 2010, 2b, 2d) and the Messina Layered Suite (Mouri et al., 2009; Zeh et al., 2010, 3b, 3d), and rare overgrowths from detrital zircons in quartzites (Zeh et al., 2008, 2010, 22b) further belong to the ca. 2.65 Ga cluster. In the western part of the CZ, zircon dates on granites of the Phikwe Complex (McCourt and Armstrong, 1998, 16; Zeh et al., 2010, 16a, 16c) complement the earlier Rb-Sr work of Hickman and Wakefield (1975). The granitoid magmas are, in most cases, intruded syntectonically (Brandl, 2002). Hf isotope data from zircons indicate that they were mainly derived from older crustal sources (Zeh et al., 2007, 2009, 2010). The dates thus document crustal anatexis in a high-grade tectonometamorphic event. In the Mahalapye Complex, magmatic zircon dates, interpreted as intrusion ages, cluster about 2.02 Ga (McCourt and Armstrong, 1998, 19; Zeh et al., 2007, 19a, 19b; Millong et al., 2010, 19d, 19f, 19h), but xenocrysts in the range 2.65–2.82 Ga occur (19c, 19g, 19e, 19i).

Four sets of zircon U-Pb dates between 2.45 and 2.52 Ga make up a set of overgrowths from the Messina Layered Suite (Mouri et al., 2009, 3b) and magmatic zircons from three widely separated gneisses and granitoids (Barton et al., 2003, 7; Kröner et al., 1999, 8; Holzer et al., 1999, 9). Given the considerable range of dates defining the ca. 2.65 Ga "event," it is not clear whether this younger cluster represents a mere tailing off of the latter, or a separate episode resulting in Pb loss or zircon regrowth. Even if the latter is true, the long duration of the ca. 2.65 Ga "event" is well documented by the rest of the data set. A possible reason for this phenomenon is discussed by Andreoli et al. (2011).

Apart from the intrusive ages in the Mahalapye Complex, the zircon U-Pb dates from the CZ clustering closely about 2.0 Ga (Fig. 2B) are from zircon rims with clearly metamorphic Th/U ratios (Buick et al., 2003, 14a, 17a; Zeh et al., 2007, 2008, 2010, 2b, 2d, 3d, 22b; Mouri et al., 2008, 2009, 3b), zircon grains in centimeter- to meter-size post-tectonic melt patches (Jaeckel et al., 1997, 18; Zeh et al., 2010, 2g) and apparently metamorphic zircons in metapelite (Jaeckel et al., 1997, 18a). Based on whole-rock Rb-Sr and zircon evidence alone, and excluding the evidence from the Mahalapye Complex, it could be argued that the ca. 2.0 Ga event in the CZ might have consisted of crustal reheating without related tectonism, leaving the dates about 2.65 Ga as time markers for a single major tectono-metamorphic event (i.e., orogeny) in the province. This view has indeed been held in a number of papers (Barton and van Reenen, 1992a; Roering et al., 1992; Treloar et al., 1992), reflecting thinking on the origin of the Limpopo Complex before many further dates on metamorphic minerals had been produced.

Many of the zircon dates from the Limpopo Complex (whether multigrain ID, SIMS, or LA-ICP-MS analyses) are less precise than zircon dates normally are. This is a consequence of extremely high U contents, particularly in quartzofeldspathic gneisses but also in the gray gneisses, which led to zircons being metamict. As a result, analytical data are frequently strongly discordant, and the apparent ages are in many cases concordia intercept dates. The lower intercepts on the concordia curves tend to reflect nonzero ages, mostly between 300 and 800 Ma. Such nonzero intercepts were in the early days of geochronology ascribed to continuous diffusion (e.g., Tilton, 1960; Wasserburg, 1963), but these models were invalidated by diffusion measurements (see Cherniak and Watson, 2001), giving rise to speculations about later, unspecified heating or hydrothermal activity, possibly associated with the widespread ca. 0.5 to ca. 0.6 Ga Pan-African orogenic events. This lower intercept phenomenon is, however, probably a side effect of the mode of radiogenic Pb loss from partly metamict zircons in deep weathering, as argued by Kramers et al. (2009) and discussed below. Note, however, that the large spread in ages that reflect the ca. 2.65 Ga event is not an artifact of these discordances: The spread is well outside error limits.

Dates from Apatite, Rutile, Monazite, and Metamorphic Silicates

In geochronological studies relating to metamorphism and tectonism, it is obviously preferable to obtain dates from major rock-forming minerals that can be characterized in terms of metamorphic reactions as well as deformation. Dates obtained in the Limpopo Complex from metamorphic minerals, after the early Rb-Sr biotite studies mentioned above, include Sm-Nd and Lu-Hf in garnet, Ar/Ar in hornblende, and Pb/Pb dates from garnet whole-rock pairs as well as those obtained by stepwise leaching (SL) (Frei and Kamber, 1995; Frei et al., 1997) mainly from garnet and titanite, but including also staurolite, kyanite, sillimanite, and orthopyroxene analyses (Fig. 3). Further, U-Pb dates on apatite, monazite, and rutile may indicate either the age of magmatic emplacement or metamorphism. The robustness of all these mineral chronometers is not in all cases well defined. The considerable number of such mineral dates in the Limpopo Complex, which in some cases confirm, and in others contradict, the zircon dates, serves not only to define metamorphic events but also to inform on the resetting behavior of the chronometers themselves.

In the NMZ (Fig. 3A) a cluster of dates about 2.6 Ga is provided by U-Pb dates from monazite (both included in cordierite, and interstitial in the matrix) and Pb/Pb SL dates from orthopyroxene (Kamber et al., 1998, 1, 2 in Fig. 3A) and garnet (Blenkinsop and Frei, 1996, 3 in Fig. 3A) in metasedimentary rocks occurring as large enclaves in, or bordering on, magmatic enderbitic bodies. The age of the metamorphism recorded by these is indistinguishable from the magmatic ages of the NMZ (Fig. 2A). The dates about 2.0 Ga in the NMZ are all related to distinct shearing; hornblende Ar/Ar results are from amphibolite-facies shear zones in the southern part of the NMZ, <20 km from the CZ (Kamber et al., 1995a, 1995b, 1996, 4–6 in Fig. 3A), whereas the Pb/Pb SL dates are from titanite recrystallized in greenschist-facies shear zones in the northern part of the NMZ (Kamber et al., 1996, 7 in Fig. 3A). These few results thus document a N-S upward metamorphic gradient at 2.0 Ga in the NMZ. They also display a spread from 2.1 to 1.95 Ga, which is much greater than that for the 2.0 Ga event in the CZ (see Fig. 2B, and below).

In the SMZ an early thin-slice Rb/Sr date defines high-grade metamorphism at 2.59 ± 0.08 Ma (Barton et al., 1983, 1 in Fig. 3D), indistinguishable in age from the tectonism as dated via the Matok Pluton (Fig. 2C). Later, a cluster of dates about 2.72 Ga to 2.58 Ga was obtained (Kreissig et al., 2001). These comprise monazite U-Pb results from metapelites from the Bandelierkop Quarry and a xenolith in the Matok Pluton (2, 3), and Pb/Pb SL dates (4) as well as Ar/Ar hornblende results (5) from the Hout River Shear Zone. All these are interpreted as ages of metamorphic growth of the minerals. As in the NMZ, there is no significant difference between syntectonic magmatic dates (Matok Pluton) and the U-Pb and Pb/Pb metamorphic ages. However, the hornblende Ar/Ar ages are ~100 m.y. younger than these.

In the Triangle Shear Zone the results of Van Breemen and Hawkesworth (1980) inspired further mineral dating activity in the same region, including an Sm-Nd mineral isochron that confirmed the dates obtained by Van Breemen and Hawkesworth (1980), and eight garnet whole-rock feldspar Pb/Pb dates (Kamber et al., 1995a, 2–4a). Three of these latter results also yielded ages close to 2.0 Ga (3), whereas one yielded 2.65 ± 0.1 Ga (4a), and three samples yielded identical intermediate dates at 2.2 Ga (4). The younger dates were interpreted as giving the age of garnet growth or regrowth during shearing in the Triangle Shear Zone. A further 2.0 Ga Pb/Pb garnet whole-rock date from the Beit Bridge area (Kamber et al., 1995a, 14 in Fig. 3C) led to a renewed interest in the metamorphic geochronology of the CZ and a large number of studies combining petrography with geochronology from metamorphic minerals, summarized below.

In the western portion of the CZ (Fig. 3B), such work was done in the Palala Shear Zone in the Koedoesrand Window near Lephalale (formerly Ellisras) in the Lose Quarry in the Mahalapye Complex and the Phikwe Complex. The only Neoarchean metamorphic mineral date from the western CZ is a garnet Pb/Pb SL date of 2.6 ± 0.028 Ga obtained from a small boudin within a greenschist facies mylonite of the Palala Shear Zone (Schaller et al., 1997, 11 in Fig. 3B). Along this shear zone, strike-slip movement occurred as well as uplift of the northern block (CZ) relative to the southern block (Kaapvaal Craton and Bushveld Complex) at ca. 2.0 Ga (Schaller et al., 1999). The mineral dates obtained directly to the north of the shear zone (on the uplifted side) are from Sm-Nd in garnet (Chavagnac et al., 2001, 8 in Fig. 3B), U-Pb in rutile (Holzer et al., 1999, 7), and Pb/Pb SL on titanite in a calc-silicate rock (Schaller et al., 1999, 10). All these give the ca. 2.0 Ga age and document at least upper amphibolite-facies metamorphism at 2.0 Ga in the CZ directly north of the Palala Shear Zone. The 2.6 ± 0.028 Ga garnet date from the mylonite thus represents a relict from the earlier, Neoarchean metamorphism, which was not recrystallized or reset during the shearing event, as temperatures in the shear zone were lower than in the uplifted block to the north at 2.0 Ga.

The mineral dates from the Mahalapye Complex include Sm-Nd and Lu-Hf garnet and U-Pb monazite dates (Chavagnac et al., 2001, 5, 6; Millonig et al., 2010, 6a–6c) as well as U-Pb from apatite (Holzer et al., 1999, 6). In the Phikwe Complex, Holzer et al. (1999) obtained Pb/Pb SL dates on titanite and garnet (2, 3) and a U-Pb apatite date (2). All except the Pb/Pb garnet date sharply define the 2.0 Ga event, again documenting at least amphibolite facies metamorphism at that time. The imprecise garnet date of 2.075 ± 0.06 from the Selebi orebody may reflect incomplete resetting at 2.0 Ga, similar to the Rb-Sr thin-slice date of Hickman and Wakefield (1975).

In the CZ east of the Koedoesrand Window (Fig. 3C), localities from which metamorphic mineral dates have been obtained (apart from the Triangle Shear Zone, mentioned above) include the Bulai Pluton and environs, the Sand River Gneiss at the Causeway locality near Musina, the Tshipise Straightening Zone, the eastern flank of the Baklykraal structure, and the Venetia Mine open pit and immediate surroundings.

In the Bulai Pluton area, similar to the Triangle Shear Zone, metamorphic mineral dates close to 2.6 Ga and at 2.0 Ga were obtained. Pb/Pb SL dating of garnet and sillimanite yielded about 2.52 Ga (Holzer et al., 1998, 8a in Fig. 3C) and U-Pb in monazite from a metapelitic xenolith in the Bulai Pluton gave 2.644 ± 0.008 Ga (Millonig et al., 2008, 8b). U-Pb monazite and Pb/Pb garnet SL dates close to 2.60 Ga (van Reenen et al., 2008, 20 in Fig. 3C) were obtained from a metapelitic xenolith in a quartzofeldspathic Singelele-type gneiss 1 km north of the Bulai intrusive contact. A further Pb/Pb garnet SL date of 2.54 ± 0.06 Ga (Holzer, 1998, 13a in Fig. 3C) was not from a xenolith but from a metapelite outcrop showing little shearing. The texture and mineralogy (garnet mantled by cordierite, with sillimanite) document high-temperature decompression. Matrix monazite from the same rock yielded a 2.0 Ga date (Holzer et al., 1996, 13b in Fig. 3C). Further mineral dates clustering about 2.0 Ga in the Musina area were a Pb/Pb SL date on a retrograde titanite from a shear zone in the Bulai Pluton itself (Holzer, 1998, 9 in Fig. 3C), an enstatite Pb/Pb SL date (Holzer, 1998, 7 in Fig. 3C), and four dates from a calc-silicate lens in a shear zone developed between the Bulai Pluton and the Ha-Tshanzi fold structure directly to the south. These are Pb/Pb SL dates from garnet and titanite (10 in Fig. 3C) and U-Pb dates from monazite and apatite (11 in Fig. 3C; both Holzer et al., 1998). The enstatite is part of the peak high-pressure granulite paragenesis (enstatite, sillimanite, cordierite, biotite, and quartz) in an Mg-Al-rich semipelite occurring as a ~2-km-long xenolithic raft within the Bulai Pluton, ~2 km N of the Ha-Tshanzi structure. Its apparent young age is therefore paradoxical, and is discussed further below.

Hornblende stepheating Ar/Ar dates from the Sand River Gneisses and the Tshipise Straightening Zone are well defined at 2.0 Ga (Holzer, 1998; 12 and 15a in Fig. 3C), as are two titanite Pb/Pb SL dates from the latter zone (15b in Fig. 3C). Some rock units within the Tshipise Shear Zone may, however, be monometamorphic (see 8 in Fig. 2B). A titanite from a calc-silicate rock in the Gumbu Group E of Tshipise, which is certainly monometamorphic, was dated at 2015 ± 7 Ma by U-Pb (Buick et al., 2003, 17 in Fig. 3C). Further, garnets from two metapelitic units in which definitely only one metamorphic event was recorded, and which probably also belong to the Gumbu Group, were dated by Pb/Pb SL. These garnets are from a shear zone in the west flank of the Baklykraal structure, ~30 km W of Alldays (Boshoff et al., 2006, 16a in Fig. 3C) and from a metapelite in the Venetia mine, showing open folding (Rigby et al., 2011, 19a in Fig. 3C) and gave well-defined dates between 2.02 and 2.04 Ga, in accord with Ar/Ar dates of 2.035 ± 0.015 Ga from fabric-forming hornblende and muscovite from the metapelite (Barton et al., 2003, 19b.. A monazite from metapelite in the more than 2.65 Ga Krone Metamorphic Terrane in the Venetia Mine area yielded 2.015 ± 0.008 Ga (Chudy et al., 2008, 18 in Fig. 3C).

Boshoff et al. (2006) also attempted Pb/Pb SL dating of garnets from metapelites in the area 30 km W of Alldays that were intruded by Singelele-type granitoids and thus are older than ca. 2.65 Ga. They also show petrological evidence of two metamorphic events (Perchuk et al., 2008). The dates are ill-defined between 2.0 and 2.2 Ga (16b and 16c in Fig. 3C), with large errors owing to scattering of data in the isochron diagrams. A garnet from the intruding Singelele-type gneiss yielded a likewise poorly defined apparent age between 2.4 and 2.5 Ga (16d in Fig. 3C), also with considerable scatter.

Summary

Zircon dates (along with Rb-Sr whole-rock isochron dates) show predominantly a period of high-grade metamorphism and anatectic magmatism, lasting from ca. 2.72 to ca. 2.55 Ga (and possibly to ca. 2.5 Ga) in all the NMZ and CZ, and 2.7–2.68 Ga in the SMZ. In addition, zircons document magmatic emplacement of precursor granitoids to the Sand River Gneiss as well as of the Messina Layered Suite at ca. 3.3 Ga, and a further high-temperature episode at ca. 3.15 Ga. In the CZ alone a small number of dates from zircons in melt patches and on zircon rims reflect a thermal event at ca. 2.0 Ga, without giving much indication on tectonism during this episode. In contrast, metamorphic mineral dates of various kinds in the CZ overwhelmingly document the ca. 2.0 Ga event, which appears to be very much more sharply defined in time (ca. 1.99–2.02 Ga) than the ca. 2.65 Ga event and which is definitely associated with tectonism (open folding and large shear zones). Among these dates in the CZ, only some Pb/Pb SL ages record the ca. 2.65 Ga event, and these are from samples within xenoliths or shielded in some way from post-Archean penetrative deformation. Several intermediate Pb/Pb SL dates occur in the CZ, and these have large errors owing to scatter of the isotope data arrays. Three of the garnet-feldspar whole-rock Pb/Pb dates from the Triangle Shear Zone form a well-defined cluster at 2.2 Ga. In the NMZ the ca. 2.0 Ga event appears less well defined in time than in the CZ, spanning ~1.93–2.1 b.y., and is recorded by minerals associated with medium- and low-grade shear zones (becoming lower in grade to the north), whereas U-Pb monazite and Pb/Pb SL dates from garnet and orthopyroxene reflect the ca. 2.65 Ga event. In the SMZ the 2.0 Ga episode is recorded only by the post-tectonic, 2.0 Ga Entabeni Pluton and by biotite Rb-Sr dates, without apparent deformation. Other metamorphic mineral dates show only the ca. 2.65 Ga episode. In all three zones the ranges of ca. 2.65 Ga metamorphic mineral dates confirm the long duration of that earlier event, as apparent from zircon dates.

In broad terms, the present geochronological database confirms the early conclusion of Van Breemen and Dodson (1972) and Hickman and Wakefield (1975) that a major tectonometamorphic event occurred at ca. 2.65 Ga and that the 2.0 Ga ages did not reflect late cooling following this event but recorded a separate episode. Major results from geochronological refinements since the early work are (1) differentiation among the three zones of the Limpopo Complex, particularly with respect to the 2.0 Ga event, (2) documentation of a long duration of the ca. 2.65 Ga period of metamorphism, (3) chronological sharpness of the 2.0 Ga metamorphic event, and (4) evidence that the latter was also associated with intense tectonism.

DISCUSSION

In this section we review, for the mineral chronometers used for the Limpopo Complex, relevant theory and experiments, and discuss how these predict their resetting or retention behavior in multiple high- to medium-grade metamorphism. We then examine the record in the light of these predictions.

The New Zirconology and the Zircon Record from the Limpopo Complex

U-Pb zircon dates are extremely robust against thermal perturbation and have traditionally been regarded as yielding the age of magmatic crystallization. The diffusion coefficient of U and Th in zircon, even at magmatic temperatures, is extremely low (~10^{-29} cm^{2}sec^{-1} at 900 °C, extrapolated, Cherniak et al., 1997). That of divalent Pb is also low, though ~3–4 orders of magnitude higher (Cherniak and Watson, 2001). However, for radiogenic Pb in zircon it might be as low as those of Th and U if this Pb is tetravalent, as argued by Kober (1987) and Kramers et al. (2009). These authors further note that tetravalent Pb would be compatible in the zircon lattice, so that it would be built into it (effectively replacing U) upon annealing of α-recoil damage at high temperature. Even if the zircon structure is metamict, high temperatures of metamorphism would cause it to be annealed before Pb out-diffusion could occur (Mezger and Krogstad, 1997). For these reasons, U-Th-Pb systematics in zircon grains could be retained for many millions of years at magmatic temperatures, and zircon dates cannot be "reset" by thermally activated volume diffusion in metamorphic events.

Nevertheless, partial to complete Pb loss from zircons is common. In metamict zircons it can occur through the interaction of hydrothermal fluids associated with metamorphism, as shown experimentally by Pidgeon et al. (1966, 1973) and Sinha et al. (1992), and also clearly demonstrated by Gerdes and Zeh (2009) from Hf isotope data. Pb loss, even from only slightly metamict zircons, also commonly occurs by interaction with groundwater in weathering (e.g., Mezger and Krogstad, 1997). The last α-decay in the ^{235}U decay chain (leading to ^{207}Pb) has a higher energy than the last one in the ^{238}U chain (leading to ^{206}Pb); therefore radiogenic ^{207}Pb resides in sites of greater lattice damage than those of ^{206}Pb. The resulting, somewhat greater leachability of ^{207}Pb over ^{206}Pb out of zircon is inferred by Kramers et al. (2009) to cause the nonzero lower concordia intercepts that are so common in discordant zircons. As the Pb loss in metamorphism is mostly partial, it can, together with subrecent Pb loss in weathering, lead to multiple discordancy, in which case much of the age information is lost. Zircon studies in polymetamorphic terranes have greatly benefited from microbeam techniques such as SIMS and LA-ICP-MS, which can produce large sets of analytical data from which the most concordant points can be selected.

These techniques, together with thermoluminescence and backscattered electron imaging (e.g., Corfu et al., 2003; Vavra et al., 1999; Zeh et al., 2007, 2008, 2010) also facilitate the selective analysis of cores and rims. Zircon rims can form as overgrowths after partial resorption of older grains in a partial melt, in a hydrous fluid, or in the interim stage between the two, a pegmatite-type hydrous melt. These can generally be distinguished by their Th/U ratios. Typical values given are 0.5 (magmatic) and 0.05 (metamorphic; Hoskin and Schaltegger, 2003; Möller et al., 2003), although intermediate values are often found (e.g., Mouri et al., 2009; Zeh et al., 2007, 2010). In each of these cases, Pb is excluded from new zircon as it crystallizes. Paired analyses of zircon cores and rims are obviously an important means for identifying and dating secondary magmatic as well as metamorphic events.

Rims do not necessarily have to be overgrowths. Zircon recrystallization has been reported to have occurred under greenschist facies conditions (Hay and Dempster, 2009) and even under <200 °C hydrothermal conditions (Geisler et al., 2003a), producing cryptocrystalline porous mantles enriched in Al and Fe. In hydrothermal experiments (at temperatures between 175 and 650 °C) on metamict zircons from Sri Lanka, Geisler et al. (2002, 2003b) produced reaction rims and zones similar to those referred to above. The boundaries between cores and rims were in most, but not in all, cases highly irregular. This is in contrast to the mostly straight or cuspate boundaries observed in zircons with magmatic or high-temperature metamorphic overgrowths. The rims produced appear to consist of cryptocrystalline to microcrystalline zircon. Radiogenic Pb was seen to be mostly lost from the rims in this process in the lower temperature experiments, but retained in some runs at 650 °C. U and Th were lost (at low temperatures) or retained (at 650 °C) to similar degrees. The implication is that at the higher temperatures, recrystallization would be rapid enough to prevent U and Th diffusion through the recrystallizing rim before it became crystalline (Geisler et al., 2002). This is logical, considering the extremely low diffusivity of U and Th in zircon (Cherniak et al., 1997). The same may apply to radiogenic Pb (Cherniak and Watson, 2001; Kramers et al., 2009). Geisler et al. (2003b) cite a number of examples from the literature of zircon rims that show the same age as their cores. Clearly it is uncertain, for this type of rim, whether the U-Th-Pb systematics would show the time of original zircon formation, or that of the rim, or some date in between. Given sufficient precision, an in-between date could be revealed by discordance, but this would be unlikely to be resolved in SIMS or LA-ICP-MS analyses if the events forming core and rim are "only" ~600 m.y. or less apart.

Questions to be addressed regarding the core-rim results from the Limpopo Complex are, therefore, whether all rims observed around zircon cores are in fact overgrowths, or whether they could also be recrystallization rims, and whether this matters for the interpretation of these data as records of a polymetamorphic history.

A comparison between U contents and Th/U ratios of cores and associated rims may be helpful to determine whether the rims are real overgrowths, and under what types of conditions they formed. In Figure 4, ^{207}Pb/^{206}Pb dates for core-rim pairs are plotted against these parameters, from the more recent studies of Limpopo Complex zircons that employed imaging and microbeam techniques, and listed Th concentrations. These are the publications by Buick et al. (2003), Mouri et al. (2008, 2009), and Zeh et al. (2007, 2008, 2010). None of the rims analyzed in these studies are microcrystalline. Note that these publications, in particular Zeh et al. (2007, 2008, 2010), contain many more dates (defining the summarized age results shown in Fig. 2), but only a small fraction of these from associated cores and rims were close to concordant.

The behavior of U-Th systematics differs greatly between the samples. In the pairs studied by Buick et al. (2003) and Mouri

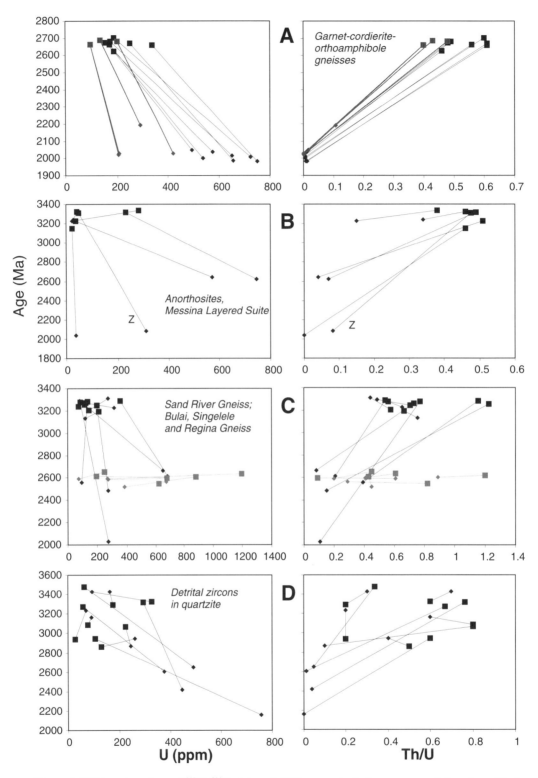

Figure 4. U-Th systematics and $^{207}Pb/^{206}Pb$ dates of 95% concordant zircon cores and rims from lithologies in the Central Zone. Squares, cores; diamonds, rims. Core-rim pairs are connected by lines. (A) Mafic granulites (garnet-cordierite-orthoamphibole gneisses) near Swartwater, 190 km E of Musina. Gray, Buick et al. (2003); black, Mouri et al. (2008); data obtained by SIMS. (B) Anorthosites from the Musina Layered Suite, Mouri et al. (2009), SIMS; and Zeh et al. (2010), Z, LA-ICP-MS. (C) Sand River Gneiss (black), and Bulai, Singelele, and Regina Gneisses (gray), Zeh et al. (2007, 2010), LA-ICP-MS. (D) Detrital zircons in quartzites from the Beit Bridge Group near Musina, Zeh et al. (2007, 2010), LA-ICP-MS.

et al. (2008) (Fig. 4A), the rims have higher U contents than the cores, have Th/U ratios of ~0.01 or lower, and give ca. 2.0 Ga ages. This is a paradigm of metamorphic zircon overgrowths crystallized from a metamorphic fluid, over primarily magmatic grains, in a single metamorphic event. An exception are two rim analyses from Buick et al. (2003), which give intermediate ages and Th/U ratios, looking almost like a mixture (which could be due to insufficient spatial resolution). In the more complex results of Mouri et al. (2009) and Zeh et al. (2010) on meta-anorthosites (Fig. 4B), U contents of rims are not invariably higher, but Th/U ratios are in all cases lower than those of the cores. Rims dating a ca. 3.2 Ga event, which have grown around earlier than 3.3 Ga cores, have Th/U ratios of 0.15 and 0.34 that are intermediate between values seen in metamorphic (hydrothermal) and magmatic zircon. In contrast, rims that date the ca. 2.6 and 2.0 Ga events have hydrothermal Th/U ratios of 0.07 and lower.

Zircon core-rim pairs from various orthogneisses in the Central Zone of the Limpopo Belt (Zeh et al., 2007, 2010; Fig. 4C) show two kinds of behavior. Those from ca. 2.6 Ga rocks show a clustering of core as well as rim dates in the same age range whereby rims can have both lower and higher U contents and Th/U ratios than cores. These rims are interpreted by Zeh et al. (2007) as being mainly magmatic overgrowths produced in the same events as the cores, or successive magmatic episodes belonging to the same extended high-temperature period. Some of the Sand River Gneiss zircon pairs, with ca. 3.3 Ga cores (Zeh et al., 2007, 2010), show the same behavior, but most have rims belonging to the ca. 2.65 or 2.0 Ga metamorphic events, which have mostly higher U contents and lower Th/U contents than the cores, although not in the typical "metamorphic" range.

In Beit Bridge quartzites, Zeh et al. (2008, 2010; Fig. 4D) found a number of zircon core-rim pairs in which the rims show normal metamorphic high U contents and low Th/U ratios. In others, the rims are similar in U content and Th/U ratios to the cores. Such rims appear up to ca. 200 Ma older or younger than the associated cores, which themselves show a considerable spread. These dates are interpreted by Zeh et al. (2008) as having resulted from partial and variable Pb loss of cores and/or rims, e.g., during the ca. 2.65 Ga event and/or in pre-quartzite weathering, whereby the precision of the U-Pb analyses was insufficient to reveal discordance. The rims appear mainly magmatic and, like the cores, are probably actually about or more than 3.2 Ga in age.

In the study of Barton and Sergeev (1997) a discordia (between ca. 3.3 and 2.04 Ga) was defined by high-precision ID-TIMS dating of five single grains, all consisting of cores and rims. Their Th/U ratios range from 0.15 to 0.45, with the lower values occurring in the grains closer to the lower intercept. From its position on the discordia, the grain with Th/U = 0.15 has ~45% rim, and it is therefore likely that the rims have Th/U <0.1 and are formed from a fluid. Among the earlier zircon studies, only Kröner et al. (1999) and McCourt and Armstrong (1998) reported Th concentrations. The Th/U ratios are mostly >0.3, with some values of <0.1 found only in very high U and strongly discordant grains.

In summary, the various zircon data from the Limpopo Complex confirm that zircons can record multiple metamorphic events and that Th/U ratios allow distinguishing between magmatic and hydrothermal zircon overgrowths. All rims of ca. 2.0 Ga observed have clearly metamorphic Th/U ratios, and mostly higher U than the associated cores, and are clearly overgrowths. Of the zircon rims produced in the ca. 2.65 Ga event or earlier, some have a hydrothermal Th/U ratio (but higher than those typical of the 2.0 Ga event), whereas most appear magmatic; many are the same age as their associated cores. No microcrystalline zircon rims have been reported, and it remains unclear (and rather irrelevant) whether there were any in situ recrystallized zones that were later annealed.

In the quartzite samples from the Musina area, where metamorphism at ca. 2.65 Ga reached high grade as documented by anatexis (producing the Singelele Gneiss), zircon rims of ca. 2.65 Ga age are rare. Only a few were found in hundreds of grains (Zeh et al., 2008, 2010). As discussed above, the formation of zircon rims, whether as overgrowths or by recrystallization, requires either a fluid or a melt. The absence of a fluid phase in quartzites of the Beit Bridge Group during the 2.65 Ga could be explained by anatexis nearby, with the melt scavenging any hydrous fluid. In more general terms, a high-grade metamorphic event need not necessarily be recorded in zircon overgrowths.

Monazite Dates, Apparent Contradictions, and Their Resolution

The diffusivity of divalent Pb in monazite has been shown to be even lower than that in zircon (Cherniak et al., 2004), and if radiogenic Pb is tetravalent in monazite, it is likely to be 2–3 orders of magnitude lower still, by analogy with U and Th. Thus the U-Th-Pb systematics can be expected to be retained even at high-grade metamorphism. However, monazite takes part in metamorphic exchange reactions. A monazite-garnet Y-exchange thermometer has been defined (Pyle et al., 2001), and Y as well as rare earth element (REE) exchange takes place between monazite, xenotime, apatite, and allanite (Pyle and Spear, 2003; Berger et al., 2005; Kelly et al., 2006; McFarlane et al., 2006; Buick et al., 2006). Pyle and Spear (2003) demonstrated three stages of monazite growth interspersed with two stages of monazite resorption, within the prograde and retrograde paths of a single high-temperature, medium-pressure metamorphic loop. These authors note that monazite dating could provide an excellent means for placing time markers on stages of pressure-temperature (P-T) loops and thus measure the speed of tectono-metamorphic cycles. On the other hand, however, memory of previous metamorphic cycles can be wiped out as monazite is reacted away and reforms. Monazite dating thus needs to be accompanied by careful petrological work to enable its interpretation. Most of the dating work on monazite in the Limpopo Complex was done before the work cited above was published, and in most cases no detailed petrographical studies were carried out.

The monazite U-Pb SIMS analyses from the NMZ (Kamber et al., 1998, 1 in Fig. 3A) were done in situ from thin sections; some of the grains occurred as inclusions in cordierite, and others in the matrix, and all gave ages of ca. 2.65 Ga with no systematic difference between the two modes of occurrence. These monazites apparently remained intact during the 2.0 Ga event, which reached lower amphibolite facies in this region of the NMZ (Kamber et al., 1995b). All other monazite dates from the Limpopo Complex were done on mineral separates. In the CZ, all but one give 2.0 Ga ages (6 in Fig. 3B; 11, 13b, 18, 21 in Fig. 3C) and thus record either first crystallization or recrystallization in the Paleoproterozoic event. The monazite date by Buick et al. (2006, 13 in Fig. 3B) was accompanied by a petrological study which makes it plausible that the monazite had grown at the expense of allanite and xenotime during prograde metamorphism. The 2.0 Ga monazite, 13b in Figure 3C (Holzer et al., 1996), is noteworthy, as garnet in the same rock documents peak metamorphism at ca. 2.6 Ga (13a in Fig. 3C, Holzer, 1998). Apparently the monazite in this rock was resorbed and newly formed in the 2.0 Ga event. In contrast, a single case of a monazite date of ca. 2.6 Ga age is from a metapelitic xenolith in the Singelele Gneiss (van Reenen et al., 2008, 20 in Fig. 3C) a few kilometers north of the Bulai Pluton, a region where the 2.0 Ga metamorphism reached high grade. This demonstrates the thermal retentivity of U-Th-Pb in monazite, provided no mineral reactions led to its recrystallization.

Thus, while monazite dates provide excellent time markers for metamorphic events, the data demonstrate that absence of a ca. 2.0 Ga or ca. 2.65 Ga monazite date in a rock within the CZ does not necessarily mean that that rock did not undergo metamorphism at the respective "age-absent" time. This is as the case with zircon, but for an entirely different reason.

Dating of Discrete Metamorphic Minerals

The high Sm/Nd and Lu/Hf ratios of garnet make this mineral highly suitable for geochronological studies of metamorphism, especially in Precambrian terranes (Van Breemen and Hawkesworth, 1980). An estimate of 620 °C for the closure temperature of Sm-Nd systematics in garnet (Mezger et al., 1992) is unrealistically low, as Nd isotope disequilibrium has been seen to persist under upper mantle conditions (Jagoutz, 1995; Roden et al., 1996). Sm-Nd garnet–whole-rock dates in metamorphic rocks are therefore expected to yield the age of garnet crystallization. A problem may be presented by heterogeneous initial ^{143}Nd/^{144}Nd ratios (Thöni and Jagoutz, 1992), but the relative importance of this decreases with age. The interpretation that the Sm-Nd garnet dates determined in the CZ and the Triangle Shear Zone, all in the 2.0 Ga cluster, date a metamorphic event of at least amphibolite facies is therefore justified.

Relatively high U/Pb ratios in some metamorphic minerals (especially titanite, but also garnet) have given rise to a number of geochronological studies using ^{207}Pb/^{206}Pb isochrons or two-point dates from analyses of garnet and other rock-forming minerals, e.g., feldspar and/or whole-rock compositions (e.g., Mezger et al., 1989; Kamber et al., 1995a). One important assumption of the dating method is that all phases in the sample had the same Pb isotope composition at the time of the metamorphism. In the case of the dates produced in the latter study on the Triangle Shear Zone, the Pb in garnets was highly radiogenic with ^{206}Pb/^{204}Pb ratios up to 120, so that the dating results are robust against small initial Pb isotope heterogeneity between the phases. Further, most of the ages are derived from isochrons, including feldspar and whole rock, giving additional confidence in the results.

The samples are high-temperature mylonites. P-T studies by Kamber et al. (1995a) on the dated rocks yielded peak metamorphic conditions between 700 and 850 °C and 5–9 kb, whereby they note that it is unsure whether these apply to a single metamorphic episode in the Paleoproterozoic, or may contain inherited signals. Four samples (3 in Fig. 3C) yielded identical ages of 2.0 Ga, which Kamber et al. (1995a) interpreted as the age of shearing and the last metamorphism. One sample (4a in Fig. 3C) yielded an imprecise age of 2.64 ± 0.11 Ga. This shows (1) that some rock units in the Triangle Shear Zone also underwent a Neoarchean metamorphism, and documents (2) that the U-Pb systematics in garnet were not reset at the high temperatures of the 2.0 Ga event. Diffusive loss of radiogenic Pb from (relatively large) garnet grains in metamorphism, even at high grade, is indeed unlikely, given the persistence of prograde chemical zonations of the similar Ca^{2+} ion in garnet through peak T of P-T loops seen in many metamorphic studies. If radiogenic Pb is tetravalent (see 3.1 and below), its diffusive loss would be even more improbable, as discussed below.

Pb/Pb Stepwise Leaching Dates

Frei and Kamber (1995) have shown that Pb/Pb dating was possible by leaching separates of single mineral phases with progressively stronger acids (e.g., 4N HNO_3, then aqua regia, then HF+HNO_3), higher temperatures (room temperature, then 100 °C) and longer leaching times (from 15 min to 24 h). Generally, early "gentle" leaching steps yielded relatively unradiogenic Pb, and the proportion of radiogenic Pb increased in subsequently more aggressive leaching steps, producing datable arrays in ^{207}Pb/^{204}Pb versus ^{206}Pb/^{204}Pb space. The perceived advantages of this method over using mineral–whole-rock or mineral-mineral pairs were that the problem of possible initial isotope equilibrium between minerals, or minerals and matrix, was circumvented, and that a greater spread of isotope ratios was achieved, giving greater precision. In a study of a Variscan staurolite (too young to be reliably dated with ^{207}Pb/^{206}Pb), it was further found that U and radiogenic Pb were leached proportionally, so that consistent U-Pb dates could be produced by progressive leaching (Frei et al., 1995). In a detailed study of a gem-quality titanite sample, Frei et al. (1997) showed that leaching transformed the mineral into a spongelike Ti-Si-oxide gel, in which radiogenic Pb was surprisingly well retained.

Although the method has given results highly consistent with other dating methods (see, e.g., Fig. 3), it is by no means clear why it works so well. Possible factors brought forward include (a) the presence of microinclusions with high U/Pb ratios, (b) greater accessibility of radiogenic Pb (located on α-recoil tracks) than common Pb to leaching, or (c) a tetravalent state of radiogenic Pb. Factor (a) indeed appears to play a part in some cases, as illustrated and discussed below. Factor (b) was initially proposed by Frei and Kamber (1995) but was not favored later, as radiogenic Pb is generally better retained in stepwise leaching (SL) than common Pb. The tetravalent radiogenic Pb hypothesis (c) (Frei et al., 1997) appears to best predict the leaching behavior of Pb in their inclusion-free titanite study. Kober (1987) proposed and argued for a tetravalent state of radiogenic Pb in zircon, and a XANES experiment on young and old titanites indeed showed a difference in the valency and/or site coordination of common and radiogenic Pb (Kramers et al., 2009).

These hypotheses clearly predict different retentivity and resetting behavior for these mineral chronometers. In the microinclusions hypothesis (a), resetting would involve transport of Pb between these inclusions and the metamorphic mineral host, a process that would be dependent not only on the diffusion coefficient and the size of the inclusions but also on the partition coefficient for Pb between the phases and possibly the presence of fluids. In the accessibility hypothesis (b), the mineral chronometer would be effectively reset if the crystal structure became annealed, annihilating the α-recoil track defects. Given that these tracks represent much less damage than fission tracks, and the latter are annealed in zircon in the 200–250 °C range, this suggests that resetting would occur under low-grade metamorphic conditions, or at even lower temperatures. For the tetravalent hypothesis (c), resetting would occur on reduction of tetravalent to divalent Pb, with or without annealing of the structure. The reduction would depend on conductivity, e.g., access by a fluid, and would not be a function of temperature alone nor subject to diffusion law. For the specific case of garnet, note further that Pb^{4+} has effective ionic radius values similar to the heaviest REEs (compatible in garnet) at similar coordination numbers (Shannon, 1976; Whittaker and Muntus, 1970).

Apart from possible partial or complete resetting of the radiometric clock in an existing crystal by any of the above mechanisms, recrystallization or the addition of new growth zones to existing mineral grains would clearly result in a changed pattern of isotope arrays, ranging from a complete "loss of memory" to mixed patterns. Because of the generally low concentrations of U and Pb in common metamorphic silicates, sample amounts subjected to leaching are normally large, 30 mg or more in the case of garnet. If garnets in the sample are of two generations, it is normally impossible to separate grains from these two in a mineral concentrate, and the robustly predicted result from an SL dating attempt on the combined population would be a data array with a large scatter in the $^{207}Pb/^{204}Pb$ versus $^{206}Pb/^{204}Pb$ diagram, yielding an age intermediate between the older and the younger population, and with wide error limits.

In Figure 5 four examples of SL Pb isotope arrays from metamorphic minerals from the CZ are shown: an enstatite (Figs. 5A, 5B) that must have undergone metamorphism at 2.65 and at 2.0 Ga, a garnet from a rock that has undergone a single metamorphism (Figs. 5C, 5D), a garnet (Figs. 5E, 5F) from a rock that has undergone both metamorphic events, and a titanite (Figs. 5G, 5H) from a calc-silicate rock in the CZ, close to the Palala Shear Zone.

The enstatite (Holzer, 1998; 7 in Fig. 3C) is part of the peak high-pressure granulite facies paragenesis of a semipelite xenolith in the Bulai Pluton. Its SL Pb isotope array in the $^{208}Pb/^{204}Pb$ versus $^{206}Pb/^{204}Pb$ indicates dominance of the radiogenic Pb by microinclusions of probable monazite (extremely high Th/U ratio) and zircon (very low Th/U ratio). The latter was seen in the last leaching step, which is the dissolution of the residue of previous steps by $HF + HNO_3$, and the chemically resistant nature of material dissolved in this last step thus fits the assumption that zircon microinclusions were involved. Unsurprisingly, no U appears to have been hosted in the enstatite itself. Similar behavior was described by Schaller et al. (1997) for a garnet from the Palala Shear Zone. In this case, the common Pb in the garnet and the radiogenic Pb from the "monazite" and "zircon" microinclusions together defined an age of 2.6 Ga (11 in Fig. 3B). In contrast, the U-Pb systematics in this enstatite with microinclusions yielded a poorly defined date close to 2.1 Ga. As metamorphic minerals in other xenoliths in the Bulai Pluton yielded dates close to 2.6 (Holzer et al., 1998, 8 in Fig. 3C) the enstatite date is most probably reset. The inferred inclusions were not seen microscopically and must be sub-micron in size. Mechanisms whereby the U-Pb systematics of such small inclusions could be reset in age could include recrystallization (in the case of monazite) or fluid access. Depending on the size of such inclusions, diffusion could also be possible.

The monometamorphic garnet sample (Figs. 5C, 5D; 16a in Fig. 3C; Boshoff et al., 2006), SL in duplicate, yielded a fairly precise date corresponding to the peak of the 2.0 Ga metamorphism. In the $^{208}Pb/^{204}Pb$ versus $^{206}Pb/^{204}Pb$ diagram, none of the steps show a high Th/U ratio ("monazite"), but the last step in each analysis showed low Th/U ("zircon"), although in a far less dominant signal than in the enstatite. The most radiogenic Pb (step 4 of analysis 2) appears to be hosted in the garnet itself, with a normal Th/U ratio. The "zircon" microinclusions appear cogenetic with the garnet, as shown by their good colinearity with the other data in the $^{207}Pb/^{204}Pb$ versus $^{206}Pb/^{204}Pb$ diagram (Fig. 5C).

Likewise, the garnet from a polymetamorphic metapelite (Figs. 5E, 5F; Boshoff et al., 2006; 16b in Fig. 3C) does not show isotopic indicators for monazite microinclusions, but a low Th/U last step provides a signal of zircon microinclusions, albeit weak. All steps together yield a date of 2173 ± 73 Ma, intermediate between the two major metamorphic episodes, with an extremely high mean square of weighted deviates (MSWD) value of 1150, expressing considerable scatter beyond the analytical uncertainty. The "zircon" step defines, regressed with unradiogenic steps 1 and 2, an older age (2418 ± 250 Ma). This example thus

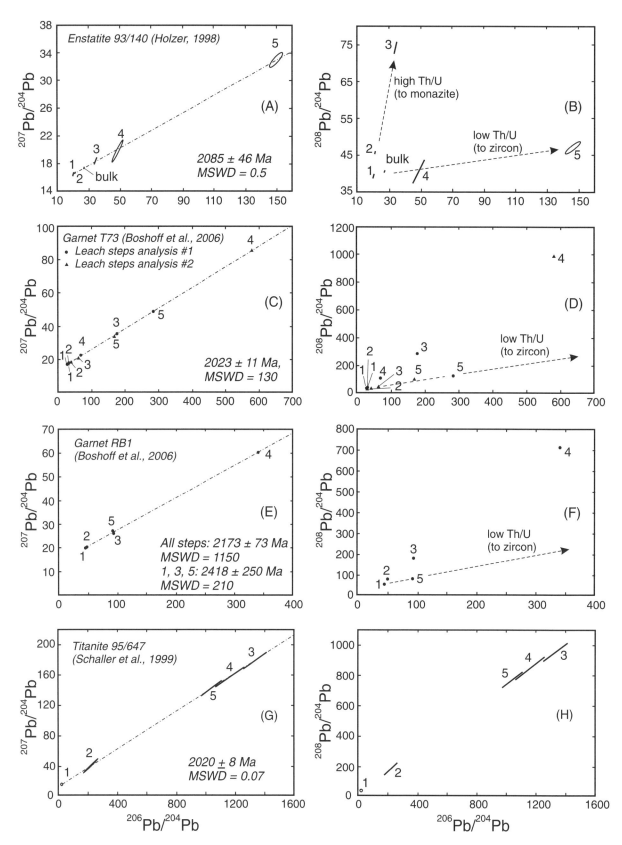

Figure 5. Four sets of results of Pb/Pb stepwise leaching (SL) dating experiments shown in $^{207}Pb/^{204}Pb$ versus $^{206}Pb/^{204}Pb$ (left) and $^{208}Pb/^{204}Pb$ versus $^{206}Pb/^{204}Pb$ (right) spaces. MSWD—mean square of weighted deviates. (A, B) Enstatite from semipelite xenolith 93/140 in Bulai Gneiss (Holzer, 1998). (C, D) Garnet from metapelite T73. (E, F) Garnet from metapelite RB1 (both from the Baklykraal structure W of Alldays; Boshoff et al., 2006). (G, H) Titanite from calc-silicate rock 95/647 just N of the Palala Shear Zone, Koedoesrand Window (Schaller et al., 1999). Discussed in text.

portrays partial resetting of the U-Pb systematics in the 2.0 Ga event, whereby the "zircon" microinclusions have retained more of a memory of the earlier metamorphism than U-Pb systematics of the hosting garnet. Similar patterns are seen in the other two garnet samples with intermediate dates (Boshoff et al., 2006; 16c, 16d in Fig. 3C).

The host rock of the titanite (10 in Fig. 3B) is not obviously polymetamorphic. The colinearity in the $^{207}Pb/^{204}Pb$ versus $^{206}Pb/^{204}Pb$ diagram is good, yielding a very low MSWD value and a relatively low error. The array in the $^{208}Pb/^{204}Pb$ versus $^{206}Pb/^{204}Pb$ plot is also closely linear, showing a homogeneous Th/U ratio for all the leaching steps and indicating that U and Th are dominantly sited in the titanite lattice itself, with inclusions playing no noticeable part. Such patterns are typical for titanite Pb/Pb SL dating.

These examples demonstrate that microinclusions of high U/Pb minerals can play a role in the systematics of Pb/Pb SL in the dating of metamorphic minerals, but that in the case of garnet and titanite, an important component or even all of the datable systematics can also reside in the host mineral. Further examination of dates obtained in the CZ sheds light on the factors affecting the resetting of such dates.

Garnet dates from the CZ that cluster about 2.0 Ga mostly have relatively small error limits, indicating a good colinearity of points in the isochron diagrams and thus a single generation of grains in the samples. The samples yielding garnet dates in this class are mainly from shear zones and other strongly tectonized domains. In some, a single metamorphic event is documented, and the grains clearly crystallized syntectonically (Boshoff et al., 2006; Figs. 5A, 5B, and 16a in Fig. 3C; Rigby et al., 2011; 19 in Fig. 3C), whereas others are probably polymetamorphic (Holzer et al., 1998; 10 in Fig. 3C).

One garnet retaining a 2.6 Ga age is from a ~4-cm-thick boudin in the Palala Shear Zone (Schaller et al., 1997; 11 in Fig. 3B), where intense shearing took place under low-grade metamorphic conditions. Channeling of deformation (and probably fluids) around the boudin at these temperatures probably helped to preserve the U-Pb systematics in this sample. The ca. 2.6 Ga or slightly younger garnet dates in the Musina area of the CZ (Holzer et al., 1998, 8a in Fig. 3C; Boshoff et al., 2006, 20 in Fig. 3C; Holzer, 1998, 13a in Fig. 3C) deserve special attention. The first is from a metapelitic xenolith in the Bulai Pluton and is accompanied by sillimanite also dated at ca. 254 Ga (8a in Fig. 3C). The second is from a metapelitic xenolith in the Singelele Gneiss and is accompanied by monazite also dated at ca. 2.6 Ga (20 in Fig. 3C). Although metamorphism up to high grade at 2.0 Ga in the area is well documented by a host of other dates, these rocks were shielded from deformation in the 2.0 Ga event by the surrounding granitoid bodies. The third sample represents a relict garnet core mantled by cordierite and associated with sillimanite. In this same rock, matrix monazite yielded 2.0 Ga as mentioned above (13b in Fig. 3C). These results provide unequivocal evidence that the U-Pb systematics revealed in the Pb/Pb SL dating of garnet are not reset by metamorphic heating alone, even up to high-grade temperatures. Of the factors enabling this dating method, discussed above, the "α-recoil track accessibility" hypothesis (b) is thereby excluded, leaving the "microinclusion" (a) and "tetravalent" (c) hypotheses.

Apart from mineral recrystallization during a second metamorphic event, fluid access is predicted by hypotheses (a) and (c) above to be a possible major factor in resetting the U-Pb systematics accessed in Pb/Pb SL dating (the "age signal"), and both processes are enhanced by deformation. Thus it is understandable that such dates, even in polymetamorphic rocks, can be reset in shear zones that were active during the youngest event, whereas in rocks that were shielded from deformation during that event the isotope systematics can be preserved, even if metamorphic temperatures were the same in each case.

Titanite dates from the CZ are all ca. 2.0 Ga (Holzer, 1998; Holzer et al., 1998, 1999; Schaller et al., 1999; 2, 10 in Fig. 3B; 10, 15b, 15c in Fig. 3C). These samples were taken in and near shear zones and from syntectonic leucosomes, but they are not in all cases recrystallized. For instance, 15c in Figure 3C consists of coarse-grained relict titanite grains in a mylonite in the Tshipise Straightening Zone, and the rather precise date of 2.001 ± 0.006 Ga is at the younger end of the ca. 2.0 Ga event range. Following the hypothesis that radiogenic Pb is tetravalent, and resetting of U-Pb systematics as accessed by Pb/Pb SL dating can occur by its reduction through fluid access, this young age, coupled with a relict texture, can be understood. Titanite Pb/Pb SL dating thus appears to be a good chronometer for synmetamorphic tectonism, even at relatively low grade.

In contrast to the CZ, the interpretation of Pb stepwise leaching dates from the NMZ and the SMZ is straightforward. In the SMZ they constitute coeval garnet, staurolite, and kyanite dates (Kreissig et al., 2001; 4 in Fig. 3D) that record metamorphic mineral growth in the ca. 2.65 Ga tectono-metamorphic event. In the NMZ, titanite dates (Kamber et al., 1996; 7 in Fig. 3A) yield the age of retrograde shear zones, related to the ca. 2.0 Ga event.

Summarizing, Pb/Pb SL dates from titanite, both in the CZ and in the NMZ, appear to record the last synmetamorphic tectonism, whether at high grade or relatively low grade. We have no evidence about possible resetting behavior of this chronometer in the absence of deformation and/or fluid access. The dates from garnet in the CZ in some cases record the earlier metamorphism (ca. 2.6 Ga), and mostly the later one (2.0 Ga). Some of the younger dates demonstrably reflect new garnet growth, whereas others are probably a result of resetting of the U-Pb systematics by fluid access. A few intermediate dates result either from mixing of older and younger generations of garnet in the analyzed sample or from incomplete resetting. The mechanism whereby fluid access can cause resetting of the U-Pb systematics in garnet, as accessed by SL, is thought to be reduction of radiogenic Pb from the tetravalent to the divalent state, accompanied by open system behavior of microinclusions. A key agent favoring both fluid access and recrystallization is deformation, and we note that the older ages particularly occur in rock units (xenoliths or boudins) that were not deformed in the 2.0 Ga event. The fluid access hypothesis is in accord with new findings following which mineral dates are mostly "hygrochronometers" rather than "thermochronometers" (e.g., Villa, 2010). Further, the record from the polymetamorpic CZ shows clearly that negative evidence from Pb/Pb SL metamorphic mineral dates (as from zircon and monazite U-Pb dates) is unreliable: In many areas of the CZ that were affected by the ca. 2.65 Ga metamorphic episode, none or very few of the available Pb/Pb SL dates reflect this event.

Amphibole Ar/Ar Dates

The β^+ decay of K^+ produces a neutral Ar atom. Although this is not electrostatically bound in a mineral lattice, its large size prevents easy diffusion and comparison with Sr diffusion as well as degassing experiments (Brabander and Giletti, 1995; Villa et al., 1996), allow us to conclude that Ar cannot diffuse out of amphibole unless recrystallization occurs, i.e., the lattice has transient holes. Thus Ar/Ar dating of amphibole can reveal several generations of amphibole growth. In mafic rocks of Val Malenco (Italian Alps) Villa et al. (2000) could discriminate four amphibole generations by stepwise heating, from relicts with a ca. 225 Ma age to overgrowths 67–73 m.y. old. They could further show that these were characterized by specific Ca/K ratios. The Alpine metamorphism reached 450 °C in the region, and although a pure thermal overprint did not cause any Ar loss from relict amphiboles of the earlier generations, a sample from an alpine shear zone yielded exclusively the youngest age.

A number of Ar/Ar hornblende stepheating dates from the CZ and the SMZ present a relatively simple picture, as they record the youngest tectono-metamorphic event for each zone (Belluso et al., 2000; 12 in Fig. 3B; Kamber et al., 1995a; Holzer, 1998; 5, 12, 15a in Fig. 3C; Kreissig et al., 2001; 5 in Fig. 3D; Barton et al., 2003; 19b in Fig. 3C). The dated amphiboles all represent fabric-forming phases, and they are mostly somewhat younger than most other dates from metamorphic minerals. For instance, two precise amphibole dates from the Sand River Gneiss (12 in Fig. 3C) and two from the Tshipise Straightening Zone (15a in Fig. 3C) are ca. 1.995 Ga, whereas Pb/Pb garnet and titanite dates in the same regions cluster about 2.02 Ga. In the Hout River Shear Zone of the SMZ the difference is greater. These differences can possibly be explained by some still remaining uncertainties in the decay constant of ^{40}K and the branching ratio of its decays to ^{40}Ca and ^{40}Ar (Begemann et al., 2001) and do not call for a geological discussion.

Ar/Ar dates in the NMZ present a more complicated picture, with some Ar/Ar amphibole dates significantly younger (similar to titanites; 7 in Fig, 3A) or older than 2.0 Ga. In this zone, the 2.0 Ga event is manifested in widely separated shear zones with a general decrease in metamorphism from amphibolite facies in the south to greenschist facies in the north. Following the notion that the Ar/Ar amphibole geochronometer dates synmetamorphic tectonism rather than cooling, the spread toward younger ages could be explained by protracted activity of these localized shear zones. A single ca. 2.1 Ga date from the NMZ ~10 km N of the Triangle Shear Zone (Kamber et al., 1995a; 6 in Fig. 3A) is problematic, as it significantly predates the 2.0 Ga event. This datum is difficult to explain as having resulted from a mixture of amphiboles from the ca. 2.65 and the 2.0 Ga events, as the stepwise heating produced a plateau well defined by four heating steps, and two amphibole generations of different ages would most likely have yielded a more complicated pattern. It is thus, together with the garnet Pb/Pb dates from the Triangle Shear Zone, an indication of possibly localized tectono-metamorphism between the two episodes.

SUMMARY AND CONCLUSIONS

Geochronological research in the Limpopo Complex over the past 50 yr has revealed three prevalent age clusters, the first at ca. 3.3 Ga, the second at ca. 2.65 Ga, and the third at 2.0 Ga. The first is revealed only by zircon U-Pb dates, it occurs in the CZ, and it is represented by a single data set from the SMZ. In the CZ this first age cluster reflects magmatic crystallization of the Sand River Gneiss and anorthosite of the Messina Layered Suite. A minor cluster of somewhat younger dates (ca. 3.15 Ga), defined by zircon rims, may signal an early metamorphic event or a second magmatic episode.

The second age cluster is represented in all three zones of the Limpopo Belt and recorded by all types of mineral chronometers used (except biotite Rb-Sr) as well as Rb-Sr whole-rock isochrons. It evidently reflects a long-lasting period, ca. 2.7 to ca. 2.55 Ga (and possibly locally to ca. 2.5 Ga) in which temperatures in the lower crust reached 900 °C and anatexis was widespread. Most of the zircon dates in this age cluster are magmatic crystallization ages. In the CZ, these date quartzofeldspathic (Singelele-type) gneisses formed by syntectonic melting, mostly occurring as conformable sheets in the metasedimentary sequence, and larger coherent plutons such as the Bulai granitoid gneiss and the so-called gray gneisses, which were all produced by crustal anatexis. In the NMZ, the zircon dates record the intrusion of numerous large enderbitic and charnockitic plutons and some postkinematic dikes and stocks, constraining the age of tectonism. In the SMZ, only the syntectonic Matok Pluton is dated by zircons in this age range. Some monazite U-Pb and Pb/Pb SL dates from metamorphic minerals, mainly garnet (but also orthopyroxene and alumosilicates), belong to the ca. 2.65 Ga cluster. In the NMZ the metamorphism they date is clearly related to the syntectonic intrusion of the charnockitic and enderbitic plutons and can be characterized as contact metamorphism. In the SMZ the minerals dated by Pb/Pb SL, as well as hornblende dated by Ar/Ar, are fabric forming in rocks remote from plutons, thus defining the age of syntectonic metamorphism. In the CZ, some (but not all) Pb/Pb SL dates from metamorphic minerals in the ca. 2.65 Ga cluster are from metapelitic xenoliths in granitoids. Yet these minerals form part of the synmetamorphic deformation fabric within the xenoliths and therefore are not products of contact metamorphism.

The great majority of metamorphic mineral dates in the CZ belong to the third, 2.0 Ga age cluster. These include most of the garnet and all of the titanite Pb/Pb SL dates, most U-Pb dates from monazite, and all Ar/Ar hornblende and biotite Rb-Sr ages. The dates with good precision define this cluster sharply in time, between 2.03 and 1.99 Ga, in contrast to the long-lasting "event" at ca. 2.65 Ga. In the SMZ this cluster is represented only by biotite Rb-Sr dates, and in the NMZ Ar/Ar hornblende ages and titanite Pb/Pb SL dates, all from shear zones, belong to it. The NMZ dates for this cluster are more spread out in time than those for the CZ.

With the exception of the biotite Rb-Sr dates, the 2.0 Ga ages in the CZ and NMZ are mostly associated with shearing, which in the CZ occurred under medium- to high-grade metamorphic conditions. The shearing appears to have mainly a strike-slip character. Exceptions are some thrust-sense shears in the NMZ dated by hornblende Ar/Ar, and open folding dated by garnet Pb/Pb SL at the Venetia Mine. In the CZ there are also a few garnet Pb/Pb SL dates intermediate between 2.65 and 2.0 Ga, but these have large errors and are thought to have been partially reset. The few Pb/Pb SL garnet dates in the ca. 2.65 Ga cluster are from samples taken far away from shear zones and/or shielded from tectonism.

The behavior of zircon U-Pb systematics in this polymetamorphic history confirms the extreme retentivity of this chronometer at high temperatures. It is further shown that overgrowths trend from a more magmatic Th/U ratio at 3.2 Ga, to intermediate values in the ca. 2.65 Ga event, and to a clearly hydrothermal character at 2.0 Ga. The rarity of ca. 2.65 Ga zircon overgrowths may be attributed to the absence of fluids owing to the widespread partial melting. Monazite, although a retentive chronometer, records mostly the 2.0 Ga metamorphic event, even in rocks that demonstrably also have been heated to high grade at ca. 2.65 Ga. This is ascribed to its possible participation in metamorphic reactions involving garnet and to consequent resorption and reforming along the prograde and retrograde paths of metamorphism.

Pb/Pb dates from garnet–whole-rock pairs (±feldspar) are shown to have been retentive at high metamorphic grade and essentially to have dated mineral growth. Pb/Pb SL results from garnet may also show ca. 2.65 Ga ages in spite of having been reheated to high grade in the 2.0 Ga event, demonstrating that the U-Pb systematics are robust against temperatures of at least 800 °C. Pb/Pb SL garnet dates younger than 2.5 Ga are of two kinds: (1) ages close to 2.0 Ga with a precision of ±25 Ma or better and relatively low MSWD values, which record mineral growth (or complete regrowth) in a single metamorphic event, and (2) ages intermediate between 2.0 and 2.5 Ga with large errors and high MSWD values. These are thought to have resulted from either partial regrowth of garnet at 2.0 Ga, or partial resetting owing to fluid access during deformation. Following the hypothesis that Pb/Pb SL dating of single phases works because of the tetravalent state of radiogenic Pb, the age signal would have been lost by its reduction to the divalent state, which could have been mediated by fluids. Pb/Pb SL dates from titanite invariably record the 2.0 Ga event, which is thought to have been due to recrystallization of this mineral, or susceptibility of the age signal to fluid access. Either way, titanite Pb/Pb SL dating emerges as a useful tool for dating Precambrian shear zones.

Hornblende Ar/Ar dating appears to record the crystallization or recrystallization of the mineral, in retrogression of charnockites or in shear zones. Like Pb/Pb SL in titanite, this method reaffirms itself as a good dating tool for amphibolite-facies shear zones.

Attempts to interpret the metamorphic mineral dates obtained in the Limpopo Complex (other than biotite Rb-Sr) in a thermochronological way would lead to numerous contradictions. Instead, it can be inferred that in the 2.0 Ga event, much of the resetting of chronometers (whether with or without recrystallization) is mediated by fluid activity, which in turn is facilitated by deformation such as shearing. There is thus abundant evidence that the 2.0 Ga event in the CZ and NMZ of the Limpopo Complex was a genuine tectono-metamorphic episode rather than a case of static reheating.

Processes during the 2.0 Ga event have, in many instances, effectively wiped out the memory of earlier events in minerals other than zircon. This has led to an apparent paradox in which these earlier events are almost exclusively recorded by magmatic zircon dates, and the last event chiefly by metamorphic mineral dates. As this is resolved in the above discussion, it becomes clear that in polymetamorphic terranes absence of evidence is not evidence of absence, and that obtaining a more or less complete record requires the use of multiple chronometers in different tectonic settings.

ACKNOWLEDGMENTS

Martin Whitehouse and an anonymous referee are thanked for their reviews, which helped to considerably improve this paper.

REFERENCES CITED

Andreoli, M.A.G., Brandl, G., Coetzee, H., Kramers, J.D., and Mouri, H., 2011, this volume, Intracrustal radioactivity as an important heat source for Neoarchean metamorphism in the Central Zone of the Limpopo Complex, in van Reenen, D.D., Kramers, J.D., McCourt, S., and Perchuk, L.L., eds., Origin and Evolution of Precambrian High-Grade Gneiss Terranes, with Special Emphasis on the Limpopo Complex of Southern Africa: Geological Society of America Memoir 207, doi: 10.1130/2011.1207(09).

Armstrong, R.A., Compston, W., Dodson, M.H., Kröner, A., and Williams, I.S., 1988, Archean crustal history in southern Africa: Canberra, Australian National University, Research School of Earth Sciences, Annual Report, v. 1987, p. 62–64.

Armstrong, R.L., Jäger, E., and Eberhardt, P., 1966, A comparison of K-Ar and Rb-Sr ages on alpine biotites: Earth and Planetary Science Letters, v. 1, p. 13–19, doi:10.1016/0012-821X(66)90097-5.

Bahnemann, K.P., 1972, A review of the structure, stratigraphy and metamorphism of the basement rocks in the Messina District, Northern Transvaal [Ph.D. thesis]: University of Pretoria, South Africa, 156 p.

Barton, J.M., Jr., 1996, The Messina Layered Intrusion, Limpopo Belt, South Africa: An example of in-situ contamination of an Archean anorthosite complex by continental crust: Precambrian Research, v. 78, p. 139–150, doi:10.1016/0301-9268(95)00074-7.

Barton, J.M., Jr., and Sergeev, S., 1997, High precision, U-Pb analyses of single grains of zircon from quartzite in the Beit Bridge group yield a discordia: South African Journal of Geology, v. 100, p. 37–41.

Barton, J.M., Jr., and van Reenen, D.D., 1992a, When was the Limpopo Orogeny?: Precambrian Research, v. 55, p. 7–16, doi:10.1016/0301-9268(92)90010-L.

Barton, J.M., Jr., and van Reenen, D.D., 1992b, The significance of Rb-Sr ages of biotite and phlogopite for the thermal history of the Central and Southern Marginal Zones of the Limpopo belt of southern Africa and the adjacent portions of the Kaapvaal Craton: Precambrian Research, v. 55, p. 7–16.

Barton, J.M., Jr., Fripp, R.E.P., Horrocks, P., and McLean, N., 1979a, The geology, age, and tectonic setting of the Messina Layered Intrusion, Limpopo Mobile Belt, southern Africa: American Journal of Science, v. 279, p. 1108–1134.

Barton, J.M., Jr., Ryan, B.D., Fripp, R.E.P., and Horrocks, P.C., 1979b, Effects of metamorphism on the Rb-Sr and U-Pb systematics of the Singelele and Bulai Gneisses, Limpopo Mobile Belt, southern Africa: Geological Society of South Africa Transactions, v. 82, p. 259–269.

Barton, J.M., Jr., Du Toit, M.C., van Reenen, D.D., and Ryan, B., 1983, Geochronologic studies in the Southern Marginal Zone of the Limpopo Mobile Belt, southern Africa, in Van Biljon, W.J., and Legg, J.H., eds., The Limpopo Belt: Geological Society of South Africa Special Publication 8, p. 55–64.

Barton, J.M., Jr., Doig, R., Smith, C.B., Bohlender, F., and van Reenen, D.D., 1992, Isotopic and REE characteristics of the intrusive charno-enderbite and enderbite geographically associated with the Matok Pluton, Limpopo Belt, southern Africa: Precambrian Research, v. 55, p. 451–467, doi:10.1016/0301-9268(92)90039-Q.

Barton, J.M., Jr., Holzer, L., Kamber, B., Doig, R., Kramers, J.D., and Nyfeler, D., 1994, Discrete metamorphic events in the Limpopo Belt, southern Africa: Implications for the application of P-T paths in complex metamorphic terranes: Geology, v. 22, p. 1035–1038, doi:10.1130/0091-7613 (1994)022<1035:DMEITL>2.3.CO;2.

Barton, J.M., Jr., Barton, E.S., and Smith, C.B., 1996, Petrography, age and origin of the Shiel alkaline complex, northern Transvaal, South Africa: Journal of African Earth Sciences, v. 22, p. 133–145, doi:10.1016/0899-5362(96)00005-X.

Barton, J.M., Jr., Barnett, J.W.R., Barton, E.S., Barnett, M., Doorgapershad, A., Twiggs, C., Klemd, R., Martin, L., Millonig, L., and Zenglein, R., 2003, The geology of the area surrounding the Venetia kimberlite pipes, Limpopo Belt, South Africa: A complex interplay of nappe tectonics and granite magmatism: South African Journal of Geology, v. 106, p. 109–128, doi:10.2113/106.2-3.109.

Begemann, F., Ludwig, K.R., Lugmair, G.W., Min, K., Nyquist, L.E., Patchett, P.J., Renne, P.R., Shih, C.-Y., Villa, I.M., and Walker, R.J., 2001, Call for an improved set of decay constants for geochronological use: Geochimica et Cosmochimica Acta, v. 65, p. 111–121, doi:10.1016/S0016-7037 (00)00512-3.

Belluso, E., Ruffini, R., Schaller, M., and Villa, I.M., 2000, Electron-microscope and Ar isotope characterization of chemically heterogeneous amphiboles from the Palala shear zone, Limpopo Belt, South Africa: European Journal of Mineralogy, v. 12, p. 45–62.

Berger, A., Scherrer, N.C., and Bussy, F., 2005, Equilibration and disequilibration between monazite and garnet: Indication from phase-composition and quantitative texture analysis: Journal of Metamorphic Geology, v. 23, p. 865–880.

Berger, M., Kramers, J.D., and Nägler, T.F., 1995, Geochemistry and geochronology of charno-enderbites in the Northern Marginal Zone of the Limpopo Belt, southern Africa and genetic models: Schweizerische Mineralogische und Petrographische Mitteilungen, v. 75, p. 17–42.

Blenkinsop, T.G., and Frei, R., 1996, Archaean and Proterozoic mineralization and tectonics at the Renco Mine (Northern Marginal Zone, Limpopo Belt, Zimbabwe): Economic Geology and the Bulletin of the Society of Economic Geologists, v. 91, p. 1225–1238.

Blenkinsop, T.G., Kröner, A., and Chiwara, V., 2004, Single stage, late Archaean exhumation of granulites in the Northern Marginal Zone, Limpopo Belt, Zimbabwe, and relevance to gold mineralization at Renco Mine: South African Journal of Geology, v. 107, p. 377–396, doi:10.2113/107.3.377.

Boshoff, R., van Reenen, D.D., Smit, C.A., Perchuk, L.L., Kramers, J.D., and Armstrong, R., 2006, Geologic history of the Central Zone of the Limpopo Complex: The West Alldays Area: Journal of Geology, v. 114, p. 699–716, doi:10.1086/507615.

Brabander, D.J., and Giletti, B.J., 1995, Strontium diffusion kinetics in amphiboles and significance to thermal history determinations: Geochimica et Cosmochimica Acta, v. 59, p. 2223–2238, doi:10.1016/0016-7037 (95)00102-6.

Brandl, G., 2002, The Geology of the Alldays Area: Pretoria, Council for Geoscience, Explanation, Sheet 2228 Alldays, 71 p.

Buick, I.S., Williams, I.S., Gibson, R.L., Cartwright, I., and Miller, J.A., 2003, Carbon and U-Pb evidence for a Palaeoproterozoic crustal component in the Central Zone of the Limpopo Belt, South Africa: Journal of the Geological Society [London], v. 160, p. 601–612, doi:10.1144/0016-764902-059.

Buick, I.S., Hermann, J., Williams, I.S., Gibson, I.L., and Rubatto, D., 2006, A SHRIMP U-Pb and LA-ICP-MS trace element study of the petrogenesis of garnet-cordierite-orthoamphibole gneisses from the Central Zone of the Limpopo Belt, South Africa: Lithos, v. 88, p. 150–172, doi:10.1016/j.lithos.2005.09.001.

Chavagnac, V., Kramers, J.D., Naegler, Th.F., and Holzer, L., 2001, The behaviour of Nd and Pb isotopes during 2.0 Ga migmatization in paragneisses of the Central Zone of the Limpopo Belt (South Africa and Botswana): Precambrian Research, v. 112, p. 51–86, doi:10.1016/S0301-9268(01)00170-X.

Cherniak, D.J., and Watson, E.B., 2001, Pb diffusion in zircon: Chemical Geology, v. 172, p. 5–24, doi:10.1016/S0009-2541(00)00233-3.

Cherniak, D.J., Hanchar, J.M., and Watson, E.B., 1997, Diffusion of tetravalent cations in zircon: Contributions to Mineralogy and Petrology, v. 127, p. 383–390, doi:10.1007/s004100050287.

Cherniak, D.J., Watson, E.B., Grove, M., and Harrison, T.M., 2004, Pb diffusion in monazite: A combined RBS/SIMS study: Geochimica et Cosmochimica Acta, v. 68, p. 829–840, doi:10.1016/j.gca.2003.07.012.

Chudy, T.C., Zeh, A., Gerdes, A., Klemd, R., and Barton, J.M., Jr., 2008, Palaeoarchaean (3.3 Ga) magmatism and Palaeoproterozoic (2.02 Ga) amphibolite facies metamorphism in the Central Zone of the Limpopo Belt: New geochronological, petrological and geochemical constraints from metabasic and metapelitic rocks from the Venetia area: South African Journal of Geology, v. 111, p. 387–408, doi:10.2113/gssajg.111.4.387.

Corfu, F., Hanchar, J.M., Hoskin, P.W.O., and Kinny, P., 2003, Atlas of zircon textures, in Hanchar, J.M., and Hoskin, P.W.O., eds., Zircon: Mineralogical Society of America, Reviews in Mineralogy and Geochemistry, v. 53, p. 468–500.

Eglington, B.M., and Armstrong, R.A., 2004, The Kaapvaal Craton and adjacent orogens, southern Africa. A geochronological database and overview of the geological development of the craton: South African Journal of Geology, v. 107, p. 13–32, doi:10.2113/107.1-2.13.

Frei, R., and Kamber, B.S., 1995, Single-mineral Pb-Pb dating: Earth and Planetary Science Letters, v. 129, p. 261–268, doi:10.1016/0012-821X (94)00248-W.

Frei, R., Biino, G.G., and Prospert, C., 1995, Dating a Variscan pressure-temperature loop with staurolite: Geology, v. 23, p. 1095–1098, doi: 10.1130/0091-7613(1995)023<1095:DAVPTL>2.3.CO;2.

Frei, R., Villa, I.M., Kramers, J.D., Nägler, T.F., Przybylowicz, W.J., Prozeski, V.M., Hofmann, B., and Kamber, B.S., 1997, Single mineral dating by the Pb-Pb step leaching method: Assessing the mechanisms: Geochimica et Cosmochimica Acta, v. 61, p. 393–414, doi:10.1016/S0016-7037 (96)00343-2.

Frei, R., Blenkinsop, T.G., and Schönberg, R., 1999, Geochronology of the late Archaean Razi and Chilimanzi suites of granites in Zimbabwe: Implications for the late Archaean tectonics of the Limpopo Belt and Zimbabwe Craton: South African Journal of Geology, v. 102, p. 55–64.

Geisler, Th., Pidgeon, R.T., van Bronswijk, W., and Kurtz, R., 2002, Transport of uranium, thorium and lead in metamict zircon under low-temperature hydrothermal conditions: Chemical Geology, v. 191, p. 141–154, doi:10.1016/S0009-2541(02)00153-5.

Geisler, T., Rashwan, A.A., Rahn, M., Poller, U., Zwingmann, H., Pidgeon, R.T., Schleicher, H., and Tomaschek, F., 2003a, Low-temperature hydrothermal alteration of natural metamict zircons from the Eastern Desert, Egypt: Mining Magazine (London), v. 67, p. 485–508, doi:10.1180/0026461036730112.

Geisler, Th., Pidgeon, R.T., Kurtz, R., van Bronswijk, W., and Schleicher, H., 2003b, Experimental hydrothermal alteration of partially metamict zircon: American Mineralogist, v. 88, p. 1496–1513.

Gerdes, A., and Zeh, A., 2009, Zircon formation versus zircon alteration—New insights from combined U-Pb and Lu-Hf in-situ LA-ICP-MS analyses, and consequences for the interpretation of Archean zircon from the Central Zone of the Limpopo Belt: Chemical Geology, v. 261, p. 230–243, doi:10.1016/j.chemgeo.2008.03.005.

Harper, C.T., 1967, On the interpretation of potassium-argon ages from Precambrian shields and Phanerozoid orogens: Earth and Planetary Science Letters, v. 3, p. 128–132, doi:10.1016/0012-821X(67)90023-4.

Hay, D.C., and Dempster, T.J., 2009, Zircon behaviour during low-temperature metamorphism: Journal of Petrology, v. 50, p. 571–589, doi:10.1093/petrology/egp011.

Hickman, M.H., 1978, Isotopic evidence for crustal reworking in the Rhodesian Archean craton, southern Africa: Geology, v. 6, p. 214–216, doi:10.1130/0091-7613(1978)6<214:IEFCRI>2.0.CO;2.

Hickman, M.H., and Wakefield, J., 1975, Tectonic implications of new geochronologic data from the Limpopo Belt at Phikwe, Botswana, southern Africa: Geological Society of America Bulletin, v. 86, p. 1468–1472, doi:10.1130/0016-7606(1975)86<1468:TIONGD>2.0.CO;2.

Holmes, A., and Cahen, L., 1957, Géochronologie africaine 1956, Resultats acquis au 1er juillet 1956: Mémoires de l'Académie Royale Scientifique Coloniale: Belge, Bruxelles, v. 1957, 169 p.

Holzer, L., 1998, The transpressive orogeny at 2 Ga in the Limpopo Belt, southern Africa [Ph.D. thesis]: Bern, Switzerland, University of Bern, 200 p.

Holzer, L., Kamber, B., Kramers, J.D., and Frei, R., 1996, The tectonometamorphic event at 2 Ga in the Limpopo belt and the resetting behaviour of chronometers at high temperature: Geological Survey of Namibia Special Publication 1, p. 127–138.

Holzer, L., Frei, R., Barton, J.M., Jr., and Kramers, J.D., 1998, Unravelling the record of successive high-grade events in the Central Zone of the Limpopo Belt using Pb single-phase dating of metamorphic minerals: Precambrian Research, v. 87, p. 87–115, doi:10.1016/S0301-9268(97)00058-2.

Holzer, L., Barton, J.M., Jr., Paya, B.K., and Kramers, J.D., 1999, Tectonothermal history in the western part of the Limpopo Belt: Test of the tectonic models and new perspectives: Journal of African Earth Sciences, v. 28, p. 383–402, doi:10.1016/S0899-5362(99)00011-1.

Hoskin, P.W.O., and Schaltegger, U., 2003, The composition of zircon and igneous and metamorphic petrogenesis, in Hanchar, J.M., and Hoskin, P.W.O., eds., Zircon: Mineralogical Society of America, Reviews in Mineralogy and Geochemistry, v. 53, p. 27–62.

Jaeckel, P., Kröner, A., Kamo, S.L., Brandl, G., and Wendt, J.I., 1997, Late Archaean to Early Proterozoic granitoid magmatism and high-grade metamorphism in the central Limpopo Belt, South Africa: Journal of the Geological Society [London], v. 154, p. 25–44, doi:10.1144/gsjgs.154.1.0025.

Jäger, E., 1967, Kritische Betrachtungen zur Interpretation der Alterswerte, in Jäger, E., Niggli, E., and Wenk, E., eds., Rb-Sr Altersbestimmungen an Glimmern der Zentralalpen: Zürich, Beiträge zur Geologischen Karte der Schweiz, NF 134, p. 38–40.

Jagoutz, E., 1995, Isotopic constraints on garnet equilibration: Strasbourg, France, European Union of Geosciences Annual Meeting, 8th, March 1995: TERRA Abstracts, v. 7, Suppl. 1, p. 339.

Kamber, B.S., Kramers, J.D., Napier, R., Cliff, R.A., and Rollinson, H.R., 1995a, The Triangle Shear Zone, Zimbabwe, revisited: New data document an important event at 2.0 Ga in the Limpopo Belt: Precambrian Research, v. 70, p. 191–213, doi:10.1016/0301-9268(94)00039-T.

Kamber, B.S., Blenkinsop, T.G., Villa, I.M., and Dahl, P.S., 1995b, Proterozoic transpressive deformation in the Northern Marginal Zone, Limpopo Belt, Zimbabwe: Journal of Geology, v. 103, p. 493–508, doi:10.1086/629772.

Kamber, B.S., Biino, G.G., Wijbrans, J.R., Davies, G.R., and Villa, I.M., 1996, Archaean granulites of the Limpopo Belt, Zimbabwe: One slow exhumation or two rapid events?: Tectonics, v. 15, p. 1414–1430, doi:10.1029/96TC00850.

Kamber, B.S., Frei, R., and Gibb, A.J., 1998, Pitfalls and new approaches in granulite chronometry: An example from the Limpopo Belt, Zimbabwe: Precambrian Research, v. 91, p. 269–285, doi:10.1016/S0301-9268(98)00053-9.

Kelly, N.M., Clarke, G.L., and Harley, S.L., 2006, Monazite behaviour and age significance in poly-metamorphic high-grade terrains: A case study from the western Musgrave Block, central Australia: Lithos, v. 88, p. 100–134, doi:10.1016/j.lithos.2005.08.007.

Kober, B., 1987, Single-zircon evaporation combined with Pb+ emitter bedding for $^{207}Pb/^{206}Pb$ age investigations using thermal ion mass spectrometry, and implications to zirconology: Contributions to Mineralogy and Petrology, v. 96, p. 63–71, doi:10.1007/BF00375526.

Kramers, J.D., Frei, R., Newville, M., Kober, B., and Villa, I.M., 2009, On the valency state of radiogenic lead in zircon and its consequences: Chemical Geology, v. 261, p. 4–11, doi:10.1016/j.chemgeo.2008.09.010.

Kreissig, K., Holzer, L., Frei, I.M., Kramers, J.D., Kröner, A., Smit, C.A., and van Reenen, D.D., 2001, Geochronology of the Hout River Shear Zone and the metamorphism in the Southern Marginal Zone of the Limpopo Belt, southern Africa: Precambrian Research, v. 109, p. 145–173, doi:10.1016/S0301-9268(01)00147-4.

Kröner, A., Jaeckel, P., Brandl, G., Nemchin, A.A., and Pidgeon, R.T., 1999, Single zircon ages for granitoid gneisses in the Central Zone of the Limpopo Belt, southern Africa and geodynamic significance: Precambrian Research, v. 93, p. 299–337, doi:10.1016/S0301-9268(98)00102-8.

Kröner, A., Jaeckel, P., and Brandl, G., 2000, Single zircon ages for felsic to intermediate rocks from the Pietersburg and Giyani greenstone belts and bordering granitoid orthogneisses, northern Kaapvaal Craton, South Africa: Journal of African Earth Sciences, v. 30, p. 773–793, doi:10.1016/S0899-5362(00)00052-X.

Ludwig, K.R., 2000, Isoplot/Ex version 2.2, Berkeley Geochronology Centre Special Publication no. 1a.

MacGregor, A.M., 1953, Precambrian formations of tropical southern Africa: Algiers, International Geological Congress, 19th, Proceedings, v. 1, p. 39–50.

McCourt, S., and Armstrong, R.A., 1998, SIMS U-Pb zircon geochronology of granites from the Central Zone, Limpopo Belt, southern Africa: Implications for the age of the Limpopo Orogeny: South African Journal of Geology, v. 101, p. 329–338.

McFarlane, C.R.M., Connelly, J.N., and Carlson, W.D., 2006, Contrasting response of monazite and zircon to a high-T thermal overprint: Lithos, v. 88, p. 135–149, doi:10.1016/j.lithos.2005.08.008.

Mezger, K., and Krogstad, E.J., 1997, Interpretation of discordant U-Pb zircon ages: An evaluation: Journal of Metamorphic Geology, v. 15, p. 127–140, doi:10.1111/j.1525-1314.1997.00008.x.

Mezger, K., Hanson, G.N., and Bohlen, S.R., 1989, U-Pb systematics of garnet: Dating the growth of garnet in the Late Archean Pikwitonei granulite domain at Cauchon and Natawahunan Lakes, Manitoba, Canada: Contributions to Mineralogy and Petrology, v. 101, p. 136–148, doi:10.1007/BF00375301.

Mezger, K., Essene, E.J., and Halliday, A.N., 1992, Closure temperatures of the Sm-Nd system in metamorphic garnets: Earth and Planetary Science Letters, v. 113, p. 397–409, doi:10.1016/0012-821X(92)90141-H.

Millonig, L., Zeh, A., Gerdes, A., and Klemd, R., 2008, Neoarchaean high-grade metamorphism in the Central Zone of the Limpopo Belt (South Africa): Combined petrological and geochronological evidence from the Bulai pluton: Lithos, v. 103, p. 333–351, doi:10.1016/j.lithos.2007.10.001.

Millonig, L., Zeh, A., Gerdes, A., Klemd, R., and Barton, J.M., Jr., 2010, Decompressional heating of the Mahalapye Complex (Limpopo Belt, Botswana): A response to Palaeoproterozoic magmatic underplating?: Journal of Petrology, v. 51, p. 703–729.

Mkweli, S., Kamber, B., and Berger, M., 1995, A westward continuation of the Zimbabwe Craton–Northern Marginal Zone tectonic break and new age constraints on the timing of the thrusting: Journal of the Geological Society [London], v. 152, p. 77–83, doi:10.1144/gsjgs.152.1.0077.

Möller, A., O'Brien, P.J., Kennedy, A., and Kröner, A., 2003, Linking growth episodes of zircon and metamorphic textures to zircon chemistry: An example from the ultrahigh temperature granulites of Rogaland (SW Norway), in Vance, D., Muller, W., and Villa, I., eds., Geochronology: Linking the Isotopic Record with Petrology and Textures: Geological Society [London] Special Publication 220, p. 65–81.

Mouri, H., Brandl, G., Whitehouse, M.J., de Waal, S., and Guiraud, M., 2008, CL-imaging and ion microprobe dating of single zircons from a high-grade rock from the Central Zone, Limpopo Belt, South Africa: Evidence for a single metamorphic event at ~2.0 Ga: Journal of African Earth Sciences, v. 50, p. 111–119, doi:10.1016/j.jafrearsci.2007.09.011.

Mouri, H., Whitehouse, M.J., Brandl, G., and Rajesh, H.M., 2009, A magmatic age and four successive metamorphic events recorded in zircons from a single meta-anorthosite sample in the Central Zone of the Limpopo Belt, South Africa: Journal of the Geological Society [London], v. 166, p. 827–830, doi:10.1144/0016-76492008-148.

Perchuk, L.L., van Reenen, D.D., Varlamov, D.A., Van Kal, S.M., Boshoff, R., and Tabatabaeimanesh, S.M., 2008, P-T record of two high-grade metamorphic events in the Central Zone of the Limpopo Complex, South Africa: Lithos, v. 103, p. 70–105, doi:10.1016/j.lithos.2007.09.011.

Pidgeon, R.T., O'Neil, J.R., and Silver, L.T., 1966, Uranium and lead isotopic stability in a metamict zircon under experimental hydrothermal conditions: Science, v. 154, p. 1538–1540, doi:10.1126/science.154.3756.1538.

Pidgeon, R.T., O'Ncil, J.R., and Silver, L.T., 1973, Observations on the crystallinity and the U-Pb system of a metamict Ceylon zircon under experimental hydrothermal conditions: Fortschritte der Mineralogie, v. 50, p. 118.

Pyle, J.M., and Spear, F.S., 2003, Four generations of accessory-phase growth in low-pressure migmatites from SW New Hampshire: American Mineralogist, v. 88, p. 338–351.

Pyle, J.M., Spear, F.S., Rudnick, R.L., and McDonough, W.E., 2001, Monazite-xenotime-garnet equilibrium in metapelites and a new monazite-garnet thermometer: Journal of Petrology, v. 42, p. 2083–2107, doi:10.1093/petrology/42.11.2083.

Rigby, M.J., Basson, I.J., Kramers, J.D., Barnett, W.P., Graser, P., and Mavimbela, P.K., 2011, The structural, metamorphic and temporal evolution of the country rocks surrounding Venetia Mine, Limpopo Belt, South Africa: Evidence for a single Palaeoproterozoic tectono-metamorphic event and implications for a tectonic model: Precambrian Research (in press).

Roden, M.F., Laźko, E.E., and Jagoutz, E., 1996, Garnet pyroxenites from the Mir kimberlite pipe; mineral compositions and Sm-Nd systematics of dikes crosscutting coarse garnet lherzolite in the Siberian lithosphere: Eos (Transactions, American Geophysical Union), v. 77, Supplement, p. 277.

Roering, C., van Reenen, D.D., Smit, C.A., Barton, J.M., Jr., De Beer, J.H., De Wit, M.J., Stettler, E.H., Van Schalkwyk, J.F., Stevens, G., and Pretorius, S.J., 1992, Tectonic model for the evolution of the Limpopo Belt: Precambrian Research, v. 55, p. 539–552, doi:10.1016/0301-9268(92)90044-O.

Schaller, M., Steiner, O., Studer, I., Frei, R., and Kramers, J.D., 1997, Pb stepwise leaching (PbSL) dating of garnet—Addressing the inclusion problem: Schweiz, Mineralogische und Petrographische Mitteilungen, v. 77, p. 113–121.

Schaller, M., Steiner, O., Studer, I., Holzer, L., Herwegh, M., and Kramers, J.D., 1999, Exhumation of Limpopo Central Zone granulites and dextral continental-scale transcurrent movement at 2.0 Ga along the Palala Shear Zone, Northern Province, South Africa: Precambrian Research, v. 96, p. 263–288, doi:10.1016/S0301-9268(99)00015-7.

Shannon, R.D., 1976, Revised effective ionic radii and systematic studies of interatomic distances in halides and chalcogenides: Acta Crystallographica. Section A, Crystal Physics, Diffraction, Theoretical and General Crystallography, v. 32, p. 751–767, doi:10.1107/S0567739476001551.

Sinha, A.K., Wayne, D.M., and Hewitt, D.A., 1992, The hydrothermal stability of zircon: Preliminary experimental and isotopic studies: Geochimica et Cosmochimica Acta, v. 56, p. 3551–3560, doi:10.1016/0016-7037(92)90398-3.

Thöni, M., and Jagoutz, E., 1992, Some new aspects of dating eclogites in orogenic belts: Sm-Nd, Rb-Sr, and Pb-Pb isotopic results from the Austroalpine Saualpe and Koralpe type-locality (Carinthia/Styria, southeastern Austria): Geochimica et Cosmochimica Acta, v. 56, p. 347–368, doi:10.1016/0016-7037(92)90138-9.

Tilton, G.R., 1960, Volume diffusion as a mechanism for discordant lead ages: Journal of Geophysical Research, v. 65, p. 2933–2945, doi:10.1029/JZ065i009p02933.

Treloar, P.J., Coward, M.P., and Harris, N.B.W., 1992, Himalayan–Tibetan analogies for the evolution of the Zimbabwe Craton and the Limpopo Belt: Precambrian Research, v. 55, p. 571–587.

Van Breemen, O., and Dodson, M.H., 1972, Metamorphic chronology of the Limpopo Belt, southern Africa: Geological Society of America Bulletin, v. 83, p. 2005–2018, doi:10.1130/0016-7606(1972)83[2005:MCOTLB]2.0.CO;2.

Van Breemen, O., and Hawkesworth, C.J., 1980, Sm-Nd isotopic study of garnets and their metamorphic host rocks: Transactions of the Royal Society of Edinburgh, v. 71, p. 97–102.

Van Breemen, O., Dodson, M.H., and Vail, J.R., 1966, Isotopic age measurements on the Limpopo orogenic belt, southern Africa: Earth and Planetary Science Letters, v. 1, p. 401–406, doi:10.1016/0012-821X(66)90036-7.

van Reenen, D.D., Boshoff, R., Smit, C.A., Perchuk, L.L., Kramers, J.D., McCourt, S., and Armstrong, R.A., 2008, Geochronological problems related to polymetamorphism in the Limpopo Complex, South Africa: Gondwana Research, v. 14, p. 644–662, doi:10.1016/j.gr.2008.01.013.

Vavra, G., Schmid, R., and Gebauer, D., 1999, Internal morphology, habit and U-Th-Pb microanalysis of amphibolite-to-granulite facies zircons: Geochronology of the Ivrea Zone (Southern Alps): Contributions to Mineralogy and Petrology, v. 134, p. 380–404, doi:10.1007/s004100050492.

Villa, I.M., 2010, Disequilibrium textures vs. equilibrium modelling: Geochronology at the Crossroads, in Spalla, M.I., Marotta, A.M., and Gosso, G., eds., Advances in Interpretation of Geological Processes: Geological Society [London] Special Publication 332, p. 1–15, doi:10.1144/SP332.1.

Villa, I.M., Grobéty, B., Kelley, S.P., Trigila, R., and Wieler, R., 1996, Assessing Ar transport paths and mechanisms in the McClure Mountains hornblende: Contributions to Mineralogy and Petrology, v. 126, p. 67–80, doi:10.1007/s004100050236.

Villa, I.M., Hermann, J., Müntener, O., and Trommsdorff, V., 2000, ^{39}Ar-^{40}Ar dating of multiply zoned amphibole generations (Malenco, Italian Alps): Contributions to Mineralogy and Petrology, v. 140, p. 363–381, doi:10.1007/s004100000197.

Wasserburg, G.J., 1963, Diffusion processes in lead-uranium systems: Journal of Geophysical Research, v. 68, p. 4832–4846.

Whittaker, E.J.W., and Muntus, R., 1970, Ionic radii for use in geochemistry: Geochimica et Cosmochimica Acta, v. 34, p. 945–956, doi:10.1016/0016-7037(70)90077-3.

Zeh, A., Gerdes, A., Klemd, R., and Barton, J.M., Jr., 2007, Archaean to Proterozoic crustal evolution in the Central Zone of the Limpopo Belt (South Africa–Botswana): Constraints from combined U-Pb and Lu-Hf isotope analyses of zircon: Journal of Petrology, v. 48, p. 1605–1639, doi:10.1093/petrology/egm032.

Zeh, A., Gerdes, A., Klemd, R., and Barton, J.R., Jr., 2008, U–Pb and Lu–Hf isotope record of detrital zircon grains from the Limpopo Belt—Evidence for crustal recycling at the Hadean to early-Archean transition: Geochimica et Cosmochimica Acta, v. 72, p. 5304–5329, doi:10.1016/j.gca.2008.07.033.

Zeh, A., Gerdes, A., and Barton, J.M., Jr., 2009, Archean accretion and crustal evolution of the Kalahari Craton—The zircon age and Hf isotope record of granitic rocks from Barberton/Swaziland to the Francistown Arc: Journal of Petrology, v. 50, p. 933–966, doi:10.1093/petrology/egp027.

Zeh, A., Gerdes, A., Barton, J.R., Jr., and Klemd, R., 2010, U–Th–Pb and Lu–Hf systematics of zircon from TTG's, leucosomes, meta-anorthosites and quartzites of the Limpopo Belt (South Africa): Constraints for the formation, recycling and metamorphism of Palaeoarchean crust: Precambrian Research, v. 179, p. 50–68, doi:10.1016/j.precamres.2010.02.012.

MANUSCRIPT ACCEPTED BY THE SOCIETY 24 MAY 2010

High-pressure and ultrahigh-temperature granulite-facies metamorphism of Precambrian high-grade terranes: Case study of the Limpopo Complex

Toshiaki Tsunogae

Graduate School of Life and Environmental Sciences (Earth Evolution Sciences), University of Tsukuba, Ibaraki 305-8572, Japan, and Department of Geology, University of Johannesburg, Auckland Park 2006, Johannesburg, Republic of South Africa

Dirk D. van Reenen

Department of Geology, University of Johannesburg, Auckland Park 2006, Johannesburg, Republic of South Africa

ABSTRACT

We report new petrological data for granulites from the Central Zone of the Limpopo Complex, southern Africa, and construct a prograde P-T path that traverses from high-pressure granulite-facies metamorphism to peak ultrahigh-temperature (UHT) metamorphism by rapid decompression, which was followed by further decompression and cooling. Mg-rich (X_{Mg} ~0.58) staurolite enclosed within poikiloblastic garnet in an Mg-Al-rich rock from the Beit Bridge area is rarely mantled by a sapphirine + quartz corona, suggesting the progress of the prograde dehydration reaction: staurolite + garnet → sapphirine + quartz + H_2O. The symplectic sapphirine + quartz developed around staurolite probably implies decompression from P >14 kbar toward the stability of sapphirine + quartz at T ~1000 °C along a clockwise P-T path. The orthopyroxene + sillimanite + quartz assemblage mantled by cordierite aggregates in a pelitic granulite from the same area also suggests extreme metamorphism and subsequent further decompression. Various corona textures such as kyanite + sapphirine, sapphirine + cordierite, and orthopyroxene + cordierite were probably formed as a result of decompression cooling events. The prograde high-pressure metamorphism and the following UHT event relate to the collisional tectonics of the Zimbabwe and Kaapvaal Cratons, which are associated with the amalgamation of microcontinents during the Neoarchean.

INTRODUCTION

The Limpopo Complex in southern Africa is known for its classic exposures of regionally metamorphosed granulite-facies rocks formed by the collision of two Mesoarchean to Neoarchean cratons; the Zimbabwe Craton to the north and the Kaapvaal Craton to the south during the Neoarchean (e.g., van Reenen et al., 1987). The complex has been internally subdivided into three zones—the Central Zone (CZ), Southern Marginal Zone (SMZ), and Northern Marginal Zone (NMZ), by major ca. 2.0 Ga strike-slip shear zones that bound the CZ (Fig. 1) (Mason, 1973; van Reenen et al., 1987, 2008). The two marginal zones are separated from the adjacent cratons by inward dipping thrust zones (Fig. 1). The NMZ and SMZ have been regarded as the high-grade equivalents of the adjacent granite-greenstone terranes (Zimbabwe and Kaapvaal Craton lithologies, respectively) (e.g., van Reenen et al., 1987, 1992; Kreissig et al., 2000, 2001) because of widespread occurrences of orthogneisses and associated metasedimentary and metavolcanic rocks, whereas the CZ is composed mainly of various supracrustal lithologies (e.g., Brandl, 1983; Watkeys et al., 1983).

Previous petrological investigations of the CZ demonstrated that the zone underwent peak granulite-facies metamorphism at ~800–900 °C and 8–10 kbar (e.g., Watkeys et al., 1983; Tsunogae and Miyano, 1989; Droop, 1989; Zeh et al., 2004; Perchuk et al., 2008). Recently Rigby (2009) performed P-T pseudosections, garnet isopleth thermobarometry, and mineral mode–isopleth modeling studies for metapelites from the CZ and obtained a high-pressure stage at 10–11 kbar and 800 °C, followed by a peak stage (8 kbar and 850 °C) and a subsequent retrograde stage (4–5 kbar at $T<650$ °C) along a single clockwise P-T path. However, earlier petrological investigations of sapphirine-bearing Mg-Al-rich rocks from the CZ demonstrated that the peak metamorphic condition of the granulites from the Zimbabwean part of the CZ is higher, 890–930 °C at 9–10 kbar (e.g., Tsunogae and van Reenen, 2006). The rocks formed through such ultrahigh-temperature (UHT) metamorphism are commonly characterized by the occurrence of unique minerals or mineral assemblages such as sapphirine + quartz, spinel + quartz, orthopyroxene + sillimanite + quartz, Al-rich orthopyroxene, inverted pigeonite, and mesoperthite (e.g., Harley, 1998a, 2008; Kelsey, 2008).

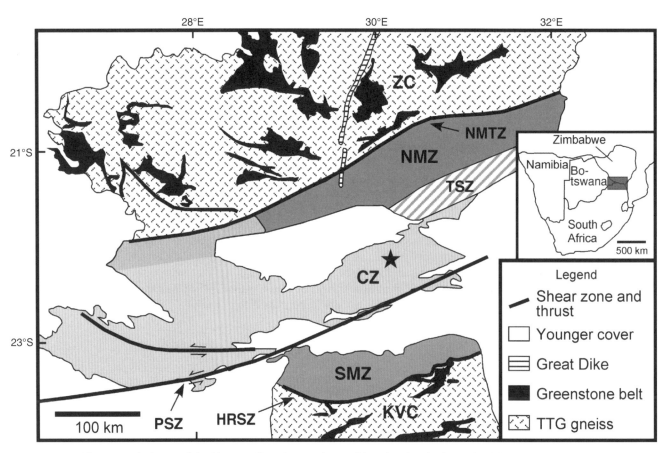

Figure 1. Simplified geological map of the Limpopo Complex, southern Africa, showing the internal subdivision into a Central Zone (CZ), separated from the Southern (SMZ) and Northern (NMZ) Marginal Zones by ca. 2.0 Ga strike-slip shear zones, the Palala Shear Zone (PSZ) in the south and the Triangle Shear Zone (TSZ) in the north. The two marginal zones are separated from the Kaapvaal Craton (KVC) and the Zimbabwe Craton (ZC) by ca. 2.65 Ga inward-dipping thrusts, the Hout River Shear Zone (HRSZ) in the south, and the North Marginal Thrust Zone (NMTZ) in the north. Also shown is the sampling locality (star) of this study. TTG—tonalite-trondhjemite-granodiorite.

Among the diagnostic assemblages of UHT metamorphism, equilibrium sapphirine + quartz has been regarded as the most robust evidence for extreme temperatures above 950 °C, and some exceeding 1000 °C (e.g., Hensen and Green, 1973; Bertrand et al., 1991; Harley, 1998a; Kelsey et al., 2004; Kelsey, 2008). In the Limpopo CZ, granulites with peak temperatures exceeding 900 °C have been found from several localities (e.g., Tsunogae and van Reenen, 2006). Although Schreyer et al. (1984) reported coexisting sapphirine and quartz from one locality in the South African part of the CZ, these authors never considered it as evidence for UHT. In contrast, the sapphirine + quartz assemblage has not yet been reported from the adjacent SMZ and NMZ, for which maximum temperature conditions of ~800–850 °C have previously been estimated (e.g., van Reenen, 1983; Stevens and van Reenen, 1992; Tsunogae et al., 1992). However, available petrological studies in the SMZ have provided evidence for an extreme thermal event at T ~900 °C on the basis of geothermobarometric considerations and mineral phase equilibria analyses (e.g., Tsunogae et al., 2004; Belyanin et al., 2010).

In this study we report a new occurrence of the equilibrium sapphirine + quartz assemblage from the CZ exposed in Zimbabwe (Fig. 1). Our detailed studies of this new locality in the CZ recorded petrographic evidence for sapphirine + quartz in direct association, which mantles Mg-rich (X_{Mg} = Mg/(Fe+Mg) ~0.58) staurolite enclosed in coarse-grained poikiloblastic garnet. The sapphirine + quartz + staurolite texture record here has been reported only rarely, for instance, from the Neoproterozoic to Cambrian Palghat-Cauvery suture zone in southern India (Nishimiya et al., 2010). We describe the detailed petrography and mineral chemistry of the sapphirine + quartz–bearing granulite from the CZ of the Limpopo Complex and interpret the various prograde and retrograde reactions for which evidence is preserved in this rock. In addition we report the rare orthopyroxene + sillimanite + quartz assemblage, also diagnostic of UHT metamorphism, from an adjacent locality as further confirmation of extreme peak metamorphic conditions at UHT followed by further decompression. We apply the available KFMASH (K_2O-FeO-MgO-Al_2O_3-SiO_2-H_2O) petrogenetic grid and experimental data to estimate the peak metamorphic conditions and P-T path of the studied samples. Our findings have important implications for the understanding of the development and exhumation of UHT orogenic belts associated with the collision of two Archean cratons.

GEOLOGICAL SETTING

Detailed discussions pertaining to the geological, structural, and geochronological evolution of the CZ of the Limpopo Complex have been published in numerous papers (e.g., van Reenen et al., 1992, and references therein; Kramers et al., 2006, and references therein), including different papers in this volume. Briefly, the CZ is mainly composed of the supracrustal Beit Bridge Complex, which comprises metaquartzite and meta–banded iron formation (meta-BIF), marble and calc-silicate rocks, leucocratic gneisses, garnet-biotite paragneisses, and mafic granulites with rare pelitic granulites. The rocks of the Beit Bridge Complex have, over an extended period of time from ca. 3.3 Ga to ca. 2.5 Ga, been affected by major igneous events and by at least three high-grade metamorphic and deformational events at ca. 3.3 Ga, ca. 2.65 Ga, and ca. 2.0 Ga. The igneous activity includes the emplacement before 3.0 Ga of meta-anorthosite and meta-gabbro of the Messina Suite (Mouri et al., 2009) and of a suite of gray gneisses that form part of the Sand River Gneiss. This was followed after 3.0 Ga by the emplacement of a second suite of gray gneisses (Alldays-Verbaard Gneisses) and by various leucocratic gneisses of which the garnet-bearing Singelele Gneiss is the most prominent. The major igneous activity in the CZ ended with the emplacement of the syn- to post-tectonic Bulai granitoid at ca. 2.61 Ga (Millonig et al., 2008).

Available age data for the NMZ and SMZ suggest a single high-grade metamorphic event that occurred between 2.6 and 2.7 Ga (e.g., Barton and van Reenen, 1992; Berger et al., 1995; Kreissig et al., 2000, 2001; Blenkinsop et al., 2004; Kramers et al., 2006). In contrast, the peak metamorphic age of the CZ is still a matter of controversy, as both the Neoarchean (2.6–2.7 Ga) (e.g., Armstrong et al., 1991; Boshoff et al., 2006; van Reenen et al., 2008) and the Paleoproterozoic (ca. 2.0 Ga) (e.g., Kamber et al., 1995; Jaeckel et al., 1997; Holzer et al., 1998; Kramers et al., 2001; Zeh et al., 2004) ages have been reported from the same metamorphic complex.

Several of the previously published P-T paths proposed for the CZ reflect peak metamorphism at 800–900 °C at 7–9 kbar, followed by near isothermal decompression to 4–5 kbar along a clockwise P-T path (Windley et al., 1984; Harris and Holland, 1984; Droop, 1989; Tsunogae and Miyano, 1989). In contrast, Perchuk et al. (2008), van Reenen et al. (2004), and Zeh et al. (2004) suggested retrograde decompression-cooling paths for the CZ. A multistage P-T history recently proposed for pelitic granulites (Perchuk et al., 2006, 2008; Perchuk and van Reenen, 2008; van Reenen et al., 2008) documents evidence for a Neoarchean (ca. 2.65 Ga) post-peak decompressional cooling event that traverses from ~8.5 kbar at ~850 °C to ~5.5 kbar at ~550 °C. The rocks of the CZ were again reworked more than 600 Ma later during a ca. 2.02 Ga Paleoproterozoic high-grade event. This event records an isobaric (P = 5.5 kbar) re-heating (550–750 °C) stage followed by decompressional cooling to ~3.3 kbar. Although there are some differences in the post-peak retrograde P-T paths, most studies support a clockwise P-T evolution for the CZ. However, the peak metamorphic conditions reached in the CZ, and the maximum pressure that the CZ lithologies underwent, are still unknown.

PETROGRAPHY

Samples of UHT granulite were collected from the supracrustal Beit Bridge Complex in the CZ north of Beitbridge in Zimbabwe (Fig. 1) by Malcohm Light and presented to the second author in the mid-1970s (sample MPL1872; Mg-Al-rich granulite). The first author collected sample C227 (pelitic granulite) from the same general area. We carried out field observation and lithological classification as well as sample collection for the latter locality. Photomicrographs of representative thin sections are shown in Figure 2. Details of the representative reaction textures are also shown in backscattered electron (BSE) images in Figure 3. Laser Raman spectroscopic analysis for kyanite was done using the JASCO NRS-1000 Raman microspectrometer at the National Institute of Polar Research (Japan) with an excitation radiation of 514.5 nm using an Ar$^+$ green laser.

The studied area north of Beitbridge in Zimbabwe is composed mainly of various supracrustal lithologies that include pelitic, mafic, and quartzofeldspathic gneisses, meta-BIF, calc-silicate gneisses, and intrusive granitoids. The rocks show obvious compositional layering with dominant N-S–trending foliation. Pelitic granulites (e.g., sample C227) discussed in this study show migmatitic textures with garnet-, biotite-, and cordierite-rich melanocratic layers and quartzofeldspathic layers. Leucocratic coarse patches composed of garnet + quartz + feldspar, a few centimeters to several centimeters in length, and probably formed by in situ partial melting, are present in biotite-rich matrix in some localities. Sample C227, discussed in this study, was collected from an outcrop without intense migmatization. The rock comprises dark patches with fine-grained orthopyroxene and cordierite, surrounded by a quartzofeldspathic matrix.

Mg-Al-RICH GRANULITE (SAMPLE MPL1872)

This rock is composed of poikiloblastic garnet (40%–50%) in a matrix of orthopyroxene (20%–30%), quartz (20%–30%), and cordierite (5%–10%) (Fig. 2A). Rare biotite and muscovite occur as retrograde phases around orthopyroxene. The poikiloblastic garnet is coarse grained (~3 cm) and contains numerous monophase or multiphase inclusions of subidioblastic to xenoblastic sapphirine, staurolite, kyanite, rutile, and ilmenite. The inclusion staurolite is fine grained (<0.7 mm in length) and yellowish in color (Fig. 2B) and is commonly mantled by xenoblastic sapphirine (Fig. 2C) or sapphirine + quartz corona (Figs. 2B, 3A). In places staurolite has been completely replaced by sapphirine + quartz symplectite (Fig. 2D), whereas fine-grained (20–30 µm) relict staurolite grains are rarely present as inclusions in the sapphirine and quartz (Figs. 3B, 3C). Kyanite occurs as isolated inclusions in garnet and is commonly intergrown with sapphirine (Fig. 2E). The occurrence of kyanite has been confirmed by laser Raman spectroscopy analysis (Fig. 4). Relict staurolite is also present in the kyanite + sapphirine intergrowth (Fig. 2E). The garnet also contains sapphirine + cordierite aggregate with kyanite (Fig. 2F). Sillimanite is absent in the garnet, but it is present as a rare mineral in the matrix. Such sillimanite occurs as aggregates of fine-grained mineral, which is a common texture of pseudomorphic sillimanite after kyanite (e.g., Kanazawa et al., 2009). The sapphirine + cordierite also occurs as symplectite around the matrix pseudomorphic sillimanite adjacent to garnet (Fig. 2G).

The coarse-grained poikiloblastic garnet is commonly mantled by a symplectic corona of cordierite (Fig. 2A) or by a cordierite + orthopyroxene symplectite (Fig. 2H). Matrix orthopyroxene is coarse grained (~2 cm), subidioblastic to xenoblastic, and surrounded by garnet and quartz (Fig. 2A).

PELITIC GRANULITE (SAMPLE C227)

Garnet- and sillimanite-bearing granulite of probable metasedimentary origin is a common supracrustal lithology in the CZ. Sample C227 is a garnet-poor orthopyroxene-sillimanite–bearing granulite, similar to pelitic granulites sampled throughout the CZ. It is composed of cordierite (15%–25%), orthopyroxene (5%–10%), sillimanite (5%–10%), biotite (5%–10%), quartz (40%–50%), K-feldspar (10%–15%), and plagioclase (2%–5%). Accessory minerals are rutile and zircon. Garnet occurs only in one thin section out of six thin sections studied. The rock is characterized by aggregates of fine-grained (0.2–0.4 mm) cordierite present as patches of ~2–4 mm in length in the matrix of coarse-grained quartz (up to 6 mm) and feldspars (up to 3.5 mm) (Fig. 2I). The foliation of the rock is defined by elongated quartz. The cordierite aggregates contain fine-grained (0.05–0.3 mm) inclusions of xenoblastic orthopyroxene, sillimanite, and biotite (Fig. 2I). Some coarse-grained subidioblastic orthopyroxene (up to 3 mm) and sillimanite (up to 0.7 mm) are also present along the margin of the patch (Fig. 2J). Rare garnet, which also occurs along the grain boundary, is xenoblastic and contains numerous inclusions of sillimanite (Fig. 2K).

Mineral Chemistry

Chemical analyses of all minerals were carried out using electron microprobe analyzer (JEOL JXA8621 and 8530F) at the Chemical Analysis Division of the Research Facility Center for Science and Technology, University of Tsukuba. The analyses were performed under conditions of 20 kV accelerating voltage and 10 nA sample current, and the data were regressed using an oxide-ZAF correction program supplied by JEOL. Below, we describe compositions of some important ferromagnesian minerals formed during prograde to peak metamorphism in the examined rock. Representative compositions of the minerals are given in Tables 1 and 2. Fe^{3+} contents of sapphirine and garnet were calculated on the basis of stoichiometry.

Garnet

Garnet in sample MPL1872 is essentially a solid solution between pyrope and almandine (X_{Mg} = 0.42–0.61) with minor grossular (<6 mol%) and spessartine (<1 mol%). Both the core and rim of the analyzed garnet grains are rich in Mg and are nearly compositionally homogeneous (Alm$_{38-42}$Pyr$_{55-58}$Grs$_{2-6}$Sps$_{0-1}$). The garnet core adjacent to the sapphirine inclusions is slightly enriched in Mg (X_{Mg} = 0.58–0.61). The garnet rim adjacent to the retrograde cordierite + orthopyroxene corona shows the highest Fe content of Alm$_{49-54}$Pyr$_{40-45}$Grs$_{5-6}$Sps$_{0-1}$.

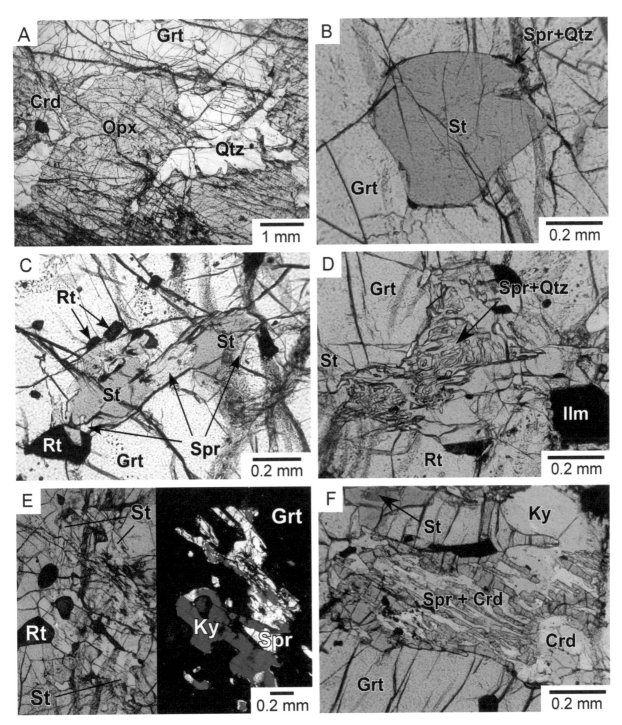

Figure 2 (*Continued on following page*). Photomicrographs showing representative rock textures discussed in this study. (A) to (H): sample MPL1872. (I) to (L): sample C227. (A) Coarse-grained garnet (Grt) + orthopyroxene (Opx) + quartz (Qtz) assemblage in sample MPL1872. Cordierite (Crd) is also present around garnet as a retrograde mineral. (B) Subidioblastic staurolite (St) inclusion in poikiloblastic garnet. Fine-grained corona of sapphirine (Spr) + staurolite is partly present between garnet and staurolite. (C) Corona of sapphirine mantling staurolite. (D) Sapphirine + quartz intergrowth included in garnet. Relict staurolite grains are present as inclusions in sapphirine and quartz, as shown in Figure 3B; Ilm—ilmenite; Rt—rutile. (E) Intergrowth of kyanite (Ky) and sapphirine within poikiloblastic garnet. Rare inclusions of staurolite are present in kyanite and sapphirine. Right: polarized light; left: crossed polars. (F) Sapphirine + cordierite intergrowth with kyanite included in garnet.

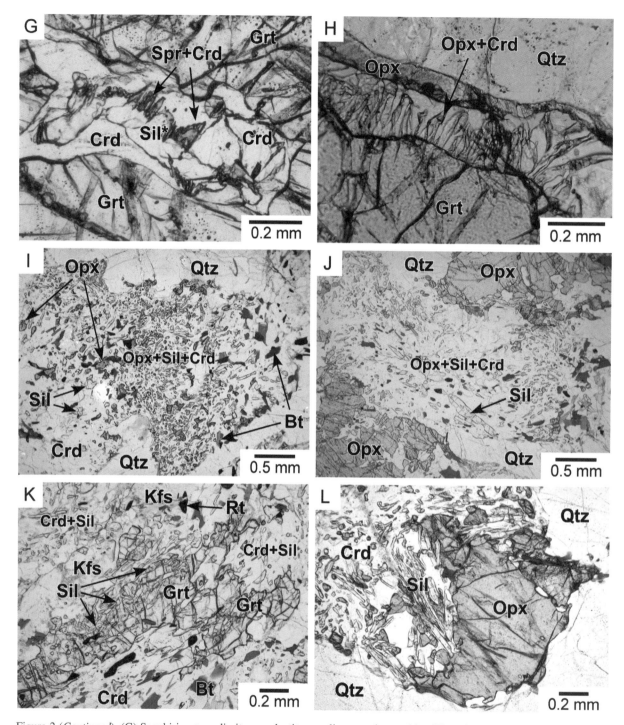

Figure 2 (*Continued*). (G) Sapphirine + cordierite symplectite mantling pseudomorphic sillimanite (Sil*) after kyanite adjacent to garnet. (H) Orthopyroxene + cordierite corona between rim of poikiloblastic garnet and matrix quartz. (I) Cordierite aggregate with inclusions of sillimanite, orthopyroxene, and biotite (Bt) in the matrix of quartz. (J) Coarse-grained orthopyroxene and sillimanite present along the grain margin of cordierite aggregate. (K) Garnet with numerous sillimanite inclusions. Garnet is present between cordierite patch and quartzofeldspathic matrix; Kfs—K-feldspar. (L) Orthopyroxene, sillimanite, and quartz separated by retrograde cordierite.

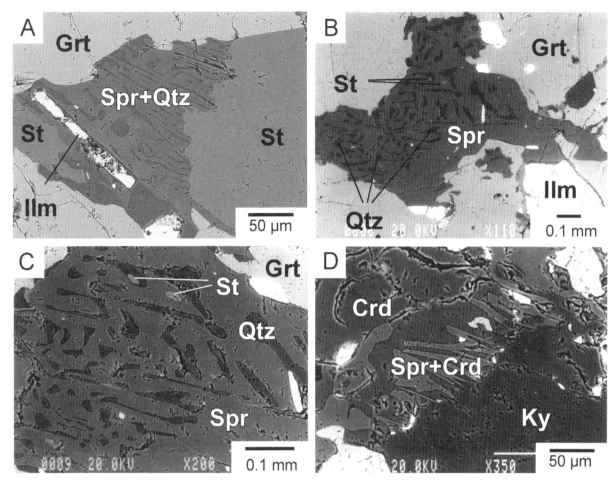

Figure 3. Backscattered electron images showing detailed textures in sample MPL1872. (A) Sapphirine + quartz symplectite between staurolite and garnet. (B) Irregular-shaped sapphirine + quartz intergrowth included in garnet. (C) Relict staurolite grains in sapphirine and quartz. (D) Sapphirine + cordierite symplectite around kyanite in garnet.

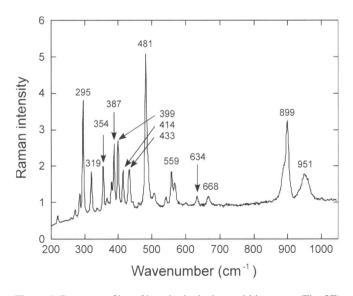

Figure 4. Raman profiles of kyanite inclusions within garnet (Fig. 2E) in sample MPL1872.

Garnet in pelitic granulite (sample C227) also shows a slight increase in Fe from core ($Alm_{53–54}Pyr_{43–45}Grs_{2–3}Sps_{0–1}$) to rim ($Alm_{54–55}Pyr_{42–43}Grs_{2–3}Sps_{0–1}$).

Sapphirine

Sapphirine in sample MPL1872 exhibits two types of occurrence: an earlier xenoblastic phase as inclusions in garnet with staurolite and/or kyanite (Fig. 3A), and a later subidioblastic phase intergrown with cordierite around kyanite (Fig. 3D). Both types have magnesian compositions (X_{Mg} = 0.83–0.89), and the (Mg, Fe)O:Al_2O_3:SiO_2 ratio is close to the 7:9:3 end-member composition (Fig. 5). The earlier sapphirine is slightly more Fe-rich (X_{Mg} = 0.83–0.87) than the later phase (X_{Mg} = 0.87–0.89). As shown in Figure 5, sapphirine with staurolite and/or quartz is relatively Si-rich and Al-poor (Si = 1.47–1.59 per formula unit, p.f.u.) compared with the symplectitic phase with cordierite (1.34–1.42). The calculated $Fe^{3+}/(Fe^{2+}+Fe^{3+})$ ratio is low: 0.01–0.28. Other components such as TiO_2, Cr_2O_3, and ZnO total <0.2 wt%.

TABLE 1. REPRESENTATIVE ELECTRON MICROPROBE ANALYSES
OF GARNET AND SAPPHIRINE IN SAMPLES MPL1872 AND C227

Rec. no.	79	174	163	5b	6	45	66	68	52
Mineral name			Garnet (O = 12)				Sapphirine (O = 20)		
Sample no.	MPL1872	MPL1872	MPL1872	C227	C227	MPL1872	MPL1872	MPL1872	MPL1872
Remarks	Core	Rim	Rim (with Crd)	Core	Rim	In Grt	With St	With Qtz	With Crd
Figures*	2A	2A	2H	2K	2K	2E	2C	2D	2D
SiO_2	40.54	38.97	39.84	39.40	39.27	12.49	12.30	12.52	11.91
Al_2O_3	22.71	22.86	22.34	22.46	22.57	63.53	63.77	63.91	64.74
TiO_2	0.07	0.05	0.01	0.25	0.02	0.06	0.17	0.05	0.02
Cr_2O_3	0.01	0.00	0.07	0.12	0.08	0.04	0.09	0.08	0.03
Fe_2O_3	1.06	1.61	0.00	0.00	0.00	1.70	0.95	0.60	1.06
FeO	18.55	22.18	25.25	25.60	25.45	4.96	4.77	5.64	4.07
MnO	0.24	0.38	0.50	0.43	0.41	0.05	0.00	0.00	0.00
MgO	16.01	12.04	10.44	11.39	11.10	16.87	16.91	16.55	16.94
ZnO	0.00	0.04	0.03	0.00	0.06	0.05	0.04	0.00	0.03
CaO	0.95	2.00	2.03	0.73	0.75	0.02	0.02	0.00	0.00
Na_2O	0.00	0.01	0.03	0.00	0.00	0.00	0.00	0.00	0.02
K_2O	0.01	0.00	0.00	0.00	0.01	0.01	0.00	0.00	0.01
Total	100.14	100.13	100.52	100.40	99.72	99.78	99.01	99.34	98.83
Si	2.982	2.937	3.015	2.982	2.991	1.480	1.465	1.488	1.417
Al	1.969	2.030	1.992	2.003	2.025	8.875	8.947	8.954	9.071
Ti	0.004	0.003	0.000	0.014	0.001	0.005	0.015	0.004	0.002
Cr	0.001	0.000	0.004	0.007	0.005	0.004	0.009	0.007	0.003
Fe^{3+}	0.059	0.091	0.000	0.000	0.000	0.151	0.085	0.054	0.095
Fe^{2+}	1.141	1.398	1.598	1.620	1.620	0.492	0.474	0.560	0.404
Mn	0.015	0.024	0.032	0.028	0.026	0.005	0.000	0.000	0.000
Mg	1.754	1.352	1.176	1.284	1.259	2.979	3.000	2.932	3.000
Zn	0.000	0.002	0.002	0.000	0.004	0.005	0.003	0.000	0.002
Ca	0.075	0.161	0.164	0.059	0.061	0.002	0.002	0.000	0.000
Na	0.000	0.001	0.005	0.001	0.001	0.000	0.000	0.000	0.005
K	0.001	0.000	0.000	0.000	0.001	0.001	0.000	0.000	0.001
Total	8.000	8.000	7.989	7.999	7.994	14.000	14.000	14.000	14.000
Mg/(Fe+Mg)	0.61	0.49	0.42	0.44	0.44	0.86	0.86	0.84	0.88
$Fe^{3+}/(Fe^{2+}+Fe^{3+})$	0.05	0.06	0.00	0.00	0.00	0.23	0.15	0.09	0.19
Alm	0.38	0.48	0.54	0.54	0.55	N.D.	N.D.	N.D.	N.D.
Pyr	0.59	0.46	0.40	0.43	0.42	N.D.	N.D.	N.D.	N.D.
Grs	0.03	0.05	0.06	0.02	0.02	N.D.	N.D.	N.D.	N.D.
Sps	0.00	0.01	0.01	0.01	0.01	N.D.	N.D.	N.D.	N.D.

Note: N.D.—not determined.
*Textures shown in Figures 2 and 3. See captions for Figures 1 and 2 for mineral abbreviations.

Staurolite

Staurolite in sample MPL1872 is unusually Mg-rich (X_{Mg} = 0.44–0.58) compared with staurolite previously reported from the CZ (Schreyer et al., 1984; X_{Mg} 0.51, Tsunogae and van Reenen, 2006; X_{Mg} = 0.47–0.53) and from other granulite terranes (Fig. 6). Staurolite is also characterized by fairly high amounts of TiO_2 (up to 1.9 wt%) and ZnO (up to 1.2 wt%). The possibility of retrograde Fe-Mg exchange between staurolite and host garnet, with modification of the respective X_{Mg} ratio, cannot be neglected. As the large staurolite grain in garnet (Fig. 2B) is more Mg-rich (X_{Mg} = 0.55–0.58) than the fine-grained relict phase included in sapphirine (0.49–0.51) and in quartz (0.44–0.45) (Fig. 3C), we regard that the effect of retrograde metamorphism may only lower the X_{Mg} ratio. The high X_{Mg}, up to 0.58, is therefore regarded as a primary nature of staurolite in the examined sample. Other components such as Cr_2O_3 and MnO are minor and total <0.1 wt%. As shown in Figure 6A, the staurolite data exhibit a consistent substitution of AlTi ↔ SiR^{2+} (e.g., Hiroi et al., 1994).

Orthopyroxene

Orthopyroxene compositions vary, depending on the mode of occurrence (Fig. 7). Orthopyroxene inclusions in garnet in sample MPL1872 exhibit the highest X_{Mg} ratio (0.77–0.79) and Al_2O_3 content (5.3–6.1 wt%), followed by coarse-grained porphyroblastic orthopyroxene (0.74–0.75, 5.9–6.0 wt%) in the matrix, and symplectitic orthopyroxene (0.71–0.75, 3.2–5.5 wt%) with cordierite. Orthopyroxene in sample C227 also shows compositional variation in X_{Mg}, with higher X_{Mg} ratios (0.71–0.72) from coarse-grained matrix orthopyroxene and lower X_{Mg} ratios (0.62–0.63) from fine-grained orthopyroxene with cordierite. However, the Al_2O_3 contents are nearly consistent (6.8–7.3 and 6.0–7.3 wt%, respectively).

Cordierite

Cordierite occurs either as symplectitic intergrowths with sapphirine in garnet or as coronas with orthopyroxene needles around garnet in sample MPL1872. Both varieties are texturally retrograde in origin. Although cordierite shows a uniform

TABLE 2. REPRESENTATIVE ELECTRON MICROPROBE ANALYSES OF STAUROLITE, ORTHOPYROXENE, AND CORDIERITE IN SAMPLES MPL1872 AND C227

Rec. no.	195	181	76	77	83	169	7	2	49	166	166
Mineral name		Staurolite (O=46)				Orthopyroxene (O=6)				Cordierite (O=18)	
Sample no.	MPL1872	MPL1872	MPL1872	MPL1872	MPL1872	MPL1872	C227	C227	MPL1872	MPL1872	C227
Remarks	In Grt	In Ky	In Spr	In Qtz	Matrix	Corona	Matrix	In Crd	With Spr	With Opx	With Opx + Sil
Figures*	2B	2E	3C	3C	2A	2H	2J	2I	2F	2H	2I
SiO_2	27.21	26.33	27.20	27.10	51.07	52.61	50.25	48.51	50.06	50.06	49.09
Al_2O_3	55.47	58.15	54.05	55.28	5.92	3.93	7.30	7.35	33.82	33.71	33.32
TiO_2	1.53	0.80	1.64	1.42	0.14	0.08	0.13	0.20	0.00	0.01	0.00
Cr_2O_3	0.05	0.08	0.00	0.08	0.00	0.03	0.11	0.06	0.00	0.00	0.00
FeO	7.42	7.91	8.72	9.78	15.67	17.13	17.42	21.83	1.74	2.58	4.17
MnO	0.04	0.00	0.05	0.02	0.08	0.07	0.07	0.17	0.01	0.03	0.05
MgO	5.86	5.55	5.16	4.35	25.82	25.67	24.32	20.72	12.52	12.10	10.55
ZnO	0.51	0.60	0.76	0.89	0.02	0.08	0.00	0.00	0.07	0.00	0.00
CaO	0.00	0.01	0.01	0.00	0.18	0.10	0.05	0.06	0.02	0.02	0.06
Na_2O	0.02	0.02	0.01	0.03	0.01	0.00	0.00	0.02	0.09	0.03	0.06
K_2O	0.02	0.00	0.00	0.00	0.00	0.00	0.00	0.02	0.01	0.00	0.02
Total	98.12	99.42	97.59	98.94	98.90	99.70	99.65	98.94	98.33	98.55	97.32
Si	7.356	7.038	7.449	7.354	1.861	1.912	1.831	1.822	5.001	5.007	5.007
Al	17.666	18.318	17.438	17.679	0.254	0.168	0.314	0.325	3.982	3.972	4.005
Ti	0.310	0.160	0.337	0.290	0.004	0.002	0.004	0.006	0.000	0.001	0.000
Cr	0.010	0.016	0.001	0.017	0.000	0.001	0.003	0.002	0.000	0.000	0.000
Fe^{2+}	1.676	1.767	1.996	2.220	0.477	0.520	0.531	0.686	0.145	0.216	0.356
Mn	0.009	0.000	0.011	0.004	0.002	0.002	0.002	0.005	0.001	0.002	0.004
Mg	2.361	2.209	2.105	1.757	1.402	1.390	1.320	1.159	1.863	1.803	1.603
Zn	0.101	0.119	0.154	0.178	0.000	0.002	0.000	0.000	0.005	0.000	0.000
Ca	0.000	0.003	0.001	0.000	0.007	0.004	0.002	0.002	0.002	0.002	0.007
Na	0.011	0.008	0.003	0.017	0.001	0.000	0.000	0.002	0.018	0.007	0.012
K	0.006	0.000	0.001	0.000	0.000	0.000	0.000	0.001	0.001	0.000	0.003
Total	29.505	29.639	29.497	29.516	4.009	4.001	4.007	4.010	11.018	11.010	10.997
Mg/(Fe + Mg)	0.58	0.56	0.51	0.44	0.75	0.73	0.71	0.63	0.93	0.89	0.82

*Textures shown in Figures 2 and 3. See captions for Figures 1 and 2 for mineral abbreviations.

magnesian composition of $X_{Mg} = 0.89–0.94$, cordierite associated with sapphirine is slightly more Mg-rich ($X_{Mg} = 0.92–0.94$) than that associated with orthopyroxene (0.87–0.89). Cordierite in sample C227 shows slightly lower X_{Mg} of 0.81–0.82. The low total of the cordierite analyses (~98.5 wt%) suggests the presence of channel-filling volatiles such as CO_2 and/or H_2O.

Other Minerals

Biotite in sample C227 is Mg-rich ($X_{Mg} = 0.75–0.76$) and is characterized by a high-TiO_2 content (~5.0 wt%). Plagioclase in the same sample is albite-rich (An22–23), whereas K-feldspar is orthoclase-rich (Or79). Kyanite (sample MPL1872) and sillimanite (C227) are close to the ideal composition (Al_2SiO_5), although they also contain small amounts of Fe_2O_3 (0.49 and 0.62 wt%, respectively). Rutile in sample MPL1872 contains up to 0.79 wt% FeO.

Metamorphic Reactions

The studied samples preserve unique mineral assemblages and textures that are useful for the construction of P-T paths. Several important mineral assemblages and reaction textures are described in this section on the basis of the (K_2O-) FeO-MgO-Al_2O_3-SiO_2 (-H_2O) [(K)FMAS(H)] system. The textures in the examined samples are divided into two major stages on the basis of the host-inclusion relationships and grain shape, and the size of the minerals in the rocks.

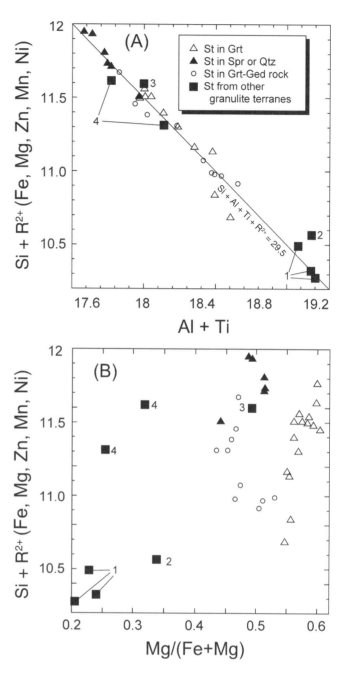

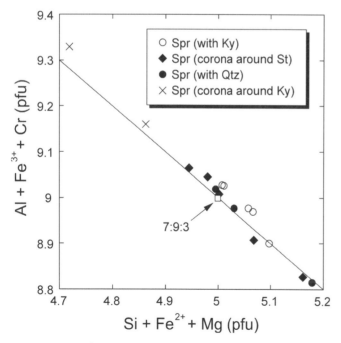

Figure 5. $Al + Fe^{3+} + Cr$ versus $Si + Fe^{2+} + Mg$ per formula unit (pfu) diagram, showing compositional variations of sapphirine in terms of Al and Si. Square with labels 7:9:3 in the figure implies a representative composition of $Mg_{3.5}Al_9Si_{1.5}O_{20}$.

Figure 6. Compositional diagrams showing staurolite chemistry. (A) $Al + Ti$ versus $Si + R^{2+}$ (Fe, Mg, Zn, Mn, Ni) (pfu [per formula unit]) diagram. (B) Mg/(Fe+Mg) versus $Si + R^{2+}$ diagram, showing the different Mg/(Fe+Mg) ratio of staurolite, depending on textures. Available staurolite compositions from granulite-facies terranes are also shown. 1—Hiroi et al. (1994); 2—Droop and Bucher-Nurminen (1984); 3—Schreyer et al. (1984); 4—Osanai et al. (1992). Staurolite data in garnet-gedrite (Ged) rock from the Limpopo Complex were taken from Tsunogae and van Reenen (2006).

Prograde to Peak Assemblages/Reactions

Staurolite + sapphirine + quartz. It is generally known that inclusion minerals in coarse-grained poikiloblastic phases such as garnet in granulites rarely preserve any trace of prograde metamorphism. Poikiloblastic garnet in sample MPL1872 contains inclusions of sapphirine, staurolite, kyanite, quartz, and rutile, which are all subidioblastic to xenoblastic (Figs. 2B–2G). The inclusion staurolite is fine grained (<0.7 mm in length) and commonly surrounded by sapphirine or sapphirine + quartz coronas (Figs. 2B, 2C), suggesting the formation of sapphirine + quartz after staurolite. The corona texture of sapphirine + quartz from staurolite in UHT terrane is relatively rare, and has previously been reported from the South African part of the Limpopo CZ (Schreyer et al., 1984) and from southern India (Nishimiya et al., 2010; Tsunogae and Santosh, 2010b). As the host garnet displays irregular grain boundaries with sapphirine (Figs. 2D, 3A, 3B), we conclude that the host garnet also reacted to form sapphirine + quartz from staurolite through the progress of the following FeO-MgO-Al_2O_3-SiO_2-H_2O (FMASH) continuous reaction (Schreyer et al., 1984; Nishimiya et al., 2010):

$$Grt + St \rightarrow Spr + Qtz + H_2O. \quad (1)$$

As this is a dehydration reaction, it probably occurred during prograde metamorphism. The progress of this reaction is supported by the occurrence of fine-grained staurolite inclusions within the product sapphirine and quartz (Fig. 3C), as well as by rutile inclusions in garnet (Figs. 2C, 2D) as the reactant staurolite contains up to 1.9 wt% TiO_2, although some coarse-grained rutile grains may not have been formed by the breakdown of staurolite. At such high temperature, a melting reaction rather than the dehydration reaction might have taken place, although we have no conclusive evidence. Lack of extensive partial melting in the rock is probably due to a low water activity condition at the peak of metamorphism, which is supported by the common presence of CO_2-rich fluid inclusions trapped in garnet in the sample (Tsunogae and van Reenen, 2007).

Sapphirine + kyanite. Intergrowth of sapphirine and kyanite and the occurrence of relict staurolite inclusions in the minerals enclosed in garnet in sample MPL1872 (Fig. 2E) also suggest the prograde hydration reaction (2) in the FMASH system:

$$St \rightarrow Spr + Ky + Grt + H_2O. \quad (2)$$

Reaction (2) probably took place at high-pressure ($P > 9$ kbar) within the stability field of kyanite.

Orthopyroxene + sillimanite + quartz. Pelitic granulite sample C227 contains the early orthopyroxene + sillimanite + quartz assemblage, which now is mantled by aggregates of retrograde cordierite (Fig. 2I). Such orthopyroxene + sillimanite + quartz assemblages have been reported from several UHT terranes such as the Napier Complex in Antarctica and the Madurai Block in southern India (e.g., Harley et al., 1990; Harley, 1998b; Tateishi et al., 2004; Tsunogae and Santosh, 2010a).

Retrograde Reactions

Various retrograde reactions have been observed in pelitic and mafic granulites and in Mg-Al-rich rocks as corona and symplectite textures that mantle early coarse-grained porphyroblastic minerals. Below, we summarize some typical examples of retrograde reactions, all of which were probably formed by decompression during the retrograde stage.

Cordierite + orthopyroxene symplectite. Several retrograde reaction textures involving garnet have so far been reported from pelitic granulites in the CZ. One typical example is a cordierite + orthopyroxene corona developed around garnet in Mg-rich varieties, which was probably formed by the following FeO-MgO-Al_2O_3-SiO_2 (FMAS) continuous reaction (Fig. 2H):

$$Grt + Qtz \rightarrow Opx + Crd. \quad (3)$$

The vermicular occurrence of orthopyroxene mantled by cordierite as a symplectite is regarded as being indicative of rapid decompression with limited time for the growth of the mineral. The progress of reaction (3) in sample MPL1872 is supported by the decreasing Mg content of the reactant garnet from the core to the rim adjacent to the corona. Similar decompression textures were also found in Mg-Al-rich rocks from the area around Beitbridge (e.g., Tsunogae and van Reenen, 2006). Similar orthopyroxene + cordierite symplectites developed around garnet have been reported from many granulite terranes in the world as evidence of rapid decompression (e.g., Harley, 1989).

Cordierite corona around garnet. In more Fe-rich varieties of pelitic granulite, subidioblastic aggregates of cordierite

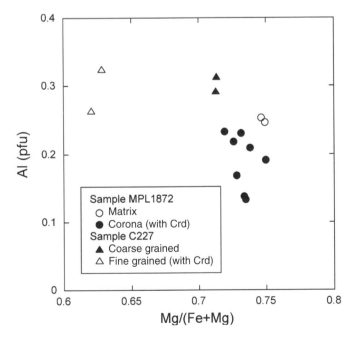

Figure 7. Compositional diagrams showing orthopyroxene chemistry. pfu—per formula unit.

coronas commonly mantle xenoblastic garnet (e.g., Fig. 2A). Sillimanite and quartz are commonly present as fine-grained inclusions in garnet or as coarse-grained matrix minerals. The microstructure suggests the progress of the following FMAS continuous reaction:

$$Grt + Sil + Qtz \rightarrow Crd. \quad (4)$$

This reaction texture is the most common retrograde reaction observed in many pelitic granulites from the CZ (e.g., Harris and Holland, 1984; Tsunogae and Miyano, 1989; van Reenen et al., 2004; Perchuk et al., 2008). Such cordierite coronas around garnet are also a common decompressional texture in granulite terranes worldwide (e.g., Harley, 1989).

Sapphirine + cordierite symplectite. In Mg-Al-rich rocks from the CZ around Beitbridge, sillimanite pseudomorphs after kyanite and garnet are separated from fine-grained intergrowth of sapphirine and cordierite or partly by spinel and cordierite (e.g., Windley et al., 1984; Tsunogae and van Reenen, 2006). A similar texture was also found in sample MPL1872 (Figs. 2G, 3D). The texture suggests the progress of the following FMAS continuous reaction:

$$Grt + Sil \rightarrow Spr + Crd. \quad (5)$$

If we adopt the ideal sapphirine chemistry of $Mg_{3.5}Al_9Si_{1.5}O_{20}$ and data set of Holland and Powell (1998), the slope of reaction (5) can be tentatively drawn as 0.6 kbar/100 °C. Sapphirine and cordierite appear in the low-pressure or high-temperature side of the reaction. As sapphirine and cordierite occur as symplectite, we regard the texture as having been formed during the decompressional stage.

Cordierite aggregate. Another common example of a retrograde metamorphic texture is aggregates of cordierite mantling orthopyroxene, sillimanite, and quartz (Fig. 2I). The texture suggests the progress of the following FMAS continuous reaction:

$$Opx + Sil + Qtz \rightarrow Crd. \quad (6)$$

Reaction (6) has been regarded as an example of decompression from the high-pressure stage at UHT conditions (Kelsey et al., 2003).

Discussion

A combination of detailed petrographic analyses of reaction textures and available data on petrogenetic grids has been employed to construct prograde and retrograde P-T trajectories for the studied granulites from the CZ of the Limpopo Belt. In this study we adopted the revised petrogenetic grids of Kelsey (2008) in the FMAS system (Fig. 8) to construct a qualitative P-T path that records evidence for the early prograde high pressure to peak UHT evolution of this crustal segment. The post-peak evolution of the CZ involves decompression cooling to P ~5.5 kbar and T ~550 °C, as described in numerous papers (e.g., Zeh et al., 2004; Kramers et al., 2006; Perchuk et al., 2008).

PROGRADE METAMORPHISM

The relict Mg-rich (X_{Mg} ~0.58) staurolite inclusions in poikiloblastic garnet in sample MPL1872, discussed in this study, demonstrate prograde high-pressure granulite-facies metamorphism of P >14 kbar (Fig. 8A), which is close to the eclogite-facies condition. In general, staurolite is known as a common mineral in metasedimentary rocks subjected to Barrovian-type metamorphism. Available experimental and thermodynamic studies of staurolite suggest that the mineral is typically Fe-rich (X_{Mg} <0.5) and becomes unstable above the amphibolite facies owing to the progress of the following (K_2O-) FeO-MgO-Al_2O_3-SiO_2-H_2O ((K)FMASH) continuous reactions at ~700 °C (e.g., Spear and Cheney, 1989):

$$St + Ms + Qtz \rightarrow Grt + Bt + Sil/Ky + H_2O; \quad (7)$$

$$St + Qtz \rightarrow Grt + Sil/Ky + H_2O. \quad (8)$$

Therefore, staurolite and sapphirine + quartz are thought to have separate stability fields. However, staurolite has also been reported from some eclogite-facies rocks (e.g., Enami and Zang, 1988; Peacock and Goodge, 1995) and high-pressure granulites (e.g., Shimpo et al., 2006; Tsunogae and Santosh, 2006a). In such examples the staurolite-bearing rocks are poor in K_2O and SiO_2 and therefore lack muscovite and quartz. Reactions (7) and (8) thus fail to take place, allowing staurolite to survive at higher temperature conditions (e.g., Nishimiya et al., 2010). Also, staurolite in high-pressure rocks is typically Mg-rich and occurs as magnesian staurolite. Staurolite with X_{Mg} >0.5 has been reported as inclusions in garnet from garnet-gedrite-orthopyroxene granulites in the CZ of the Limpopo Complex by Tsunogae and van Reenen (2006) as well as from the Gondwana suture zone in southern India (e.g., Collins et al., 2007; Tsunogae et al., 2008; Kanazawa et al., 2009; Nishimiya et al., 2010), and is taken as evidence for prograde high-pressure metamorphism. The occurrence of Mg-rich staurolite is therefore regarded as a unique feature of granulites in Precambrian collisional orogens.

Experimental investigation by Schreyer (1988) suggests that pure Mg-staurolite is stable at P >14 kbar and T >710–760 °C in the MASH system. Fockenberg (1998) confirmed the high-pressure stability of magnesian staurolite, but obtained higher P–T conditions of 12–66 kbar and 608–918 °C. However, as these experiments were based on pure Mg-staurolite, the experimental results cannot be directly applicable to the natural rocks (Kelsey et al., 2006). Recently Sato et al. (2010) performed high-P-T experiments of staurolite with moderate X_{Mg} (= 0.7–0.5) at 12–19 kbar and 850–1050 °C. According to these results, staurolite with X_{Mg} = 0.7 decomposed to orthopyroxene + corundum + melt, whereas staurolite with X_{Mg} = 0.5–0.6

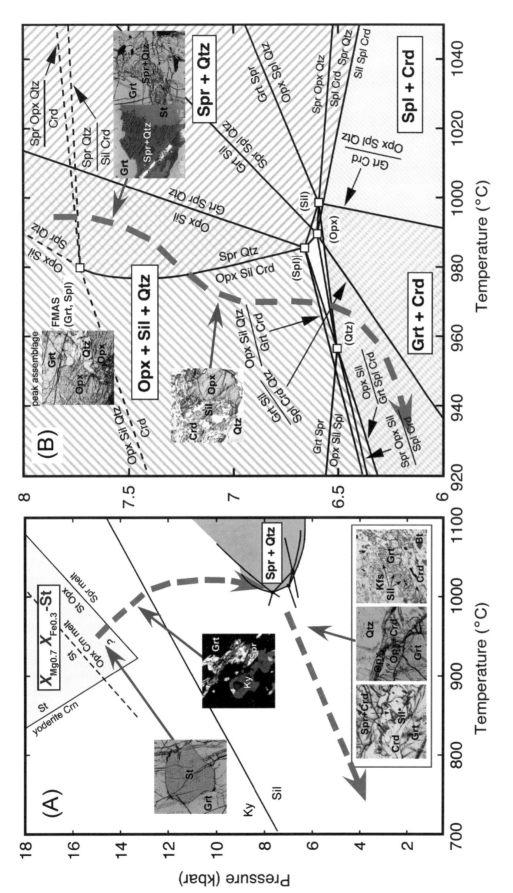

Figure 8. P-T diagram showing pressure-temperature path (broken arrows) for the Central Zone of the Limpopo Complex. (A) High-pressure and ultrahigh-temperature evolution of the study area. The stability fields of sapphirine + quartz and $X_{Mg0.7}X_{Fe0.3}$-staurolite are after Kelsey et al. (2004) and $X_{Mg0.7}X_{Fe0.3}$-staurolite are after van Reenen et al. (2004) and Sato et al. (2010), respectively. The retrograde P-T path below 7 kbar was taken after van Reenen et al. (2004) and Perchuk et al. (2008). (B) Petrogenetic grid in the FeO-MgO-Al_2O_3-SiO_2 (FMAS) system (modified from Kelsey, 2008), showing the P-T trajectory around the peak metamorphism of the CZ. Arrows and photomicrographs indicate approximate positions of the formation of the textures. See text for further discussion. Sil—sillimanite; Spl—spinel.

decomposed with orthopyroxene to sapphirine + melt between 14 and 16 kbar at 950 °C (Fig. 8A). These data correspond to the lower pressure stability limit of the mineral. Based on the results of Sato et al. (2010), the Mg-rich staurolite with X_{Mg} ~0.58 from the Limpopo CZ might have been stable at P >14 kbar. This result is also consistent with observations of Mg-rich staurolite from the Palghat-Cauvery suture zone in southern India by Nishimiya et al. (2010) and Tsunogae and Santosh (2010b) as evidence of prograde high-pressure metamorphism (P >14 kbar). These authors based their observations on the occurrence of Mg-rich staurolite and kyanite as inclusions in garnet, similar to the textures described in sample MPL1872 of this study. Significantly, magnesian staurolite (X_{Mg} = 0.53–0.57), synthesized experimentally by Hellman and Green (1979) at P = 24–26 kbar and T = 740–760 °C, has a composition close to that of the staurolite (X_{Mg} = 0.47–0.53) in sample MPL1872. We thus interpret the Mg-rich staurolite reported from the CZ of the Limpopo Complex as a possible indicator of prograde high-pressure (P > 14 kbar) metamorphism (Fig. 8A).

Similar Mg-rich staurolites were previously reported from two localities in the CZ as relict minerals enclosed in poikiloblastic garnet. Schreyer et al. (1984) first discovered staurolite included in garnet in Mg-Al-rich rocks from the CZ in South Africa. This staurolite has an Mg/(Fe+Mg) ratio of 0.51 and coexists with orthopyroxene, kyanite, sapphirine, and quartz. Tsunogae and van Reenen (2006) reported on Mg-rich staurolite (X_{Mg} = 0.47–0.52) in a garnet-gedrite-plagioclase rock from the Beitbridge area in the CZ. In their sample (MPL867), staurolite is mantled by sapphirine, and the assemblage is in turn included in poikiloblastic garnet. Tsunogae and van Reenen (2006) regarded the magnesian staurolite as a relict phase and inferred the following dehydration reaction that possibly took place during prograde metamorphism:

$$St \rightarrow Grt + Spr + Sil/Ky + H_2O. \quad (9)$$

Such prograde high-pressure metamorphism at P >14 kbar is consistent with previous reports on the occurrence of prograde kyanite from the CZ (e.g., Chinner and Sweatman, 1968; Watkeys et al., 1983; Windley et al., 1984; Tsunogae and Miyano, 1989). Chinner and Sweatman (1968) first reported on kyanite associated with orthopyroxene and quartz. As the kyanite is coarse grained and subidioblastic, they regarded the mineral as a prograde phase. Tsunogae and Miyano (1989) reported on relict kyanite surrounded by sillimanite in a pelitic granulite from the Musina area as evidence of decompression from high pressure (P >10 kbar), and inferred the progress of the following reaction:

$$kyanite \rightarrow sillimanite. \quad (10)$$

Sample MPL1872, discussed in the present study, also contains relict kyanite inclusions in poikiloblastic garnet (Fig. 2E), which has been confirmed by laser Raman spectroscopic analysis (Fig. 4). All the available petrological data therefore suggest that the CZ underwent prograde high-pressure metamorphism within the stability field of kyanite, probably at P >14 kbar (Fig. 8A).

PEAK UHT METAMORPHISM

Mg-rich staurolite inclusions within porphyroblastic garnet are rarely mantled by sapphirine + quartz symplectite (Figs. 2D, 3A). These two minerals show a direct contact relationship without any reaction texture developed between them. This microstructure is the second report of equilibrium sapphirine + quartz assemblage from the CZ, and the first from the Zimbabwean part of the Beit Bridge Complex. Although sapphirine has been reported from several localities within the CZ (e.g., Horrocks, 1983; Droop, 1989; Hisada and Miyano, 1996; Tsunogae and van Reenen, 2006), it occurs mostly in quartz-free Mg-Al-rich rocks, therefore lacking the diagnostic sapphirine + quartz assemblage of UHT metamorphism.

Available phase equilibrium data suggest that sapphirine + quartz is stable at T >1030 °C at 9.5 kbar (Hensen and Green, 1973) or T >1050 °C at 11 kbar (Bertrand et al., 1991). Recent thermodynamic calculations in the KFMASH system also support the high-temperature stability of this assemblage (T >1005 °C, Fig. 8; Kelsey et al., 2004; Kelsey, 2008). The peak temperature condition inferred from the examined rocks is therefore ~1000 °C, although the sapphirine + quartz stability field might shift slightly to a lower temperature because staurolite is not included in the construction of the FMAS petrogenetic grid of Figure 8B. We thus attempted to estimate metamorphic temperatures for poikiloblastic garnet and orthopyroxene in sample MPL1872 using the garnet-orthopyroxene geothermometer of Lee and Ganguly (1988). The calculated temperature range is 880–890 °C, which is slightly lower than the stability of sapphirine + quartz. This is probably because of retrograde Fe-Mg exchange between the two minerals.

Our report of equilibrium sapphirine + quartz from the CZ is taken as robust evidence of UHT metamorphism associated with the Neoarchean Limpopo orogeny. The fine-grained intergrowth of sapphirine and quartz enclosed within garnet in sample MPL1872 (e.g., Fig. 3A) is similar to the occurrence reported from several granulite terranes in other parts of the world, such as southern India (Tateishi et al., 2004; Tsunogae and Santosh, 2006b, 2010a), Ouzzal (Ouzegane and Boumaza, 1996), and the Highland Complex (Sajeev and Osanai, 2004). In all these cases, the assemblage has been regarded as clear evidence for peak metamorphism at UHT conditions (T ~1000 °C). Previous P-T investigations of the peak metamorphism in the CZ based on geothermometry (e.g., Tsunogae and van Reenen, 2006) are consistent with high to ultrahigh temperatures >900 °C. The results thus provide unequivocal evidence for the formation and exhumation of an ultra-hot orogen (Santosh et al., 2009a).

The orthopyroxene + sillimanite + quartz assemblage (Fig. 2L) is regarded as stable at T >900 °C and at relatively high pressure at P >6 kbar in the FMAS system (Fig. 8B) and P >8–9 kbar in the KFMASH system (Kelsey et al., 2003, 2004), and thus it is also a diagnostic assemblage of UHT metamorphism (e.g., Hensen and Green, 1973; Carrington and Harley, 1995; Harley, 1998a; Kelsey, 2008). The assemblage is therefore reported in many Mg-Al-rich rocks and pelitic granulites from most UHT terranes (see Kelsey et al., 2003; Kelsey, 2008).

RETROGRADE METAMORPHISM

The peak metamorphic stage (Fig. 8) was followed by decompression or decompression cooling that resulted in the formation of various corona textures such as sapphirine + cordierite after kyanite/sillimanite + garnet (Figs. 2F, 2G), cordierite + orthopyroxene around garnet (Fig. 2H), and cordierite after orthopyroxene + sillimanite + quartz (Fig. 2I). Plagioclase + orthopyroxene symplectite developed around garnet in mafic granulite (e.g., Watkeys et al., 1983; Tsunogae and Miyano, 1989), and sapphirine + spinel + cordierite symplectite developed between sillimanite and garnet-gedrite in Mg-Al-rich rocks (Tsunogae and van Reenen, 2006), are also reported from the CZ. We consider that all the corona textures probably formed during the post-peak decompression and/or decompressional cooling stage (e.g., Perchuk et al., 2008). This is consistent with the occurrence of similar textures in numerous granulites that underwent HP and UHT metamorphism (e.g., Harley, 2008; Kelsey, 2008, and references therein).

P-T TRAJECTORY

The microstructures in the studied Mg-Al-rich rock differ in important aspects from previous reports of similar rocks in the CZ. This is due to the identification of Mg-rich staurolite and the replacement of this mineral by sapphirine + quartz symplectite, all of which are presently included as relict minerals in porphyroblastic garnet. Although staurolite with X_{Mg} >0.5 included in garnet was reported by Schreyer et al. (1984), the staurolite of the present study has the highest X_{Mg} value (~0.58) yet reported from the Limpopo Complex. This high value is recorded from isolated staurolite inclusions in garnet, which are not associated with retrograde corona textures, suggesting that the composition of primary staurolite might have been more magnesian. Shimpo et al. (2006) first suggested that the occurrence of such Mg-rich staurolite is indicative of prograde high-pressure metamorphism at P >12 kbar. Recently Sato et al. (2010) performed high-P-T experiments of staurolite with moderate X_{Mg} (= 0.7–0.5) at 12–19 kbar and 850–1050 °C. These authors demonstrated that at 950 °C, while pressures between 14 and 16 kbar staurolite with X_{Mg} = 0.7 decomposed to orthopyroxene + corundum + melt, that staurolite with X_{Mg} = 0.5–0.6 decomposed with orthopyroxene to sapphirine + melt (Fig. 8A). These data correspond to the lower pressure stability limit of the staurolite. Based on these results the Mg-rich staurolite with X_{Mg} = 0.58 thus can be stable at a higher pressure (P >14 kbar, Fig. 8A) than that inferred by Shimpo et al. (2006). The occurrence of relict kyanite from several localities in the CZ as well as in sample MPL1872 is also consistent with the proposed prograde high-pressure event. Thus, the prograde P-T path for the CZ inferred from the composition of magnesian staurolite (Fig. 2B) suggests that maximum pressure conditions of P >14 kbar were reached at T ~950 °C (Fig. 8A).

This peak high-pressure event was followed by rapid decompression to ~9 kbar at the peak UHT metamorphism of T ~1000 °C (Fig. 8B). This is inferred from the presence of sapphirine + quartz coronas (Fig. 2D) developed around Mg-rich staurolite as already discussed. The inferred rapid decompressional event is supported by the symplectic nature of the sapphirine and quartz (Figs. 2D, 3C). Kyanite + sapphirine (Fig. 2E) and sapphirine + cordierite intergrowths enclosed in garnet (Figs. 2F, 2G) might also have formed at this stage.

The peak temperature conditions within the stability field of sapphirine + quartz might have been followed by possible isobaric cooling toward the stability field of orthopyroxene + sillimanite + quartz, which is present in pelitic granulites (sample C227). However, as sapphirine + quartz and orthopyroxene + sillimanite + quartz were observed in different samples, we do not exclude an alternative interpretation that might be due to different bulk-rock chemistries. The two diagnostic UHT assemblages thus might have formed together within their overlapping stability fields, at ~1020 °C and 8.5 kbar (Fig. 8B). Various corona textures that developed around coarse-grained poikiloblastic minerals—such as orthopyroxene + cordierite symplectites, cordierite coronas around garnet, and cordierite aggregates around orthopyroxene, sillimanite, and quartz—were probably formed by subsequent decompressional cooling (Fig. 8A), as was previously described by various authors (e.g., Perchuk et al., 2008). Available petrologic data discussed in this study are therefore consistent with a clockwise P-T evolution of this region of the CZ that involved an early high-pressure event that culminated with rapid decompression to UHT conditions, followed finally by decompressional cooling down to the closure temperature of Fe-Mg exchange between minerals (600 °C and 3 kbar, van Reenen et al., 2004; Fig. 8A).

The P-T path estimated in this study is nearly consistent with that reported from the Palghat-Cauvery suture zone in India, a classic example of late Neoproterozoic to Early Cambrian high-grade metamorphic orogens formed by collisional orogeny during the final stage of assembly of the Gondwana supercontinent (e.g., Santosh et al., 2009a, 2009b). Prograde high-pressure (P >14 kbar) metamorphism and subsequent decompression and peak UHT (T ~1000 °C) metamorphism, inferred from sapphirine + quartz + staurolite association similar to that in sample MPL1872, was recently reported by Nishimiya et al. (2010) and Tsunogae and Santosh (2010b). The results of this study are therefore consistent with the proposal that prograde high-pressure metamorphism followed by an UHT event can be related to the collisional tectonics of the Zimbabwe and Kaapvaal Cratons associated with the amalgamation of supercontinents during the Neoarchean.

ACKNOWLEDGMENTS

The first author would like to thank the late Takashi Miyano for introducing him to the Limpopo Complex. We pay our homage to him, and dedicate this contribution to his memory. We also thank Takashi Kashima, Ai Nabara, Tomomi Fukui, Katsuhiko Nasu, and the staff at the Geological Survey of Zimbabwe and the University of Zimbabwe for their field support and thin-section preparation. We also thank all participants at the Limpopo Field Workshop in 2008 for their helpful discussions and comments, and N. Nishida and T. Hokada for their assistance with microprobe and laser Raman analyses. Alexei Perchuk and H.M. Rajesh are acknowledged for their constructive and helpful reviews of this chapter. E. Grew and D.J. Ellis are also acknowledged for their valuable comments and suggestions. Partial funding for this project was supported by a grant-in-aid for Scientific Research (B) from the Japanese Ministry of Education, Culture, Sports, Science and Technology (MEXT) (nos. 20340148 and 22403017) and the JSPS-INSA joint research program to T. Tsunogae and no. 21253008 to Y. Osanai. DDVR acknowledges NRF grant no. 68288 and financial support from the University of Johannesburg.

REFERENCES CITED

Armstrong, R.A., Compston, W., Retief, E.A., Williams, I.S., and Welke, H.J., 1991, Zircon ion microprobe studies bearing on the age and evolution of the Witwatersrand triad: Precambrian Research, v. 53, p. 243–266, doi:10.1016/0301-9268(91)90074-K.

Barton, J.M., Jr., and van Reenen, D.D., 1992, When was the Limpopo Orogeny?: Precambrian Research, v. 55, p. 7–16, doi:10.1016/0301-9268(92)90010-L.

Belyanin, G.A., Rajesh, H.M., van Reenen, D.D., and Mouri, H., 2010, Corundum + orthopyroxene ± spinel intergrowths in an ultrahigh temperature Al-Mg granulite from the Southern Marginal Zone, Limpopo Belt, South Africa: American Mineralogist, v. 95, p. 196–199, doi:10.2138/am.2010.3383.

Berger, M., Kramers, J.D., and Nägler, F.T., 1995, Geochemistry and geochronology of charnoenderbites in the Northern Marginal Zone of the Limpopo Belt, Southern Africa, and genetic models: Schweizerische Mineralogische und Petrographische Mitteilungen, v. 75, p. 17–42.

Bertrand, P., Ellis, D.J., and Green, D.H., 1991, The stability of sapphirine-quartz and hypersthene-sillimanite-quartz assemblages: An experimental investigation in the system FeO-MgO-Al$_2$O$_3$-SiO$_2$ under H$_2$O and CO$_2$ conditions: Contributions to Mineralogy and Petrology, v. 108, p. 55–71, doi:10.1007/BF00307326.

Blenkinsop, T.G., Kröner, A., and Chiwara, V., 2004, Single stage, late Archean exhumation of granulites in the Northern Marginal Zone, Limpopo Belt, Zimbabwe, and relevance of gold mineralization at Renco Mine: South African Journal of Geology, v. 107, p. 377–396, doi:10.2113/107.3.377.

Boshoff, R., van Reenen, D.D., Smit, C.A., Perchuk, L.L., Kramers, J.D., and Armstrong, R.A., 2006, Geologic history of the Central Zone of the Limpopo Complex: The West Alldays Area: Journal of Geology, v. 114, p. 699–716, doi:10.1086/507615.

Brandl, G., 1983, Geology and geochemistry of various supracrustal rocks of the Beit Bridge Complex east of Messina: Geological Society of South Africa Special Publication 8, p. 103–112.

Carrington, D.P., and Harley, S.L., 1995, Partial melting and phase relations in high-grade metapelites: An experimental petrogenetic grid in the KFMASH system: Contributions to Mineralogy and Petrology, v. 120, p. 270–291, doi:10.1007/BF00306508.

Chinner, G.A., and Sweatman, T.R., 1968, A former association of enstatite and kyanite: Mineralogical Magazine, v. 36, p. 1052–1060, doi:10.1180/minmag.1968.036.284.03.

Collins, A.S., Clark, C., Sajeev, K., Santosh, M., Kelsey, D.E., and Hand, M., 2007, Passage through India: The Mozambique ocean suture, high-pressure granulites and the Palghat-Cauvery Shear System: Terra Nova, v. 19, p. 141–147, doi:10.1111/j.1365-3121.2007.00729.x.

Droop, G.T.R., 1989, Reaction history of garnet-sapphirine granulites and conditions of Archaean high-pressure granulite-facies metamorphism in the central Limpopo Mobile Belt, Zimbabwe: Journal of Metamorphic Geology, v. 7, p. 383–403, doi:10.1111/j.1525-1314.1989.tb00604.x.

Droop, G.T.R., and Bucher-Nurminen, K., 1984, Reaction textures and metamorphic evolution of sapphirine-bearing granulites from the Gruf Complex, Italian Central Alps: Journal of Petrology, v. 25, p. 766–803.

Enami, M., and Zang, Q., 1988, Magnesian staurolite in garnet-corundum rocks and eclogite from the Donghai district, Jiangsu province, east China: American Mineralogist, v. 73, p. 48–56.

Fockenberg, T., 1998, An experimental investigation on the P-T stability of Mg-staurolite in the system MgO-Al$_2$O$_3$-SiO$_2$-H$_2$O: Contributions to Mineralogy and Petrology, v. 130, p. 187–198, doi:10.1007/s004100050359.

Harley, S.L., 1989, The origin of granulites: A metamorphic perspective: Geological Magazine, v. 126, p. 215–247, doi:10.1017/S0016756800022330.

Harley, S.L., 1998a, On the occurrence and characterization of ultrahigh-temperature crustal metamorphism: Geological Society [London] Special Publication 138, p. 81–107, doi:10.1144/GSL.SP.1996.138.01.06.

Harley, S.L., 1998b, Ultrahigh temperature granulite metamorphism (1050 °C, 12 kbar) and decompression in garnet (Mg70)-orthopyroxene-sillimanite gneisses from the Rauer Group, East Antarctica: Journal of Metamorphic Geology, v. 16, p. 541–562, doi:10.1111/j.1525-1314.1998.00155.x.

Harley, S.L., 2008, Refining the P-T records of UHT crustal metamorphism: Journal of Metamorphic Geology, v. 26, p. 125–154, doi:10.1111/j.1525-1314.2008.00765.x.

Harley, S.L., Hensen, B.J., and Sheraton, J.W., 1990, Two-stage decompression in orthopyroxene-sillimanite granulites from Forefinger Point, Enderby Land, Antarctica: Implications for the evolution of the Archaean Napier Complex: Journal of Metamorphic Geology, v. 8, p. 591–613, doi:10.1111/j.1525-1314.1990.tb00490.x.

Harris, N.B.W., and Holland, T.J.B., 1984, The significance of cordierite-hypersthene assemblages from the Beitbridge region of the Central Limpopo Belt: Evidence for rapid decompression in the Archaean?: American Mineralogist, v. 69, p. 1036–1049.

Hellman, P.L., and Green, T.H., 1979, The high-pressure experimental crystallization of staurolite in hydrous mafic compositions: Contributions to Mineralogy and Petrology, v. 68, p. 369–372, doi:10.1007/BF01164521.

Hensen, B.J., and Green, D.H., 1973, Experimental study of the stability of cordierite and garnet in pelitic compositions at high pressures and temperatures. III. Synthesis of experimental data and geological applications: Contributions to Mineralogy and Petrology, v. 38, p. 151–166, doi:10.1007/BF00373879.

Hiroi, Y., Ogo, Y., and Namba, K., 1994, Evidence for prograde metamorphic evolution of Sri Lankan pelitic granulites, and implications for the development of continental crust: Precambrian Research, v. 66, p. 245–263, doi:10.1016/0301-9268(94)90053-1.

Hisada, K., and Miyano, T., 1996, Petrology and microthermometry of aluminous rocks in the Botswanan Limpopo Central Zone: Evidence for isothermal decompression and isobaric cooling: Journal of Metamorphic Geology, v. 14, p. 183–197, doi:10.1046/j.1525-1314.1996.05857.x.

Holland, T.J.B., and Powell, R., 1998, An internally consistent thermodynamic data set for phases of petrological interest: Journal of Metamorphic Geology, v. 16, p. 309–343, doi:10.1111/j.1525-1314.1998.00140.x.

Holzer, L., Frey, R., Barton, J.M., Jr., and Kramers, J.D., 1998, Unravelling the record of successive high-grade events in the Central Zone of the Limpopo belt using Pb single phase dating of metamorphic minerals: Precambrian Research, v. 87, p. 87–115, doi:10.1016/S0301-9268(97)00058-2.

Horrocks, P.C., 1983, A corundum and sapphirine paragenesis from the Limpopo Mobile Belt, southern Africa: Journal of Metamorphic Geology, v. 1, p. 13–23, doi:10.1111/j.1525-1314.1983.tb00262.x.

Jaeckel, P., Kröner, A., Kamo, S.L., Brandl, G., and Wendt, J.I., 1997, Late Archaean to early Proterozoic granitoid magmatism and high-grade metamorphism in the central Limpopo belt, South Africa: Journal of the Geological Society [London], v. 154, p. 25–44, doi:10.1144/gsjgs.154.1.0025.

Kamber, B.S., Kramers, J.D., Napier, R., Cliff, R.A., and Rollinson, H.R., 1995, The Triangle Shearzone, Zimbabwe, revisited: New data document an important event at 2.0 Ga in the Limpopo Belt: Precambrian Research, v. 70, p. 191–213, doi:10.1016/0301-9268(94)00039-T.

Kanazawa, T., Tsunogae, T., Sato, K., and Santosh, M., 2009, The stability and origin of sodicgedrite in ultrahigh-temperature Mg-Al granulites: A case study from the Gondwana suture in southern India: Contributions to Mineralogy and Petrology, v. 157, p. 95–110, doi:10.1007/s00410-008-0322-0.

Kelsey, D.E., 2008, On ultrahigh-temperature crustal metamorphism: Gondwana Research, v. 13, p. 1–29, doi:10.1016/j.gr.2007.06.001.

Kelsey, D.E., White, R.W., and Powell, R., 2003, Orthopyroxene-sillimanite-quartz assemblages: Distribution, petrology, quantitative P-T-X constraints and P-T paths: Journal of Metamorphic Geology, v. 21, p. 439–453, doi:10.1046/j.1525-1314.2003.00456.x.

Kelsey, D.E., White, R.W., Holland, T.J.B., and Powell, R., 2004, Calculated phase equilibria in K_2O-FeO-MgO-Al_2O_3-SiO_2-H_2O for sapphirine-quartz-bearing mineral assemblages: Journal of Metamorphic Geology, v. 22, p. 559–578, doi:10.1111/j.1525-1314.2004.00533.x.

Kelsey, D.E., Clark, C., Hand, M., and Collins, A.S., 2006, Comment on "First report of garnet-corundum rocks from southern India: Implications for prograde high-pressure (eclogite-facies?) metamorphism": Earth and Planetary Science Letters, v. 249, p. 529–534, doi:10.1016/j.epsl.2006.07.048.

Kramers, J.D., Kreissig, K., and Jones, M.Q.W., 2001, Crustal heat production and style of metamorphism: A comparison between two Archean high grade provinces in the Limpopo Belt, southern Africa: Precambrian Research, v. 112, p. 149–163, doi:10.1016/S0301-9268(01)00173-5.

Kramers, J.D., McCourt, S., and van Reenen, D.D., 2006, The Limpopo Belt, *in* Johnson, M.R., Anhaeusser, C.R., and Thomas, R.J., eds., The Geology of South Africa: Johannesburg, Geological Society of South Africa/Pretoria, Council for Geoscience, p. 209–236.

Kreissig, K., Nägler, T.F., Kramers, J.D., van Reenen, D.D., and Smit, C.A., 2000, An isotopic and geochemical study of the northern Kaapvaal Craton and the Southern Marginal Zone of the Limpopo Belt: Are they juxtaposed terranes?: Lithos, v. 50, p. 1–25, doi:10.1016/S0024-4937(99)00037-7.

Kreissig, K., Holzer, L., Frei, R., Villa, I.M., Kramers, J.D., Kröner, A., Smit, C.A., and van Reenen, D.D., 2001, Geochronology of the Hout River Shear Zone and the metamorphism in the Southern Marginal Zone of the Limpopo Belt, Southern Africa: Precambrian Research, v. 109, p. 145–173, doi:10.1016/S0301-9268(01)00147-4.

Lee, H.Y., and Ganguly, J., 1988, Equilibrium compositions of coexisting garnet and orthopyroxene: Experimental determinations in the system FeO-MgO-Al_2O_3-SiO_2, and applications: Journal of Petrology, v. 29, p. 93–113.

Mason, R., 1973, The Limpopo Mobile Belt—Southern Africa: Philosophical Transactions of the Royal Society of London, ser. A, v. 273, p. 463–485.

Millonig, L., Zeh, A., Gerdes, A., and Klemd, R., 2008, Neoarchaean high-grade metamorphism in the Central Zone of the Limpopo Belt (South Africa): Combined petrological and geochronological evidence from the Bulai pluton: Lithos, v. 103, p. 333–351, doi:10.1016/j.lithos.2007.10.001.

Mouri, H., Whitehouse, M.J., Brandl, G., and Rajesh, H.M., 2009, A magmatic age and four successive metamorphic events recorded in zircons from a single meta-anorthosite sample in the Central Zone of the Limpopo Belt, South Africa: Journal of the Geological Society [London], v. 166, p. 827–830, doi:10.1144/0016-76492008-148.

Nishimiya, Y., Tsunogae, T., and Santosh, M., 2010, Sapphirine + quartz corona around magnesian (X_{Mg} ~0.58) staurolite from the Palghat-Cauvery Suture Zone, southern India: Evidence for high-pressure and ultrahigh-temperature metamorphism within the Gondwana suture: Lithos, v. 114, p. 490–502, doi:10.1016/j.lithos.2009.10.012.

Osanai, Y., Owada, M., and Kawasaki, T., 1992, Tertiary deep crustal ultrametamorphism in the Hidaka metamorphic belt, northern Japan: Journal of Metamorphic Geology, v. 10, p. 401–414, doi:10.1111/j.1525-1314.1992.tb00092.x.

Ouzegane, K., and Boumaza, S., 1996, An example of ultrahigh-temperature metamorphism: Orthopyroxene-sillimanite-garnet, sapphirine-quartz and spinel-quartz parageneses in Al-Mg granulites from In Hihaou, In Ouzzal, Hoggar: Journal of Metamorphic Geology, v. 14, p. 693–708, doi:10.1111/j.1525-1314.1996.00049.x.

Peacock, S.M., and Goodge, J.W., 1995, Eclogite-facies metamorphism preserved in tectonic blocks from a lower crustal shear zone, central Transantarctic Mountains, Antarctica: Lithos, v. 36, p. 1–13, doi:10.1016/0024-4937(95)00006-2.

Perchuk, L.L., and van Reenen, D.D., 2008, Reply to "Comments on 'P–T record of two high-grade metamorphic events in the Central Zone of the Limpopo Complex, South Africa,'" by A. Zeh and R. Klemd: Lithos, v. 106, p. 403–410, doi:10.1016/j.lithos.2008.07.011.

Perchuk, L.L., Varla Mov, D.A., and van Reenen, D.D., 2006, Unique record of P-T history of high-grade polymetamorphism: Doklady Academy of Sciences, Earth Science, v. 409A, p. 958–962.

Perchuk, L.L., van Reenen, D.D., Varlamov, D.A., and van Kal, S.M., Tabatabaeimanesh, and Boshoff, R., 2008, P-T record of two high-grade metamorphic events in the Central Zone of the Limpopo Complex, South Africa: Lithos, v. 103, p. 70–105.

Rigby, M.J., 2009, Conflicting P-T paths within the Central Zone of the Limpopo Belt: A consequence of different thermobarometric methods?: Journal of African Earth Sciences, v. 54, p. 111–126.

Sajeev, K., and Osanai, Y., 2004, Ultrahigh-temperature metamorphism (1150 °C, 12 kbar) and multistage evolution of Mg-, Al-rich granulites from the Central Highland Complex, Sri Lanka: Journal of Petrology, v. 45, p. 1821–1844.

Santosh, M., Maruyama, S., and Yamamoto, S., 2009a, The making and breaking of supercontinents: Some speculations based on superplumes, superdownwelling and the role of tectosphere: Gondwana Research, v. 15, p. 324–341, doi:10.1016/j.gr.2008.11.004.

Santosh, M., Maruyama, S., and Sato, K., 2009b, Anatomy of a Cambrian suture in Gondwana: Pacific-type orogeny in southern India?: Gondwana Research, v. 16, p. 321–341, doi:10.1016/j.gr.2008.12.012.

Sato, K., Santosh, M., and Tsunogae, T., 2010, High P-T phase relation of magnesian staurolite in the system FeO-MgO-Al_2O_3-SiO_2-H_2O: Evidence for high-pressure metamorphism in collisional orogens: American Mineralogist, v. 95, p. 177–184, doi:10.2138/am.2010.3170.

Schreyer, W., 1988, Experimental studies on metamorphism of crustal rocks under mantle pressures: Mineralogical Magazine, v. 52, p. 1–26, doi:10.1180/minmag.1988.052.364.01.

Schreyer, W., Horrocks, P.C., and Abraham, K., 1984, High-magnesium staurolite in a sapphirine-garnet rock from the Limpopo Belt: Southern Africa: Contributions to Mineralogy and Petrology, v. 86, p. 200–207, doi:10.1007/BF00381847.

Shimpo, M., Tsunogae, T., and Santosh, M., 2006, First report of garnet-corundum rocks from southern India: Implications for prograde high-pressure (eclogite-facies?) metamorphism: Earth and Planetary Science Letters, v. 242, p. 111–129, doi:10.1016/j.epsl.2005.11.042.

Spear, F.S., and Cheney, J.T., 1989, A petrogenetic grid for pelitic schists in the system SiO_2-Al_2O_3-FeO-MgO-K_2O-H_2O: Contributions to Mineralogy and Petrology, v. 101, p. 149–164, doi:10.1007/BF00375302.

Stevens, G., and van Reenen, D.D., 1992, Partial melting and the origin of metapelitic granulites in the Southern Marginal Zone of the Limpopo Belt, South Africa: Precambrian Research, v. 55, p. 303–319, doi:10.1016/0301-9268(92)90030-R.

Tateishi, K., Tsunogae, T., Santosh, M., and Janardhan, A.S., 2004, First report of sapphirine + quartz assemblage from southern India: Implications for ultrahigh-temperature metamorphism: Gondwana Research, v. 7, p. 899–912, doi:10.1016/S1342-937X(05)71073-X.

Tsunogae, T., and Miyano, T., 1989, Granulite facies metamorphism in the Central and Southern Marginal Zones of the Limpopo Belt, South Africa: Journal of the Geological Society of Japan, v. 95, p. 1–16 [in Japanese with English abstract].

Tsunogae, T., and Santosh, M., 2006a, Reply to Comment on "First report of garnet–corundum rocks from Southern India: Implications for prograde high-pressure (eclogite–facies?) metamorphism" by D.E. Kelsey, C. Clark, M. Hand, A.S. Collins: Earth and Planetary Science Letters, v. 249, p. 535–540, doi:10.1016/j.epsl.2006.07.049.

Tsunogae, T., and Santosh, M., 2006b, Spinel-sapphirine-quartz bearing composite inclusion within garnet from an ultrahigh-temperature pelitic granulite: Implications for metamorphic history and P-T path: Lithos, v. 92, p. 524–536, doi:10.1016/j.lithos.2006.03.060.

Tsunogae, T., and Santosh, M., 2010a, Ultrahigh-temperature metamorphism and decompression history of sapphirine granulites from Rajapalaiyam, southern India: Implications for the formation of hot orogens during Gondwana assembly: Geological Magazine, v. 147, p. 42–58, doi:10.1017/S0016756809990100.

Tsunogae, T., and Santosh, M., 2010b, Sapphirine + quartz assemblage from the Southern Indian Granulite Terrane: Diagnostic evidence for ultrahigh-temperature metamorphism within a Gondwana collisional orogen: Geological Journal, doi:10.1002/gj.1244 (in press).

Tsunogae, T., and van Reenen, D.D., 2006, Corundum + quartz and Mg-staurolite bearing granulite from the Limpopo Belt, southern Africa: Implications for a P-T path: Lithos, v. 92, p. 576–587, doi:10.1016/j.lithos.2006.03.052.

Tsunogae, T., and van Reenen, D.D., 2007, Carbonic fluid inclusions in sapphirine + quartz bearing garnet granulite from the Limpopo Belt, southern Africa: Journal of Mineralogical and Petrological Sciences, v. 102, p. 57–60, doi:10.2465/jmps.060829.

Tsunogae, T., Miyano, T., and Ridley, J., 1992, Metamorphic P-T profiles from the Zimbabwe Craton to the Limpopo Belt, Zimbabwe: Precambrian Research, v. 55, p. 259–277, doi:10.1016/0301-9268(92)90027-L.

Tsunogae, T., Miyano, T., van Reenen, D.D., and Smit, C.A., 2004, Ultrahigh-temperature metamorphism of the Southern Marginal Zone of the Archean Limpopo Belt, South Africa: Journal of Mineralogical and Petrological Sciences, v. 99, p. 213–224, doi:10.2465/jmps.99.213.

Tsunogae, T., Santosh, M., Ohyama, H., and Sato, K., 2008, High-pressure and ultrahigh-temperature metamorphism at Komateri, northern Madurai Block, southern India: Journal of Asian Earth Sciences, v. 33, p. 395–413, doi:10.1016/j.jseaes.2008.02.004.

van Reenen, D.D., 1983, Cordierite + garnet + hypersthene + biotite-bearing assemblages as a function of changing metamorphic conditions in the Southern Marginal Zone of the Limpopo Metamorphic Complex, South Africa: Geological Society of South Africa Special Publication 8, p. 143–167.

van Reenen, D.D., Barton, J.M., Jr., Roering, C., Smit, C.A., and van Schalkwyk, J.F., 1987, Deep crustal response to continental collision: The Limpopo belt of southern Africa: Geology, v. 15, p. 11–14, doi:10.1130/0091-7613(1987)15<11:DCRTCC>2.0.CO;2.

van Reenen, D.D., Roering, C., Ashwal, L.D., and De Wit, M.J., 1992, The Archaean Limpopo Granulite Belt: Tectonics and Deep Crustal Processes: Precambrian Research, v. 55, 587 p.

van Reenen, D.D., Perchuk, L.L., Smit, C.A., Varlamov, D.A., Boshoff, R., Huizenga, J.M., and Gerya, T.V., 2004, Structural and P-T evolution of a major cross fold in the Central Zone of the Limpopo high-grade terrain, South Africa: Journal of Petrology, v. 45, p. 1413–1439, doi:10.1093/petrology/egh028.

van Reenen, D.D., Boshoff, R., Smit, C.A., Perchuk, L.L., Kramers, J.D., McCourt, S., and Armstrong, R.A., 2008, Geochronological problems related to polymetamorphism in the Limpopo Complex, South Africa: Gondwana Research, v. 14, p. 644–662, doi:10.1016/j.gr.2008.01.013.

Watkeys, M.K., Light, M.P.R., and Broderick, T.J., 1983, A retrospective view of the Central Zone of the Limpopo Belt, Zimbabwe: Geological Society of South Africa Special Publication 8, p. 65–80.

Windley, B.F., Ackermand, D., and Herd, R.K., 1984, Sapphirine/kornerupine-bearing rocks and crustal uplift history of the Limpopo belt, southern Africa: Contributions to Mineralogy and Petrology, v. 86, p. 342–358, doi:10.1007/BF01187139.

Zeh, A., Klemd, R., Buhlmann, S., and Barton, J.M., Jr., 2004, Pro- and retrograde PT evolution of granulites of the Beit Bridge Complex (Limpopo Belt, South Africa): Constraints from quantitative phase diagrams and geotectonic implications: Journal of Metamorphic Geology, v. 22, p. 79–95, doi:10.1111/j.1525-1314.2004.00501.x.

Manuscript Accepted by the Society 24 May 2010

Granite emplacement and the retrograde P-T-fluid evolution of Neoarchean granulites from the Central Zone of the Limpopo Complex

Jan-Marten Huizenga*
Department of Geology, University of Johannesburg, P.O. Box 524, Auckland Park, 2006, South Africa

Leonid L. Perchuk[†]
*Department of Petrology, Geological Faculty, Moscow State University, Leninskie Gory, Moscow, 119192, Russia;
Institute of Experimental Mineralogy, Russian Academy of Sciences, Chernogolovka, Moscow District, 142432, Russia; and
Department of Geology, University of Johannesburg, P.O. Box 524, Auckland Park, 2006, Johannesburg, South Africa*

Dirk D. van Reenen
Yvonne Flattery
Department of Geology, University of Johannesburg, P.O. Box 524, Auckland Park, 2006, South Africa

Dimitri A. Varlamov
Department of Petrology, Geological Faculty, Moscow State University, Vorobievy Gory, Moscow, 119899, Russia

C. André Smit
Department of Geology, University of Johannesburg, P.O. Box 524, Auckland Park, 2006, South Africa

Taras V. Gerya
*Department of Earth Sciences, Swiss Federal Institute of Technology (ETH-Zurich), CH-8092 Zurich, Switzerland, and
Adjunct Professor of Geological Faculty, Moscow State University, Vorobievy Gory, Moscow, 119899, Russia*

ABSTRACT

Petrological and fluid-inclusion data of high-grade metapelitic gneisses that occur as enclaves and in the immediate surroundings of the 2.612 Ga old Bulai granitoid intrusive are presented in this chapter. The Bulai intrusive is an important time marker in the tectono-metamorphic evolution of the Central Zone of the Limpopo Complex. The host-rock gneisses show one generation of garnet, cordierite, and sillimanite, whereas the enclave gneisses show two different generations of

---

*Present address: Department of Geology, School of Environmental Science and Development, North-West University, Potchefstroom, South Africa; jan.huizenga@nwu.ac.za.
[†]posthumous

garnet ($Grt_{1,2}$), cordierite ($Crd_{1,2}$), and sillimanite ($Sil_{1,2}$). The first generation defines a gneissic texture, whereas the second generation shows a random mineral orientation. Grt_1 and Crd_1 show a higher Mg content compared with Grt_2 and Crd_2. Host rock garnet and Grt_1 show K-feldspar micro-veins at the contact with quartz as a result of high-temperature metasomatism. Host rock garnet, Grt_1, and Grt_2 are zoned and participate in two simultaneously operating reactions: sillimanite + garnet + quartz = cordierite and garnet + K-feldspar + H_2O = biotite + sillimanite + quartz. The combination of petrographic, geothermobarometric, and fluid-inclusion results shows evidence of two different pressure-temperature (P-T) paths in the enclave and a single P-T path in the host rocks. The decompressional cooling P-T path in the host rock is typical of the country rocks throughout the Central Zone. The high-pressure part of the host-rock P-T path overlaps with the Grt_1-Crd_1-Sil_1 P-T path found in the enclave rocks. The second P-T path is calculated from the Grt_2-Crd_2-Sil_2 assemblage and is found only in the enclave rocks. The two P-T paths in the enclave rocks can be connected by a sub-isobaric heating event of ~50 °C at 5.5 kbar. This increase in temperature is followed by decompressional cooling but with a lower P-T gradient compared with that of the country rocks caused by the emplacement of the Bulai Pluton. Fluids present during granulite metamorphism include CO_2 and brines. Retrograde infiltration of water in graphite-bearing country rocks under relatively reduced conditions resulted in the formation of a methane-rich fluid.

INTRODUCTION

The Neoarchean Limpopo High-Grade Terrane of southern Africa is an excellent terrane for studying geologic problems that are unique to high-grade terranes worldwide. These complex problems refer to virtually all aspects of this high-grade terrane, including fluid-rock interaction (e.g., van Reenen, 1986; Van den Berg and Huizenga, 2001) and the timing, style, and number of high-grade deformational and metamorphic events (e.g., Kramers et al., 2007). In particular the number and timing of metamorphic events is of importance for the Limpopo High-Grade Terrane. For example, different workers disagree on whether the Central Zone was affected by a single high-grade tectono-metamorphic event at ca. 2 Ga that postdates the emplacement of the Bulai Pluton (e.g., Barton et al., 1994; Millonig et al., 2008), or whether the Bulai Pluton is an important time marker for a major Neoarchean regional tectono-metamorphic event in the Central Zone (e.g., Barton and van Reenen, 1992; van Reenen et al., 2008). Furthermore, the behavior of the fluid phase during retrograde metamorphism in the Central Zone has not been the subject of any detailed study so far.

In this study we integrate fluid-inclusion data with classical geothermobarometry obtained from metapelitic granulites that respectively occur as enclaves and as lit-par-lit layers (here referred to as country rocks) within the main porphyritic granite phase of the Bulai Pluton to determine (1) whether the emplacement of the Bulai granite at 2.612 Ga (Zeh et al., 2005) has had any noticeable contact metamorphic overprint onto the enclaves and country rock; and, if this is the case, (2) the nature of this effect on the retrograde P-T-fluid evolution of these high-grade rocks; and (3) the composition of the fluids present during peak and retrograde conditions.

In addition, geothermobarometric data from this study will be compared with data obtained from metapelites that occur throughout the Central Zone to establish whether the Bulai Pluton did indeed intrude supercrustal rocks that have already undergone a regional high-grade Neoarchean tectono-metamorphic event (van Reenen et al., 2008).

REGIONAL GEOLOGY

The Limpopo High-Grade Terrane is subdivided into three subzones (Fig. 1): the Southern Marginal Zone, the Central Zone, and the Northern Marginal Zone (e.g., van Reenen et al., 1990). The Southern and Northern Marginal Zones are the high-grade metamorphic equivalents of the greenstone-gneiss terranes of the adjacent Kaapvaal and Zimbabwe Cratons, respectively (e.g., van Reenen et al., 1990).

The Central Zone is dominated by a shelf-type supracrustal sequence, the Beit Bridge Complex, a variety of Mesoarchean (Sand River gneisses) and Neoarchean (Alldays gneisses) orthogneisses of different ages that intrude and are folded with the supracrustal sequence, Mesoarchean meta-gabbroic and meta-anorthositic rocks of the Messina Layered Suite, Neoarchean quartzofeldspathic intrusive rocks of varying compositions (Singelele-type gneisses), and the Bulai Pluton (Barton et al., 1994; Holzer et al., 1998; Kröner et al., 1999). The Beit Bridge Complex comprises mainly quartzofeldspathic and metasedimentary gneisses, which include metapelitic gneisses, mafic gneisses, marbles, calc-silicate rocks, and (magnetic) quartzites.

The Southern and Northern Marginal Zones are bounded in the south and north by the Neoarchean inward-dipping dip-slip Hout River Shear Zone and the North Marginal Thrust Zone, respectively (Fig. 1). The Central Zone is separated from the

Northern Marginal Zone by the southerly dipping strike-slip ca. 2.0 Ga Tuli-Sabi–Triangle Shear Zone (Kamber et al., 1995) and from the Southern Marginal Zone by the northerly dipping ca. 2.0 Ga strike-slip Palala Shear Zone (Schaller et al., 1999) (Fig. 1).

The entire Central Zone is characterized by a complex deformational pattern consisting of two domains: the Tshipise Straightening Zone in the south and the Cross Folded Zone in the Musina area (e.g., van Reenen et al., 2004) (Fig. 2). The Cross Folded Zone comprises large-scale NNW-SSE–trending cross isoclinal folds and WSW plunging sheath folds (Roering et al., 1992; van Reenen et al., 2008). Boshoff et al. (2006) demonstrated that major structural features mapped as cross folds in the west Alldays area are in fact structures that developed as the result of superposition of N-S–trending Paleoproterozoic high-grade shear zones onto early Neoarchean folds. This is supported by the fact that P-T paths for metapelitic gneisses from closely associated cross folds and sheath folds within the Cross Folded Zone indicate two distinctly different retrograde decompression cooling histories: The sheath folds are indicative of relatively high pressure during decompressional cooling, whereas the cross folds indicate a relatively low pressure during decompressional cooling (Boshoff et al., 2006; van Reenen et al., 2008). The low-pressure P-T path is linked to the high-pressure P-T path by an isobaric (5.5 kbar) heating path that occurred at ca. 2.02 Ga. This isobaric heating event resulted in the widespread formation of polymetamorphic granulites in the Central Zone (Boshoff et al., 2006; van Reenen et al., 2008).

The Bulai Pluton

The Bulai Pluton, recently dated at 2612 ± 7 Ma (Millonig et al., 2008) mainly crops out in the area northwest of Musina (Fig. 1) and is an important time marker for interpreting the structural and P-T-fluid evolution of the Messina–Beitbridge area (Holzer et al., 1998). It mainly comprises undeformed porphyritic granitoids with geological and structural features that are of particular importance for this study: (1) The granitoids intrude supracrustal rocks (paragneisses of the Beit Bridge Complex) that occur either as enclaves or as lit-par-lit layers within the granites. These rocks were already metamorphosed, migmatized, and deformed before 2644 ± 8 Ma (Boshoff et al., 2006; Millonig et al., 2008; van Reenen et al., 2008). This tectono-metamorphic event in the Central Zone represents the uplift of the high-grade rocks to mid-crustal levels (Boshoff et al., 2006; van Reenen et al., 2008). (2) The metapelitic enclaves that occur within the Bulai porphyritic granites show granulite-facies mineral assemblages and a gneissic migmatitic texture, similar to that of the country rocks that occur as lit-par-lit layers within the outcrop area of the pluton as well as within the country rocks surrounding the pluton. This implies that deformation under granulite facies conditions occurred prior to, or at the very least, during the emplacement of the Bulai Pluton (e.g., van Reenen et al., 2008). (3) There is no evidence of a fold deformation event after the emplacement of the Bulai Pluton. Narrow, late magmatic dikes with the same mineralogical composition as the Bulai granite, are completely undeformed, like the main body of the Bulai Pluton.

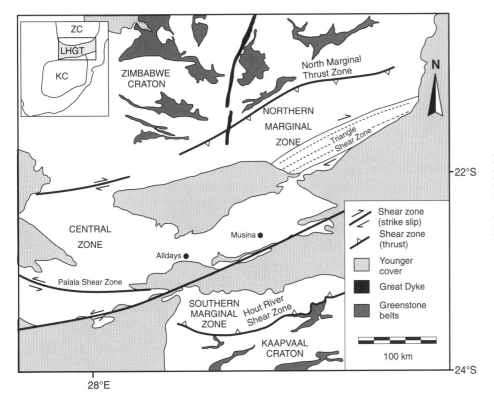

Figure 1. Geological map of the Limpopo High-Grade Terrane (LHGT), showing the bounding shear zones and the subdivision into the Northern Marginal Zone, the Central Zone, and the Southern Marginal Zone. ZC—Zimbabwe Craton; KC—Kaapvaal Craton.

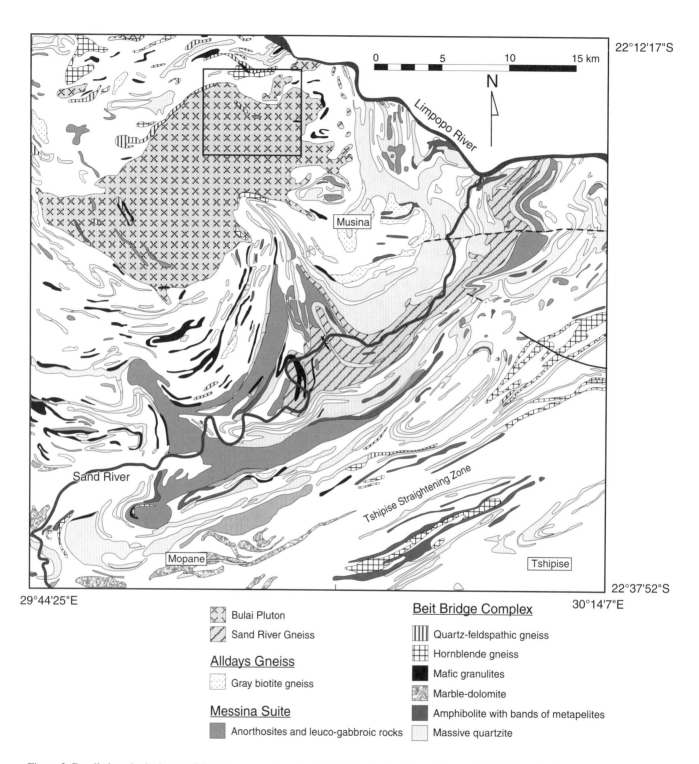

Figure 2. Detailed geological map of the area surrounding the Bulai Pluton in the Central Zone of the high-grade Limpopo terrane (modified from van Reenen et al., 2004). Box shows study area (Three Sisters locality).

(4) Evidence for post-emplacement deformation is restricted to narrow (less than a meter) high-temperature shear zones that cut both the Beit Bridge Complex rocks and the Matok Pluton (van Reenen et al., 2008).

PETROGRAPHY AND MINERAL COMPOSITIONS

Samples (Table 1) were collected from metapelitic granulites from the enclaves in the main porphyritic phase of the Bulai Pluton at the Three Sisters locality on the Boston farm northwest of Musina, and from the metapelitic country rocks that occur lit-par-lit with the intrusive porphyritic phase (van Reenen et al., 2008), and sampled ~100 m northwest of the Three Sisters locality (Fig. 2). Both varieties of metapelitic gneisses contain large, up to 10–20 mm, garnet porphyroblasts (Fig. 3A) in a cordierite-biotite-plagioclase-quartz matrix. The orientation of matrix sillimanite is consistent with the southwest-plunging quartz and feldspar lineation in the country rocks. These lineations are oriented parallel to the lineations associated with sheath folds throughout the Central Zone. This observation provides supporting evidence for a major Neoarchean tectono-metamorphic event in the Central Zone prior to the emplacement of the Bulai Pluton (van Reenen et al., 2008).

Minerals were analyzed at the Department of Petrology of Moscow State University using a Cameca SX50 electron microprobe, and at the Institute of Experimental Mineralogy (Russian Academy of Sciences) using a Tescan CamScan MV2300 equipped with an EDX analyzer with a semiconductor Si (Li) INCA Energy detector. Data obtained with both instruments are consistent. Mineral abbreviations are after Kretz (1983). Typical petrographic and compositional characteristics of the different generations of garnet (Grt), cordierite (Crd), and biotite (Bt) are summarized in Table 2.

Garnet crystallized in two generations. The first generation of garnet (Grt_1) occurs in both the enclave and country rocks and is represented by large (5–25 mm) equigranular poikiloblastic (Fig. 3A) and elongated grains. Some of the elongated garnets show an inclusion-free core, whereas the rim contains abundant quartz inclusions that are commonly rimmed by K-feldspar. In many cases a K-feldspar rim is also present where the garnet is in contact with quartz (Fig. 3B). Grt_1 is commonly surrounded by large (0.5–1 mm) sillimanite needles, defining a schistose fabric. Grt_1 shows two types of systematic chemical zoning. The first type relates to the Crd-free assemblages and shows X_{Mg} mol ratios, Mg/(Mg+Fe), varying between 0.47 and 0.44 from core to rim (Fig. 3D). The highest X_{Mg} value of 0.47 was measured in the core of the pencil-shaped garnet with rutile inclusions. The second type of Grt_1 is typical for Crd-bearing rocks, showing X_{Mg} values ranging from 0.40 (core) to 0.36 (rim).

The second type of garnet, Grt_2, occurs only in Crd-bearing rocks from the enclave (Fig. 3C) and shows lower X_{Mg} values compared with Grt_1 (Fig. 3E, 4A). At the contact with cordierite, Grt_1 is in places overgrown by Grt_2. Individual Grt_2 grains have an irregular (≤1 mm) shape and contain abundant, randomly oriented inclusions of sillimanite. In places ~5-μm-sized euhedral Grt_2 is included in a medium- to fine-grained Crd-Sil matrix (Fig. 3C). X_{Mg} decreases from 0.37 toward 0.26, where it is in direct contact with cordierite (Fig. 3E).

Similar to garnet, cordierite also occurs in two generations with distinct X_{Mg} values (Fig. 3F). The first generation, Crd_1, occurs in both the enclave and the country rocks as large (~3 mm) grains and aggregates and is associated with Grt_1 (Fig. 3A). X_{Mg} values range between 0.78 and 0.81. The second generation, Crd_2, is present only in rocks from the enclave and typically occurs in the matrix together with a fine-grained intergrowth of oriented Sil and Bt. This cordierite has lower X_{Mg} values, ranging between 0.71 and 0.76, compared with Crd_1 (Fig. 4B) and is associated with Grt_2. At the contact with Grt_2, X_{Mg} increases to values between 0.82 and 0.86. Cordierite is replaced by secondary chlorite and white mica.

Three generations of biotite show a wide variation in X_{Mg}. The first generation (Bt_1) is present in the enclave and country rocks, and occurs in the rock matrix and as inclusions in Grt_1. This biotite is characterized by a positive correlation of X_{Mg} (0.59–0.68) with titanium that varies from 0.40 to 0.73 cations per formula unit (c.p.f.u.). The second generation of biotite (Bt_2) is associated with Grt_2 and Crd_2 and occurs in the enclave only. The X_{Mg} of this biotite varies between 0.54 and 0.62 (Fig. 5A). Secondary Ti-free biotite (Bt_3) was found only in one sample (Lim165-2) from the country rock as a rim between Grt_1 and Crd_1. This biotite shows a X_{Mg} variation between 0.57 and 0.62, and increases to 0.82 toward the contact with garnet (Fig. 5A). The highest Al concentration is ~1.5 c.p.f.u. for Bt_3 in the country rock (sample Lim165-2, Fig. 5A).

High-Temperature Metasomatism

Both the perthitic and antiperthitic feldspars occur at the contact of the matrix quartz with Grt_1 (Fig. 3B) and as rims around quartz inclusions within the Grt_1 porphyroblasts (Fig. 3B) in the enclave and country rocks. The composition of the perthitic feldspars (Fig. 5B) varies from $An_8Ab_{68}Or_{24}$ to $An_2Ab_{22}Or_{76}$ (Fig. 5B). Biotite (Bt_1) associated with the micro-veins is characterized by

TABLE 1. MINERALOGY OF STUDIED SAMPLES

Sample	Rock type	Mineral assemblage
Bx1-4		$Grt_{1,2}+Crd_{1,2}+Bt_{1,2}+Spl+Sil+Pl+Qtz+Kfs$
Bx1-5		$Grt_{1,2}+Crd_{1,2}+Bt_{1,2}+Spl+Sil+Pl+Qtz+Kfs$
Lim164-2	Enclave	$Grt_{1,2}+Crd_{1,2}+Bt_{1,2}+Spl+Sil+Pl+Qtz+Kfs$
Lim164-3		$Grt_{1,2}+Crd_{1,2}+Bt_{1,2}+Spl+Sil+Pl+Qtz+Kfs$
Lim164-6		$Grt_{1,2}+Crd_{1,2}+Bt_{1,2}+Spl+Sil+Pl+Qtz+Kfs$
Lim165-2	Country rock	$Grt_1+Crd_1+Bt_{1,3}+Spl+Sil+Pl+Qtz$
Lim165-3		$Grt_1+Crd_1+Bt_1+Spl+Sil+Pl+Qtz+Kfs$

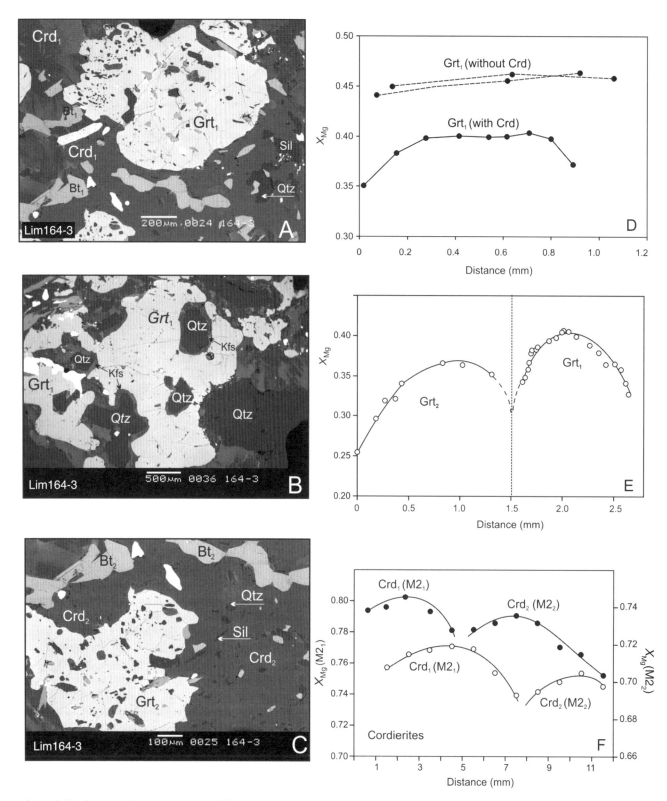

Figure 3. Backscattered electron images (BSI) and chemical (X_{Mg}) compositions, showing the different generations of garnet (Grt) and cordierite (Crd) in metapelites from the enclave within the Bulai Pluton. (A) BSI of Grt_1 with Crd_1, Sil (sillimanite), and Qtz (quartz). Bt—biotite. (B) BSI image of Grt_1 in the Crd + Sil free portion of the rock. This garnet is separated from quartz by Kfs (K-feldspar) rims that resulted from interaction of the rock with a potassium-rich metasomatic fluid. (C) BSI of Grt_2 in paragenesis with Crd_2, Sil, and Qtz. (D) Chemistry of two types of Grt_1 related to the Crd-free paragenesis (see B) and to the Crd abundant assemblage (see A). (E) Compositional relationships between Grt_1 and Grt_2 (Bx1-5). (F) Chemical compositions of cordierite related to stages $M2_1$ and $M2_2$ (Lim164-3).

TABLE 2. SUMMARY OF MINERAL CHEMISTRY

Mineral	Occurrence	Composition: $X_{Mg}/(X_{Mg} + X_{Fe})$	Description
Grt_1	Enclave and country rocks	Without Crd: 0.47–0.44 (core to rim)	Rounded and elongated poikiloblasts (5–25 mm), surrounded by Sil defining a schistose fabric
		With Crd: 0.40–0.36 (core to rim)	
Crd_1	Enclave and country rocks	0.78–0.81	Crd_1 occurs as grains and aggregates together with Grt_1
Bt_1	Enclave and country rocks	0.60–0.68	Ti-rich biotite occurring in the matrix and as inclusions in Grt_1
Grt_2	Enclave only	0.37–0.26 at the contact with Crd_2	Grt_2 contains randomly orientated Sil inclusions and occurs at the contact between Grt_1 and Crd_1. Grt_2 has lower X_{Mg} values compared to Grt_1
Crd_2	Enclave only	0.76–0.71	Crd_2 is associated with Grt_2 and has lower X_{Mg} compared with Crd_1
Bt_2	Enclave only	0.54–0.62	Ti-rich biotite occurring in the matrix and as inclusions in Grt_2
Bt_3	Country rocks only	0.57–0.62 at the contact with Grt_1	Ti-free biotite occurring between Grt_1 and Crd_1

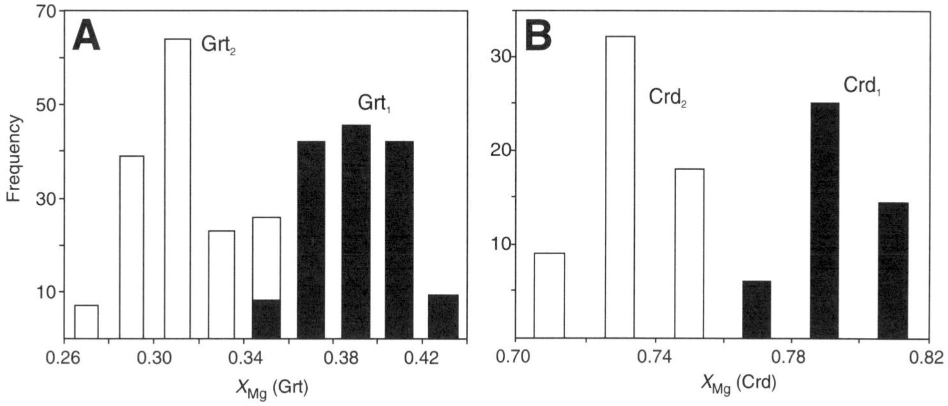

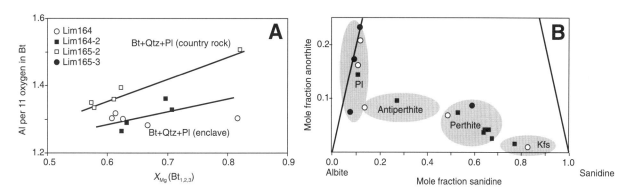

Figure 4. Histograms demonstrating two distinct generations of garnet and cordierite in samples Bx1-5 and Bx1-4, collected from Al-rich portions in the enclave. The first generations of Grt_1 and Crd_1 were formed at peak P-T conditions and during exhumation to mid-crustal levels. However, Crd_1 is preserved in sample Bx1-5 only, whereas sample Bx1-5 contains both Grt_1 and Crd_1. The second generations of Grt_2 and Crd_2 occur in both samples and may reflect the influence of the Bulai intrusion.

Figure 5. Compositions of biotite (A) and feldspar (B) associated with K-feldspar (Kfs) rims between quartz and garnet. Al—alvite; Pl—plagioclase.

TABLE 3. SELECTED GEOTHERMOBAROMETRIC RESULTS FOR THE ENCLAVE

Mg numbers			$\alpha_{H_2O}^\dagger$	T (°C)		P (kbar)	Mg numbers			α_{H_2O}	T (°C)		P (kbar)
Grt	Crd	Bt*		Crd-Grt	Bt-Grt		Grt	Crd	Bt		Crd-Grt	Bt-Grt	
Sample Bx1-5 (Crd$_1$+Grt$_1$+Bt$_1$+Kfs+Sil)							Sample Bx1-4 (Crd$_2$+Grt$_2$+Bt$_2$+Kfs+Sil)						
0.406	0.771	0.594	0.239	821	819	7.25	0.284	0.740	*0.577*	0.119	682	684	4.85
0.392	0.771	0.597	0,225	797	797	6.91	0.279	0.737	*0.574*	0.118	680	681	4.78
0.441	0.807	0.647	0.224	796	795	7.24	0.292	0.750	*0.591*	0.118	678	680	4.88
0.406	0.790	0.619	0.214	780	787	6.84	0.280	0.740	*0.579*	0.118	676	677	4.76
0.396	0.787	0.619	0.209	771	774	6.68	0.275	0.737	*0.576*	0.117	674	675	4.70
0.392	0.785	0.617	0.208	770	772	6.63	0.288	0.750	*0.592*	0.117	673	674	4.80
0.393	0.787	0.628	0.206	767	761	6.61	0.284	0.748	0.589	0.116	670	672	4.73
0.437	0.822	0.677	0.198	753	751	6.77	0.285	0.757	*0.605*	0.114	658	659	4.64
0.384	0.790	0.632	0.194	746	745	6.34	0.282	0.757	*0.606*	0.113	654	655	4.58
0.372	0.784	0.619	0.190	740	745	6.18	0.281	0.759	0.611	0.113	650	649	4.54
0.356	0.778	0.617	0.183	727	729	5.95	0.285	0.765	0.618	0.112	646	647	4.54
0.406	0.815	0.660	0.181	725	737	6.29	Sample Bx1-5 (Crd$_2$+Grt$_2$+Bt$_2$+Kfs+Sil)						
0.358	0.782	0.628	0.180	723	719	5.92	0.304	0.751	0.583	0.168	704	707	5.28
0.358	0.787	0.632	0.175	714	715	5.83	0.306	0.746	0.580	0.175	715	712	5.40
0.343	0.780	0.628	0.169	705	703	5.63	0.310	0.754	0.597	0.171	707	700	5.36
0.340	0.780	0.628	0.167	701	700	5.57	0.280	0.728	0.560	0.169	705	702	5.08
0.384	0.815	0.678	0.163	695	693	5.84	0.277	0.728	0.560	0.166	700	698	4.99
0.388	0.787	0.628	0.202	759	755	6.50	0.281	0.736	0.578	0.163	694	685	4.98
0.382	0.787	0.628	0.196	750	747	6.36	0.281	0.719	0.545	0.178	719	717	5.23
0.385	0.794	0.638	0.190	740	740	6.29	0.274	0.719	0.547	0.171	709	707	5.05
0.376	0.790	0.629	0.188	736	740	6.18	0.274	0.735	0.580	0.157	685	675	4.82
0.390	0.803	0.647	0.183	728	734	6.20	0.274	0.719	0.550	0.171	709	704	5.05
Sample Lim164-3 (Crd$_1$+Grt$_1$+Bt$_1$+Kfs+Sil)							0.266	0.751	0.595	0.137	652	653	4.42
0.385	0.792	*0.633*	0.193	744	744	6.33							
0.385	0.794	*0.636*	0.190	740	741	6.29							
0.376	0.790	*0.633*	0.187	735	734	6.18							
0.381	0.794	*0.637*	0.186	733	735	6.18							
0.390	0.803	*0.652*	0.182	727	728	6.20							
0.376	0.797	*0.645*	0.178	720	721	6.03							
Sample Bx1-5 (Crd$_2$+Grt$_2$+Bt$_2$+Kfs+Sil)													
0.296	0.715	0.536	0.167	745	743	5.40							
0.319	0.738	0.560	0.166	743	747	5.57							
0.302	0.731	0.558	0.163	727	727	5.28							
0.304	0.734	0.561	0.162	725	726	5.27							
0.288	0.727	0.550	0.160	710	717	5.00							
0.321	0.731	0.547	0.169	758	763	5.73							

*Values in italic for Bt are calculated at a given T (°C) using the Crd-Grt thermometer.
$^\dagger \alpha_{H_2O}$ is taken as 0.2 as an average value for M2$_1$ stage.
Note: Bt—biotite; Crd—cordierite; Grt—garnet; Kfs—K-feldspar; Sil—sillimanite.

a low Al content. Similar K-feldspar reaction textures have been described in other granulite facies terranes (Perchuk and Gerya, 1993; Harlov et al., 1998; Harlov and Wirth, 2000; Harlov and Förster, 2002) and also in the Southern Marginal Zone of the Limpopo High-Grade Terrane (Van den Berg and Huizenga, 2001). Taking into account the low-Al content in Bt$_1$ associated with Kfs, the following reaction is suggested for the formation of the Kfs micro-veins between Qtz and Grt$_1$:

$$Grt_1 + Qtz + (K, Na)_{fluid} = Kfs + Ab + Bt. \quad (1)$$

Geothermobarometry and P-T Paths

An internally consistent thermodynamic data set, based on experimental data for the chemical equilibria involving garnet, biotite, cordierite, quartz, and sillimanite (Perchuk and Lavrent'eva, 1983; Perchuk et al., 1985; Aranovich and Podlesskii, 1989), was used for calculation of local temperatures and pressures in the studied metapelites using the following reactions:

$$Mg\text{-}Bt + Fe\text{-}Grt = Mg\text{-}Grt + Fe\text{-}Bt; \quad (2)$$

$$Mg\text{-}Crd + Fe\text{-}Grt = Mg\text{-}Grt + Fe\text{-}Crd; \quad (3)$$

$$Mg\text{-}Crd = Mg\text{-}Grt + Sil + Qtz. \quad (4)$$

For the derivation of P-T paths we used the procedures described by Perchuk as summarized in this issue (Perchuk, this volume). Representative compositional data and calculated P-T values are presented in Tables 3 and 4.

H$_2$O activities in the fluid phase (a_{H_2O}) for each stage of the retrograde stage of the metamorphic evolution were calculated using the reaction:

TABLE 4. SELECTED GEOTHERMOBAROMETRIC RESULTS FOR THE COUNTRY ROCK

Mg numbers			a_{H_2O}†	T (°C)		P (kbar)
Grt	Crd	Bt*		Crd-Grt	Bt-Grt	
			Sample Lim165-2			
0.423	0.736	*0.529*	0.200	936	934	8.37
0.417	0.736	*0.530*	0.200	924	924	8.22
0.422	0.747	*0.545*	0.200	909	910	8.10
0.404	0.752	*0.559*	0.200	866	865	7.57
0.389	0.767	*0.585*	0.200	811	813	6.94
0.378	0.768	*0.589*	0.200	792	794	6.69
0.350	0.779	*0.615*	0.200	731	732	5.92
0.344	0.779	*0.615*	0.200	724	726	5.81
0.337	0.778	*0.615*	0.200	716	718	5.68
0.325	0.782	*0.625*	0.200	693	695	5.39
0.320	0.790	*0.639*	0.200	673	675	5.17
0.314	0.806	*0.673*	0.200	639	637	4.84
0.300	0.803	*0.669*	0.200	627	627	4.62
			Lim165-3			
0.475	0.777	*0.584*	0.200	930	930	8.64
0.474	0.793	*0.612*	0.200	887	887	8.23
0.460	0.800	*0.627*	0.200	848	848	7.77
0.466	0.806	*0.637*	0.200	841	842	7.75
0.423	0.789	*0.617*	0.200	815	815	7.22
0.443	0.815	*0.659*	0.200	786	786	7.10
0.398	0.795	*0.632*	0.200	767	767	6.61
0.354	0.772	*0.604*	0.200	749	749	6.12
0.345	0.789	*0.634*	0.200	707	707	5.67
0.365	0.808	*0.662*	0.200	698	698	5.76
0.372	0.822	*0.686*	0.200	680	680	5.66
0.350	0.827	*0.700*	0.200	644	644	5.19
0.374	0.852	*0.741*	0.200	623	624	5.21

*Values in italic for Bt are calculated at a given T (°C) using the Crd-Grt thermometer.
†a_{H_2O} is taken as 0.2 as an average value for $M2_1$ stage.

$$Grt + Kfs + H_2O = Bt + Sil + Qtz. \quad (5)$$

The a_{H_2O} depends on temperature only (Fig. 6), which can be approximated by the following equations:

$$a_{H_2O} (M2_1) = 0.0006\ T\ (°C) - 0.2538; \quad (6)$$

$$a_{H_2O} (M2_2) = 0.0002\ T\ (°C) + 0.0175. \quad (7)$$

These equations were used for the calculation of the isopleths (in Figs. 8A and 8B) and for the pressure calculation of each local equilibrium (Tables 3, 4). For some local assemblages we were unable to measure distinct compositions of biotite in local equilibrium. In those cases, X_{Mg} of the biotite was calculated using the Bt-Grt, Fe-Mg distribution coefficient that was calculated using a temperature obtained from the Crd-Grt thermometer.

The two generations of garnet must have coexisted with two generations of other Fe-Mg minerals, reflecting, therefore, different physicochemical conditions of formation. Figure 7 clearly shows the systematic difference in X_{Mg} for the coexisting minerals involved in reactions (2) and (3). The data points for the assemblages Crd-Grt and Bt-Grt form two separate fields in the diagrams: 700–860 °C for $M2_1$, and 650–780 °C for $M2_2$ (where M2 refers to the retrograde metamorphic stage) at different pressures, suggesting the existence of two different P-T paths (Perchuk et al., 2008).

The calculated P-T paths are shown in Figure 8. The compositions of Grt_1, Crd_1, and Bt_1 in the metapelites from both the country rocks and the enclave represent the same $M2_1$ stage of the P-T path. The maximum P-T values estimated for the $M2_1$

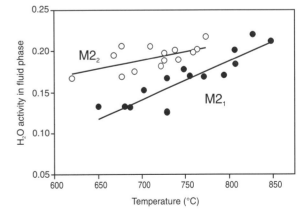

Figure 6. Temperature dependence of H_2O activity in a fluid, calculated on the basis of data from Table 3 and the compositions of coexisting Kfs related to the reaction (4) for metamorphic stages $M2_1$ and $M2_2$ for the Bulai metapelites.

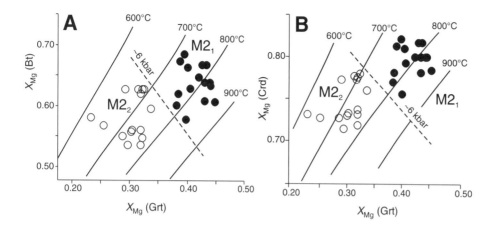

Figure 7. Distribution of Mg and Fe between biotite (Bt) and garnet (Grt) (A) and cordierite (Crd) and garnet (B), coexisting in sample Bx1-5. Isotherms are calculated with the Bt-Grt and Crd-Grt thermometers (Perchuk and Lavrent'eva, 1983), whereas key compositions of locally equilibrated minerals are taken from Table 3. Black circles reflect the distribution for stage $M2_1$ (regional metamorphism), and open circles belong to stage $M2_2$ (reheating from the Bulai intrusion).

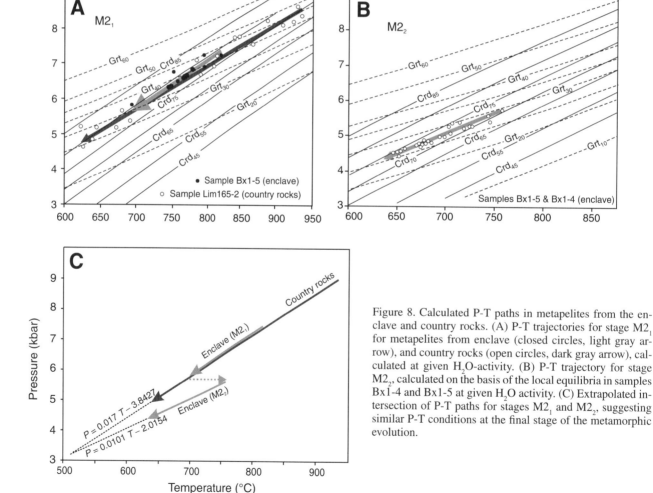

Figure 8. Calculated P-T paths in metapelites from the enclave and country rocks. (A) P-T trajectories for stage $M2_1$ for metapelites from enclave (closed circles, light gray arrow), and country rocks (open circles, dark gray arrow), calculated at given H_2O-activity. (B) P-T trajectory for stage $M2_2$, calculated on the basis of the local equilibria in samples Bx1-4 and Bx1-5 at given H_2O activity. (C) Extrapolated intersection of P-T paths for stages $M2_1$ and $M2_2$, suggesting similar P-T conditions at the final stage of the metamorphic evolution.

stage of the country rock (930 °C and 8.6 kbar; see Table 4 and Fig. 8A) are significantly higher than those of the M2$_1$ stage estimated for the enclave, which shows lower maximum P-T conditions (820 °C and 7.3 kbar, Fig. 8A) and is also shorter. The second mineral assemblage of Grt$_2$, Crd$_2$, and Bt$_2$, which is present only in the enclave, clearly yields a significantly lower M2$_2$, P-T path (Fig. 8B). The two different P-T paths can be connected by an isobaric heating event of ~50 °C (Fig. 8C).

FLUID INCLUSIONS

Fluid inclusions were studied in five samples (four from the enclave, one from the country rock) from the same outcrops that were used for the metamorphic study, comprising identical mineral assemblages and mineral chemistry. Sample Bx1-4 (Table 1) from the enclave was the only sample suitable for both fluid-inclusion studies and thermobarometry. The main purpose of the fluid inclusion study is (1) to identify which fluids are present during the peak and retrograde part of the P-T path, and (2) to determine how the retrograde P-T evaluation has affected fluid densities.

Microthermometric measurements were carried out on doubly polished sections, ~200 μm thick, using a Linkam TMH-600 heating-freezing stage, which is connected to a TP93 programmable thermal control unit (temperature range between −190 and +600 °C). The following temperatures were measured: (1) final melting and homogenization of CO_2 (±CH_4) carbonic inclusions: Tm_{CAR} (sol$_{CO_2}$ + liq$_{CAR}$ + vap = vap + liq$_{CAR}$) and Th_{CAR} (liq$_{CAR}$ + vap = liq$_{CAR}$), respectively; (2) final melting temperature of hydrohalite in NaCl saturated aqueous inclusions: Tm_{HH} (hydrohalite + liq$_{AQ}$ + vap = liq$_{AQ}$ + vap); (3) homogenization temperatures of aqueous inclusions: Th_{AQ} (liq$_{AQ}$ + vap = liq$_{AQ}$). Precise measurements of homogenization into the vapor phase were not done, as these are unreliable. Pure CO_2 and H_2O inclusions were used for calibration at the triple points (−56.6 and 0 °C). Measurements of Tm_{CAR}, Tm_{HH}, and Th_{CAR} have an accuracy of ~0.5 °C, whereas Th_{AQ} measurements have an accuracy of ~5 °C. Raman microprobe analyses were carried out at the Vrije Universiteit Amsterdam, using a multichannel Microdil-28® Raman laser probe with a 514 nm Ar-ion laser excitation source (Burke and Lustenhouwer, 1987). Fluid compositions are calculated using Raman cross sections (Dubessy et al., 1989) and by calibrating against standard gas samples. The relative error for the minor component is in the range of ~15%.

Four fluid-inclusion types (Table 5) were identified on the basis of mode of occurrence (isolated-clustered or in trails) and fluid composition, including isolated CO_2 (±CH_4) carbonic inclusions, trail-bound CO_2-rich (±CH_4) inclusions, isolated CH_4-rich (±CO_2) inclusions, and isolated aqueous brine inclusions.

Isolated and Clustered CO_2-Rich Inclusions

This is the most abundant type of carbonic inclusions found in Grt$_1$, in quartz blebs in Grt$_1$, and in matrix quartz (both country rock and enclaves) (Table 5). The inclusions (10–30 μm) occur either isolated or clustered and have irregular shapes, comprising a liquid and a vapor phase at room temperature. The inclusions that occur in garnet are primary with respect to the growth of their host mineral (Touret, 2001). The timing of the fluid inclusions in matrix quartz is difficult to tell owing to the possibility of recrystallization of quartz during the entire metamorphic history.

Tm_{CAR} ranges between −61.7 and −56.6 °C. The depression of Tm_{CAR} is due to the presence of CH_4 (up to 20 mol%), as confirmed by Raman microspectrometry. Th_{CAR} into the liquid phase ranges between −13.8 and +30.2 °C (Figs. 9A–9D). Inclusions in matrix quartz also show homogenization in the vapor phase (Fig. 9C). Notable is the large spread in Th_{CAR} in matrix quartz compared with the inclusions in Grt$_1$ and in quartz blebs in Grt$_1$. The reasonable correlation between Tm_{CAR} and Th_{CAR} for inclusions in matrix quartz can be explained by the variation of the mole fraction CH_4 in the carbonic phase (Fig. 10) (Thiéry et al., 1994; Van den Kerkhof and Thiéry, 2001).

Trail-Bound CO_2-Rich Inclusions

These inclusions define trails of rounded CO_2 inclusions (10–30 μm) in Grt$_1$ (enclave) and in matrix quartz (country rock) (Table 5). Trails found in Grt$_1$ in the enclave show a Th_{CAR} between +2 and +19 °C (Fig. 9E) and Tm_{CAR} varying between −57.2 and −56.8 °C, indicating an almost pure CO_2 carbonic fluid phase. Two measurable trails were identified in matrix quartz in the country rock. The first trail inclusions show a Tm_{CAR} between −56.9 and −56.6 °C, and all homogenize into the vapor phase. The second trail inclusions form a cluster of en-echelon trails. These inclusions show Tm_{CAR} between −62.5 and −56.7 °C and Th_{CAR} between −19 and +30 °C (Fig. 9F). The large variation of both Tm_{CAR} and Th_{CAR} can be explained by compositional variations between CO_2 and CH_4 (mol fraction CH_4 <0.3) (Fig. 11).

Isolated Methane Inclusions

These isolated inclusions are rare and restricted to the country rock metapelite where they occur in matrix quartz. They have irregular shapes and sizes varying from <10 μm to ~30 μm. At room temperature, most inclusions contain a carbonic liquid and a vapor phase. Some inclusions contain only a vapor phase, and bubble formation occurred only after subsequent cooling. These inclusions appear very dark, making it difficult, and in some cases impossible, to measure any phase transitions accurately. All inclusions homogenize into the vapor phase, indicating very low density. Tm_{CAR} ranges from −74.1 to −68.8 °C. Raman analyses identified CH_4 as the major constituent (>80 mol%) and CO_2. Other gaseous fluid species were not identified. Some water (<20 mol%), which is invisible, may be present in these inclusions.

Aqueous Brine Inclusions

Aqueous brine inclusions occur dominantly in matrix quartz in both the country rocks and enclaves. Two isolated inclusions

TABLE 5. SUMMARY OF FLUID-INCLUSION TYPE CHARACTERISTICS

Description	Rock type	Host mineral	Th/Tm (°C)	Molar volume	Composition*
		Carbonic inclusions			
Isolated CO_2-rich inclusions	Country rock	Matrix Qtz	Th_{CAR}: $-14 \sim$ into vap.† Tm_{CAR}: $-61.7 \sim -56.6$	$> \sim 48$ cm³/mol	CO_2-CH_4 $X_{CH_4} < 0.2$
		Qtz blebs in Grt_1	Th_{CAR}: $+1 \sim$ into vap. Tm_{CAR}: $-58.1 \sim 56.7$	$> \sim 48$	CO_2-CH_4 $X_{CH_4} < 0.05$
	Enclave	Matrix Qtz	Th_{CAR}: $+2 \sim$ into vap. Tm_{CAR}: $-57.4 \sim -56.7$	$> \sim 48$	CO_2-CH_4 $X_{CH_4} < 0.1$
		Qtz blebs in Grt_1	Th_{CAR}: $+9 \sim +21$ Tm_{CAR}: $-57.4 \sim -56.9$	~ 51–~ 58	
		Grt_1	Th_{CAR}: $-14 \sim +22$ Tm_{CAR}: $-62.0 \sim -56.9$	~ 46–~ 58	
Trail-bound CO_2-rich inclusions	Country rock	Matrix Qtz	Th_{CAR}: $-19.4 \sim +29.5$ Tm_{CAR}: $-62.5 \sim -56.7$	~ 51	CO_2-CH_4 $X_{CH_4} < 0.3$
	Enclave	Grt_1	Th_{CAR}: $-3 \sim +19$ Tm_{CAR}: $-57.2 \sim -56.8$	~ 48–~ 56	CO_2-CH_4 $X_{CH_4} < 0.02$
Isolated CH_4-rich inclusions	Country rock	Matrix Qtz	Th_{CAR}: into vap.‡ Tm_{CAR}: $-74.1 \sim -68.8$	$> \sim 85$	CH_4-CO_2 $X_{CH_4} > 0.8$
		Aqueous inclusions			
Isolated brines (±halite)	Country rock	Matrix Qtz	Th_{AQ}: not measured Tm_{AQ}: metastable	N/A	H_2O-NaCl-$CaCl_2$
	Enclave	Matrix Qtz and Qtz blebs in Grt_1	Th_{AQ}: $+100 \sim +400$ Tm_{AQ}: metastable	N/A	

*Composition of carbonic inclusions was determined using Th-Tm data (Thiéry et al., 1994; Van den Kerkhof and Thiéry, 2001), and Raman microspectrometry.
†Critical molar volume of CO_2 is 94.4 cm³/mol.
‡Critical molar volume of CH_4 is 99.2 cm³/mol.
Note: Grt—garnet; Qtz—quartz; Vap.—vapor.

were found in quartz blebs in garnet (Grt_1) in metapelite from the enclave. Unfortunately, collecting reliable microthermometric data from these two inclusions was not possible owing to their small size (10–20 μm) and irregular shape. They do contain sporadically a halite crystal, indicating saturation of the solution in NaCl. Tm_{HH} of the inclusions in quartz shows a large variation between −24 and +26 °C. Eutectic melting could not be measured because of observation difficulties. The large variation in Tm_{HH} can be explained by (1) metastable melting, and (2) compositional variations in the H_2O-NaCl-$CaCl_2$ system. Metastable melting of halite at temperatures >0 °C typically occurs in NaCl-saturated fluid inclusions (Goldstein and Reynolds, 1994). Tm_{HH} values below the eutectic melting point of the H_2O-NaCl system (−21.2 °C) can be explained only with an additional species, of which $CaCl_2$ is the most likely. Low Tm_{HH} can be expected in inclusions of which the $CaCl_2/(NaCl + CaCl_2)$ mass ratio is $> \sim 0.5$. The homogenization temperatures show a spread between 100 and 380 °C.

DISCUSSION

Petrological Data

Petrological evidence suggests the following scheme for the retrograde tectono-metamorphic (exhumation) history of the studied area. High-grade metamorphism, reflected by the $M2_1$ P-T path is present in both the country rocks and the enclave, but peak conditions at ~950 °C and ~8.5 kbar are preserved only in the country rocks (Fig. 8A). Peak conditions were followed by decompression and cooling during which exhumation of granulites took place to mid-crustal levels (~700 °C, 5.5 kbar), prior to the emplacement of the Bulai Pluton. During this early event of exhumation the granulites were deformed and metasomatized. This was followed by the emplacement of the 2.612 Ga Bulai Pluton at ~5.5 kbar and the reheating of the enclave by the granite magma (stage $M2_2$, Figs. 8B, 8C). This heating event is not reflected in the mineral chemistry of the country rocks, indicating that it was a localized event restricted to the enclaves. Subsequent exhumation led to decompression and cooling of the entire Bulai Pluton, including the enclave and the country rocks. The extrapolated intersection of the two P-T paths at ~510 °C and ~3.2 kbar represents the point at which both the enclave and the country rocks reached the same temperature (Fig. 8C). Oriented matrix sillimanite surrounding Grt_1 porphyroblasts (Fig. 4D), and the K-feldspar micro-veins in the enclave, are postdated by the growth of unorientated Grt_2 + Sil aggregates during $M2_2$. The static nature of this second metamorphic stage is consistent with field observations that no regional deformation occurred after emplacement of the Bulai Pluton. Low-grade hydration (<500 °C, <3 kbar) is responsible for widespread hydration of preexisting mineral assemblages.

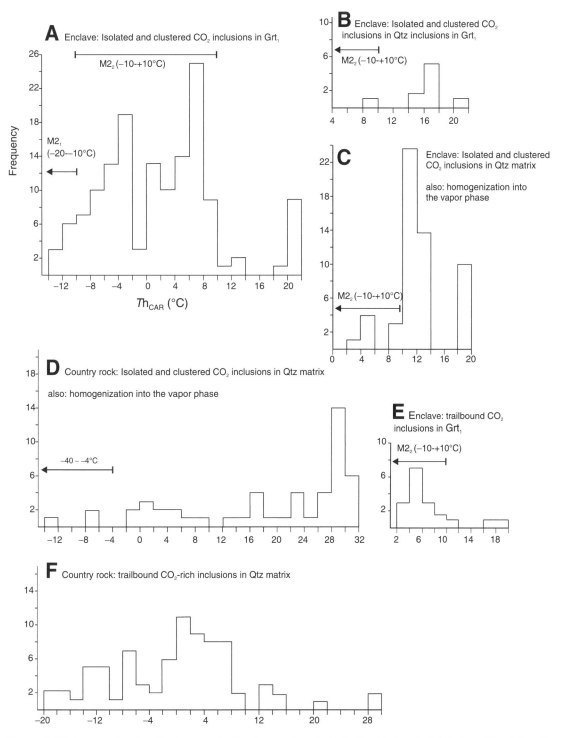

Figure 9. Histograms showing the homogenization temperature (into the liquid phase) distribution (Th_{CAR}) for all inclusion types in both the enclave and country rock. For the enclave, the $M2_1$ and $M2_2$ homogenization temperatures related to the retrograde P-T path are indicated (Figs. 8A, 8B). For the country rock, homogenization temperatures are indicated for the entire retrograde P-T path (Fig. 8A). See text for further explanation.

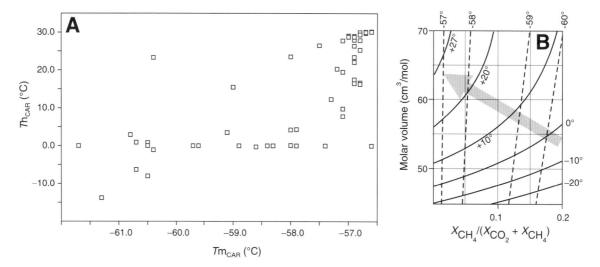

Figure 10. Th_{CAR} (into the liquid phase) versus Tm_{CAR}, illustrating the positive correlation for isolated and clustered CO_2-rich inclusions in the country rock as a result of variation in the CH_4 content (Thiéry et al., 1994; Van den Kerkhof and Thiéry, 2001). Arrow shows an increase in Th_{CAR} and Tm_{CAR} (A) as a result of a decrease in CH_4 in the fluid phase and an increase in molar volume (B).

Interpretation of Fluid-Inclusion Results

High-Grade Metamorphic Fluids

The isolated CO_2 fluid inclusions that occur in the Grt_1 represent remnants of a fluid that was present during the peak of metamorphism. The calculated H_2O activity of ~0.2 at the peak of metamorphism (see Tables 3 and 4, and Fig. 6) would correspond to fluid inclusions with ~10 vol% of water (corresponding to ~20 mol% H_2O), which should have been present in these inclusions. However, the fact that clathrate melting was not observed implies that water is not present and that water leakage must have taken place.

The exact timing of trapping of the high-salinity fluid inclusions is difficult to establish. Although establishing the time is tempting, one cannot conclude that the presence of brine inclusions in quartz blebs in Grt_1 represent a peak metamorphic

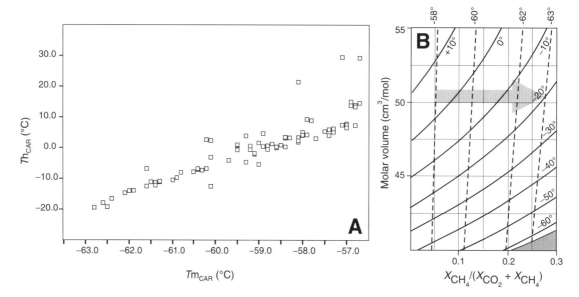

Figure 11. Th_{CAR} (into the liquid phase) versus Tm_{CAR}, illustrating the positive correlation for trail-bound CO_2-rich inclusions in one trail in matrix quartz in the country rock (A) as a result of a variation in the CH_4 content (Thiéry et al., 1994; Van den Kerkhof and Thiéry, 2001). Arrow shows an increase in Th_{CAR} and Tm_{CAR} (A) as a result of a decrease in CH_4 in the fluid phase at constant molar volume (B).

fluid, as it is impossible to establish when the quartz blebs were formed with respect to the garnet (Vernon et al., 2008). Also, to determine the exact composition of the brine inclusions is difficult, if not impossible, owing to their small size and rare occurrence. The best evidence, however, of the widespread presence of a high-grade brine fluid is the presence of the metasomatic K-feldspar veining around quartz that is in contact with Grt_1. The rare presence of brine inclusions can be explained by the high mobility of these fluids (Watson and Brennan, 1987); i.e., the fluids are simply not trapped. Furthermore, the steep slope of the isochores of aqueous systems means that during retrograde metamorphism, brine inclusions will be subjected to a severe underpressure, resulting in implosion and destruction of the inclusions (e.g., Touret and Huizenga, 1999).

Fluid Densities

It has been well established in experiments that fluid-inclusion densities may be adjusted during decompression and/or cooling, resulting in a spreading of the homogenization temperatures of one single fluid generation (e.g., Vityk and Bodnar, 1998). The expected CO_2 homogenization temperature range for the retrograde P-T paths were calculated using the software supplied by Bakker (2003) and are indicated in Figure 9. The histograms show clearly that the homogenization temperatures of the inclusions in Grt_1 in the enclave overlap largely with the $M2_1$ and $M2_2$ retrograde stage (Fig. 9A). CO_2 inclusions that occur in matrix quartz in the country rocks (Figs. 9C, 9D), on the other hand, show homogenization temperatures that can only be explained by re-equilibration at very low pressure and temperature.

In order to relate the retrograde P-T path more precisely to the fluid-inclusion densities, we calculated the fluid-inclusion densities and the corresponding homogenization temperatures and compared these with the observed homogenization temperatures using mineral and fluid-inclusion data of sample Bx1-4. The procedure is as follows: (1) Grt_1-Crd-Sil assemblages in sample Bx1-4 and Bx1-5 were used to calculate pressure and temperature (Fig. 12A). (2) These P-T data are correlated with molar volumes for pure CO_2 (Fig. 12A), and corresponding homogenization temperatures are calculated (Fig. 12B). (3) A histogram is constructed from the calculated homogenization temperatures (Fig. 12C), which is compared with measured homogenization temperatures of primary and clustered CO_2 inclusions in Grt_1 (Fig. 12D). Figures 12C and 12D show that the calculated histogram has an almost identical shape as the histogram of the measured inclusions in sample Bx1-4, confirming that these CO_2 inclusions most likely formed or re-equilibrated during $M2_1$ and $M2_2$.

Retrograde Fluids: CH_4

The methane-bearing inclusions occur in matrix quartz and were observed only in the country rocks. The main difference, apart from the presence of different generations of the same minerals, between the country rock and enclaves is the degree

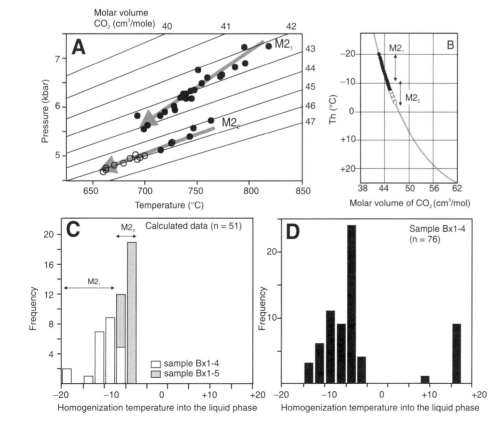

Figure 12. (A) Comparison of isochores for pure CO_2 with P-T compositions determined from geothermobarometry in this study. Data points (51) related to samples Bx1-4 (open circles) and Bx1-5 (filled circles) are indicated. Black lines with values of the CO_2 molar volumes represent isochores for pure CO_2 (Gerya and Perchuk, 1997). (B) CO_2 molar volume versus homogenization temperature diagram (modified from Tomilenko and Chupin, 1983) with 51 data points for which the CO_2 homogenization temperatures were determined. (C) Calculated histogram. (D) Histogram of homogenization temperatures from isolated and clustered CO_2-rich fluid inclusions in Grt_1 in sample Bx1-4.

of retrograde alteration. The country rock underwent severe low-grade hydration, whereas the enclaves show only minor low-grade alteration. It is therefore likely that the presence of CH_4 is associated with the infiltration of meteoric water at low-grade conditions into graphite-bearing rock, resulting in the reaction:

$$C + 2 H_2O \rightarrow CH_4 + O_2. \quad (8)$$

This infiltration must have occurred under redox conditions of $\log_{10} f_{O_2}$ <FMQ+0.34 for pressure and temperature conditions of ~1 kbar and ~400 °C, respectively (Figs. 13A, 13B); otherwise the system would have been too oxidized, resulting in the formation of a low-density CO_2 fluid. The ability of graphite in the rocks to buffer the fluid composition is directly related to the water/graphite ratio. For a low water/graphite ratio it is likely that graphite has an unlimited buffering capacity, even when there is little graphite present in the rock. In this case, the activity of carbon, a_{carbon}, will be 1. On the other hand, if the water/graphite ratio is high, a_{carbon} will be <1. From the above, it is clear that a decrease in the carbon activity reflects an increase in the water/graphite ratio. Model calculations were therefore carried out for a variable carbon activity ($0 < a_{carbon} \leq 1$) in order to evaluate the effect of different water/graphite ratios on the fluid composition (Fig. 13C). The calculations demonstrate that a CH_4-H_2O fluid will form as a result of water-graphite interaction for a low temperature-pressure system that is reduced relative to FMQ (Fayalite-Magnetite-Quartz) (Fig. 13C). A CH_4-rich fluid can only be produced if the fluid system is close to carbon saturation, i.e., a low water/graphite ratio. The final fluid composition, after reaction (8) has reached an equilibrium state, may still have some water (Fig. 13C). However, this amount of water (<20 mol%) is invisible in fluid inclusions.

CONCLUSIONS

An important conclusion that follows from this study is that the enclaves within the Bulai Pluton represent true xenoliths. This is supported by the observation that the enclaves preserve evidence for a local contact metamorphic overprint reflected by the $M2_1$ and $M2_2$ P-T paths in these rocks (Fig. 8C). In contrast, the country rocks do not show any evidence for this overprint and should, therefore, not be interpreted as xenoliths (Millonig et al., 2008) but rather as layers that are part of the regional Neoarchean structural pattern of the Central Zone (van Reenen et al.,

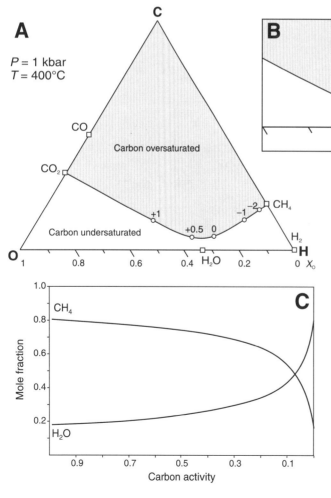

Figure 13. (A) Calculated carbon saturation surface (black line) in a C-O-H diagram, showing the compositional variation of carbon-saturated fluid with changing oxygen fugacity (expressed in $\log_{10}$ units relative to the Fayalite-Magnetite-Quartz [FMQ] oxygen buffer). (B) Re-equilibration of a water-rich fluid with graphite toward an H_2O-CH_4 fluid (oxygen fugacity buffered by FMQ−2) for varying carbon activities ranging from 0 (pure water) to 1 (on the carbon saturation surface). (C) Fluid compositions of water re-equilibrating with graphite at an oxygen fugacity of FMQ−2($\log_{10}$ units) for different carbon activities. Diagrams were calculated using an updated version of the spreadsheet COH (Huizenga, 2005). See text for further explanation.

2008). The fact that the Bulai Pluton intrudes already deformed, metamorphosed, and migmatized pelites from the Beit Bridge Complex implies that high-grade metamorphic conditions in the Central Zone prevailed during a late (Neoarchean) tectonometamorphic event that not only affected the area around the Bulai Pluton but also the entire Central Zone.

The presence of high-grade CO_2 and saline fluids is similar to the observations by other researchers (e.g., Newton et al., 1998; Touret and Huizenga, 1999). The widespread occurrence of metasomatic K-feldspar in the Central Zone and also in the Southern Marginal Zone (Van den Berg and Huizenga, 2001) implies that brine fluids occurred at a large scale. The rare occurrence of brine inclusions can be explained by the high mobility of these fluids in the lower crust (Watson and Brennan, 1987) and by the destruction of brine inclusions by implosion during exhumation (Touret and Huizenga, 1999).

ACKNOWLEDGMENTS

This work was carried out as part of the Russian–South African scientific collaboration. Financial support was given by the National Research Foundation (South Africa) to DDVR, and RFBR grants to LLP (project 06-05-64196) and to DAV (project 06-05-64098). Raman analyses were performed by Ernst Burke at the Vrije Universiteit Amsterdam with financial support of the Netherlands Organization for the Advancement of Pure Research (NWO). The authors would like to thank A.M. Van den Kerkhof, P. Pitra, and J.L.R. Touret for valuable comments on this manuscript. Reviews of the manuscript by A. Schoch and J. Reinhardt are highly appreciated.

REFERENCES CITED

Aranovich, L.Y., and Podlesskii, K.K., 1989, Geothermobarometry of high-grade metapelites: Simultaneously operating reactions, in Daly, J.S., Yardley, B.W.D., and Cliff, B.R., eds., Evolution of Metamorphic Belts: Geological Society [London] Special Publication 42, p. 41–65.

Bakker, R.J., 2003, Package FLUIDS 1: Computer programs for analysis of fluid inclusion data and for modelling of bulk fluid properties: Chemical Geology, v. 194, p. 3–23, doi:10.1016/S0009-2541(02)00268-1.

Barton, J.M., and van Reenen, D.D., 1992, When was the Limpopo orogeny?: Precambrian Research, v. 55, p. 7–16, doi:10.1016/0301-9268(92)90010-L.

Barton, J.M., Holzer, L., Kamber, B., Doig, R., Kramers, J.D., and Nyfeler, D., 1994, Discrete metamorphic events In the Limpopo belt, southern Africa: Implications for the application of P-T paths in complex metamorphic terranes: Geology, v. 22, p. 1035–1038, doi:10.1130/0091-7613(1994)022<1035:DMEITL>2.3.CO;2.

Boshoff, R., van Reenen, D.D., Smit, C.A., Perchuck, L.L., Kramers, J.D., and Armstrong, R.A., 2006, Geological history of the Central Zone of the Limpopo Complex: The west Alldays area: Journal of Geology, v. 114, p. 699–716, doi:10.1086/507615.

Burke, E.A.J., and Lustenhouwer, W.J., 1987, The application of a multichannel laser Raman microprobe (Microdil-28) to the analysis of fluid inclusions: Chemical Geology, v. 61, p. 11–17, doi:10.1016/0009-2541(87)90021-0.

Dubessy, J., Poty, B., and Ramboz, C., 1989, Advances in C-O-H-N-S fluid geochemistry based on micro Raman spectrometric analysis of fluid inclusions: European Journal of Mineralogy, v. 1, p. 517–534.

Gerya, T.V., and Perchuk, L.L., 1997, Equations of state of compressed gasses for thermodynamic databases used in petrology: Petrology, v. 5, p. 366–380.

Goldstein, R.H., and Reynolds, T.J., 1994, Systematics of Fluid Inclusions in Diagenetic Minerals: SEPM (Society for Sedimentary Geology) Short Course, v. 31, 199 p.

Harlov, D.E., and Förster, H.-J., 2002, High-grade fluid metasomatism on both a local and regional scale: The Seward Peninsula, Alaska, and the Val Strona di Omegna, Ivrea-Verbano Zone, northern Italy, Part I: Petrography and silicate mineral chemistry: Journal of Petrology, v. 43, p. 769–799, doi:10.1093/petrology/43.5.769.

Harlov, D.E., and Wirth, R., 2000, K-feldspar-quartz and K-feldspar-plagioclase phase boundary interactions in garnet-orthopyroxene gneisses from the Val Strona di Omegna, Ivrea-Verbano Zone, northern Italy: Contributions to Mineralogy and Petrology, v. 140, p. 148–162, doi:10.1007/s004100000185.

Harlov, D.E., Hansen, E.C., and Bigler, C., 1998, Petrologic evidence for K-feldspar metasomatism in granulite facies rocks: Chemical Geology, v. 151, p. 373–386, doi:10.1016/S0009-2541(98)00090-4.

Holzer, L., Frei, R., Barton, J.M., and Kramers, J.D., 1998, Unraveling the record of successive high-grade events in the Central Zone of the Limpopo Belt using Pb single phase dating of metamorphic minerals: Precambrian Research, v. 87, p. 87–115, doi:10.1016/S0301-9268(97)00058-2.

Huizenga, J.M., 2005, COH, an Excel spreadsheet for composition calculations in the C-O-H fluid system: Computers & Geosciences, v. 31, p. 797–800, doi:10.1016/j.cageo.2005.03.003.

Kamber, B.S., Blenkinsop, T.G., Villa, I.M., and Dahl, P.S., 1995, Proterozoic transpressive deformation in the Northern Marginal Zone, Limpopo Belt, Zimbabwe: Journal of Geology, v. 103, p. 493–508, doi:10.1086/629772.

Kramers, J.D., McCourt, S., and van Reenen, D.D., 2007, The Limpopo Belt, in Johnson, M.R., Anheausser, C.R., and Thomas, R.J., eds., The Geology of South Africa: Geological Society of South Africa Publication: Pretoria, Council for Geoscience, p. 209–236.

Kretz, R., 1983, Symbols for rock forming minerals: American Mineralogist, v. 68, p. 277–278.

Kröner, A., Jaeckel, P., Brand, G., Nemchin, A.A., and Pidgeon, R.T., 1999, Single zircon ages from granitoid gneisses in the Central Zone of the Limpopo belt, southern Africa and geodynamic significance: Precambrian Research, v. 93, p. 299–337, doi:10.1016/S0301-9268(98)00102-8.

Millonig, L., Zeh, A., Gerdes, A., and Klemd, R., 2008, Neoarchean high-grade metamorphism in the Central Zone of the Limpopo Belt (South Africa): Combined petrological and geochronological evidence from the Bulai Pluton: Lithos, v. 103, p. 333–351, doi:10.1016/j.lithos.2007.10.001.

Newton, R.C., Aranovich, L.Ya., Hansen, E.C., and Vandenheuvel, B.A., 1998, Hypersaline fluids in Precambrian deep-crustal metamorphism: Precambrian Research, v. 91, p. 41–63, doi:10.1016/S0301-9268(98)00038-2.

Perchuk, L.L., 2011, this volume, Local mineral equilibria and P-T paths: Fundamental principles and applications for high-grade metamorphic terranes, in van Reenen, D.D., Kramers, J.D., McCourt, S., and Perchuk, L.L., eds., Origin and Evolution of Precambrian High-Grade Gneiss Terranes, with Special Emphasis on the Limpopo Complex of Southern Africa: Geological Society of America Memoir 207, doi:10.1130/2011.1207(05).

Perchuk, L.L., and Gerya, T.V., 1993, Fluid control of charnockitization: Chemical Geology, v. 108, p. 175–186, doi:10.1016/0009-2541(93)90323-B.

Perchuk, L.L., and Lavrent'eva, I.V., 1983, Experimental investigation of exchange equilibria in the system cordierite-garnet-biotite, in Saxena S.K., ed., Kinetics and Equilibrium in Mineral Reactions: Advances in Physical Geochemistry 3: New York, Springer-Verlag, p. 199–239.

Perchuk, L.L., Aranovich, L.Ya., Podlesskii, K.K., Lavrent'eva, I.V., Gerasimov, V.Yu., Fed'kin, V.V., Kitsul, V.N., and Karsakov, L.P., 1985, Precambrian granulites of the Aldan shield, eastern Siberia, USSR: Journal of Metamorphic Geology, v. 3, p. 265–310, doi:10.1111/j.1525-1314.1985.tb00321.x.

Perchuk, L.L., Gerya, T.V., van Reenen, D.D., and Smit, C.A., 2006, P-T paths and problems of high-temperature polymetamorphism: Petrology, v. 14, p. 117–153, doi:10.1134/S0869591106020019.

Perchuk, L.L., van Reenen, D. D., Varlamov, D.A., Van Kal, S.M., Boshoff, R., and Tabatabaeimanesh, 2008, P-T record of two high-grade metamorphic events in the Central Zone of the Limpopo Complex, South Africa: Lithos, v. 103, p. 70–105, doi:10.1016/j.lithos.2007.09.011.

Roering, C., van Reenen, D.D., Smit, C.A., Barton, J.M., De Beer, J.H., De Wit, M.J., Stettler, E.H., Van Schalkwyk, J.F., Stevens, G., and Pretorius, S., 1992, Tectonic model for the evolution of the Limpopo Belt: Precambrian Research, v. 55, p. 539–552, doi:10.1016/0301-9268(92)90044-O.

Schaller, M., Steiner, O., Studer, I., Holzer, L., Herweg, M., and Kramers, J.D., 1999, Exhumation of Limpopo Central Zone granulites and dextral continental-scale transcurrent movement at 2.0 Ga along the Palala Shear Zone, Northern Province, South Africa: Precambrian Research, v. 96, p. 263–288, doi:10.1016/S0301-9268(99)00015-7.

Thiéry, R., Van den Kerkhof, A.M., and Dubessy, J., 1994, vX properties of CH_4-CO_2 and CO_2-N_2 fluid inclusions: Modelling for T < 31°C and P < 400 bars: European Journal of Mineralogy, v. 6, p. 753–771.

Tomilenko, A.A., and Chupin, V.P., 1983, Thermobarogeochemistry of metamorphic complexes: Moscow, Nauka Press, 201 p. [in Russian].

Touret, J.L.R., 2001, Fluids in metamorphic rocks: Lithos, v. 55, p. 1–25, doi:10.1016/S0024-4937(00)00036-0.

Touret, J.L.R., and Huizenga, J.M., 1999, Precambrian intraplate magmatism: High temperature, low pressure crustal granulites: Journal of African Earth Sciences, v. 28, p. 367–382, doi:10.1016/S0899-5362(99)00010-X.

Van den Berg, R., and Huizenga, J.M., 2001, Fluids in granulites of the Southern Marginal Zone of the Limpopo Belt, South Africa: Contributions to Mineralogy and Petrology, v. 141, p. 529–545.

Van den Kerkhof, A.M., and Thiéry, R., 2001, Carbonic inclusions: Lithos, v. 55, p. 49–68, doi:10.1016/S0024-4937(00)00038-4.

van Reenen, D.D., 1986, Hydration of cordierite and hypersthene and a description of the retrograde orthoamphibole isograd in the Limpopo belt, South Africa: American Mineralogist, v. 71, p. 900–915.

van Reenen, D.D., Roering, C., Brandl, G., Smit, C.A., and Barton, J.M., 1990, The granulite facies rocks of the Limpopo Belt, Southern Africa, *in* Vielzeuf, D., and Vidal, P., eds., Granulites and Crustal Evolution: Dordrecht, Kluwer Academic, NATO-ASI Ser. C311, p. 257–289.

van Reenen, D.D., Perchuk, L.L., Smit, C.A., Varlamov, D.A., Boshoff, R., Huizenga, J.M., and Gerya, T.V., 2004, Structural and P-T evolution of a major cross fold in the Central Zone of the Limpopo High-Grade Terrain, South Africa: Journal of Petrology, v. 45, p. 1413–1439, doi:10.1093/petrology/egh028.

van Reenen, D.D., Boshoff, R., Smit, C.A., Perchuk, L.L., Kramers, J.D., McCourt, S., and Armstrong, R.A., 2008, Geochronological problems related to polymetamorphism in the Limpopo Complex, South Africa: Gondwana Research, v. 14, p. 644–662, doi:10.1016/j.gr.2008.01.013.

Vernon, R.H., White, R.W., and Clarke, G.L., 2008, False metamorphic events inferred from misinterpretation of microstructural evidence and P-T data: Journal of Metamorphic Geology, v. 26, p. 437–449, doi:10.1111/j.1525-1314.2008.00762.x.

Vityk, M.O., and Bodnar, R.J., 1998, Statistical microthermometry of synthetic fluid inclusions in quartz during decompression reequilibration: Contributions to Mineralogy and Petrology, v. 132, p. 149–162, doi:10.1007/s004100050413.

Watson, E.B., and Brennan, J.M., 1987, Fluids in the lithosphere I: Experimentally determined wetting characteristics of H_2O-CO_2 fluids and their implications for fluid transport, host rock physical properties and fluid inclusion formation: Earth and Planetary Science Letters, v. 125, p. 55–70.

Zeh, A., Holland, T.J.B., and Klemd, R., 2005, Phase relationships in grunerite-garnet-bearing amphibolites in the system CFMASH, with applications to metamorphic rocks from the Central Zone of the Limpopo Belt: Journal of Metamorphic Geology, v. 23, p. 1–17, doi:10.1111/j.1525-1314.2005.00554.x.

Manuscript Accepted by the Society 24 May 2010

Intracrustal radioactivity as an important heat source for Neoarchean metamorphism in the Central Zone of the Limpopo Complex

Marco A.G. Andreoli
*South African Nuclear Energy Corporation, P.O. Box 582, Pretoria 0001, South Africa and
School of Geosciences, University of the Witwatersrand, Private Bag 3, Wits 2050, South Africa*

Günther Brandl
Henk Coetzee
Council for Geoscience, Private Bag X112, Pretoria 0001, South Africa

Jan D. Kramers*
*School of Geosciences, University of the Witwatersrand, Private Bag 3, Wits 2050, South Africa, and
Department of Geology, University of Johannesburg, P.O. Box 524, Auckland Park 2006, Johannesburg, South Africa*

Hassina Mouri
Department of Geology, University of Johannesburg, P.O. Box 524, Auckland Park 2006, Johannesburg, South Africa

ABSTRACT

The major periods of metamorphism in the Central Zone (CZ) of the Limpopo Complex occurred at 2.0 Ga and in the time range between ca. 2.7 and ca. 2.55 Ga. We investigate intracrustal radioactivity as a possible heat source for the earlier of these episodes. Available airborne radiometric surveys that cover the South African part of the CZ, combined with rock analyses, yield 2.15 µg/g U, 12.3 µg/g Th, and 12,650 µg/g K as a weighted regional average. The corresponding heat production rate at 2.65 Ga is 2.6 µW × m^{-3}. A steady-state geotherm, calculated assuming uniform [U], [Th], and [K] throughout the crustal column and its thickening to 45 km during the ca. 2.65 Ga event (both arguable on the basis of peak metamorphic pressure-temperature [P-T] data), surpasses temperatures of the peak metamorphism at middle and lower crustal levels, which cluster around the fluid-absent biotite dehydration solidus. Intracrustal radioactivity thus provided a sufficient heat source to account for the metamorphism at ca. 2.65 Ga, and partial melting acted as a lower crustal thermostat. After crustal thickening, up to more than 100 m.y. (dependent on U, Th, and K concentrations)

*jkramers@uj.ac.za

would be needed to approach a new steady state. Predicted regional variations thus account for the long duration of the ca. 2.65 Ga metamorphism. Lower crustal partial melting could have led to diapirism, yielding the steep structures in the CZ, which are not aligned to a regional fabric. Metamorphism ceased after crustal thinning to a normal 30 km. The metamorphic event at 2.0 Ga cannot be explained by this type of process.

INTRODUCTION

After well over 30 yr of intensive research in the Central Zone (CZ) of the Limpopo Complex, it is now established that it is a polymetamorphic province, in which major high-grade events occurred at ca. 2.65 and 2.02 Ga (see Kramers and Mouri, this volume; Perchuk et al., 2008; Smit et al., this volume, for reviews). In this paper we examine the possibility that intracrustal radioactive heating, after crustal thickening, might have been the main cause for the metamorphism at ca. 2.65 Ga.

It is now possible to distinguish structures and metamorphism of the 2.02 Ga event from those of the ca. 2.65 Ga period (Holzer et al., 1998, 1999; Schaller et al., 1999; Boshoff et al., 2006; van Reenen et al., 2008; Perchuk et al., 2008; Smit et al., this volume). The Paleoproterozoic structures are large shear zones with mainly subhorizontal displacement; those exposed at the present surface were formed at mid-crustal (e.g., Tshipise Straightening Zone and Baklykraal Shear Zone) or at mid- to upper crustal levels (Palala Shear Zone). Where there is a single metamorphism and deformation, dated at 2.02 Ga (Boshoff et al., 2006; Mouri et al., 2008), this can be seen from the style of deformation, and from retrograde pressure-temperature (P-T) paths within them. Also, where two separate P-T paths at different pressures have been derived from a single rock, those with higher pressures appear to reflect the earlier metamorphism (van Reenen et al., 2004, 2008; Boshoff et al., 2006; Perchuk et al., 2008; Smit et al., this volume). Further, the quartzofeldspathic gneisses of the Singelele type, which are widespread in the CZ, were formed by anatexis in the Neoarchean, between 2550 and 2700 Ma (Jaeckel et al., 1997; Kröner et al., 1999; Zeh et al., 2007; Kramers and Mouri, this volume). With this knowledge it is now possible, to some extent, to "strip away" the effects of the Paleoproterozoic event from those of the Neoarchean event, and to study the latter in isolation.

It is particularly difficult to find analogues in Phanerozoic orogenic belts that would help to provide a context for the generation of many of the characteristics of the Central Zone Neoarchean episode, which reached high-grade metamorphism in many areas of the province. Five points, in particular, are not readily explained. First, there is a lack of planar features (or linear ones on the map), e.g., well-defined thrust planes, such as are commonly found in younger orogens. Instead, structures at the map scale are curved and have random strike directions instead of being elongated along the WSW-ENE axis of the province itself (Fig. 1), whereby their planar features and lineations are steeply dipping. Second, there are a number of conspicuous circular structures that have been interpreted as sheath folds (Roering et al., 1992), but with diameters up to 5 km or more. Rocks in the interior of these structures show a strong, steep stretching lineation. If these were true sheath folds, their size and the high aspect ratio normally associated with sheath folds (Cobbold and Quinquis, 1980) would project their base to well below the base of the crust. Therefore, they cannot be true sheath folds. Third, there is a considerable spread in zircon U-Pb dates for magmatic rocks formed by anatexis in this event, from ca. 2.57–2.68 Ga and well outside analytical uncertainties (Kröner et al., 1999; Jaeckel et al., 1997; van Reenen et al., 2008; Zeh et al., 2007; Kramers and Mouri, this volume). This is in stark contrast to the Paleoproterozoic event, for which zircon dates are tightly clustered within a 30 m.y. range (see summary in Kramers and Mouri, this volume). Fourth, most retrograde P-T path segments that can be clearly attributed to the Neoarchean event end at mid-crustal level, even if extrapolated to lower temperatures. The rocks at the surface today were therefore mainly exhumed in later events. Fifth, in accord with the previous point, siliciclastic sediments are all but absent in the lower portion of the Transvaal Supergroup stratigraphy, which prior to ca. 2.5 Ga was dominated by carbonate platform rocks (see summary in Eriksson et al., 2006), and no detrital zircons of Neoarchean age occur (Beukes et al., 2004).

A number of different scenarios for the main cause of the metamorphism may be examined for compatibility with these features. In most episodes of tectono-metamorphism, the P-T conditions determined are much hotter than those of normal continental crustal geotherms. This could have been caused by delamination, slab breakoff (Davies and von Blanckenburg, 1995, 1998) leading to anomalous mantle heat flow, by advection of heat by magma intrusion or magmatic underplating, or by tectonic crustal thickening. The latter process increases both the blanketing effect of the crust and the total amount of radioactive heat generated within it; and in horizontal crustal stacking the hot base of the upper stack heats the crust below it (Oxburgh and Turcotte, 1974; England and Richardson, 1977; Ashwal et al., 1992). This study focuses on intracrustal radioactive heating (Sandiford, 1989; McLaren et al., 1999, 2005; Kramers et al., 2001; Andreoli et al., 2006). This clearly must have been more important in early Earth history than today, as the amount of all radioactive isotopes has declined exponentially. The decrease is particularly dramatic for the relatively short-lived ^{235}U (half life 704 Ma) and ^{40}K (half life 1270 Ma). Thus intracrustal radioactive heating is non-actualistic, and examining its effects is useful in cases where modern analogues are absent.

The concentration of heat-producing elements in the crust can be regionally highly variable and has been shown to correlate with present-day regional heat-flow patterns, in Australia (Neumann et al., 2000) as well as in southern Africa (Kramers et al., 2001). Unusually high concentrations of U and particularly Th were also noted in the Central Zone of the Limpopo Complex, and in other granulite facies belts once contiguous in a Gondwana continental assembly, including the Namaqualand-Natal belt (Andreoli and Hart, 1990; Andreoli et al., 2003).

Estimates of regional U, Th, and K concentrations can be derived from airborne γ-spectrometric surveys, combined with rock analyses from a number of regions. In the present paper we compile these data for the portion of the CZ in South Africa, and calculate heat production values for rock units presently at the surface, for 2.0 and 2.65 Ga. Many P-T studies clearly show that the rocks now at the surface in the CZ underwent high-grade metamorphism at lower crustal levels (P, 8–11 kb; depth, 30–40 km; Chinner and Sweatman, 1968; Droop, 1989; Holzer, 1995; Horrocks, 1983; Schreyer et al., 1984; Watkeys et al., 1983; Perchuk et al., 2008; Zeh et al., 2004; van Reenen et al., 2004; Hisada and Miyano, 1996). Therefore their U, Th, and K concentrations and regional averages allow estimation of the lower crustal heat production at the time of metamorphism. Following the approach of Kramers et al. (2001) and McLaren et al. (1999, 2005), we calculate steady-state geotherms as first-order "infinity conditions" (which may never be reached in the case of tectonic thickening activity). Further, we estimate the time scales of heating in the event of thickening and/or redistribution of radioactive rocks to depth. The results are compared with measured P-T conditions, and we show that intracrustal radioactive heat production after crustal thickening can indeed account for many features of Neoarchean metamorphism. Further, it appears that the concept may also help in understanding salient tectonic features.

DATA SOURCES FOR RADIOACTIVE ELEMENT CONCENTRATIONS IN THE CENTRAL ZONE OF THE LIMPOPO COMPLEX AND ESTIMATED HEAT PRODUCTION IN THE CRUSTAL COLUMN

Estimates of U, Th, and K Concentrations

The data sources are radioactivity maps from airborne surveys, and geochemical analyses of rocks. The latter can help to

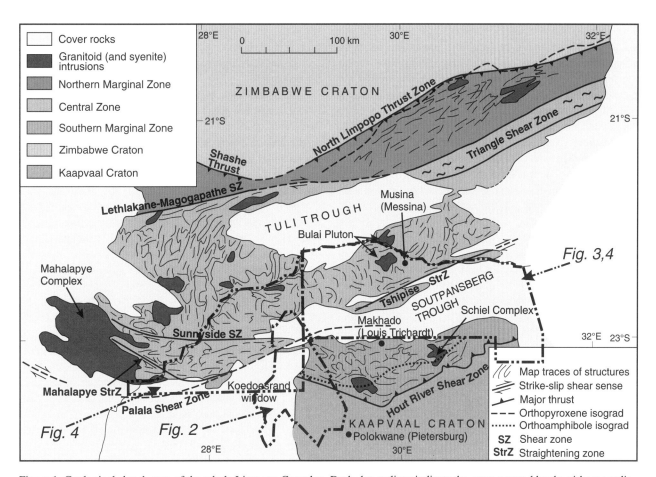

Figure 1. Geological sketch map of the whole Limpopo Complex. Dash-dot outlines indicate the areas covered by the airborne radiometric survey shown in Figures 2, 3, and 4 as marked.

calibrate the former (Bodorkos et al., 2004; see below) throughout the area covered by data. Averages and ranges of available U, Th, and K concentration data are listed in Table 1 with sources. Most data are directly analyzed concentrations, and for a number of units we have calculated the U and Th concentrations from Pb isotope data originally obtained for dating, and Pb concentrations (see notes to Table 1). These data compare well with directly measured U and Th concentrations. It thus appears that U concentrations in the rock units studied are not strongly affected by weathering, a point that is also important in interpreting the regional pattern from the airborne surveys.

One airborne gamma-spectrometric survey, conducted in 1990 by Geodass covers the area NW of Polokwane (formerly Pietersburg), reaching to the Central Zone of the Limpopo Complex in the Alldays area (Figs. 1, 2). This survey was calibrated for U, Th, and K. The areal coverage of the Central Zone in this survey is small, but it clearly shows the Central Zone surface to have elevated levels of Th and particularly U (>6 and >2 ppm, respectively), relative to that in the northern Kaapvaal Craton and the Southern Marginal Zone of the Limpopo Belt. Low U and Th concentrations were also noted in those latter regions by Kramers et al. (2001).

Further, a noncalibrated γ-spectrometric survey, conducted by the Council of Geosciences, covers the area from Alldays to the Mozambique border (Fig. 3). Interpretation at the time was limited to the identification of localized anomalies that were of interest for uranium exploration (Minty, 1977). This survey clearly shows high levels of U, Th, and K in the area around and NW of the town of Musina, including the Bulai intrusion and the Tshipise Straightening Zone. The contrast between the Central Zone rocks and the extremely U-, Th-, and K-poor sediments of the Soutpansberg Trough is very clear. A small SE section of the surveyed area includes units of the Southern Marginal Zone, and this again confirms the low level of radioactive heat-producing elements in that zone. The data collected in the survey were not quantitatively calibrated to radioelement concentrations, but the stripped count-rates can be semiquantitatively back-calibrated (International Atomic Energy Agency, 2003) via the data in Table 1 as well as via an area adjacent to the northernmost part of the first survey (compare Figs. 1, 2, and 3).

For the Central Zone west of 29°E, only total-count γ-ray intensity data are available (Fig. 4), also produced by the Council for Geosciences. A remarkable feature detected in this survey is an E-W–trending strip of strongly enhanced radioactivity ~20 km wide and 80 km long just south of the Limpopo River (extending 22°30′–22°45′ S and 28°10′–29° E). This area is dominated by quartzofeldspathic gneisses. Analyses from Brandl (1983), Boryta and Condie (1990), and this work show similar U, Th, and K concentration values to those found in the high-radioactivity area around Musina and the Bulai Pluton. Th concentrations are particularly elevated, although the U content is also higher than normal (Table 1). The total γ radioactivity in this area appears higher on the map than that in the area around Musina and the Bulai Pluton, but it should be noted that the map shown in Figure 4 was produced in two separate surveys that may not have been fully intercalibrated.

The question of whether surface γ-emanation (i.e., mainly from soils) reflects the element concentrations in the rocks is important if the airborne surveys are to be used to estimate intracrustal heat production. Soils are not, per se, expected to reflect the U, Th, and K concentrations of their bedrock. Certainly lateritic soils are often strongly depleted in U and possibly enriched in Th. However, soils in the Central Zone are typically sandy or sandy-loamy, not indicating the extreme chemical weathering that produces laterites. In a careful comparative study of the northern Pilbara Craton, where the climate is currently arid, Bodorkos et al. (2004) found that airborne γ-ray surveys generally very well reflect the K, Th, and U concentrations in extensive outcrop areas. Over soils formed in situ, [Th] of the underlying bedrock is still determined accurately. [K] is underestimated by up to about a factor of 2 over felsic rocks, whereas [U] is frequently overestimated by about the same amount. This may be due to strong secular disequilibrium between intermediate daughter nuclides: The proxy used for ^{238}U in γ-spectrometry is the 1.76 MeV γ-emission with the β-decay of ^{214}Bi to ^{214}Po, which is very late in the decay chain, so that chemical fractionation effects such as retention of ^{230}Th or ^{226}Ra in the soil could influence the measurement (Bodorkos et al., 2004). Further, overestimation may be caused by the concentration of zircon and other radioactive heavy minerals in soils and stream channels. In general, however, it may be expected that the presence of elevated levels of U and Th progeny in a soil profile represent elevated levels within the bedrock.

A direct check of the calibrated survey (Fig. 2) with analytical data is not possible, as no samples from the small Central Zone area covered by it were analyzed. The eastern extension of the area in the northernmost part of this survey can be followed eastward into Figure 3, where high Th values are shown, similar to the very high values of the Musina area, but somewhat lower U values. Perusing Table 1 for the Singelele and Bulai Gneisses near Musina, this comparison is not in conflict with the calibrated values in Figure 2. A westward extension from a second sliver of the Central Zone in Figure 2 continues to an area just north of the Palala Shear Zone in the Koedoesrand Window, where metapelites were analyzed that gave somewhat lower Th and, particularly, lower U concentrations than the calibrated values of Figure 2. The difference may be similar to that reported by Bodorkos et al. (2004). Although the other two surveys are only qualitative to semiquantitative, there is a coincidence of the "hottest" areas with those of U- and Th-rich analyzed samples, notably the areas around Musina and the strip extending 22°30′–22°45′ S and 28°10′–29° E, just south of the Limpopo River. The strip of lower radioactivity seen in Figure 3 between the "hot" areas around Tshipise and Musina is mainly occupied by the Sand River Gneiss and mafic rocks of the Messina (Musina) Suite, which are both relatively low in U and Th. Thus we conclude that the aerial surveys provide broad, semiquantitative but usable constraints on the U, Th, and K contents of rocks in regions of the province.

TABLE 1. SUMMARY OF K, Th, AND U ANALYTICAL DATA FOR ROCK UNITS FROM THE CENTRAL ZONE OF THE LIMPOPO COMPLEX

Rock unit, region	K_2O (%) Average	K (%) Average	K (%) Range (N)	Th (µg/g) Average	Th (µg/g) Range (N)	U (µg/g) Average	U (µg/g) Range (N)	Reference
Sand River Gneiss, Causeway locality near Musina[†]	1.45	1.20	0.97–1.71 (19)	6.56	1.9–24.9 (26)	0.72	0.4–2.0 (26)	Barton et al. (1983)
Singelele Gneiss, type locality near Musina[†]	N.D.	N.D.	N.D.	34.25	23.8–45.8 (17)	4.23	2.71–9.07 (17)	Barton et al. (1979)
Bulai Granite, type locality near Musina[†]	N.D.	N.D.	N.D.	29.14	9.2–135 (9)	3.75	3.0–5.0 (9)	Barton et al. (1979)
Quartzofeldspathic (QF) gneisses, Artonvilla Mine, Musina	3.04	2.53	0.95–4.6 (11)	N.D.	N.D.	N.D.	N.D.	Ryan et al. (1983)
Garnetiferous QF gneiss, Mount Dowe Group near Musina	2.52	2.09	2.6–4.1 (4)	N.D.	N.D.	N.D.	N.D.	Brandl (1983)
Hornblende-bearing QF gneiss, Malala Drift Group, near Musina	3.75	3.11	2.3–4.1 (7)	19.29	17–30 (7)	N.D.	N.D.	Brandl (1983)
Calc-silicate rocks and marbles, Gumbu Group, near Tshipise	0.97	0.81	0.1–1.6 (6)	3.8	3–8 (6)	N.D.	N.D.	Brandl (1983)
Migmatitic metapelite, Koedoesrand Window, just N of Palala Shear Zone	2.85	2.36	N.D.	5.65	1.05–7.2 (18)	1.68	0.5–3 (18)	Chavagnac et al. (2001)
Leucocratic quartzofeldspathic gneiss, Alldays area	5.03	4.18	(5)	14	(5)	2	(5)	Boryta and Condie (1992)
Singelele-type gneisses, in quadrant 22°30′–22°45′ S, 28°10′–29° E	3.37	2.81	0.97–4.6 (8)	35.5	2.5–85 (8)	4.1	1.5–14.9 (8)	This work[§]
Alldays gneisses, in quadrant 22°30′–22°45′ S, 28°10′–29° E	3.83	3.19	1.3–4.6 (6)	28	2.5–51.6 (6)	3	1.5–8 (6)	This work[§]

Notes: Quartzites of the Mount Dowe Group are very poor in K, Th, and U, except in sedimentary iron ore deposits in which U increases to 10–20 ppm, and Th is up to 10 ppm (Brandl, 2002). Amphibolites of the Malala Drift Group have generally <1 ppm for both U and Th, and <0.5% K (Brandl, 2002). Anorthositic gneisses of the Messina Suite (Table 8 in Brandl, 2002) all have U and Th of <1 ppm and K_2O generally below 1%. N.D.—not determined.
[†]Th and U concentrations derived from Pb isotope ratios and Pb concentrations.
[§]Council for Geosciences, Pretoria. Analyses by XRF.

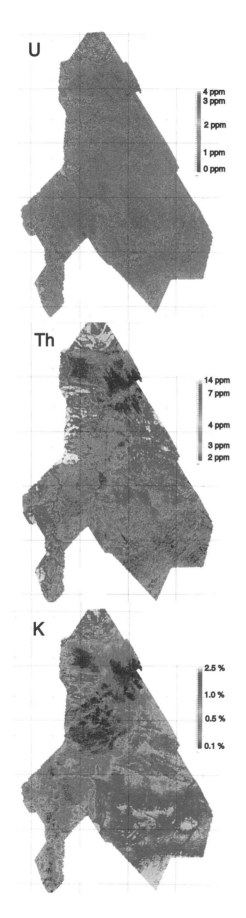

Some of the rock units of the CZ of the Limpopo Complex did not exist at 2.65 Ga, at least not in their present form. Whereas such units (e.g., the Bulai Pluton, as mentioned above) have been useful in relating airborne γ-spectrometric data to the U, Th, and K content of bedrock, they must obviously be excluded from estimates of radioactive heat production prior to 2.65 Ga (or leading up to the 2.65 Ga episode).

There are many indications that the units of the Gumbu Group were only deposited after the ca. 2.65 Ga metamorphic events. In areas mapped as the Gumbu Group by the Council of Geoscience (Brandl, 2002; Musina and Alldays Sheets), U-Pb dates for zircons (considered to be detrital) are intermediate between the two metamorphic events (Buick et al., 2003; Barton et al., 2003), and $\delta^{13}C$ values found in marbles portray the 2.3–2.1 Ga Lomagundi-Jatulian carbon isotope excursion (Buick et al., 2003). Further, some metamorphic studies show only one episode, for which zircon (Mouri et al., 2008), garnet Pb/Pb (Boshoff et al., 2006), and muscovite as well as hornblende Ar/Ar (Barton et al., 2003) dates yield ages of ca. 2.02 Ga. Such samples show peak conditions at ~800 °C and ~6.5 kb, and retrogression to ~600 °C and ~3 kb (van Reenen et al., 2004, 2008).

Singelele-type quartzofeldspathic gneisses as well as the Bulai Pluton and most gray gneisses (such as the Alldays Gneiss) range between 2.7 and 2.55 Ga. The Hf isotope data on zircons from the Singelele gneisses show that they were derived by intracrustal anatexis (Zeh et al., 2007). They still occur interlayered with sedimentary units, particularly metapelites, in a lit-par-lit relation. The field evidence, particularly the distribution of xenoliths, suggests that they are not far removed from the source rocks. A study of metapelitic migmatites (Chavagnac et al., 2001) shows no strong enrichment of U and Th in the leucosome. Therefore, average U, Th, and K contents of the regions of Singelele-type quartzofeldspathic gneisses and metapelites, etc. (i.e., the Malala Drift Group, Brandl, 2002) may be broadly regarded as representative of the pre–2.65 Ga sedimentary terrane. The Bulai granitoid and gray gneisses in part show the presence of a juvenile ca. 2.65 Ga crustal component (Kramers and Zeh, this volume) and intruded as medium-size to large discrete plutons. While their U, Th, and K concentrations may hold some information on those of lower crustal rocks at the time of their formation, enrichment in partial melting and magma differentiation depends on unknown parameters, so that these data cannot be used to calculate past crustal heat production.

Heat Production in the Crustal Column

On the basis of the rock analyses and the radiometric maps, we have defined three lithological domains within the Central

Figure 2. Results of calibrated airborne γ-spectrometric survey carried out in 1990 by Geodass, reaching from just W of Polokwane on the Kaapvaal Craton to the Central Zone of the Limpopo Complex (see Fig. 1).

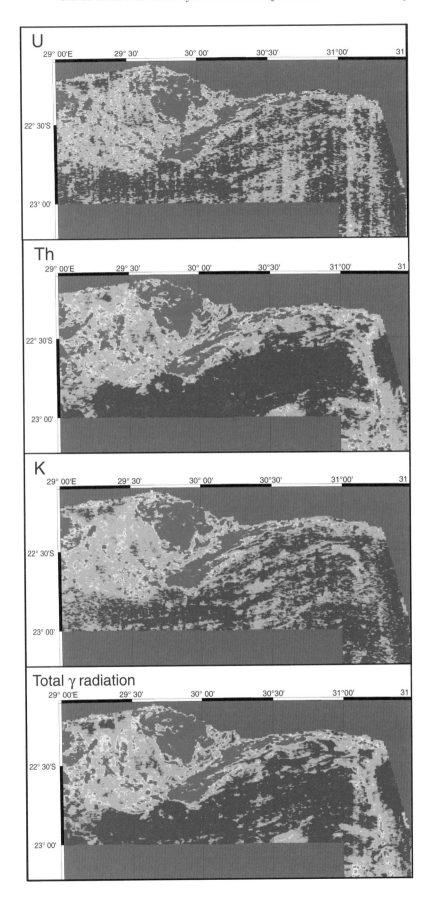

Figure 3. Results of airborne uncalibrated γ-spectrometric surveys carried out in 1971 and 1973 by the Compagnie Générale de Géophysique for the council of Geosciences (Ledwaba et al., 2009), stretching from Alldays to the Mozambique border along the northern border of South Africa (see Fig. 1). γ-ray intensity increases over the optical spectrum from violet to red. Data collected at a flight-line spacing of 1 km, a tie-line spacing of 10–15 km, a nominal altitude of 150 m above ground level, and a nominal airspeed of 240 km/h. Sodium iodide detectors were used, and data were corrected for Compton interactions to produce stripped count rates in radioelement channels for K (based on γ radiation from ^{40}K), U (based on γ radiation from ^{214}Bi), and Th (based on γ radiation from ^{208}Tl). These survey specifications were expected to identify regional-scale anomalies, although not necessarily having sufficient spatial resolution for the mapping of fine-scale features.

Figure 4. Results of two airborne uncalibrated total γ-radiation surveys, carried out in 1971 by the council of Geosciences, juxtaposed to cover the northernmost Limpopo Province from Stockpoort in the west to the Mozambique border (see Fig. 1); γ-ray intensity increases over the optical spectrum from violet to red.

Zone, each with their own characteristic average U, Th, and K contents, and taking into account that [U] may be overestimated in the airborne surveys. These domains are listed in Table 2 and mapped in Figure 5. This also shows the areas occupied by units of the Gumbu Group, the Bulai Pluton, and the ca. 2.6 Ga gray gneisses, where no estimate of heat production at 2.65 Ga is possible. For domain 3, estimates of U and particularly Th, are difficult to make because the domain is heterogeneous and contains rock units that are extremely enriched in these elements: U (-Th) mineralizations were prospected in a quartzite-metapelite association SW of Tshipise (Andreoli et al., 2003, 2006), and Th values up to ~90 ppm were measured in patchy enderbite of uncertain stratigraphic position west of Alldays, close to the Botswana border (this study). The values listed in Table 2 are conservative estimates.

Table 2 also shows the heat production corresponding to the relevant U, Th, and K concentrations for 0, 2.0, and 2.65 Ga. These were calculated using the heat-production constants of Rybach (1976), and the decay constants for the relevant isotopes:

$$H(\tau) = \rho[3.48 \times 10^{-6} \times C_K \times e^{0.554\tau} + 0.0256 \times C_{Th} \times e^{0.0495\tau} + C_U \times (0.9929 \times 0.0918 \times e^{0.1551\tau} + 0.0071 \times 0.576 \times e^{0.98485\tau})]. \quad (1)$$

Here H is the heat production in μWm^{-3}, ρ is the density (2.7 g × cm^{-3} is assumed), C are concentrations of the elements K, Th, and U in μg/g, and τ is time in the past (age) in Ga (see Kramers et al., 2001; Kamber et al., 2005).

In modeling crustal heat production and heat flow patterns, it is normally assumed that the concentration of heat-producing elements decreases exponentially with depth in the crust (Bickle, 1978; Oxburgh and Turcotte, 1974; England and Richardson, 1977; Ashwal et al., 1992). In the case of the CZ, however, medium- to high-pressure peak conditions were determined (on metasediments) for the Neoarchean metamorphism (P, 8–11 kb; i.e., depth, 30–40 km; see Introduction) and the Paleoproterozoic event at ca. 2.0 Ga does not appear to have led to widespread syn-tectonic anatexis and subsequent vertical redistribution of matter (see Boshoff et al., 2006; van Reenen et al., 2004, 2008; Smit et al., this volume). Therefore the U, Th, and K concentrations and

TABLE 2. THREE DOMAINS WITHIN THE PRE-2.65 Ga CENTRAL ZONE, THEIR REGIONAL AVERAGE U, TH, AND K CONTENTS, AND PAST AND PRESENT HEAT PRODUCTION

	μg/g	μg/g	μg/g	μW·m^{-3} at time:		
	U	Th	K	2.65 Ga	2.0 Ga	Today
Domain 1 (mainly Sand River Gneiss, mafic rocks of Messina Suite)	1	4	10,000	1.24	1.00	0.63
Domain 2 (Palala-Sunnyside; mainly metasediments of Mount Dowe Group)	2	10	12,000	2.32	1.93	1.32
Domain 3 (mainly quartzofeldspathic gneisses and metapelites of Malala Drift Group)	3	20	15,000	3.75	3.20	2.29
Weighted average of the three domains: 1, 20%; 2, 45%; 3, 35%	2.15	12.3	12,650	2.60	2.19	1.52

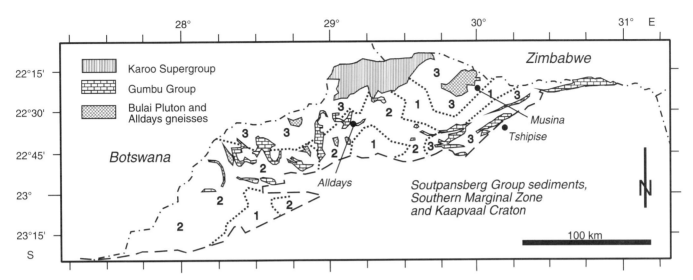

Figure 5. Subdivision of the South African portion of the Central Zone of the Limpopo Complex in domains corresponding to Table 2, based on data of Table 1 and Figures 2, 3, and 4. Also shown are units that did not exist prior to 2.65 Ga: the Bulai Pluton and gray gneisses (Alldays Gneiss in particular), the sediments mapped as the Gumbu Group, and sediments and volcanic rocks of the Karoo Supergroup in the Tuli-Sabi Trough. Dash-dot lines are international boundaries, and dashed line is the southern limit of exposure of the Central Zone.

heat-production rates found apply to deep levels in a thickened crust at ca. 2.65 Ga. Applying an exponential model for the distribution of these elements in the crustal column to these data would lead to improbably high concentrations and heat production rates in the Neoarchean CZ upper crust. We must conclude that the thickened crust in the CZ at ca. 2.65 Ga did not conform to the rule of decreasing radioactivity with depth. This can be understood in the light of the abundance of metasediments in the CZ, which were probably emplaced in the lower crust through imbrication or stacking (Ashwal et al., 1992) in the process of crustal shortening and thickening. The resulting high radioactivity in the lower crust is not an isolated case. In a study of the Namaqua belt, Andreoli et al. (2006) noted an apparent geochemical paradox: The upper granulite facies rocks are the most enriched in U and Th, the amphibolite facies gneisses the more depleted, and the lower granulite facies rocks somewhere in between. For the above reasons, we have in our calculations assumed a uniform abundance of U, Th, and K throughout the upper 45 km of the crustal column of the CZ.

Estimates of the total crustal thickness during the ca. 2.65 Ga metamorphic period are difficult to make. From teleseismic surveys, the crust of the CZ appears to be, anomalously, up to 52 km thick, with a complex Moho structure (Nguuri et al., 2001; Gore et al., 2009). In order to reconcile this finding with apparent isostatic equilibrium, these authors postulated a thin (~10 km) felsic crust underlain by a ~40-km-thick mafic crust. This is in agreement with findings from kimberlite xenoliths (Pretorius and Barton, 2002). Magmatic underplating, related either to the 2.06 Ga Bushveld Complex or to the Mesozoic Karoo magmatism, has been suggested by the above authors as a possible solution. Irrespective of these details, the implication of a thin present-day felsic crust in the CZ means that there is no requirement to postulate for the Neoarchean metamorphic event a crustal thickening much beyond the ~40 km indicated by the peak metamorphic P-T data. As a conservative estimate, we have therefore chosen 45 km.

GEOTHERM CALCULATIONS AND TIME SCALES

Using the heat production data in Table 2, we have calculated steady-state geotherms, for a one-dimensional crustal column, for relevant conditions (Figs. 6, 7). In a steady-state situation, the heat capacity of rocks is canceled out, and integrating the heat production over depth yields T as a function of depth (as used by Kramers et al., 2001; Kamber et al., 2005):

$$T(z) = T_{top} + (Q_{base} + LH) \times z/K - Hz^2/(2K), \qquad (2)$$

where T_{top} is the temperature at the surface (set at 0 °C), Q_{base} the heat flow at the base of the crust (mW × m^{-2}), L the thickness of the crust (m), K its thermal conductivity (Wm^{-1} × °C^{-1}), and H its heat production (μW × m^{-3}). We used a K value of 2.5 Wm^{-1} × °C^{-1} (see England and Thompson, 1984; Sandiford et al., 2002).

In Figure 6, geotherms for a crustal thickness of 30 km and U, Th, and K concentrations of domains 1, 2, and 3 and their weighted averages are shown for 0, 2.0, and 2.65 Ga. The value used for the basal heat flow is 20 mW × m^{-2} in all cases. This is the present-day value of Archean cratons that are underlain by a thick subcontinental lithospheric keel (Jones, 1998; Nyblade et al., 1990; Nyblade and Pollack, 1993). The basal heat flow can attain much higher values even today, e.g., through delamination of the lithospheric mantle (England and Houseman, 1989) or slab breakoff (Davies and von Blanckenburg, 1995), but such increases would be transient. In the Archean, the heat production of the mantle was at least twice as high as today. On the

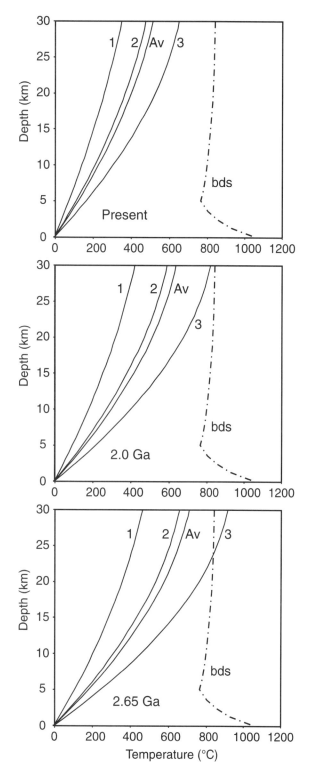

Figure 6. Steady-state geotherms calculated using equation (2) for a 30-km-thick crust made up of a single layer, with heat production rates for domains 1, 2, and 3 and the weighted average (labels at curves) from Table 2, for 0 Ga, 2.0 Ga, and 2.65 Ga. Boundary conditions were a surface temperature of 0 °C and a basal heat flow of 20 mWm^{-2} in all cases. The biotite dehydration solidus (bds, Wyllie, 1977) has been drawn in for comparison.

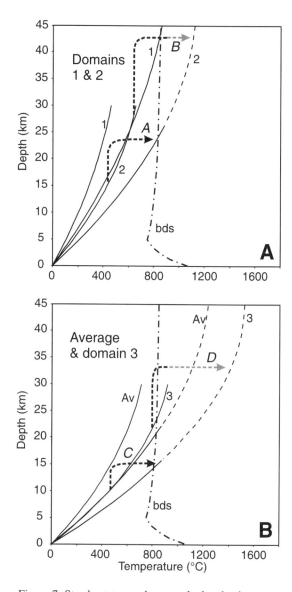

Figure 7. Steady-state geotherms calculated using equation (2) for homogeneous crust with the heat production rates of domains 1, 2, 3, and the weighted average (Av) from Table 2 for 2.65 Ga for crustal thicknesses of 30 and 45 km (distinguished by length of curves). Surface temperature assumed to be 0 °C. For the 30 km curves, the basal heat flux was assumed to be 20 mWm^{-2}; 45 km curves are shown for values of 20 mWm^{-2} for domain 1, and 0 mWm^{-2} for domains 2, 3, and average, as basal crust temperatures would be close to those of the mantle. Biotite dehydration solidus (bds, Wyllie, 1977) included for comparison. Supersolidus parts of geotherms, dashed, are unrealistic. Stippled arrows A, B, C, and D are possible trajectories of compression and heating at two different levels, following homogeneous thickening of a domain 2 (A, B) or 3 (C, D) crust from 30 to 45 km. Gray parts are supersolidus and thus are unrealistic.

other hand, the lower viscosity in the Archean mantle compared to today might have enabled more efficient channeling of heat to spreading centers, leaving continental keels cool (Richter, 1985; Nyblade and Pollack, 1993; Ballard and Pollack, 1987). It can be seen that, up to 2.0 Ga, all domains remain subsolidus, and that the solidus is barely surpassed at the base of the 30 km crust for domain 3 at 2.65 Ga.

In Figure 7, steady-state model geotherms for 2.65 Ga, for the three domains defined in Table 2, and the weighted average, are shown for crustal thicknesses of 30 and 45 km. It is clear that for domains 2, 3, and the average CZ, even with zero heat input from the mantle, the "steady state" geotherm could never be reached because the lower crust would be extensively molten long before it became established, and the thermal structure of the crust would be completely reorganized by magmatic heat advection.

The crustal thickening documented by the P-T data for the metamorphism at ca. 2.65 Ga in the Limpopo Complex Central Zone must have been caused by tectonic shortening. After the thickening, the geotherm adjusted itself to the new, thicker crust during a transitional period. Important questions concern the time scale for such a transition, and whether the thermal adjustment could be completed. For well-understood tectonic situations this can be modeled in some detail (Oxburgh and Turcotte, 1974; England and Richardson, 1977; England and Thompson, 1984; Sandiford et al., 2002). For Precambrian high-grade terranes there are rarely enough boundary conditions available for such modeling. There are, however, simple considerations that may yield a first-order estimate for the time scale at which a crustal domain is heated up after its heat production rate is increased or after it is brought to a deeper level by thickening of the crust.

The first scenario would occur during tectonic imbrication and stacking, whereby relatively U-, Th-, and K-rich rock units from the upper crust, e.g., metapelites, are emplaced at lower crustal levels. Excess heating power ΔH, the difference between the new heat production rate of the domain and the old one, is then available for heating the lower crustal domain, in principle to the temperature given by a steady-state geotherm for the new U, Th, and K distribution in the crust. In the second case, the heat production rate of a rock domain is not changed, but the integrated thermal conductivity of the crust above it is reduced by the crustal thickening, and therefore there is heat available to raise the temperature until a new steady-state geotherm is reached. The increased effective heating power is given by $\Delta H_{eff} = H(1 - z_1/z_2)$ where z_1 and z_2 are the depth of the rock unit in the crust before and after crustal thickening.

The initial heating rate for a crustal rock domain at depth, in which the internal heat production rate available for temperature increase has been increased as discussed above, is readily calculated from its heat capacity. The heat capacity C of rocks increases with temperature, and for both granitoids and pelitic rocks it increases from ~0.96 J °C^{-1}g^{-1} at 200 °C to ~1.25 J °C^{-1}g^{-1} at temperatures ~600 °C. For example, crustal thickening from 30 to 45 km for the average heat production of the CZ at 2.65 Ga (see Table 2) results in ΔH_{eff} = 0.87 µWm^{-3}. Using this value, a heat capacity of 1.25 J °C^{-1}g^{-1} and a rock density of 2.7 would yield an initial temperature increase of 8.1 °C per million years.

However, as the domain heats up, it will also give off heat to its surroundings, whereby ultimately the new "steady state" geotherm would be established. Thereby, the amount of heat available for temperature increase becomes less with time, and heating to the temperature of a new "steady state" geotherm is slowed down. The rate of initial temperature increase (from a previous steady-state geotherm) without heat transport would be $dT/dt = \Delta H_{eff}/C$. If some heat is being lost by conduction, then this becomes $dT/dt = \Delta H_{eff,i}(1-L)/C$, where $\Delta H_{eff,i}$ is the initially available excess heating power, and L is the fraction of heat that is being lost. In the case of adjustment from a previous "steady state" geotherm, it can be assumed that, on a first approximation, the amount of heat lost to the surroundings (to establish the new geotherm) is proportional to the temperature excess over the old geotherm. The new geotherm then represents an equilibrium where $L = 1$, and the temperature is no longer increased. Thus the fraction L of extra heat generated (ΔH_{eff}) that is not used for local temperature increase is approximated by $L(t) = (T(t) - T_1)/(T_2 - T_1)$, where T, T_1, and T_2 are the actual temperatures of the rock domain, the "old" geotherm, and the future geotherm, respectively. The temperature increase is now described by

$$dT/dt = \Delta H_{eff,i} [1 - (T(t) - T_1)/(T_2 - T_1)]/C. \qquad (3)$$

The solutions for this equation are exponential expressions:

$$dT/dt = \Delta H_{eff,i}/C \times e^{-\psi t} \text{ and } T(t) = T_2 - (T_2 - T_1)e^{-\psi t}, \qquad (4)$$

where ψ is a parameter describing the attenuation of the amount of heat available for temperature increase. It is a linear function of $(T_2 - T_1)$ and of ΔH_i. The equations describe an exponential "decay" of the heating rate and an asymptotic approach to the future geotherm. Analogous to radioactive decay, there is a half-time $t_{1/2}$ that is the time interval in which the heat effectively available for further temperature increase is halved: $\psi \times t_{1/2} = -\ln(0.5)$.

The interesting feature of these simple equations and the parameters ψ and $t_{1/2}$ is that they are not scaled to geometric factors, diffusion, etc., and thus allow estimation of time scales for the heating of crustal domains without need of further details. If $t_{1/2}$ is in millions of years, then it is given by

$$t_{1/2} = 0.0207 \times \rho \times C \times (T_2 - T_1)/\Delta H_{eff}, \qquad (5)$$

where ρ(density) is in gcm^{-3}, C is in Jg^{-1}°C^{-1}, and ΔH_{eff} is in µWm^{-3}. Table 3 shows some values of $t_{1/2}$ relevant to the Central Zone situation, as portrayed in Figure 5.

For example, in case (A), shown in Figure 7A, rock units with U, Th, and K contents of domain 2 are brought to a 50% deeper crustal level. Here, $\Delta H_{eff} = 2.32 \times (1 - 2/3) = 0.773$ µWm^{-3}, and the temperature differences between the old and the new geotherms are ~360 °C. The half-time for heating would thus

TABLE 3. VALUES FOR THE HALF-TIME OF HEATING AS A FUNCTION OF THE EFFECTIVE AVAILABLE HEATING POWER DIFFERENCE, ΔH_{EFF}, AND BOUNDARY TEMPERATURE DIFFERENCE

ΔH_{eff}	$T_2 - T_1$ (°C)							
	50	100	150	200	250	300	350	400
	$t_{\frac{1}{2}}$ (Ma)							
0.5	7.0	14.0	21.0	28.0	35.0	42.0	49.0	56.0
1	3.5	7.0	10.5	14.0	17.5	21.0	24.5	28.8
2	1.75	3.5	5.3	7.0	8.8	10.5	12.3	14.0
3	1.17	2.33	3.5	4.7	5.8	7.0	8.2	9.3

be 32.4 Ma (equation 5). Two to three half-times, thus ~80 m.y., must have elapsed before the temperature could be said to have approached the value of the new geotherm. By comparison, the time scale for the relaxation of a thermal anomaly generated by stacking two crust layers, each 30 km thick, is ~30 m.y. (e.g., Ashwal et al., 1992). Therefore, radioactive heating, not thermal conductivity, appears to be rate-limiting.

The long time span for radioactive heating of the crust is highly sensitive to regional differences in ΔH_{eff}, which in turn (at least in the case of crustal thickening) is a linear function of

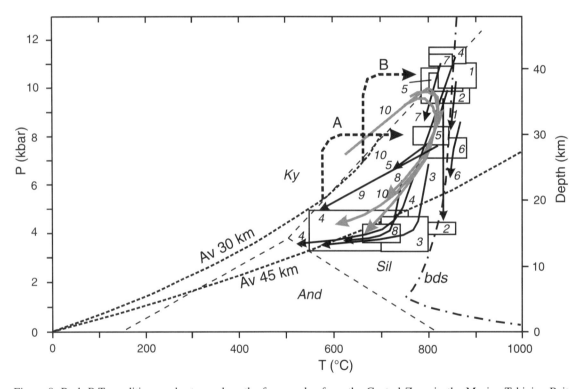

Figure 8. Peak P-T conditions and retrograde paths for samples from the Central Zone in the Musina–Tshipise–Beit Bridge area, for the 2.65 Ga metamorphism. Sources, areas, and parageneses: (1) Chinner and Sweatman (1968), Beit Bridge area, ky/sill + opx + qtz. (2) Droop (1989), Europe Claim, grt + cor + spr + phlog + rtl. (3) Harris and Holland (1984), Beit Bridge area, crd + spl + opx from grt + sill + qtz. (4) Holzer (1995), NW of Musina, grt + opx + cpx + hbl + pl. (5) Horrocks (1983), S of Musina, grt + cor. (6) Schreyer et al. (1984), S of Musina, grt + mg + str. (7) Watkeys et al. (1983), Beit Bridge, ky + qtz + grt or opx. (8) Windley et al. (1984), Beit Bridge, spr + krn + spl + opx + crd + phlog. (9) Perchuk et al. (2008), low strain metapelitic sample from the Ha-Tshansi closed structure, grt-crd equilibria. (10) (thicker gray arrows) Zeh et al. (2004), prograde and retrograde paths from pseudosections, Sand River Gneiss and two metapelites. Superimposed: steady-state geotherms for the weighted average of the CZ at 2650 Ga for 30 and 45 km crustal thickness (dashed curves; zero basal heat flux assumed in the 45-km case), and related theoretical prograde paths (thick dashed arrows A and B) that would reflect rapid crustal thickening by 50%, followed by heating toward the new steady-state geotherm, as in Figure 7. Ky—kyanite; Sil—sillimanite; And—andalusite; hbl—hornblende; bds—biotite dehydration solidus.

the internal radioactive heat production H. If radioactive heating of the crust was a main factor leading to the Neoarchean metamorphism of the Central Zone, then it is expected that the metamorphism would be diachronous from area to area.

After a solidus temperature is surpassed, most of the effective excess heat will be used for melting reactions instead of further heating, so that temperatures will be buffered in the melting interval (Completely melting a rock requires as much heat as would be needed to raise its temperature by ~100 °C). Further, partial melting would result in both a lowering of the density and a reduction in viscosity by several orders of magnitude (e.g., Wickham, 1987; Sawyer, 2001), so that vertical movement of some sort would disturb the geotherm, as discussed further below. For these reasons temperatures are not expected to exceed the lower part of the melting interval, and the strongly supersolidus parts of calculated "steady state" geotherms in Figures 6 and 7 are unrealistic.

DISCUSSION

Heat Production and P-T-t Data for the 2.65 Ga Metamorphism of the Central Zone

In Figures 8 and 9 the calculated "steady state" geotherms for the weighted average CZ crust, 30 km thick and thickened to 45 km, are compared with measured peak P-T conditions and retrograde P-T paths that are all thought to pertain to the 2.65 Ga metamorphism (van Reenen et al., 2004; Boshoff et al., 2006; see Perchuk et al., 2008; Smit et al., this volume, for a discussion). The geotherm for the thickened crust extends to much higher temperatures than the peak P-T data at mid- to lower crustal levels. This also applies to thickened-crust geotherms for domains 2 and especially 3 (compare Fig. 7). Thus the question raised in the introduction, whether intracrustal heating would be sufficient to have caused the metamorphism at 2.65 Ga, is answered in the affirmative, at least for domains 2 and 3 and the weighted average CZ.

Peak P-T conditions cluster closely around the biotite dehydration solidus (bds). There is indeed evidence for widespread anatexis from the Singelele-type quartzofeldspathic gneisses. A ready explanation for the clustering of P-T peak data around the solidus is, then, that partial melting acted as a thermostat, preventing a further major temperature increase as long as melting progressed. Retrograde paths take two forms: decompression-dominated ones along the solidus, and subsolidus cooling with minor decompression at mid-crustal level. The trajectories of the subsolidus retrogressive P-T paths range from being subparallel to geotherms in the 6–7 kb range, to almost isobaric in the 3–5 kb range. It is noteworthy that there are no decompression-dominated retrograde paths at subsolidus temperatures.

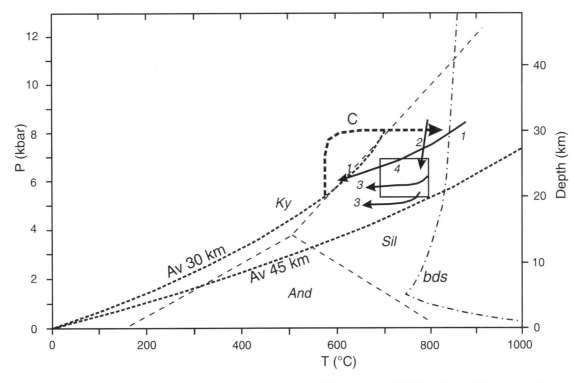

Figure 9. Peak P-T conditions and retrograde P-T paths from the Alldays–Baines Drift–Zanzibar and Koedoesrand areas for the 2.65 Ga metamorphism. Sources, areas, and parageneses: (1) Perchuk et al. (2008), bt + grt + crd thermobarometry from a low strain area 10 km W of Alldays. (2) Hisada and Miyano (1996), Baines Drift, opx + sill + spr + crd. (3) Hisada and Miyano (1996), W of Zanzibar, gedrite formation in metapelites. (4) Schaller et al. (1999), Central Zone, N Palala Shear Zone in Koedoesrand Window, opx + cpx + hbl + pl. Superimposed: steady-state geotherms and related arbitrary theoretical prograde path C, as in Figure 8; pl—plagioclase; other abbreviations as in Figure 8 caption.

The calculated "steady state" 45 km geotherm for domain 1 does not reach the solidus (Fig. 7) or the P-T conditions measured. However, all existing P-T measurements relating to the ca. 2.65 Ga metamorphism were carried out on rock units from domains 2 and 3, or regions very close to them, such as the thin slice of domain 1 between the Tshipise and Musina "hot" areas. Such neighboring regions may have received heat via conduction. The probable reason why no P-T studies have been done in extensive domain 1 areas is that the required mineral parageneses have not been found there. This may be due to the rock types or to a lower grade of metamorphism in these regions, as predicted by the geotherms.

Dating the Neoarchean metamorphism in the Central Zone has been an enduring effort over the past 40 yr. There are some dates ca. 2.65 Ga that were obtained directly from metamorphic major phases such as garnet (Fig. 3 of Kramers and Mouri, this volume; van Reenen et al., 2008; Holzer et al., 1998). However, their precision is generally poor (error limits are ±30 m.y. or larger) so that they do not allow resolution as to whether the metamorphic episode was short-lived or long-lasting. This is in contrast to the Northern and Southern Marginal Zones of the Limpopo Complex, where metamorphic mineral ages of sufficient precision are available to define considerable age ranges (Barton et al., 1992; Kamber et al., 1996; Kreissig et al., 2001). In the CZ, ages of peak metamorphism for the 2.65 Ga event have been chiefly obtained as U-Pb dates from metamorphic zircon overgrowths (Mouri et al., 2009; Zeh et al., 2007, 2010) or zircons in granitoids that are the products of crustal anatexis (Kröner et al., 1999; Boshoff et al., 2006; Jaeckel et al., 1997; van Reenen et al., 2008; Zeh et al., 2007; see Figure 2 of Kramers and Mouri, this volume). These zircon dates vary in an age range between 2.7 and 2.55 Ga., well outside analytical precision and accuracy (generally less than 5 m.y.), clearly documenting a long duration of the 2.65 Ga event. As discussed above, the time scale for heating by intracrustal radioactivity can vary, depending on heat production and crustal thickening, in the range 10 to more than 100 m.y., and is longer than that for conductive relaxation of a thermal anomaly brought about by, e.g., crustal stacking, delamination of the subcontinental mantle lithosphere, slab breakoff, or underplating. Also, the dependency of the heating time scale on the intracrustal heat production means that the rate of heating would be regionally variable (Table 3). Thus the large age range of Neoarchean metamorphism in the Central Zone is more in accord with the predictions from an intracrustal heating scenario than with one of anomalous heat flow from below.

In summary, it appears that the hypothesis of intracrustal radioactive heating, with the assumptions of a homogeneous vertical distribution of the concentration of heat-producing elements and a moderate crustal thickening from 30 to 40 km (both assumptions being supported by data), can account for the observed peak metamorphic conditions of the Neoarchean metamorphism in the Central Zone and for the large spread in metamorphic ages. Important features that have not yet been addressed here are the diverging nature of retrogressive paths and possible reasons for the cessation of metamorphism. These are discussed below in the context of tectonics.

Heat Production at 2.65 Ga, Tectonic Styles of the Central Zone, and the End of Metamorphism

In many areas of the Central Zone, uncertainty remains as to whether structures were produced in the ca. 2.65 Ga or in the 2.02 Ga event. For the purpose of this study, it is necessary only to have a generalized picture characterizing the main features of the ca. 2.65 Ga structures. Toward this aim, the structures that belong to the 2.02 Ga episode must be identified. These are, first of all, the shear zones delimiting the Central Zone to the north and south (McCourt and Vearncombe, 1987; Kamber et al., 1995; Schaller et al., 1999; Holzer et al., 1999), as well as dominant shear zones within the Central Zone that run subparallel to these boundaries, the Sunnyside Shear Zone and the Mahalapye Straightening Zone (Holzer et al., 1999; Chavagnac et al., 2001). Further, prominent north-south–trending shear zones, previously termed *cross folds*, such as the Baklykraal structure, belong to this category (Boshoff et al., 2006; van Reenen et al., 2008). All these shear zones have subhorizontal movement indicators. Where peak P-T conditions and retrograde paths reflecting a single metamorphism have been measured (or where intense shearing has obliterated pre-2.02 Ga mineral parageneses), these reflect pressures in the 3–6 kb range (Boshoff et al., 2006; Smit et al., this volume).

If these 2.0 Ga structures are stripped away, blocks with steep structures of random strike, with fold axes dipping ~60° to the SW, and circular structures with stretching lineations of the same dip, remain as dominant features of the Central Zone. Notably, there is no structural grain parallel to the geographical WSW-ENE elongation of the zone, and its boundaries that follow the same trend. There is a growing body of evidence that these structures were generated in the 2.65 Ga episode. The 2.7–2.55 Ga zircon dates from Singelele-type quartzofeldspathic gneisses, which form sheetlike intrusions with this type of orientation, have already been mentioned. Further, a granite with a very strong mineral lineation in the center of the Avoca closed structure yielded a zircon U-Pb date of 2.594 ± 0.038 Ga (no younger overgrowths are recorded); this is interpreted as the age of its syntectonic intrusion (Boshoff et al., 2006; van Reenen et al., 2008).

It can be argued that the hypothesis of intracrustal heating, as considered in this study, could possibly predict the emergence of such structures at ca. 2.65 Ga. A density inversion and drastic reduction in viscosity by several orders of magnitude (e.g., Wickham, 1987; Sawyer, 2001), both resulting from partial melting, could have led to diapirism in the lower and middle crust. Lower crustal doming in the solid state has been demonstrated and modeled (Parphenuk et al., 1994; Whitney et al., 2004; Gerya et al., 2001, 2004; Perchuk and Gerya, this volume). Even lower viscosities than envisaged in these studies could have resulted from partial melting, leading to diapirism on the smaller horizontal scale of several kilometers as seen in the CZ (and also in the NMZ, Blenkinsop, this volume). Partially molten parcels of lower crust could have intruded in diapiric fashion into mid-crustal levels, where the surrounding rocks would have partly sunk. This hypothetical process is sketched in Figure 10: At mid-crustal level, this would have led to steep structures and local compressional tectonics, without a regional lateral compression being necessary to have caused it. The process would have involved some horizontal mass flow (toward zones of upward movement) within the lower crust, and here subhorizontal structures would be predicted. As illustrated in Figure 10, this lowest crustal level with horizontal structures would have become attenuated, and the level with vertical structures would have been thick and more likely to have been seen at the surface after uplift and erosion.

The retrogressive P-T paths that first follow the solidus, and then reflect almost isobaric cooling as seen prominently in Figure 8, could have been generated by this type of diapiric reorganization mechanism: The decompression seen at perisolidus temperatures may reflect intrusion of migmatitic parcels from lower to mid-crustal levels. Here, the parcels' temperature would be above the prevailing "average" geotherm, and they would therefore cool relatively rapidly to adjust to it; no crustal thinning would be required for this cooling. The subsolidus cooling-decompression paths at higher pressures require crustal thinning and are discussed below. In areas particularly rich in heat-producing elements, the degree of partial melting could have become high enough to generate larger scale granitoid intrusions such as the Bulai Pluton and the Alldays Gneiss. These areas are mostly spatially associated with areas of high crustal radioactivity (domain 3; see Figs. 5, 8, 9), and the Hf-isotopes of their zircons show that they are mainly crustally derived (Zeh et al., 2007; Kramers and Zeh, this volume).

Intracrustal radioactive heating, after crustal thickening, would not be a unique cause for the diapir-like scenario speculatively proposed here. It could also occur as a consequence of greatly increased basal heat flow, e.g., from mantle keel delamination or slab breakoff. Strong arguments in favor of intracrustal heating remain (1) the demonstrated availability of sufficient heat generation in the lower crust at 2.65 Ga, and (2) the observed long time span of heat-related processes (metamorphism and intrusion), predicted by the radioactive heating scenario. However, an important question that specifically arises for the intracrustal radioactive heating hypothesis is: How could a metamorphic episode of this nature cease? With heat production continuing, there are only two mechanisms for this: first, magmatic transport of radioactive elements to higher levels in the crust (e.g., Sandiford et al., 2002; McLaren et al., 2005), or second, thinning of the crust. The latter could happen either by erosion or by extension.

From the available P-T data it is evident that the rocks now at the surface in the CZ were mainly at mid- to deep-crustal levels at 2.65 Ga. As they still contain heat-producing elements at above average crustal concentrations, it is unlikely that these elements were transported by magmas to shallow crustal levels. Also, the

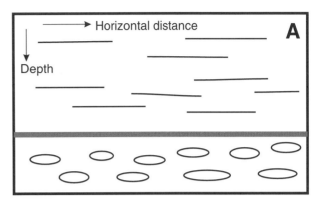

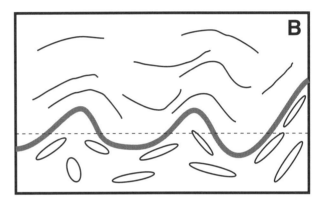

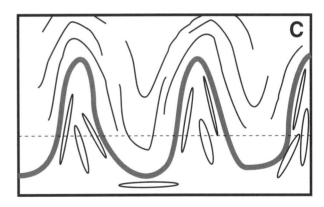

Figure 10. Cartoon illustrating the concept of diapir-like gravitational reorganization following partial melting in the lower crust. Thick gray line is a hypothetical boundary between "more partially molten" lower crust (with ellipses) and "less partially molten" middle crust (with lines showing the structural trends). (A) All structures are still horizontal. (B) Gravitational instability (inversion) begins to lead to localized doming of lower crust and sinking of middle crust. (C) The reorganization has become more extreme, and steep structures dominate. Note that the area of horizontal flow toward domes in the lower crust has become highly attenuated. Note also that higher viscosity in the middle crust than in the lower crust qualitatively expresses itself in sinking zones being wider than domes.

quartzofeldspathic gneisses presently observed occur mostly in relatively thin (up to several 100-m-thick) sheets that appear to have been "frozen" in mid-crust. Another argument against high-level Neoarchean intrusions in the CZ is the absence of zircons of that age range in coeval sediments of the Transvaal Supergroup (Beukes et al., 2004).

Thus, thinning of the crust remains as a possible mechanism for cooling. The effectiveness of this scenario can be seen from Figure 7 if a transition from a 45-km- to a 30-km-thick crust is envisaged. Crustal thinning is documented by some of the sub-solidus retrogressive P-T paths seen in Figures 8 and 9: Arrows 9 and 5 in Figure 8 and 1 in Figure 9 are compatible with a gradual reduction in crust thickness of ~30% (from the pressure decrease) and coeval cooling to an average CZ geotherm for 30 km thickness. Either uplift and erosion, or extension, could have produced this result. The dispersed occurrence of Gumbu Group metasediments, between 2.6 and 2.0 Ga in the CZ, suggests that some parts of the ca. 2.65 Ga high-grade metamorphic terrane were exhumed to the surface prior to their deposition. However, siliciclastic sediments are absent in the in the Neoarchean part of the Transvaal Supergroup stratigraphic column (Eriksson et al., 2006). On the other hand, there are no ca. 2.65 Ga structures in the CZ that point to crustal extension. Thus, geological observations do not provide an answer to the question of what caused crustal thinning following the ca. 2.65 Ga event in the CZ.

Some Comments on the 2.02 Ga Metamorphism

From the above, it appears that the notion of intracrustal radioactive heating, after crustal thickening, as the major cause for the ca. 2.65 Ga metamorphism (and some of the tectonism) in the CZ, is a valid working hypothesis. We now consider whether this could be true for the 2.0 Ga metamorphism as well. There are several reasons why this is unlikely:

1. At 2.0 Ga, the heat production of rock units with given concentrations of U, Th, and K is already ~20% less than at 2.65 Ga, owing to the relatively short half-lives of ^{40}K and particularly ^{235}U. For domain 3, 50% crustal thickening could still have produced the T and P recorded in the rocks (up to 800 °C at ~6 kb, Boshoff et al., 2006; Perchuk et al., 2008). For domain 2 and the average CZ, greater crustal thickening (at least 70%) would be required.

2. The pressures recorded from rock units that show only the 2.02 Ga metamorphism are lower than for those of the 2.65 Ga metamorphism. For the shear zone in the Baklykraal structure the peak pressure is ~6 kb, and the heating appears to have been near-isobaric (Boshoff et al., 2006; Perchuk et al., 2008). These peak pressures do not reflect the large crustal thickness that would be needed to heat up the mid-crust to the temperatures observed if intracrustal radioactive heating were the heat source.

3. The 2.02 Ga metamorphism in the CZ is characterized by a sharp isochronism of peak conditions (2.02 ± 0.02 Ga; Boshoff et al., 2006; Holzer et al., 1998, 1999; Jaeckel et al., 1997; Kröner et al., 1999; Mouri et al., 2008, 2009; Schaller et al., 1999; Zeh et al., 2007; see summary in Kramers and Mouri, this volume). As shown above, intracrustal radioactive heating is a slow process that would normally lead to a large spread in the ages of metamorphic peak conditions (as observed in the ca. 2.65 Ga metamorphism). The heating becomes ever slower as time progresses from the Archean to the present. The isochronism of the 2.02 Ga metamorphism is thus a strong argument against radioactive intracrustal heating as a cause.

Intracrustal radioactive heating is always present, but for the above reasons it can clearly be ruled out as the major mechanism responsible for the 2.02 Ga metamorphism in the Limpopo Belt. Therefore, in the case of that later metamorphic episode, it is necessary to invoke a rapid, transient increase in basal heat flux into the crust. In this context we note that a very thick mafic layer, possibly of Bushveld age (2.04 Ga), in the crust underneath the CZ is inferred from teleseismic data (Nguuri et al., 2001; Gore et al., 2009). Speculations about its possible effects are, however, outside the scope of this chapter.

SUMMARY AND CONCLUSIONS

The steady-state geotherms calculated in this study were based on the combined data on concentrations of U, Th, and K in rock units of the Central Zone of the Limpopo Complex, coupled with the observation that the rocks presently at the surface were mostly at mid- to lower crustal levels at 2.65 Ga. This led to the assumption that the concentrations observed would be uniform over the thickness of the crust. Further, we assumed a crustal thickening to 45 km during the 2.65 Ga metamorphic episode. This is supported by petrological P-T studies.

Mid- and lower crustal temperatures found from the steady-state geotherm calculations exceed the temperatures of peak metamorphism found in the CZ for the given depths for the two U and Th richer domains and the weighted average. The steady-state geotherms are unrealistic for two reasons: First, intracrustal radioactive heating is a slow process, taking tens to more than a hundred million years to approach a new steady-state geotherm after crustal thickening, and second, partial melting acts as a thermostat in the lower crust as heat is consumed in melting reactions. It is unsurprising that the peak P-T conditions found in petrological studies cluster mainly around the vapor-absent biotite dehydration solidus.

The concentration data used in this study are not sharply constrained, strong assumptions were made, and the approach of calculating steady-state geotherms is a considerable simplification. However, the results indicate sufficient intracrustal radioactive heat for the observed metamorphic conditions by a considerable margin. This provides confidence in the main finding of this study, that intracrustal radioactive heating could indeed have caused the Neoarchean high-grade metamorphism observed in the CZ.

The intracrustal radioactive heating scenario also helps to explain the long duration of the metamorphism, from 2.7 to 2.55 Ga. Further, it can be plausibly argued that the typical structural patterns observed in the CZ can have been caused by it, via a density inversion and a strong contrast in viscosity in the lower crust. Only crustal thinning (indicated by some retrograde P-T paths) could have terminated the metamorphic episode, but there is no direct evidence for the mechanism (extension or uplift and erosion).

The 2.0 Ga metamorphic event in the CZ cannot have been caused by intracrustal radioactive heating alone. The reasons are (1) a strong reduction in radioactive heat production in the intervening ca. 0.65 Ga; (2) the 2.0 Ga event appears to be characterized by lower pressures than the ca. 2.65 Ga event; therefore, there is no indication for strong crustal thickening as required to generate the observed temperatures at 2.0 Ga; and (3) the 2.0 Ga event was sharply isochronous over the entire CZ and lasted only ~30 m.y., whereas an intracrustal radioactive heating hypothesis would predict a much longer duration.

ACKNOWLEDGMENTS

We thank Jackie van Schalkwyk for making available the calibrated airborne radiometric survey data of Figure 2. Lew Ashwal and John Ridley are thanked for their thoughtful and constructive reviews, which helped greatly to improve the manuscript.

REFERENCES CITED

Andreoli, M.A.G., and Hart, R.J., 1990, Metasomatized granulites and eclogites of the Mozambique belt: Implications for mantle devolatilization, *in* Herbert, H.K., and Ho, S.E., eds., Proceedings of the Conference on Stable Isotopes and Fluid Processes in Mineralization: Nedlands, University of Western Australia, Geology Department & University Extension Publication 23, p. 121–140.

Andreoli, M.A.G., Coetzee, H., Hart, R., Ashwal, L.D., Huizenga, J.-M., Smit, A., Brandl, G., and Grantham, G., 2003, The Erlank paradox: High-U granulites, low-U amphibolite facies gneisses in the mobile belts of southern Africa, *in* Cuney, M., ed., Proceedings of International Conference on Uranium Geochemistry: Nancy, France, 13–16 April 2003, University of Nancy, p. 47–50.

Andreoli, M.A.G., Hart, R.J., Ashwal, L.D., and Coetzee, H., 2006, Correlation between U, Th content and metamorphic grade in the western Namaqualand belt, South Africa, with implications for radioactive heating of the crust: Journal of Petrology, v. 47, p. 1095–1118, doi:10.1093/petrology/egl004.

Ashwal, L.D., Morgan, P., and Hoisch, Th.D., 1992, Tectonics and heat sources for granulite metamorphism of supracrustal-bearing terrains: Precambrian Research, v. 55, p. 525–538, doi:10.1016/0301-9268(92)90043-N.

Ballard, S., and Pollack, H.N., 1987, Diversion of heat by Archean cratons: A model for southern Africa: Earth and Planetary Science Letters, v. 85, p. 253–264, doi:10.1016/0012-821X(87)90036-7.

Barton, J.M., Jr., Ryan, B.D., Fripp, R.E.P., and Horrocks, P.C., 1979, Effects of metamorphism on the Rb-Sr and U-Pb systematics of the Singelele and Bulai Gneisses, Limpopo Mobile Belt, southern Africa: Geological Society of South Africa Transactions, v. 82, p. 259–269.

Barton, J.M., Jr., Ryan, B.D., and Fripp, R.E.P., 1983, Rb-Sr and U-Th-Pb isotope studies of the Sand River Gneisses, Limpopo Mobile Belt, *in* Van Biljon, W.J., and Legg, J.H., eds., The Limpopo Belt: Geological Society of South Africa Special Publication, no. 8: Johannesburg, Geological Society of South Africa, p. 9–18.

Barton, J.M., Doig, R., Smith, C.B., Bohlender, F., and van Reenen, D.D., 1992, Isotopic and REE characteristics of the intrusive charnoenderbite and enderbite geographically associated with the Matok Pluton, Limpopo Belt, Southern Africa: Precambrian Research, v. 55, p. 451–467, doi:10.1016/0301-9268(92)90039-Q.

Barton, J.M., Jr., Barnett, J.W.R., Barton, E.S., Barnett, M., Doorgapershad, A., Twiggs, C., Klemd, R., Martin, L., Mellonig, L., and Zenglein, R., 2003, The geology of the area surrounding the Venetia kimberlite pipes, Limpopo Belt, South Africa: A complex interplay of nappe tectonics and granite magmatism: South African Journal of Geology, v. 106, p. 109–128, doi:10.2113/106.2-3.109.

Beukes, N.J., Dorland, H.C., Gutzmer, J., Evans, D.A.D., and Armstrong, R.A., 2004, Timing and provenance of Neoarchean-Paleoproterozoic unconformity bounded sequences on the Kaapvaal Craton: Geological Society of America Abstracts with Programs, v. 36, no. 5, p. 255.

Bickle, M.J., 1978, Heat loss from the Earth: A constraint on Archean tectonics from the relation between geothermal gradients and the rate of plate production: Earth and Planetary Science Letters, v. 40, p. 301–315, doi:10.1016/0012-821X(78)90155-3.

Blenkinsop, T.G., 2011, this volume, Archean magmatic granulites, diapirism, and Proterozoic reworking in the Northern Marginal Zone of the Limpopo Belt, *in* van Reenen, D.D., Kramers, J.D., McCourt, S., and Perchuk, L.L., eds., Origin and Evolution of Precambrian High-Grade Gneiss Terranes, with Special Emphasis on the Limpopo Complex of Southern Africa: Geological Society of America Memoir 207, doi: 10.1130/2011.1207(13).

Bodorkos, S., Sandiford, M., Minty, B.R.S., and Blewett, R.S., 2004, A high-resolution, calibrated airborne radiometric dataset applied to the estimation of crustal heat production in the Archaean northern Pilbara Craton, Western Australia: Precambrian Research, v. 128, p. 57–82, doi:10.1016/j.precamres.2003.08.008.

Boryta, M., and Condie, K.C., 1990, Geochemistry and origin of the Archaean Beit Bridge Complex, Limpopo Belt, South Africa: Journal of the Geological Society [London], v. 147, p. 229–239, doi:10.1144/gsjgs.147.2.0229.

Boshoff, R., van Reenen, D.D., Smit, C.A., Perchuk, L.L., Kramers, J.D., and Armstrong, R.A., 2006, Geologic history of the Central Zone of the Limpopo Complex: The west Alldays area: Journal of Geology, v. 114, p. 699–716, doi:10.1086/507615.

Brandl, G., 1983, Geology and geochemistry of various supracrustal rocks of the Beit Bridge Complex east of Messina, *in* Van Biljon, W.J., and Legg, J.H., eds., The Limpopo Belt: Geological Society of South Africa Special Publication, no. 8: Johannesburg, Geological Society of South Africa, p. 103–112.

Brandl, G., 2002, The Geology of the Alldays Area: Pretoria, Council for Geoscience, Sheet 2228 (Alldays), Explanation, 71 p.

Buick, I.S., Williams, I.S., Gibson, R.L., Cartwright, I., and Miller, J.A., 2003, Carbon and U-Pb evidence for a Palaeoproterozoic crustal component in the Central Zone of the Limpopo Belt, South Africa: Journal of the Geological Society [London], v. 160, p. 601–612, doi:10.1144/0016-764902-059.

Chavagnac, V., Kramers, J.D., Naegler, Th.F., and Holzer, L., 2001, The behaviour of Nd and Pb isotopes during 2.0 Ga migmatization in paragneisses of the Central Zone of the Limpopo Belt (South Africa and Botswana): Precambrian Research, v. 112, p. 51–86, doi:10.1016/S0301-9268(01)00170-X.

Chinner, G.A., and Sweatman, T.R., 1968, A former association of enstatite and kyanite: Mineralogical Magazine, v. 36, p. 1052–1060, doi:10.1180/minmag.1968.036.284.03.

Cobbold, P.R., and Quinquis, H., 1980, Development of sheath folds in shear regimes: Journal of Structural Geology, v. 2, p. 119–126, doi:10.1016/0191-8141(80)90041-3.

Davies, J.H., and von Blanckenburg, F., 1995, Slab breakoff: A model of lithospheric detachment and its test in the magmatism and deformation of collisional orogens: Earth and Planetary Science Letters, v. 129, p. 85–102, doi:10.1016/0012-821X(94)00237-S.

Davies, J.H., and von Blanckenburg, F., 1998, Thermal controls on slab breakoff and the rise of high pressure rocks during continental collisions, *in* Hacker, B.R., and Liou, J.G., eds., When Continents Collide: Geodynamics and Geochemistry of Ultrahigh-Pressure Rocks: Dordrecht, Kluwer Academic, p. 97–115.

Droop, G.T.R., 1989, Reaction history of garnet-sapphirine granulites and conditions of Archaean high-pressure granulite-facies metamorphism in the central Limpopo mobile belt, Zimbabwe: Journal of Metamorphic Geology, v. 7, p. 383–403, doi:10.1111/j.1525-1314.1989.tb00604.x.

England, P.C., and Houseman, G.A., 1989, Extension during continental convergence, with application to the Tibetan plateau: Journal of Geophysical Research, v. 94, p. 17,561–17,579, doi:10.1029/JB094iB12p17561.

England, P.C., and Richardson, S.W., 1977, The influence of erosion upon the mineral facies of rocks from different metamorphic environments: Journal of the Geological Society [London], v. 134, p. 201–213, doi:10.1144/gsjgs.134.2.0201.

England, P.C., and Thompson, A.B., 1984, Pressure-temperature-time paths of regional metamorphism: I. Heat transfer during the evolution of regions of thickened continental crust: Journal of Petrology, v. 25, p. 894–928.

Eriksson, P.G., Altermann, W., and Hartzer, F.J., 2006, The Transvaal Supergroup and its precursors, *in* Johnson, M.R., Anhaeusser, C., and Thomas, R.J., eds., The Geology of South Africa: Johannesburg, Geological Society of South Africa, and Pretoria, South Africa, Council for Geoscience, p. 237–260.

Gerya, T.V., Maresch, W.V., Willner, A.P., van Reenen, D.D., and Smit, C.A., 2001, Inherent gravitational instability of thickened continental crust with regionally developed low- to medium pressure granulite facies metamorphism: Earth and Planetary Science Letters, v. 190, p. 221–235, doi:10.1016/S0012-821X(01)00394-6.

Gerya, T.V., Perchuk, L.L., Maresch, W.V., and Willner, A.P., 2004, Inherent gravitational instability of hot continental crust: Implications for doming and diapirism, *in* Whitney, D.L., Teyssier, C., and Siddoway, C.S., eds., Gneiss Domes in Orogeny: Geological Society of America Special Paper 380, p. 97–115.

Gore, J., James, D.E., Zengeni, T.G., and Gwavava, O., 2009, Crustal structure of the Zimbabwe Craton and the Limpopo Belt of Southern Africa: New constraints from seismic data and implications for its evolution: South African Journal of Geology, v. 112, p. 213–228, doi:10.2113/gssajg.112.3-4.213.

Harris, N.B.W., and Holland, T.J.B., 1984, The significance of cordierite-hypersthene assemblages from the Beitbridge region of the central Limpopo belt: Evidence for rapid decompression in the Archaean?: American Mineralogist, v. 69, p. 1036–1049.

Hisada, K., and Miyano, T., 1996, Petrology and microthermometry of aluminous rocks in the Botswanan Limpopo Central Zone: Evidence for isothermal decompression and isobaric cooling: Journal of Metamorphic Geology, v. 14, p. 183–197, doi:10.1046/j.1525-1314.1996.05857.x.

Holzer, L., 1995, The magmatic petrology of the Bulai Pluton and the tectonometamorphic overprint at 2.0 Ga in the Central Zone of the Limpopo Belt, Southern Africa [M.S. thesis]: Bern, Switzerland, University of Bern, 201 p.

Holzer, L., Frei, R., Barton, J.M., and Kramers, J.D., 1998, Unraveling the record of successive high grade events in the Central Zone of the Limpopo Belt using single phase dating of metamorphic minerals: Precambrian Research, v. 87, p. 87–115, doi:10.1016/S0301-9268(97)00058-2.

Holzer, L., Barton, J.M., Jr., Paya, B.K., and Kramers, J.D., 1999, Tectonothermal history in the western part of the Limpopo Belt: Test of tectonic models and new perspectives: Journal of African Earth Sciences, v. 28, p. 383–402, doi:10.1016/S0899-5362(99)00011-1.

Horrocks, P.C., 1983, A corundum and sapphirine paragenesis from the Limpopo belt, southern Africa: Journal of Metamorphic Geology, v. 1, p. 13–23, doi:10.1111/j.1525-1314.1983.tb00262.x.

International Atomic Energy Agency, 2003, Guidelines for Radioelement Mapping Using Gamma Ray Spectrometry: Vienna, International Atomic Energy Agency, 173 p.

Jaeckel, P., Kröner, A., Kamo, S.L., Brandl, G., and Wendt, J.I., 1997, Late Archaean to Early Proterozoic granitoid magmatism and high-grade metamorphism in the central Limpopo Belt, South Africa: Journal of the Geological Society [London], v. 154, p. 25–44, doi:10.1144/gsjgs.154.1.0025.

Jones, M.Q.W., 1998, A review of heat flow in southern Africa and the thermal structure of the lithosphere: Southern African Geophysical Review, v. 2, p. 115–122.

Kamber, B.S., Kramers, J.D., Napier, R., Cliff, R.A., and Rollinson, H., 1995, The Triangle Shearzone, Zimbabwe, revisited: New data document an important event at 2.0 Ga in the Limpopo Belt: Precambrian Research, v. 70, p. 191–213, doi:10.1016/0301-9268(94)00039-T.

Kamber, B.S., Biino, G.G., Wijbrans, J.R., Davies, G.R., and Villa, I.M., 1996, Archean granulites of the Limpopo Belt, Zimbabwe: One slow exhumation or two rapid events?: Tectonics, v. 15, p. 1414–1430, doi:10.1029/96TC00850.

Kamber, B.S., Whitehouse, M.J., Bolhar, R., and Moorbath, S., 2005, Volcanic resurfacing and the early terrestrial crust: Zircon U–Pb and REE constraints from the Isua Greenstone Belt, southern West Greenland: Earth and Planetary Science Letters, v. 240, p. 276–290, doi:10.1016/j.epsl.2005.09.037.

Kramers, J.D., and Mouri, H., 2011, this volume, The geochronology of the Limpopo Complex: A controversy solved, *in* van Reenen, D.D., Kramers, J.D., McCourt, S., and Perchuk, L.L., eds., Origin and Evolution of Precambrian High-Grade Gneiss Terranes, with Special Emphasis on the Limpopo Complex of Southern Africa: Geological Society of America Memoir 207, doi:10.1130/2011.1207(06).

Kramers, J.D., and Zeh, A., 2011, this volume, A review of Sm-Nd and Lu-Hf isotope studies in the Limpopo Complex and adjoining cratonic areas, and their bearing on models of crustal evolution and tectonism, *in* van Reenen, D.D., Kramers, J.D., McCourt, S., and Perchuk, L.L., eds., Origin and Evolution of Precambrian High-Grade Gneiss Terranes, with Special Emphasis on the Limpopo Complex of Southern Africa: Geological Society of America Memoir 207, doi: 10.1130/2011.1207(10).

Kramers, J.D., Kreissig, K., and Jones, M.Q.W., 2001, Crustal heat production and style of metamorphism: A comparison between two Archean high grade provinces in the Limpopo Belt, Southern Africa: Precambrian Research, v. 112, p. 149–163, doi:10.1016/S0301-9268(01)00173-5.

Kreissig, K., Holzer, L., Frei, R., Villa, I.M., Kramers, J.D., Kröner, A., Smit, C.A., and van Reenen, D.D., 2001, Geochronology of the Hout River Shear Zone and the metamorphism in the Southern Marginal Zone of the Limpopo Belt, Southern Africa: Precambrian Research, v. 109, p. 145–173, doi:10.1016/S0301-9268(01)00147-4.

Kröner, A., Jaeckel, P., Brandl, G., Nemchin, A.A., and Pidgeon, R.T., 1999, Single zircon ages for granitoid gneisses in the Central Zone of the Limpopo Belt, southern Africa and geodynamic significance: Precambrian Research, v. 93, p. 299–337, doi:10.1016/S0301-9268(98)00102-8.

Ledwaba, L., Dingoko, O., Cole, P., and Havenga, M., 2009, Compilation of Survey Specifications for All the Regional Airborne Geophysical Surveys Conducted over South Africa: Pretoria, Council for Geoscience, Report 2009-0130, 49 p.

McCourt, S., and Vearncombe, J.R., 1987, Shear zones bounding the Central Zone of the Limpopo Mobile Belt, southern Africa: Journal of Structural Geology, v. 9, p. 127–137, doi:10.1016/0191-8141(87)90021-6.

McLaren, S., Sandiford, M., and Hand, M., 1999, High radiogenic heat producing granites and metamorphism: An example from the western Mount Isa Inlier, Australia: Geology, v. 27, p. 679–682, doi:10.1130/0091-7613(1999)027<0679:HRHPGA>2.3.CO;2.

McLaren, S., Sandiford, M., and Powell, R., 2005, Contrasting styles of Proterozoic crustal evolution; a hot-plate tectonic model for Australian terranes: Geology, v. 33, p. 673–676, doi:10.1130/G21544.1.

Minty, B.R.S., 1977, Assessment of Uranium Anomalies from Airborne Radiometric Data of the Areas 7/69, 8/69, 9/71, 10/71, 11/71, 12/71, 13/71, 14/73, 15/73 and 21/74: Pretoria, Geological Survey of South Africa Report 1977-0154, 29 p.

Mouri, H., Brandl, G., Whitehouse, M., de Waal, S., and Guiraud, M., 2008, CL-imaging and ion microprobe dating of single zircons from a high-grade rock from the Central Zone, Limpopo Belt, South Africa: Journal of African Earth Sciences, v. 50, p. 111–119, doi:10.1016/j.jafrearsci.2007.09.011.

Mouri, H., Whitehouse, M., Rajesh, H.M., and Brandl, G., 2009, A magmatic age and four successive metamorphic events recorded in zircons from a single meta-anorthosite sample in the Central Zone of the Limpopo belt, South Africa: Journal of the Geological Society [London], v. 166, p. 827–830, doi:10.1144/0016-76492008-148.

Neumann, N., Sandiford, M., and Foden, J., 2000, Regional geochemistry and continental heat flow; implications for the origin of the South Australian heat flow anomaly: Earth and Planetary Science Letters, v. 183, p. 107–120, doi:10.1016/S0012-821X(00)00268-5.

Nguuri, T.K., Gore, J., James, D.E., Webb, S.J., Wright, C., Zengeni, T.G., Gwavava, O., Snoke, J.A., and Kaapvaal Seismic Group, 2001, Crustal structure beneath southern Africa and its implications for the formation of the Kaapvaal and Zimbabwe Cratons: Geophysical Research Letters, v. 28, p. 2501–2504, doi:10.1029/2000GL012587.

Nyblade, A.A., and Pollack, H.N., 1993, A global analysis of heat flow from Precambrian terrains: Implications for the thermal structure of Archean and Proterozoic lithosphere: Journal of Geophysical Research, v. 98, p. 12,207–12,218, doi:10.1029/93JB00521.

Nyblade, A.A., Pollack, H.N., Jones, D.L., Podmore, F., and Mushayendebvu, M., 1990, Terrestrial heat flow in east and southern Africa: Journal of Geophysical Research, v. 95, p. 17,371–17,384, doi:10.1029/JB095iB11p17371.

Oxburgh, E.R., and Turcotte, D.L., 1974, Thermal gradients and regional metamorphism in overthrust terrains with special reference to the Eastern Alps: Schweiz: Mineralogische und Petrographische Mitteilungen, v. 54, p. 641–662.

Parphenuk, O.I., Dechoux, V., and Mareschal, J.-C., 1994, Finite-element models of evolution for the Kapuskasing structural zone: Canadian Journal of Earth Sciences, v. 31, p. 1227–1234, doi:10.1139/e94-108.

Perchuk, L.L., and Gerya, T.V., 2011, this volume, Formation and evolution of Precambrian granulite terranes: A gravitational redistribution model, in van Reenen, D.D., Kramers, J.D., McCourt, S., and Perchuk, L.L., eds., Origin and Evolution of Precambrian High-Grade Gneiss Terranes, with Special Emphasis on the Limpopo Complex of Southern Africa: Geological Society of America Memoir 207, doi:10.1130/2011.1207(15).

Perchuk, L.L., van Reenen, D.D., Varlamov, D.A., Van Kal, S.M., Boshoff, R., and Tabatabaeimanesh, S.M., 2008, P-T record of two high-grade metamorphic events in the Central Zone of the Limpopo Complex, South Africa: Lithos, v. 103, p. 70–105, doi:10.1016/j.lithos.2007.09.011.

Pretorius, W., and Barton, J.M., Jr., 2002, Measured and calculated compressional wave velocities of crustal and upper mantle rocks in the Central Zone of the Limpopo Belt, South Africa—Implications for lithospheric structure: South African Journal of Geology, v. 105, p. 303–310.

Richter, F.M., 1985, Models for the Archean thermal regime: Earth and Planetary Science Letters, v. 73, p. 350–360, doi:10.1016/0012-821X(85)90083-4.

Roering, C., van Reenen, D.D., Smit, C.A., Barton, J.M., Jr., De Beer, J.H., De Wit, M.J., Stettler, E.H., Van Schalkwyk, J.F., Stevens, G., and Pretorius, S.J., 1992, Tectonic model for the evolution of the Limpopo Belt: Precambrian Research, v. 55, p. 539–552, doi:10.1016/0301-9268(92)90044-O.

Rybach, L., 1976, Radioactive heat production; a physical property determined by the chemistry of rocks, in Strens, R.G.J., ed., The Physics and Chemistry of Minerals and Rocks: New York, Wiley Interscience, p. 309–318.

Sandiford, M., 1989, Secular trends in the thermal evolution of metamorphic terrains: Earth and Planetary Science Letters, v. 95, p. 85–96, doi:10.1016/0012-821X(89)90169-6.

Sandiford, M., McLaren, S., and Neumann, N., 2002, Long-term thermal consequences of the redistribution of heat-producing elements associated with large-scale granitic complexes: Journal of Metamorphic Geology, v. 20, p. 87–98, doi:10.1046/j.0263-4929.2001.00359.x.

Sawyer, E.W., 2001, Melt segregation in the continental crust: Distribution and movement of melt in anatectic rocks: Journal of Metamorphic Geology, v. 19, p. 291–309, doi:10.1046/j.0263-4929.2000.00312.x.

Schaller, M., Steiner, O., Studer, I., Holzer, L., Herwegh, M., and Kramers, J.D., 1999, The Palala Shear Zone, Transvaal: Exhumation of the Limpopo Central Zone granulites and continent-scale transcurrent movement at 2.0 Ga: Precambrian Research, v. 96, p. 263–288, doi:10.1016/S0301-9268(99)00015-7.

Schreyer, W., Horrocks, P.C., and Abraham, K., 1984, High magnesium staurolite in a sapphirine garnet rock from the Limpopo Belt, southern Africa: Contributions to Mineralogy and Petrology, v. 86, p. 200–207, doi:10.1007/BF00381847.

Smit, C.A., van Reenen, D.D., Roering, C., Boshoff, R., and Perchuk, L.L., 2011, this volume, Neoarchean to Paleoproterozoic evolution of the polymetamorphic Central Zone of the Limpopo Complex, in van Reenen, D.D., Kramers, J.D., McCourt, S., and Perchuk, L.L., eds., Origin and Evolution of Precambrian High-Grade Gneiss Terranes, with Special Emphasis on the Limpopo Complex of Southern Africa: Geological Society of America Memoir 207, doi:10.1130/2011.1207(12).

van Reenen, D.D., Perchuk, L.L., Smit, C.A., Varlamov, D.A., Boshoff, R., Huizenga, J.M., and Gerya, T.V., 2004, Structural and P-T Evolution of a Major Cross Fold in the Central Zone of the Limpopo High-Grade Terrain, South Africa: Journal of Petrology, v. 45, p. 1413–1439, doi:10.1093/petrology/egh028.

van Reenen, D.D., Boshoff, R., Smit, C.A., Perchuk, L.L., Kramers, J.D., McCourt, S., and Armstrong, R.A., 2008, Geochronological problems related to polymetamorphism in the Limpopo Complex, South Africa: Gondwana Research, v. 14, p. 644–662, doi:10.1016/j.gr.2008.01.013.

Watkeys, M.K., Light, M.P.R., and Broderick, T.J., 1983, A retrospective view of the Central Zone of the Limpopo Belt, Zimbabwe, in Van Biljon, W.J., and Legg, J.H., eds., The Limpopo Belt: Geological Society of South Africa Special Publication 8, p. 65–80.

Whitney, D.L., Teyssier, C., and Vanderhaeghe, O., 2004, Gneiss domes and crustal flow, in Whitney, D.L., Teyssier, C., and Siddoway, C.S., eds., Gneiss Domes in Orogeny: Geological Society of America Special Paper 380, p. 15–33.

Wickham, S.M., 1987, The segregation and emplacement of granitic magmas: Journal of the Geological Society [London], v. 144, p. 281–297, doi:10.1144/gsjgs.144.2.0281.

Windley, B.F., Ackermann, D., and Herd, R.K., 1984, Sapphirine/kornerupine bearing rocks and crustal uplift history of the Limpopo belt, southern Africa: Contributions to Mineralogy and Petrology, v. 86, p. 342–358, doi:10.1007/BF01187139.

Wyllie, P.J., 1977, Crustal anatexis: An experimental review: Tectonophysics, v. 43, p. 41–71, doi:10.1016/0040-1951(77)90005-1.

Zeh, A., Klemd, R., Buhlmann, S., and Barton, J.M., Jr., 2004, Pro- and retrograde P-T evolution of granulites of the Beit Bridge Complex (Limpopo Belt, South Africa): Constraints from quantitative phase diagrams and geotectonic implications: Journal of Metamorphic Geology, v. 22, p. 79–95, doi:10.1111/j.1525-1314.2004.00501.x.

Zeh, A., Gerdes, A., Klemd, R., and Barton, J.M., Jr., 2007, Archaean to Proterozoic crustal evolution in the Central Zone of the Limpopo Belt (South Africa–Botswana): Constraints from combined U-Pb and Lu-Hf isotope analyses of zircon: Journal of Petrology, v. 48, p. 1605–1639, doi:10.1093/petrology/egm032.

Zeh, A., Gerdes, A., Barton, J.R., Jr., and Klemd, R., 2010, U–Th–Pb and Lu–Hf systematics of zircon from TTG's, leucosomes, meta-anorthosites and quartzites of the Limpopo Belt (South Africa): Constraints for the formation, recycling and metamorphism of Palaeoarchaean crust: Precambrian Research, v. 179, p. 50–68, doi:10.1016/j.precamres.2010.02.012.

Manuscript Accepted by the Society 24 May 2010

A review of Sm-Nd and Lu-Hf isotope studies in the Limpopo Complex and adjoining cratonic areas, and their bearing on models of crustal evolution and tectonism

Jan D. Kramers*
Department of Geology, University of Johannesburg, P.O. Box 524, Auckland Park, 2006, Johannesburg, South Africa, and School of Geosciences, University of the Witwatersrand, Private Bag 3, Witwatersrand 2050, South Africa

Armin Zeh
Lehrstuhl für Geodynamik und Geomaterialforschung, Am Hubland, D-97074 Würzburg, Germany

ABSTRACT

Published whole-rock Sm-Nd and zircon Lu-Hf data from the Limpopo Complex and adjoining areas of the Zimbabwe and Kaapvaal Cratons provide insight into the regional crustal evolution and tectonic processes that shaped the complex.

The Northern Marginal Zone of the complex, and the Francistown area of the Zimbabwe craton, represent an accretionary margin (active at 2.6–2.7 Ga) at the southern edge of that craton, at deep and shallow crustal levels, respectively. The Southern Marginal Zone represents a deep crustal level of the northern Kaapvaal Craton and was not an accretionary margin at the time of high-grade metamorphism (2.72–2.65 Ga). The syntectonic Matok granite was produced by crustal anatexis.

In the Central Zone, the presence of ca. 3.5–3.3 Ga crust is indicated throughout its E-W extent by $T_{Nd,DM}$ model ages of metapelites and by zircon xenocrysts and their $T_{Hf,DM}$ model ages. The ca. 2.65 Ga granitoids in the Central Zone (the Singelele-type quartzofeldspathic gneisses in the Musina area, granitoids in the Phikwe Complex, Botswana, the so-called gray gneisses, and the Bulai charnockite) were formed by anatexis of such old crust, whereas 2.6 Ga juvenile (arc-related?) magmatism produced the Bulai enderbite, and may be a component in the Zanzibar gneiss. The Mahalapye granitoid complex in Botswana was formed by crustal anatexis at 2.0 Ga, but mafic and hybrid rocks of this age have a mantle-derived component.

The data do not prohibit a collisional model for the Neoarchean high-grade metamorphic event in the Central Zone and Southern Marginal Zone of the Limpopo Complex.

---

*jkramers@uj.ac.za

Kramers, J.D., and Zeh, A., 2011, A review of Sm-Nd and Lu-Hf isotope studies in the Limpopo Complex and adjoining cratonic areas, and their bearing on models of crustal evolution and tectonism, *in* van Reenen, D.D., Kramers, J.D., McCourt, S., and Perchuk, L.L., eds., Origin and Evolution of Precambrian High-Grade Gneiss Terranes, with Special Emphasis on the Limpopo Complex of Southern Africa: Geological Society of America Memoir 207, p. 163–188, doi: 10.1130/2011.1207(10). For permission to copy, contact editing@geosociety.org. © 2011 The Geological Society of America. All rights reserved.

INTRODUCTION

In studies of orogenic provinces, questions on the crustal evolution are usually as important as those concerning the age of the tectonic and metamorphic events themselves. Traditionally crust formation ages are derived from samarium-neodymium (Sm-Nd) systematics, which allow in principle estimation at what time the (averaged) material in a crustal province was derived from the CHondritic Uniform Reservoir (CHUR) or from the depleted mantle (DM). The model ages are denoted $T_{Nd,CHUR}$ or $T_{Nd,DM}$ ages (McCulloch and Wasserburg, 1978; DePaolo, 1980, 1988). The method is based on the fractionation of the Sm/Nd ratio in mantle melting, whereby Nd is more incompatible than Sm. This leads to Sm/Nd ratios in the crust being lower than in the mantle (~60%), with correspondingly slower ingrowth of radiogenic ^{143}Nd in the crust. Further, as a result of the preferential removal of Nd over Sm from the mantle, this depleted mantle acquires a higher Sm/Nd ratio.

Lutetium-hafnium (Lu-Hf) systematics behave similarly to Sm-Nd in mantle melting, with Hf being more incompatible than Lu. The fractionation is stronger than for Sm/Nd, and the average continental crust has a Lu/Hf ratio only ~30% of that of the mantle source. In the past 10 years Lu-Hf isotope data have greatly enhanced the study of crustal growth and mantle evolution through time. The extremely strong partitioning of Hf in zircon, leading to Hf concentrations of ~1%, is a disadvantage in whole-rock Hf isotope studies of felsic rocks because of the resulting nugget effect. On the other hand it enables work on single zircons, dated by U-Pb (e.g., Amelin et al., 1999, 2000). Since laser-ablation multicollector ICP mass spectrometry made in situ measurements and thus combined U-Pb and Lu-Hf (and oxygen isotope studies) possible (e.g., Kemp et al., 2006; Valley et al., 2005; Gerdes and Zeh, 2006, 2009), the potential of this method for crust history studies has become even greater.

The Nd isotope approach can, in addition to intracrustal rocks, be applied to metapelites, as in erosion and clastic sedimentation the Sm/Nd fractionation is normally not highly significant (DePaolo, 1988). A problem with the Nd method is, however, that high-grade metamorphism, with partial melting and/or high-temperature fluid transport, can change Sm/Nd ratios of rock units, causing scatter or even systematic error of $T_{Nd,CHUR}$ or $T_{Nd,DM}$ ages (Moorbath et al., 1997). Hf isotope work on U-Pb dated zircons suffers less from this problem. The Hf isotope signature in zircons is robust even where U-Pb systematics have been disturbed in metamorphism or subrecent lead loss (e.g., Zeh et al., 2009, 2010). The method can be applied to detrital as well as to magmatic zircons and provides a superb window on the provenance of their source magmas. The high Hf in zircons entails an extremely low Lu/Hf ratio, so that in situ age corrections of the ^{176}Hf/^{177}Hf ratios are usually quite small, although for interpretation, the zircon age is nevertheless critical, as discussed further below. The ^{176}Hf/^{177}Hf ratios at the time of zircon crystallization then yield information on crust or mantle derivation of the parent magma (e.g., Davis et al., 2005; Flowerdew et al., 2007).

If this is derived from older continental crust, $T_{Hf,CHUR}$ or $T_{Hf,DM}$ model ages can be derived using an assumed Lu/Hf ratio for this crustal source.

For the Limpopo Complex, a number of collision-type models have been proposed (e.g., Coward et al., 1976; Robertson and DuToit, 1981; Roering et al., 1992; Treloar et al., 1992; Holzer et al., 1998; Smit et al., this volume; van Reenen et al., this volume). On the other hand, differences in geological features between the different zones of the complex have also inspired the proposal of terrane accretion models (Rollinson, 1993; Barton et al., 2006). Nd and Hf isotope studies hold the promise of testing such models. For instance, collisional models imply ocean closure, and if ophiolites are absent, accretionary margins (providing evidence for a leading continental edge) can be detected by Nd and/or Hf isotopic data, indicating the addition of juvenile crust at the time of the orogeny or slightly earlier. Further, in collisional origins, crustal thickening and vertical reorganization, with upper crustal units rich in U and Th placed deep within the crust, might promote intracrustal melting (Ashwal et al., 1992; Andreoli et al., this volume). Again, such isotopic studies can reveal this by showing crustal ages far in excess of the emplacement age. Conversely, terrane accretion models can be tested by gauging Nd and Hf model age differences between adjacent provinces, whereby large contrasts could point to juxtaposed terranes.

In the three zones of the Limpopo Complex and the adjoining cratonic areas, Nd isotope studies have been carried out since the 1980s (Harris et al., 1987; Taylor et al., 1991; Berger et al., 1995; Barton, 1996; Kröner et al., 1999; Kreissig et al., 2000; Zhai et al., 2006). Hf studies of zircon have been done much more recently in the Central Zone (Zeh et al., 2007, 2008, 2010; Millonig et al., 2010) as well as in the Southern Marginal Zone and the Kaapvaal and Zimbabwe Cratons (Zeh et al., 2009). In this contribution the results of these studies are reviewed. It is found that the Nd and Hf isotope data complement each other very well. In addition to providing a broad regional outline of crustal evolution, they indeed impose some constraints on tectonic models.

THE DATABASE, REFERENCE MODELS, AND GENERAL PROBLEMS

In Figure 1, the localities of samples studied for Nd and Hf isotopes are shown for the Limpopo Complex and adjacent regions of the Zimbabwe and Kaapvaal Cratons. In this chapter, a general introduction of the regional geology and the geological setting of samples taken is omitted, and the reader is referred to contributions to this volume by Blenkinsop (this volume), Smit et al. (this volume), and van Reenen et al. (this volume) in which the regional geology is described in detail. Figures 2–14 show the Sm-Nd and Lu-Hf systematics in ε_{143Nd}-versus-age and ε_{176Hf}-versus-age diagrams. All Sm-Nd results are from whole-rock analyses done by isotope dilution thermal ionization mass spectrometry (ID-TIMS). The Hf isotope data were all obtained by laser ablation, inductively coupled plasma mass spectrometry

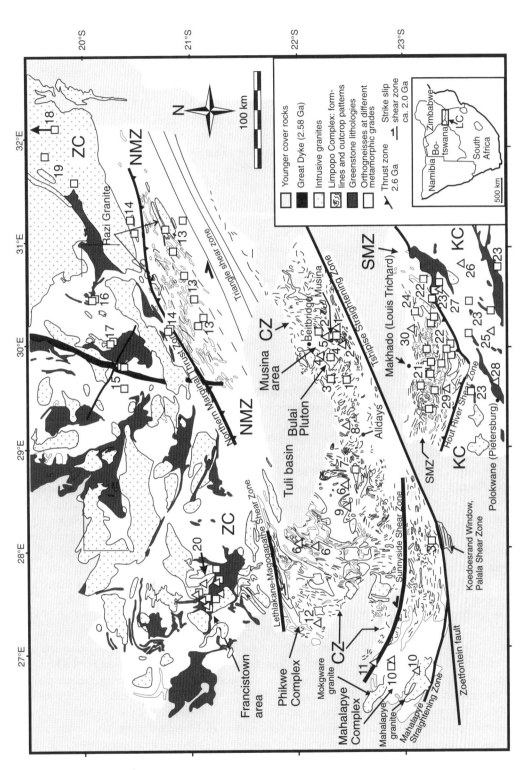

Figure 1. Localities in the Limpopo Belt and adjoining cratonic areas sampled for Sm-Nd and Lu-Hf studies. CZ—Central Zone; NMZ—Northern Marginal Zone; SMZ—Southern Marginal Zone; ZC—Zimbabwe Craton; KC—Kaapvaal Craton; squares—Sm-Nd whole-rock samples; triangles—Lu-Hf in zircon. (1) Sand River Gneiss (Harris et al., 1987; Kröner et al., 1999; Zeh et al., 2007, 2010). (2) Musina Layered Suite (Barton, 1996; Zeh et al., 2010). (3) Metapelites (Chavagnac et al., 2001; Harris et al., 1987). (4) Bulai Pluton (Harris et al., 1987; Zeh et al., 2007). (5) Singelele Gneiss (Zeh et al., 2007). (6) Granitoids, Phikwe Complex (Zeh et al., 2009). (7) Alldays Gneiss (Kröner et al., 1999). (8) Regina Gneiss (Zeh et al., 2007). (9) Zanzibar Gneiss (Zeh et al., 2007). (10) Lose quarry (Chavagnac et al., 2001; Zeh et al., 2007; Millonig et al., 2010 and Mahalapye granite (Millonig et al., 2010). (11) Mokgware granite and mafic rock (Zeh et al., 2007, 2009). (12) Granitoid in westernmost Phikwe Complex (Zeh et al., 2009). (13) Enderbites, Northern Marginal Zone (Berger et al., 1995). (14) Razi Granite, NMZ (Berger et al., 1995; Schmidt Mumm and Kreissig, 1999; Schmidt Mumm, 1999). (15) Chingezi gneiss. (16) Tokwe gneiss. (17) Shabani gneiss (Taylor et al., 1991). (18) Wedza gneiss. (19) Chilimanzi granite (Schmidt Mumm and Kreissig, 1999; Schmidt Mumm, 1999). (20) Granitoids in Francistown area, analyzed for Sm-Nd (Zhai et al., 2006) and zircon Lu-Hf (Zeh et al., 2009). (21) Gneisses and metapelites analyzed for Sm-Nd, granulite facies area of the SMZ. (22) Gneisses and metapelites analyzed for Sm-Nd, orthoamphibole zone of the SMZ. (23) Gneisses and metapelites analyzed for Sm-Nd, northern Kaapvaal Craton (Kreissig et al., 2000). (24) Schiel alkaline complex (Barton et al., 1996). (25–30) Samples analyzed for zircon Lu-Hf (Zeh et al., 2009): N Kaapvaal Craton: (25) Groot Letaba tonalite gneiss. (26) Groot Letaba granodiorite gneiss. (27) Meriri granite (Giyani Greenstone Belt). (28) Turfloop granite. Southern Marginal Zone: (29) Matok Pluton. (30) Entabeni Pluton.

(LA-ICP-MS) on zircons dated by the same method. Results are shown only for zircons with U-Pb systematics that are 5% or less discordant. ε values represent a normalization of isotope ratios to those of the CHondritic Uniform Reservoir (CHUR; Jacobsen and Wasserburg, 1980; Patchett et al., 2004; Blichert-Toft and Albarède, 1997; Bouvier et al., 2008). The results are defined as follows: $\varepsilon_{143Nd} = [(^{143}Nd/^{144}Nd_{sample(t)})/(^{143}Nd/^{144}Nd_{CHUR(t)}) - 1] \times 10,000$ and analogous for ε_{176Hf}. Compared with plotting actual $^{143}Nd/^{144}Nd$ and $^{176}Hf/^{177}Hf$ ratios, the ε notations have the advantage that small differences in the isotope ratios are better resolved, as the graph does not have to be expanded vertically to show the large increase of CHUR Nd and Hf isotope ratios over time. The parameters used for CHUR are (Bouvier et al., 2008) $^{147}Sm/^{144}Nd_{today} = 0.1960 \pm 0.0004$, $^{143}Nd/^{144}Nd_{today} = 0.512630 \pm 0.000011$, $^{176}Lu/^{177}Hf_{today} = 0.0336 \pm 0.0001$, $^{176}Hf/^{177}Hf_{today} = 0.282785 \pm 0.000011$. For Sm-Nd, the differences from earlier published values are insignificant. For Lu-Hf, corrections in the ε value of samples relative to those based on the next most recently published CHUR parameters (Blichert-Toft and Albarède, 1997) range from −0.46 for the present to +0.79 for 4.5 Ga.

Sm-Nd crustal history studies may be affected by fractionation in magmatic events and also by metamorphism, in cases where this involves anatexis or important fluid transport. This was demonstrated by Moorbath et al. (1997) for the Acasta gneisses in the Slave Craton, Canada, and the Amîtsoq Gneisses in West Greenland, where ε_{143Nd} values had been extrapolated to the age of the oldest zircons in the rocks to apparently reveal huge heterogeneities in ε_{143Nd} of the mantle in the earliest Archean (Bennett et al., 1993; Bowring and Houch, 1995). $^{143}Nd/^{144}Nd$ versus $^{147}Sm/^{144}Nd$ plots revealed near-isochrons of mid-Archean age, indicating partial homogenization and Sm/Nd differentiation in later events. In the ε_{143Nd} versus age diagram, such behavior is betrayed by development lines fanning out from a single point; an example from the Limpopo Complex is shown in Figure 7A and inset. Another process that can hamper the interpretation of Sm-Nd data for crustal history is mixing of components of different crustal age in a magmatic event; here the ε_{143Nd} versus age evolution lines may be subparallel (portraying similar Sm/Nd ratios), but they still yield differing apparent $T_{Nd,DM}$ ages. Several examples of this behavior are seen in the data set (e.g., Figs. 4A, 11, 13A, 14A).

Age corrections for the Sm-Nd data, calculated from present-day ε_{143Nd} values and $^{147}Sm/^{144}Nd$ ratios, are shown for the individual samples up to the approximate ages of emplacement or metamorphism. Possible Nd and Sm mobilization during these events prohibits the projection of these individual sample lines further back in time, and gray bands indicating the hypothetical earlier evolution for the suite as a whole are shown instead.

While the Lu-Hf systematics in zircons are hardly affected by metamorphic events causing isotopic homogenization or fractionation (except for cases in which zircons are resorbed and new ones formed; see Gerdes and Zeh, 2009), the interpretation of the data is as susceptible to source mixing as in the case of Sm-Nd. Further, interpretations in terms of T_{CHUR} or T_{DM} model ages are strongly dependent on the zircon U-Pb age used for correction. As the Lu/Hf ratios of zircons are generally about an order of magnitude lower than those of the average continental crust, the effect of zircon age on the initial $^{176}Hf/^{177}Hf$ ratio is non-critical. However, $\varepsilon_{176Hf(t)}$, $T_{Hf,CHUR}$, and $T_{Hf,DM}$ are affected, as the $^{176}Lu/^{177}Hf$ ratios of CHUR and DM are high, and the $^{176}Hf/^{177}Hf$ ratios of these reservoirs increase rapidly with time (or decrease with age). Although the Lu-Hf data in Figures 2–14 have been filtered to include those from zircons 5% or less discordant only, there is still an apparent age spread in some sample suites that could be at least in part due to Pb loss in a later metamorphic event. For instance, a 10% Pb loss at 2.0 Ga from a 2.65 Ga zircon causes only ~3% discordancy, whereas the apparent age shifts to 0.065 b.y. younger. This in turn would cause a shift of −1.5 in the $\varepsilon_{176Hf(t)}$ value, and increases of 0.03 b.y. in the apparent T_{CHUR} and T_{DM} ages.

The Lu-Hf data reviewed are all from zircons, and their age-corrected ε_{Hf} values are plotted against the U-Pb ages. Error limits are shown at the 2 standard error (95% confidence) level. Factored in are measurement uncertainties of $^{176}Hf/^{177}Hf$ and $^{176}Lu/^{177}Hf$ ratios, as well as uncertainties of ε_{176Hf} resulting from the $^{207}Pb/^{206}Pb$ age error as discussed above. Possible correction trends for zircons that appear younger than main populations and might be discordant are shown by gray arrows marked C. Hypothetical crustal evolution before zircon crystallization is mostly shown as gray bands for each population. Published estimates of average values for $^{176}Lu/^{177}Hf$ ratios of the total continental crust range from 0.010 (Wedepohl, 1995) to 0.015 for the bulk Archaean crust (Taylor and McLennan, 1995; Rudnick and Gao, 2003). A best estimate from sediments is 0.0117 (Vervoort et al., 1999) and from global modeling, 0.012 (Tolstikhin et al., 2006). The gray crustal evolution bands therefore widen with increasing age (before the zircon date), corresponding to the range 0.010–0.015. The range of apparent mantle derivation ages for a zircon population is then given by the intersection of the gray crustal evolution band with the mantle development line (see below). "Apparent" means that the $\varepsilon_{176Hf,t}$ values of zircons may result from mixing of mantle- and crust-derived melts, in which case the gray bands would not truly portray crustal evolution.

A further source of uncertainty for mantle derivation ages, for both Sm-Nd and Lu-Hf systematics, lies in the mantle model used. Melts leading to the formation of continental crust are considered to have been (albeit indirectly) derived from a homogeneous upper mantle, depleted by the previous extraction of crust. Depleted mantle ε_{143Nd} and ε_{176Hf} versus time curves therefore depend critically on the crustal mass as a function of time in the past and the history of crustal recycling into the mantle, for which a number of models exist. As the average age of the continental crust today is ca. 2 Ga (Jacobsen and Wasserburg, 1979), models in which the crustal mass has been more or less constant over geologic time (e.g., Armstrong, 1981, Bowring and Houch, 1995) imply a great deal of crust recycling into the mantle. This would produce convex-up ε_{143Nd} and ε_{176Hf} versus time curves for the resulting depleted mantle. Models in which the crustal mass has increased over time (e.g., Taylor and McLennan, 1995) produce strongly concave-up curves if the amount of crust recycling into the mantle is relatively small, and an increased recycling flux leads to a less concave-up shape. Relatively low total recycling models satisfy Pb and Nd isotope systematics of crust and mantle (Kramers and Tolstikhin, 1997; Nägler and Kramers, 1998): However, strong indications of an early-formed, relatively isolated and enriched reservoir now residing at the core-mantle boundary as the D″ layer (Tolstikhin and Hofmann, 2005; Boyet and Carlson, 2005, 2006) further complicate matters. A model in which the whole mantle is depleted, first by the early segregation of a large mafic crust that became this D″ reservoir, and subsequently by the growth of continental crust over time, provided an almost linear ε_{143Nd} versus time curve (Tolstikhin et al., 2006). Note that the different scenarios in all cases result in similar shaped curves of ε_{143Nd} and ε_{176Hf} versus time, as magmatic processes affect parent/daughter ratios in the same way. For this review, linear increases of ε_{143Nd} and ε_{176Hf} from zero at 4.5 Ga to 9.0 (Blichert-Toft and Albarède, 1994) and 16.4 (Chauvel and Blichert-Toft, 2001), respectively, are used. This linear shape (Galer et al., 1989) has been shown to be close to consistent with large data sets for much of Earth's history. However, particularly for the mid- and early Archean, it is uncertain how homogeneous the upper mantle was, and how models intermediate between CHUR and the linear DM may apply.

DISCUSSION OF INDIVIDUAL DATA SETS

Central Zone (CZ)

The data from this Zone are reviewed first, as they provide the best opportunity for direct comparisons between Sm-Nd and Lu-Hf data sets. The data include the Sand River gneisses, the Musina layered mafic suite (Sm-Nd and Lu-Hf; Figs. 2A–2D), metapelites from various localities (Sm-Nd only; Fig. 3), the Bulai Pluton (Sm-Nd and Lu-Hf; Fig. 4), the Singelele-type quartzofeldspathic gneisses (Lu-Hf only; Fig. 5A), granite and gneisses from the Phikwe Complex, Botswana (Lu-Hf only; Figs. 5B, 9), the "gray" Regina and Zanzibar Gneisses (Lu-Hf; Fig. 6), and intrusive rocks in the Mahalapye Complex (Sm-Nd and Lu-Hf, Figs. 7, 8) at the western exposed end of the CZ.

Sand River Gneisses

Five Sm-Nd results from this unit, four in the Musina area and one some 50 km SW of Musina in the bed of the Sand River (Fig. 1) yield a range for possible $T_{Nd,CHUR}$ model ages between 2.8 and 3.25 Ga, or $T_{Nd,DM}$ between 3.05 and 3.50 Ga (solid lines in Fig. 2A). Zircon ID-TIMS dates for two of the samples are 3.314 ± 0.005 and 3.297 ± 0.006 Ga (Kröner et al., 1999). The Lu-Hf data (Fig. 2B) are from five rock samples (Zeh et al., 2007, 2010), and the zircon U-Pb dates form a main cluster between 3.24 and 3.3 Ga. A second cluster is seen at 3.2 Ga. Overgrowths in the age range from 2.5 to 2.7 Ga and at 2.02 Ga were found. $\varepsilon_{176Hf(t)}$ values for the main group of zircons are between 1 and −3, showing heterogeneity outside the error limits. The ranges of $T_{Hf,CHUR}$ and $T_{Hf,DM}$ model ages are 3.25–3.55 and 3.45–3.75 Ga, respectively. Assuming that a CHUR model would imply the Sand River gneisses could represent juvenile ca. 3.3 Ga crust, the DM model could suggest that they originated by reworking of crust up to at least 3.75 Ga. ε_{176Hf} vs age plots for zircons of two leucosomes in the Sand River gneisses (Zeh et al., 2010) are shown in Figure 2C. The tonalitic ca. 2.0 Ga melt patch appears to have been formed by remelting of mid-Archean crust. The granitic leucosome intruded between 3.27 and 3.07 Ga and contains inherited (xenocryst) zircon cores formed at 3.40 Ga and altered during intrusion. These cores yielded $T_{Hf,DM}$ model ages up to 3.9 Ga.

Zeh et al. (2008, 2010) found many detrital zircon grains in the age range 3.5–3.6 Ga in quartzites from the Beit Bridge Complex, and two at ca. 3.9 Ga. Their $\varepsilon_{176Hf(t)}$ values are variable but almost all below CHUR, down to −7. Therefore, they could have been derived from Hadean crust similar to that represented by the detrital zircons from the Jack Hills quartzite in the Australian Yilgarn Craton (Amelin et al., 1999, 2000; Kemp et al., 2010). It is suggested (Zeh et al., 2008, 2010) that such Hadean crust could have been reworked in the formation of the Sand River gneisses.

Musina Layered Suite

This mafic igneous suite is intruded into the oldest Central Zone metasediments. Two samples analyzed by B. Eglington (Barton, 1996) yield (with the caveat of metamorphic effects as above) apparent $T_{Nd,CHUR}$ model ages of 2.75 and 3.2 Ga, and $T_{Nd,DM}$ ages of 3.39 and 3.5 Ga (Fig. 2A). ε_{176Hf} versus age plots for zircons from anorthosites (Zeh et al., 2010) are shown in Figure 2D. The zircon dates ca. 3.2 Ga may be due to Pb loss, and the corresponding $\varepsilon_{176Hf,t}$ values could be corrected upward as shown by the gray arrow C and discussed above. The oldest zircon population of ca. 3.3 Ga (in accord with 3.344 ± 0.004 Ga, Mouri et al., 2009), with U contents <70 ppm, typical for zircons in mafic rocks, has ε_{176Hf} from −2 to +2 outside analytical error. This suggests variable crustal contamination of the mantle-derived mafic magma. Overgrowths formed at ca. 2.6 Ga also have ε_{176Hf} values close to CHUR, which may be a chance result of the whole-rock anorthosite Lu/Hf ratio having been close to chondritic; 2.0 Ga zircon overgrowths (also found by Mouri et al., 2009) have negative ε_{176Hf} values, which probably result from partial dissolution of old zircon grains followed by new (over)growth.

Metapelites of the Central Zone

Figure 3 shows four metapelite Sm-Nd results from the Musina area (Harris et al., 1987) and data from an Sm-Nd study on a single migmatitic metapelite outcrop just north of the Palala Shear Zone in the Koedoesrand Window (Chavagnac et al., 2001) (Fig. 1). The Musina area samples yielded apparent $T_{Nd,CHUR}$ and $T_{Nd,DM}$ ages in the ranges 2.65–2.8 and 2.86–3.0 Ga, respectively. Of the Koedoesrand Window set, the four whole-rock samples gave $T_{Nd,CHUR}$ and $T_{Nd,DM}$ model ages in the ranges 2.8–3.2 and 3.14–3.36 Ga, respectively. Subsamples of leucosomes and paleosomes were taken to study the effect of local anatexis on the Sm-Nd systematics. There is a poorly defined convergence of the ε_{143Nd} versus time lines for most of the subsamples in the 2.0–2.5 Ga age range, suggesting differentiation in the 2.0 Ga or 2.65 Ga metamorphism or in both events. Further, the lines for two leucosome samples are markedly displaced upward, suggesting that these leucosomes may not be entirely in situ. This set of results further illustrates the points made by Moorbath et al. (1997) on disturbance of Sm-Nd systematics in high-grade metamorphism.

Nevertheless, tentative conclusions can be drawn from the whole-rock data. The Palala metapelite Sm-Nd data suggest the previous presence of continental crust of Sand River Gneiss age as far west as 28° E, ~200 km WSW of Musina. However, the samples are significantly different from the Musina area metapelites studied by Harris et al. (1987), not only in their older $T_{Nd,CHUR}$ and $T_{Nd,DM}$ model ages but also in their $^{147}Sm/^{144}Nd$ ratios (0.13–0.14 for Palala, compared with values of ~0.10 for Musina). The higher $^{147}Sm/^{144}Nd$ ratios for the Palala metapelites suggest a mafic component in their source region, which could have been volcanics related to the Musina Layered Suite. The younger model age range for the Musina metapelites suggests a mixed-age source region and a deposition age much younger than the Sand River gneisses and Musina layered suite. This

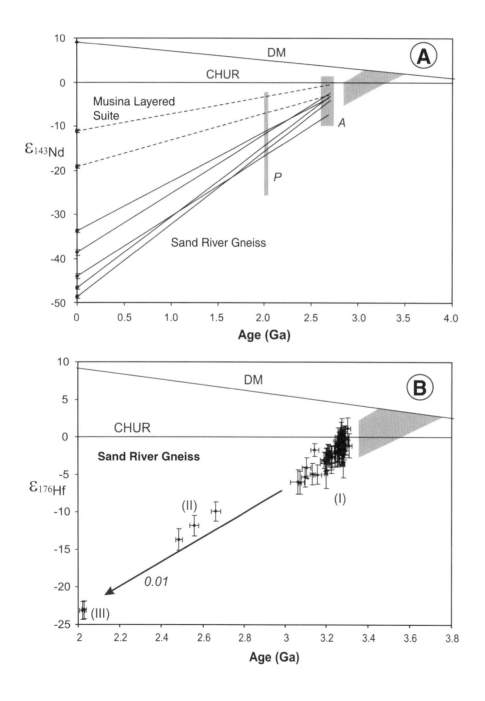

Figure 2 (*Continued on following page*). (A) ε_{143Nd} versus age for whole-rock samples from Sand River Gneiss (solid lines, Harris et al., 1987; Kröner et al., 1999) and anorthosites of the Musina Layered Suite (dashed lines, Barton, 1996). Gray bars *P* and *A* are Paleoproterozoic and Neoarchean periods of metamorphism, respectively (see Kramers and Mouri, this volume, for database). Sample lines are not projected beyond the latter period. Gray trapezoid indicates likely earlier evolution path. (B) ε_{176Hf} versus age for zircons from five tonalitic Sand River Gneiss samples (Zeh et al., 2007, 2010). Error limits as discussed in the text. Only zircons with U-Pb data 5% or less discordant were considered. (I) Magmatic zircon cores, giving emplacement age. (II and III) Overgrowths. Gray trapezoid beyond zircon (I) ages shows possible crust evolution path for $0.01 < {}^{176}Lu/{}^{177}Hf < 0.015$ from depleted mantle. Note relative homogeneity of ε_{176Hf} values of (I), with error limits overlapping. Arrow 0.01 shows ε_{176Hf} evolution in crust with $^{176}Lu/^{177}Hf$ of that value.

observation raises an apparent age paradox. The Musina Suite has an intrusive age of 3.34 Ga (Mouri et al., 2009; Zeh et al., 2010), and it intruded into metasediments, as demonstrated by xenoliths (Barton et al., 1979). This same paradox is found in the study of Zeh et al. (2008) of a quartzite from the Musina area, which showed the main detrital zircon population to be between 3.3 and 3.1 Ga, thus showing that the sedimentary series to which this quartzite belonged could not have been intruded by the Musina Layered Suite. The subdivision of sediments of the Beit Bridge Complex in three groups of distinct ages (Brandl, 1983, 2001, 2002) may provide a way out of this problem, although owing to the intense tectonism and high-grade metamorphism, it is mostly impossible to interpret the field relationships in terms of a time sequence.

Bulai Pluton

This intrusion just W of Musina exists as enderbitic and porphyritic granitic-charnockitic varieties. Both were intruded at 2.61 ± 0.05 Ga (Barton et al., 1994; Zeh et al., 2007), and the complex has been characterized as calc-alkaline (Millonig et al., 2008). Four separate whole-rock enderbite samples (Fig. 4A) yielded $T_{Nd,CHUR}$ and $T_{Nd,DM}$ model ages in the ranges 2.22–2.54

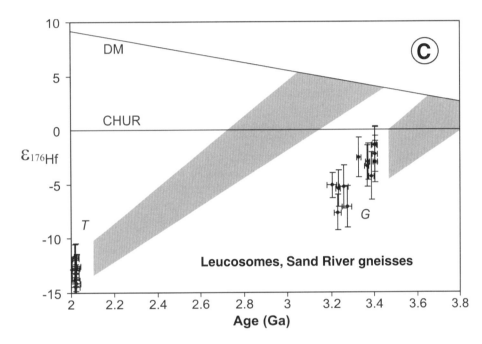

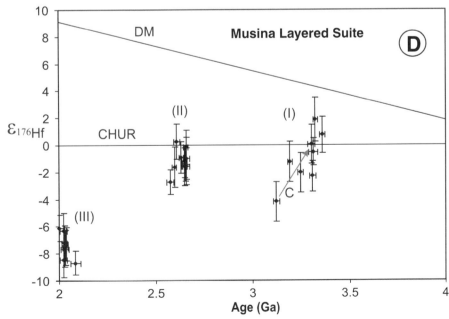

Figure 2 (*Continued*). (C) ε_{176Hf} versus age for zircons from granitic *(G)* and tonalitic *(T)* leucosomes in Sand River gneisses (Zeh et al., 2010). Gray trapezoids as in Figure 2A. (D) ε_{176Hf} versus age for zircons from two anorthosite samples of the Musina Layered Suite (Zeh et al., 2010). Gray arrow (C) shows the direction in which ε_{176Hf} values change as a result of correcting U-Pb age that may have been measured too young. (I) First generation (magmatic) zircons. (II and III) Overgrowths. DM—depleted mantle; CHUR—CHondritic Uniform Reservoir.

and 2.50–2.80 Ga, respectively (Harris et al., 1987). Unlike the case of the Sand River gneisses, there is no convergence of the ε_{Nd} versus time lines at any intermediate age. Deformation of the Bulai Pluton at 2.0 Ga is limited to shear zones (Holzer et al., 1998; Huizenga et al., this volume), and a disturbance of the systematics in the 2.0 Ga event appears unlikely to have affected the data. An intrusion age cannot be older than a mantle derivation model age, and this allows the conclusion that the CHUR model is inappropriate for mantle magma sources in the CZ at ca. 2.6 Ga. Also, a strongly concave-up depleted mantle model would yield model ages in part significantly younger than the emplacement age. The linear DM model fits much better with the zircon date, portraying a magmatic suite with a juvenile end member and variable contamination by older crust indicated by the apparent model age range extending to 2.8 Ga.

A zircon population from a charnockitic sample (Fig. 4B) yielded $\varepsilon_{176Hf(t)}$ values between –2.6 and –4.4 (Zeh et al., 2007), significant beyond analytical errors, corresponding to a range of apparent $T_{Hf,DM}$ model ages from 3.08 to 3.28 Ga. These ranges are not affected by some zircons that apparently lost some Pb at ca. 2.0 Ga. In contrast to the conclusion from the Sm-Nd data on the enderbitic Bulai body, the zircon Lu-Hf results for the charnockite thus indicate that a mid-Archean crustal source was dominant. A minor juvenile crust component is, however, not ruled out.

It can be concluded that the Bulai Pluton is composite. The enderbitic part appears to represent predominantly new crust with some older crustal contamination, whereas the charnockite body is mainly a product of crustal anatexis. The Bulai data illustrate the advantages of a combined approach in which whole-rock Sm-Nd and zircon Lu-Hf data complement each other.

Singelele Gneiss

This widespread meta-igneous rock suite consists mainly of garnet-bearing quartzofeldspathic gneisses that generally occur in up to several 100-m-thick sheets, interleaved with metasediments of the Beit Bridge Complex. The intrusion ages vary between localities in the range 2.58–2.72 Ga (Jaeckel et al., 1997; Kröner et al., 1999; see summary in Kramers and Mouri, this volume). No Sm-Nd data exist for the suite. The Lu-Hf systematics (Zeh et al., 2007) for zircons from a single sample taken from the type locality near Musina are shown in Figure 5A. The two lowest apparent $\varepsilon_{176Hf(t)}$ values (–6.2 and –7.1; see Fig. 5A) are associated with zircon U-Pb ages of ca. 2.55 Ga, which are probably due to Pb loss at ca. 2 Ga from the 2.65 Ga population, causing minimal discordancy. The gray arrows (C) in Figure 5A show the correction of corresponding $\varepsilon_{176Hf(t)}$ values.

Disregarding these data, the scatter in $\varepsilon_{176Hf(t)}$ for the remaining concordant grains, all close to 2.65 Ga in age, is still rather large, between –2.2 and –4.9, and significant outside analytical error. Thus the magma was heterogeneous in ε_{176Hf} during emplacement and crystallization. The corresponding $T_{Hf,DM}$ model age range 3.14–3.25 Ga, younger than the Sand River gneisses but within the range for T_{Nd} model ages of metapelites discussed above. The model age range is thus in accord with the view that the protoliths of the Singelele gneisses are the products of anatexis of metasediments of the Beit Bridge Complex, formed during the high-grade metamorphic event at ca. 2.65 Ga. In this scenario, the observed $\varepsilon_{176Hf(t)}$ heterogeneity could have been inherited from the sediment source. Although a juvenile crust component is not excluded by these data, it is not required, and the component derived by anatexis of older crust is clearly dominant.

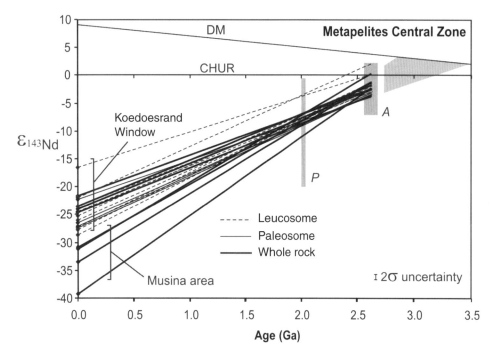

Figure 3. ε_{143Nd} versus age for whole-rock metapelite samples from the Musina area (Harris et al., 1987) and whole-rock as well as paleosome and leucosome portions from the Koedoesrand Window just N of the Palala Shear Zone (Chavagnac et al., 2001). Both areas are within the CZ (Fig. 1). P and A as in Figure 2A. Note convergence of sample lines between P and A. They are not projected beyond A, and the gray trapezoid, as in Figure 2A, gives tentative range of $T_{Nd,DM}$ model ages between 2.85 and 3.5 Ga.

Phikwe Complex Granitoids

An interesting comparison with the Singelele Gneiss data is provided by zircons from three samples from the Phikwe Complex in the western part of the CZ, studied by Zeh et al. (2009). These are granite gneisses, in part migmatitic, and interleaved on a kilometer scale with quartzofeldspathic gneisses of the Singelele type. They are considered anatectic on the basis of their field relationships (e.g., Key, 1976) and are similar in age range to the Singelele-type gneisses (McCourt and Armstrong, 1998; Zeh et al., 2009; see summary in Kramers and Mouri, this volume). There are no Sm-Nd data for rocks from this complex. The Lu-Hf data (Zeh et al., 2009) are shown in Figure 5B. Zircons mainly have U-Pb ages between 2.58 and 2.66 Ga. One 2.22 Ga zircon in the population has a low Th/U ratio (0.17), verging on the values for metamorphic zircons (Hoskin and Schaltegger, 2003). This, therefore, does not contradict a conclusion that these granitoids were emplaced at ca. 2.65 Ga. The 2.58 and 2.66 Ga zircons mostly have $\varepsilon_{176Hf,t}$ values in the wide range between +0.3 and −6, which, remarkably, is shown by zircons from the single Phikwe Town sample. It corresponds to apparent $T_{Hf,DM}$ model ages in the range 2.99–3.4 Ga, with one grain (perhaps a xenocryst zircon core with reset U-Pb age) having $\varepsilon_{176Hf,t} = -9$ and $T_{Hf,DM}$ up to 3.6 Ga. An unambiguous zircon xenocryst with an age of 2.87 Ga yielded a slightly younger $T_{Hf,DM}$ model age of 3.33–3.44 Ga. The

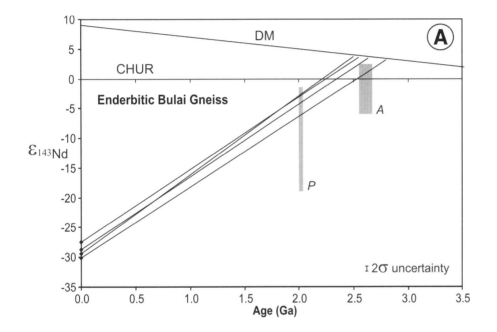

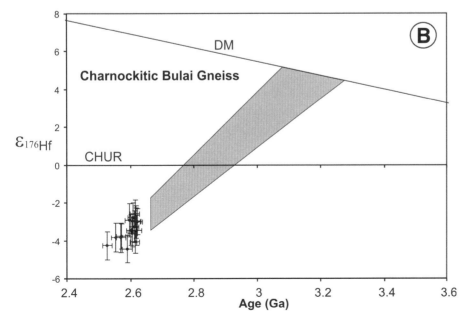

Figure 4. (A) ε_{143Nd} versus age for whole-rock samples of enderbitic Bulai Gneiss (Harris et al., 1987). (B) ε_{176Hf} versus age for zircons (5% or less discordant) from a single sample of charnockitic Bulai Gneiss (Zeh et al., 2007), both W Musina. In A, gray bars *P* and *A* as in Figure 2A. DM—depleted mantle; CHUR—CHondritic Uniform Reservoir.

Lu-Hf data sets of each of the three individual samples are even more heterogeneous than those for the Singelele Gneiss type locality discussed above.

This set of results appears to portray, in all three cases, poorly mixed magmas from a heterogeneous source, in which zircons preserved an image of this heterogeneity as they crystallized. This situation is common in S-type granites, as shown by Hf (and O) isotope studies of zircons from granites of several orogens (e.g., Kemp et al., 2007; Bolhar et al., 2008). The data confirm that a mid-Archean (3.3–3.4 Ga) continental crust province extended, pre-2.65 Ga, to the westernmost part of the CZ. The range of $\varepsilon_{176Hf,t}$ values extending up to 0.3 and $T_{Hf,DM}$ ages down to 2.96 Ga, makes a small juvenile crust component (with a mantle signature at the time of emplacement) more likely than in the case of the Singelele type locality discussed above. However, older crustal sources appear dominant.

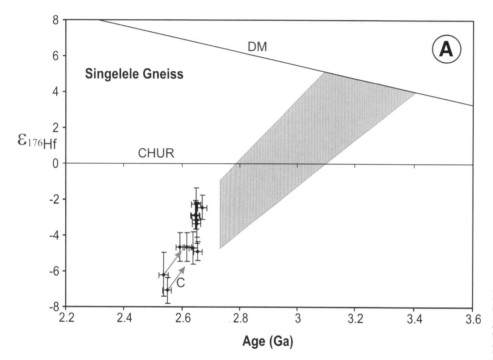

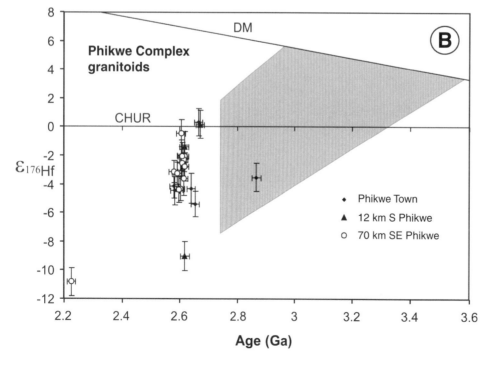

Figure 5. ε_{176Hf} versus age for zircons (5% or less discordant) from (A) a single sample of Singelele Gneiss at the type locality near Musina (Zeh et al., 2007) and (B) three granitoid samples from the Phikwe Complex (Zeh et al., 2009). Gray arrows (C) in panel A show corrections of ε_{176Hf} values as in Figure 2D. Note within-sample heterogeneity in ε_{176Hf} values significantly outside uncertainty limits.

Gray Gneisses

In the CZ, this term is used mainly for tonalitic to granodioritic gneisses that appear to form distinct intrusive bodies within the general mix of metasediments, mafic rocks of the Musina suite, and anatectic (Singelele-type) quartzofeldspathic gneisses that make up the regional geologic framework. Among these, the Alldays, Regina, and Zanzibar Gneisses have been studied with relevance to crustal history (Fig. 1). All of these have zircon dates between 2.6 and 2.65 Ga without older xenocrysts or younger metamorphic rims (Jaeckel et al., 1997; Kröner et al., 1999; Zeh et al., 2007). For the Alldays Gneiss, a single Sm-Nd datum is available (Kröner et al., 1999). This yields a $T_{Nd,DM}$ model age of 2.98 Ga, interpreted by these authors as evidence for an important older crust component in the magma. For the Regina and Zanzibar Gneisses, zircon Lu-Hf data (Zeh et al., 2007) are shown in Figure 6.

For the Regina Gneiss, four of the six near-concordant grains yielded 2.646–2.653 Ga zircon dates, identical within error, giving the intrusion age, whereas two younger grains with ca. 2.60 Ga dates may have lost radiogenic Pb, as discussed above and by Zeh et al. (2007). The Lu-Hf systematics are similar to those of the Bulai charnockite, with $\varepsilon_{176Hf,t}$ values in the narrow range between −3.8 and −5.4, and $T_{Hf,DM}$ model ages from 3.15 to 3.37 Ga. The Regina Gneiss protolith was thus formed by anatexis of older crust. Further, in this case almost perfect homogenization was achieved in anatexis.

Of the Zanzibar Gneiss, Zeh et al. (2007) studied a granodioritic and a granitic variety. Both show a spread in apparent zircon dates. The oldest, 2.631 and 2.625 Ga, respectively (and identical within error), most likely give the intrusion age of both varieties, and younger ones (down to 2.54 Ga) indicate minor radiogenic lead loss, probably at 2.0 Ga, without becoming resolvably discordant. Even taking this younging effect into account, the heterogeneity in $\varepsilon_{176Hf,t}$ values and $T_{Hf,DM}$ model ages is greater than for the Regina Gneiss, with higher $\varepsilon_{176Hf,t}$ values (+0.3 to −3) and younger apparent $T_{Hf,DM}$ model ages (2.94–3.18 Ga). This is a younger average than for the Regina Gneiss, and a component of new crust formed at ca. 2.6 Ga could be present (suggested in Fig. 6).

Mahalapye Complex

In the Mahalapye-Mokgware area in the western extremity of the CZ, extensive magmatism occurred during the 2.0 Ga event (McCourt and Armstrong, 1998; Zeh et al., 2007, 2009; Millonig et al., 2010). Further, syntectonic metamorphism up to high grade at this time has been well documented (Hisada and Miyano, 1996; Chavagnac et al., 2001; Millonig et al., 2010). Some reconstruction of the crustal history is possible through Sm-Nd and Lu-Hf studies, in part on the same outcrop.

Sm-Nd systematics (Chavagnac et al., 2001) of whole-rock portions of a migmatite sample from a quarry at Lose, on the Mahalapye-Palapye main road, are shown in Figure 7A. The migmatite consists of a garnet-biotite gneiss host rock, described as a metagraywacke (probably erroneously, see below) by Chavagnac et al. (2001) and a garnet-bearing leucosome occurring as veins on diverse scales. The ε_{143Nd} versus time lines for most samples of both the garnet-biotite gneiss and the leucosome converge clearly at ca. 2.0 Ga, demonstrating extensive Nd isotope homogenization and some Sm/Nd fractionation at this time. For the garnet-biotite gneiss, this resulted in five of the six samples defining a near-isochron yielding 2.02 ± 0.17 Ga (mean square of weighted deviates [MSWD] = 27; Fig. 7A inset). Further, apparent

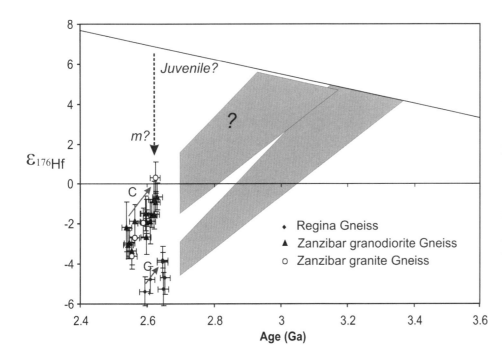

Figure 6. ε_{176Hf} versus age for zircons (5% or less discordant) from the Regina Gneiss near Venetia Mine, and two samples representing different varieties of the Zanzibar Gneiss (Fig. 1). Correction arrow (C) as in Figure 2D. For the Zanzibar Gneiss, the question mark in the gray trapezoid yielding apparent $T_{Hf,DM}$ model ages indicates uncertainty about this being real. Alternatively, addition of juvenile mantle-derived magma and mixing (dashed arrow, m?) is suggested.

$T_{Nd,DM}$ model ages are spread out as a result (similar to the situation discussed by Moorbath et al., 1997), but they mainly cluster between 2.5 and 2.7 Ga. The leucosome data bear no apparent relation to the garnet-biotite gneiss data. Their apparent $T_{Nd,DM}$ model ages range from 2.80 to 3.07 Ga.

The zircon Lu-Hf data for the same lithologies from the same outcrop (Zeh et al., 2007) are shown in Figure 7B. For the garnet-biotite gneiss, the Th/U ratios of the zircons and their internal zoning indicate magmatic crystallization (Zeh et al., 2007), characterizing the protolith as a magmatic rock. The zircon data cluster tightly, with U-Pb dates at 2.046–2.064 Ga and $\varepsilon_{176Hf,t}$ values at −0.03 to −1.23, showing strong Hf isotope homogenization. The apparent $T_{Hf,DM}$ model ages lie in the narrow range 2.48–2.66 Ga (not shown in Fig. 7B). As in the case of Sm-Nd, the leucosome Lu-Hf data bear no relation to those of the garnet-biotite gneiss. The zircon U-Pb dates range from 2.47 to 2.71 Ga, the $\varepsilon_{176Hf,t}$ values from +4 to −4.5 and the apparent $T_{Hf,DM}$ model ages from 2.8 to 3.26 Ga.

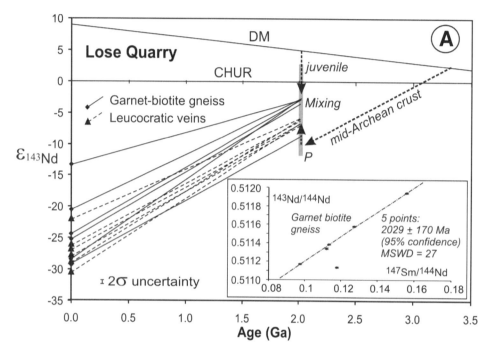

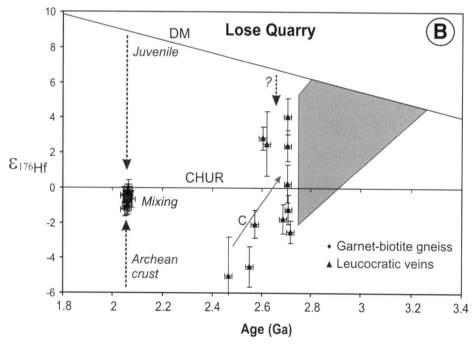

Figure 7. Data for garnet biotite gneiss and a leucocratic granitic rock occurring in veins in the quarry at Lose, Botswana, in the Mahalapye Complex. (A) ε_{143Nd} versus age (Chavagnac et al., 2001); inset, $^{143}Nd/^{144}Nd$ versus $^{147}Sm/^{144}Nd$ diagram for garnet biotite gneiss. (B) ε_{176Hf} versus age for zircons 5% or less discordant (Zeh et al., 2007). Gray bar P in panel A as in Figure 2A. Dashed arrows discussed in the text. In B, correction arrow (C) as in Figure 2D. MSWD—mean square of weighted deviates.

The Sm-Nd and Lu-Hf results are thus in very good agreement. Both show a near-complete rehomogenization of daughter isotope ratios in the 2.0 Ga event in the garnet-biotite gneiss. For the Lu-Hf data, this implies resorption and new crystallization of zircons in a magmatic event. Further, the leucosome appears not to have been produced by local anatexis but was intrusive (Zeh et al., 2007). At first sight it appears paradoxical that the zircons in the intruding leucosome are older than those in the intruded garnet-biotite gneiss. The zircons in the leucosome can, however, be xenocrysts, derived from its heterogeneous source. We note that Zeh et al. (2007) found zircon overgrowths on some of the grains, which they assumed were 2 Ga, but which were not dated.

Lu-Hf results from samples of Mokgware granite (Zeh et al., 2007) and a mafic unit occurring as a raft within it (Zeh et al., 2009), both taken 20 km N of Lose, are shown in Figure 8A. They present a similar situation to that of Lose, except that in this case the emplacement age of the granite (2.03 Ga) is clearly shown by three out of the four near-concordant zircons, with

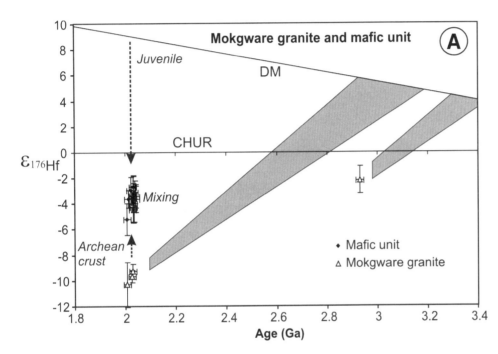

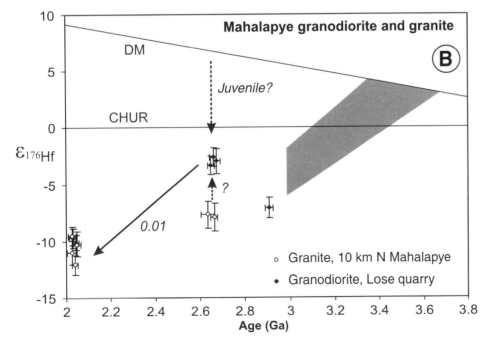

Figure 8. ε_{176Hf} versus age for zircons 5% or less discordant for granitoids from the Mahalapye Complex. (A) Mokgware granite (Zeh et al., 2007) and enclosed mafic rock (Zeh et al., 2009), both taken 20 km N of Lose in the Mahalapye Complex (Fig. 1). (B) Granite at Lose quarry, and granodiorite 10 km N of Mahalapye (Millonig et al., 2010).

one xenocryst at 2.93 Ga also in evidence. The mafic rock has identical zircon ages to the granite, with $\varepsilon_{176Hf,t}$ values between and −2.7 and −5.2, giving apparent $T_{Hf,Nd}$ model ages in the range 2.64–2.76 Ga (not shown). The granite is clearly the product of remelting of much older crust. Its 2.03 Ga zircons and the 2.93 Ga xenocryst give $T_{Hf,DM}$ model ages in the ranges 2.93–3.2 and 3.33–3.44 Ga, respectively. A zircon xenocryst in the latter age range was also found in the Mahalapye Complex by McCourt and Armstrong (1998).

Zircon Lu-Hf results from two further granitoids in the Mahalapye Complex (Millonig et al., 2010) are shown in Figure 8B. The granite and granodiorite were both apparently emplaced at ca. 2.02 Ga. Both units, at 40 km distance from each other, have the same narrow range of $\varepsilon_{176Hf,t}$ values between −9.5 and −12.5. Both have ca. 2.65 Ga zircon xenocrysts; a 2.92 Ga xenocryst was further found in the granodiorite. The three oldest and lowest $\varepsilon_{176Hf,t}$ xenocrysts yielded a possible apparent $T_{Hf,DM}$ model age range of 3.34–3.68 Ga.

Thus mid- to early Archean crust is clearly in evidence in the Mahalapye Complex. The strong heterogeneity in $\varepsilon_{176Hf,t}$ values (−10 to +4) shown by Neoarchean zircon xenocrysts may indicate that new crust (with a mantle Hf isotopic signature) was produced at that time and that magmas were variably contaminated by the older crust, as seen in many accretionary margins (e.g., Kemp et al., 2006, 2007).

The homogeneity of the zircon populations within the garnet-biotite gneiss at Lose Quarry and the mafic rock of the Mokgware set, both in age and in $\varepsilon_{176Hf,t}$ values, suggest that these are 2.05 and 2.03 Ga magmatic rocks (Zeh et al., 2007, 2009). The composition of the Lose garnet-biotite gneiss (e.g., 14%–19% Al_2O_3, 1.5%–2.7% MgO, 1.5%–5.5% CaO; Chavagnac et al., 2001) shows that the magma would have contained an important metapelite-derived component, also evident from the abundant garnet.

The $\varepsilon_{176Hf,t}$ values of the ~2.02 magmatic zircons in the Mahalapye Complex are vastly different from one unit to the next. Those shown in Figure 8B could have been produced from anatexis of the crust portrayed by some of the Neoarchean zircons, as shown by the evolution arrow for a crust with $^{176}Lu/^{177}Hf$ = 0.01. "Pure" crustal anatexis probably also produced the Mokgware granite and the leucosome at Lose Quarry at ca. 2.02 Ga. However, for the $\varepsilon_{176Hf,t}$ values of the mafic unit at Mokgware and the garnet-biotite gneiss at Lose, and the Nd isotope systematics of the latter, the most likely explanation is that they resulted from the intrusion, at ca. 2.02 Ga, of mantle-derived magmas that were, to regionally varying degrees, contaminated by old crust. This hypothesis is sketched in Figures 7A, 7B, and 8A. An alternative explanation would be that they originated by selective melting of mafic units in older crust, and were emplaced without being contaminated by felsic crust. This is highly unlikely from a petrological perspective.

Westernmost Phikwe Complex

In a granite gneiss sample taken in the westernmost exposed area of the Phikwe complex (Zeh et al., 2009; Figs. 1, 9) four concordant zircon overgrowths define U-Pb ages of

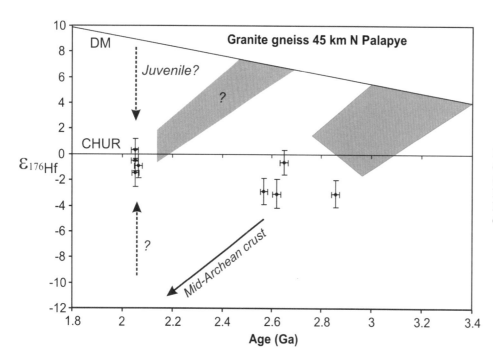

Figure 9. ε_{176Hf} versus age for zircons 5% or less discordant for a granitoid gneiss (Zeh et al., 2009) in the westernmost Phikwe Complex, 45 km N of Palapye, Botswana (Fig. 1). Sketched hypothetical interpretations discussed in the text.

2.05–2.06 Ga. They have $\varepsilon_{176Hf,t}$ values of ~0 and apparent $T_{Hf,DM}$ ages between 2.5 and 2.7 Ga. Of four zircon cores, three have ages of ca. 2.6 Ga, and one of 2.86 Ga; their apparent $T_{Hf,DM}$ ages are between 3.0 and 3.4 Ga, documenting contamination by mid-Archean crust. The zircon overgrowths have low U contents (70–100 ppm) and high Th/U ratios (0.5–0.7). These features preclude a metamorphic origin of the overgrowths (Hoskin and Schaltegger, 2003) and granitoid magmatism at 2.05–2.06 Ga, intruding 2.6–3.4 Ga crust, is therefore indicated. Derivation of this magma by remelting of crust formed at ca. 2.6 Ga (gray trapezoid with question mark) would have excluded contamination by older crust and therefore was not possible. In another hypothesis, a mantle-derived magma could have been contaminated with old crust (shown by dashed arrows in Fig. 9). The resulting rock would then be mafic, unless the setting was an accretionary margin, where a granitoid magma with a mantle isotope signature could exist.

Summary of CZ Data

The assembled Sm-Nd and zircon Lu-Hf data sets from diverse rock units in the CZ show that crustal material between 3.0 and 3.5 Ga exists throughout the zone. Widespread magmatism during the ca. 2.65 Ga orogeny was chiefly the result of intracrustal melting, with clear evidence of newly formed crust found only in the enderbitic portion of the Bulai Pluton, and in a few zircon cores from the Lose quarry of the Mahalapye Complex. During the 2.0 Ga event, mantle-derived magma intruded in the Mahalapye Complex and was contaminated by older crust, whereas the ca. 2.0 Ga magmas of the same unit were formed by crustal anatexis. It can be speculated that the extensive plutonism in the Mahalapye Complex, and the 2.0 Ga metamorphism at least in the western portion of the CZ, were the result of heat advection by mantle-derived magma, perhaps related to the Bushveld Complex intrusion and/or to the Magondi orogeny.

Northern Marginal Zone (NMZ) and Southern Zimbabwe Craton

From the Northern Marginal Zone, no zircon Lu-Hf data are available; Sm-Nd data (Berger et al., 1995) with zircon U-Pb age ranges are shown in Figure 10. Apart from a comparison with the CZ, one with regions of the Zimbabwe Craton is also useful. Sm-Nd data exist for most of the width of this craton adjacent to the Northern Marginal Zone (Taylor et al., 1991; Schmidt Mumm and Kreissig, 1999; Schmidt Mumm, 1999; Zhai et al., 2006) and are shown in Figures 11 and 12A. Figure 12B shows zircon Lu-Hf data for the westernmost part of the Zimbabwe Craton (Zeh et al., 2009).

Enderbites and Razi Granites of the NMZ

Enderbitic magmatic granulites are the dominant lithology of the Northern Marginal Zone. Their magmatic character is generally clearly shown by xenoliths on the meter to kilometer scale that are enclosed in them. Although quasi-continuous throughout, they can be subdivided into a number of plutonic domains separated by metasedimentary and metavolcanic septa (Berger et al., 1995; Blenkinsop, this volume). In addition, porphyritic charnockites and granites occur, known as the Razi suite. These are mainly concentrated along the North Limpopo Thrust Zone that marks the boundary between the Northern Marginal Zone and the Zimbabwe Craton, and generally coincides with the northern boundary of the occurrence of metamorphic orthopyroxene

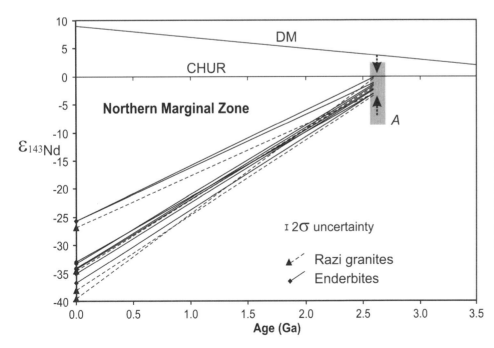

Figure 10. ε_{143Nd} versus age for whole-rock samples of enderbites (Berger et al., 1995) and Razi granites and charnockites (Berger et al., 1995; Schmidt-Mumm and Kreissig, 1999; Schmidt Mumm, 1999) of the Northern Marginal Zone. Gray vertical band *A* represents the range of intrusion ages, corresponding to the ca. 2.65 Ga event. Dashed vertical arrows depict hypothesis of mixing during this event.

(Mkweli et al., 1995; Blenkinsop and Frei, 1996; Frei et al., 1999). The Razi granites are seen to intrude enderbites sheared along the North Limpopo Thrust Zone, but are also locally sheared by it. Their intrusion is thus syntectonic.

The enderbite samples analyzed for Sm-Nd were in part dated by zircon multigrain ID-TIMS (Berger et al., 1995) and a single Rb-Sr whole-rock isochron (Hickman, 1978). The zircon populations are mostly highly discordant owing to high U contents. Strongly discordant results mimic a bias toward younger apparent ages, possibly as a result of multiple discordance. The most concordant and precise dates are between 2.6 and 2.7 Ga (Berger et al., 1995; see Fig. 10) and are regarded as magmatic emplacement ages. The $\varepsilon_{143Nd,t}$ values for the enderbites, sampled from a wide area (Fig. 1), range from 0 to −3, and would correspond to apparent $T_{Nd,DM}$ model ages 2.9–3.1 Ga if projected beyond the intrusion ages. This pattern can be interpreted in two ways: (1) It portrays a crustal province with mantle derivation ages on average between 2.9 and 3.1 Ga throughout, in which the enderbites of the NMZ formed by crustal anatexis in the ca. 2.65 Ga event; (2) it results from mixing, at ca. 2.65 Ga, of magma with a

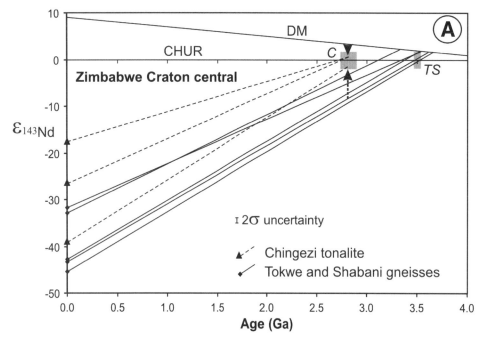

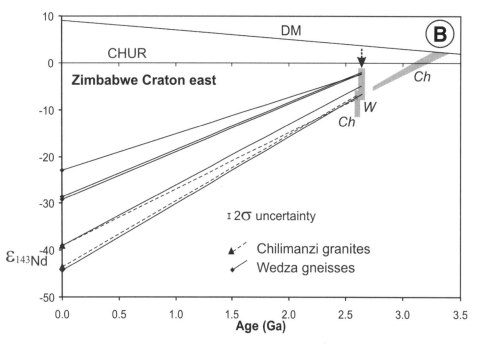

Figure 11. ε_{143Nd} versus age for whole-rock samples from the Zimbabwe Craton E of 29° E. (A) Tokwe and Shabani gneisses and Chingezi tonalite (Fig. 1). Gray rectangles: C, range of Pb/Pb whole-rock dates for Chingezi tonalite (Taylor et al., 1991); TS, range of U-Pb zircon dates for Tokwe and Shabani gneisses (Horstwood et al., 1999). (B) Eastern Zimbabwe Craton (Schmidt Mumm and Kreissig, 1999; Schmidt Mumm, 1999). Gray rectangles Ch and W show emplacement ages for the Chilimanzi granites and Wedza gneisses, respectively. Gray trapezoid Ch extends beyond emplacement age for Chilimanzi granites only; in both A and B, dashed arrows indicate mixing hypothesis, as discussed in the text.

depleted mantle Hf isotope composition with crustal matter older than ca. 3.1 Ga. The ubiquitous enderbitic magmatism indicates high lower crustal temperatures with large amounts of melt being produced, thus facilitating mixing. Since the NMZ enderbites are uniformly felsic rocks, this mixing hypothesis implies an accretionary margin setting, where felsic magmas with mantle isotopic compositions can exist.

In either case, it is impossible to assess how much Sm/Nd fractionation occurred during anatexis and emplacement, as homogenization, producing a convergence of ε_{143Nd} versus time

(t) lines, is prohibited by the large sample spacing. An Rb-Sr whole-rock isochron date of 2.88 ± 0.05 Ga for an enderbite body at Bangala Dam (Hickman, 1978) is significantly older than the range of zircon dates. The result is meaningless as an age, but it may reflect incomplete Sr isotope homogenization on the scale of the sampled area (~10 km) at the time of emplacement.

Like the enderbites, the zircon dates on the Razi charnockites and granites suffer from considerable discordance, and also in this case this appears to produce a bias toward younger apparent upper intercept ages (Frei et al., 1999). The more concordant

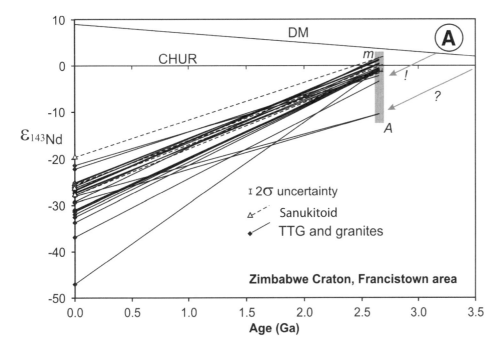

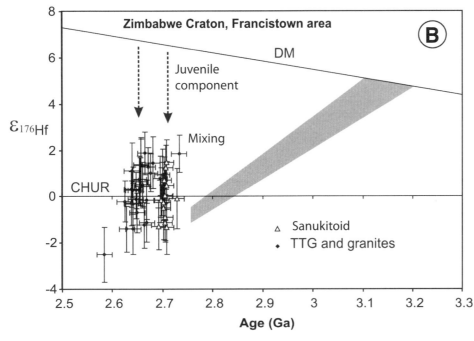

Figure 12. Data from the Tati granite-greenstone terrane in the Francistown area, western Zimbabwe Craton (Fig. 1). (A) ε_{143Nd} versus age for whole-rock samples (Zhai et al., 2006). Gray rectangle A gives the range of intrusion ages (Bagai et al., 2002; Zeh et al., 2009). Gray arrows with question and exclamation marks show possible back projections for the sample with lowest ε_{143Nd} value at 2.65 Ga (a K-rich granite measured in duplicate), and the lowest value of the main population, respectively; m denotes probable mixing of mantle-derived melts with older crust. (B) ε_{176Hf} versus age for zircons 5% or less discordant (Zeh et al., 2009). Gray trapezoid shows possible back projection from the zircon with lowest $\varepsilon_{176Hf,t}$ value to give minimum $T_{Hf,DM}$ model age for the pre-existing crust if $\varepsilon_{176Hf,t}$ spread is the result of mixing of mantle-derived magma and crust (dashed arrows). TTG—tonalite-trondhjemite-granodiorite.

and precise dates converge to 2.6 Ga. The apparent projected $T_{Nd,DM}$ model ages would scatter between 2.8 and 3.2, slightly more than those of the enderbites. The ε_{143Nd} versus t lines for the granites-charnockites do not converge on the emplacement age. Instead, $\varepsilon_{143Nd,t}$ values range from 0 to −5. The two possible interpretations are the same as those for the enderbites, whereby lower melt volumes for the Razi Suite would result in less Nd isotope homogenization in either case, so that a greater apparent range of $T_{Nd,DM}$ model ages would be predicted than for the enderbites.

Berger and Rollinson (1997) could show, mainly on the basis of consistently high $^{207}Pb/^{204}Pb$ ratios of the enderbites and Razi granites and charnockites, that crustal material significantly older than 3 Ga makes up an important component of the NMZ throughout. Thus they provided an important argument for the mixing hypothesis, implying an accretionary margin setting at ca. 2.65 Ga. This hypothesis is thus preferred. This margin would have been active during, or just preceding, the major ca. 2.65 Ga tectono-metamorphic event in the Limpopo Central Zone.

In accretionary margins studied to date in this way, a component with an almost pure mantle Hf isotope composition is observed along with the mixed signature. The mantle isotope signature is mainly seen in the arc-proximal regions but can also appear elsewhere during periods of extension (Davis et al., 2005; Kemp et al., 2006, 2009). The absence of a pure juvenile component in the NMZ at ca. 2.65 Ga could mean that not much extension occurred. Further, the NMZ is characterized by unusually high concentrations of the heat-producing elements U and Th, even at lower crustal levels (as documented by the enderbites), which could have promoted widespread lower crustal melting (Kramers et al., 2001) and thus extensive mixing.

Zimbabwe Craton East of 29° E

Between 30° and 30°45′ E, the NMZ borders on the oldest province of the Zimbabwe Craton, known as the Tokwe Segment. Zircon dates close to 3.5 Ga (Horstwood et al., 1999) for the Tokwe and Shabani tonalite-trondhjemite-granodiorite (TTG) gneisses in this region (Fig. 1) are in accord with $T_{Nd,DM}$ ages in the range 3.4–3.6 Ga (Fig. 11A; Taylor et al., 1991). Horstwood et al. (1999) showed that this ancient province extends northward to ~18°40′ S. This agreement between T_{DM} and emplacement ages records an early crust formation episode not involving the partial reworking of yet earlier crust. W of the Belingwe Greenstone Belt, between 29°20′ and 29°50′ E, tonalitic gneisses of the Chingezi suite (Fig. 1) have Pb/Pb whole-rock dates between 2.75 and 2.87 Ga, contrasting with apparent $T_{Nd,DM}$ model ages in the range 3.05–3.13 (Taylor et al., 1991; Fig. 11A; model age range not shown). This indicates crustal reworking, with a possible contribution of a juvenile crust component as shown by the dashed arrow in Figure 11A. A similar situation is seen east of 31°30′ E, where gneisses of the Wedza TTG suite have zircon dates ca. 2.65 Ga (Jelsma et al., 1996; Wilson et al., 1995). Samples of this suite taken on the southern edge of the Odzi Greenstone Belt have apparent $T_{Nd,DM}$ model ages between 3.09 and 3.27 Ga (Schmidt Mumm and Kreissig, 1999; Schmidt Mumm, 1999; Fig. 11B, model ages not shown). In this case the range of $\varepsilon_{143Nd,t}$ values for the time of emplacement is quite large, and the dashed arrow in Figure 11B illustrates the hypothesis of variable admixture of a juvenile component. Two samples of the ca. 2.6 Ga Chilimanzi granite suite from the eastern craton have older $T_{Nd,DM}$ model ages of 3.26 and 3.4 Ga. Although Sm/Nd fractionation could have occurred during their formation, these may point to "pure" anatexis of crust in the 3.3–3.4 Ga range.

Francistown Area

At the western extremity of the Zimbabwe Craton, the Francistown granitoid-greenstone terrane comprises the Tati, Vumba, and Matsitama Greenstone Belts (including a subvolcanic sanukitoid member) and a number of granitoid plutons. This terrane has been interpreted as an ancient accretionary margin (Kampunzu et al., 2003; Zhai et al. 2006; Zeh et al., 2009). Zircon U-Pb dates for granitoid plutons range from 2.71 to 2.645 Ga (Bagai et al., 2002; Zeh et al., 2009). Sm-Nd (Zhai et al., 2006) and zircon Lu-Hf (Zeh et al., 2009) data are shown in Figures 12A and 12B and broadly agree. The main range of apparent $T_{Nd,DM}$ model ages of a range of lithologies, including the sanukitoid, high and low Al TTGs, and high-K granites, would be 2.82–3.13 Ga with an average of 2.97 Ga, if projected back to the DM line. The $\varepsilon_{176Hf,t}$ values of zircons from the same range of rock types are heterogeneous, ranging from 1.8 to −1.4, well outside analytical error, and would correspond to apparent $T_{Hf,DM}$ model ages between 2.85 and 3.20 Ga. The $\varepsilon_{176Hf,t}$ values of the zircons from the sanukitoid, with ages ca. 2.70 Ga, are in the same range as the values for the younger granitoids with U-Pb ages ca. 2.65 Ga. The $\varepsilon_{143Nd,t}$ values at the time of intrusion are also in the same range for all rock types, with the exception of one high-K granite sample (analyzed in duplicate) with an extremely low $\varepsilon_{143Nd,t}$ value at 2.65 Ga, which would yield a $T_{Nd,DM}$ model age older than 3.5 Ga (gray arrow with question mark in Fig. 12A).

None of the data contradict the hypothesis that this province represents an accretionary margin. Assuming variable contamination of juvenile magma with old crust (Zhai et al., 2006; Zeh et al., 2009), this would account for the large spread of $\varepsilon_{176Hf,t}$ and $\varepsilon_{143Nd,t}$ values. Although the least radiogenic $\varepsilon_{143Nd,t}$ value and its attendant extremely old $T_{Nd,DM}$ model age is difficult to explain, a minimum age of ca. 3.1–3.2 Ga for the preexisting crust (see gray trapezoid in Figure 12B and upper gray arrow [!] in Fig. 12A) appears to be a safe assumption. It is interesting that this margin, dated by the zircon U-Pb results, was (like that proposed for the NMZ) active at the time of the ca. 2.65 Ga orogeny in the Limpopo Complex. Blenkinsop (this volume) proposes that the Francistown province is the western extension of the NMZ, at a higher crustal level. Thus the Francistown Arc and the NMZ could represent the same southern accretionary margin to the Zimbabwe Craton.

In summary, the Tokwe Segment in the Zimbabwe Craton is the only crustal province in the combined NMZ and southern Zimbabwe Craton where zircon U-Pb dates are identical to mantle derivation model ages (within the inherent uncertainty of the latter). In all other areas, including the NMZ, there are differences of several hundred millions of years between the two. The accretionary margin hypothesis is most well founded in the Francistown area, but it is not contradicted by data in either the eastern Zimbabwe Craton or the NMZ.

Southern Marginal Zone and Northern Kaapvaal Craton

As part of a comparative geochemical study of the northern Kaapvaal Craton and the Southern Marginal Zone (SMZ) of the Limpopo Complex, Kreissig et al. (2000) carried out Sm-Nd isotope work on gneisses and metapelites (Figs. 13A, 14A). Further Sm-Nd work was done by Barton et al. (1996) on the ca. 2 Ga Schiel Alkaline Complex within the SMZ (Fig. 13B). Zeh et al. (2009), in the framework of a large study aimed at defining the regional crustal history and constraining terrane accretion scenarios, analyzed a large number of zircons from plutonic rocks from both the SMZ and the Kaapvaal Craton for Lu-Hf. The results from the SMZ and from the cratonic area adjacent to it are reviewed here (Figs. 1, 13C, 14B).

Southern Marginal Zone

The enderbitic gneisses of the SMZ are of general tonalite-trondhjemite-granodiorite (TTG) composition. Together with metapelites, they underwent syntectonic high-grade metamorphism at 2.72–2.65 Ga (see Kramers and Mouri, this volume, for a summary). Zircons from gneisses in the granulite and the ortho-amphibole subzones in the eastern SMZ have yielded 2.816 ± 0.016 and 3.28 ± 0.02 Ga, respectively (Kröner et al., 2000; Fig. 13A). Thus, in contrast to the NMZ, the emplacement of gneiss protoliths significantly predated tectono-metamorphism. Gneisses from the granulite subzone of the SMZ, as well as all metapelites analyzed, have $\varepsilon_{143Nd,t}$ values between −1 and −4 at ca. 2.65 Ga and apparent $T_{Nd,DM}$ model ages in the range 3.02–3.23 Ga. Four out of five gneiss samples from the retrogressed orthoamphibolite subzone yielded apparent $T_{Nd,DM}$ model ages between 3.15 and 3.62 Ga, with one exception (measured in duplicate) as high as 3.82 Ga (Fig. 13A; Kreissig et al., 2000). This sample presents no anomaly in the Pb isotope array (Kreissig et al., 2000). It has a relatively high Sm/Nd ratio; its $\varepsilon_{143Nd,t}$ value at ca. 2.65 Ga is not significantly below that of the other four samples, and metamorphic Sm/Nd fractionation cannot be excluded as a cause for this apparent model age anomaly. Disregarding this sample, the data still clearly indicate the presence of mid- to early Archean crust in this subzone. As in the NMZ, the spread of $\varepsilon_{143Nd,t}$ values for the (poorly defined) range of possible times of protolith emplacement (2.8–3.3 Ga, see above, and $\varepsilon_{143Nd,t}$ between −8 and DM) can be interpreted as indicating an accretionary margin in that time range, with mixing of juvenile and older crust.

The Lu-Hf systematics of zircons from the 2.69 Ga syntectonic and synmetamorphic Matok granite give apparent $T_{Hf,DM}$ model ages in the range 3.13–3.4 Ga, and those from the 2.02 Ga post-tectonic Entabeni Granite yield 3.02–3.4 Ga (Fig. 13C). Within the inherent uncertainties, these ranges are identical to the apparent $T_{Nd,DM}$ ranges for the SMZ given above. This shows that these plutons could have been wholly derived by crustal anatexis. In contrast, the Sm-Nd data from syenites and granites of the ca. 2 Ga Schiel Complex, which is described as an anorogenic mantle-derived intrusion (Barton et al., 1996; Fig. 13B), yielded apparent model ages from 2.70 to 2.96 Ga, somewhat younger than the above range for the geological framework of the SMZ. Also, the convergence of $\varepsilon_{143Nd,t}$ versus time lines at ca. 2 Ga in Figure 13B is evidence for Nd isotope homogenization and Sm/Nd fractionation at this time. Dashed arrows in Figure 13B illustrate a possible scenario of contamination of differentiated mantle-derived melts by older crust. The coeval and adjacent Entabeni Granite, although crustally derived, may have been related to the accompanying thermal pulse. This situation appears somewhat reminiscent of the data from the Mahalapye Complex (Figs. 7, 8), except that in the SMZ the ca. 2.0 Ga intrusions are not coeval with a tectono-metamorphic event.

Northern Kaapvaal Craton

Gneisses sampled in the northern Kaapvaal Craton for Sm-Nd work by Kreissig et al. (2000) have zircon U-Pb ages between 2.75 and 2.95 Ga (Kröner et al., 2000; Zeh et al., 2009). These, and metapelites from intercalated greenstone belts (Fig. 14A), have apparent $T_{Nd,DM}$ model age values between 3.04 and 3.24 Ga. One exception is a metapelite from the Giyani Greenstone Belt (analyzed in duplicate) that has $T_{Nd,DM}$ equal to 3.5 Ga. The Giyani Belt contains a Mesoarchean supracrustal succession: A meta-andesite has yielded a zircon U-Pb date of 3.20 ± 0.01 Ga, and zircon dates of 3.17 ± 0.02 and 2.95 ± 0.06 Ga were obtained from gneisses just south of the Giyani Greenstone Belt (Kröner et al., 2000). The latter are equal to and slightly younger than the gneiss $T_{Nd,DM}$ model ages. The zircon Lu-Hf systematics (Zeh et al. 2009) from four gneiss and plutonic rock samples, ranging in zircon U-Pb age from 2.68 to 2.95 Ga, give $\varepsilon_{176Hf,t}$ values between −2 and +2 (Fig. 14B) and would define apparent $T_{Hf,DM}$ model ages between 3.0 and 3.27 Ga (not shown in Fig. 14B), essentially an identical range to that given above for Nd model ages. Three xenocrysts with ca. 3.2 Ga U-Pb dates give $T_{Hf,DM}$ model ages between 3.33 and 3.53 Ga. The two zircons with the oldest model ages come from the Meriri granite adjacent to the Giyani Greenstone Belt, whereas the third is from the Groot Letaba tonalitic gneiss sample taken 80 km farther SW (Fig. 1).

As in the Zimbabwe Craton outside the Tokwe segment, the granitoid or gneiss protolith emplacement ages in the northern Kaapvaal Craton are all at least 0.2 m.y. younger than the apparent T_{DM} model ages of the same units. For the Kaapvaal Craton this is made more striking by the large spread of $\varepsilon_{176Hf,t}$ values in the zircon populations of single samples. As in the case of the Zimbabwe Craton, and for the same reason, we favor the hypothesis that much of this province was formed at the time given by the zircon U-Pb emplacement ages, between 2.75 and 2.95 Ga

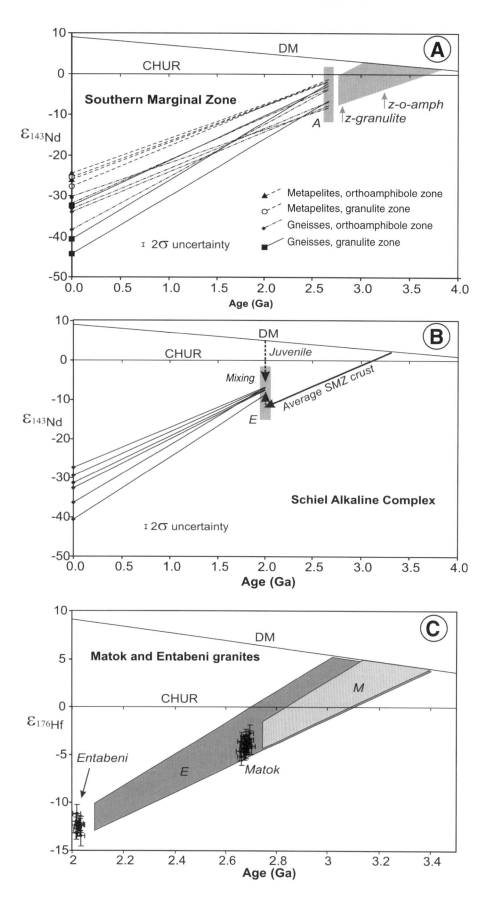

Figure 13. Sm-Nd and Lu-Hf data from the Southern Marginal Zone. (A) ε_{143Nd} versus age for gneisses and metapelites (Kreissig et al., 2000). Gray rectangle A gives the age range of the Neoarchean high-grade metamorphism; gray trapezoid gives range of possible crustal evolution lines prior to this time. Gray arrows *z-granulite* and *z-o-amph* are individual zircon U-Pb dates of SMZ gneisses in the granulite and orthoamphibolite zone, respectively (Kröner et al., 2000). (B) ε_{143Nd} versus age for samples from the ca. 2 Ga Schiel alkaline complex (Barton et al., 1996). Gray rectangle E shows the emplacement age. Solid arrow shows average crustal development in the region, panel A, and dashed arrows indicate contamination of (juvenile) mantle-derived magma by this component. (C) ε_{176Hf} versus age for zircons 5% or less discordant from the 2.69 Ga Matok and the 2.02 Ga Entabeni Plutons in the SMZ (Zeh et al., 2009). Gray trapezoids M (Matok) and E (Entabeni) show possible crustal evolution paths prior to emplacement, limited by $^{176}Lu/^{177}Hf$ ratios of 0.010 and 0.015.

in an accretionary margin with extensive assimilation of older crust. This scenario is sketched by vertical "mixing" *(m)* arrows in Figs. 14A, 14B. The accretion scenario is in accord with the conclusions of work encompassing other or larger parts of the Kaapvaal Craton (de Wit et al., 1992; Schöne et al., 2009; Zeh et al., 2009). A mid- to early Archean continental nucleus representing this older crust probably still exists in the surroundings of the Giyani Greenstone Belt and the adjacent region of the SMZ.

This brief summary presents no contradiction between the results obtained from Lu-Hf in zircons from plutonic complexes and Sm-Nd systematics of the gneissic framework and metapelites. Further, there are no significant differences between values from the SMZ and the northern Kaapvaal Craton. Kreissig et al. (2000) noted this observation from the Nd isotope systematics as well as Pb isotopes and trace element geochemistry. With the work of Zeh et al. (2009) these results are now extended

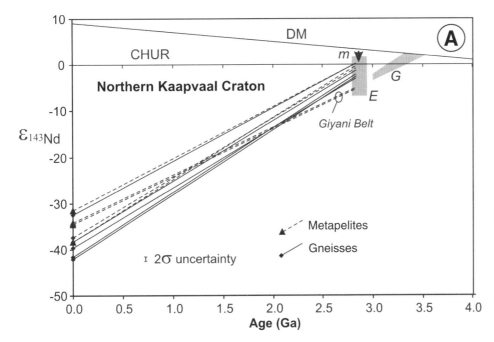

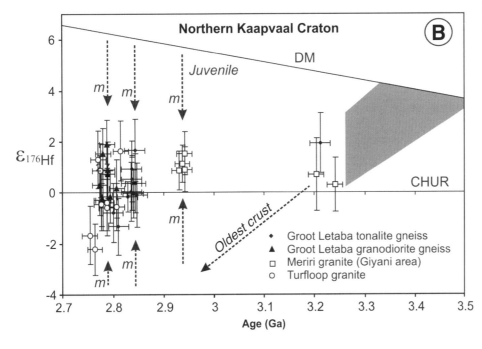

Figure 14. Sm-Nd and Lu-Hf data from the northern Kaapvaal Craton. (A) ε_{143Nd} versus age for gneisses and metapelites (Kreissig et al., 2000). Gray rectangle *E* gives the emplacement age range of gneisses sampled (Kröner et al., 2000; Zeh et al., 2009), which is also that of metamorphism in the greenstone belts. Gray trapezoid *G* gives possible pre-metamorphic crustal evolution path for precursor rocks of the Giyani Greenstone Belt metapelite (analyzed in duplicate), which has lowest ε_{143Nd} at 2.8 Ga. Arrow *m* illustrates mixing hypothesis. (B) ε_{176Hf} versus age for zircons 5% or less discordant from granite and gneiss samples (Zeh et al., 2009). Gray trapezoid indicates possible crustal evolution path for xenocryst zircons. Dashed arrows illustrate further evolution and mixing hypothesis *(m)*.

to zircon Lu-Hf systematics. Thus the data do not contradict the notion that the SMZ represents a high-grade metamorphic equivalent of the Kaapvaal Craton (e.g., Roering et al., 1992; van Reenen et al., this volume). Along with the northern Kaapvaal Craton, the SMZ may have been part of an accretionary margin at ca. 2.8 Ga, but not (in contrast to the case of the NMZ) during the ca. 2.65 Ga tectono-metamorphic event.

SUMMARY AND NOTES ON TECTONIC SCENARIOS

It is worthwhile to consider to what extent the combined Lu-Hf and Sm-Nd data set reviewed here is compatible with aspects of scenarios that have been proposed to explain the observed geology of the Limpopo Complex. Three main questions are central for testing models: (1) How heterogeneous are the provinces internally? (2) How strongly do they differ from each other in their crustal history? (3) How do the relative amounts of anatexis of older, and generation of new, crust compare in the different periods of tectonism and metamorphism?

The southern Zimbabwe Craton displays the classical pattern characteristic of many Archean cratons. It contains an early Archean nucleus ca. 3.5 Ga (the Tokwe Segment), bordered on the west and east by areas of TTG gneisses with younger apparent $T_{Nd,DM}$ model ages (3.0–3.25 Ga). The respective protolith emplacement ages are still younger (2.7–2.8 Ga), with no evidence of any rocks as old as the apparent $T_{Nd,DM}$ model ages. The western extremity of the craton has been described as 2.65–2.7 Ga accretionary margin (Francistown Arc), and the Nd and Hf isotope data are in accord with this. The pattern of emplacement ages versus $T_{Nd,DM}$ model ages in all parts of the southern Zimbabwe Craton, except for the Tokwe Segment, best fits a mixing model in an accretionary margin scenario, where arc magmas with depleted-mantle Nd and Hf isotope signatures are contaminated by older crust at the time given by the protolith emplacement ages. Thus the notion that the southern Zimbabwe Craton consists of an ancient nucleus surrounded by accretionary margins, successively younger toward the edges of the present craton, is not in conflict with the data. The ca. 2.6 Ga Chilimanzi suite of granites may then represent the product of later intracrustal melting.

In the NMZ of the Limpopo Complex, the relation between emplacement ages of the dominant TTG units and their apparent $T_{Nd,DM}$ model ages is similar to that in the eastern and westernmost Zimbabwe Craton, so the paradigm of an accretionary margin (e.g., an Andean-type arc) would also fit this zone, with juvenile magma having assimilated older crust at 2.6–2.7 Ga, resulting in the 2.95–3.09 Ga range of apparent $T_{Nd,DM}$ model ages. The Razi charnockite suite appears equivalent in age and origin to the Chilimanzi suite. The proposal that the NMZ represents a section of the Zimbabwe Craton at a deeper level (Blenkinsop, this volume; Robertson and DuToit, 1981) is not in conflict with these observations, and there is no compelling reason to postulate that the NMZ represents a separate terrane. The crustal growth in the NMZ was approximately coeval with, or could have slightly preceded, the ca. 2.65 Ga metamorphism in the CZ and the SMZ, and is also coeval with the accretion in the Francistown Arc. The latter could be the NMZ's western extension, exposed at a higher crustal level (Blenkinsop, 2010). One consequence of the hypothesis that the NMZ represents an accretionary margin of the Zimbabwe craton at ca. 2.65 Ga is that the associated subduction zone should dip to the north and not to the south, as suggested in most collisional models (e.g., Roering et al., 1992).

In the Kaapvaal Craton, the oldest cratonic nucleus is situated in the province south of the Barberton Greenstone Belt (de Wit et al., 1992; Schöne et al., 2009; Zeh et al., 2009), with accretionary margins and accreted terranes making up the craton to the north. The data reviewed here for the northernmost part of the Craton do not contradict this model, as the ensemble of zircon U-Pb dates and Sm-Nd systematics is mostly consistent with an accretionary margin involving intrusion of juvenile magma at times between 2.75 and 2.95 Ga with significant assimilation of older crust (which is also betrayed by zircon xenocrysts). The Giyani Greenstone Belt region appears to be an exception, with emplacement and model ages in the 3.2–3.5 Ga range. This may be an additional old continental nucleus, possibly a terrane.

The Southern Marginal Zone of the Limpopo Complex is similar to the northern Kaapvaal Craton in Sm-Nd isotope systematics, and the hypothesis that it represents a high-grade metamorphic equivalent of the northern Kaapvaal Craton (e.g., Kreissig et al., 2000; Roering et al. 1992; van Reenen et al., this volume) is not contradicted by the data. There is no reason to postulate a separate SMZ terrane.

In the CZ, evidence of 3.3–3.5 Ga crust is manifest throughout its length, from the Phikwe and Mahalapye Complexes in the west to the Musina area in the east. Apart from the Sand River gneisses in the latter locality, this evidence consists of $T_{Hf,DM}$ model ages yielded by xenocrystic and some emplacement-age zircons, as well as $T_{Nd,DM}$ model ages of some metapelites. Of the widespread magmatism seen for ca. 2.6 to ca. 2.7 Ga, only the enderbitic part of the Bulai Pluton has a mantle Nd isotopic signature (from Hf isotopes, however, its charnockitic portion appears dominated by crustal anatexis). Further indications of admixture of juvenile magma at that time are heterogeneities of $\varepsilon_{176Hf,t}$ in the Zanzibar gneisses and in the granitoids of the Phikwe Complex, with ranges of apparent $T_{Hf,DM}$ model ages extending to less than 3 Ga. In these cases the old crustal component is dominant. Other units, such as the Singelele Gneiss at its type locality and the Regina Gneiss near the Venetia Mine, all show the signature of remelting of more than 3 Ga crust at ca. 2.65 Ga.

In the Mahalapye Complex, granitic magmatism at ca. 2.0 Ga is a product of crustal anatexis, but the $\varepsilon_{176Hf,t}$ values of magmatic ca. 2.0 Ga zircons in a mafic and hybrid gneiss point to mantle-derived magma, with variable crustal assimilation. The magmatism is coeval with the ca. 2.0 Ga tectonism and metamorphism, or precedes it slightly, as shown by deformation and metamorphic mineralogy.

In spite of this occurrence of ca. 2.0 Ga magmatism only in the western part of the CZ, the remarkable features summarized above that are observed throughout the zone from east to west and across its width do not appear to require the discrimination of separate terranes within it, as suggested by Barton et al. (2006). The same features set the CZ apart from the SMZ. It is, further, distinct from the SMZ and the Kaapvaal Craton in having, on average, anomalously high U and Th concentrations (Barton et al., 2006; Andreoli et al., this volume), but it shares this feature with the NMZ and parts of the Zimbabwe Craton (Barton et al., 2006; Kramers et al., 2001). The hypothesis that the CZ represents a separate terrane from both the SMZ and NMZ (Rollinson, 1993; McCourt and Vearncombe, 1992; Barton et al., 2006) cannot be tested on the basis of the available isotopic data.

Extensive magmatism in the ca. 2.65 Ga event characterizes both the NMZ and the CZ. In the NMZ, it appears that this occurred in an arc environment, with significant contribution of juvenile magma. For the CZ, it occurred almost solely by crustal anatexis, and the metamorphic P-T data (Smit et al., this volume; Tsunogae and van Reenen, this volume; Millonig et al., 2008) indicate crustal thickening. These features are characteristic of collisional tectonics and thus support a continental collisional model (e.g., Roering et al., 1992) for the ca. 2.65 Ga event as recorded in the CZ. The unusual amount of crustal remelting in both the NMZ and the CZ has been explained by a high content of U and Th in both provinces, which at ca. 2.65 Ga could have sufficed to heat the lower crust (at normal thickness) to close to solidus temperatures, so that either magmatic heat advection (in the NMZ) or crustal thickening (in the CZ) could trigger extensive melting (Kramers et al., 2001; Andreoli et al., this volume).

Like the NMZ and the CZ, the SMZ underwent high-grade metamorphism during the ca. 2.65 Ga Limpopo event. However, very little magmatism (other than migmatization) appears to have occurred here. The Matok Pluton is the only magmatic unit in this age range studied in the SMZ, and the Lu-Hf systematics of its zircons show that it is fully crustally derived. Thus the SMZ was probably not an active accretionary margin at ca. 2.65 Ga (but it shares the indications for earlier, ca. 2.8 Ga marginal accretion with the northern Kaapvaal Craton). The ca. 2.65 Ga metamorphism and accompanying anatexis appear to have been caused by crustal thickening, and thus collisional tectonics.

Magmatism in the SMZ at ca. 2 Ga, represented by the Schiel Alkaline Complex and the Entabeni Granite, is anorogenic. This is in contrast to the Mahalapye Complex of the CZ. However, neither in this complex nor in the SMZ do regional tectonics suggest the proximity of an accretionary margin at ca. 2 Ga. On the other hand, the well-dated Mahalapye and Entabeni magmatism rather precisely coincides in time with the intrusion of the Bushveld Complex (2052 ± 2 Ma; Walraven et al., 1990; Walraven and Hattingh, 1993), and a connection with this complex remains a valid hypothesis to explain at least some of the ca. 2.0 Ga magmatism and metamorphism in the CZ of the Limpopo Complex.

ACKNOWLEDGMENTS

Highly constructive and thoughtful reviews by Lew Ashwal and Tony Kemp have helped a great deal to improve the manuscript.

REFERENCES CITED

Amelin, Y., Lee, D.-C., Halliday, A.N., and Pidgeon, R.T., 1999, Nature of the Earth's earliest crust from hafnium isotopes in single detrital zircons: Nature, v. 399, p. 252–255, doi:10.1038/20426.

Amelin, Y., Lee, D.-C., and Halliday, A.N., 2000, Early-middle Archaean crustal evolution deduced from Lu-Hf and U-Pb isotopic studies of single zircon grains: Geochimica et Cosmochimica Acta, v. 64, p. 4205–4225, doi:10.1016/S0016-7037(00)00493-2.

Andreoli, M.A.G., Brandl, G., Coetzee, H., Kramers, J.D., and Mouri, H., 2011, this volume, Intracrustal radioactivity as an important heat source for Neoarchean metamorphism in the Central Zone of the Limpopo Complex, in van Reenen, D.D., Kramers, J.D., McCourt, S., and Perchuk, L.L., eds., Origin and Evolution of Precambrian High-Grade Gneiss Terranes, with Special Emphasis on the Limpopo Complex of Southern Africa: Geological Society of America Memoir 207, doi:10.1130/2011.1207(09).

Armstrong, R.L., 1981, Radiogenic isotopes: The case for crustal recycling on a near-steady-state no-continental growth Earth: Philosophical Transactions of the Royal Society of London, ser. A, v. 301, p. 443–472, doi:10.1098/rsta.1981.0122.

Ashwal, L.D., Morgan, P., and Hoisch, T.D., 1992, Tectonics and heat sources for granulite metamorphism of supracrustal-bearing terrains: Precambrian Research, v. 55, p. 525–538, doi:10.1016/0301-9268(92)90043-N.

Bagai, Z., Armstrong, R., and Kampunzu, A.B., 2002, U–Pb single zircon geochronology of granitoids in the Vumba granite–greenstone terrain (NE Botswana): Implication for the Archaean Zimbabwe craton: Precambrian Research, v. 118, p. 149–168, doi:10.1016/S0301-9268(02)00074-8.

Barton, J.M., Jr., 1996, The Messina Layered Intrusion, Limpopo Belt, South Africa: An example of in-situ contamination of an Archean anorthosite complex by continental crust: Precambrian Research, v. 78, p. 139–150, doi:10.1016/0301-9268(95)00074-7.

Barton, J.M., Jr., Fripp, R.E.P., Horrocks, P., and McLean, N., 1979, The geology, age, and tectonic setting of the Messina Layered Intrusion, Limpopo Mobile Belt, southern Africa: American Journal of Science, v. 279, p. 1108–1134.

Barton, J.M., Jr., Holzer, L., Kamber, B.S., Doig, R., Kramers, J.D., and Nyfeler, D., 1994, Discrete metamorphic events in the Limpopo Belt, southern Africa: Implications for the interpretation of P-T-time paths in metamorphic terrains: Geology, v. 22, p. 1035–1038, doi:10.1130/0091-7613(1994)022<1035:DMEITL>2.3.CO;2.

Barton, J.M., Jr., Barton, E.S., and Smith, C.B., 1996, Petrography, age and origin of the Schiel alkaline complex, northern Transvaal, South Africa: Journal of African Earth Sciences, v. 22, p. 133–145, doi:10.1016/0899-5362(96)00005-X.

Barton, J.M., Jr., Klemd, R., and Zeh, A., 2006, The Limpopo Belt: A result of Archean to Proterozoic, Turkic-type orogenesis?, in Reimold, W.U., and Gibson, R.L., eds., Processes on the Early Earth: Geological Society of America Special Paper 405, p. 315–331.

Bennett, V.C., Nutman, A.P., and McCulloch, M.T., 1993, Nd isotopic evidence for transient, highly depleted mantle reservoirs in the early history of the Earth: Earth and Planetary Science Letters, v. 119, p. 299–317, doi:10.1016/0012-821X(93)90140-5.

Berger, M., and Rollinson, R., 1997, Isotopic and geochemical evidence for crust-mantle interaction during late Archean crustal growth: Geochimica et Cosmochimica Acta, v. 61, p. 4809–4829, doi:10.1016/S0016-7037(97)00271-8.

Berger, M., Kramers, J.D., and Nägler, Th., 1995, Geochemistry and geochronology of charnoenderbites in the Northern Marginal Zone of the Limpopo Belt, Southern Africa, and genetic models: Schweiz: Mineralogische und Petrographische Mitteilungen, v. 75, p. 17–42.

Blenkinsop, T.G., 2011, this volume, Archean magmatic granulites, diapirism, and Proterozoic reworking in the Northern Marginal Zone of the Limpopo Belt, in van Reenen, D.D., Kramers, J.D., McCourt, S., and Perchuk, L.L., eds., Origin and Evolution of Precambrian High-Grade Gneiss Terranes, with Special Emphasis on the Limpopo Complex of Southern Africa: Geological Society of America Memoir 207, doi:10.1130/2011.1207(13).

Blenkinsop, T.G., and Frei, R., 1996, Archean and Proterozoic mineralization and tectonics at Renco Mine (Northern Marginal Zone, Limpopo Belt, Zimbabwe): Economic Geology and the Bulletin of the Society of Economic Geologists, v. 91, p. 1225–1238.

Blichert-Toft, J., and Albarède, F., 1994, Short-lived chemical heterogeneities in the Archean mantle with implications for mantle convection: Science, v. 263, p. 1593–1596, doi:10.1126/science.263.5153.1593.

Blichert-Toft, J., and Albarède, F., 1997, The Lu-Hf isotope geochemistry of chondrites and the evolution of the mantle-crust system: Earth and Planetary Science Letters, v. 148, p. 243–258, doi:10.1016/S0012-821X(97)00040-X.

Bolhar, R., Weaver, S.D., Whitehouse, M.J., Palin, J.M., Woodhead, J.D., and Cole, J.W., 2008, Sources and evolution of arc magmas inferred from coupled O and Hf isotope systematics of plutonic zircons from the Cretaceous Separation Point Suite (New Zealand): Earth and Planetary Science Letters, v. 268, p. 312–324, doi:10.1016/j.epsl.2008.01.022.

Bouvier, A., Vervoort, J.D., and Patchett, P.J., 2008, The Lu–Hf and Sm–Nd isotopic composition of CHUR: Constraints from unequilibrated chondrites and implications for the bulk composition of terrestrial planets: Earth and Planetary Science Letters, v. 273, p. 48–57, doi:10.1016/j.epsl.2008.06.010.

Bowring, S.A., and Housh, T., 1995, The Earth's early evolution: Science, v. 269, p. 1535–1540, doi:10.1126/science.7667634.

Boyet, M., and Carlson, R.W., 2005, ^{142}Nd Evidence for Early (>4.53 Ga) Global Differentiation of the Silicate Earth: Science, v. 309, p. 576–581, doi:10.1126/science.1113634.

Boyet, M., and Carlson, R.W., 2006, A new geochemical model for the Earth's mantle inferred from ^{146}Sm–^{142}Nd systematics: Earth and Planetary Science Letters, v. 250, p. 254–268, doi:10.1016/j.epsl.2006.07.046.

Brandl, G., 1983, Geology and geochemistry of various supracrustal rocks of the Beit Bridge Complex east of Messina, in Van Biljon, W.J., and Legg, J.H., eds., The Limpopo Belt: Geological Society of South Africa Special Publication 8, p. 103–112.

Brandl, G., 2001, Explanation: Geological Map of the Limpopo Belt and Its Environs (1:500,000): Pretoria, Council for Geoscience, 46 p.

Brandl, G., 2002, The Geology of the Alldays Area. Explanation, Sheet 2228 Alldays: Pretoria, Council for Geoscience, 71 p.

Chauvel, C., and Blichert-Toft, J., 2001, A hafnium isotope and trace element perspective on melting of the depleted mantle: Earth and Planetary Science Letters, v. 190, p. 137–151, doi:10.1016/S0012-821X(01)00379-X.

Chavagnac, V., Kramers, J.D., Nägler, Th.F., and Holzer, L., 2001, The behaviour of Nd and Pb isotopes during 2.0 Ga migmatization in paragneisses of the Central Zone of the Limpopo Belt (South Africa and Botswana): Precambrian Research, v. 112, p. 51–86, doi:10.1016/S0301-9268(01)00170-X.

Coward, M.P., James, P.R., and Wright, L., 1976, Northern Margin of the Limpopo Belt, southern Africa: Geological Society of America Bulletin, v. 87, p. 601–611, doi:10.1130/0016-7606(1976)87<601:NMOTLM>2.0.CO;2.

Davis, D.W., Amelin, Y., Nowel, G.M., and Parrish, R.R., 2005, Hf isotopes in zircon from the western Superior province, Canada: Implications for Archean crustal development and evolution of the depleted mantle reservoir: Precambrian Research, v. 140, p. 132–156, doi:10.1016/j.precamres.2005.07.005.

DePaolo, D.J., 1980, Crustal growth and mantle evolution: Inferences from models of element transport and Nd and Sr isotopes: Geochimica et Cosmochimica Acta, v. 44, p. 1185–1196, doi:10.1016/0016-7037(80)90072-1.

DePaolo, D.J., 1988, Neodymium Isotope Geochemistry: Berlin, Springer-Verlag, 187 p.

de Wit, M.J., Roering, C., Hart, R.J., Armstrong, R.A., de Ronde, C.E.J., Green, R.W.E., Tredoux, M., Pederby, E., and Hart, R.A., 1992, Formation of an Archean continent: Nature, v. 357, p. 553–562, doi:10.1038/357553a0.

Flowerdew, M.J., Millar, I.L., Curtis, M.L., Vaughan, A.P.M., Horstwood, M.S.A., Whitehouse, M.J., and Fanning, C.M., 2007, Combined U-Pb isotopic age and Hf isotope geochemistry of detrital zircons from early Paleozoic sedimentary rocks, Ellsworth-Whitmore Mountains block, Antarctica: Geological Society of America Bulletin, v. 119, p. 275–288, doi:10.1130/B25891.1.

Frei, R., Blenkinsop, T.G., and Schönberg, R., 1999, Geochronology of the late Archaean Razi and Chilimanzi suites of granites in Zimbabwe: Implications for the late Archean tectonics of the Limpopo Belt and Zimbabwe Craton: South African Journal of Geology, v. 102, p. 55–63.

Galer, S.J.G., Goldstein, S.L., and O'Nions, R.K., 1989, Limits on chemical and convective isolation in the Earth's interior: Chemical Geology, v. 75, p. 257–290, doi:10.1016/0009-2541(89)90001-6.

Gerdes, A., and Zeh, A., 2006, Combined U–Pb and Hf isotope LA-(MC-)ICP-MS analyses of detrital zircons: Comparison with SHRIMP and new constraints for the provenance and age of an Armorican metasediment in Central Germany: Earth and Planetary Sciences Letters, v. 249, p. 47–61.

Gerdes, A., and Zeh, A., 2009, Zircon formation versus zircon alteration—New insights from combined U-Pb and Lu-Hf in-situ LA-ICP-MS analyses, and consequences for the interpretation of Archean zircon from the Limpopo Belt: Chemical Geology, v. 261, 230–243, doi:10.1016/j.chemgeo.2008.03.005.

Harris, N.B.W., Hawkesworth, C.J., van Casteren, P., and McDermott, F., 1987, Evolution of continental crust in southern Africa: Earth and Planetary Science Letters, v. 83, p. 85–93, doi:10.1016/0012-821X(87)90053-7.

Hickman, M.H., 1978, Isotopic evidence for crustal reworking in the Rhodesian Archean craton, southern Africa: Geology, v. 6, p. 214–216, doi:10.1130/0091-7613(1978)6<214:IEFCRI>2.0.CO;2.

Hisada, K., and Miyano, T., 1996, Petrology and microthermometry of aluminous rocks in the Botswanan Limpopo Central Zone: Evidence for isothermal decompression and isobaric cooling: Journal of Metamorphic Geology, v. 14, p. 183–197, doi:10.1046/j.1525-1314.1996.05857.x.

Holzer, L., Frei, R., Barton, J.M., Jr., and Kramers, J.D., 1998, Unravelling the record of successive high-grade events in the Central Zone of the Limpopo Belt using Pb single-phase dating of metamorphic minerals: Precambrian Research, v. 87, p. 87–115, doi:10.1016/S0301-9268(97)00058-2.

Horstwood, M.S.A., Nesbitt, R.W., Noble, S.R., and Wilson, J.F., 1999, U-Pb zircon evidence for an extensive early Archean craton in Zimbabwe: A reassessment of the timing of craton formation, stabilization, and growth: Geology, v. 27, p. 707–710, doi:10.1130/0091-7613(1999)027<0707:UPZEFA>2.3.CO;2.

Hoskin, P.W.O., and Schaltegger, U., 2003, The composition of zircon and igneous and metamorphic petrogenesis, in Hanchar, J.M., and Hoskin, P.W.O., eds., Zircon: Mineralogical Society of America, Reviews in Mineralogy and Geochemistry, v. 53, p. 27–62.

Huizenga, J.-M., Perchuk, L.L., van Reenen, D.D., Flattery, Y., Varlamov, D.A., Smit, C.A., and Gerya, T.V., 2011, this volume, Granite emplacement and the retrograde P-T-fluid evolution of Neoarchean granulites in the Central Zone of the Limpopo Complex, in van Reenen, D.D., Kramers, J.D., McCourt, S., and Perchuk, L.L., eds., Origin and Evolution of Precambrian High-Grade Gneiss Terranes, with Special Emphasis on the Limpopo Complex of Southern Africa: Geological Society of America Memoir 207, doi:10.1130/2011.1207(08).

Jacobsen, S.B., and Wasserburg, G.J., 1979, The mean age of mantle and crustal reservoirs: Journal of Geophysical Research, v. 84, p. 7411–7427.

Jacobsen, S.B., and Wasserburg, G.J., 1980, Sm-Nd isotopic evolution of chondrites: Earth and Planetary Science Letters, v. 50, p. 139–155, doi:10.1016/0012-821X(80)90125-9.

Jaeckel, P., Kröner, A., Kamo, S.L., Brandl, G., and Wendt, J.I., 1997, Late Archaean to Early Proterozoic granitoid magmatism and high-grade metamorphism in the central Limpopo Belt, South Africa: Journal of the Geological Society [London], v. 154, p. 25–44, doi:10.1144/gsjgs.154.1.0025.

Jelsma, H.A., Vinju, M.L., Valbracht, P.J., Davies, G.R., Wijbrans, J.R., and Verdurmen, E.A.T., 1996, Constraints on Archean crustal evolution of the Zimbabwe craton: A U-Pb zircon, Sm-Nd and Pb-Pb whole rock isotope study: Contributions to Mineralogy and Petrology, v. 124, p. 55–70, doi:10.1007/s004100050173.

Kampunzu, A.B., Tombale, A.R., Zhai, M., Majaule, T., and Modisi, M.P., 2003, Major and trace element geochemistry of plutonic rocks from Francistown, NE Botswana: Evidence for a Neoarchaean continental active margin in the Zimbabwe craton: Lithos, v. 71, p. 431–460, doi:10.1016/S0024-4937(03)00125-7.

Kemp, A.I.S., Hawkesworth, C.J., Paterson, B.A., and Kinny, P.D., 2006, Episodic growth of the Gondwana supercontinent from hafnium and oxygen isotopes in zircon: Nature, v. 439, p. 580–583, doi:10.1038/nature04505.

Kemp, A.I.S., Hawkesworth, C.J., Foster, G.L., Paterson, B.A., Woodhead, J.D., Hergt, J.M., Gray, C.M., and Whitehouse, M.J., 2007, Magmatic and crustal differentiation history of granitic rocks from Hf-O isotopes in zircon: Science, v. 315, p. 980–983, doi:10.1126/science.1136154.

Kemp, A.I.S., Hawkesworth, C.J., Collins W.J., Gray, C.M., Blevin P.L., and EIMF, 2009, Isotopic evidence for rapid continental growth in an extensional accretionary orogen: The Tasmanides, eastern Australia: Earth and Planetary Science Letters, v. 284, p. 455–466, doi:10.1016/j.epsl.2009.05.011.

Kemp, A.I.S., Wilde, S.A., Hawkesworth, C.J., Coath, C.D., Nemchin, A., Pidgeon, R.T., Vervoort, J.D., and DuFrane, A., 2010, Hadean crustal evolution revisited: New constraints from Pb-Hf isotope systematics of the Jack Hills zircons: Earth and Planetary Science Letters, v. 296, p. 45–56, doi:10.1016/j.epsl.2010.04.043.

Key, R.M., 1976, Quarter Degree Sheet 2127D, Pikwe: Lobatse, Botswana, Geological Survey of Botswana, scale 1:125,000.

Kramers, J.D., and Mouri, H., 2011, this volume, The geochronology of the Limpopo Complex: A controversy solved, in van Reenen, D.D., Kramers, J.D., McCourt, S., and Perchuk, L.L., eds., Origin and Evolution of Precambrian High-Grade Gneiss Terranes, with Special Emphasis on the Limpopo Complex of Southern Africa: Geological Society of America Memoir 207, doi:10.1130/2011.1207(06).

Kramers, J.D., and Tolstikhin, I.N., 1997, Two major terrestrial Pb isotope paradoxes, forward transport modelling, core formation and the history of the continental crust: Chemical Geology, v. 139, p. 75–110, doi:10.1016/S0009-2541(97)00027-2.

Kramers, J.D., Kreissig, K., and Jones, M.Q.W., 2001, Crustal heat production and style of metamorphism: A comparison between two Archean high grade provinces in the Limpopo Belt, Southern Africa: Precambrian Research, v. 112, p. 149–163, doi:10.1016/S0301-9268(01)00173-5.

Kreissig, K., Nägler, Th.F., Kramers, J.D., van Reenen, D.D., and Smit, C.A., 2000, An isotopic and geochemical study of the northern Kaapvaal Craton and the Southern Marginal Zone of the Limpopo Belt: Are they juxtaposed terranes?: Lithos, v. 50, p. 1–25, doi:10.1016/S0024-4937(99)00037-7.

Kröner, A., Jaeckel, P., Brandl, G., Nemchin, A.A., and Pidgeon, R.T., 1999, Single zircon ages for granitoid gneisses in the Central Zone of the Limpopo Belt, southern Africa and geodynamic significance: Precambrian Research, v. 93, p. 299–337, doi:10.1016/S0301-9268(98)00102-8.

Kröner, A., Jaeckel, P., and Brandl, G., 2000, Single zircon ages for felsic to intermediate rocks from the Pietersburg and Giyani greenstone belts and bordering granitoid orthogneisses, northern Kaapvaal Craton, South Africa: Journal of African Earth Sciences, v. 30, p. 773–793, doi:10.1016/S0899-5362(00)00052-X.

McCourt, S., and Armstrong, R.A., 1998, SIMS U-Pb zircon geochronology of granites from the Central Zone, Limpopo Belt, southern Africa: Implications for the age of the Limpopo Orogeny: South African Journal of Geology, v. 101, p. 329–338.

McCourt, S., and Vearncombe, J.R., 1992, Shear zones of the Limpopo Belt and adjacent granitoid-greenstone terranes: Implications for late Archaean collision tectonics in southern Africa: Precambrian Research, v. 55, p. 553–570, doi:10.1016/0301-9268(92)90045-P.

McCulloch, M.T., and Wasserburg, G.J., 1978, Sm-Nd and Rb-Sr chronology of continental crust formation: Science, v. 200, p. 1003–1011, doi:10.1126/science.200.4345.1003.

Millonig, L., Zeh, A., Gerdes, A., and Klemd, R., 2008, Neoarchaean high-grade metamorphism in the Central Zone of the Limpopo Belt (South Africa): Combined petrological and geochronological evidence from the Bulai pluton: Lithos, v. 103, p. 333–351, doi:10.1016/j.lithos.2007.10.001.

Millonig, L., Zeh, A., Gerdes, A., Klemd, R., and Barton, J.M., Jr., 2010, Decompressional heating of the Mahalapye Complex (Limpopo Belt, Botswana): A response to Palaeoproterozoic magmatic underplating?: Journal of Petrology, v. 51, p. 703–729, doi:10.1093/petrology/egp097.

Mkweli, S., Kamber, B., and Berger, M., 1995, Westward continuation of the Craton–Limpopo Belt tectonic break in Zimbabwe and new age constraints on the timing of the thrusting: Journal of the Geological Society [London], v. 152, p. 77–83, doi:10.1144/gsjgs.152.1.0077.

Moorbath, S., Whitehouse, M.J., and Kamber, B.S., 1997, Extreme Nd-isotope heterogeneity in the early Archaean—Fact or fiction? Case histories from northern Canada and West Greenland: Chemical Geology, v. 135, p. 213–231, doi:10.1016/S0009-2541(96)00117-9.

Mouri, H., Whitehouse, M.J., Brandl, G., and Rajesh, H.M., 2009, A magmatic age and four successive metamorphic events recorded in zircons from a single meta-anorthosite sample in the Central Zone of the Limpopo Belt, South Africa: Journal of the Geological Society [London], v. 166, p. 827–830, doi:10.1144/0016-76492008-148.

Nägler, Th.F., and Kramers, J.D., 1998, Nd isotopic evolution of the upper mantle during the Precambrian: Models, data and the uncertainty of both: Precambrian Research, v. 91, p. 233–252, doi:10.1016/S0301-9268(98)00051-5.

Patchett, P.J., Vervoort, J.D., Soderlund, U., and Salters, V.J.M., 2004, Lu-Hf and Sm-Nd isotopic systematics in chondrites and their constraints on the Lu-Hf properties of the Earth: Earth and Planetary Science Letters, v. 222, p. 29–41, doi:10.1016/j.epsl.2004.02.030.

Robertson, I.D.M., and DuToit, M.C., 1981, Mobile belts: A. The Limpopo belt, in Hunter, D.R., ed., The Precambrian of the Southern Hemisphere: Amsterdam, Elsevier, p. 641–671.

Roering, C., van Reenen, D.D., Smit, C.A., Barton, J.M., Jr., De Beer, J.H., De Wit, M.J., Stettler, E.H., Van Schalkwyk, J.F., Stevens, G., and Pretorius, S.J., 1992, Tectonic model for the evolution of the Limpopo Belt: Precambrian Research, v. 55, p. 539–552, doi:10.1016/0301-9268(92)90044-O.

Rollinson, H.R., 1993, A terrane interpretation of the Archaean Limpopo Belt: Geological Magazine, v. 130, p. 755–765, doi:10.1017/S001675680002313X.

Rudnick, R.L., and Gao, S., 2003, Composition of the continental crust, in Rudnick, R.L., ed., The Crust, Treatise on Geochemistry, Vol. 3: Amsterdam, Elsevier, p. 1–64.

Schmidt Mumm, A., 1999, Regional aktive Fluidsysteme während der Kratonisierung im späten Archaikum Simbabwes und Paläoproterozoikum Ghanas [Habilitation thesis]: Halle-Wittenberg, Germany, University of Halle-Wittenberg, 147 p.

Schmidt Mumm, A., and Kreissig, K., 1999, Crustal evolution in the SE-Zimbabwe Craton: European Union of Geosciences Conference Abstracts: EUG 10, Journal of Conference Abstracts, v. 4, no. 141, p. 1999.

Schöne, B., Dudas, F.O.L., Bowring, S.A., and de Wit, M., 2009, Sm-Nd isotopic mapping of lithospheric growth and stabilization in the eastern Kaapvaal craton: Terra Nova, v. 21, p. 219–228, doi:10.1111/j.1365-3121.2009.00877.x.

Smit, C.A., van Reenen, D.D., Roering, C., Perchuk, L.L., and Boshoff, R., 2011, this volume, Neoarchaean to Paleoproterozoic evolution of the polymetamorphic Central Zone of the Limpopo Complex, in van Reenen, D.D., Kramers, J.D., McCourt, S., and Perchuk, L.L., eds., Origin and Evolution of Precambrian High-Grade Gneiss Terranes, with Special Emphasis on the Limpopo Complex of Southern Africa: Geological Society of America Memoir 207, doi:10.1130/2011.1207(12).

Taylor, P.N., Kramers, J.D., Moorbath, S., Wilson, J.F., Orpen, J.L., and Martin, A., 1991, Pb/Pb, Sm-Nd and Rb-Sr geochronology in the Archean Craton of Zimbabwe: Chemical Geology, v. 87, p. 175–196.

Taylor, S.R., and McLennan, S., 1995, The geochemical evolution of the continental crust: Reviews of Geophysics, v. 33, p. 241–265, doi:10.1029/95RG00262.

Tolstikhin, I.N., and Hofmann, A.W., 2005, Early crust on top of the Earth's core: Physics of the Earth and Planetary Interiors, v. 148, p. 109–130, doi:10.1016/j.pepi.2004.05.011.

Tolstikhin, I.N., Kramers, J.D., and Hofmann, A.W., 2006, A chemical Earth model with whole mantle convection: The importance of a core-mantle boundary layer (D″) and its early formation: Chemical Geology, v. 226, p. 79–99, doi:10.1016/j.chemgeo.2005.09.015.

Treloar, P.J., Coward, M.P., and Harris, N.B.W., 1992, Himalayan-Tibetan analogies for the evolution of the Zimbabwe Craton and the Limpopo Belt: Precambrian Research, v. 55, p. 571–587, doi:10.1016/0301-9268(92)90046-Q.

Tsunogae, T., and van Reenen, D.D., 2011, this volume, High-pressure and ultrahigh-temperature granulite-facies metamorphism of Precambrian high-grade terranes: Case study of the Limpopo Complex, in van Reenen, D.D., Kramers, J.D., McCourt, S., and Perchuk, L.L., eds., Origin and Evolution of Precambrian High-Grade Gneiss Terranes, with Special Emphasis on the Limpopo Complex of Southern Africa: Geological Society of America Memoir 207, doi:10.1130/2011.1207(07).

Valley, J.W., Lackey, J.S., Cavosie, A.J., Clechenko, C.C., Spicuzza, M.J., Basei, M.A.S., Bindeman, I.N., Ferreira, V.P., Sial, A.N., King, E.M., Peck, W.H., Sinha, A.K., and Wei, C.S., 2005, 4.4 billion years of crustal maturation: Oxygen isotope ratios of magmatic zircon: Contributions to Mineralogy and Petrology, v. 150, p. 561–580, doi:10.1007/s00410-005-0025-8.

van Reenen, D.D., Smit, C.A., Perchuk, L.L., Roering, C., and Boshoff, R., 2011, this volume, Thrust exhumation of the Neoarchean UHT Southern Marginal Zone, Limpopo Complex: Convergence of decompression-cooling paths in the hanging wall and prograde P-T paths in the footwall, in van Reenen, D.D., Kramers, J.D., McCourt, S., and Perchuk, L.L., eds., Origin and Evolution of Precambrian High-Grade Gneiss Terranes, with Special Emphasis on the Limpopo Complex of Southern Africa: Geological Society of America Memoir 207, doi:10.1130/2011.1207(11).

Vervoort, J.D., Patchett, P.J., Blichert-Toft, J., and Albarède, F., 1999, Relationships between Lu–Hf and Sm–Nd isotopic systems in the global sedimentary system: Earth and Planetary Science Letters, v. 168, p. 79–99, doi:10.1016/S0012-821X(99)00047-3.

Walraven, F., and Hattingh, E., 1993, Geochronology of the Nebo Granite, Bushveld Complex, South Africa: Chemical Geology, v. 72, p. 17–28.

Walraven, F., Armstrong, R.A., and Kruger, F.J., 1990, A chronostratigraphic framework for the north-central Kaapvaal Craton, the Bushveld

Complex and Vredefort structure: Tectonophysics, v. 171, p. 23–48, doi:10.1016/0040-1951(90)90088-P.

Wedepohl, K.H., 1995, The compositions of the continental crust: Geochimica et Cosmochimica Acta, v. 59, p. 1217–1232, doi:10.1016/0016-7037 (95)00038-2.

Wilson, J.F., Nesbitt, R.W., and Fanning, C.M., 1995, Zircon geochronology of Archean felsic sequences in the Zimbabwe craton: A revision of greenstone stratigraphy and a model for crustal growth, in Coward, M.P., and Ries, A.C., eds., Early Precambrian Processes: Geological Society [London] Special Publication 95, p. 109–126.

Zeh, A., Gerdes, A., Klemd, R., and Barton, J.M., Jr., 2007, Archaean to Proterozoic Crustal Evolution in the Central Zone of the Limpopo Belt (South Africa–Botswana): Constraints from combined U-Pb and Lu-Hf isotope analyses of zircon: Journal of Petrology, v. 48, p. 1605–1639, doi:10.1093/petrology/egm032.

Zeh, A., Gerdes, A., Klemd, R., and Barton, J.R., Jr., 2008, U–Pb and Lu–Hf isotope record of detrital zircon grains from the Limpopo Belt—Evidence for crustal recycling at the Hadean to early-Archean transition: Geochimica et Cosmochimica Acta, v. 72, p. 5304–5329, doi:10.1016/j.gca.2008.07.033.

Zeh, A., Gerdes, A., and Barton, J.M., Jr., 2009, Archean accretion and crustal evolution of the Kalahari Craton—The zircon age and Hf isotope record of granitic rocks from Barberton/Swaziland to the Francistown Arc: Journal of Petrology, v. 50, p. 933–966, doi:10.1093/petrology/egp027.

Zeh, A., Gerdes, A., Barton, J.M., Jr., and Klemd, R., 2010, U–Th–Pb and Lu–Hf systematics of zircon from TTG's, leucosomes, meta-anorthosites and quartzites of the Limpopo Belt (South Africa): Constraints for the formation, recycling and metamorphism of Palaeoarchaean crust: Precambrian Research, v. 179, p. 50–68, doi:10.1016/j.precamres.2010.02.012.

Zhai, M., Kampunzu, A.B., Modisi, M.P., and Bagai, Z., 2006, Sr and Nd isotope systematics of Francistown plutonic rocks, Botswana: Implications for Neoarchaean crustal evolution of the Zimbabwe craton: International Journal of Earth Sciences, v. 95, p. 355–369, doi:10.1007/s00531-005-0054-6.

MANUSCRIPT ACCEPTED BY THE SOCIETY 24 MAY 2010

The Geological Society of America
Memoir 207
2011

Thrust exhumation of the Neoarchean ultrahigh-temperature Southern Marginal Zone, Limpopo Complex: Convergence of decompression-cooling paths in the hanging wall and prograde P-T paths in the footwall

Dirk D. van Reenen
C. Andre Smit
Department of Geology, University of Johannesburg, P.O. Box 524, Auckland Park, 2006, Johannesburg, South Africa

Leonid L. Perchuk*
Department of Geology, University of Johannesburg, P.O. Box 524, Auckland Park, 2006, Johannesburg, South Africa;
Department of Petrology, Geological Faculty, Moscow State University, Leninskie Gory, Moscow, 119192, Russia; and
Institute of Experimental Mineralogy, Russian Academy of Sciences, Chernogolovka, Moscow District, 142432, Russia

Chris Roering
Department of Geology, University of Johannesburg, P.O. Box 524, Auckland Park, 2006, Johannesburg, South Africa

René Boshoff
Department of Geology, University of Johannesburg, P.O. Box 524, Auckland Park, 2006, Johannesburg, South Africa, and
Department of Earth Science, Stellenbosch University, Private Bag X1, Matieland, 7602, Stellenbosch, South Africa

ABSTRACT

Integrated structural, metamorphic, and geochronological data indicate that the evolution of the Southern Marginal Zone (SMZ) of the Limpopo Complex of southern Africa was controlled by a single Neoarchean high-grade tectono-metamorphic event. The exhumation history reflected by the high-grade rocks is determined by their location relative to the contact with the low-grade rocks of the Kaapvaal Craton. Exhumation of granulites far north from this contact is recorded by a decompression-cooling (DC) pressure-temperature (P-T) path linked to steep southward-verging thrusts related to the Hout River Shear Zone. This P-T path traverses from P ~8 kbar, T ~825 °C to P ~5 kbar, T ~550 °C and reflects exhumation of the SMZ in the interval ca. 2.68–2.64 Ga. P-T paths for granulites close to this contact are characterized by a distinct inflection at P ~6 kbar, T ~700 °C that exhibits near-isobaric cooling (IC) to

---

*posthumous

van Reenen, D.D., Smit, C.A., Perchuk, L.L., Roering, C., and Boshoff, R., 2011, Thrust exhumation of the Neoarchean ultrahigh-temperature Southern Marginal Zone, Limpopo Complex: Convergence of decompression-cooling paths in the hanging wall and prograde P-T paths in the footwall, *in* van Reenen, D.D., Kramers, J.D., McCourt, S., and Perchuk, L.L., eds., Origin and Evolution of Precambrian High-Grade Gneiss Terranes, with Special Emphasis on the Limpopo Complex of Southern Africa: Geological Society of America Memoir 207, p. 189–212, doi:10.1130/2011.1207(11). For permission to copy, contact editing@geosociety.org. © 2011 The Geological Society of America. All rights reserved.

T ~580 °C. The IC stage is linked to low-angle, out-of-sequence, southward-verging thrusts that developed in the interval 2.63–2.6 Ga. The thrust-controlled exhumation of the SMZ furthermore is demonstrated by the convergence at P ~6 kbar, T ~700 °C of DC P-T paths in the hanging wall with prograde P-T loops in the footwall of the steeply southward-verging Hout River Shear Zone, and by the establishment of a retrograde isograd and zone of rehydrated granulites in the hanging wall derived from the dehydration of the low-grade rocks in the footwall. A composite deformation-pressure-temperature-time (D-P-T-t) diagram provides evidence in support of a tectonic model for the evolution of the Limpopo Complex that involves early crustal thickening and peak metamorphic conditions followed by doming and diapirism related to gravitational redistribution mechanisms.

INTRODUCTION

The Limpopo Complex of southern Africa (Fig. 1), between the Kaapvaal and Zimbabwe Cratons, is a SW-NE–trending high-grade gneiss terrane that has traditionally been subdivided into three distinct terranes on the basis of lithological and structural differences, namely a Southern Marginal Zone (SMZ), a Central Zone (CZ), and a Northern Marginal Zone (NMZ) (e.g., van Reenen et al., 1992, and references therein). The SMZ and NMZ are separated from the adjacent granite-greenstone cratons by ca. 2.65 Ga inward-dipping dip-slip shear zones, termed the *Hout River Shear Zone* in the south and the *North Marginal Thrust Zone* in the north (van Reenen et al., 1987, 1988, 1990; Roering, et al., 1992a, 1992b; Blenkinsop et al., 2004; Blenkinsop, this volume). The CZ is separated from the marginal zones by ca. 2.0 Ga inward-dipping strike-slip shear zones, termed the *Palala Shear Zone* in the south and the *Triangle Shear Zone* in the north (e.g., McCourt and Vearncombe, 1987, 1992; Schaller et al., 1999).

The SMZ, entirely exposed within South Africa, highlights the relationship of the high-grade gneissic Limpopo Complex in the north with the juxtaposed low-grade granite-greenstone Kaapvaal Craton in the south (Fig. 2). This relationship documents evidence for a single Neoarchean tectono-metamorphic event. This is in contrast to the CZ, where the superposition of a major Paleoproterozoic high-grade event complicates the interpretation of data (van Reenen et al., 2008; Smit et al., this volume).

In dealing with high-grade rocks that are gneissic, as is the case throughout the SMZ, conventional geological mapping and field techniques are not readily applicable. Fieldwork thus relied heavily on lithological and structural observations, which then were combined with metamorphic and geochronologic data. The results of these studies highlight the essential compressive tectonic style of the SMZ, which is responsible for the juxtaposition of ca. 2.72 Ga ultrahigh-temperature (UHT) gneisses with more than 3 Ga low-grade greenstones (Kaapvaal Craton) along a major NE-dipping tectonic boundary (Smit et al., 1992). Compressive tectonism is responsible for the southward emplacement (spreading) of high-grade rocks of the Limpopo Complex onto the cooler granite-greenstone terrane of the Kaapvaal Craton. Detailed metamorphic petrologic investigations were undertaken to track in detail the response of strategically selected rock samples to the changing tectonic-metamorphic conditions.

The published geological data for the SMZ and rocks of the Kaapvaal Craton can be logically synthesized with reference to two distinct time periods: data compiled before and tested during the first Limpopo international field workshop held in the Limpopo Complex in 1990 (van Reenen et al., 1992), whereas data published and accumulated after 1992 were presented at a second Limpopo international field workshop held in 2008.

GEOLOGICAL DATA ACCUMULATED UP TO 1992

Major Rock Types and Setting of the SMZ

Data generated before 1992 already highlighted aspects of the geology of the Limpopo Complex, namely, the change in the style of deformation and grade of metamorphism across well-defined tectonic boundaries that separate the high-grade SMZ and Northern Marginal Zone (NMZ), respectively, from the Kaapvaal and Zimbabwe Cratons (Figs. 1, 2) (van Reenen et al., 1992, and references therein).

The SMZ (Fig. 2; Plate 1 [on loose insert accompanying this volume and in the Data Repository[1]]) comprises intensely deformed high-grade supracrustal migmatitic gneisses of mafic, ultramafic, and metapelitic composition and rare banded iron formation (BIF), collectively termed the *Bandelierkop Formation* by Du Toit and van Reenen (1977) and Du Toit et al. (1983). These rocks are tectonically intermixed with tonalitic-enderbitic gray gneisses locally referred to as the *Baviaanskloof Gneiss* (Du Toit and van Reenen, 1977; Du Toit et al., 1983). Du Toit et al. (1983) interpreted these rocks on the basis of limited geochemical data to represent the high-grade equivalents of the more than 3.0 Ga

---
[1]Plate 1 is available as GSA Data Repository Item 2011062, online at www.geosociety.org/pubs/ft2011.htm, or on request from editing@geosociety.org, Documents Secretary, GSA, P.O. Box 9140, Boulder, CO 80301-9140, USA.

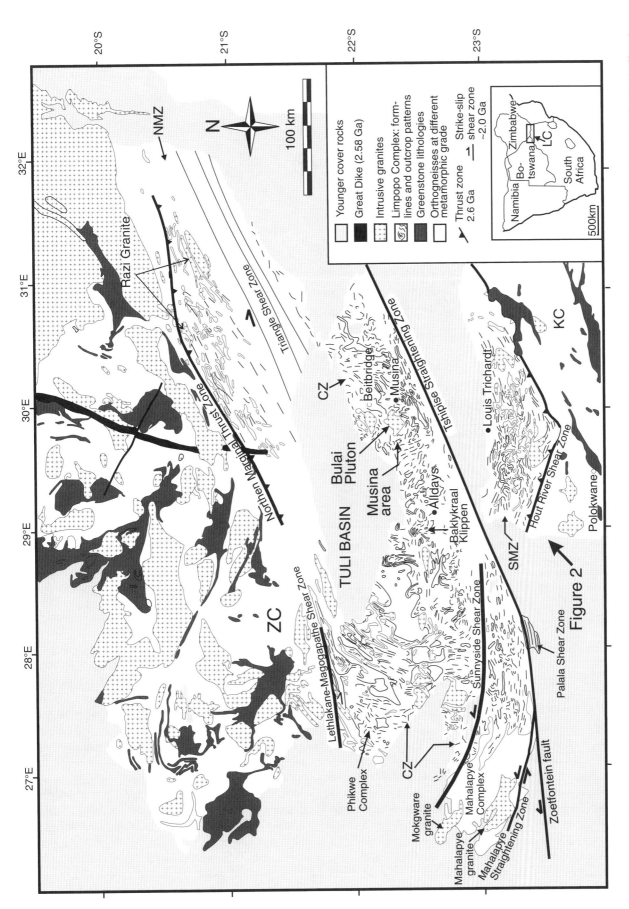

Figure 1. Map of the Limpopo Complex (LC in inset map), located between the Zimbabwe Craton (ZC) in the north and the Kaapvaal Craton (KC) in the south (after Boshoff et al., 2006). Shown is the internal subdivision into a Central Zone (CZ), a Southern Marginal Zone (SMZ), and a Northern Marginal Zone (NMZ), which are separated from one another and from the adjacent cratons by SW-NE-trending inward-dipping dip-slip and strike-slip shear zones. Also shown is ca. 2.6 Ga granitic magmatism that occurred throughout the Limpopo Complex. The location of Figure 2 is indicated.

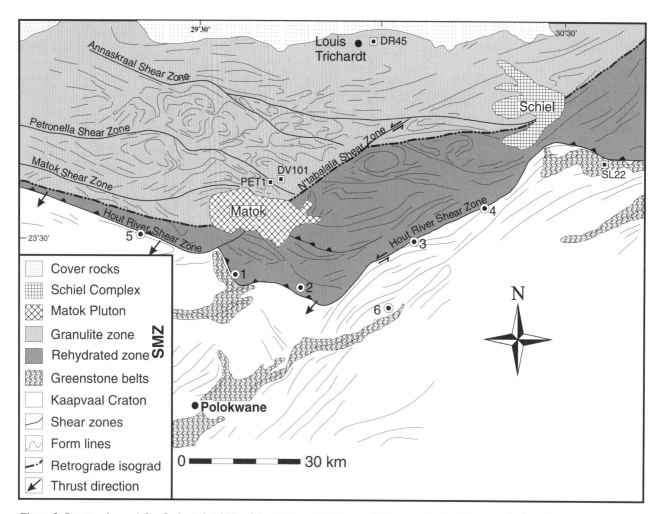

Figure 2. Structural map (after Smit et al., 1992) of the SMZ and the Kaapvaal Craton, showing $D2_A$ crustal blocks bounded by steeply SW-verging $D2_B$ shear zones, which include the Hout River Shear Zone that bounds the SMZ in the south. Also shown is the retrograde isograd that subdivides the SMZ into a northern granulite zone and a southern zone of rehydrated granulites. Black arrows show the SW direction of exhumation (thrusting) of the SMZ. Sample localities (PET1, etc.), as well as other localities (1–6) discussed in the text, are shown, including the position of the major tectonic klippen on the Kaapvaal Craton (loc. 6).

low-grade Pietersburg-, Giyani-, and Rhenosterkoppies Greenstone Belts of the Kaapvaal Craton located within tonalitic to trondhjemitic gneisses (Fig. 2; Plate 1).

Granitoids intrusive into the high-grade SMZ gneisses include the ca. 2.671–2.664 Ga late- to post-kinematic Matok Pluton, leucocratic anatectic gneisses with a crystallization age of ca. 2.643 Ga, the ca. 2.46 Ga post-tectonic Palmietfontein Granite, and rocks of the Schiel alkaline complex (Fig. 2; Plate 1) (Barton et al., 1983; Barton and van Reenen, 1992a). The Matok Pluton comprises a variety of mainly undeformed granitic rocks that include the ca. 2.671 Ga charnoenderbites (Barton et al., 1992; Barton and van Reenen, 1992a), a main ca. 2.667 Ga porphyritic granitic phase, and a coarse-grained ca. 2.664 Ga granodioritic phase (Bohlender et al., 1992). The granodioritic phase intrudes the main porphyritic phase as well as the steeply SW-verging Matok Shear Zone (Fig. 3) in the ramp section of the Hout River Shear Zone. All phases of the Matok Pluton, as well as the ca. 2.46 Ga Palmietfontein Granite (Barton et al., 1983; Barton and van Reenen, 1992a), were affected by the reactivation of the N'Tabalala Shear Zone, which resulted in the development of a near-vertical, narrow, mylonitic shear structure (Fig. 4D) that cuts through the Matok Pluton (Fig. 2; Plate 1).

Subvolcanic dikes (with groundmass volcanic glass) related to the ca. 2.46 Ga post-tectonic Palmietfontein Granite (Barton et al., 1983) intrude the Baviaanskloof gneisses at the type Goudplaas locality (Fig. 2, loc. 2). It follows therefore that the SMZ must have been at the near surface before ca. 2.46 Ga.

Late- to post-tectonic ca. 2.6 Ga granitic intrusions similar to the Matok Pluton also occur in the Central Zone (Bulai Pluton) and in the Northern Marginal Zone (the Razi granites) (Fig. 1) and have been considered as important time markers for the high-grade tectono-metamorphic event that affected the Limpopo Complex in the Neoarchean (Barton and van Reenen, 1992a).

Structural Geological Setting

Structural Geology of the Juxtaposed Granite Greenstone Terrane

The internal structure of all the greenstone belts on the northern Kaapvaal Craton (Plate 1) shows similarities (e.g., de Beer and Stettler, 1992; de Wit et al., 1992a, 1992c; McCourt and van Reenen, 1992; Roering et al., 1992a, 1992b): all have regional ENE-WSW–trending schistosity, here termed $D1_A$ structures, with a dominant shallow northward vergence associated with considerable crustal shortening (de Wit et al., 1992a). The $D1_A$ schistosity is overprinted by low-grade and low-angle northward-verging $D1_B$ shear zones that abut against the southward-verging Hout River Shear Zone (see Fig. 5) (de Wit et al., 1992c). Rocks of distinctly different structural and metamorphic histories occupy the hanging wall and footwall of the Hout River Shear Zone (de Wit et al., 1992c; McCourt and van Reenen, 1992).

Structural Geology of the SMZ

The structural framework of the SMZ was based on a photo geological interpretation (Smit et al., 1992), which showed the presence of different generations of gneissic fabrics (Fig. 2; Plate 1). At the time, two major deformational elements were identified: (1) large crustal blocks with steeply dipping and strongly foliated and isoclinally folded high-grade rocks termed $D2_A$, bounded by (2) large-scale, steeply dipping shear zones defined as $D2_B$, and collectively termed the Hout River Shear System.

The Hout River Shear System (Fig. 2; Plate 1) is defined by a regionally interconnecting pattern of reverse faults that include the Hout River, Petronella, Matok, and Annaskraal Shear Zones within a well-developed SW-verging structural ramp with NW-SE–trending frontal (dip-slip) and SW-NE–trending lateral ramp sections. These shear zones form an anastomosing pattern (Fig. 2; Plate 1) that resulted in large crustal blocks characterized by a strong $D2_A$ gneissic fold pattern (Fig. 4A). The outcrop pattern of some of the folds resembles large-scale sheath folds. Individual shear zones are kilometer-wide, steeply north-dipping (>70°) structures with a dip-slip to steeply oblique SW-verging sense (Smit et al., 1992). They display strong gneissic shear fabrics (Figs. 4B–4D) (Smit and van Reenen, 1997; Smit et al., 2001) related to deep crustal movement, and reflect the exhumation of the high-grade SMZ rocks along a steeply oriented SW trajectory to the present-day erosional level. The vertical $D2_B$ N'Tabalala Shear Zone (Fig. 2) shows evidence for late reactivation, which is manifested as transpressional movement, termed $D3_A$, that cuts through the Matok Pluton (Fig. 2; Plate 1). This minor event is reflected by narrow mylonitic shear zones (Fig. 4D) and is not critical to the formation of the P-T paths discussed here.

The structural character of the northern Kaapvaal Craton and the SMZ was further refined on the basis of published geophysical data that include deep electrical soundings, gravity data, and the results of a regional vibroseismic survey undertaken in the early 1990s by the National Geophysics Programme and the then Geological Survey of South Africa (de Beer and Stettler, 1992; Roering et al., 1992b). The crustal section (Fig. 5) based on this survey suggested that one of the most important structural elements on the Kaapvaal Craton is a shallow, thin-skinned thrust with a northward vergence. A much larger system of reflectors also indicates a similar style of deformation in the deep crust, resulting in a system of northward-verging faults (here termed $D1_B$) in the Kaapvaal Craton. The sole $D1_B$ thrust has been interpreted to account for the shallow nature of the greenstone belts and the suggestion that these belts are allochthonous (de Beer and Stettler, 1992). Within the SMZ the suggested large-scale, steeply SW-verging shear zones indicate a large duplex of which the Hout River Shear Zone forms the sole thrust (here termed $D2_B$). Geophysical data (de Beer and Stettler, 1992; Roering et al., 1992b) furthermore showed that granulite facies rocks, which are restricted to the area north of the Hout River Shear Zone,

Figure 3. Ca. 2.664 Ga granodioritic phase of the Matok Pluton (right) intrudes a $D2_B$ shear zone developed in metapelitic gneiss (left). The $D3_A$ overprint, which is demonstrated by the mylonitic fabric developed in the granodiorite, is linked to upper crustal reactivation of the $D2_B$ N'Tabalala Shear Zone (Fig. 2). Scale is provided by camera case.

do not underlie the Kaapvaal Craton. Fault-bounded outcrops of granulite-facies rocks on the Kaapvaal Craton far south of the Hout River Shear Zone (Fig. 2, loc. 6) (van Reenen et al., 1988) are small, rootless erosional remnants (klippen) of a thrust sheet. The hanging wall of the terrane boundary was at the time assumed to consist of two major structural features: the steeply SW-verging Hout River Shear System, and an as yet unidentified flat, SW-verging thrust that was called upon to explain the thrusting of granulites onto the Kaapvaal Craton.

It is important to note that the structural evolution of the SMZ at the time was entirely based on the regional distribution of large-scale linear structures and geophysical data. Subsequent

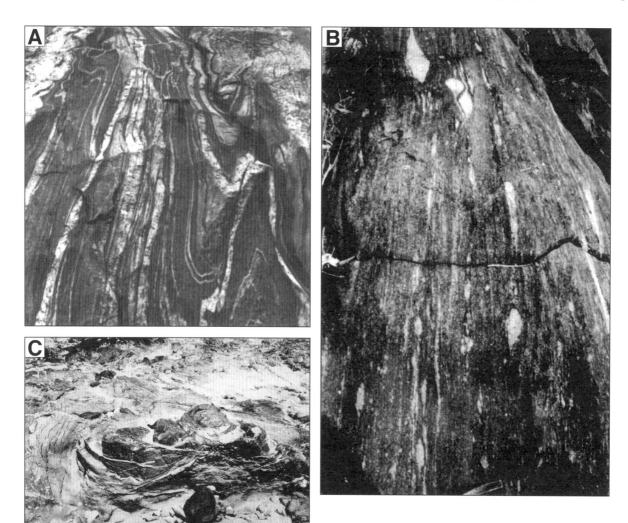

Figure 4. (A) Example of a typical $D2_A$ low-shear-strain gneissic fabric that characterizes $D2_A$ crustal blocks in the SMZ (Fig. 2, loc. 2). (B) Typical example of a near-vertical $D2_B$ high-shear-strain fabric that characterizes metapelitic straight gneisses from the Petronella Shear Zone (Fig. 2, loc. PET1). (C) Steep NE-plunging sheath fold at the type locality (Fig. 2, loc. 5) of the Hout River Shear Zone. The fold axis is oriented parallel to other $D2_B$ movement indicators. (D) $D3_A$ mylonitic shear fabric, which in this photograph completely overprints the early high-grade $D2_B$ shear fabric of the N'Tabalala Shear Zone (Fig. 2).

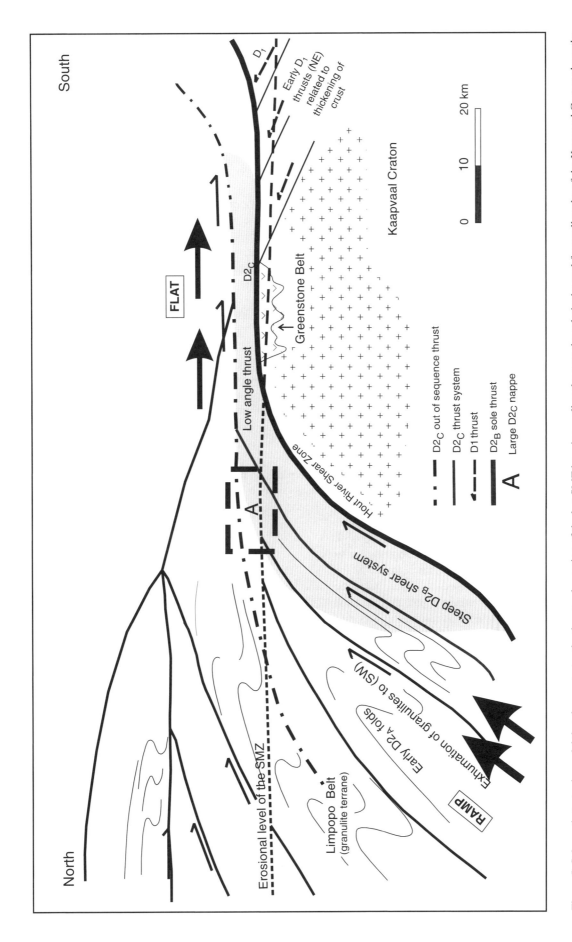

Figure 5. Schematic tectonic model that demonstrates the thrust exhumation of the hot SMZ hanging-wall rocks over the relatively cool footwall rocks of the Kaapvaal Craton along the Hout River Shear Zone. The initial exhumation history at ca. 2.69–2.664 Ga was controlled by steep, SW-verging $D2_B$ shear zones (Figs. 2, 4B, 4C) in the ramp section, whereas the final thrusting of the hot SMZ onto the cool Kaapvaal Craton was controlled by near-horizontal out-of-sequence SW-verging $D2_C$ thrusts (see Figs. 10A–10D) in the flat section. These thrusts were active in the interval 2.63–2.6 Ga. The zone of hydrated granulites (Fig. 2) is shown in gray, and the position of the large structural nappe is shown at A. See text for discussion.

structural, metamorphic, and radiometric studies have focused mainly on testing the validity of these interpretations.

Metamorphic Petrology

Metamorphic Setting of the SMZ and Juxtaposed Kaapvaal Craton

Van Reenen and Du Toit (1977) showed that the SMZ is subdivided into a northern granulite zone and a southern zone of rehydrated granulites (Fig. 2; Plate 1). Both zones are characterized mainly by mafic, ultramafic, metapelitic, and gray tonalitic gneisses of similar bulk chemical compositions (Du Toit et al., 1983; van Reenen, 1983). Van Reenen (1986) subsequently demonstrated that the transition from the granulite zone to the zone of rehydrated granulites is a well-defined retrograde isograd (Fig. 2; Plate 1). Low-grade rocks from the juxtaposed greenstone belts thus occupy the footwall of this major tectonic boundary, and rehydrated granulites the hanging wall (Fig. 2; Plate 1) (de Wit et al., 1992a; McCourt and van Reenen, 1992; Roering et al., 1992a).

Prograde Metamorphic Zonation of Greenschists in the Footwall of the Hout River Shear Zone

Several authors (van Reenen et al., 1988; de Wit et al., 1992c; McCourt and van Reenen, 1992; Miyano et al., 1992; Van Schalkwyk and van Reenen, 1992) have documented evidence for the telescopic prograde metamorphic zonation of rocks from the Giyani and Rhenosterkoppies Greenstone Belts that are located in the footwall of the Hout River Shear Zone (Fig. 2; Plate 1). In the Giyani Greenstone Belt (Fig. 2, loc. SL22) the prograde metamorphic zonation from greenschist through epidote-amphibolite to garnet amphibolite can be observed over a few hundred meters from south to north as the contact with the SMZ is approached (McCourt and van Reenen, 1992). Quartzitic schists immediately adjacent to the terrane bounding shear zone demonstrate the replacement of kyanite by sillimanite (McCourt and van Reenen, 1992). Miyano et al. (1992) also showed that prograde garnet-kyanite–bearing assemblages in the Rhenosterkoppies Greenstone Belt (Fig. 2, loc. 1) can be linked to thrusting of high- over low-grade rocks.

Roering et al. (1992a; Fig. 3) ascribed the prograde zonation of greenstone belts in the footwall of the Hout River Shear Zone to a "hot-iron" effect (isobaric heating) owing to thrusting of hot SMZ rocks over cold cratonic rocks.

Post-Peak Metamorphic History of the SMZ

The granulite zone. Metapelitic granulites in the granulite zone (Fig. 2; Plate 1) are coarse-grained migmatitic gneisses composed of orthopyroxene + biotite + plagioclase + quartz ± cordierite ± garnet ± K-feldspar with minor kyanite-sillimanite and spinel (van Reenen and Du Toit, 1977, 1978). The presence or absence of garnet and/or cordierite in these rocks is controlled by systematic variations in the bulk MgO/(MgO+FeO) (X_{MgO}) ratio, which explain the presence of three main mineralogical groups in the SMZ (van Reenen, 1983): (1) X_{MgO} <0.6: Fe-rich garnet + orthopyroxene + biotite + quartz + plagioclase (Fig. 6A); (2) X_{MgO} = 0.6–0.7: garnet + orthopyroxene + cordierite + biotite + quartz + K-feldspar + plagioclase ± kyanite-sillimanite and spinel (Fig. 6B); (3) X_{MgO} >0.7: Mg-rich orthopyroxene + cordierite + biotite + quartz + plagioclase. The Mg-rich cordierite-bearing varieties are always characterized by cordierite-orthopyroxene symplectite (Fig. 6B), interpreted to reflect the progress of the garnet-breakdown reaction 1, which is preserved in rocks with 0.6 < X_{MgO} < 0.7 (van Reenen, 1983; Stevens and van Reenen, 1992):

$$\text{Grt} + \text{Qtz} \rightarrow \text{Crd} + \text{Opx}. \quad (1)$$

The observation that reaction 1 is complete in Mg-rich metapelitic compositions, has not run to completion in intermediate X_{MgO} compositions (Fig. 6B), and did not start in Fe-rich compositions (Fig. 6A), was taken as evidence (van Reenen, 1983;

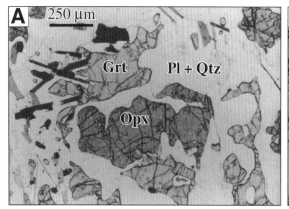 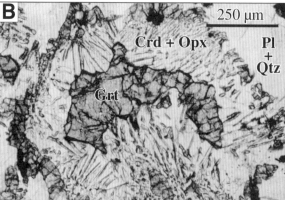

Figure 6. (A) Typical granoblastic texture of cordierite-free (X_{MgO} <0.60) garnet-orthopyroxene-bearing metapelitic granulite. (B) Opx-Crd symplectite replacing garnet in metapelitic granulite with 0.6 < X_{MgO} < 0.7, after the reaction (1) Grt + Qtz → Crd + Opx. Grt—garnet; Qtz—quartz; Pl—plagioclase; Opx—orthopyroxene; Crd—cordierite.

Stevens and van Reenen, 1992) that a decompression-cooling P-T path (Fig. 7B) rather than an isothermal-decompression path characterized much of the post-peak metamorphic evolution of the SMZ. This is because the four-phase field for coexisting garnet + orthopyroxene + cordierite + quartz oriented subparallel to the T-axis shifts to lower pressure in Fe-rich bulk compositions (Fig. 7A) (Hensen and Green, 1973). The observation that Fe-rich garnet is still stable in metapelite with X_{MgO} <0.6 (Fig. 6A) suggests that the P-T path during exhumation did not enter the stability field of the Fe-rich garnet, which should have been the case with isothermal decompression (Fig. 7A).

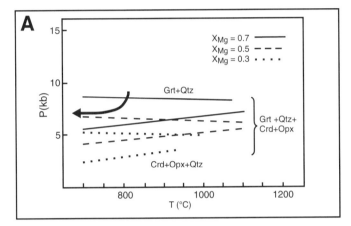

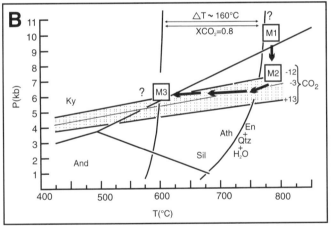

Figure 7. (A) Whole-rock X_{MgO} control on the relative position in P-T space of the divariant stability field of Grt + Crd + Opx + Qtz (after Hensen and Green, 1973; van Reenen, 1983). Isothermal decompression is excluded because whereas Mg-rich garnet is unstable in metapelitic gneisses with X_{MgO} >0.6, Fe-rich garnet still remain stable in rocks with X_{MgO} <0.6 (see P-T path with arrow). See text for discussion. (B) P-T diagram illustrating the metamorphic evolution of the SMZ (van Reenen, 1986; van Reenen and Hollister, 1988). M1—inferred early prograde conditions; M2—post-peak decompression-cooling history; M3—rehydration and establishment of retrograde isograd. Fluid inclusion data (–12, etc.) refer to T_H of CO_2-rich inclusions. Also shown is the displacement to lower temperature of the dehydration reaction enstatite (En) + Qtz + H_2O = anthophyllite (Ath) owing to extensive dilution of the pure-water fluid phase by CO_2. Ky—kyanite. See text for discussion.

Mafic granulites closely associated with metapelitic gneisses in the granulite zone (Plate 1) are coarse-grained granoblastic rocks composed of plagioclase (An40–61) + orthopyroxene + clinopyroxene + olive-brown hornblende + magnetite-ilmenite + minor quartz (Fig. 8A) (van Reenen, 1983).

Ultramafic granulites occur within the regional high-grade foliation as podlike bodies (Plate 1). The coarse-grained granoblastic rocks are composed mainly of forsterite + spinel + enstatite + pargasitic hornblende (Fig. 8C) (Van Schalkwyk and van Reenen, 1992).

The retrograde isograd. The metamorphic transition from the granulite zone in the north to the zone of rehydrated granulites in the south (Fig. 2; Plate 1) is reflected by retrograde reactions that characterize all granulite-facies rocks belonging to the Bandelierkop Formation, also including the gray tonalitic Baviaanskloof gneisses. The retrograde isograd has been mapped over a distance of >200 km from west to east across the entire SMZ (Fig. 2; Plate 1), and the position in the field is defined by the orthopyroxene-breakdown reaction 2, which is recorded by all three compositional groups of metapelitic gneisses (Fig. 9A) (van Reenen, 1986).

$$7(Mg, Fe)SiO_3 + SiO_2 + H_2O \rightarrow (Mg, Fe)_7Si_8O_{22}(OH)_2 \quad (2)$$

or $\quad 7Opx + Qtz + H_2O \rightarrow Ath.$

Reaction 2 occurs over a very restricted interval in the field owing to the fact that it is univariant in the subsystem (MgO-FeO)-SiO_2-H_2O (f = c+2-p = 3+2-4 = 1) because orthopyroxene that coexists with anthophyllite in FeO- and MgO-rich metapelitic gneisses (Fig. 9A) along the retrograde isograd have identical X_{MgO} = 0.62 (van Reenen, 1986).

On the other hand, the cordierite-breakdown reaction 5 (Fig. 9B),

$$5(Mg, Fe)_2Al_4Si_5O_{18} + H_2O \rightarrow$$
$$2(Mg, Fe)5Al_4Si_6O_{22}(OH)_2 + 6Al_2SiO_5 + 7SiO_2 \quad (3)$$

or $\quad 5Crd + H_2O \rightarrow 2Ged + 6Ky/Sil + 7Qtz$

in Mg-rich metapelitic gneisses straddles the position of the retrograde isograd over a much wider interval in the field (Fig. 2; Plate 1) (van Reenen, 1986). The divariant nature of reaction 3 (f = c+2-p = 5+2-5 = 2) within the subsystem MgO-FeO-Al_2O_3-SiO_2-H_2O (Fig. 9B) is due to the distinctly different X_{MgO} ratios of coexisting cordierite (0.86) and gedrite (0.68) (van Reenen, 1986).

Published field, petrologic, and whole-rock–mineral O-isotope fractionation data suggest that the retrograde isograd is a metamorphic feature that was established in metapelitic gneisses by a water-bearing CO_2-rich fluid that infiltrated hot SMZ rocks along a north-to-south decreasing P-T gradient during exhumation (van Reenen, 1986; van Reenen and Hollister 1988; Baker et al., 1992). This is demonstrated by the observation that the retrograde isograd cuts through the high-grade fold pattern of the

SMZ (Fig. 2; Plate 1). The only exception is a restricted area directly east of the Matok Pluton, where the isograd follows the reactivated N'Tabalala Shear Zone for a short distance before it swerves east, again cutting through the fold pattern. Garnet-biotite thermometry, obtained from garnet cores and from matrix biotite in metapelitic gneisses sampled throughout the SMZ (van Reenen, 1986), shows a systematic decrease in temperature from the granulite zone in the north (T >700–800 °C) across the isograd (T ~620 °C) to the zone of rehydrated granulites in the south (T <~600 °C). The presence of kyanite as a product of reaction 3 (Figs. 9B, 9C) allows the minimum pressure of the orthopyroxene-breakdown reaction 2 to be estimated as P ~6kbar at T ~600 °C (Fig. 7B) (van Reenen, 1986). The low-temperature conditions of hydration of orthopyroxene in the presence of quartz (reaction 2) show that the isograd was established by a water-bearing fluid with $X_{CO2} = 0.8$ (Fig. 7B) (van Reenen, 1986). The results of a fluid-inclusion study (van Reenen and Hollister, 1988) subsequently confirmed the CO_2-rich nature of the infiltrating fluid.

Additional research confirmed both the metamorphic and suggested P-T-X conditions of the retrograde isograd based on the studied phase equilibrium relationships of orthopyroxene + cordierite + anthophyllite + kyanite-sillimanite + quartz in the FMASH system at P = 6 kbar, using the data set and THERMOCALC program of Holland and Powell (1990). Baker et al. (1992; Fig. 2) interpreted the results of this study to indicate that the equilibrium hydration of orthopyroxene to anthophyllite through reaction 2 occurred at T = 450–650 °C for P = 5–7 kbar at water activities between 0.6 and 0.1. Van Schalkwyk and van Reenen (1992, Figs. 12–14), also assuming equilibrium, showed that the olivine-breakdown reaction 4:

$$Mg_2SiO_4 + CO_2 \rightarrow Mg(SiO_3)^2 + MgCO_3 \qquad (4)$$

observed in partially rehydrated ultramafic granulites occurred at T ~670 °C and X_{CO_2} >0.67. These results followed from the phase equilibrium analyses of the isobaric (P ~6 kbar) invariant assemblage of olivine + orthopyroxene + calcic-amphibole + spinel + magnesite + chlorite + dolomite in the system $CaO-MgO-SiO_2-Al_2O_3-CO_2-H_2O$ (Van Schalkwyk and van Reenen, 1992). The calculations were done with the GEOCALC program (Berman et al., 1985) and the thermodynamic data set of Berman et al. (1985).

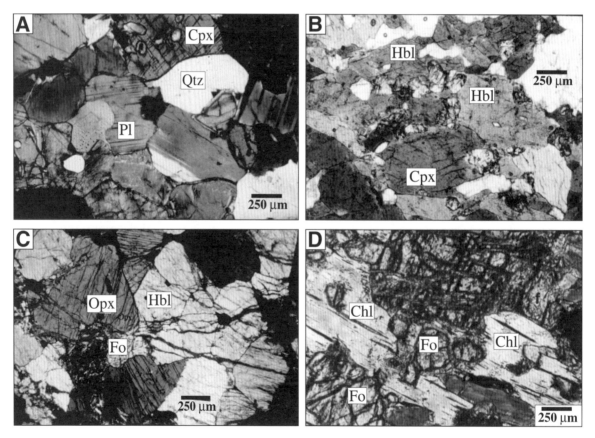

Figure 8. (A) Mafic granulite comprising orthopyroxene (Opx) + clinopyroxene (Cpx) + hornblende (Hbl) + plagioclase (Pl) + quartz (Qtz). Note the granoblastic texture. (B) Partially hydrated mafic granulite in which pyroxene is partially or completely replaced by hornblende. (C) Ultramafic granulite comprising olivine (Fo) + enstatite (En) + pargasitic hornblende (Hbl) + spinel (black). Note the granoblastic texture. (D) Partially hydrated ultramafic granulite showing olivine (Fo) replaced by chlorite (Chl).

Subsequently, the results of a whole-rock–mineral O-isotope fractionation study of a variety of granulites and their rehydrated equivalents by Hoernes et al. (1995) constrained the "peak" temperature conditions reached in the granulite zone to a range of 800–870 °C. Partial O-isotopic equilibria furthermore suggested that orthoamphibole formation occurred at temperatures close to 600 °C. Hoernes et al. (1995) concluded that retrogression of SMZ granulites was triggered by the influx of an externally derived fluid rich in CO_2 into an existing north-to-south temperature gradient across the retrograde isograd (Fig. 2; Plate 1).

Data published by different authors (Baker et al., 1992; Van Schalkwyk and van Reenen, 1992; Hoernes et al., 1995) thus are in accordance with retrogression from the introduction of a CO_2-rich H_2O-bearing fluid into rocks still at high grade (T ~630 °C, P ~6 kbar) that resulted in a retrograde metamorphic isograd being established in the hanging wall of the Hout River Shear Zone. Van Reenen and Hollister (1988) suggested that the infiltrating fluid could have been derived from devolatilization reactions related to the thrusting of hot granulites in the hanging wall over cool greenschists in the footwall of the Hout River Shear Zone (Fig. 2; Plate 1).

Stevens (1997), on the other hand, attributed the retrograde isograd to a fault boundary between two high-grade SMZ blocks with different H_2O contents, which resulted from different degrees of melt loss during the crystallization of anatectic granites.

Zone of rehydrated granulites. Metapelitic gneisses with X_{MgO} >0.6 in the zone of rehydrated granulites (van Reenen, 1986) are characterized by coexisting anthophyllite + gedrite + kyanite-sillimanite + biotite + Ca-rich plagioclase ± garnet (Figs. 9C, 9D). The presence or absence of garnet is determined by the X_{MgO} ratios of the rocks as described for the dehydrated equivalents. Rehydrated granulites with X_{MgO} <0.6 comprise garnet + anthophyllite + biotite + quartz + plagioclase (van Reenen, 1986).

Rehydrated mafic granulites characteristically contain relict orthopyroxene and clinopyroxene, which are invariably rimmed and partly replaced by green hornblende (Fig. 8B). The peak granulite assemblage of forsterite + orthopyroxene + spinel + pargasitic hornblende (Fig. 8C) of ultramafic granulites is replaced mainly by magnesite and chlorite (Fig. 8D) (Van Schalkwyk and van Reenen, 1992). The replacement of forsterite by a second generation of orthopyroxene and magnesite according to forsterite-breakdown reaction 4 is convincing evidence for

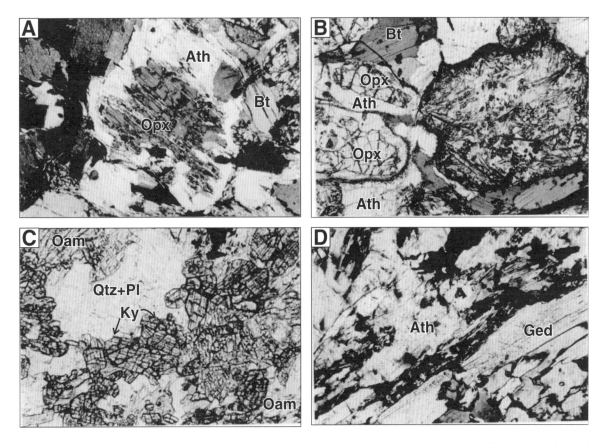

Figure 9. (A) Orthopyroxene (Opx) mantled by anthophyllite (Ath) in Fe-rich metapelitic gneiss along the retrograde isograd (Fig. 2; Plate 1). (B) Orthopyroxene mantled by anthophyllite, and pseudomorphic cordierite (right) being replaced by an intergrowth of gedrite and kyanite in Mg-rich metapelitic gneisses along the retrograde isograd. (C) Coexisting orthoamphibole (Oam) and kyanite (Ky) in the zone of rehydrated granulite (Fig. 2; Plate 1). (D) Coexisting gedrite (Ged) and anthophyllite (Ath) in rehydrated granulite (after van Reenen, 1986).

the post-peak infiltration under high-temperature retrograde conditions of an externally derived fluid rich in CO_2 (van Reenen et al., 1988; Van Schalkwyk and van Reenen, 1992, Figs. 12–14).

The results of the early metamorphic studies showed that the granulite zone (Fig. 2; Plate 1) prior to the establishment of the retrograde isograd and associated zone of rehydrated granulites extended all the way south to the present position of the Hout River Shear Zone. This is suggested by the observed distribution of relict orthopyroxene in mafic and ultramafic granulites. Geophysical data (de Beer and Stettler, 1992) furthermore showed that granulite facies rocks do not underlie the Kaapvaal Craton south of the Hout River Shear Zone.

GEOCHRONOLOGY

Barton and van Reenen (1992a, 1992b) summarized the then available geochronologic data for the SMZ (and also for the entire Limpopo Complex).

The virtually undeformed Matok granitic pluton that intruded the high-grade gneisses of the SMZ, as well as the steeply NE-dipping Matok Shear Zone (Figs. 2, 3) is a most important time marker that provided a minimum age of ca. 2.671 Ga (age of the charnoenderbitic phase) for high-grade metamorphism and associated deformation of the SMZ (Barton and van Reenen, 1992a). A minimum age of ca. 2.65 Ga for the high-grade event is supported by Rb-Sr muscovite ages obtained from a variety of pegmatites, which intrude high-grade shear zones associated with the Hout River Shear Zone (Barton and van Reenen, 1992a).

The important conclusion, that the SMZ was emplaced at the near surface before ca. 2.46 Ga, follows from the observation that subvolcanic dikes with volcanic glass in the groundmass related to the ca. 2.46 Ga post-tectonic Palmietfontein Granite (Barton et al., 1983) intrude the Baviaanskloof gneisses at the Goudplaas locality (Fig. 2, loc. 2).

Evidence that the high-grade SMZ rocks have been reworked in the Paleoproterozoic is suggested by the reactivation of the N'Tabalala Shear Zone (Fig. 4D) that cuts through both the Matok Pluton (Fig. 2; Plate 1) and the Palmietfontein Granite (Du Toit et al., 1983). Further evidence is given by Rb/Sr biotite ages of ca. 2.0 Ga obtained from a variety of rocks from the SMZ and granite-greenstone Kaapvaal Craton (Barton and van Reenen, 1992b). However, Barton and van Reenen (1992b) linked the ca. 2.0 Ga biotite ages not to a distinct tectonic event but to the final uplift and cooling of the high-grade SMZ.

A Neoarchean Pop-Up Model for the Evolution of the SMZ

The Neoarchean tectonic pop-up model suggested for the evolution of the SMZ (and the entire Limpopo Complex) by van Reenen et al. (1987) and Roering et al. (1992b) was proposed on the basis of the available geophysical data (Fig. 5) and the observation that the evolution of the SMZ was basically controlled by exhumation during an early tectonic event that led to crustal thickening and large-scale instability. Any discussion about the early pre-exhumation stage of the Limpopo orogenic event at the time was speculative and depended entirely on sketchy evidence provided by the early northward verging thrusts then recognized within the Pietersburg Greenstone Belt (Fig. 2; Plate 1) (de Wit et al., 1992b). However, deformational features throughout the Limpopo Complex provided evidence for continued horizontal compression for the duration of the Neoarchean Limpopo orogeny (de Wit et al., 1992c; Roering et al., 1992b). High-grade rocks of the Limpopo Complex that were thrust onto the adjacent cratons are considered as the major response to the deep burial of shallow crustal material (Light, 1982; van Reenen et al., 1987; Roering et al., 1992b; de Wit et al., 1992c). In the SMZ (Fig. 5) exhumation from deep to shallow crustal levels occurred along southward-verging shear zones compatible with a single exhumation event in the Neoarchean. The observed prograde metamorphic zonation of greenschists in the footwall of the Hout River Shear Zone at the time was considered to reflect evidence for a hot-iron effect (isobaric heating) owing to the thrusting of hot over cool rocks.

Major outstanding issues in 1992 include the following: (1) detailed age data to accurately constrain the time of the post-peak retrograde metamorphic event in the hanging wall and of the prograde event in the footwall of the Hout River Shear Zone; (2) structural data to support the proposed near-horizontal SW-verging thrusts necessary to explain the high-grade tectonic klippen on the Kaapvaal Craton far south of the SMZ; these structures, if present, also need to be dated; (3) detailed metamorphic and age data to construct well-constrained retrograde P-T-t paths for the hanging wall and prograde P-T-t paths for the footwall; (4) detailed geochemical data to confirm the suggestion that the SMZ rocks are the high-grade equivalents of low-grade granite-greenstones of the Kaapvaal Craton; (5) finally, and most important, structural, metamorphic, and age data need to be integrated for the construction of D-P-T-t diagrams necessary to study the evolution of the SMZ.

GEOLOGICAL STATUS QUO IN 2010

Structural Geology

After 1992, different researchers (Perchuk et al., 1996, 2000a, 2000b, 2001; van Reenen et al., 1995; Smit and van Reenen, 1997; Gerya et al., 2000; Kreissig et al., 2000, 2001; Smit et al., 2001) were involved in detailed studies of the SMZ and adjacent granite-greenstone terrane in an effort to refine existing knowledge on the distribution and nature of shear zones, the timing of different geological events, and the construction of well-constrained P-T paths. The geological database thus produced was used as the basis for a detailed geological map of the SMZ (Plate 1) at a scale of ~1:500,000. The map shows the distribution of lithological units both within the SMZ and adjacent granite-greenstone terrane of the Kaapvaal Craton. It also shows the regional structural fabric pattern as form lines, the location of shear zones and distribution of large crustal blocks within the SMZ, the distribution of a large nappe in the hanging wall of the

Hout River Shear Zone, and the position of the retrograde isograd that subdivides the SMZ into a northern granulite zone and a southern zone of rehydrated granulites.

Mapping in the Rhenosterkoppies Greenstone Belt (Fig. 2; Plate 1) (Passeraub et al., 1999) established the presence of a system of shallow N-verging thrust faults in the footwall of the steeply SW-verging Hout River Shear Zone (Fig. 2, loc.1; Plate 1). These structures, dated at 2729 ± 19 Ma on the basis of the syntectonic growth of titanite in sheared mafic rocks (Passeraub et al., 1999) are labeled $D1_B$ and represent the earliest identifiable structures that can be related to the Limpopo orogeny. The steeply SW-verging $D2_B$ Hout River Shear Zone terminates the shallow NE-verging D_{1B} structures within the Rhenosterkoppies Greenstone Belt. Kreissig et al. (2001) subsequently dated the time of shear deformation associated with the Hout River Shear Zone at ca. 2.71–2.67 Ga on the basis of prograde staurolite, garnet, and kyanite from sheared mica schists in the footwall of this structure in the Giyani Greenstone Belt (Fig. 2, loc. SL22). The growth of these prograde minerals has been linked to thrusting of hot over cool rocks.

Recent geometric and kinematic studies of the $D2_B$ Hout River Shear Zone and the associated SW-verging shear zones developed within the hanging-wall section showed that this major tectonic boundary consists of distinct structural features: First, a steeply oriented $D2_B$ structural duplex with steep SW-verging crustal movement that represents the ramp portion of the structure (Figs. 2, 5). Second, a flat section (Figs. 5, 10D) represented by a major SW-verging $D2_C$ structural nappe that occupies large

Figure 10. Reworked gray tonalitic gneisses along a $D2_C$ sole thrust that separates the hot high-grade gneisses in the hanging wall of the Hout River Shear Zone from the cool reworked schists of the Kaapvaal Craton in the footwall. (A) Thrust contact between steep $D2_B$ (left) and near horizontal $D2_C$ (right) shear fabrics in the Middle Letaba River on the Goudplaas farm at locality 2 (Fig. 2). (B) Small structural duplex developed in quartzofeldspathic gneisses in the Sand River on the Breipaal farm (Fig. 2, loc. 1). The shallow SW-verging $D2_C$ thrust (left) cuts through the near vertical shear fabric of a steep SW-verging $D2_B$ thrust. (C) Near-horizontal SW-verging intersection lineations developed in quartzofeldspathic gneisses in the Sand River on the Breipaal farm (Fig. 2, loc. 1). (D) Near-horizontal $D2_C$ fabric characterizes quartzofeldspathic gneisses within the large nappe in the hanging wall of the Hout River Shear Zone (Fig. 2, locs. 1–4; Plate 1).

parts of the SMZ in the hanging wall of the shear zone (Plate 1). The flat-lying gneisses (Fig. 10D) of the nappe overprint the steep fabric of the ramp portion (Figs. 10A, 10B), suggesting that the flat structure is out-of-sequence with its steeper counterpart. This is well demonstrated in the Sand River on the Breipaal farm (Fig. 1, loc. 1), where over a distance of ~500 m from north to south the near vertical $S2_B$ shear foliation of quartzofeldspathic gneisses is progressively overprinted by the shallow, NE-dipping $S2_C$ shear fabric (Fig. 10B). Ages between 2.63 and 2.6 Ga (Kreissig et al., 2001), obtained from syntectonic Ca-amphibole that define an intersection lineation (Fig. 10C) associated with the development of these flat fabrics, support this suggestion. The existence of the major nappe would also account for the presence of high-grade outliers (klippen) that occur on the Kaapvaal Craton south of the Hout River Shear Zone (Fig. 2, loc. 6) (van Reenen et al., 1988), confirming the presence of flat structures in the SMZ predicted many years earlier on the basis of geophysical data (Fig. 5) (Roering et al., 1992a, 1992b; Percival et al., 1997).

At the same time research in the SMZ shifted toward understanding the geological processes that were operational during the geodynamic growth of the high-grade terrane (Gerya et al., 2000; Perchuk et al., 2000a, 2000b, 2001; Smit et al., 2001; Perchuk and Gerya, this volume). Distinct fabric development linked to specific mineral assemblages and large structural features (Smit and van Reenen, 1997; Smit et al., 2001) formed the basis for a new research approach in the high-grade gneiss terrane of the SMZ. This approach follows the fact that changes in pressure, temperature, fluid activity, and stress conditions within a given time span control the rheological properties of rocks, the growth and re-equilibration of metamorphic mineral assemblages, and thus the development of a particular fabric related to large-scale structures, either large-scale folds or deep crustal shear zones. Data thus generated formed the basis for the construction of D-P-T-t diagrams, which proved to be an essential tool for studying the evolution of the high-grade SMZ.

The same technique was applied to the study of the geodynamic evolution of the much more complex polymetamorphic CZ of the Limpopo Complex (Smit et al., this volume), and also proved to be invaluable for geodynamic models for the evolution of high-grade terranes in general (Mahan et al., this volume).

Metamorphic Petrology

Quantitative metamorphic data generated for the SMZ and the juxtaposed granite-greenstone terrane of the Kaapvaal Craton since 1992 formed the basis for the construction of well-constrained P-T paths. These paths provide evidence for the thermal and dynamic interaction of hanging-wall granulite (retrograde P-T paths) and footwall mica schist (prograde P-T paths) during the thrust exhumation of the SMZ onto and over the adjacent Kaapvaal Craton (Perchuk et al., 1996, 2000a, 2000b, 2001; Gerya et al., 2000; Kreissig et al., 2001; Smit et al., 2001). Detailed petrologic studies (Tsunogae et al., 2004; Belyanin et al., 2010) furthermore provided the first evidence for UHT conditions in the SMZ that have important consequences for tectonic models for the Limpopo Complex (Kramers et al., this volume; Tsunogae and van Reenen, this volume).

Non-Isobaric P-T Evolution of Mica Schists in the Footwall of the Hout River Shear Zone

Perchuk et al. (1996, 2000b) constructed a narrow non-isobaric P-T loop (Fig. 11A; sample SL22) for strongly sheared mica schists from the Giyani Greenstone Belt in the footwall of the Hout River Shear Zone (Fig. 2, loc. SL22). This is the same locality for which previous authors (McCourt and van Reenen, 1992; Roering et al., 1992a) first described evidence for prograde metamorphic zonation, which they interpreted to reflect isobaric heating (hot-iron effect) owing to thrusting of the hot SMZ rocks over cool rocks from the Kaapvaal Craton. Mica schists (e.g., sample SL22, Fig. 2) at this locality are composed of quartz, garnet, biotite, muscovite, staurolite, kyanite, Ca-amphibole, and plagioclase. The prograde and retrograde stages of the metamorphic evolution are documented by garnet porphyroblasts with prograde (core) and retrograde (rim) zonation, and two distinct generations of prograde and retrograde biotite. Perchuk et al. (1996, 2000b) calibrated the geobarometric reaction

$$\text{Prp} + 2\text{Ms} + \text{Phl} \rightarrow 6 \text{ Qtz} + 3 \text{ East}, \qquad (5)$$

which also depends on temperature to construct the narrow non-isobaric prograde P-T loop (Fig. 11A). This loop reflects the heating and burial of the low-grade cratonic rocks from a depth of ~12 km, and T = 500 °C to a depth of ~16 km at T ~570 °C as the result of the dynamic and thermal interaction of footwall schists with hanging-wall granulites during thrusting of hot SMZ rocks over the cold cratonic rocks in the interval ca. 2.69–2.60 Ga. The subsequent joint exhumation of the SMZ granulites and the juxtaposed greenstones to the upper crustal level (Fig. 11A) is recorded only in mineral equilibria preserved in mica schists (Perchuk et al., 2000b). The situation described here is similar to the non-isobaric P-T loop that involves the formation of snowball garnet porphyroblasts in sheared rocks in the footwall of the Lapland granulite terrane (Perchuk and Krotov, 1998; Perchuk et al., 2000b, 2001).

Metamorphic Evolution of Granulites in the Hanging Wall of the Hout River Shear Zone

Detailed pressure-temperature-$a^{fl}_{H_2O}$ information necessary for the construction of P-T paths for metapelitic granulite are derived from local mineral equilibria (core-core → rim-rim method) coupled with conventional thermobarometry. The application of this method to metapelitic rocks was recently explained in great detail (Perchuk, this volume) (see also Perchuk, 1977, 1989; Perchuk et al., 2008; Smit et al., 2001; Perchuk and van Reenen, 2008; van Reenen et al., 2004). However, useful background information is that each individual point on a P-T path (e.g., Fig. 12A) reflects a specific P-T-$a^{fl}_{H_2O}$ value, which was calculated from the corresponding equilibrium mineral compositions based on phase correspondence. The particular set of

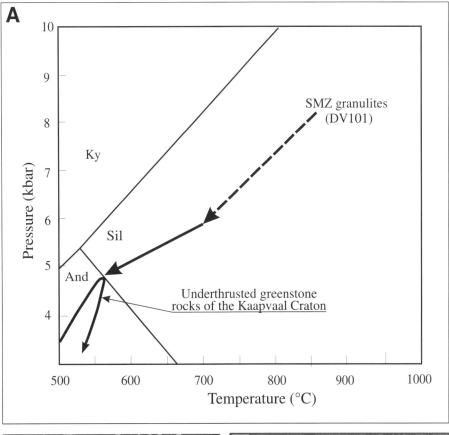
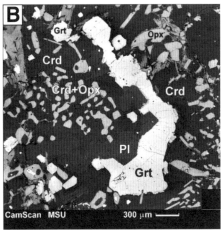
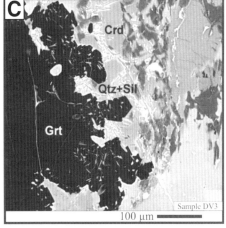
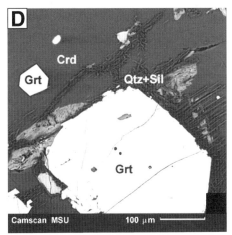

Figure 11. (A) Convergence at ca. 2.69 Ga (Kreissig et al., 2001) of near-isobaric-cooling (IC) P-T path in footwall granulite (DV101) with prograde P-T loop in hanging-wall mica schist (SL22) along the Hout River Shear Zone (after Perchuk et al., 1996, 2000a; Smit et al., 2001). The inferred decompression-cooling (DC) segment of the P-T path (taken from Fig. 12A) is reflected by Crd-Opx symplectite (see B) in which early Mg-rich garnet has been completely consumed (after the reaction Grt + Qtz = Crd + Opx). The near-IC segment is calculated from garnet-producing reactions (B, C). (B) Replacement of early Opx-Crd symplectite (see A) by Fe-rich atoll garnet (after the reaction Opx + Crd = Grt + Qtz). (C) Fe-rich Grt-Sil-Qtz intergrowth at the contact with symplectic Crd (after the reaction Crd = Grt + Sil + Qtz). (D) Small idioblastic grains of Fe-rich garnet. Sil—sillimanite. See Figure 6 caption for abbreviations explanation and text for discussion.

equilibrium mineral compositions, which are linked to a specific P-T-a^{fl}_{H2O} value (Fig. 12A) were identified from detailed microprobe zoning profiles obtained from the relevant mineral assemblage. The following exchange and net-transfer reactions

$$Grt_{Mg} + Bt_{Fe} = Grt_{Fe} + Bt_{Mg}, \qquad (6)$$

$$Grt_{Mg} + Crd_{Fe} = Grt_{Fe} + Crd_{Mg}, \qquad (7)$$

$$Grt + Sil + Qtz = Crd, \qquad (8)$$

$$Grt + Qtz = Crd + Opx, \text{ and} \qquad (9)$$

$$Grt + Kfs + H_2O = Bt + Sil + Qtz \qquad (10)$$

are the basis for the calculation of individual P-T-a^{fl}_{H2O} values and of the corresponding P-T path. The application of an internally consistent thermodynamic data set (e.g., Gerya and Perchuk, 1990) to metapelitic gneisses of similar bulk chemical and mineralogical compositions furthermore allow the direct comparison of calculated P-T paths for rocks sampled from different major structural features.

The paper by Smit et al. (2001) provides a detailed description of the methodology used to construct P-T paths for the metapelitic granulites under discussion (DR45, DV101, PET1, and CAPR7) and also demonstrate how robust the inferred slopes in P-T space are. Furthermore, the two types of P-T paths, decompression cooling (DC) and near-isobaric cooling (IC), were modeled using two-dimensional (2D) numerical experiments (Gerya et al., 2000) to test the consistency between the shapes of the calculated and modeled paths. The high accuracy of the thermometers studied experimentally follows from linear regression of kd = f (T) with a correlation coefficient of 0.999 (Perchuk and Lavrent'eva, 1983):

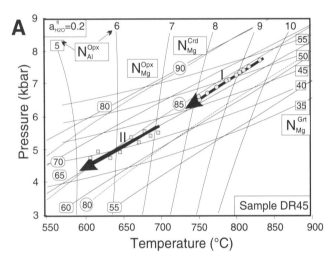

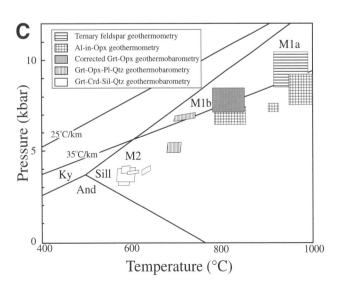

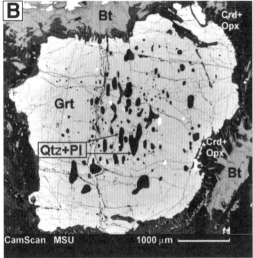

Figure 12. (A) Post-peak decompression-cooling (DC) P-T path and compositional isopleths calculated for D2$_A$ sample DR45 located far from the boundary with the Kaapvaal Craton (Fig. 1, loc. DR45) (Perchuk et al., 2000a; Smit et al., 2001). Note the distinct gap in the P-T path that separates the early and late DC segments. P-T parameters for the early post-peak DC segment (before the gap; open circles) are calculated from the inclusion-rich cores of porphyroblasts of Grt + Opx + Crd (see B), whereas P-T parameters (open squares) for the second segment (after the gap) are calculated from the inclusion-free rim of garnet surrounded by the Opx-Crd symplectite (see B). (B) Photomicrograph of garnet porphyroblast surrounded by a narrow Crd-Opx symplectite. Note the oriented inclusions in the MgO-rich Grt core overgrown by an inclusion-free Fe-rich Grt rim. Bt—biotite. (C) UHT P-T path constructed for SMZ metapelitic granulite (Tsunogae et al., 2004) shows peak T to be at least 100 °C higher than the maximum T of the DC P-T path in A. Note the similar shape and P-T conditions reflected by two P-T paths (A, C), which were constructed using completely different methodologies. See text for discussion.

Bt-Grt thermometer: ± 13.49 °C (p. 234);

Crd-Grt thermometer: ± 11 °C (p. 230).

The accuracy of the cordierite-garnet-sillimanite-quartz geobarometer (Aranovich and Podlesskii (1983)) is as follows: P (max): ± 1 kbar, T: ± 15 °C (p. 191).

Perchuk et al. (1996, 2000a) showed that a large number of metapelitic gneisses in the granulite zone of the SMZ reflect evidence for two distinct post-peak P-T paths that are determined by the position of the samples relative to the tectonic boundary with the cool rocks of the Kaapvaal Craton. Rocks sampled far from this boundary (Fig. 2, loc. DR45) reflect evidence for a decompression-cooling P-T path (Fig. 12A), whereas rocks sampled close to this boundary (Fig. 2, loc. DV101, PET1, CAPR7) are characterized by a distinct kink in the P-T path (Figs. 11A, 13A, 13B) that reflects evidence for near-IC following the early DC stage. The DC stage of exhumation is reflected by reaction textures that resulted from garnet consumption (reaction 1), whereas the IC stage is linked to reaction textures resulting from garnet growth (reactions 11 and 12).

A DC P-T path (e.g., Fig. 12A, sample DR45) characterizes metapelitic gneisses sampled from low-shear-strain $D2_A$ crustal blocks far to the north of the Hout River Shear Zone (Fig. 2, loc. DR 45). Rocks with X_{MgO} >0.6 <0.7 that reflect evidence for the garnet-breakdown reaction (1) (Grt + Qtz → Crd + Opx) (Fig. 6B) are characterized by two distinct generations of garnet, cordierite, and orthopyroxene (van Reenen, 1983). These different generations of the same minerals were respectively used to construct different segments of the DC P-T path (Fig. 12A). The early post-peak DC segment (Fig. 12A), which traverses from P ~8 kbar, T ~825 °C, and ends at P ~6.5 kbar, T ~750 °C, is calculated from the inclusion-rich cores of porphyroblastic Mg-rich garnet (Fig. 12B), which is interpreted to have coexisted with matrix porphyroblastic cordierite and orthopyroxene. The second segment of this DC path, which traverses from P ~6 kbar, T ~700 °C to P ~4.5 kbar and T ~620 °C (Fig. 12A), is calculated from the inclusion-free Fe-rich garnet rim in contact with symplectic cordierite and orthopyroxene (Fig. 12B) (Perchuk et al., 2000a; Smit et al., 2001). The two segments of the P-T path are not continuous but are separated by a distinct gap (P ~6.5–6 kbar, T ~750–700 °C) that coincides with the contact between the inclusion-rich core and the inclusion-free rim of porphyroblastic garnet.

DC paths followed by near-IC paths (Figs. 11A, 13A, 13B) are constructed for rocks located much closer to the tectonic boundary with the cool rocks of the Kaapvaal Craton (Fig. 2, loc. DV101, PET1, CAPR7). Such rocks preserve only indirect evidence for the early DC segment, which is reflected by symplectic textures (Fig. 11B) in which Mg-rich garnet has been completely consumed owing to the complete shift of reaction (1) to the right. The subsequent near-IC segment is recorded by reaction textures that reflect the growth of Fe-rich garnet (Figs. 11B–11D) (Perchuk et al., 1996, 2000a; Smit et al., 2001):

Opx + Crd (products of reaction 1) → Grt + Qtz;   (11)

Crd (product of reaction 1) → Grt + Sil + Qtz.   (12)

Sample DV101 from a low-shear-strain crustal block close to the boundary with the Kaapvaal Craton (Fig. 2, loc. DV101) documents clear evidence for a DC → IC P-T path. This evidence is provided by Crd-Opx symplectite (Fig. 11B) that formed after Mg-rich garnet (reaction 1), which was completely consumed during the early DC stage of the evolution from the post-peak conditions to T = 675 °C and P = 5.7 kbar (inferred broken P-T path in Fig. 11A). The early-formed Crd-Opx symplectite is partly replaced by more Fe-rich atoll garnet that resulted from reaction 5 (Fig. 11B). Delicate garnet-quartz-sillimanite intergrowths (Fig. 11C) were also formed by decomposition of early symplectic cordierite according to reaction 12, which proceeded simultaneously with reaction 11. Newly formed tiny euhedral Fe-rich garnet characterizes this reaction texture (Fig. 11D). The near-IC segment of the P-T path of sample DV101 (solid line in Fig. 11A) shows T decreasing from ~650 °C to 550 °C, whereas P changes from ~6 to 5 kbar.

High-strain $D2_B$ straight gneisses (PET1 and CAPR7) (Smit and van Reenen, 1997; Smit et al., 2001) from the Petronella Shear Zone (Fig. 2, loc. PET1) reflect reaction textures (Figs. 13C, 13D) and P-T histories (Figs. 13A, 13B) that are identical to those preserved in the closely associated low-shear-strain $D2_A$ gneiss (sample DV101). However, in contrast to the observation that symplectite in low-shear-strain rocks is characterized by randomly oriented orthopyroxene needles (Figs. 6B, 12B), reaction textures in high-shear-strain gneisses are oriented (Figs. 13C, 13D), reflecting the syn-kinematic growth of minerals during high-temperature, high-strain shear deformation (Smit et al., 2001). The high-shear-strain gneisses (Figs. 13C, 13D) also preserve indirect evidence for the early DC segment (inferred broken paths in Figs. 13A, 13B) owing to the complete consumption of Mg-rich garnet after reaction 1.

The differences in the shapes of DC paths and DC paths followed by IC paths, and the link respectively with ca. 2.69–2.664 Ga steeply SW-verging $D2_B$ and ca. 2.63–2.6 Ga near-horizontal SW-verging $D2_C$ thrusts, are discussed with reference to an integrated D-P-T-t diagram (Fig. 14).

UHT Metamorphism of the SMZ

Tsunogae et al. (2004) first published evidence (Fig. 12C) that the actual peak T metamorphic conditions reached in the SMZ are ~100 °C higher than the previously published "peak" conditions of ~870 °C (Fig. 12A) (Perchuk et al., 1996, 2000a; Smit et al., 2001). Different approaches were used to calculate UHT conditions (Fig. 12C): first, a ternary feldspar geothermometer as applied to antiperthitic feldspars in a leucocratic granulite gave T = 920–980 °C; second, Al solubility in orthopyroxene in a metapelitic granulite gave T = 970–1020 °C; and third, a revised (Pattison et al., 2003) garnet-orthopyroxene geothermometer corrected for retrograde Fe-Mg exchange gave T = 920–990 °C.

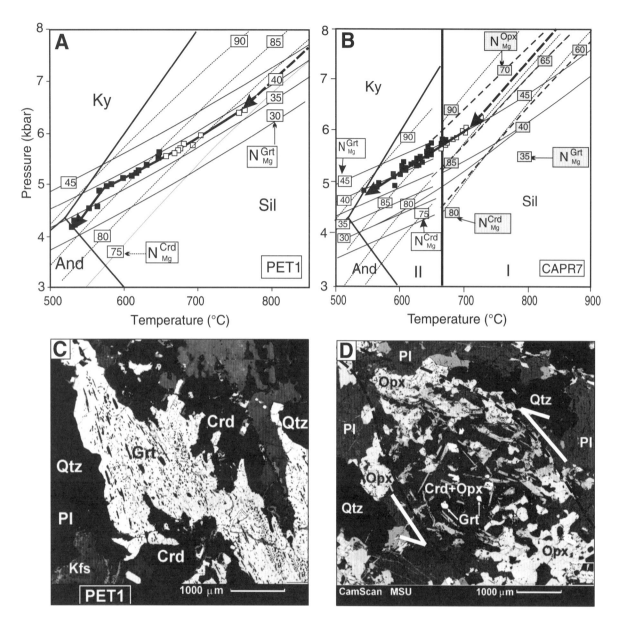

Figure 13. (A, B) P-T paths and compositional isopleths (after Smit et al., 2001) constructed for sheared $D2_B$ metapelitic samples PET1 (A) and CAPR7 (B) (Fig. 1, loc. PET1). Isopleths shown in fields I and II (P-T path for sample CAPR7) respectively refer to the assemblage Grt + Opx + Crd + Qtz, and Grt + Crd + Sil + Qtz. DC P-T paths in A and B reflect the inferred early history (taken from Fig. 12A), whereas near-IC paths reflect the subsequent near-isobaric evolution. Open and filled symbols respectively indicate P-T parameters calculated from core and rim compositions. (C, D) Backscattered electron images, showing the syntectonic growth of reaction textures preserved in sheared $D2_B$ samples PET1 (C) and CAPR7 (D) (after Smit et al., 2001). (C) Sample PET1 illustrates a large, oriented, pencil-shaped porphyroblast of Fe-rich garnet with oriented inclusions of sillimanite and surrounded by cordierite. The texture reflects evidence for the reaction Crd = Grt + Sil + Qtz. (D) Sample CAPR7 illustrates a sigmoid-shaped aggregate in which early Mg-rich Grt has been completely replaced by oriented wormlike Crd-Opx symplectite. The intergrowth contains small idioblastic Fe-rich garnet. The pencil-shaped garnet and sigmoid-shaped aggregates are oriented in the transport direction of the SW-verging $D2_B$ Petronella Shear Zone (Smit et al., 2001). See text for discussion.

Two different approaches (Perchuk et al., 1996, 2000a; Tsunogae et al., 2004) applied for the construction of P-T paths for SMZ granulites thus produced similar P-T paths (Figs. 12A, 12C). Belyanin et al. (2010) recently reported on an unusual corundum lamellar intergrowth with orthopyroxene from similar rocks in support of the UHT signature of SMZ rocks.

The timing of this peak UHT event is indirectly constrained to ca. 2.72 Ga on the basis of the age of shallow NE-verging $D1_B$ thrusts in the Rhenosterkoppies Greenstone Belt, which are terminated by the steeply SW-verging $D2_B$ Hout River Shear Zone (Fig. 5) (Passeraub et al., 1999). These $D1_B$ structures are interpreted to reflect evidence for early N-verging thrusts within the SMZ that were associated with the crustal thickening event that initially resulted in the formation of UHT granulites in the SMZ.

GEOCHRONOLOGY

Post-peak granulite conditions in the SMZ have been constrained to 2691 ± 7 Ma (Kreissig et al., 2001; Kramers et al., 2006) by a U-Pb age for monazite from a metapelitic granulite in the Bandelierkop Quarry (Fig. 2, loc. quarry). A 1-m-wide leucocratic vein, which intrudes and is folded with metapelitic granulites at this type locality, has been interpreted as the product of in situ fluid-absent melting during high-grade metamorphism (Stevens and van Reenen, 1992). A zircon Pb-evaporation age of 2643 ± 1 Ma obtained from this vein constrains the time of crystallization of leucocratic anatectic melts during exhumation of the SMZ, and also highlights the long-lasting post-peak metamorphism (ca. 2.69 Ga) (Kreissig et al., 2001; Kramers et al., 2006).

Kreissig et al. (2001) applied the Pb-Pb step leaching method to date syn-kinematic staurolite, garnet, and kyanite in mica schists from the footwall section of the Hout River Shear Zone in the Giyani Greenstone Belt (Fig. 2, loc. SL22), which respectively yielded ages of 2712 ± 37 Ma, 2691 ± 20 Ma, and 2672 ± 51 Ma. These ages, on the one hand, are much younger than those of the more than 3.0 Ga greenstones, but on the other hand, are within error of the Matok Pluton U-Pb zircon dates (Barton and van Reenen, 1992a; Kramers et al., 2006) and the U-Pb monazite "peak" metamorphic age of the SMZ (Kreissig et al., 2001). Thus, they are interpreted to reflect the time of peak prograde metamorphism that affected the footwall rocks during the early stage of thrusting of the high-grade SMZ over and onto the Kaapvaal Craton along the steeply northerly dipping $D2_B$ Hout River Shear Zone (Fig. 5). The final stage of thrusting is reflected by the near-horizontal top-to-the-SW $D2_C$ thrusts (Figs. 5, 10) that are restricted to the southern boundary of the SMZ (Plate 1). These flat structures controlled the final thrusting of the high-grade SMZ over the cool Kaapvaal Craton. This also accounts for the presence of the 2-km-long, shear-bounded high-grade SMZ klippen that occur on the Vreedzaam farm on the Kaapvaal Craton northeast of the Pietersburg Greenstone Belt (Fig. 2, loc. 6) (van Reenen et al., 1988). Ar-Ar dating of syn-kinematic hornblende that defines the near-horizontal mineral intersection lineation developed in sheared Baviaanskloof gneiss (Fig. 10C) in the out-of-sequence low-angle thrust section of the Hout River Shear Zone at the Breipaal locality in the Sand River (Fig. 2, loc. 1) yielded ages between ca. 2.63 Ga and ca. 2.6 Ga (Kreissig et al., 2001; Kramers et al., 2006) for this event, emphasizing the long-lasting exhumation history of the SMZ that commenced at ca. 2.69 Ga. The SMZ rocks were finally emplaced at the upper crustal level before ca. 2.46 Ga, as discussed.

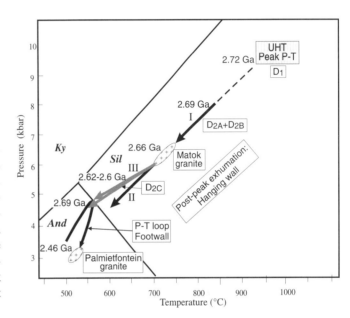

Figure 14. Integrated D-P-T-t diagram, showing the convergence at ca. 2.69–2.6 Ga of retrograde P-T paths constructed for hanging-wall granulites (DR45, DV101, PET1, CAPR7) and prograde P-T loops constructed for footwall mica schist (SL22). SMZ rocks document evidence for a two-stage post-peak exhumation history. The first stage of exhumation (I) is documented by early DC segments of DC → DC P-T paths that end at the gap in the P-T path. This stage resulted in the exhumation and emplacement of the rocks at ~21 km crustal depth controlled by steep, SW-verging $D2_B$ shear zones dated at ca. 2.69–2.664 Ga. The second stage of exhumation and emplacement at ca. 15 km depth (after the gap in the P-T path, II and III), controlled by shallow SW-verging $D2_C$ thrusts, was dated at ca. 2.62–2.6 Ga. The DC or IC P-T evolution of rocks is determined by the location of the samples relative to the contact with the cool rocks of the Kaapvaal Craton. DC → DC P-T paths are constructed for rocks sampled far from this contact (sample DR45, Fig. 12A), whereas DC → IC P-T paths are documented by rocks sampled in the granulite zone close to the boundary with the cool cratonic rocks. The inflection in the P-T path (III, Fig. 13) occurs at P-T conditions that correspond with the gap in the P-T path of sample DR45. The P-T loop (SL22) reflects the response of the footwall mica schist to the thrusting of hot hanging-wall granulite onto cool footwall mica schist. The UHT history of the SMZ (Fig. 14) (Tsunogae et al., 2004; Belyanin et al., 2010) is shown as a broken path. The D-P-T-t history (Fig. 14) of the SMZ and the juxtaposed granite-greenstones is well reflected in metapelitic rocks (Perchuk et al., 1996, 2000a, 2000b; Smit et al., 2001). See text for further explanation.

Geochemical Link of the SMZ Granulites and Juxtaposed Greenstones

On the basis of their similar trace-element geochemistry, Nd model ages between 2.9 and 3.1 Ga, and contents of the heat-producing elements U and Th, Kreissig et al. (2000) showed that rocks from the Bandelierkop Formation and Baviaanskloof gneisses in the SMZ are the high-grade equivalents of low-grade rocks from the adjacent granite-greenstone terrane of the Kaapvaal Craton. Geochemical data thus support earlier suggestions (van Reenen et al., 1987, 1988, 1990; Roering et al., 1992a, 1992b) that these two distinct geological terranes were a geotectonic entity during the Neoarchean Limpopo orogeny.

D-P-T-t EVOLUTION OF THE SMZ AND JUXTAPOSED MICA SCHISTS OF THE KAAPVAAL CRATON

Integration of field, structural, metamorphic, magmatic, and age data allow the construction of integrated D-P-T-t paths (Fig. 14), which demonstrate the early exhumation of the SMZ and coeval convergence at ca. 2.69–2.67 Ga of retrograde P-T paths in the hanging wall and prograde P-T paths in the footwall of the steeply NE-dipping Hout River Shear Zone. That this thrust controlled the exhumation history of the SMZ is best demonstrated by reference to the link of distinct deformational events (D1, $D2_A$, $D2_B$, and $D2_C$) with P-T-t paths constructed for five well-documented samples collected from four carefully selected localities. Three localities are from the SMZ (hanging-wall section) and one from the juxtaposed Giyani Greenstone Belt (footwall section). The post-peak evolution of samples from these four localities is discussed with reference to the D-P-T-t diagram (Fig. 14):

Retrograde P-T paths characterized by two distinct DC stages. (DC → DC paths, sample DR45). Low-shear-strain samples collected in $D2_A$ crustal blocks in the SMZ far north of the contact with the cool cratonic rocks are characterized by a P-T path in which two distinct DC stages are separated by a gap in the path at P ~6.5–6 kbar, T ~750–700 °C (Fig. 14). The first DC stage (I in Fig. 14) of this low-shear-strain rock, which is constructed on the basis of core compositions of porphyroblastic minerals (Fig. 12B) commences at the post-peak conditions of P ~8 kbar, T ~830 °C, and ends at the position of the gap in the P-T path. The second DC stage (II in Fig. 14) is calculated from garnet-rim compositions in contact with symplectic cordierite and orthopyroxene (Fig. 12B). It commences after the gap in the P-T path and ends at P ~4 kbar, T ~600 °C.

P-T paths with a distinct inflection (DC → IC P-T paths). P-T paths with a distinct inflection (near IC segment) (II in Fig. 14) characterize rocks from both low- and high-shear-strain domains close to the contact with the cool cratonic rocks (Fig. 2, loc. DV101, PET1, CAPR7). Although none of the studied rocks reflect direct evidence for the early DC stage of evolution, this stage is indirectly indicated by the presence of reaction textures (i.e., Crd-Opx symplectite; Fig. 11B) in which early Mg-rich garnet has been completely consumed owing to the shift of reaction 1 (Grt + Qtz → Crd + Opx) to the right (Smit et al., 2001). The near-IC stage of evolution (Fig. 14) traverses from P ~6 kbar, T ~700 °C to P ~4.5 kbar, T ~550 °C, and is directly calculated from garnet-producing reaction textures that reflect the subsequent growth of more Fe-rich garnet at the expense of early symplectic cordierite (Figs. 11B–11D) (Smit et al., 2001).

Unsheared rocks from low-shear strain $D2_A$ crustal blocks (Fig. 2, loc. DV101) and highly sheared rocks from steeply SW-verging $D2_B$ shear zones (Fig. 2, loc. PET1 and CAPR7) close to the contact with the cool cratonic rocks thus are characterized by identical reaction assemblages and near-IC P-T paths but by different reaction textures. Whereas sheared samples (PET1 and CAPR7) composed of kinematic reaction textures (Figs. 13C, 13D) that reflect the syn-kinematic crystallization of reaction products during high-temperature, high-shear-strain conditions, their unsheared equivalents (DV101, DR45) are characterized by randomly oriented reaction textures that reflect the static conditions of mineral growth (Fig. 6B) (Smit et al., 2001).

P-T loop of footwall mica schist. Evidence for the dynamic and thermal interaction of hot SMZ rocks and cool greenstones during the $D2_C$ thrust exhumation of the high-grade rocks is demonstrated by the convergence at a crustal depth of ~5 kbar (~17 km) of IC paths preserved in hanging-wall granulite and prograde P-T loops preserved in footwall mica schist (Fig. 11A) along the SW-verging Hout River Shear Zone. Coeval ages (Kreissig et al., 2001) of metamorphism (ca. 2.69–2.67 Ga) obtained from mica schist (Fig. 2, loc. SL22), and from granulite (Fig. 2, loc. quarry) provide further evidence in support of this process. In addition, the documented widespread retrograde metamorphism of the hanging-wall granulite by water-bearing CO_2-rich fluids derived by dehydration of low-grade footwall schist provides another convincing link between the contrasting metamorphic evolutions (Fig. 2; Plate 1).

UHT conditions in the SMZ. The D-P-T-t diagram (Fig. 14) was constructed on the basis of the common mineral assemblage (Grt + Opx/Sil + Crd + Bt + Kfs + Pl + Qtz), which presently characterizes metapelitic granulites in the SMZ. On the other hand, Tsunogae et al. (2004) and Belyanin et al. (2010) published evidence in support of an early UHT event (T >900 °C at P > 10 kbar), which preceded the post-peak exhumation history of the SMZ (Figs. 12C, 14). This UHT peak event is linked to early N-verging structures on the Kaapvaal Craton that were dated at ca. 2.72 Ga (Passeraub et al., 1999).

The D-P-T-t diagram (Fig. 14) thus allows the following conclusions:

1. The early DC stage of evolution of the SMZ (I in Fig. 13) is linked to steeply SW-verging D2B shear zones, which were active between ca. 2.67 Ga (peak metamorphic age of the P-T loop in the footwall of the Hout River Shear Zone) and ca. 2.664 Ga, the time of emplacement of the granodioritic phase of the Matok Pluton (Fig. 3) (Kreissig et al., 2001). These shear zones controlled the initial emplacement of the rocks at the crustal level of ~20 km

(P ~6 kbar). The subsequent emplacement of the rocks to a crustal level of ~15 km (P ~4.5 kbar) is reflected by the near-IC stage (III in Fig. 14), which is linked to flat SW-verging D2C thrusts dated at 2.63–2.6 Ga (Kreissig et al., 2001).

2. The gap in the P-T path of sample DR45 (Figs. 12A, 14) coincides with the "peak" P-T conditions reflected by the near-IC P-T paths (P ~6 kbar, T ~700 °C) of a large number of samples studied in the SMZ (see Perchuk et al., 1996, 2000a). This gap also coincides with the conditions (P ~6.5 kbar, T ~650 °C) at which the retrograde isograd was established (Figs. 7A, 7B) (van Reenen, 1986).

Perchuk et al. (1996, 2000a, 2000b) and Gerya et al. (2001) explained the existence of two groups of P-T paths preserved in SMZ granulites by differences in the geodynamic behavior of different crustal blocks during their exhumation and emplacement in the crust. Blocks far from the contact with the cool underthrusted greenstones in the footwall of the Hout River Shear Zone (Fig. 2, loc. DR45) were simply exhumed along a common DC trajectory (Figs. 12A, 14), allowing the adjustment to the prevailing thermal gradient of the crust at that point during exhumation, which involved a relatively large decrease in pressure. Crustal blocks close to the contact with the cool greenstones were initially emplaced at the crustal level of 15–18 km, where the rocks were rapidly cooled with a relatively small decrease in pressure because of the large thermal gradient that was established between the hot granulites and the cool underthrusted greenstones. This is reflected in these rocks by the near-IC stage of exhumation (III in Fig. 14).

DISCUSSION

D-P-T-t diagrams (Fig. 14) constructed for metapelitic granulites and juxtaposed mica schist (Fig. 2; Plate 1) demonstrate the convergence of decompression-cooling paths in the hanging wall and prograde P-T paths in the footwall of the Hout River Shear Zone, interpreted to reflect evidence for the thrust-controlled exhumation of the SMZ. The data presented by the D-P-T-t diagram is interpreted to reflect the reaction or response of the rocks to an event that produced tectonically unstable over-thickened crust related to the collision of the Kaapvaal and Zimbabwe Cratons in the Neoarchean before 2.7 Ga (van Reenen et al., 1987; Treloar et al., 1992; Roering et al., 1992b). This collisional event was responsible for low-grade crustal material being transported from shallow depths to depths >~30 km (Tsunogae and van Reenen, this volume; Belyanin et al., 2010). Evidence in support of a crustal thickening event is suggested by the presence of early (ca. 2.72 Ga; Passeraub et al., 1999), shallowing northward, verging thrusts developed in the footwall of the Hout River Shear Zone (Fig. 5). Deformation features developed throughout the entire Limpopo Complex furthermore provide evidence of continued horizontal compression for the duration of the Neoarchean Limpopo orogeny within which the different high-grade shear systems are interpreted to have developed. High-grade rocks of the Limpopo Complex that were thrusted over the two adjoining cratons (Fig. 5) in the interval ca. 2.69–2.6 Ga are the major response of the deep burial of shallow crustal material during this event. The SMZ exhumation occurred along the southward-verging Hout River Shear System (Fig. 5), which is compatible with a single exhumation event from deep to shallow crustal levels.

The suggestion (van Reenen et al., 1995, 2008; Kramers et al., this volume) that the Neoarchean exhumation history of the SMZ and NMZ of the Limpopo Complex (Fig. 1) should be similar is supported by metamorphic (Tsunogae et al., 1992, 2001) as well as by published geochronological and structural data for the NMZ (Blenkinsop et al., 2004, this volume; van Reenen et al., 1995), and by the fact that high-grade klippen occur on both the Kaapvaal Craton south of the Hout River Shear Zone (this paper) and on the Zimbabwe Craton north of the North Marginal Thrust Zone (Fig. 1) (Light, 1982). The SMZ and NMZ of the Limpopo Complex therefore were both exhumed along large ductile thrusts from deep- through mid- and to upper crustal levels during a single Neoarchean event. Smit et al. (this volume) discuss evidence that the CZ of the Limpopo Complex initially evolved as the result of the same Neoarchean event. Available data thus tend to favor a Himalayan model for the evolution of the Limpopo Complex (Kramers et al., this volume) in which early crustal thickening and the establishment of peak metamorphism are related to the collision of the Kaapvaal and Zimbabwe Cratons in the Neoarchean. The post-peak evolution of the Limpopo Complex is the response to the gravitational instability created by crustal thickening and UHT conditions. This is reflected by the thrust-controlled exhumation of the SMZ and NMZ ("pop-up" tectonic model of van Reenen et al., 1987; Roering et al., 1992b; Kramers et al., this volume; Smit et al., this volume). The alternative mechanism for the Neoarchean evolution of the Limpopo Complex is that of gravitational redistribution as suggested by Gerya et al. (2000) and Perchuk and Gerya (this volume). However, Perchuk and Gerya (this volume) also show that collisional and gravitational mechanisms of rock deformation are not mutually exclusive. They highlight the possibility that whereas collisional mechanisms probably operated during the early prograde stage of the tectono-metamorphic cycle, gravitational mechanisms should dominate during the post-peak retrograde stages through regional doming and diapirism (also Smit et al., this volume).

CONCLUSIONS

The post-peak segment of the P-T path for UHT gneisses in the Southern Marginal Zone of the Limpopo Complex situated in the hanging wall of the Hout River Shear Zone is integrated with the prograde P-T path for greenschist- to amphibolite-facies rocks of the Kaapvaal Craton in its footwall. Together these data provide evidence for a single D-P-T-t cycle of convergent tectono-metamorphic evolution in the interval ca. 2.69–2.6 Ga, with the high-grade rocks in the hanging wall characterized by decompressional and isothermal cooling paths with or without retrogression from incursion of hydrous and CO_2-bearing fluids, and the low-grade rocks in the footwall characterized by a prograde P-T path. The two contrasting paths converge at P-T conditions of ~5 kbar and 550 °C.

Post-peak decompression-cooling paths from P ~8 kbar, T ~825 °C to P ~6.5, T ~725 °C reflect the transport of high-grade rocks up the ramp section of SW-verging $D2_B$ shear zones dated at ca. 2.69–2.664 Ga. On the other hand, near-isobaric cooling paths from P ~6.5 kbar, T ~750 °C to P ~5 kbar, T ~550 °C reflect the subsequent transport of the high-grade rocks along the flat section of SW-verging $D2_C$ thrusts dated at ca. 2.63–2.6 Ga. The SMZ was finally emplaced at the near surface before 2.46 Ga, as is demonstrated by subvolcanic dikes of the Palmietfontein Granite that intrude the high-grade gneisses. The subsequent reactivation of early high-grade shear zones with the formation of upper crustal mylonites signifies the end of tectonic activity in the SMZ. This minor event is probably related to the same event that controlled the development of the ca. 2.0 Ga Palala mylonitic strike-slip shear zone that separates the SMZ from the CZ in the north. Direct evidence that the high-grade SMZ might have been reworked in the Paleoproterozoic is suggested by ca. 2.0 Ga Rb-Sr biotite ages obtained from a variety of rock types from the SMZ and adjacent Kaapvaal Craton (Barton and van Reenen, 1992b).

UHT (P >10 kbar, T >900 °C) metamorphism preceded exhumation by ~30 m.y. and is linked to shallow ca. 2.72 Ga N-verging structures for which evidence is restricted to the northern Kaapvaal Craton. These structures, truncated by the steeply SW-verging ca. 2.69–2.66 Ga Hout River Shear Zone, are interpreted as early N-verging structures related to the initial burial of the SMZ rocks. New structural data also confirmed the existence of huge ca. 2.63–2.6 Ga southward-verging high-grade gneissic nappes that outcrop over large areas along the southern boundary of the SMZ. All data are indicative of a long period of compressive tectonism in Archean times that in essence is similar to modern-day orogenic environments. Evidence in favor of a tectonic model in which crustal thickening and peak metamorphism, followed by post-peak doming and diapirism (gravitational redistribution) related to the Neoarchean collision of the Kaapvaal and Zimbabwe Cratons, is further discussed in this volume by Smit et al. (this volume) and Kramers et al. (this volume).

ACKNOWLEDGMENTS

Since 1995 this research was carried out as a part of the RF-RSA collaboration supported by National Research Foundation grant 68288 to DDvR, and grants 08-05-00351 and 1949.2008, respectively, from the RFBR and the RF President Program entitled "Leading Research Schools of Russia" to LLP. Sadly, Leonid Perchuk passed away on 19 June 2009 in Johannesburg, South Africa, while working on an earlier version of this paper, but his contribution is acknowledged. DDvR also wishes to acknowledge the consistent support from the Faculty of Science, University of Johannesburg (previously the Rand Afrikaans University), without which this research could not have been successfully completed. Michael Chakuparira is thanked for his assistance with the drafting of many of the figures in the text, and especially for producing the geological map of Plate 1. The authors gladly acknowledge comments on a previous version of the paper by Peter Treloar and Toby Rivers. We are especially grateful to Toby for his detailed and highly instructive review, and also for his comments on the revised version of the paper.

REFERENCES CITED

Aranovich, L.Y., and Podlesskii, K.K., 1983, Geothermobarometry of high-grade metapelites: Simultaneously operating reactions, in Daly, J.S., Jardley, B.W.D., and Cliff, B.R., eds., Evolution of Metamorphic Belts: Geological Society [London] Special Publication 42, p. 41–65.

Baker, J., van Reenen, D.D., Van Schalkwyk, J.F., and Newton, R.C., 1992, Constraints on the composition of fluids involved in retrograde anthophyllite formation in the Limpopo Belt, South Africa: Precambrian Research, v. 55, p. 327–336, doi:10.1016/0301-9268(92)90032-J.

Barton, J.M., Jr., and van Reenen, D.D., 1992a, When was the Limpopo Orogeny?: Precambrian Research, v. 55, p. 7–16, doi:10.1016/0301-9268(92)90010-L.

Barton, J.M., Jr., and van Reenen, D.D., 1992b, The significance of RB-Sr ages of biotite and phlogopite for the thermal history of the Central and Southern Marginal Zones of the Limpopo Belt of southern Africa and the adjacent portions of the Kaapvaal Craton: Precambrian Research, v. 55, p. 17–31, doi:10.1016/0301-9268(92)90011-C.

Barton, J.M., Jr., Du Toit, M.C., van Reenen, D.D., and Ryan, B., 1983, Geochronologic studies in the southern marginal zone of the Limpopo mobile belt, southern Africa, in Van Biljon, W.J., and Legg, J.G., eds., The Limpopo Mobile Belt: Geological Society of South Africa Special Publication 8, p. 55–64.

Barton, J.M., Jr., Doig, R., Smith, C.B., Bohlender, F., and van Reenen, D.D., 1992, Isotopic and REE characteristics of the intrusive charnoenderbite and enderbite geographically associated with the Matok Complex: Precambrian Research, v. 55, p. 451–467, doi:10.1016/0301-9268(92)90039-Q.

Belyanin, G.A., Rajesh, H.M., van Reenen, D.D., and Mouri, H., 2010, Corundum + orthopyroxene ± spinel intergrowths in an ultrahigh temperature Al-Mg granulite from the Southern Marginal Zone, Limpopo Belt, South Africa: American Mineralogist, v. 95, p. 196–199, doi:10.2138/am.2010.3383.

Berman, R.G., Brown, T.H., and Greenwood, H.J., 1985, An internally Consistent Thermodynamic Data Base for Minerals in the System Na_2O-K_2O-CaO-MgO-FeO-Fe_2O_3-Al_2O3-SiO_2-H_2O-CO_2: Atomic Energy of Canada Ltd., Technical Report 377, 62 p.

Blenkinsop, T., 2011, this volume, Archean magmatic granulites, diapirism, and Proterozoic reworking in the Northern Marginal Zone of the Limpopo Belt, in van Reenen, D.D., Kramers, J.D., McCourt, S., and Perchuk, L.L., eds., Origin and Evolution of Precambrian High-Grade Gneiss Terranes, with Special Emphasis on the Limpopo Complex of Southern Africa: Geological Society of America Memoir 207, doi:10.1130/2011.1207(13).

Blenkinsop, T.G., Kröner, A., and Chiwara, V., 2004, Single stage, late Archaean exhumation of granulites in the Northern Marginal Zone, Limpopo Belt, Zimbabwe, and relevance to gold mineralization at Renco Mine: South African Journal of Geology, v. 107, p. 377–396, doi:10.2113/107.3.377.

Bohlender, F., van Reenen, D.D., and Barton, J.M., Jr., 1992, Evidence for metamorphic and igneous charnockites in the Southern Marginal Zone of the Limpopo Belt: Precambrian Research, v. 55, p. 429–449, doi:10.1016/0301-9268(92)90038-P.

Boshoff, R., van Reenen, D.D., Smit, C.A., Perchuk, L.L., Kramers, J.D., and Armstrong, R., 2006, Geologic history of the Central Zone of the Limpopo Complex: The West Alldays Area: Journal of Geology, v. 114, p. 699–716, doi:10.1086/507615.

de Beer, J.H., and Stettler, E.H., 1992, The deep structure of the Limpopo Belt: Precambrian Research, v. 55, p. 173–186, doi:10.1016/0301-9268(92)90022-G.

de Wit, M.J., van Reenen, D.D., and Roering, C., 1992a, Geologic observations across a tectono-metamorphic boundary in the Babangu area, Giyani (Sutherland) greenstone belt, South Africa: Precambrian Research, v. 55, p. 111–122, doi:10.1016/0301-9268(92)90018-J.

de Wit, M.J., Jones, M.G., and Buchanan, D.L., 1992b, The geology and tectonic evolution of the Pietersburg greenstone belt, South Africa: Precambrian Research, v. 55, p. 123–153, doi:10.1016/0301-9268(92)90019-K.

de Wit, M.J., Roering, C., Hart, R.J., Armstrong, R.A., de Ronde, C.E.J., Green, R.W.E., Tredoux, M., Peberdy, E., and Hart, R.A., 1992c, Formation of an Archean continent: Nature, v. 357, p. 553–562, doi:10.1038/357553a0.

Du Toit, M.C., and van Reenen, D.D., 1977, The southern margin of the Limpopo Belt, Northern Transvaal, with special reference to metamorphism and structure: Geological Survey of Botswana Bulletin, v. 12, p. 83–97.

Du Toit, M.C., van Reenen, D.D., and Roering, C., 1983, Some aspects of the geology, structure and metamorphism of the Southern Marginal Zone of the Limpopo Metamorphic Complex: Geological Society of South Africa Special Publication 8, p. 89–102.

Gerya, T.V., and Perchuk, L.L., 1990, GEOPATH—A thermodynamic database for geothermobarometry and related calculations with the IBM PC computer: Calgary, Canada, Calgary University Press, Program and Abstracts, p. 59–61.

Gerya, T.V., Perchuk, L.L., van Reenen, D.D., and Smit, C.A., 2000, Two-dimensional numerical modeling of pressure-temperature-time paths for the exhumation of some granulite facies terranes in the Precambrian: Journal of Geodynamics, v. 30, p. 17–35, doi:10.1016/S0264-3707(99)00025-3.

Gerya, T.V., Maresch, W.V., Willner A.P., van Reenen, D.D., and Smit, C.A., 2001, Inherent gravitational instability of thickened continental crust with regionally developed low- to medium-pressure granulite facies metamorphism: Earth and Planetary Science Letters, v. 120, p. 221–235.

Hensen, B.J., and Green, D.H., 1973, Experimental study of the stability of cordierite and garnet in pelitic compositions at high pressure and temperature. III. Synthesis of experimental data and geological applications: Contributions to Mineralogy and Petrology, v. 38, p. 151–166, doi:10.1007/BF00373879.

Hoernes, S., Lichtenstein, U., van Reenen, D.D., and Mokgatla, K.P., 1995, Whole rock/mineral O-isotope fractionations as a tool to model fluid-rock interaction: South African Journal of Geology, v. 98, p. 488–497.

Holland, T.J.B., and Powell, R., 1990, An enlarged and updated internally consistent thermodynamic dataset with uncertainties and correlations: The system $K_2O–Na_2O–CaO–MgO–MnO–FeO–Fe_2O_3–Al_2O_3–TiO_2–SiO_2–C–H_2–O_2$: Journal of Metamorphic Geology, v. 8, p. 89–124, doi:10.1111/j.1525-1314.1990.tb00458.x.

Kramers, J.D., McCourt, S., and van Reenen, D.D., 2006, The Limpopo Belt, in Anhaeusser, C.R., and Thomas, R.J., eds., Geology of South Africa: Johannesburg, Geological Society of South Africa, and Pretoria, the Council for Geosciences, p. 209–236.

Kramers, J.D., McCourt, S., Roering, C., Smit, C.A., and van Reenen, D.D., 2011, this volume, Tectonic models proposed for the Limpopo Complex: Mutual compatibilities and constraints, in van Reenen, D.D., Kramers, J.D., McCourt, S., and Perchuk, L.L., eds., Origin and Evolution of Precambrian High-Grade Gneiss Terranes, with Special Emphasis on the Limpopo Complex of Southern Africa: Geological Society of America Memoir 207, doi:10.1130/2011.1207(16).

Kreissig, K., Nägler, T.F., Kramers, J.D., van Reenen, D.A., and Smit, C.A., 2000, An isotopic and geochemical study of the northern Kaapvaal craton and the Southern Marginal Zone of the Limpopo Belt: Are they juxtaposed terranes?: Lithos, v. 50, p. 1–25, doi:10.1016/S0024-4937(99)00037-7.

Kreissig, K., Holzer, L., Frei, I.M., Villa, J.D., Kramers, J.D., Kröner, A., Smit, C.A., and van Reenen, D.D., 2001, Geochronology of the Hout River Shear Zone and the metamorphism in the Southern Marginal Zone of the Limpopo belt, Southern Africa: Precambrian Research, v. 109, p. 145–173, doi:10.1016/S0301-9268(01)00147-4.

Light, M.P.R., 1982, The Limpopo Belt, Southern Africa: A result of continental collision: Tectonics, v. 1, p. 325–342, doi:10.1029/TC001i004p00325.

Mahan, K.H., Smit, C.A., Williams, M.L., Dumond, G., and van Reenen, D.D., 2011, this volume, Scales of heterogeneous strain and polymetamorphism in high-grade terranes: Insight from the Athabasca Granulite Terrane, western Canada, and the Limpopo Complex, in van Reenen, D.D., Kramers, J.D., McCourt, S., and Perchuk, L.L., eds., Origin and Evolution of Precambrian High-Grade Gneiss Terranes, with Special Emphasis on the Limpopo Complex of Southern Africa: Geological Society of America Memoir 207, doi:10.1130/2011.1207(14).

McCourt, S., and van Reenen, D.D., 1992, Structural geology and tectonic setting of the Sutherland greenstone belt, Kaapvaal Craton, South Africa: Precambrian Research, v. 55, p. 93–110, doi:10.1016/0301-9268(92)90017-I.

McCourt, S., and Vearncombe, J.R., 1987, Shear zones bounding the central zone of the Limpopo mobile belt, southern Africa: Journal of Structural Geology, v. 9, p. 127–137, doi:10.1016/0191-8141(87)90021-6.

McCourt, S., and Vearncombe, J.R., 1992, Shear zones of the Limpopo Belt and adjacent granitoid-greenstone terranes: Implications for late Archaean collision tectonics in southern Africa: Precambrian Research, v. 55, p. 553–570, doi:10.1016/0301-9268(92)90045-P.

Miyano, T., Ogata, H., van Reenen, D.D., Van Schalkwyk, H.F., and Arawaka, Y., 1992, Peak metamorphic conditions of sapphirine-bearing rocks in the Rhenosterkoppies greenstone belt, Northern Kaapvaal Craton, South Africa, in Glover, J.E, and Ho, S.E., eds., The Archean Terranes, Processes and Metalogeny: 31A Symposium Volume: University of Western Australia, Geology Department (Key Centre), Publication 22, p. 73–87.

Passeraub, M., Wuest, T., Kreissig, K., Smit, C.A., and Kramers, J.D., 1999, Structure, metamorphism, and geochronology of the Rhenosterkoppies greenstone belt, South Africa: South African Journal of Geology, v. 102, p. 323–334.

Pattison, D.R.M., Chacko, T., Farquhar, J., and McFarlane, C.R.M., 2003, Temperatures of granulite-facies metamorphism; constraints from experimental phase equilibria and thermobarometry corrected for retrograde exchange: Journal of Petrology, v. 44, p. 867–900, doi:10.1093/petrology/44.5.867.

Perchuk, L.L., 1977, Thermodynamic control of metamorphic processes, in Saxena, S.K., and Bhattacharji, S., eds., Energetics of Geological Processes: New York, Springer-Verlag, p. 285–352.

Perchuk, L.L., 1989, P-T-fluid regimes of metamorphism and related magmatism with specific reference to the Baikal Lake granulites, in Daly, S., Yardley, D.W.D., and Cliff, B., eds., Evolution of Metamorphic Belts: Geological Society [London] Special Publication 2, p. 275–291.

Perchuk, L.L., 2011, this volume, Local mineral equilibria and P-T paths: Fundamental principles and their application for high-grade metamorphic terranes, in van Reenen, D.D., Kramers, J.D., McCourt, S., and Perchuk, L.L., eds., Origin and Evolution of Precambrian High-Grade Gneiss Terranes, with Special Emphasis on the Limpopo Complex of Southern Africa: Geological Society of America Memoir 207, doi:10.1130/2011.1207(05).

Perchuk, L.L., and Gerya, T.V., 2011, this volume, Formation and evolution of Precambrian granulite terranes, in van Reenen, D.D., Kramers, J.D., McCourt, S., and Perchuk, L.L., eds., Origin and Evolution of Precambrian High-Grade Gneiss Terranes, with Special Emphasis on the Limpopo Complex of Southern Africa: Geological Society of America Memoir 207, doi:10.1130/2011.1207(15).

Perchuk, L.L., and Krotov, A.V., 1998, Petrology of the mica schist of the Tanaelv belt in the southern tectonic framing of the Lapland granulite complex: Petrology, v. 6, p. 149–179.

Perchuk, L.L., and Lavrent'eva, I.V., 1983, Experimental investigation of exchange equilibria in the system cordierite-garnet-biotite: Kinetics and equilibrium in mineral reactions: New York, Springer-Verlag, Advances in Physical Geochemistry, v. 3, p. 199–239.

Perchuk, L.L., and van Reenen, D.D., 2008, Reply to comments on "P-T record of two high-grade metamorphic events in the Central Zone of the Limpopo Complex, South Africa" by Armin Zeh and Reiner Klemd: Lithos, v. 106, p. 403–410, doi:10.1016/j.lithos.2008.07.011.

Perchuk, L.L., Gerya, T.V., van Reenen, D.D., Safanov, O.G., and Smit, C.A., 1996, The Limpopo metamorphic complex, South Africa: 2. Decompression/cooling regimes of granulites and adjacent rocks of the Kaapvaal Craton: Petrology, v. 4, p. 571–599.

Perchuk, L.L., Gerya, T.V., van Reenen, D.D., Krotov, A.V., Safonov, O.G., Smit, C.A., and Shur, M.Yu., 2000a, Comparative petrology and metamorphic evolution of the Limpopo (South Africa) and Lapland (Fennoscandia) high-grade terrains: Mineralogy and Petrology, v. 69, p. 69–107, doi:10.1007/s007100050019.

Perchuk, L.L., Gerya, T.V., van Reenen, D.D., Smit, C.A., and Krotov, A.V., 2000b, P-T paths and tectonic evolution of shear zones separating high-grade terrains from cratons; examples from Kola Peninsula (Russia) and Limpopo region (South Africa): Mineralogy and Petrology, v. 69, p. 109–142, doi:10.1007/s007100050020.

Perchuk, L.L., Gerya, T.V., van Reenen, D.D., and Smit, C.A., 2001, Formation and dynamics of granulite complexes within cratons: Gondwana Research, v. 4, p. 729–732, doi:10.1016/S1342-937X(05)70524-4.

Perchuk, L.L., van Reenen, D.D., Varlamov, D.A., Van Kal, S.M., Boshoff, R., and Tabaebaeimanesh, 2008, P-T record of two high-grade metamorphic events in the Central Zone of the Limpopo Complex, South Africa: Lithos, v 103, p 70–105.

Percival, J.A., Roering, C., van Reenen, D.D., and Smit, C.A., 1997, Tectonic evolution of associated greenstone belts and high-grade terranes, *in* de Wit, M.J., and Ashwal, L.D., eds., Greenstone Belts: Oxford, UK, Oxford University Press, p. 398–421.

Roering, C., van Reenen, D.D., de Wit, M.J., Smit, C.A., de Beer, J.H., and Van Schalkwyk, J.F., 1992a, Structural geological and metamorphic significance of the Kaapvaal Craton/Limpopo Belt contact: Precambrian Research, v. 55, p. 69–80, doi:10.1016/0301-9268(92)90015-G.

Roering, C., van Reenen, D.D., Smit, C.A., Barton, J.M., Jr., de Beer, J.H., de Wit, M.J., Stettler, E.H., Van Schalkwyk, J.F., Stevens, G., and Pretorius, S., 1992b, Tectonic model for the evolution of the Limpopo Belt: Precambrian Research, v. 55, p. 539–552, doi:10.1016/0301-9268(92)90044-O.

Schaller, M., Steiner, O., Studer, I., Holzer, L., Herwegh, M., and Kramers, J.D., 1999, Exhumation of Limpopo Central Zone granulites and dextral continent-scale transcurrent movement at 2.0 Ga along the Palala Shear Zone, Northern Province, South Africa: Precambrian Research, v. 96, p. 263–288, doi:10.1016/S0301-9268(99)00015-7.

Smit, C.A., and van Reenen, D.D., 1997, Deep crustal shear zones, high-grade tectonites and associated alteration in the Limpopo belt, South Africa: Implication for deep crustal processes: Journal of Geology, v. 105, p. 37–57, doi:10.1086/606146.

Smit, C.A., Roering, C., and van Reenen, D.D., 1992, The structural framework of the southern margin of the Limpopo Belt, South Africa: Precambrian Research, v. 55, p. 51–67, doi:10.1016/0301-9268(92)90014-F.

Smit, C.A., van Reenen, D.D., Gerya, T.V., and Perchuk, L.L., 2001, P-T conditions of decompression of the Limpopo high-grade terrain: Record from shear zones: Journal of Metamorphic Geology, v. 19, p. 249–268, doi:10.1046/j.0263-4929.2000.00310.x.

Smit, C.A., van Reenen, D.D., Roering, C., Boshoff, R., and Perchuk, L.L., 2011, this volume, Neoarchean to Paleoproterozoic evolution of the polymetamorphic Central Zone of the Limpopo Complex, *in* van Reenen, D.D., Kramers, J.D., McCourt, S., and Perchuk, L.L., eds., Origin and Evolution of Precambrian High-Grade Gneiss Terranes, with Special Emphasis on the Limpopo Complex of Southern Africa: Geological Society of America Memoir 207, doi:10.1130/2011.1207(12).

Stevens, G., 1997, Melting carbonic fluids and water recycling in the deep crust, an example from the Limpopo Belt, South Africa: Journal of Metamorphic Geology, v. 15, p. 141–154, doi:10.1111/j.1525-1314.1997.00010.x.

Stevens, G., and van Reenen, D.D., 1992, Constraints on the form of the P-T loop in the Southern Marginal Zone of the Limpopo Belt, South Africa: Precambrian Research, v. 55, p. 51–67.

Treloar, P.J., Coward, M.P., and Harris, N.B.W., 1992, Himalyan-Tibetan analogies for the evolution of the Zimbabwe Craton and Limpopo Belt: Precambrian Research, v. 55, p. 571–587, doi:10.1016/0301-9268(92)90046-Q.

Tsunogae, T., and van Reenen, D.D., 2011, this volume, High-pressure and ultrahigh-temperature granulite-facies metamorphism of Precambrian high-grade terranes: Case study of the Limpopo Complex, *in* van Reenen, D.D., Kramers, J.D., McCourt, S., and Perchuk, L.L., eds., Origin and Evolution of Precambrian High-Grade Gneiss Terranes, with Special Emphasis on the Limpopo Complex of Southern Africa: Geological Society of America Memoir 207, doi:10.1130/2011.1207(07).

Tsunogae, T., Miyano, T., and Ridley, J., 1992, P-T profiles from the Zimbabwe Craton to the Limpopo Belt, Zimbabwe: Precambrian Research, v. 55, p. 259–277.

Tsunogae, T., Nabara, A., Fukui, T., Harada, H., Mzvanga, W., and Mugumbate, F., 2001, Ultrahigh-temperature metamorphism of the Archean Limpopo Belt and its thermal effect on the adjacent low-grade Zimbabwe Craton, Southern Africa: Record—Australian Geological Survey Organisation, p. 362–364.

Tsunogae, T., Miyano, T., van Reenen, D.D., and Smit, C.A., 2004, Ultrahigh-temperature metamorphism of the southern marginal zone of the Archean Limpopo Belt, South Africa: Journal of Mineralogical and Petrological Sciences, v. 99, p. 213–224, doi:10.2465/jmps.99.213.

van Reenen, D.D., 1983, Cordierite+garnet+hypersthene+biotite–bearing assemblages as a function of changing metamorphic conditions in the Southern Marginal Zone of the Limpopo metamorphic complex, South Africa, *in* Van Biljon, W.J., and Leg, J.H., eds., The Limpopo Belt: Geological Society of South Africa Special Publication 8, p. 143–167.

van Reenen, D.D., 1986, Hydration of cordierite and hypersthenes and a description of the retrograde orthoamphibole isograd in the Limpopo Belt, South Africa: American Mineralogist, v. 71, p. 900–915.

van Reenen, D.D., and Du Toit, M.C., 1977, Mineral reactions and the timing of metamorphic events in the Limpopo Metamorphic Complex south of the Soutpansberg: Botswana Geological Survey Bulletin, v. 12, p. 107–128.

van Reenen, D.D., and Du Toit, M.C., 1978, The reaction garnet + quartz = cordierite + hypersthene in granulites of the Limpopo Complex in Northern Transvaal: Geological Society of South Africa Special Publication 4, p. 149–177.

van Reenen, D.D., and Hollister, L.S., 1988, Fluid inclusions in hydrated granulite facies rocks, southern marginal zone of the Limpopo Belt, South Africa: Geochimica et Cosmochimica Acta, v. 52, p. 1057–1064, doi:10.1016/0016-7037(88)90260-8.

van Reenen, D.D., Barton, J.M., Jr., Roering, C., Smit, C.A., and Van Schalkwyk, J.R., 1987, Deep crustal response to continental collision: The Limpopo Belt of South Africa: Geology, v. 15, p. 11–14, doi:10.1130/0091-7613(1987)15<11:DCRTCC>2.0.CO;2.

van Reenen, D.D., Roering, C., Smit, C.A., Van Schalkwyk, J.F., and Barton, J.M., Jr., 1988, Evolution of the northern high-grade margin of the Kaapvaal craton: South African Journal of Geology, v. 9, p. 549–560.

van Reenen, D.D., Roering, C., Brandl, G., Smit, C.A., and Barton, J.M., Jr., 1990, The granulite-facies rocks of the Limpopo Belt, southern Africa, *in* Vielzeuf, D., and Vidal Ph., eds., Granulites and Crustal Evolution: Dordrecht, the Netherlands, Kluwer, p. 257–289.

van Reenen, D.D., Roering, C., Ashwal, L.D., and de Wit, M.J., eds., 1992, The Archaean Limpopo granulite belt: Tectonics and deep crustal processes: Precambrian Research, v. 55, p. 587.

van Reenen, D.D., McCourt, S., and Smit, C.A., 1995, Are the Southern and Northern Marginal Zones of the Limpopo Belt related to a single continental collisional event?: South African Journal of Geology, v. 98, p. 489–504.

van Reenen, D.D., Perchuk, L.L., Smit, C.A., Varlamov, D.A., Boshoff, R., Huizenga, J.M., and Gerya, T.V., 2004, Structural and P-T evolution of a major cross fold in the Central Zone of the Limpopo High-Grade Terrain, South Africa: Journal of Petrology, v. 45, p. 1413–1439, doi:10.1093/petrology/egh028.

van Reenen, D.D., Boshoff, R., Smit, C.A., Perchuk, L.L., Kramers, J.D., McCourt, S.M., and Armstrong, R.A., 2008, Geochronological problems related to polymetamorphism in the Limpopo Complex, South Africa: Gondwana Research, v. 14, p. 644–662, doi:10.1016/j.gr.2008.01.013.

Van Schalkwyk, J.F., and van Reenen, D.D., 1992, High temperature hydration of ultramafic granulites from the Southern Marginal Zone of the Limpopo Belt by infiltration of CO_2-rich fluid: Precambrian Research, v. 55, p. 337–352, doi:10.1016/0301-9268(92)90033-K.

MANUSCRIPT ACCEPTED BY THE SOCIETY 24 MAY 2010

Neoarchean to Paleoproterozoic evolution of the polymetamorphic Central Zone of the Limpopo Complex

C. André Smit
Dirk D. van Reenen
Chris Roering
René Boshoff
Department of Geology, University of Johannesburg, P.O. Box 524, Auckland Park, 2006, Johannesburg, South Africa

Leonid L. Perchuk
Department of Geology, University of Johannesburg, P.O. Box 524, Auckland Park, 2006, Johannesburg, South Africa; Department of Petrology, Geological Faculty, Moscow State University, Leninskie Gory, Moscow, 119192, Russia; and Institute of Experimental Mineralogy, Russian Academy of Sciences, Chernogolovka, Moscow District, 142432, Russia

ABSTRACT

Integrated geological studies in the Central Zone of the Limpopo Complex formed the basis for the construction of a composite deformation (D)–pressure (P)–temperature (T)–time (t) (D-P-T-t) diagram that shows the following: First, in the Neoarchean the Central Zone probably underwent high-pressure (HP) (P >14 kbar, T ~950 °C) conditions followed by near isothermal decompression to ultrahigh-temperature conditions (UHT) (T ~1000 °C, P ~10 kbar), before ca. 2.68 Ga. Second, the post-peak exhumation history linked to two distinct decompression cooling stages commenced at ca. 2.68 Ga and ended before the emplacement of the Bulai Pluton at ca. 2.61 Ga. Stage 1 started at P ~9 kbar, T = 900 °C, and culminated with the emplacement of leucocratic anatectic granitoids at ca. 2.65 Ga. Stage 2, linked to the development of major SW-plunging sheath folds and related shear zones, started at P ~6 kbar, T ~700 °C and ended at P ~5 kbar, T ~550 °C, before ca. 2.61 Ga. The rocks resided at the mid-crustal level for more than 600 m.y. before they were again reworked at ca. 2.02 Ga by a Paleoproterozoic event. This event commenced with isobaric (P ~5 kbar) reheating (T ~150 °C) of the rocks related to the emplacement at ca. 2.05 Ga of magma linked to the Bushveld Igneous Complex. This was followed by final exhumation of the Central Zone. The Neoarchean high-grade event that affected the Limpopo Complex is linked to a Himalayan-type collision of the Kaapvaal and Zimbabwe Cratons that resulted in over-thickened unstable crust and the establishment of HP and UHT conditions. This unstable crust initially responded to the compressional event by thrust-driven uplift and spreading of the marginal zones onto the two adjacent granite-greenstone cratons. The post-peak exhumation history was probably driven by a doming-diapiric mechanism (gravitational redistribution).

INTRODUCTION

The high-grade Limpopo Complex of southern Africa (Fig. 1) is a ~750-km-long southwest-northeast–trending zone of complexly deformed and metamorphosed Neoarchean high-grade rocks traditionally subdivided into a Central Zone (CZ) flanked by two marginal zones, the Southern Marginal Zone (SMZ) and the Northern Marginal Zone (NMZ). The CZ is presently separated from the marginal zones by reactivated crustal-scale SW-NE–trending shear zones, in which the final movement is expressed by ca. 2.0 Ga mylonitic strike-slip displacement: the Palala Shear Zone in the south, and the Triangle Shear Zone in the north. The marginal zones are separated from the Zimbabwe granite-greenstone craton in the north and the Kaapvaal granite-greenstone craton in the south by ca. 2.65 Ga inward-dipping gneissic dip-slip shear zones, referred to as the *Hout River Shear Zone* in the south and the *Northern Marginal Thrust Zone* in the north (Fig. 1).

The NMZ, which persists throughout most of southern Zimbabwe (Fig. 1), consists essentially of charnockites and enderbites in which an ENE-WSW–striking south-dipping fabric varies from intense to weak (Ridley, 1992; Rollinson and Blenkinsop, 1995; Blenkinsop et al., 2004; Blenkinsop, this volume). Intercalated mafic-ultramafic enclaves with rare associated metasedimentary rocks occur and also strike ENE. The SMZ, which is not continuous along the length of the belt (Fig. 1), comprises mainly gray enderbitic gneisses with a strongly developed E-W to ENE-WSW fabric, in which metasedimentary (mainly metapelitic) gneisses and mafic-ultramafic units are common (Du Toit et al., 1983; van Reenen et al., this volume). The rocks composing the marginal zones have been shown to represent the high-grade metamorphic equivalents of the adjacent cratons (Kreissig et al., 2000, 2001). The CZ occupies the largest area of the three domains (Fig. 1) and contains a number of rock types not recognized in either of the marginal zones (details below).

The high-grade gneiss terrane of the Limpopo Complex has been the focus of research by numerous individuals and teams of researchers for close to 50 yr (e.g., van Reenen et al., 1992; Barton et al., 2006; Kramers et al., 2006; and references therein). The SMZ and NMZ, with their strong granite-greenstone affinities (Kreissig et al., 2000), have been studied (van Reenen et al., this volume; Blenkinsop, this volume, and references therein) with the specific objective of establishing the tectono-metamorphic relationship of the high-grade terrane to the juxtaposed low-grade granite-greenstone cratons (Fig. 1). Research in the CZ, on the other hand, has focused on the geodynamic evolution of the core portion of the Limpopo Complex (Fig. 1, inset map), which is juxtaposed on both sides against the high-grade marginal zones.

Most researchers (Kramers et al., 2006, and references therein) agree that the SMZ and NMZ have been affected by single ca. 2.65 Ga Neoarchean high-grade tectono-metamorphic events, but they disagree on the nature of the three distinct events that affected the much more complex CZ in the Mesoarchean at ca. 3.3 Ga, in the Neoarchean at ca. 2.65 Ga, and in the Paleoproterozoic at ca. 2Ga. In fact, different researchers not only disagree as to whether there is convincing evidence for a high-grade event in the CZ at ca. 2.65 Ga, but also disagree on virtually all aspects of the geology of the CZ, including the deformational history, the pressure-temperature (P-T) evolution, the interpretation of a large geochronological database, and the different tectonic models that have been suggested for the evolution of the Limpopo Complex after 3 Ga (Kramers et al., 2006; Barton et al., 2006; van Reenen et al., 2008; and references therein). The major issues are summarized below.

Deformational History

Different researchers disagree as to whether the complex deformational pattern of the CZ (Fig. 2) is linked to single Neoarchean (e.g., Roering et al., 1992a) or to Paleoproterozoic high-grade events (Holzer et al., 1998, 1999), whereas others argue that the deformational pattern is due to the superposition of a Paleoproterozoic shear deformational event at high grade onto the Neoarchean high-grade event (Boshoff et al., 2006; van Reenen et al., 2008).

Metamorphic History

Based mainly on work in the eastern part of the CZ, a variety of distinctly different P-T paths have been proposed for its metamorphic evolution: (1) near-isothermal decompression P-T paths (Windley et al., 1984; Droop, 1989; Tsunogae et al., 1989), (2) near-isothermal decompression paths followed by near-isobaric cooling paths (Watkeys, 1984), (3) counterclockwise P-T paths (Holzer et al., 1998), (4) clockwise P-T loops (Zeh et al., 2004), (5) high-pressure decompression cooling P-T paths (van Reenen et al., 2004; Perchuk et al., 2008a), and (6) isobaric heating P-T paths followed by decompression cooling (Perchuk et al., 2008a).

In addition, Tsunogae and van Reenen (2006, this volume) recently documented rare evidence for an early HP (P >14 kbar, T ~950 °C) event followed by near isothermal decompression to peak metamorphic conditions (UHT) at P ~10 kbar, T ~1000 °C. Further decompression cooling followed.

Post-peak metamorphic data published by various researchers thus define two distinct P-T groups independent of the shapes of the paths. The one group includes P-T paths that reflect maximum P-T conditions >8.5 kbar and 800 °C (Droop, 1989; Zeh et al., 2004; Perchuk et al., 2006a, 2006b, 2008a), whereas the other group includes P-T paths that reflect significantly lower maximum P-T conditions of P ~6 kbar and T ~750 °C (Harris and Holland, 1984; van Reenen et al., 2004; Hisada et al., 2005). Published P-T data therefore directly suggested that the high-grade rocks of the CZ in the Musina area were affected by at least two distinct high-grade metamorphic events after 3 Ga, notwithstanding arguments to the contrary (e.g., Zeh et al., 2004, 2007; Rigby et al., 2008; Gerdes and Zeh, 2009; Rigby, 2009).

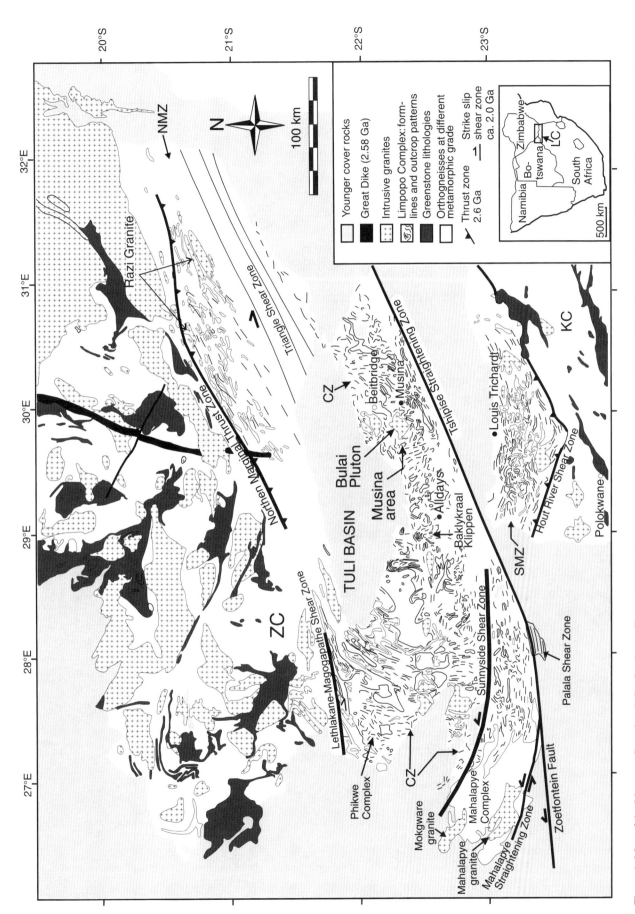

Figure 1. Map of the Limpopo Complex (LC, after Boshoff et al., 2006), showing the subdivision into a Central Zone (CZ), a Southern Marginal Zone (SMZ), and a Northern Marginal Zone (NMZ). Also shown are intrusive Neoarchean granitoids (Matok, Bulai, and Razi) and the ca. 2.58 Ga Great Dike of Zimbabwe that intrudes the NMZ. ZC—Zimbabwe Craton; KC—Kaapvaal Craton.

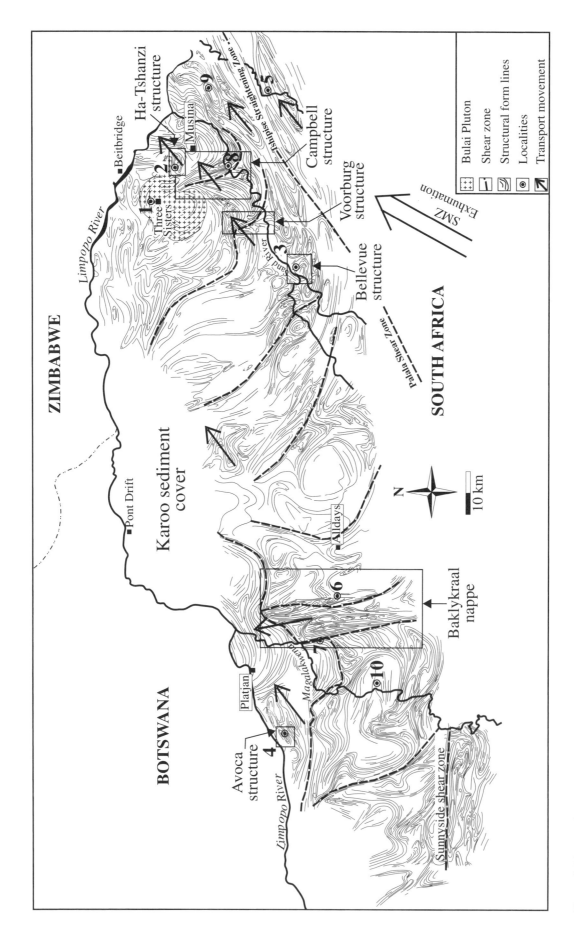

Figure 2. Photogeological interpretation map of deformational features of the CZ of the Limpopo Complex in southern Africa. The structural pattern mainly displays interference of D_2 isoclinal folds, large D_2 sheath folds, and the D_2 Tshipise Straightening Zone, which together form a complex but coherent deformational pattern. Arrows show the direction of transport along all major D_2 structures, suggesting consistent transport of high-grade material toward the NE both within the CZ and along the Tshipise Straightening Zone. It is important to note that younger N-S-trending D_3 shear zones (e.g., the Voorburg structure) show a similar transport direction as is suggested for the D_2 structures. Numbers refer to samples discussed in the text: 1 (Lim165-2), 2 (TOV13), 3 (RB47), 4 (Avoca), 5 (RB65), 6 (JC1), 7 (T73), 8 (O6-19), 9 (RB55), and 10 (T66). See text for further discussion.

Age Data

Different interpretations of the large published geochronological database for the CZ (Kamber et al., 1995a, 1995b; Jaeckel et al., 1997; Holzer et al., 1998; Kröner et al., 1999; Zeh et al., 2007; Rigby et al., 2008; Gerdes and Zeh, 2009; Rigby, 2009; van Reenen et al., 2008) formed the foundation for the ongoing debate in the literature on the number of distinct high-grade events that might have affected the CZ after 3 Ga. Different researchers interpret U-Pb zircon-monazite ages between ca. 2.62 and 2.68 Ga, obtained from anatectic (Singelele-type) granitoids that occur throughout the CZ to be associated with a protracted Neoarchean magmatic event that was not linked to a related high-grade metamorphic event (Kröner et al., 1999; Zeh et al., 2004, 2007; Buick et al., 2006; Zeh and Klemd, 2008; Millonig et al., 2008). This interpretation was proposed, notwithstanding the fact that the variably deformed Singelele-type granitoids can clearly be demonstrated to have intruded the high-grade gneissic fabric of paragneisses at numerous localities throughout the CZ (Bahnemann, 1972; Fripp et al., 1979; Watkeys et al., 1983; Horrocks, 1983; Hofmann et al., 1998; Kröner et al., 1999; Boshoff et al., 2006; van Reenen et al., 2008).

Arguments in favor of a mainly Paleoproterozoic high-grade event were strongly influenced by a large Pb-Pb age database obtained from a variety of high-grade minerals (e.g., garnet, sillimanite, titanite, and clinopyroxene) separated from different paragneisses (Holzer et al., 1998; Schaller et al., 1999; Kramers et al., 2006). These ages, which cluster around ca. 2.0 Ga (within error), were interpreted to reflect a high-grade metamorphic event at ca. 2.0 Ga (Holzer et al., 1998). However, the correlation by different authors (Kramers et al., 2006, and references therein) of this ca. 2.0 Ga high-grade event to an orogeny in the Zimbabwe Craton at ca. 2.0 Ga is at variance with unequivocal field evidence that Singelele-type granitoids and the Bulai granitic pluton intruded the regional high-grade gneissic fabric of the CZ at various localities before 2.61 Ga (e.g., van Reenen et al., 2008; Kramers and Mouri, this volume).

Tectonic Models

The lack of agreement among different researchers on different aspects of CZ geology is furthermore reflected by the variety of tectonic models that have been proposed to explain the complex evolution of the Limpopo Complex. Published models include Himalayan-type collisional models (Light, 1982; de Wit et al., 1992; McCourt and Vearncombe, 1992; Roering et al., 1992a, 1992b; Treloar et al., 1992), gravitational redistribution–diapiric models (Gerya et al., 2000; Blenkinsop, this volume; Perchuk and Gerya, this volume), a transpressive tectonic model (Holzer et al., 1998, 1999; Schaller et al., 1999), and both simple (Rollinson, 1993) and highly complex (Barton et al., 2006) accretion-type models.

GOALS OF THIS CHAPTER

The main goal of this chapter is to discuss the results of a collaborative multidisciplinary research project that was established in 2001 with the explicit goal of resolving the ongoing debate in the literature on the geological evolution of the CZ. This project is based at the Department of Geology, University of Johannesburg (formerly Rand Afrikaans University), with collaborators from the University of KwaZulu-Natal; Moscow State University and the Institute of Experimental Mineralogy at Chernokalovka, Moscow District, Russia; the University of Tsukuba in Japan; the Australian National University in Canberra; and the Isotope Geology Laboratory, University of Bern, in Switzerland. A field workshop was recently organized to discuss and evaluate the published (van Reenen et al., 2004, 2008; Boshoff et al., 2006; Hisada et al., 2005; Kramers et al., 2006; Perchuk, 2005; Perchuk et al., 2006a, 2006b, 2008a, 2008b; Perchuk and van Reenen, 2008; Tsunogae and van Reenen, 2006, this volume) and unpublished results of this collaborative project (Boshoff, 2004, 2008; Van Kal, 2004; van Reenen and Boshoff, 2008).

Our multidisciplinary research focused on the integration of field, structural, petrological, and geochronological data to construct deformational (D)–pressure (P)–temperature (T)–time (t) (D-P-T-t) diagrams. The construction of these diagrams depends heavily on the recognition of the critical role played by heterogeneous deformation in the formation and preservation of evidence for distinct granulite facies events at scales that range from the regional through the outcrop to the thin section. These diagrams are essential tools that have been employed to evaluate the various tectonic models for complexly deformed high-grade polymetamorphic terranes (e.g., Mahan et al., 2008, this volume; Kramers et al., this volume).

This chapter is an attempt to reinterpret (where necessary) and fully integrate published data (van Reenen et al., 2004, 2008; Perchuk, 2005; Boshoff et al., 2006; Perchuk et al., 2006a, 2006b, 2008a, 2008b; Perchuk and van Reenen, 2008) pertaining to our own structural, geochronologic, and metamorphic studies in the CZ. The purpose is to utilize the integrated data to construct a composite D-P-T-t diagram that demonstrates the protracted high-grade Neoarchean to Paleoproterozoic evolution of the CZ. It is, however, not our intention to review the published structural, metamorphic, or age data for the CZ, or to discuss and explain in detail the methodology involved in the construction of individual P-T paths for metapelitic gneisses that form the basis of this study. This is well documented in a number of recent publications (van Reenen et al., 2004; Perchuk et al., 2006a, 2008a; Perchuk and van Reenen, 2008; Boshoff, 2008; van Reenen et al., 2008; Perchuk and Gerya, this volume). However, pertinent details of the metamorphic and geochronological studies are useful background to this chapter and are therefore summarized in the section below.

METHODOLOGY

Construction of P-T Paths for Metapelitic Gneisses of This Study

Detailed P-T-a^{fl}_{H2O} information necessary for the construction of P-T paths (and D-P-T-t diagrams) for high-grade gneisses is, in our experience, best derived from metapelitic gneisses by combining conventional geothermobarometry with the principle of local mineral equilibria (the core → core, rim → rim method) (Perchuk, 1977, 1989; Perchuk and Gerya, this volume; Perchuk et al., 2006a, 2006b, 2008a; Smit et al., 2001; Perchuk and van Reenen, 2008; van Reenen et al., 2004). The cardinal aspects of this approach can be highlighted. First, it focuses on the common (major) mineral assemblage (garnet-cordierite-sillimanite-orthopyroxene-biotite-K-feldspar-plagioclase-quartz) that respectively characterizes Fe-rich (orthopyroxene-free) and Mg-rich (orthopyroxene-bearing) metapelitic gneisses in the CZ. Second, the early post-peak metamorphic parameters are calculated from zoning profiles preserved in inclusion-rich cores of porphyroblastic garnet in association with coarse-grained cordierite ± orthopyroxene in the matrix. Third, the final stage of the same event is calculated from zoning profiles preserved in the inclusion-free rims of the porphyroblastic garnet in direct contact with symplectic reaction products (also Vernon et al., 2008). Fourth, a specific P-T-a^{fl}_{H2O} value (reflected by a specific point on a P-T diagram) is calculated from a specific set of phase compositions considered to have been in local equilibrium. This specific set is identified on the basis of detailed microprobe zonation profiles and the concept of local (mosaic) equilibria. The following exchange and net-transfer reactions are the basis for the calculation of individual P-T-a^{fl}_{H2O} values and of the corresponding P-T path:

$$Grt_{Mg} + Bt_{Fe} = Grt_{Fe} + Bt_{Mg}, \quad (1)$$

$$Grt_{Mg} + Crd_{Fe} = Grt_{Fe} + Crd_{Mg}, \quad (2)$$

$$Grt + Sil + Qtz = Crd, \quad (3)$$

$$Grt + Qtz = Crd + Opx, \quad (4)$$

$$Grt + Kfs + H_2O = Bt + Sil + Qtz. \quad (5)$$

Thermodynamics of the reactions, including activities of components in corresponding solid solutions (Gerya and Perchuk, 1994), are based on thermodynamic treatment of numerical experimental data for a high range of pressure (P = 2–10 kbar) and temperature (T = 500–1000 °C). The application of an internally consistent thermodynamic data set (e.g., Gerya and Perchuk, 1994) to metapelitic gneisses of similar bulk chemical and mineralogical compositions thus allows the direct comparison of calculated P-T paths for rocks sampled from different localities throughout the CZ. The high accuracy of the thermometers studied experimentally follows from linear regression of kd = f(T) with a correlation coefficient of 0.999 (Perchuk and Lavrent'eva, 1983), a *Bt-Grt* thermometer ± 13.49 °C (p. 234), and a *Crd-Grt* thermometer ± 11 °C (p. 230). The accuracy of the cordierite-garnet-sillimanite-quartz geobarometer (Aranovich and Podlesskii, 1989) is P (max) ± 1 kbar, T ± 15 °C (p. 191). Reviewers thus have never requested error bars for P-T diagrams based on published data.

Timing of Discrete Events

Two different approaches to date well-constrained P-T paths constructed for metapelitic gneisses of this study can be summarized (Boshoff, 2008; Boshoff et al., 2006; van Reenen et al., 2008; Kramers and Mouri, this volume): First, U-Pb SHRIMP zircon-monazite ages were used to constrain the time of crystallization of anatectic granite melts closely associated with metapelitic gneisses at all studied localities. These anatectic veins mainly crosscut the gneissic fabric, the development of which has already been linked to a distinct P-T path, or the veins crystallized syntectonically with reference to a specific gneissic fabric-forming event. Age data thus obtained commonly provide minimum ages for the different fabric-forming events (and thus of a specific P-T path), or in the rare case of a true syntectonic vein (sample O6-19L) directly date a specific fabric-forming event linked to a specific P-T path. Second, because garnet participated in the majority of exchange and net-transfer reactions in metapelitic gneisses, PbSL data obtained from this mineral could in the rare case (sample T73) of a completely reworked rock be used to directly date a specific fabric-forming event linked to a specific P-T path. In all other cases this method produced imprecise ages between ca. 2.6 and ca. 2 Ga that are best interpreted to reflect evidence for superimposed high-grade events and thus support petrological evidence for the presence of polymetamorphic granulites in the CZ (Boshoff et al., 2006; Perchuk and van Reenen, 2008; van Reenen et al., 2008; Kramers and Mouri, this volume).

REGIONAL GEOLOGICAL SETTING OF THE CENTRAL ZONE

Major Rock Types and Their Field and Age Relationships

The CZ (e.g., Hofmann et al., 1998; Kröner et al., 1999) is dominated by a variety of supracrustal rocks of which the more than 3.3 Ga Beit Bridge Complex is the most dominant unit. It comprises pure meta-quartzite, marble, and calc-silicate gneisses, mafic gneisses, various garnet-biotite paragneisses of which true metapelitic gneisses are a minor component, banded iron formation (BIF), and magnetite quartzite, with interfolded quartzofeldspathic gneisses. The Beit Bridge Complex rocks are intruded by a variety of gray orthogneisses that include the ca. 3.3 Ga banded and migmatitic Sand River Gneiss, the ca. 2.65 Ga gray Alldays, Verbaard, and Zanzibar Gneisses that are seldom banded, and by anorthositic-gabbroic rocks belonging to the more than 3 Ga Messina Layered Intrusion (Mouri et al., 2009). All of these deformed and metamorphosed high-grade rocks were subsequently intruded between ca. 2.68–2.62 Ga by a variety of homogeneous nonbanded quartzofeldspathic gneisses considered to be the product of crustal anatexis (Fripp et al., 1979; Watkeys et al., 1983). Prominent among these are the garnetiferous, brownish-weathering Singelele Gneiss first described by Söhnge (1946) from the type outcrop at Singelele Hill within the town of Musina. A rock from the type locality was recently dated at ca. 2.64 Ga (Zeh et al., 2007). The weakly deformed Bulai Pluton with a ca. 2.61 Ga main porphyroblastic granitic phase (Millonig et al., 2008) is an important time marker that intrudes migmatized gneisses of the Beit Bridge Complex (e.g., Barton and van Reenen, 1992; Kramers et al., 2006), thus signifying the end of the Neoarchean high-grade tectono-metamorphic event that affected the CZ (Boshoff, 2008; van Reenen et al., 2008; Kramers and Mouri, this volume). The final magmatic event that affected the CZ at ca. 2.01 Ga (Jaeckel et al., 1997) is reflected by completely undeformed granitic melt patches and veins that obliterate the earlier gneissic fabric of the rocks from within which they were generated. This event is also represented by undeformed intrusive granitic dikes and granite pegmatites. These bodies are well exposed in road cuts in the Musina area and in the Sand River on the Verbaard farm SSW of Musina.

Regional Structural Framework of the CZ

The CZ of the Limpopo Complex is a high-grade gneiss terrane with a complex structural architecture, based on published data (Kramers et al., 2006, and references therein; van Reenen et al., 2008), is the product of two distinct, regional-scale, tectonometamorphic events: an older Neoarchean event and a younger Paleoproterozoic event.

The Neoarchean Event

Large parts of the CZ are underlain by high-grade rocks that display large- and small-scale SW-NE–trending isoclinal folds and moderately SW-plunging closed folds with elliptical to circular map patterns of different size that together produce the complex structural fold pattern of the CZ (Fig. 3). The large circular structures are interpreted as sheath folds (Roering et al., 1992a; Van Kal, 2004; van Reenen et al., 2008) that overprint the early isoclinal folds in different localities throughout the CZ. The area directly north of the Bulai Pluton (Fig. 2) is a typical locality for the study of large isoclinal folds, whereas several well-developed sheath folds (Avoca, Bellevue, and Ha-Tshanzi) occur distributed throughout the CZ (Figs. 1, 2).

The CZ is also affected by crustal scale, ductile straightening zones that include the SW-NE–trending and steeply SE-dipping oblique-slip Tshipise Straightening Zone in the south (Bahnemann, 1972) (Figs. 1–3), the SW-NE–trending and shallow south-dipping Triangle Shear Zone in the north (Ridley, 1992) (Fig. 1), and a system of east-dipping thrust faults in the extreme west in Botswana (McCourt and Armstrong, 1998) (Fig. 1). Traditionally, the Tshipise Straightening Zone (e.g., Bahnemann, 1972; Holzer, 1995; Holzer et al., 1998; Watkeys et al., 1983) refers to the major SW-NE–trending shear zone that bounds the CZ in the area south of Musina (Figs. 2, 3). The western extension (Fig. 1) of this steeply SE-dipping deep crustal Neoarchean shear zone, however, is a ~30-km-wide, steeply north-dipping upper crustal strike-slip mylonite belt, the Palala Shear Zone (e.g., McCourt and Vearncombe, 1987, 1992; Schaller et al., 1999), which probably reflects the reactivation of the Neoarchean Tshipise Straightening Zone in the Paleoproterozoic.

The Paleoproterozoic Event

Paleoproterozoic deformation in the CZ resulted in the reactivation of Neoarchean shear zones (Fig. 2) at both mid- and shallow crustal levels. As a result, the Tshipise Straightening Zone reflects evidence for three distinct shear events: First, an early mid-crustal event related to Neoarchean crustal thickening resulted in the development of the SW-NE– and N-S–trending shear pattern that characterizes the CZ (Roering et al., 1992a). Second, reactivation at the mid-crustal level of the early shear zones during a Paleoproterozoic high-grade shear event resulted in the formation of shear zones both subparallel to the Tshipise Straightening Zone and also oriented N-S. Third, structures within the Tshipise Straightening Zone were reactivated again at the upper crustal level subsequent to the intrusion of the ca. 2.05 Ga Bushveld Complex. This resulted in the development of ENE-trending and steeply north-dipping mylonite oriented parallel to the Palala strike-slip shear zone that forms the southern boundary of the western CZ (Figs. 1, 2). In addition to these reactivated structures, several researchers (e.g., Rollinson, 1993; Kamber et al., 1995a, 1995b; McCourt and Armstrong, 1998; Holzer et al., 1998, 1999) documented Paleoproterozoic shear zones along both the northern and western boundaries of the CZ.

Different authors (e.g., Holzer et al., 1998; Schaller et al., 1999) interpreted the Paleoproterozoic-age shear zones that define the boundaries of the CZ to represent a large transpressive shear system that displaced the entire CZ to the west or the east. Westward displacement conforms better to the metamorphic data, which suggest that shallower crustal levels are exposed in the western part of the CZ.

D-P-T-t EVOLUTION OF ROCKS ASSOCIATED WITH MAJOR DEFORMATIONAL FEATURES IN THE CZ

Introduction

The aim of this section is to discuss the deformation (D)–pressure (P)–temperature (T)–time (t) (D-P-T-t) evolution of gneissic fabrics in rocks associated with different major structures in the CZ. The purpose is to link distinct fabric-forming events to the geodynamic evolution of the CZ. We discuss the relative timing of these structures with particular emphasis on their fabric and field relations, and link the data to P-T-t paths constructed for associated metapelitic gneisses. Major structural features commonly display interference that resulted in the development of superimposed fabrics of different age that can be linked to different mechanisms of formation. Owing to the critical role that heterogeneous deformation has played in the development and preservation of multiple granulite events, it was possible to study the D-P-T-t evolution of these superimposed fabrics within a single outcrop or even within a single thin section (also Mahan et al., 2008, this volume).

In the discussion that follows we focus on specific localities that have been selected for detailed studies because they demonstrate evidence for a specific event within the tectonic evolution of the CZ, or for a sequence of events that affected the CZ at different times. We focus on localities where rocks of suitable bulk chemical compositions (metapelitic gneisses) are available to construct reliable P-T paths.

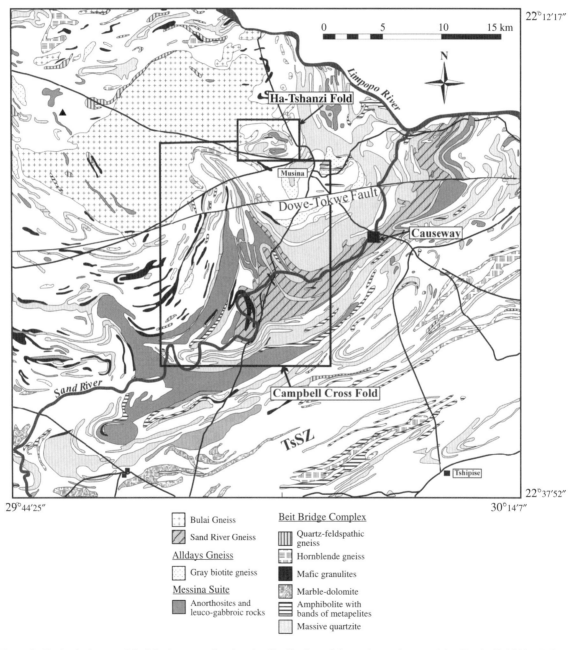

Figure 3. Geological map of the Musina area, showing the distribution of the major rock types (after Boshoff, 2008). TsSZ—Tshipise Straightening Zone.

The D$_2$ Deformational Event

The Peak Metamorphic Event That Affected the CZ before ca. 2.68 Ga

The early peak metamorphic event in the CZ is difficult to recognize, but we suggest that the steep and strongly annealed regional gneissic fabric and associated tight to isoclinal folds (Fig. 4) that characterize large areas in the CZ are the manifestations of this early event. Isoclinal folds are particularly well developed in the area directly west of Alldays and in the vicinity of the main outcrop area of the Bulai Pluton northwest of Musina (Figs. 2, 3). P-T constraints on this earlier event come from recently published petrological data (Tsunogae and van Reenen, 2006, this volume) for an Mg-Al-rich metapelitic gneiss from the Beitbridge area of the CZ in Zimbabwe that preserves evidence for an early ultrahigh-pressure (P >14 kbar, T ~950 °C) metamorphic event followed by near isothermal decompression to ultrahigh-temperature conditions reached at P ~10 kbar, T ~1000 °C. HP conditions are documented by rare Mg-rich (X_{Mg} ~0.58) staurolite enclosed within poikiloblastic garnet, whereas symplectic sapphirine + quartz developed around the Mg-staurolite implies rapid decompression along a clockwise P-T path from P >14 kbar toward the stability field of sapphirine + quartz at P ~10 kbar, T ~1000 °C. Orthopyroxene + sillimanite + quartz assemblages mantled by cordierite aggregates in a pelitic granulite from the same area provide supporting evidence for extreme metamorphism, which was followed by further decompression and cooling, as is suggested by various corona textures such as kyanite + sapphirine, sapphirine + cordierite, and orthopyroxene + cordierite.

Early Post-Peak D-P-T-t Evolution Related to Regional High-Grade Gneissic Fabric

Steeply oriented and strongly annealed high-grade banded and migmatized gneisses of different composition are linked to the early deep crustal event. The fabric of these gneisses is typically granoblastic in hand specimen and under the microscope, reflecting evidence for pervasive recrystallization and annealing of the rock during exhumation to the mid-crustal level.

The age of this fabric is well constrained, as migmatized metapelitic gneisses are intruded by Singelele gneiss throughout the CZ (Figs. 5A, 5B) and locally by the Bulai granite. A precise 2627 ± 2.6 Ma U-Pb monazite SHRIMP age (van Reenen et al., 2008) obtained for the Singelele Gneiss (Fig. 5B) from a locality near the Three Sisters locality (Fig. 2) and the 2.61 Ga age for the porphyritic phase of the mainly undeformed Bulai granitic pluton (Millonig et al., 2008) constrain the minimum age of the regional high-grade gneissic fabric to the Neoarchean.

The D-P-T-t evolution of the regional granoblastic gneissic fabric of the CZ can be demonstrated with reference to two localities that are >100 km apart. The first locality (Fig. 2, loc. 1) is

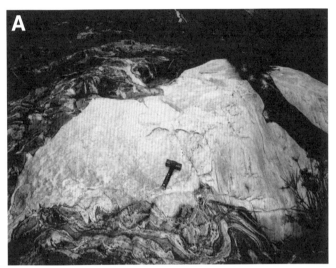

Figure 5. Intrusive relationship of the ca. 2.63 Ga, slightly deformed, garnet-bearing Singelele-type quartzofeldspathic gneiss with deformed high-grade paragneiss of the Beit Bridge Complex. (A) Causeway locality, SE of Musina (Fig. 2). (B) Rafts of paragneiss in the Singelele Gneiss (Fig. 6, loc. 3) (after Boshoff, 2008; van Reenen et al., 2008).

Figure 4. Complex deformation pattern that characterizes much of the CZ (Causeway locality, Figs. 2, 3).

northwest of Musina, and the second just east of the Baklykraal structure (Fig. 2, loc. 6).

Locality 1. The Bulai granite at the Three Sisters locality (Fig. 2, loc. 1) carries enclaves of migmatized metapelitic gneisses (Watkeys, 1984; Holzer, 1995; Huizenga et al., this volume). Field evidence (Fig. 6A) (Holzer, 1995; van Reenen et al., 2008; Huizenga et al., this volume) suggests that the Bulai Pluton intruded passively into isoclinally folded and migmatitic gneisses of the Beit Bridge Complex such that the regional fold pattern was not reoriented. This observation is supported by mineral stretching lineations measured from the metapelitic gneisses at the Three Sisters locality (Fig. 6B), which are still oriented subparallel to the general SW-plunging trend of all major structural features in the CZ (van Reenen et al., 2008). Metapelitic gneisses from the enclave at the Three Sisters (Fig. 2) and from the country rock to the Bulai granite are characterized by garnet porphyroblasts that occur in a matrix consisting of cordierite-sillimanite-biotite-K-feldspar-plagioclase-quartz (Huizenga et al., this volume). A relatively high-pressure decompression-cooling (DC) P-T path calculated for sample Lim165-2 (Fig. 6C) from the country rock commences at the maximum P-T parameters of P ~8.6 kbar, T ~930 °C, and ends at P ~5 kbar, T ~630 °C (Huizenga et al., this volume), before the emplacement of the Bulai Pluton at ca. 2.61 Ga (Huizenga et al., this volume). The P-T path calculated for sample BX1–5 from the enclave overlaps with the P-T path of sample Lim165-2.

Locality 2. The area just east of the Baklykraal structure (Fig. 2, loc. 6) that was suitable for metamorphic studies (Fig. 7A) is characterized by steeply dipping and intensely deformed garnet-biotite paragneisses (Fig. 7B) that locally occur as enclaves in less deformed garnet-bearing Singelele-type quartzofeldspathic gneiss (Boshoff et al., 2006). Sample JC1 from a small (meter-sized) boudin in this area is a leucosome-free and completely non-foliated Mg-rich granoblastic rock characterized by spectacular symplectites developed around garnet (Fig. 7C) (Perchuk et al., 2000, 2008a). The symplectitic texture comprises unoriented needles of orthopyroxene intergrown with cordierite, whereas the granoblastic matrix comprises porphyroblasts of garnet, cordierite, and orthopyroxene associated with quartz, plagioclase, a little biotite, and no K-feldspar. The cores of garnet porphyroblasts commonly preserve oriented inclusions consisting mainly of quartz, sillimanite, biotite, and plagioclase (Perchuk et al., 2008a). This is taken as rare evidence for an early high-strain regime.

Two distinct stages characterize the P-T evolution of sample JC1: an early DC stage, followed by an isobaric heating stage (Fig. 7A) (Perchuk et al., 2006a, 2006b, 2008a). The early post-peak DC P-T path (solid line in Fig. 7A) traverses from P ~9 kbar and T ~900 °C to P ~7 kbar and 750 °C and is constructed from zoning profiles preserved in the cores of garnet porphyroblasts coexisting with porphyroblasts of orthopyroxene and cordierite (Perchuk et al., 2008a). The isobaric (P ~6 kbar) heating (T ~100 °C) P-T path, which is linked to the superimposed Paleoproterozoic event (Perchuk et al., 2008a; van Reenen et al., 2008), is calculated from the inclusion-free rim of garnet in contact with symplectitic cordierite and orthopyroxene (Fig. 7C). Sample JC1 therefore records evidence for the early DC P-T history (solid line in Fig. 7A), which evolved during post-peak exhumation and which overprints the gneissic fabric and associated isoclinal folds. This P-T path has no memory of the P-T history (broken line in Fig. 7A) that followed after the emplacement of the rock to the crustal level of 23 km, and before the onset of isobaric heating (IH).

The time of formation of the gneissic fabric-forming event at this locality is constrained to the Neoarchean by the fact that garnet-biotite paragneiss occurs as enclaves in slightly deformed Singelele-type gneisses (Boshoff et al., 2006). PbSL garnet data obtained from garnet in sample JC1 and from a garnet-biotite gneiss (sample RB1) both gave "mixed" ages between ca. 2.0 and ca. 2.6 Ga (Boshoff et al., 2006; van Reenen et al., 2008). Similar mixed ages were also obtained for the Singelele-type gneiss (RB38) at the same locality (Boshoff et al., 2006; van Reenen et al., 2008). The mixed PbSL age data and the composite P-T path (Fig. 7A) show that samples JC1 and RB1 are polymetamorphic granulites that were affected by superimposed high-grade events, an early DC event and a younger IH event (Boshoff et al., 2006; Perchuk et al., 2008a). The PbSL method properly accounts for the presence of small inclusions (e.g., zircon) in garnet (e.g., Boshoff et al., 2006; Tolstikhin and Kramers, 2008), and the inferred presence of such inclusions cannot be called upon to explain the mixed Pb-Pb garnet ages, as was suggested by Zeh and Klemd (2008).

D-P-T-t Evolution of the High-Grade Shear Fabric of Rocks Associated with Major D_2 Sheath Folds

Mapping in the CZ established the fact that cylindrical shaped structures, which on various scales throughout the CZ define eye-shaped or elliptical outcrop patterns (Figs. 2, 3), represent sheath folds (Roering et al., 1992a; Van Kal, 2004; van Reenen et al., 2008). The term *sheath fold* as applied here describes a structure characterized by uniformly oriented populations of linear structures, and the fact that all foliations are cylindrically related with respect to central fold axes or linear structural orientations. All the studied sheath folds have rims defined by strongly sheared high-strain gneisses, whereas the cores are occupied either by younger intrusive granitoids (Avoca sheath fold, Fig. 2), indicating diapiric affinities (Perchuk et al., 2008b), or by gneiss including both Beit Bridge Complex gneisses (the Ha-Tshanzi sheath fold, Fig. 2) and Sand River-type gneisses (the Bellevue sheath fold, Fig. 2). Kinematic indicators are abundantly developed in the sheared rocks associated with the sheath folds and include linear structures such as rods, elongated pods, minor fold axes, and mineral stretching lineations. Similar linear elements are rarely preserved in the granoblastic gneisses that occur outside the sheath folds. On the basis of the geometry of measured fabrics, it is clear that map-scale sheath folds at distances of >100 km apart (Fig. 2) have identical orientations (van Reenen et al., 2008), suggesting that the same deformational

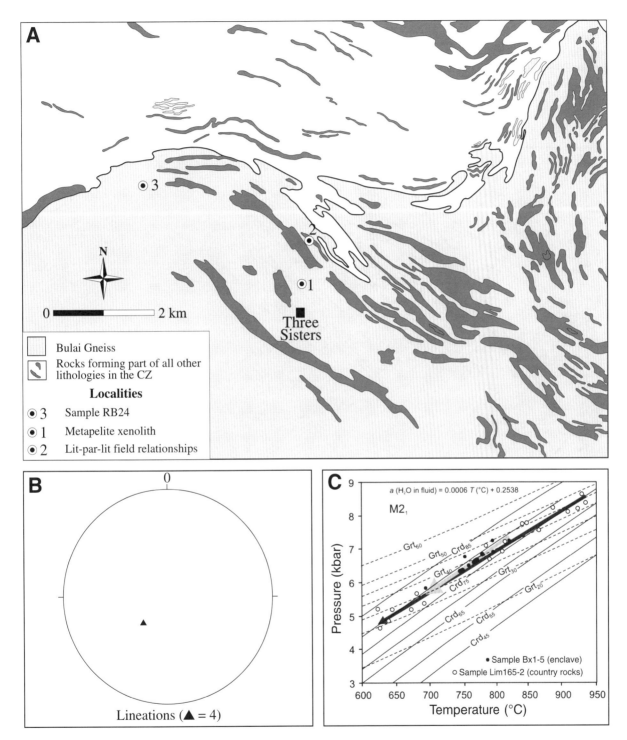

Figure 6. (A) Map of the ca. 2.61 Ga Bulai Pluton (Millonig et al., 2008) and surrounding area (Fig. 2) (after Holzer, 1995). The emplacement of this major and mainly undeformed granitic body did not disrupt the existing regional fold pattern of the surrounding area. (B) Stereographic plot of mineral stretching lineations (4× measurements) from the same outcrop (loc. 2) for which the P-T path (Lim165-2) (C) was constructed (Huizenga et al., this volume).

regime was present over most of this deep crustal region at the time of their formation.

Wherever sheath folds are developed in the CZ they can be shown to overprint the early regional granoblastic gneissic fabric and the associated folds (Fig. 8). Deformation in the rims of the folds transforms the granoblastic fabric into a strong shear fabric. Traverses across the rims of such sheath folds thus demonstrate a systematic increase in the modification of the early fabric through transposition and dynamic crystallization (Figs. 8A–8D) to produce "straight gneisses" (Davidson, 1984; Smit and van Reenen, 1997; Smit et al., 2001). Heterogeneous strain thus provides the opportunity to study the metamorphic evolution of distinctly different deformation fabrics preserved in a single outcrop, and even in a single thin section of a studied sheath fold.

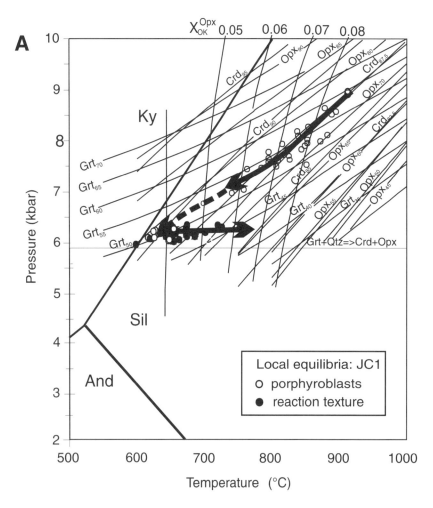

Figure 7. (A) Polymetamorphic sample JC1 characterized by two distinct P-T paths. (B) Outcrop of deformed garnet-biotite paragneiss on the Prinslust farm, directly east of the Baklykraal structure (Fig. 13A). Sample JC1 is from the same locality (Fig. 2, Loc. 6; Fig. 13A). (C) Symplectic reaction texture around garnet (Grt) in sample JC1 that reflects evidence for isobaric re-heating. See text for discussion. And—andalusite; Crd—cordierite; Grt—garnet; Ky—kyanite; Opx—orthopyroxene; Sil—sillimanite.

The term *sheath fold* has a genetic connotation that would link it with transport of material within or close to domains of simple shear. The kilometer size of the sheath folds encountered in the CZ suggests a major shear event in the CZ during the Neoarchean. The identical axial orientation (van Reenen et al., 2008) of the folds across hundreds of kilometers in the CZ thus is indicative of regional-scale simple shear (ductile flow) that affected the CZ during this time. Integrated D-P-T-t studies should provide important information on the exhumation history of these intensely sheared rocks at localities far apart. To test this hypothesis, D-P-T-t diagrams were constructed for samples from three major sheath folds (Avoca, Bellevue, and Ha-Tshanzi; Fig. 2).

Ha-Tshanzi Sheath Fold

The SW-plunging Ha-Tshanzi sheath fold (Figs. 2, 3, 9) is NW of Musina and in close proximity to the Bulai Pluton and the N-S-trending Campbell fold. This large sheath fold is mainly developed in rocks of the Beit Bridge Complex, Sand River Gneiss, Messina Layered Suite, and Singelele-type quartzofeldspathic gneisses (Fig. 9A). Well-developed mineral stretching lineations measured from a variety of rock types within the fold show a large spread of data (Fig. 9B) because the Ha-Tshanzi sheath fold is strongly overprinted by discrete N-S–trending high-grade D_3 shear zones (Holzer, 1995; Holzer et al., 1998; Van Kal, 2004; van Reenen et al., 2008).

Metapelitic sample TOV13, collected from the western rim of the fold (Fig. 9A), is a banded migmatitic gneiss consisting of garnet-cordierite-sillimanite-biotite-spinel-K-feldspar-plagioclase-quartz. The sample is unique in that it preserves evidence for much of the protracted Neoarchean to Paleoproterozoic history of the CZ in a single thin section. This is due to the critical role that heterogeneous deformation has played at the thin-section scale (Fig. 10) to develop and preserve distinct fabric-forming events that can be linked to distinct high-grade metamorphic events (Fig. 9C). TOV13 is characterized by a penetratively developed shear fabric that wraps centimeter-sized mineral aggregates that preserve evidence for the early granoblastic fabric (Fig. 10). Millimeter-sized, fabric-parallel micro-shears (Fig. 10C) overprint

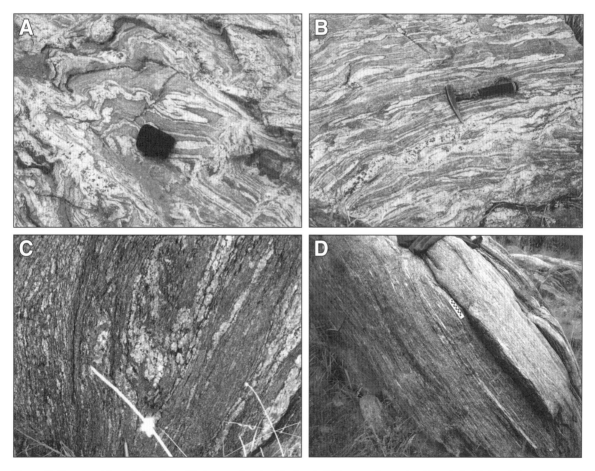

Figure 8. Superposition of the shear fabric associated with sheath folds onto the regional gneissic fabric of the Sand River Gneiss. This locality is within the rim of the Bellevue sheath fold (Fig. 11A). (A) No evidence of a superimposed shear fabric. (B) Early stage of the superimposed shear fabric. (C) More advanced stage of shear deformation. Preservation of the relict gneissic fabric wrapped by the shear fabric in the center of the photograph is explained by heterogeneous deformation. (D) Intensely sheared straight gneiss with no "memory" of the early gneissic fabric.

both these foliations. In addition the micro-shears cut garnet porphyroblasts, which in different parts of the same thin section can be linked to two different fabrics—the older granoblastic fabric, and the shear fabric related to formation of the sheath fold.

Sample TOV13 therefore is characterized by three distinct generations of $Grt_{1,2,3} + Crd_{1,2,3} + Sil_{1,2,3} + Bt_{1,2,3}$ that respectively are linked to the formation of the relict granoblastic fabric, the near penetrative sheath fold shear fabric, and the superimposed micro-shears. This protracted evolution is demonstrated by three distinct P-T paths in the composite D-P-T-t diagram (Fig. 9C) that respectively are constructed from data obtained from the three different generations of $Grt + Crd + Sil + Bt$ (Perchuk et al.,

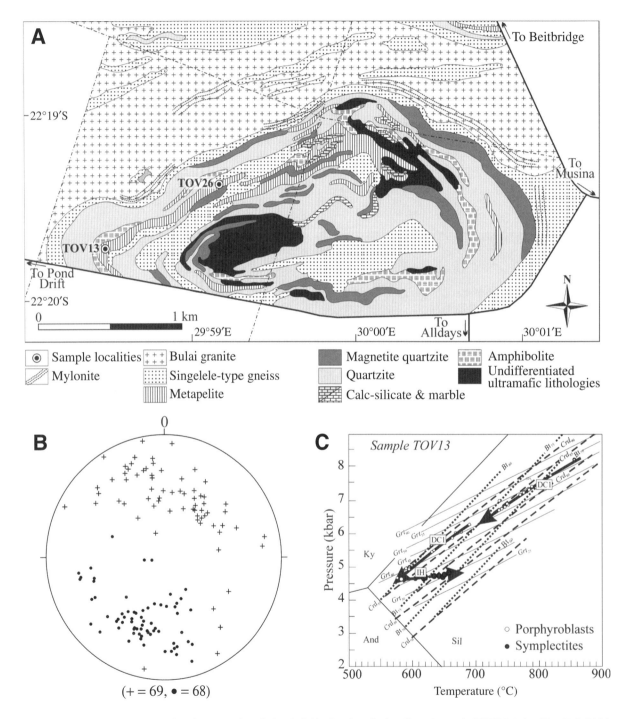

Figure 9. (A) Map of the SW-plunging Ha-Tshanzi sheath fold, showing the locality of sample TOV13 (after Van Kal, 2004; Boshoff, 2008). (B) Stereographic plot of kinematic data. + = poles to foliations; dots = lineations. (C) Composite P-T path constructed for polymetamorphic sample TOV13. Bt—biotite. See text for discussion.

2006a, 2008a; Mahan et al., this volume). The D-P-T-t evolution of sample TOV13 can be summarized as follows:

Stage 1. The first decompression cooling stage (DC1) (Fig. 9C) is calculated from zoning profiles preserved in the inclusion-rich core compositions of Grt_1 porphyroblasts (Fig. 10C) in contact with large grains of Crd_1 and Sil_1 that occur in the relict granoblastic portions of the thin section (Fig. 10C). The DC1 P-T path commences at P ~8 kbar and T ~850 °C and ends at ~6.3 kbar and T ~730 °C. The distinct gap in the P-T path defines the end of the DC1 path (Fig. 9C). Oriented mineral inclusions trapped within the cores of Grt_1 porphyroblasts (Fig. 10C) preserve evidence for the regional gneissic fabric, which initially must have developed in the deep crust under high stress conditions.

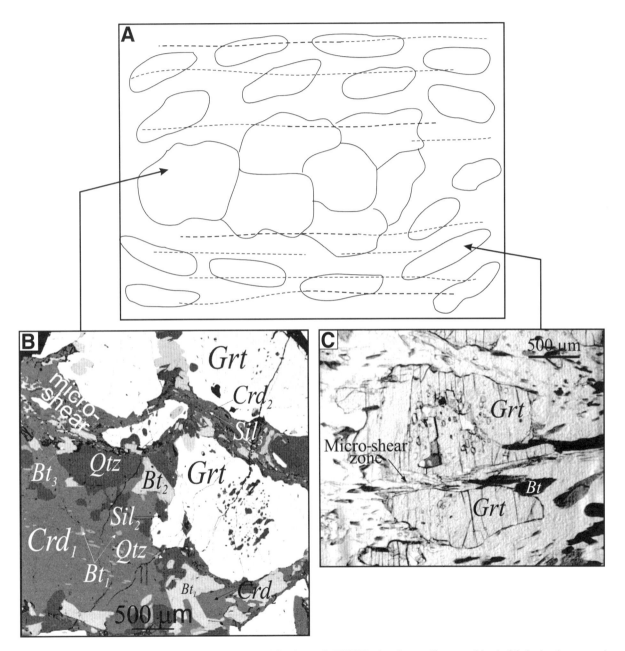

Figure 10. (A) Line sketch prepared from a polished slab of sample TOV13, showing a relict granoblastic fabric that is wrapped by the almost penetrative-developed shear fabric related to the sheath fold. Overprinted onto the shear fabric are numerous fabric-parallel micro-shears that also cut early garnet porphyroblasts. (B) Photomicrograph of the relict granoblastic fabric of sample TOV13, characterized by porphyroblasts of Grt_1 and Crd_1. The inclusion-rich cores of Grt_1 porphyroblasts are rimmed by inclusion-free Grt_2. (C) Photomicrograph showing the shear fabric of sample TOV13, which is defined by pencil-shaped Grt_2 porphyroblasts being replaced by Crd_2. Micro-shears with $Crd_3 + Bt_3 + Sil_3$ + rare Grt_3 cut Grt_2 porphyroblasts. Sil—sillimanite. See text for discussion.

Stage 2. The second decompression cooling stage (DC2) (Fig. 9C) is constructed from oriented pencil shaped Grt_2 porphyroblasts + Sil_2 + Qtz that was being replaced by Crd_2 within the superimposed shear fabric that wraps the relict granoblastic portions of sample TOV13 (Fig. 10C). The composition of the elongated Grt_2 porphyroblasts is similar to that of the retrograde rim that overgrows Grt_1 porphyroblasts in the granoblastic portion of the thin section (Fig. 10B). The DC2 P-T path commenced after the gap in the P-T path at P ~6.2 kbar and T ~680 °C and ends at P ~4.3 kbar and T ~575 °C (Fig. 9C). The distinct gap (Fig. 9C) at 750–680 °C for which no P-T data are preserved in the mineral assemblages separates the two P-T paths. The noticeable change in the slope of the two P-T paths furthermore is reflected by a distinct decrease in the water activity from a^{fl}_{H2O} = 0.115 (DC1 path) to a^{fl}_{H2O} = 0.095 (DC2 path) (Perchuk et al., 2008a). The emplacement at ca. 2.61 Ga (Millonig et al., 2008) of the largely undeformed Bulai Pluton marks the end of the D_2 stage of shear deformation and metamorphism that accompanied the formation of the Ha-Tshanzi sheath fold.

Stage 3. The isobaric heating (IH) stage of the D-P-T-t evolution of sample TOV13 is reflected by an isobaric (P ~4.5 kbar) heating (T ~150 °C) P-T path that is constructed from $Crd_3 + Bt_3 + Sil_3$ + rare tiny grains of Grt_3 from the fabric-parallel micro-shears that overprint the early fabric of sample TOV13 (Figs. 10A, 10C).

The time of formation of the D_2 Ha-Tshanzi sheath fold is poorly constrained. A garnet separated from sample TOV13 gave an imprecise PbSL age of 1956 ± 190 Ma (MSWD [mean square of weighted deviates] = 1450) (van Reenen et al., 2008). This is a mixed age that reflects the presence of different generations of garnet owing to polymetamorphism (van Reenen et al., 2008; Kramers and Mouri, this volume). The time of formation of the N-S–oriented micro-shears is indirectly limited to ca. 2.02 Ga by PbSL age data obtained from a variety of high-grade metamorphic minerals from N-S–trending D_3 shears developed at the eastern contact of the Ha-Tshanzi sheath fold with the Bulai Pluton (Fig. 9A) (Holzer et al., 1998; van Reenen et al., 2008).

Bellevue Sheath Fold

The Bellevue sheath fold (Fig. 11A) is southwest of Musina (Fig. 2) in an area that is characterized by the presence of several large sheath folds of similar size. The rim of the fold is developed in Beit Bridge Complex rocks that include marble and calc-silicate gneisses, metaquartzite, garnet-biotite paragneiss, rare metapelitic gneiss, leucogneiss, and meta-anorthosite of the Messina Layered Suite. The core is occupied by ca. 3.3 Ga gray Sand River–type gneisses (Kröner et al., 1999). The Bellevue sheath fold (Fig. 2, loc. 3) plunges gently to the SW (Fig. 11B), indicating a similar geometric orientation as the Ha-Tshanzi fold farther to the NE (Fig. 2, loc. 2). The age of the Bellevue sheath fold is not well established, but the similarity in its superimposed relationship with the regional gneissic fabric (Fig. 8) and its identical geometric orientation (Fig. 11B), compared with that of the Ha-Tshanzi (Fig. 9B) and Avoca sheath folds (see Fig. 12B), would suggest that this structure also developed during the Neoarchean sheath fold event. This conclusion is supported by the fact that Singelele-type gneisses intrude Beit Bridge Complex rocks in the northern rim of the fold (van Reenen and Boshoff, 2008).

The relationship between the early gneissic banding outside the fold (Fig. 8A) and the superimposed shear fabric developed within the rim of the fold (Figs. 8B–8D) is well exposed in the Sand River where it cuts through the northern rim of the fold (Fig. 11A). A P-T path constructed for a completely reworked gneiss in the rim of the fold should therefore only reflect the evolution of the shear event associated with the formation of the sheath fold. That this is true is demonstrated by a P-T path (Fig. 11C) derived from garnet-cordierite-sillimanite–bearing metapelitic gneiss (RB47) from the western rim portion of the Bellevue sheath fold (Perchuk et al., 2008b). The path commences at P ~6 kbar and T ~750 °C and ends at P ~3.7 kbar, T ~570 °C, and only documents the evolution of the fabric-forming event associated with the development of the sheath fold. This suggests that, in contrast to the situation described for sample TOV13 from the Ha-Tshanzi sheath fold, sample RB47 from the Bellevue fold was completely reworked (e.g., Fig. 8D) during the superimposed high-temperature shear event.

Avoca Sheath Fold

The Avoca sheath fold (Fig. 2, loc. 4; Fig. 12A) is developed mainly within Singelele-type quartzofeldspathic gneisses termed *Avoca gneiss* in the area northwest of Alldays. This area is characterized by large isoclinal folds that trend in the same SW-NE direction as that of the elliptically shaped sheath fold.

The Avoca sheath fold (Fig. 12A) maps out as an elliptical-shaped structure with a long axis of 2.5 km. The fold plunges to the SW (Fig. 12B) and is characterized by a few-hundred-meters-wide rim of intensely sheared high-strain S-L tectonite, the Avoca gneiss (Fig. 12C), whereas the core comprises mainly a much less foliated but strongly lineated leucocratic granitoid, the Avoca L-tectonite (Fig. 12D). The L-tectonite carries rare enclaves of the Avoca gneiss (van Reenen et al., 2008). In outcrop, rocks associated with the sheath fold have a well-developed shear fabric that overprints the regional gneissic fabric that defines the early isoclinal folds in the area surrounding the sheath fold (van Reenen and Boshoff, 2008). U-Pb zircon ages (van Reenen et al., 2008) of 2651 ± 8 Ma and 2627 ± 2.6 Ma, respectively, obtained for the Avoca gneiss and the L-tectonite provide support for the age relationship of the two rock types inferred from field evidence. The Avoca gneiss thus is interpreted to have developed during

Figure 11. (A) Map of the steeply SW-plunging Bellevue sheath fold (after Roering et al., 1992a) with the locality for Figure 8 shown. (B) Stereographic plot of structural data. + = poles to foliations; dots = lineations. (C) P-T path constructed for metapelitic sample RB47 from the western rim of the fold (Perchuk et al., 2008b). Avoca P-T path taken from Figure 12E, Bulai P-T path taken from Figure 6C. BBC—Beit Bridge Complex; Ky—kyanite; And—andalusite.

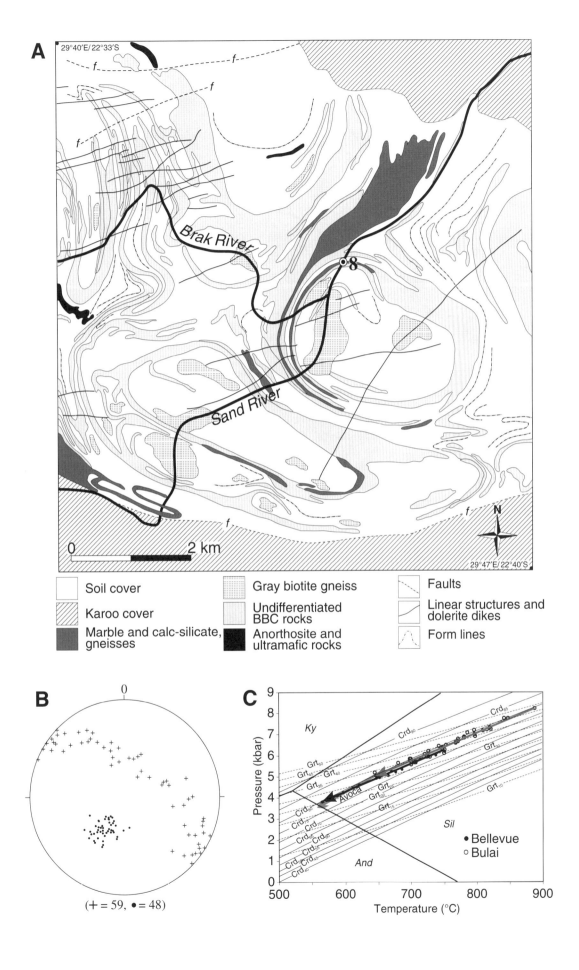

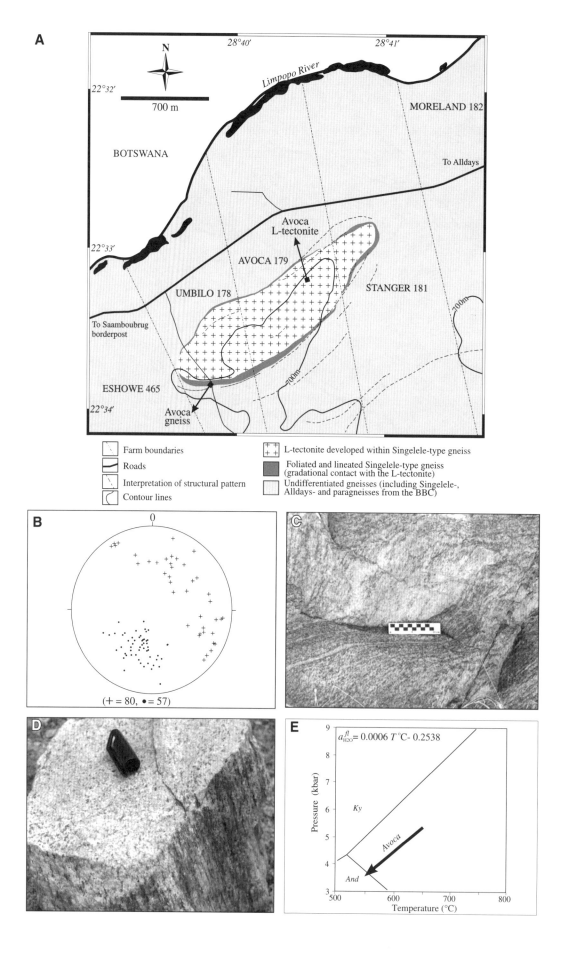

the main sheath-fold, fabric-forming stage, whereas the younger L-tectonite probably intruded only during the waning stage of this major deformational event. Perchuk et al. (2008b) regard the Avoca sheath fold as an excellent example of a granitic diapiric structure linked to the exhumation history of the CZ. The study of sheared rocks from this sheath fold therefore should reflect the changing D-P-T-t conditions that prevailed in the crust in the interval between 2.65 and 2.63 Ga, which is the time of formation of this sheath fold.

Perchuk et al. (2008b) showed that useful P-T information (Fig. 12E) might be obtained from a study of two-pyroxene mafic granulites that occur intercalated with the Avoca gneiss in the rim of the fold (van Reenen et al., 2008). The metamorphic assemblages of the metabasites *(Pl+Opx+Cpx+Hbl+Mag+Ilm+Ap)* provide temperature data derived from various versions of the two-pyroxene geothermometer, whereas data derived from fluid inclusions trapped in quartz in closely associated Avoca gneiss provide pressure estimates (Perchuk et al., 2008b). Based on this approach, an average temperature close to 675 °C was obtained from application of the two-pyroxene geothermometer, whereas the lowest P-T parameters (Fig. 12E) are based on the lowest T_h value of −8 and the fact that andalusite has not been reported from CZ rocks in South Africa. The calculated DC P-T path (Fig. 12E) thus commences at P ~5.3 kbar and T ~675 °C and ends at P ~3.6 kbar and T ~560 °C. Orthogneisses making up the rim of the Avoca sheath fold thus only document evidence for the post–ca. 2.65 Ga evolution of the shear fabric associated with the sheath fold (Perchuk et al., 2008b). This is confirmation of the situation described for sample RB47 from the Bellevue sheath fold and the DC2 P-T path of sample TOV13 from the Ha-Tshanzi sheath fold.

The U-Pb zircon age of 2626 ± 5.2 Ma obtained for the Avoca L-tectonite is statistically indistinguishable from the U-Pb monazite age of 2627 ± 2.6 Ma obtained for the Singelele gneiss that cuts the regional gneissic fabric of garnet-biotite paragneisses (Fig. 5B) NE of the Three Sisters locality (Fig. 2, loc. 1) >100 km to the east. The systematic control of the regional D_2 shear deformational event on the orientation of the SW-plunging kinematic elements throughout the CZ is well demonstrated by these two examples.

Baklykraal Structure

Söhnge (1946) described several large N-S–oriented foldlike structures that occur sporadically throughout the CZ (Fig. 2) as "cross folds" owing to the fact that they trend perpendicular to the general SW-NE orientation of the LC. The 40-km-long Baklykraal fold (Figs. 2, 13A) is the largest of these structures and occurs as a N-S–trending synformal fold with a near-horizontal fold axis (Fig. 13B) developed in Beit Bridge Complex rocks that include metaquartzite, marble, and calc-silicate gneisses, mafic gneisses, garnet-biotite paragneisses, rare metapelitic gneisses, and quartzofeldspathic gneisses. The core portion of this shallow S-plunging structure is marked by large kilometer-sized blocks of metaquartzite and fabric-parallel boudins of mafic gneiss that display the same early gneissic fabric as that observed outside the fold (Fig. 13), emphasizing the fact that the area in which the fold developed was affected by an earlier deformational history. Within this central portion of the fold (Fig. 13A) the quartzite blocks are wrapped by a younger anastomosing foliation that merges with the rim portion of the fold, collectively defining the general foldlike geometry of the Baklykraal structure. The younger fabric is developed mainly in ductile marble and calc-silicate gneisses and accompanied by well-oriented kilometer-sized boudins of mafic gneisses that carry the same younger N-S–oriented fabric (Fig. 13B); both features are considered to have developed during the younger sheath fold event.

An abundance of linear elements occur as boudinaged calc-silicate gneisses, cigar-shaped granitic boudins, quartz rods, minor fold axes, and mineral stretching lineations, and are particularly well developed within calc-silicate. Feldtmann (1996) produced a large structural database (Fig. 13B) of such kinematic indicators from the Baklykraal quarry, developed in marble and calc-silicate gneisses in the western limb of the fold. This database was subsequently supported by additional structural measurements taken from the entire structure (Boshoff et al., 2006). These kinematic indicators (Fig. 13B) show that the Baklykraal structure is characterized by a near-vertical N-S–oriented shear fabric, accompanied by a consistent population of near-horizontal lineations that represent the fold axis of the structure. Toward the center of the fold this consistent N-S–oriented fabric, however, wraps around the large quartzite blocks and produces a strongly scattered pattern. It is important, however, to note that this younger fabric, regardless of its orientation (N-S or E-W), carries near-horizontal N-S–oriented kinematic indicators (Fig. 13B).

Different authors (Pienaar, 1985; Feldtmann, 1996; van Reenen et al., 2004; Boshoff et al., 2006) interpreted this structure differently. Pienaar (1985) suggested that the structure developed as a result of interference between two superimposed fold events. Feldtmann (1996) suggested that it represents a major low-angle sheath fold, whereas van Reenen et al. (2004) and Boshoff et al. (2006) suggested that it is not a fold but the result of early folds truncated by large, younger N-S D_3 shear zones. The suggestion by Feldtmann (1996) that the Baklykraal fold is a low-angle sheath fold is probably the most correct interpretation. The Baklykraal structure will therefore be referred to as a sheath fold characterized by rocks with a shear fabric, similar to the situation described for the more steeply plunging Avoca and Bellevue sheath folds.

Along both the western and eastern boundary areas of the Baklykraal fold, large 500-m-wide, N-S–oriented D_3 shear zones

Figure 12. (A) Map of the SW-plunging Avoca sheath fold (Fig. 2). (B) Stereographic plot of kinematic data (after Roering et al., 1992a; Boshoff, 2004; van Reenen et al., 2008). + = poles to foliations; dots = lineations. (C) Foliated and lineated Avoca gneiss in the rim of the fold. (D) L-tectonite that occupies the core of the fold. (E) P-T path that reflects the evolution of the fold (Perchuk et al., 2008b). BBC— Beit Bridge Complex. See text for discussion.

(Fig. 13A) that overprint both the early gneissic and the younger sheath fold–related fabric have been recognized (Boshoff et al., 2006). The western shear zone changes toward the north into an E-W–trending, low-angle, southward-dipping shear zone with thrust geometry that clearly cuts a D_2 sheath fold (Fig. 13A, loc. A). This boundary shear zone was dated as a Paleoproterozoic structure (Boshoff et al., 2006) and will be further discussed.

The significance of this large fold structure is that it preserves fabrics related to three consecutive deformational events: the annealed gneissic fabric preserved within D_2 fragments, the shear fabric linked to the formation of the sheath fold, and finally the fabric in the D_3 shear zones. Studying rocks from these different structural domains should reveal evidence for the different metamorphic conditions under which each fabric developed and evolved.

D-P-T-t Evolution of the Shear Fabric Associated with High-Grade D_2 Shear Zones

Tshipise Straightening Zone. The 30-km-wide SW-NE–trending and steeply SE-dipping deep crustal Tshipise Straightening Zone (Bahnemann, 1972; Horrocks, 1983; Figs. 1–3)

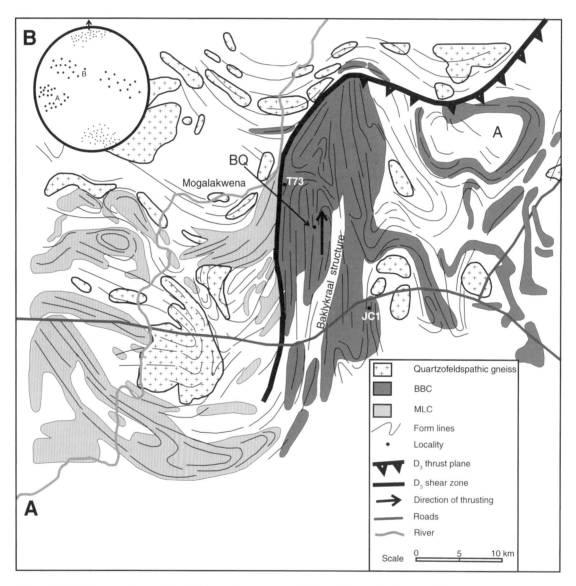

Figure 13. (A) Structural map of the Baklykraal area west of Alldays, illustrating the complex structural pattern of the area developed in Beit Bridge Complex (BBC) rocks, Messina Layered Complex (MLC) rocks, and quartzofeldspathic Alldays- and Singelele-type gneisses. The Baklykraal structure is developed only in Beit Bridge Complex rocks and is bounded along its western and northern sides by younger D_3 shear zones. Sample (T73) is from a D_3 shear zone, whereas sample JC1 is from unsheared D_2 gneisses wrapped by D_2 shear zones within the structure. The large structural database was collected mainly from the Baklykraal quarry (BQ). (B) Kinematic structural elements (after Feldtmann, 1996; Boshoff et al., 2006). Larger solid circles represent poles to foliation planes, and smaller solid circles, linear elements. A—poles to foliations measured for N-S–trending structures; B—poles to foliation measured for E-W–trending structures.

crops out only in the area south of Musina, where it defines the southern boundary of the CZ (Figs. 2, 3). Published maps of the Limpopo Complex (e.g., Fig. 1) imply that the mylonitic Palala Shear Zone, traditionally defined as the southern boundary of the CZ (e.g., McCourt and Vearncombe, 1987; Kramers et al., 2006), represents the strike-slip extension of the Tshipise Straightening Zone (Fig. 1). However, the character and geometry of these structures do not allow them to be part of a single shear zone. The Tshipise Straightening Zone is a deep crustal, steeply SE-dipping gneissic shear zone with an oblique top-to-the-NE sense of movement, whereas the Palala Shear Zone is a steeply north-dipping upper-crustal mylonitic strike-slip shear zone (McCourt and Vearncombe, 1987, 1992; Schaller et al., 1999; Smit et al., 2001). The possibility that the Tshipise Straightening Zone did extend farther to the west earlier is suggested by two observations: First, evidence that the Tshipise Straightening Zone in the type locality south of Musina was reactivated during at least two superimposed shear events. The first event resulted in the formation of fabric-parallel, mid-crustal, high-grade shear zones, whereas the final event resulted in the formation of upper-crustal mylonites similar to those of the Palala Shear Zone (Holzer et al., 1998; Boshoff, 2008). Second, McCourt and Vearncombe (1987) describe relicts of sheared gneisses preserved within intensely mylonitized rocks from the northern margin of the Palala strike-slip shear zone. These high-grade relicts still preserve evidence for oblique-slip lineations that point to an earlier shear event. If true, these data suggest that the high-grade Tshipise Straightening Zone in the west was almost completely reworked during the superimposed mylonite event.

The Tshipise Straightening Zone is a complex structure characterized by a variety of deformational features that range from medium- to large-scale SW-NE–trending shear zones of different age, large SW-NE–oriented sheath folds, and kilometer-scale rafts of low-strain Beit Bridge Complex gneisses that preserve the regional high-grade gneissic fabric of the CZ. The intense ductile shear fabric of the Tshipise Straightening Zone is best developed in metapelitic gneisses and Singelele-type garnet-bearing quartzofeldspathic gneisses. Kinematic indicators that include mineral elongation lineations, minor fold axes, and major sheath folds (Fig. 2) all plunge at moderate angles to the SW. These data (van Reenen et al., 2008) are consistent with an oblique (sinistral-reverse) displacement to the NE. The age of this displacement is not directly constrained, but the subparallelism of linear structures in the Tshipise Straightening Zone with the plunge of map-scale sheath folds both within the outcrop of the Tshipise Straightening Zone (Fig. 2) and in the CZ to the north are taken to indicate that this zone developed synchronously with the sheath folds and is thus a Neoarchean structure. However, other researchers (e.g., Holzer, et al., 1998; Schaller et al., 1999) linked the development of the Tshipise Straightening Zone to a Paleoproterozoic transpressive event. These different interpretations follow from the presence of discrete smaller shear zones that developed subparallel to the main shear foliation in the Tshipise Straightening Zone. Holzer et al. (1998) obtained a PbSL age of ca. 2.02 Ga on zircon from a synkinematic granitic vein in intensely sheared metapelitic gneiss from such a shear zone on the Kopjesfontein farm (Fig. 2, loc. 9) and interpreted the age to date the time of formation of the Tshipise Straightening Zone. This issue will be further discussed.

The type locality of the Tshipise Straightening Zone south of Musina (Figs. 1, 2) has been studied with the aim of establishing whether it can, on the basis of P-T data (Fig. 14), be linked to the regional D_2 shear event reflected by sheath folds in the CZ. A DC P-T path (Fig. 14B), derived from highly sheared D_2 $Grt + Crd + Sil + Bt + Qtz + Kfs + Pl$–bearing metapelitic gneiss (sample RB65) from a SW-NE–trending and steeply SE-dipping

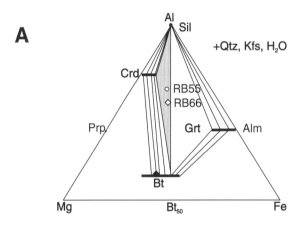

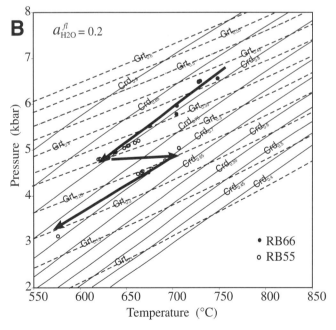

Figure 14. (A) AFM diagram showing the similar bulk chemical compositions of samples RB65 and RB55, respectively, from D_2 (Fig. 2, loc. 5) and D_3 (Fig. 2, loc. 9) shear zones within the Tshipise Straightening Zone (Perchuk et al., 2008b). (B) Distinctly different P-T paths constructed for samples RB65 and RB55 (Perchuk et al., 2008b). See text for discussion.

shear zone exposed on the Skullpoint farm (Fig. 2, loc. 5) records P-T conditions that range from maximum values of P ~6.5 kbar, T ~750 °C to minimum values of P ~4.5 kbar, T ~610 °C. The DC P-T path not only overlaps with other D_2 P-T paths (Figs. 6C, 7A, 9C, 11C, 12E), but as was the case with P-T paths constructed for highly sheared rocks associated with D_2 sheath folds, sample RB65 also records-preserves evidence only for the D_2 shear event and has no "memory" of the early part of the P-T evolution, which is linked to the development of the D_2 regional gneissic fabric and isoclinal folds (e.g., Figs. 6C, 7A).

D-P-T-t Evolution of the D_3 Shear Fabric

D_3 Shear Zones That Overprint the TsSZ

The study of sheared rocks from the TsSZ demonstrates how difficult, in the absence of crosscutting relationships, it can be to distinguish D_2 and D_3 shear deformational events in the field. In cases where field evidence is not conclusive (e.g., the Tshipise Straightening Zone), a distinction can be made only on the basis of geochronological data supported by P-T data.

A P-T path constructed for metapelitic sample RB55 from a shear zone exposed on the Kopjesfontein farm (Fig. 2, loc. 9) reflects maximum conditions of ~5.5 kbar and ~700 °C and minimum conditions of ~3.3 kbar and ~557 °C. These pressures are significantly lower than those calculated for sample RB65 (Fig. 14B) collected from a D_2 shear zone on the Skullpoint farm. Boshoff et al. (2006) interpreted this difference to indicate that sample RB55 was from a D_3 shear zone. The suggested link of the low-pressure P-T path of sample RB55 with the D_3 Paleoproterozoic shear deformational event is supported by the PbSL age of ca. 2.02 Ga on zircon from the same locality (Holzer et al., 1998).

D_3 Shear Zones That Overprint the Baklykraal Structure

The N-S–trending Baklykraal structure (Fig. 13A) is bounded in the west by a N-S–trending, near-vertical D_3 shear zone that overprints the D_2 shear fabric of this major structure. Traced northward, the shear zone changes into an E-W–trending, low-angle (~25°), southward-dipping shear zone, suggesting an overall thrust geometry.

Three highly sheared Fe-rich metapelitic gneisses (T18, T20, and T73) from two different localities along the N-S section of the D_3 shear zone (Fig. 13A) were studied in detail to determine the D-P-T-t evolution of the shear fabric (van Reenen et al., 2004). Sample T73 (Fig. 15), an Fe-rich tectonite composed of garnet-cordierite-sillimanite-biotite-quartz-K-feldspar-plagioclase, documents rare evidence in the CZ of a metapelitic granulite that was completely reworked during the younger superimposed high-temperature, shear-deformation event (van Reenen, et al., 2004). The fact that this rock has no "memory" of the early D_2 history is demonstrated by a single DC P-T path (Fig. 15A) that was calculated from minerals associated with the reaction garnet + sillimanite + quartz = cordierite (van Reenen et al., 2004). This well-constrained P-T path (Fig. 15A) traverses from P ~5.8 kbar and T ~790 °C to P ~3 kbar and T ~600 °C, indicating that the fabric in T73 formed at significantly lower pressure conditions than the fabric in rocks associated with the Neoarchean D_2 deformational event (e.g., sample JC1, Fig. 7A).

The D_3 shear fabric within the Baklykraal structure is accurately constrained to 2023 ± 11 Ma by PbSL age data obtained from garnet in sample T73 that was used to construct the single DC P-T path (Fig. 15A) (Boshoff et al., 2006; van Reenen et al., 2008).

D_3 Shear Zones That Overprint the D_2 Campbell Fold

The large N-S–trending moderately WSW-plunging Campbell fold SW of Musina (Figs. 2, 3) is overprinted by N-S–oriented D_3 shear zones (Fig. 3) (Van Kal, 2004; Boshoff, 2008; van Reenen et al., 2008) with dip-slip mineral stretching lineations and movement indicators suggesting top-to-the-NE shear (thrust) movement (van Reenen et al., 2008). The studied sample (O6-19, Fig. 2, loc. 8) is from a D_3 shear zone.

Sample O6-19 is a highly sheared Fe-rich $Grt + Crd + Sil + Bt + Kfs + Qtz + Pl + Sp$–bearing metapelitic gneiss that comprises a highly sheared metapelitic portion characterized by thin, fabric-parallel leucosomes (Fig. 16B, sample O6-19M), and syn- to late tectonic melt patches (Fig. 16C, sample O6-19L) characterized by newly grown garnet (Van Kal, 2004; Perchuk et al., 2008a). The P-T path (Fig. 16D) (Perchuk et al., 2008a) was derived from two distinctly different generations of $Grt + Crd + Sil + Bt$ and shows an isobaric (P ~5 kbar) heating (T ~130 °C) (IH) trajectory followed by a DC trajectory. The DC path commences at the peak conditions of P ~5 kbar and T ~725 °C and ends at P ~3.5 kbar and T ~575 °C, in accordance with uplift achieved during the D_3 thrust movement.

The time of formation of the D_3 shear fabric is accurately constrained by a precise U-Pb SHRIMP crystallization age of 2017 ± 2.8 Ma (Boshoff, 2008; van Reenen et al., 2008) obtained from monazite separated from the syntectonic melt patch (Fig. 16C, sample O6-19L). A cluster of four zircon grains also defines a concordant U-Pb age of 2011 ± 11 Ma. However, another cluster of four zircon grains also defines a concordant age of 2610 ± 8.3 Ma, which reflects evidence for the Neoarchean event (Perchuk and van Reenen, 2008; van Reenen et al., 2008). Highly inaccurate PbSL ages between ca. 2 and 2.6 Ga were also obtained from garnet separated from both the metapelitic portion (Fig. 16B, sample O6-19M) and from the syntectonic melt patch (Fig. 16C, sample O6-19L). These "mixed" ages (van Reenen et al., 2008; Kramers and Mouri, this volume) suggest the presence of different generations of garnet in the polymetamorphic sample O6-19, in accordance with the P-T data.

The precise U-Pb monazite age of 2017 ± 2.8 Ma obtained from the syntectonic melt patch of sample O6-19 (Fig. 16C) is indistinguishable from the precise PbSL garnet age of 2023 ± 11Ma obtained for metapelite T73 from the D_3 shear zone that overprints the Baklykraal structure >120 km to the west (Fig. 2, loc. 7) (Boshoff et al., 2006).

Precise U/Pb zircon SHRIMP ages of 2005.6 ± 4.4 Ma (Jaeckel et al., 1997) obtained from completely undeformed granitic melt patches that destroy the gneissic fabric of all rocks within which they are developed are a common feature of the CZ and provide a minimum age for the high-grade D_3 event (van Reenen et al., 2008).

High-Grade Polymetamorphism in the CZ

The crucial role that heterogeneous strain has played in the development and preservation of rare examples of multiple granulite facies events within single thin sections formed the basis for the construction of composite D-P-T-t diagrams that allowed superimposed high-grade Neoarchean and Paleoproterozoic events to be characterized (see also Mahan et al., 2008, this volume). Without this fortuitous set of circumstances, the apparent reaction history of the CZ rocks would lead incorrectly to an interpretation involving a single high-grade event at ca. 2.02 Ga (e.g., Zeh et al., 2004; Buick et al., 2006; Zeh and Klemd, 2008; Rigby, 2009).

D-P-T-t diagrams constructed for metapelitic gneisses sampled in this study indicate the following:

First (Fig. 17A), the entire CZ was affected by a single Neoarchean high-grade tectono-metamorphic event in which an early (ca. 2.68 Ga) DC1 stage is linked to the development of the regional D_2 gneissic fabric, whereas the second DC2 stage is linked to the formation of D_2 sheath folds and high-grade shear zones that developed in the interval ca. 2.65–2.63 Ga. The D_2 deformational event culminated with the emplacement of the rocks at the mid-crustal level before ca. 2.61 Ga (age of the Bulai Pluton).

Second (Fig. 17B), the rocks resided at the mid-crustal level for more than 600 m.y. before they were again reworked by a high-grade event in the Paleoproterozoic at ca. 2.02 Ga. The fact that maximum P-T conditions recorded by D_3 gneisses are the same as the minimum P-T conditions recorded by the D_2 gneisses (Fig. 18) suggest significant isobaric reheating of the rocks as a

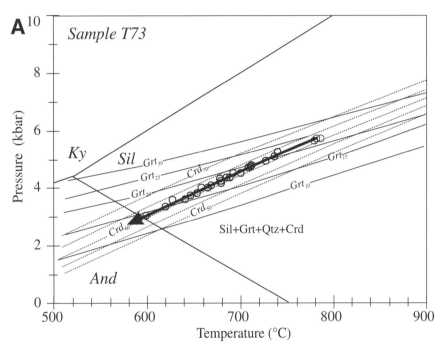

Figure 15. (A) P-T path for sample T73 (van Reenen et al., 2004). (B) Polished slab of sheared sample T73.

result of the superimposed event. This is reflected in the rocks by the isobaric (P ~5 kbar) reheating (T ~150 °C) path (Fig. 18).

Third, the suggested D_3 exhumation of the CZ toward the west is supported by the observation that an increasingly shallower crustal section is exposed from east to west (compare P-T paths for samples O6-19 (Musina area), T73 (Baklykraal area), and B5–5A2 (extreme western area in Botswana, Fig. 17).

Isobaric heating thus resulted in the widespread formation of polymetamorphic gneisses in the CZ (Perchuk et al., 2008a; van Reenen et al., 2008). High-grade polymetamorphism in the CZ furthermore is well demonstrated by comparing the P-T paths (Fig. 18) constructed for sample pairs TOV13 (D_2) and O6-19 (D_3) (Fig. 2, locs. 2 and 8), and RB65 (D_2) and RB55 (D_3) (Fig. 2, loc. 5, and Fig. 9C) from closely associated D_2 and D_3 structural features. The individual sample pairs have near-identical bulk chemical compositions (Fig. 14A; Perchuk et al., 2008a) and exemplify the same mineral parageneses *Grt + Crd + Sil + Bt* and the same reaction texture that reflects the replacement of garnet by cordierite *(Grt+Sil+Qtz => Crd)*. However, D_3 samples O6-19 and RB55 have much more modal cordierite and less modal garnet compared with D_2 samples TOV13 and RB65, and the Mg-numbers of *Grt*, *Crd*, and *Bt* are also significantly less than those of samples TOV13 and RB65 (Perchuk et al., 2008a). These differences are reflected by distinctly different P-T paths (Figs. 18A, 18B) in which the maximum P-T conditions documented by D_3 samples O6-19 and RB55 are equivalent to the minimum P-T conditions recorded by D_2 samples TOV13 and RB65. A similar situation (Fig. 18A) is also true in the case of the Mg-rich sample JC1 and the Fe-rich sample T73 (Fig. 2, locs. 6 and 7).

Integrated D-P-T-t data obtained from different localities in the CZ indicate high-grade polymetamorphism and thus do not support arguments in favor of a single high-grade Paleoproterozoic event in the CZ as has been suggested (e.g., Zeh et al., 2004, 2007; Buick et al., 2006; Zeh et al., 2007; Zeh and Klemd, 2008). Neither does it support a Turkic-type model (Barton et al., 2006) for the evolution of the CZ because such models require the CZ

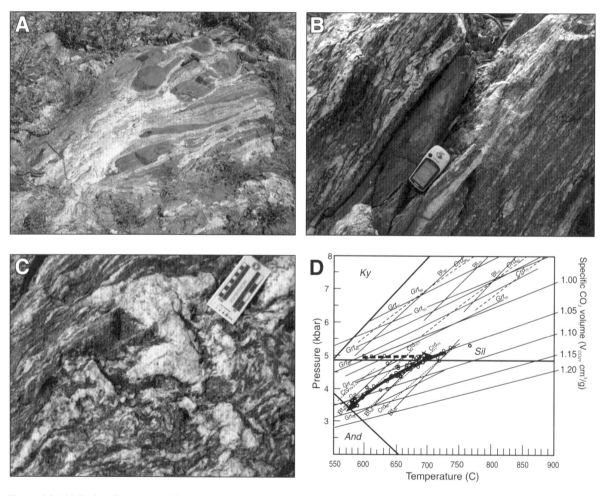

Figure 16. (A) D_2 boudins wrapped by D_3 shear zones at the Verbaard locality in the Sand River SW of Musina. (B) Sheared outcrop of sample O6-19M (Fig. 2, loc. 8) from a N-S–trending and moderately west-dipping D_3 shear zone that overprints the Campbell fold. (C) Garnet-bearing syntectonic melt patch (sample O6-19L) from the same outcrop. (D) Composite P-T path constructed for polymetamorphic sample O6-19M (Perchuk et al., 2008a).

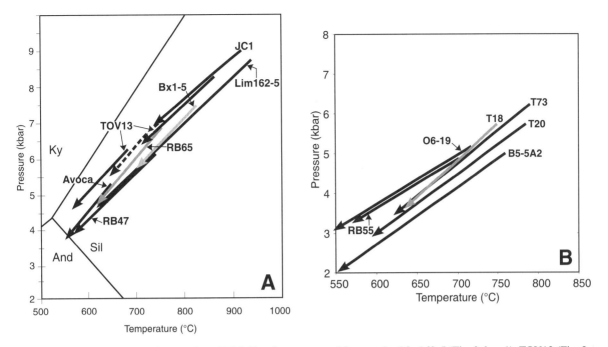

Figure 17. (A) Compilation of Neoarchean DC P-T paths constructed for samples Lim162-5 (Fig. 2, loc. 1), TOV13 (Fig. 2, loc. 2), RB65 (Fig. 2, loc. 5), RB47 (Fig. 2, loc. 3), Avoca (Fig. 2, loc. 4), JC1 (Fig. 2, loc. 6), and Bx1-5 (Fig. 2, loc. 1) (Perchuk et al., 2006a). (B) Compilation of Paleoproterozoic P-T paths constructed for samples O6-19 (Fig. 2, loc. 8), RB55 (Fig. 2, loc. 9), T73 (Fig. 2, loc. 7), T18 and T20 (Fig. 2, north of loc. 7), B5-5A2 (extreme western margin of the CZ, Hisada et al., 2005).

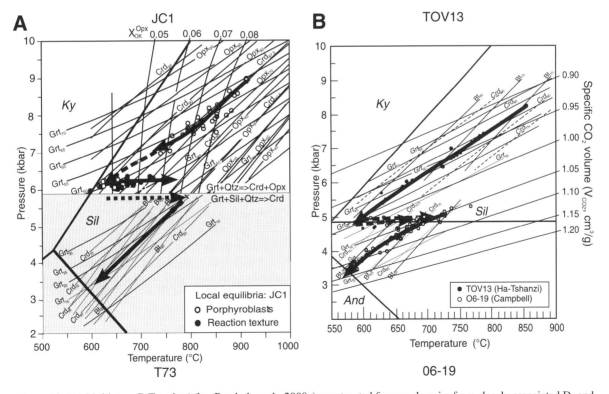

Figure 18. (A) Multistage P-T paths (after Perchuk et al., 2008a) constructed for sample pairs from closely associated D_2 and D_3 structures in the CZ. D_2 Sample JC1 (Fig. 2, loc. 6) and D_3 sample T73 (Fig. 2, loc. 7). (B) D_2 sample TOV13 (Fig. 2, loc. 2) and D_3 sample O6-19 (Fig. 2, loc. 8). Maximum P-T conditions recorded by D_3 samples T73 (A) and O6-19 (B) are respectively identical to the minimum conditions recorded by D_2 samples JC1 and TOV13. See text for further discussion.

to comprise terranes with distinctly different metamorphic and geochronological histories (see Kramers and Zeh, this volume).

Of interest is the fact that Millonig et al. (2008) recently published a ca. 2.0 Ga P-T path for the evolution of the Mahalapye Complex in the extreme western part of the CZ in Botswana (Fig. 1). This path includes a near-isobaric heating stage followed by decompression cooling, nearly identical to the ca. 2.0 Ga P-T path proposed for the Paleoproterozoic evolution of the CZ in South Africa (Fig. 18B). Millonig et al. (2008) also link the isobaric-heating stage of the P-T path to magmatic underplating related to the emplacement of the Bushveld Igneous Complex at ca. 2.05 Ga.

GEOLOGICAL IMPLICATIONS

Granulites that formed under medium to high pressure (P ~9 kbar) and temperature (T ~850 °C) conditions are commonly linked to processes of homogeneous thickening of a rheologically weak crust that resulted from continental collision. High-pressure (HP) and ultrahigh-temperature (UHT) conditions (>12 kbar, T >900 °C) are expected to develop in the basement of such over-thickened continental crust (Duchêne et al., 1997; O'Brien and Rötzler, 2003; O'Brien, 2008; Harley, 2008). However, current concepts of the exhumation of such deep-seated HP and UHT rocks (e.g., Duchêne et al., 1997; Schulmann et al., 2008) are not that clear. Gravity-driven mechanisms of exhumation of orogenic material within thickened orogenic root systems involve convective removal of the tectospheric root (England and Houseman, 1998; Andersen et al., 1991), a process considered to drive rapid exhumation from HP peak to mid-crustal conditions (P ~8–9 kbar). Gerya et al. (2004) furthermore showed that prograde changes in metamorphic mineral assemblages during thermal relaxation after collisional thickening of the crust provide a viable gravitational redistribution mechanism that leads to regional doming and diapirism.

Alternative models for exhumation of over-thickened crust focus on vertical displacement of HP rocks. Štípská et al. (2004) and Schulmann et al. (2008), for instance, suggest that early exhumation in the lower crust, as was observed in the eastern sector of the Variscan front of the Bohemian Massif, occurred during intra-crustal folding, followed by vertical extrusion of the lower crustal rocks. Perchuk and Gerya (this volume) show that collisional and gravitational mechanisms of rock deformation are not mutually exclusive and highlight the fact that collisional mechanisms should operate during the early prograde stage of the tectono-metamorphic cycle, whereas gravitational mechanisms should dominate during the post-peak retrograde stages through regional doming and diapirism.

Current concepts of exhumation of deep-seated rocks in convergent orogens are furthermore based on the style of the P-T-t paths retrieved from HP and UHT rocks (e.g., Duchêne et al., 1997). These paths commonly involve decompression heating (DH) or isothermal decompression (ID) paths. DH or ID paths are commonly followed by milder metamorphic conditions characterized by decompression-cooling paths (DC) that reflect conditions within the pressure range 9–4 kbar. Such conditions may or may not change into isobaric cooling (IC) paths when hot exhumed rocks are thrusted over cooler adjacent cratons, as was the case with the SMZ of the Limpopo Complex (van Reenen et al., this volume).

With this as background, the exhumation of the over-thickened crust that underlies the Limpopo Complex can be discussed with reference to a composite D-P-T-t diagram (Fig. 19) constructed for samples TOV 13 and O6-19 from the eastern CZ. This protracted evolutionary history commenced before ca. 2.7 Ga and ended at ca. 2.0 Ga with the emplacement of the high-grade CZ rocks at the upper crustal level. It involves seven distinct stages of evolution that can be linked to discrete P-T paths and deformational processes (Fig. 19).

An early HP (P >15 kbar, T ~950 °C) stage related to crustal shortening resulted in the development of the steep and strongly annealed regional gneissic fabric in the CZ at ca. 2.74 Ga. This poorly recorded event was followed by near-isothermal decompression when extreme UHT conditions (P ~9 kbar, T ~1000 °C) were reached (Fig. 18; Tsunogae and van Reenen, this volume) that marked the onset of exhumation and the possible development of large cuspate folds as a result of ductile flow in the deep-seated crust before 2.68 Ga. This is in accordance with suggestions by Štípská et al. (2004) and Schulmann et al. (2008) for the development of large cuspate folds in the lower crust and the onset of exhumation as reflected by isothermal decompression P-T paths (Fig. 19).

The UHT stage is normally followed by decompression cooling (DC) stages, commonly documented for the post-peak portions of the exhumational history of such terranes (e.g., Harley, 2008; O'Brien and Rötzler, 2003; Duchêne et al., 1997; O'Brien, 2008; Racek et al., 2006). The post-peak decompression cooling history of the CZ rocks within the interval ca. 2.68–2.61 Ga has been thoroughly investigated and documented in areas that have undergone heterogeneous deformation. The first stage (DC1) of this DC path (P ~9–6.5 kbar, T ~900–700 °C) (Fig. 19) in the interval ca. 2.68 to more than 2.62 Ga signifies the end of the isoclinal fold event. This is expressed in the associated rocks by an annealed fabric and on the P-T diagram (Fig. 19) by a distinct gap that is linked to the emplacement of Singelele-type granitoids in the interval 2.65–2.63 Ga (Fig. 19). The third stage (DC2, Fig. 19) (P ~6.5–5 kbar, T ~700–600 °C) is linked to the development of large shear zones and sheath folds that accommodated exhumation of high-grade rocks to the mid-crustal level before the emplacement of the Bulai Pluton at ca. 2.61 Ga. This event was probably connected with granitoid diapirism (Perchuk et al., 2008b) as observed in sheath fold formation. The CZ rocks resided at this level for more than 600 m.y. before they were finally reworked at ca. 2.0 Ga during the Paleoproterozoic high-grade event. This Paleoproterozoic event was initiated by isobaric (P ~5 kbar) reheating (T ~150 °C) IH of the rocks (Fig. 19), followed by the third DC stage during which P decreased from ~5–3 kbar with the final exhumation of the CZ to upper crustal levels

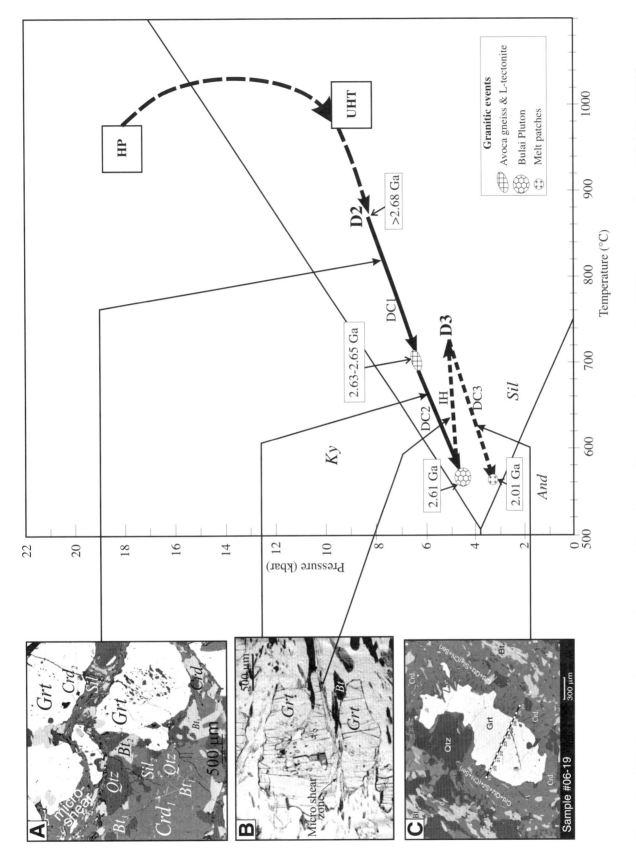

Figure 19. D-P-T-t diagram that demonstrates the protracted Neoarchean → Paleoproterozoic evolution of the CZ. The early (more than 2.68 Ga) HP and UHT events are taken from Tsunogae and van Reenen (2006, this volume). The post-peak $DC_1 \rightarrow DC_2 \rightarrow IH \rightarrow DC_3$ evolution in the interval ca. 2.68–2 Ga is from this study: (A) Relict granoblastic fabric (TOV13). (B) Younger D_2 shear fabric overprinted by D_3 micro-shear zones (TOV13). (C) D_3 shear fabric (06-19). See text for further discussion.

by a transpressive D_3 shear system. We suggest that reheating can be linked to magmatic underplating owing to the emplacement of magma related to the closely associated ca. 2.05 Ga Bushveld Igneous Complex (Scoates and Friedman, 2006; Millonig et al., 2008). The event ended with the development of large mylonitic shear zones that cut Bushveld-age granites (McCourt and Vearncombe, 1987).

MODEL FOR THE EVOLUTION OF THE LIMPOPO COMPLEX

We have always preferred to discuss the evolution of the Limpopo Complex as the *response* to a crustal-thickening event that resulted in the uplift and exhumation of high-grade rocks that presently are juxtaposed with the shallow-level crustal material of the adjacent granite-greenstone cratons (van Reenen et al., this volume). We can only speculate as to what caused the initial imbalance that responded by uplift, but several authors (de Wit et al., 1992; Roering et al., 1992a, 1992b; Treloar et al., 1992) suggested that the Zimbabwe and Kaapvaal Cratons were juxtaposed during a Himalayan-type collision in which the Kaapvaal Craton was forced over the Zimbabwe proto-Craton, leading to over-thickened crust. The response was probably controlled by a combination of processes that included a mechanism of gravitational redistribution that operated mainly at the mid-crustal level (Gerya et al., 2000, 2004; Perchuk and Gerya, this volume).

A tectonic model for the Neoarchean evolution of the Limpopo Complex (Fig. 19) should be constrained by the data discussed in this paper. Such a model furthermore should reflect the reaction or response of the Limpopo Complex to some preexisting event that led to large-scale crustal-gravitational instability when shallow crustal material was forced down to depths of the order of ~40 km (Fig. 19) (Tsunogae and van Reenen, this volume).

The crustal section (Fig. 20) suggests that the LC was thickened along an early (more than 2.7 Ga) system of north-verging thrusts that presently can be recognized within both cratons as under-thrusted crust (van Reenen et al., this volume; Blenkinsop, this volume). The thrusting and exhumation of high-grade rocks of the Marginal Zones over the adjacent cratons (e.g., van Reenen et al., this volume; Blenkinsop, this volume) are the major response to crustal thickening (van Reenen et al., 1987; Roering et al., 1992a, 1992b). In the SMZ, exhumation occurred along southward (Hout River Shear Zone) and northward (Tshipise Straightening Zone) verging shear zones compatible with a single exhumation event from deep to shallow crustal levels in the interval ca. 2.68–2.6 Ga (van Reenen et al., this volume) (Fig. 20).

The contractional event that affected the CZ in the Neoarchean is reflected by the fabrics preserved in the Tshipise Straightening Zone and the thrust-sense gneissic shear zones in the adjacent marginal zones (Fig. 20). The CZ, however, is further characterized by major sheath folds–diapiric structures that clearly relate to regional crustal movement. Gerya et al. (2004) showed that exponential lowering of viscosity with increasing temperature in thickened crust, as is suggested for the CZ (Fig. 19), provides a viable gravitational redistribution mechanism that leads to regional doming and diapirism. Perchuk and Gerya (this volume) also highlighted the fact that collisional (driven by external forces) and gravitational (driven by internal forces) mechanisms of rock deformation are not mutually exclusive. Gravitationally unstable crust is expected to result from collisional events that create crustal thickening, whereas gravitational mechanisms should dominate during the post-peak retrograde stages, thus providing an important factor for regional doming and diapirism. It is therefore suggested that the CZ was exhumed in the Neoarchean by a combination of uplift along dip and oblique-slip shear zones and diapiric uplift accompanied by sheath fold formation. It is important to note, however, that the gravitational response was synchronous with the dynamic compressional shear deformational event associated with the development of the TsSZ. This explains the general uniformity of all linear structures, large and small, within the TsSZ, other Neoarchean folds, early isoclinal folds, and related shear zones throughout the CZ. Had gravity been the sole driving force, the sheath folds would have been vertically oriented and not have a consistent, moderate SW plunge.

D-P-T-t studies of rocks from the CZ furthermore suggest that exhumation of the CZ rocks stopped at the mid-crustal level after the intrusion of the Bulai Pluton at ca. 2.61 Ga, which therefore marks the end of the Neoarchean event in the CZ. The rocks resided at this level for more than 600 m.y. before they were reworked at ca. 2.02 Ga by a major isobaric (P ~5 kbar) reheating (T ~150 °C) event (Fig. 18) that can be linked to the emplacement at 2.05 Ga of huge volumes of magma related to the Bushveld Complex (see also Millonig et al., 2008). Uplift of the CZ during this Paleoproterozoic high-grade event was accomplished by the development of a large oblique transpressive shear system (Holzer et al., 1998) that is reflected by DC3 P-T paths calculated for sheared metapelitic gneisses (Fig. 18). The end of the high-grade event was signified by the development at ca. 2.0 Ga of undeformed granitic melt patches (Jaeckel et al., 1997) that destroyed the gneissic–shear fabric of the rocks within which they were developed. The CZ rocks were finally overprinted at 2 Ga by large-scale upper-crustal strike-slip mylonitic belts such as the Palala and Triangle Shear Zones, which at present exposure levels define the boundaries to the CZ.

ACKNOWLEDGMENTS

DDvR acknowledges financial support of the Limpopo project from the University of Johannesburg and the National Science Foundation through grant 68288. LLP acknowledges RFBR grants (08-05-00354 and 09-05-00991) and grant 1949.2008.5 from the Program of the RF President "Leading Research Schools of Russia." We thank Steve McCourt for many discussions in the field and in the laboratory that contributed to our understanding of the structural evolution of the Central Zone of the Limpopo Complex. A special word of thanks to Johan

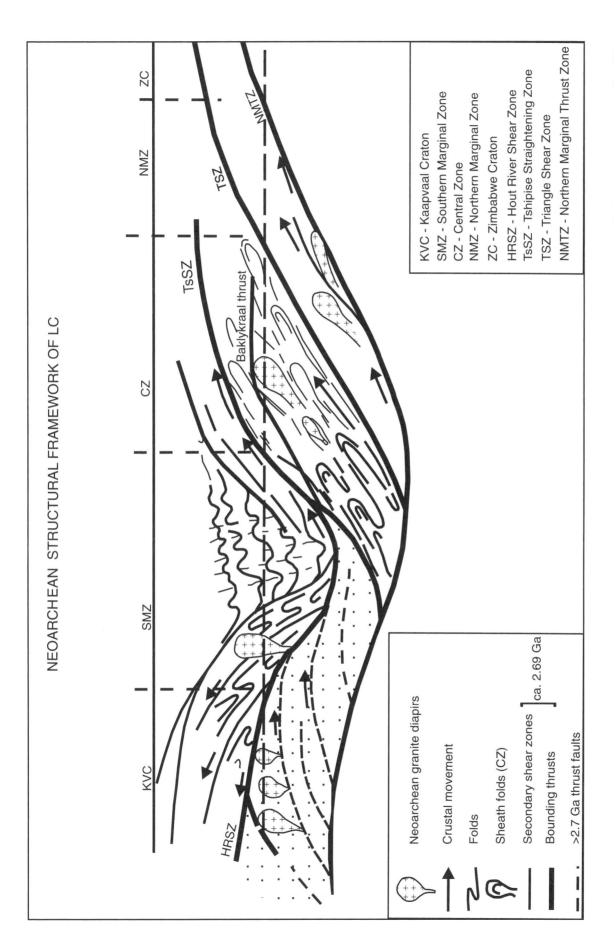

Figure 20. Schematic N-S crustal section of the Limpopo Complex and adjacent cratons. Thrust tectonics were accomplished by major inward-dipping shear zones that dominated the Neoarchean exhumation of the high-grade Limpopo Complex. Note the underthrusted Kaapvaal Craton crust (dotted), the emplacement of Neoarchean granitic diapirs, which in the CZ are linked to sheath fold formation, and the overall outward transport of high-grade material (arrows) onto the adjacent cratons within a crustal-scale pop-up structure. Exhumation within the SMZ took place along crosscutting (back-thrust) shear zones with respect to the initial transport, whereas exhumation along the NMZ developed parallel to the early crustal-thickening transport direction. Paleoproterozoic shear zones are not shown on this section. LC—Limpopo Complex.

Wolfaardt (Sand River Safaris, Musina) and Neels van Wyk (Farm Kilimanjaro, NW of Alldays) for their hospitality and free accommodations during fieldwork in the CZ. The final manuscript profited largely from critical but objective reviews by John Percival and an anonymous reviewer. The authors also acknowledge detailed comments on a previous version of the paper from Steve McCourt as the responsible editor.

REFERENCES CITED

Andersen, T.B., Jamtveit, B., Dewy, J.F., and Swensson, E., 1991, Subduction and eduction of continental crust: Major mechanism during continent-continent collision and orogenic extensional collapse, a model based on the south Norwegian Caledonides: Terra Nova, v. 3, p. 303–310, doi:10.1111/j.1365-3121.1991.tb00148.x.

Aranovich, L.Y., and Podlesskii, K.K., 1989, Geothermobarometry of high-grade metapelites: Simultaneously operating reactions, in Daly, J.S., Yardley, B.W.D. and Cliff, B.R., eds., Evolution of Metamorphic Belts: Geological Society [London] Special Publication 42, p. 41–65.

Bahnemann, K.P., 1972, A Review of the Structure, the Stratigraphy and the Metamorphism of the Basement Rocks of the Messina District, Northern Transvaal [D.Sc. thesis]: Pretoria, South Africa, University of Pretoria, 156 p.

Barton, J.M., Jr., and van Reenen, D.D., 1992, When was the Limpopo Orogeny?: Precambrian Research, v. 55, p. 7–16, doi:10.1016/0301-9268(92)90010-L.

Barton, J.M., Jr., Klemd, R., and Zeh, A., 2006, The Limpopo Belt: A result of Archean to Proterozoic, Turkic-type orogenesis?, in Reimold, W.U., and Gibson, R.L., eds., Processes on the Early Earth: Geological Society of America Special Paper 405, p. 315–331.

Blenkinsop, T.G., 2011, this volume, Archean magmatic granulites, diapirism, and Proterozoic reworking in the Northern Marginal Zone of the Limpopo Belt, in van Reenen, D.D., Kramers, J.D., McCourt, S., and Perchuk, L.L., eds., Origin and Evolution of Precambrian High-Grade Gneiss Terranes, with Special Emphasis on the Limpopo Complex of Southern Africa: Geological Society of America Memoir 207, doi:10.1130/2011.1207(13).

Blenkinsop, T.G., Kröner, A., and Chiwara, V., 2004, Single stage, late Archean exhumation of granulites in the Northern Marginal Zone, Limpopo Belt, Zimbabwe, and relevance of gold mineralization at Renco Mine: South African Journal of Geology, v. 107, p. 377–396, doi:10.2113/107.3.377.

Boshoff, R., 2004, Formation of Major Fold Types during Distinct Geological Events in the Central Zone of the Limpopo Belt, South Africa: New Structural, Metamorphic and Geochronologic data [M.S. thesis]: Johannesburg, Rand Afrikaans University, 121 p.

Boshoff, R., 2008, The Neoarchaean to Palaeoproterozoic Evolution of the Polymetamorphic Central Zone of the Limpopo High-Grade Terrain in South Africa [Ph.D. thesis]: Johannesburg, University of Johannesburg, 216 p.

Boshoff, R., van Reenen, D.D., Smit, C.A., Perchuk, L.L., Kramers, J.D., and Armstrong, R., 2006, Geologic history of the Central Zone of the Limpopo Complex: The West Alldays Area: Journal of Geology, v. 114, p. 699–716, doi:10.1086/507615.

Buick, I.S., Hermann, J., Williams, I.S., Gibson, R.L., and Rubatto, D., 2006, A SHRIMP U-Pb and LA-ICP-MS trace element study of the petrogenesis of garnet-cordierite-orthoamphibole gneisses from the Central Zone of the Limpopo Belt, South Africa: Lithos, v. 88, p. 150–172, doi:10.1016/j.lithos.2005.09.001.

Davidson, A., 1984, Identification of ductile shear zones in the southwestern Grenville Province of the Canadian Shield, in Kröner, A., and Greiling, R., eds., Precambrian Tectonics Illustrated: Stuttgart, E. Schweizerbart'sche Verlagsbuchhandelung (Nägele und Obermiller), p. 263–279.

de Wit, M.J., Roering, C., Hart, R.J., Armstrong, R.A., De Ronde, C.E., Green, R.W.E., Peberdy, E., and Hart, R.A., 1992, Formation of an Archaean continent: Nature, v. 357, p. 553–562, doi:10.1038/357553a0.

Droop, G.T.R., 1989, Reaction history of garnet-sapphirine granulites and conditions of Archaean high-pressure metamorphism in the Central Limpopo Mobile Belt, Zimbabwe: Journal of Metamorphic Geology, v. 7, p. 383–403, doi:10.1111/j.1525-1314.1989.tb00604.x.

Duchêne, S., Lardeaux, J.M., and Albarede, F., 1997, Exhumation of eclogites: Insights from depth-time path analyses: Tectonophysics v. 280, p. 125–140, doi:10.1016/S0040-1951(97)00143-1.

Du Toit, M.C., van Reenen, D.D., and Roering, C., 1983, Some aspects of the geology, structure and metamorphism of the Southern Marginal Zone of the Limpopo Metamorphic Complex: Geological Society of South Africa Special Publication 8, p. 121–142.

England, P.C., and Houseman, G.A., 1998, The mechanics of the Tibetan Plateau: Philosophical Transactions of the Royal Society of London, ser. A, p. 301–320.

Feldtmann, F., 1996, The Structural-Metamorphic Evolution of the Marble and Calc-Silicate Rocks of the Baklykraal Quarry near Alldays, Central Zone, Limpopo Belt, South Africa [M.S. thesis]: Johannesburg, Rand Afrikaans University, 132 p.

Fripp, R.E.P., Lilly, P.A., and Barton, J.M., Jr., 1979, The structure and origin of the Singelele Gneiss, Limpopo mobile belt, South Africa: Geological Society of South Africa Transactions, v. 82, p. 161–167.

Gerdes, A., and Zeh, A., 2009, Zircon formation versus zircon alteration—New insights from combined U-Pb and Lu-Hf in-situ LA-ICP-MS analyses, and consequences for the interpretation of Archean zircon from the Central Zone of the Limpopo Belt: Chemical Geology, v. 261, p. 230–243.

Gerya, T.V., and Perchuk, L.L.P., 1994, A new thermodynamic database for thermometry: Pisa, Italy, International Mineralogical Association General Meeting, 16th, p. 142.

Gerya, T.V., Perchuk, L.L., van Reenen, D.D, and Smit, C.A., 2000, Two-dimensional numerical modeling of pressure-temperature-time paths for the exhumation of some granulite facies terrains: Precambrian Journal of Geodynamics, v. 30, p. 17–35.

Gerya, T.V., Perchuk, L.L., Maresch, W.V., and Willner, A.P., 2004, Inherent gravitational instability of hot continental crust: Implication for doming and diapirism in granulite facies terrains, in Whitney, D., Teyssier, C., and Siddoway, C.S., eds., Gneiss Domes in Orogeny: Geological Society of America Special Paper 380, p. 97–115.

Harley, S., 2008, Refining the P-T records of UHT crustal metamorphism: Journal of Metamorphic Geology, v. 26, p. 125–154, doi:10.1111/j.1525-1314.2008.00765.x.

Harris, N.B.W., and Holland, T.J.B., 1984, The significance of cordierite-hypersthene assemblages from the Beitbridge region of the central Limpopo belt; evidence for rapid decompression in the Archaean?: American Mineralogist, v. 69, p. 1036–1049.

Hisada, K., Perchuk, L.L., Gerya, T.V., van Reenen, D.D., and Paya, B.K., 2005, Fluid infiltration and LP-HT event in the Mahalapye Complex of the Limpopo high-grade terrain, eastern Botswana: Johannesburg, University of the Witwatersrand, Geoscience Africa, Extended Abstracts, p. 276–277.

Hofmann, A., Kröner, A., and Brandl, G., 1998, Field relationships of mid- to late Archaean high-grade gneisses of igneous and sedimentary parentage in the Sand River, Central Zone of the Limpopo Belt, South Africa: South African Journal of Geology, v. 101, p. 185–200.

Holzer, L., 1995, The Magmatic Petrology of the Bulai Pluton and the Tectono-Metamorphic Overprint at 2.0 Ga in the Central Zone of the Limpopo Belt (Musina-Beitbridge Area, Southern Africa) [Diploma thesis]: University of Bern, Switzerland, 157 p.

Holzer, L., Frei, R., Barton, J.M., Jr., and Kramers, J.D., 1998, Unraveling the record of successive high grade events in the Central Zone of the Limpopo belt using Pb single phase dating of metamorphic minerals: Precambrian Research, v. 87, p. 87–115, doi:10.1016/S0301-9268(97)00058-2.

Holzer, L., Barton, J.M., Paya, B.K., and Kramers, J.D., 1999, Tectonothermal history of the western part of the Limpopo Belt: Tectonic models and new perspectives: Journal of African Earth Sciences, v. 28, p. 383–402, doi:10.1016/S0899-5362(99)00011-1.

Horrocks, P.C., 1983, The Precambrian geology of an area between Messina and Tshipise, Limpopo mobile belt, in Van Biljon, W.J., and Legg, J.H., eds., The Limpopo Mobile Belt: Geological Society of South Africa Special Publication 8, p. 81–88.

Huizenga, J.M., Perchuk, L.L., van Reenen, D.D., Flattery, Y., Warlamov, D., Gerya, T.V., and Smit, C.A., 2011, this volume, Emplacement of the Bulai Pluton and the retrograde P-T-fluid evolution of Neoarchean granulites in the Central Zone of the Limpopo Complex, in van Reenen, D.D., Kramers, J.D., McCourt, S., and Perchuk, L.L., eds., Origin and Evolution of Precambrian High-Grade Gneiss Terranes, with Special Emphasis on the Limpopo Complex of Southern Africa: Geological Society of America Memoir 207, p. doi:10.1130/2011.1207(08).

Jaeckel, P., Kröner, A., Kamo, S.L., Brandl, G., and Wendt, J.I., 1997, Late Archean to Early Proterozoic granitoid magmatism and high grade metamorphism in the central Limpopo belt, South Africa: Journal of the Geological Society [London], v. 154, p. 25–44, doi:10.1144/gsjgs.154.1.0025.

Kamber, B.S., Kramers, J.D., Napier, R., Cliff, R.A., and Rollinson, H.R., 1995a, The Triangle Shearzone, Zimbabwe, revisited: New data document an important event at 2.0 Ga in the Limpopo Belt: Precambrian Research, v. 70, p. 191–213, doi:10.1016/0301-9268(94)00039-T.

Kamber, B.S., Blenkinsop, T.G., Villa, I.M., and Dahl, P.S., 1995b, Proterozoic strike-slip deformation and uplift of the Central Zone, Limpopo Belt, Zimbabwe: Journal of Geology, v. 103, p. 493–508, doi:10.1086/629772.

Kramers, J.D., and Mouri, H., 2011, this volume, The geochronology of the Limpopo Complex: A controversy solved, in van Reenen, D.D., Kramers, J.D., McCourt, S., and Perchuk, L.L., eds., Origin and Evolution of Precambrian High-Grade Gneiss Terranes, with Special Emphasis on the Limpopo Complex of Southern Africa: Geological Society of America Memoir 207, doi:10.1130/2011.1207(06).

Kramers, J.D., and Zeh, A., 2011, this volume, A review of Sm-Nd and Lu-Hf isotope studies in the Limpopo Complex and adjoining cratonic areas, and their bearing on models of crustal evolution and tectonism, in van Reenen, D.D., Kramers, J.D., McCourt, S., and Perchuk, L.L., eds., Origin and Evolution of Precambrian High-Grade Gneiss Terranes, with Special Emphasis on the Limpopo Complex of Southern Africa: Geological Society of America Memoir 207, doi:10.1130/2011.1207(10).

Kramers, J.D., McCourt, S., and van Reenen, D.D., 2006, The Limpopo Belt, in Anhaeusser, C.R., and Thomas, R.J., eds., Geology of South Africa: Johannesburg, South African Geological Society, and Pretoria, Council for Geoscience, p. 209–236.

Kramers, J.D., McCourt, S., Roering, C., Smit, C.A., and van Reenen, D.D., 2011, this volume, Tectonic models proposed for the Limpopo Complex: Mutual compatibilities and constraints, in van Reenen, D.D., Kramers, J.D., McCourt, S., and Perchuk, L.L., eds., Origin and Evolution of Precambrian High-Grade Gneiss Terranes, with Special Emphasis on the Limpopo Complex of Southern Africa: Geological Society of America Memoir 207, doi:10.1130/2011.1207(16).

Kreissig, K., Thomas, F.N., Kramers, J.D., van Reenen, D.D., and Smit, C.A., 2000, An isotopic and geochemical study of the northern Kaapvaal Craton and the Southern Marginal Zone of the Limpopo Belt: Are they juxtaposed terranes?: Lithos, v. 50, p. 1–25, doi:10.1016/S0024-4937(99)00037-7.

Kreissig, K., Holzer, L., Frei, R., Ville, I.M., Kramers, J.D., Kröner, A., Smit, C.A., and van Reenen, D.D., 2001, Geochronology of the Hout River Shear Zone and the metamorphism in the Southern Marginal Zone of the Limpopo belt, southern Africa: Precambrian Research, v. 109, p. 145–173, doi:10.1016/S0301-9268(01)00147-4.

Kröner, A., Jaechel, P., Brandl, G., Nemchin, A.A., and Pidgeon, R.T., 1999, Single zircon ages for granitoid gneisses in the Central Zone of the Limpopo belt, southern Africa and geodynamic significance: Precambrian Research, v. 93, p. 299–337, doi:10.1016/S0301-9268(98)00102-8.

Light, M.P.R., 1982, The Limpopo Mobile Belt—A result of continental collision: Tectonics, v. 1, p. 325–342, doi:10.1029/TC001i004p00325.

Mahan, K.H., Goncalves, P., Flowers, R.M., Williams, M.L., and Hoffman-Setka, D., 2008, The role of heterogeneous strain in the development and preservation of a polymetamorphic record in high-P granulites, western Canadian shield: Journal of Metamorphic Geology, v. 26, p. 669–694, doi:10.1111/j.1525-1314.2008.00783.x.

Mahan, K.H., Smit, C.A., Williams, M.L., Dumond, G., and van Reenen, D.D., 2011, this volume, Heterogeneous strain and polymetamorphism in high-grade terranes: Insight into crustal processes from the Athabasca Granulite Terrane, western Canada, and the Limpopo Complex, in van Reenen, D.D., Kramers, J.D., McCourt, S., and Perchuk, L.L., eds., Origin and Evolution of Precambrian High-Grade Gneiss Terranes, with Special Emphasis on the Limpopo Complex of Southern Africa: Geological Society of America Memoir 207, doi:10.1130/2011.1207(14).

McCourt, S., and Armstrong, R.A., 1998, SHRIMP U-Pb zircon geochronology of granites from the Central Zone, Limpopo Belt, southern Africa: Implications for the age of the Limpopo Orogeny: South African Journal of Geology, v. 101, p. 329–338.

McCourt, S., and Vearncombe, J.R., 1987, Shear zones bounding the central zone of the Limpopo mobile belt, southern Africa: Journal of Structural Geology, v. 9, p. 127–137, doi:10.1016/0191-8141(87)90021-6.

McCourt, S., and Vearncombe, J.R., 1992, Shear zones of the Limpopo Belt and adjacent granitoids-greenstone terranes: Implications for late Archaean collision tectonics in southern Africa: Precambrian Research, v. 55, p. 553–570, doi:10.1016/0301-9268(92)90045-P.

Millonig, L., Zeh, A., Gerdes, A., and Klemd, R., 2008, Neoarchaean high-grade metamorphism in the Central Zone of the Limpopo Belt (South Africa): Combined petrological and geochronological evidence from the Bulai pluton: Lithos, v. 103, p. 333–351.

Mouri, H., Whitehead, M.J., Brandl, G., and Rajesh, H.M., 2009, A magmatic age and four successive metamorphic events recorded in a single meta-anorthosite sample in the Central Zone of the Limpopo Belt, South Africa: Journal of the Geological Society [London], v. 166, p. 827–830, doi:10.1144/0016-76492008-148.

O'Brien, P.J., 2008, Changes in pressure granulite metamorphism in the era of pseudosections: Reaction textures, compositional zoning and tectonic interpretation with examples from the Bohemian Massif: Journal of Metamorphic Geology, v. 26, p. 235–251, doi:10.1111/j.1525-1314.2007.00758.x.

O'Brien, P.J., and Rötzler, J., 2003, High-pressure granulites: Formation, recovery of peak conditions and implications for tectonics: Journal of Metamorphic Geology, v. 21, p. 3–20, doi:10.1046/j.1525-1314.2003.00420.x.

Perchuk, L.L., 1977, Thermodynamic control of metamorphic processes, in Saxena, S.K., and Bhattacharji, S., eds., Energetics of Geological Processes: New York, Springer-Verlag, p. 285–352.

Perchuk, L.L., 1989, P-T fluid regimes of metamorphism and related magmatism with special reference to the Baikal Lake granulites, in Day, S., Yardley, D.W.D., and Cliff, B., eds., Evolution of Metamorphic Belts: Geological Society [London] Special Publication 2, p. 275–291.

Perchuk, L.L., 2005, Configuration of P-T trends as a record of high-temperature polymetamorphism: Moscow, Doklady Russian Academy of Science, Earth Science, v. 401, p. 311–314.

Perchuk, L.L., and Gerya, T.V., 2011, this volume, Formation and evolution of Precambrian granulite terranes: A gravitational redistribution model, in van Reenen, D.D., Kramers, J.D., McCourt, S., and Perchuk, L.L., eds., Origin and Evolution of Precambrian High-Grade Gneiss Terranes, with Special Emphasis on the Limpopo Complex of Southern Africa: Geological Society of America Memoir 207, doi:10.1130/2011.1207(15).

Perchuk, L.L., and Lavrent'eva, I.V., 1983, Experimental investigation of exchange equilibria in the system cordierite-garnet-biotite: Advances in Physical Geochemistry, v. 3, p. 199–239.

Perchuk, L.L., and van Reenen, D.D., 2008, Comments on "P-T record of two high-grade metamorphic events in the Central Zone of the Limpopo Complex, South Africa": Reply: Lithos, v. 106, p. 403–410, doi:10.1016/j.lithos.2008.07.011.

Perchuk, L.L., Gerya, T.V., van Reenen, D.D., Krotov, A.V., Safonov, O.G., Smit, C.A., and Shur, M.Yu., 2000, Comparative petrology and metamorphic evolution of the Limpopo (South Africa) and Lapland (Fennoscandia) high-grade terrains: Mineralogy and Petrology, v. 69, p. 69–107, doi:10.1007/s007100050019.

Perchuk, L.L., Gerya, T.V., van Reenen, D.D., and Smit, C.A., 2006a, P-T paths and problems of high-temperature polymetamorphism: Petrology, v. 14, p. 117–153, doi:10.1134/S0869591106020019.

Perchuk, L.L., Varlamov, D.A., and van Reenen, D.D., 2006b, A unique record of P-T history of high-grade polymetamorphism: Moscow, Doklady Earth Sciences, v. 409A, p. 958–962, doi:10.1134/S1028334X06060286.

Perchuk, L.L., van Reenen, D.D., Varlamov, D.A., Van Kal, S.M., Boshoff, R., and Tabatabaeimanesh, S.M., 2008a, P-T record of two high-grade metamorphic events in the Central Zone of the Limpopo Complex, South Africa: Lithos, v. 103, p. 70–105, doi:10.1016/j.lithos.2007.09.011.

Perchuk, L.L., van Reenen, D.D., Smit, C.A., Boshoff, R., Belyanin, G.A., and Yapaskurt, V.O., 2008b, Role of granite intrusions for the formation of ring structures in granulite complexes: Examples from the Limpopo Belt, South Africa: Petrology, v. 16, p. 652–678, doi:10.1134/S0869591108070023.

Pienaar, J.C., 1985, The Geology of an Area in the Vicinity of Alldays in the Northern Transvaal [M.S. thesis]: Johannesburg, Rand Afrikaans University, 158 p. [in Afrikaans].

Racek, M., Štípská, P., Pitra, P., Schulmann, K., and Lexa, O., 2006, Metamorphic record of burial and exhumation of orogenic lower and middle crust: A new tectonothermal model for the Drosendorf window (Bohemian Massif, Austria): Mineralogy and Petrology, v. 86, p. 221–251, doi:10.1007/s00710-005-0111-7.

Ridley, J., 1992, On the origin and tectonic significance of the Charnockite Suite of the Archean Limpopo Belt, Northern Marginal Zone, Zimbabwe: Precambrian Research, v. 55, p. 407–427, doi:10.1016/0301-9268(92)90037-O.

Rigby, M.J., 2009, Conflicting P-T paths within the Central Zone of the Limpopo Belt: A consequence of different thermobarometric methods?:

Journal of African Earth Sciences, v. 54, p. 111–126, doi:10.1016/j.jafrearsci.2009.03.005.

Rigby, M.J., Mouri, H., and Brandl, G., 2008, A review of the P-T-t evolution of the Limpopo Belt: Constraints for a tectonic model: Journal of African Earth Sciences, v. 50, p. 120–132, doi:10.1016/j.jafrearsci.2007.09.010.

Roering, C., van Reenen, D.D., Smit, C.A., Barton, J.M., Jr., De Beer, J.H., de Wit, M.J., Stettler, E.H., Van Schalkwyk, J.F., Stevens, G., and Pretorius, S., 1992a, Tectonic model for the evolution of the Limpopo Belt: Precambrian Research, v. 55, p. 539–552, doi:10.1016/0301-9268(92)90044-O.

Roering, C., van Reenen, D.D., de Wit, M.J., Smit, C.A., De Beer, J.H., and Van Schalkwyk, J.F., 1992b, Structural geological and metamorphic significance of the Kaapvaal Craton–Limpopo Belt contact: Precambrian Research, v. 55, p. 69–80, doi:10.1016/0301-9268(92)90015-G.

Rollinson, H.R., 1993, A terrane interpretation of the Archaean Limpopo Belt: Geological Magazine, v. 130, p. 755–765, doi:10.1017/S001675680002313X.

Rollinson, H.R., and Blenkinsop, T., 1995, The magmatic, metamorphic, and tectonic evolution of the Northern Marginal Zone of the Limpopo Belt in Zimbabwe: Journal of the Geological Society [London], v. 152, p. 65–75, doi:10.1144/gsjgs.152.1.0065.

Schaller, M., Steiner, O., Studer, I., Holzer, L., Herwegh, M., and Kramers, J.D., 1999, Exhumation of Limpopo Central Zone granulites and dextral continent-scale transcurrent movement at 2.0 Ga along the Palala Shear Zone, Northern Province, South Africa: Precambrian Research, v. 96, p. 263–288, doi:10.1016/S0301-9268(99)00015-7.

Schulmann, K., Lexa, O., Štípská, P., Racek, M., Tajcmanová, L., Konopásek, J., Edel, J.-B., Peschler, A., and Lehmann, J., 2008, Vertical extrusion and horizontal channel flow of orogenic lower crust: Key exhumation mechanisms in large hot orogens?: Journal of Metamorphic Geology, v. 26, p. 273–297, doi:10.1111/j.1525-1314.2007.00755.x.

Scoates, J.S., and Friedman, R.M., 2006, Precise crystallization age of the Bushveld Complex, South Africa: Direct dating of the platiniferous Merensky Reef using the zircon U-Pb chemical abrasion ID-TIMS technique: American Geophysical Union Fall Meeting, San Francisco, abstract V31D-0611.

Smit, C.A., and van Reenen, D.D., 1997, Deep crustal shear zones, high-grade tectonites, and associated alteration in the Limpopo belt, South Africa: Implications for deep crustal processes: Journal of Geology, v. 105, p. 37–57, doi:10.1086/606146.

Smit, C.A., van Reenen, D.D., Gerya, T.V., and Perchuk, L.L., 2001, P-T conditions of decompression of the Limpopo high-grade terrain: Record from shear zones: Journal of Metamorphic Geology, v. 19, p. 249–268, doi:10.1046/j.0263-4929.2000.00310.x.

Söhnge, P.G., 1946, The Geology of the Messina Copper Mines and Surrounding Country: South African Geological Survey Memoirs, v. 40, 280 p.

Štípská, P., Schulmann, K., and Kröner, A., 2004, Vertical extrusion and middle crustal spreading of umphacite granulite: A model of syn-convergent exhumation (Bohemian Massif, Czech Republic): Journal of Metamorphic Geology, v. 22, p. 179–198, doi:10.1111/j.1525-1314.2004.00508.x.

Tolstikhin, I.N., and Kramers, J.D., 2008, The Evolution of Matter: From the Big Bang to the Present Day: Cambridge, UK, Cambridge University Press, 521 p.

Treloar, P.J., Coward, M.P., and Harris, N.B.W., 1992, Himalayan-Tibetan analogies for the evolution of the Zimbabwe Craton and Limpopo Belt: Precambrian Research, v. 55, p. 571–587, doi:10.1016/0301-9268(92)90046-Q.

Tsunogae, T., and van Reenen, D.D., 2006, Corundum+quartz and Mg-staurolite bearing granulite from the Limpopo Belt, southern Africa: Implications for a P-T path: Lithos, v. 92, p. 576–587, doi:10.1016/j.lithos.2006.03.052.

Tsunogae, T., and van Reenen, D.D., 2011, this volume, High-pressure and ultrahigh-temperature granulite-facies metamorphism of Precambrian high-grade terranes: Case study of the Limpopo Complex, in van Reenen, D.D., Kramers, J.D., McCourt, S., and Perchuk, L.L., eds., Origin and Evolution of Precambrian High-Grade Gneiss Terranes, with Special Emphasis on the Limpopo Complex of Southern Africa: Geological Society of America Memoir 207, doi:10.1130/2011.1207(07).

Tsunogae, T., Miyano, T., and Machacha, T.P., 1989, Retrograde orthoamphiboles in metapelites near Maratele, southwest Selebi-Phikwe, Botswana: Tsukuba, Japan, University of Tsukuba, Annual Report of the Institute of Geoscience, v. 15, p. 259–277.

Van Kal, S., 2004, Two Distinct Tectono-Metamorphic Events in the Central Zone of the Limpopo Complex, South Africa: Evidence from the Mt Shanzi Sheath Fold and the Campbell Cross Fold near Musina [M.S. thesis]: Johannesburg, Rand Afrikaans University, 158 p.

van Reenen, D.D., and Boshoff, R., 2008, Limpopo International Field Workshop 2008, unpublished Field Guide, July 13–17: Johannesburg, University of Johannesburg, Department of Geology, 130 p.

van Reenen, D.D., Barton, J.M., Jr., Roering, C., Smit, C.A., and Van Schalkwyk, J.F., 1987, Deep crustal response to continental collision: The Limpopo belt of southern Africa: Geology, v. 15, p. 11–14, doi:10.1130/0091-7613(1987)15<11:DCRTCC>2.0.CO;2.

van Reenen, D.D., Roering, C., Ashwal, L.D., and de Wit, M.J., 1992, Regional geological setting of the Limpopo Belt: Precambrian Research, v. 55, p. 1–5, doi:10.1016/0301-9268(92)90009-D.

van Reenen, D.D., Perchuk, L.L., Smit, C.A., Varlamov, D.A., Boshoff, R., Huizenga, J.M., and Gerya, T.V., 2004, Structural and P-T evolution of a major cross fold in the Central Zone of the Limpopo high-grade terrain, South Africa: Journal of Petrology, v. 45, p. 1413–1439, doi:10.1093/petrology/egh028.

van Reenen, D.D., Boshoff, R., Smit, C.A., Perchuk, L.L., Kramers, J.D., McCourt, S.M., and Armstrong, R.A., 2008, Geochronological problems in the Limpopo Complex, South Africa: Journal of Gondwana Research, v. 14, p. 644–662, doi:10.1016/j.gr.2008.01.013.

van Reenen, D.D., Smit, C.A., Perchuk, L.L., Boshoff, R., and Roering, C., 2011, this volume, Thrust exhumation of the Neoarchean UHT Southern Marginal Zone, Limpopo Complex: Convergence of decompression-cooling paths in the hanging wall and prograde P-T paths in the footwall, in van Reenen, D.D., Kramers, J.D., McCourt, S., and Perchuk, L.L., eds., Origin and Evolution of Precambrian High-Grade Gneiss Terranes, with Special Emphasis on the Limpopo Complex of Southern Africa: Geological Society of America Memoir 207, doi:10.1130/2011.1207(11).

Vernon, R.H., White, R.W., and Clarke, G.L., 2008, False metamorphic events inferred from misinterpretation of microstructural evidence and P-T data: Journal of Metamorphic Geology, v. 26, p. 437–449, doi:10.1111/j.1525-1314.2008.00762.x.

Watkeys, M.K., 1984, The Precambrian Geology of the Limpopo Belt North and West of Messina [Ph.D. thesis]: Johannesburg, University of the Witwatersrand, 349 p.

Watkeys, M.K., Light, M.P.R., and Broderick, R.J., 1983, A retrospective view of the Central Zone of the Limpopo Belt, Zimbabwe, in van Biljon, W.J., and Legg, J.H., eds., The Limpopo Belt: Geological Society of South Africa Special Publication 8, p. 65–80.

Windley, B.F., Ackermand, D., and Herd, R.K., 1984, Sapphirine/kornerupine-bearing rocks and crustal uplift history of the Limpopo belt, Southern Africa: Contributions to Mineralogy and Petrology, v. 86, p. 342–358, doi:10.1007/BF01187139.

Zeh, A., and Klemd, R., 2008, Comments on "P-T record of two high-grade metamorphic events in the Central Zone of the Limpopo Complex, South Africa": Comment: Lithos, v. 106, p. 403–410.

Zeh, A., Klemd, R., Buhlmann, S., and Barton, J.M., Jr., 2004, Pro- and retrograde P-T evolution of granulites of the Beit Bridge Complex (Limpopo Belt, South Africa): Constraints from quantitative phase diagrams and geotectonic implications: Journal of Metamorphic Geology, v. 22, p. 79–95, doi:10.1111/j.1525-1314.2004.00501.x.

Zeh, A., Gerdes, A., Klemd, R., and Barton, J.M., Jr., 2007, Archaean to Proterozoic crustal evolution in the Central Zone of the Limpopo Belt (South Africa–Botswana): Constraints from combined U-Pb and Lu-Hf isotope analyses of zircon: Journal of Petrology, v. 48, p. 1605–1639, doi:10.1093/petrology/egm032.

MANUSCRIPT ACCEPTED BY THE SOCIETY 24 MAY 2010

Archean magmatic granulites, diapirism, and Proterozoic reworking in the Northern Marginal Zone of the Limpopo Belt

T.G. Blenkinsop
School of Earth and Environmental Science, James Cook University, Townsville QLD 4811, Australia

ABSTRACT

The Northern Marginal Zone (NMZ) of the Limpopo Belt, southern Africa, is a high-grade gneiss belt dominated by magmatic granulites of the charnoenderbite suite, which intruded minor mafic-ultramafic and metasedimentary rocks between 2.74 and 2.57 Ga. The intrusive rocks have crustal and mantle components, and occur as elliptical bodies interpreted as diapirs. Peak metamorphism (P ≤800 MPa, T = 800–850 °C) occurred at ca. 2.59 Ga. The highly radiogenic nature of the rocks in the NMZ, supplemented by heat from mantle melts, led to heating and diapirism, culminating in the intrusion of distinctive porphyritic charnockites and granites. Horizontal shortening and steep extrusion of the NMZ, during which crustal thickening was limited by high geothermal gradients, contrast with overthickening and gravitational collapse observed particularly in more recent orogens. The granulites were exhumed by the end of the Archean. The pervasive late Archean shortening over the whole of the NMZ contrasts with limited deformation on the Zimbabwe Craton, possibly owing to the strengthening effect of early crust in the craton. In the southeast of the NMZ, strike-slip kinematic indicators occur within the Transition Zone and the Triangle Shear Zone, where dextral shearing reworked the Archean crust at ca. 1.97 Ga.

INTRODUCTION

Exposed high-grade gneiss terranes afford direct insights into the nature of the mid- and lower crust. Archean examples are particularly fascinating because of the clues they contain about early Earth tectonics, which complement perspectives gained from the lower grade granite-greenstone terranes of Archean cratons (e.g., Fedo et al., 1995; Kreissig et al., 2000; Kisters et al., 2003). The issues of whether plate tectonics operated in the Archean (e.g., Hamilton, 1998, 2003), when heat flow was 4–6 times present day values, and if so, how different it may have been, have been provoked by Limpopo studies (e.g., van Reenen et al., 1987; de Wit et al., 1992).

The Northern Marginal Zone (NMZ) is a high-grade gneiss terrane forming the northern part of the Limpopo Belt in southern Africa that illustrates several aspects of high-grade metamorphism. The terrane is 250 km long, with a maximum width of 70 km, and located largely within southern Zimbabwe (Fig. 1). To the north lies the Archean Zimbabwe Craton, and to the south, the Central Zone of the Limpopo Belt, containing older

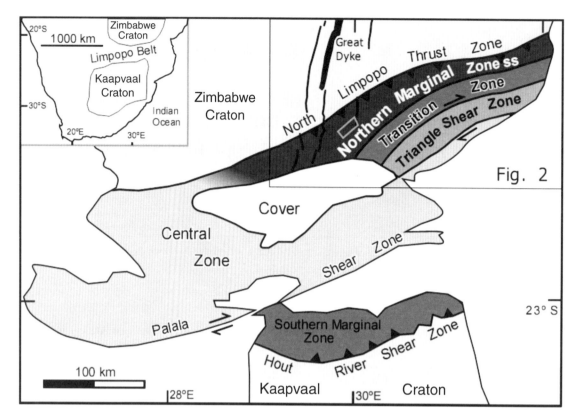

Figure 1. The Northern Marginal Zone in the context of the rest of the Limpopo Belt and the Zimbabwe and Kaapvaal Cratons (adapted from Blenkinsop and Kisters, 2005). Box in NMZ indicates position of Figure 6.

and higher grade gneisses (e.g., Barton et al., 1994; Holzer et al., 1998; Kroener et al., 1998). In the southeast corner of Zimbabwe, metasedimentary rocks of the Umkondo Group (ca. 1100 Ma) and sedimentary and volcanic rocks of the Karoo System (Jurassic) overlie the gneisses. The NMZ is contiguous with the Matsitama-Motloutse Complex to the west in Botswana (McCourt et al., 2004), but this complex has some different characteristics from the rest of the zone, and so is not included in this study.

The NMZ has been divided into four major structural zones, as shown in Figure 1 (Blenkinsop et al., 1995; Kamber and Biino, 1995; Mkweli et al., 1995; Rollinson and Blenkinsop, 1995): the North Limpopo Thrust Zone at the contact with the Zimbabwe Craton, the Northern Marginal Zone sensu stricto (NMZ s.s.), the Transition Zone, and the Triangle Shear Zone at the contact with the Central Zone of the Limpopo Belt. These zones are defined mainly on the basis of tectonic fabrics, but they also have some distinctive petrological characteristics.

A brief review of early geological knowledge of the Limpopo Belt is given by Mason (1973). MacGregor (1953) identified the belt by name, and Cox et al. (1965) defined the marginal and central zones. The Geological Survey of Zimbabwe (formerly Rhodesia) played a major role in early studies of the NMZ, followed by researchers from the Universities of Leeds, Zimbabwe, and Bern. Despite some intensive pieces of research, large areas in the NMZ are known no better than at reconnaissance level: It is probably the least well studied part of the Limpopo Belt that is reasonably well exposed. Many studies generalize over the whole zone, and little attention has been paid to heterogeneity.

The aim of this chapter is to review the geology of the NMZ, emphasizing four inter-connected themes: the relationship of the NMZ to the Zimbabwe Craton, the formation of granulites, the role of Archean versus Proterozoic tectonics, and the exhumation of granulites. The review combines petrographic, structural, geochronological, and geochemical data acquired over the past 15 yr with recent geophysical data, allowing a new understanding of high-grade metamorphic processes in this gneiss terrane, particularly in the late Archean.

ROCK TYPES AND FIELD RELATIONSHIPS

The rock types of the NMZ can be divided into a Supracrustal Assemblage that is intruded by a Plutonic Assemblage (Fig. 2; Rollinson and Blenkinsop, 1995). These assemblages are cut by discordant mafic dikes in several orientations, including the southern extensions of the Great Dyke and its satellites.

The Supracrustal Assemblage

The Supracrustal Assemblage consists of two main rock types: mafic granulites and metasedimentary rocks. The mafic

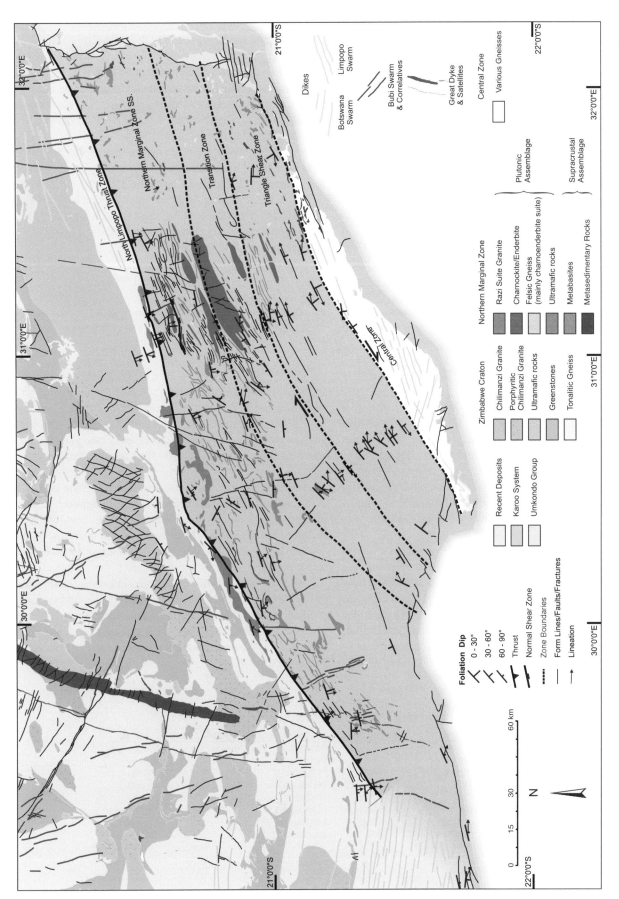

Figure 2. Detailed map of the Northern Marginal Zone, based on Brandl (1995), Pitfield (1996), and Rollinson and Blenkinsop (1995). The normal-sense shear zone is the Mtilikwe Shear Zone.

granulites comprise Opx, Pl, Hbl, Mag/Ilm ± Bt, Cpx, Qtz (see Table 1 for mineral abbreviations), with variable grain sizes from less than one to several millimeters, and variable fabrics from massive to strong gneissic banding. The metasedimentary rocks are mainly banded iron formation sensu lato (s.l.), but include quartzite, magnetite quartzite, metapelite, calc-silicate, and marble (Kamber and Biino, 1995). Metapelite consists of Qtz, Bt, Opx, Grt, Sil ± Crd, Ksp and shows evidence of partial melting. A single example of a coarse-grained marble (Cal + Cpx) has been described (Mwatahwa, 1992). Calc-silicate rocks consist of Di, Pl, Qtz, Am, Ttn ± Grt. All metasedimentary rocks show evidence of strong deformation and low- to medium-pressure granulite facies metamorphism. Highly aluminous magnesium-rich granulite was described by Robertson (1973a) from the Mount Towla area, consisting of Opx, Ath, Spl, Spr, Crd, and Bt. The mafic granulites and the metasedimentary rocks typically form sheets a few tens of meters wide, subparallel to the dominant fabric of the NMZ (Fig. 2). In places they define spectacular kilometric isoclinal folds (Fig. 2). Metabasic rocks also form meter-scale inclusions within the Plutonic Assemblage rocks.

The Plutonic Assemblage

The vast majority of rocks in the NMZ are gray, green, or brown felsic granitoids and gneisses with grain sizes of millimeters to centimeters. Their mineralogy is Fsp, Qtz, Opx ± Bi, Hbl, Cpx, Grt, with accessory Ap, Zrn, Fe-Ti oxides, and Py. These rocks have been almost always referred to as charnockite, charnoenderbite, or enderbite (e.g., Berger et al., 1995). The term charnoenderbite suite will be used to refer to the whole spectrum of compositions, which includes the specific type charnoenderbite.

Fabrics in rocks of the Plutonic Assemblage vary from undeformed, granular igneous textures to strong gneissic banding on a millimeter to centimeter scale. In places, strong fabrics are associated with grain size reduction, giving rise to protomylonites to mylonites (e.g., Blenkinsop and Kisters, 2005). Much of the Plutonic Assemblage occurs in oval bodies up to a few kilometers in size, partly to completely surrounded by supracrustal rocks, or in the cores of kilometer-scale, tight to isoclinal folds (Fig. 2: e.g., Blenkinsop et al., 2004). Various degrees of migmatization are recognized within the Plutonic Assemblage, resulting in either a distinctive massive white granitic rock with Grt, Mag, Sil, and Spl, which was called white granite by Rollinson and Blenkinsop (1995), or a coarser grained variety that occurs in obvious leucosomes parallel to the dominant east-northeast–trending fabric of the NMZ (Kamber and Biino, 1995).

An important debate concerns the interpretation of textures in these rocks. A magmatic origin for Opx is generally accepted (e.g., Berger et al., 1995), but opinions differ about the interpretation of some Bt textures. Bt is commonly found intergrown with Qtz. Ridley (1992) and Rollinson and Blenkinsop (1995) argued for magmatic crystallization of biotite (hydration crystallization) represented by:

$$Opx + melt \rightarrow Bt + Qtz \qquad (1)$$

as opposed to a retrograde hydration metamorphic reaction such as:

$$Opx + Kfs + H_2O \rightarrow Bt + Qtz \qquad (2)$$

because of the following features: (1) lack of preferential growth of Bt and Qtz along Opx-Kfs boundaries, (2) lack of resorption of Kfs, (3) uniformity of texture, (4) vermicular Bt-Qtz texture, and (5) subhedral feldspars. Kamber and Biino (1995) disputed some of these interpretations on the grounds that Bt-Qtz intergrowths occur in strongly retrogressed rocks. Blenkinsop et al., (2004) recognized that both reactions can be present and that the first reaction could be distinguished because it produced delicate intergrowths of Bt and Qtz compared with the random textures owing to retrogression. The presence of the intergrowths in the absence of adjacent Kfs could also be taken as evidence for the first reaction.

At least two types of retrogression affect the felsic members of the Plutonic Assemblage to variable degrees. Zones of coarser grained, lighter colored rock ("granodioritic zones," Berger et al., 1995) were formed where Pl is replaced by Kfs from fluid ingress. In the second type, Opx is replaced by narrow zones of Hbl or Bt.

TABLE 1. MINERAL ABBREVIATIONS*

Abbreviation	Mineral	Abbreviation	Mineral	Abbreviation	Mineral
Am	Amphibole	Fo	Forsterite	Pl	Plagioclase
Ap	Apatite	Grt	Garnet	Px	Pyroxene
Ath	Anthophyllite	Hem	Hematite	Py	Pyrite
Bt	Biotite	Hbl	Hornblende	Qtz	Quartz
Cal	Calcite	Ilm	Ilmenite	Rt	Rutile
Chl	Chlorite	Ksp	K-feldspar	Spr	Sapphirine
Cpx	Clinopyroxene	Mag	Magnetite	Ser	Sericite
Crd	Cordierite	Mc	Microcline	Sil	Sillimanite
Di	Diopside	Ol	Olivine	Spl	Spinel
Ep	Epidote	Or	Orthoclase	Ttn	Titanite
Fsp	Feldspar	Opx	Orthopyroxene	Zrn	Zircon

*After Siivola and Schmid (2007).

The youngest felsic member of the Plutonic Assemblage is a suite of late to post-tectonic granites and charnockites that are distinctive because of their large Ksp crystals. These porphyritic granites were described as occurring in the Razi subprovince (around the contact between the craton and the NMZ), and in the Kyle subprovince (slightly to the north in the Zimbabwe Craton: Robertson, 1973a, 1973b, 1974), but following Rollinson and Blenkinsop (1995) they will be described here as the *Razi suite*. Fabrics vary from random to strong; the latter are magmatic fabrics commonly parallel to the general east-northeast trend throughout the NMZ (Fig. 2). These granites appear similar to the Chilimanzi granites of the Zimbabwe Craton. Their mineralogy varies from Bt + Hbl near the Zimbabwe Craton (as in the Chilimanzi granites) to Opx dominant in the NMZ s.s., where they have a distinctive brown color. They are most common at the contact between the NMZ and the Zimbabwe Craton (the North Limpopo Thrust Zone), but there are also large bodies within the NMZ. The Razi suite is associated with pegmatites that are both parallel to and crosscut fabrics in the suite (e.g., Blenkinsop et al., 2004). Mafic minerals are generally not seen in the pegmatites, but up to 3% Bt can occur.

Ultramafic rocks in the NMZ occur in association with mafic rocks (Fig. 2), for example, in the Towla, Neshuru, Sizire, Inyala, Crown, and Chingwa-Ma-Karoro complexes (Worst, 1962; Robertson, 1974; Odell, 1975). The ultramafic rocks consist of serpentinite, dunite, and pyroxenite, and chromite layers in some complexes were mined. The complexes appear to be at least partly intrusive on the basis of relict igneous textures (banding, cumulate textures).

Mafic Dikes

A variety of mafic dikes in at least four younger generations cuts other rocks in the NMZ (Fig. 2).

The Great Dyke and Satellites

The southern extension of the Great Dyke crosses the North Limpopo Thrust Zone continuously from the craton into the NMZ at the Umlali River (Robertson, 1973a; Figs. 1, 2). A satellite to the Great Dyke to the west (the Umvimeela Dyke) is also continuous across the craton-NMZ boundary, and a mafic dike to the east in the NMZ aligns with the East Dyke, a major satellite to the east of the Great Dyke on the craton (Figs. 1, 2). A north-south dike in the east of the NMZ (Fig. 2) that crosses into the Zimbabwe Craton (Fig. 2), and has a total strike length of >100 km, may be a satellite of the Great Dyke (Pitfield, 1996), with several other subparallel dikes.

The Bubi Swarm

Northwest- to north-northwest–trending dikes occur throughout the NMZ and the southern part of the Zimbabwe craton (Fig. 2). These have been called the *Bubi swarm* in the NMZ (Robertson, 1973a), and they are parallel to the Sebanga Poort Dykes, considered to be coeval with the Mashonaland dolerites of the craton, intruded at ca. 2.0 Ga (Wilson et al., 1987).

The Limpopo and Botswana Swarms

East-northeast–trending dikes are abundant in the Triangle Shear Zone and the Transition Zone: These form the Limpopo Swarm (Fig. 2, Wilson et al., 1987). West-northwest–trending dikes in the southwest of the NMZ constitute the Botswana Swarm (Fig. 2), which intruded during the Jurassic Karoo igneous event (ca. 180 Ma), penecontemporaneously with the Limpopo Swarm (Reeves, 1978; Wilson et al., 1987).

GEOCHEMISTRY

Metabasites of the supracrustal assemblage range from basalt to trachyandesite in composition, and are olivine tholeiites by normative composition (Rollinson and Lowry, 1992). Rollinson and Lowry distinguished three types of metabasite from their REE (rare earth element) patterns: flat REE (A), enriched REE (B) and super-enriched REE (C: Fig. 3). A and B have similar patterns to tholeiites and calc-alkaline basaltic andesites of the Zimbabwe Craton, respectively. Three explanations were suggested for REE enrichment: crustal contamination, small degrees of partial melting, and source region heterogeneity. The first explanation was rejected because contamination with known local crustal compositions does not yield compatible major element compositions. The extreme enrichment of the group C dikes probably indicates an enriched mantle source (H. Rollinson, 2009, personal commun.). It is interesting that the Archean mafic dike geochemistry is similar to that of the Jurassic basalts of the Karoo igneous province in southern Africa (Rollinson and Lowry, 1992), which are also considered to be derived from an enriched mantle source (Duncan et al., 1984).

The major element geochemistry of the charnoenderbite suite classifies them mostly as tonalite (Fig. 4A), and there is a continuous compositional variation from these rocks to more felsic ones, including the granites of the Razi suite (Rollinson and Blenkinsop, 1995; Berger et al., 1995). Two features of the charnoenderbite suite geochemistry distinguish them from other Archean TTG (tonalite-trondhjemite-granodiorite) gneisses. They do not have depleted heat-producing elements (Th, U, Rb, K, Fig. 4B), and they are less sodic: Very few members of the charnoenderbite suite have trondhjemitic compositions (Berger and Rollinson, 1997; Fig. 4A).

Like the mafic rocks of the NMZ, the felsic rocks show fractionated light REE patterns (Fig. 4C), with fractionation increasing with SiO_2 content (Rollinson and Blenkinsop, 1995). The fractionation is consistent with either melting of a garnet-bearing source rock, or garnet fractionation in the source. Element variation diagrams suggest that the members of the charnoenderbite suite are related by fractionation of Cpx/Opx, and Hbl or Plag ± Cpx/Hbl for the more felsic compositions (Berger et al., 1995). The Cpx fractionation trend indicates that the charnoenderbite

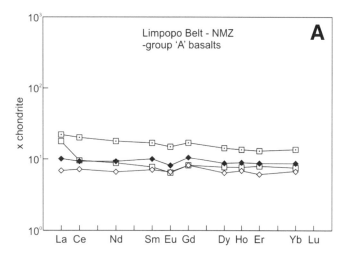
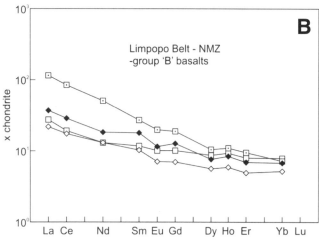
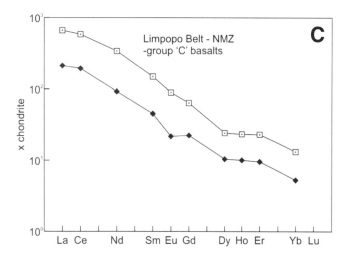

Figure 3. Rare earth element (REE) diagrams for three groups of metabasites from the NMZ (from Rollinson and Lowry, 1992). (A) Flat REE. (B) Enriched REE. (C) Super-enriched REE. REE enrichment may come from small degrees of partial melting or source region heterogeneity.

suite is magmatic. Large ion lithophile (LIL) patterns of the Razi charnockites and granites fit trends of partial melting from enderbite compositions.

The geochemical trends of the plutonic assemblage have strong affinities to various gneisses and granites of the Zimbabwe Craton (Berger et al., 1995). Chingezi and Mashaba tonalites (Luais and Hawkesworth, 1994) have similar geochemical behavior to the charnoenderbite suite. The Chingezi and Mashaba type I rocks differ slightly from the charnoenderbite suite in having Hbl fractionation trends instead of Px (Berger et al., 1995). Chilimanzi suite granites from the craton follow Kfs fractionation trends from charnoenderbite compositions, which are compatible with their derivation by partial melting of enderbites, and higher level fractionation. This is consistent with the late to post-tectonic timing of these intrusives.

The geochemistry of chromites from ultramafic rocks in the NMZ was examined by Rollinson (1995a, 1995b, 1997). Chromites from the Rhonda ultramafic complex were compared to other chromite compositions to demonstrate that they may have preserved liquidus characteristics in the chromite cores, and a trend toward chromium enrichment from cores to rims of individual grains was attributed to the gain of Cr from igneous pyroxenes during granulite facies metamorphism (Rollinson, 1995a). The liquidus chromite compositions in the Rhonda and Inyala deposits indicate an ultramafic liquid source, similar to the source of komatiites in the Belingwe Greenstone Belt on the Zimbabwe Craton, and different from chromitites of the Bushveld and some other layered mafic intrusions (Rollinson, 1997). Olivine Fo–NiO trends from the Inyala rocks are also continuous with Belingwe komatiites.

One of the most notable features of NMZ rock geochemistry as a whole are the very high Th and U contents (Berger et al., 1995; Berger and Rollinson, 1997). Kramers et al. (2001) calculated that the resulting heat production for the NMZ is comparable to, or exceeds by a factor of 2, various estimates of heat production from the average upper crust. Rocks of the southern Zimbabwe Craton also have high Th and U contents, and would have generated slightly less heat (80%) than the NMZ in the late Archean.

METAMORPHISM

The North Limpopo Thrust Zone

Metabasites in the southwest part of the North Limpopo Thrust Zone have a mineralogy of Cpx, Am, Pl, Bt ± Ttn, Ep, Ilm, Mag (Mkweli, 1997). An early high-grade assemblage of Cpx, Pl, Am is overgrown by Am, Bt, Mag, suggesting that the early Cpx has been replaced by Am and Bt in a retrogade event. This assemblage is in turn overgrown by Ep and Ttn. Both the early and the second assemblages are in the dominant foliation and appear to be syn-kinematic, but the Ep-Ttn retrogression is post-kinematic. Pressures have been estimated at 570–600 MPa and temperatures of 720–760 °C (Table 2).

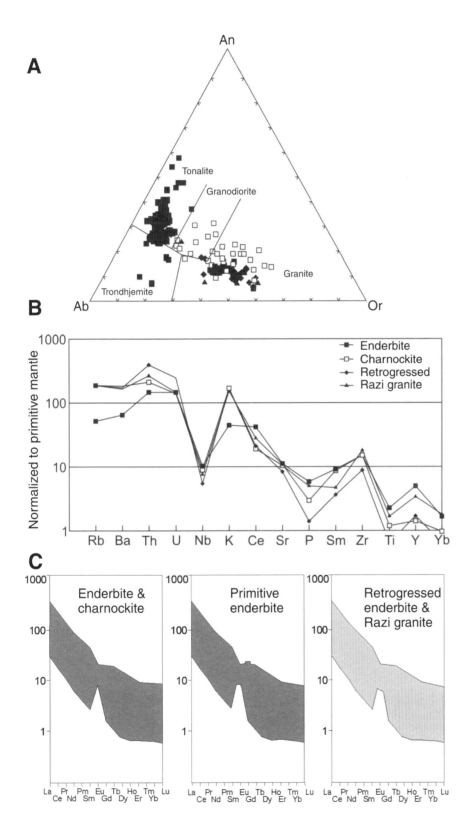

Figure 4. Geochemistry of the charnoenderbite suite, retrogressed rocks, and Razi granite suite (from Berger and Rollinson, 1997). (A) Normative An-Ab-Or classification diagram. The charnoenderbite suite is mainly tonalite. See B for key. (B) Mantle-normalized trace-element plots. Heat-producing elements Th, U, Rb, K are not depleted, unlike other TTG (tonalite-trondhjemite-granodiorite) suites. (C) Chondrite-normalized REE plots. The charnoenderbite suite has fractionated light REE patterns, like mafic rocks of the NMZ.

The Northern Marginal Zone sensu stricto

The detailed petrographic study of Kamber and Biino (1995) defined a four-stage metamorphic history for the NMZ s.s. Stages 1 and 2 are identified in metabasite enclosed within rocks of the charnoenderbite suite. Stage 1 has a mineralogy of Pl, Opx, Cpx, Mag, and/or Ilm ± Hbl, Qtz and may have bands of alternating Cpx and Opx-rich compositions. Overgrowths of Hbl and Mag and/or Ilm on the pyroxenes defines the retrogressive Stage 2. Kamber and Biino (1995) suggest that metasedimentary rocks in the NMZ did not undergo Stages 1 and 2. However, banded iron formation is found in thin layers parallel to mafic granulite, a strong argument that these two rock types were formed at the same time and share a common metamorphic history.

A critical textural observation in these mafic granulites are prograde reactions around Stage 2 Hbl, which may be expressed as Opx and Pl symplectite on Hbl:

$$Hbl + Qtz \rightarrow Pl + Opx + H_2O, \quad (3)$$

or:

$$Hbl \rightarrow Opx \pm Cpx \pm Pl + Qtz \pm Mag/Hem + H_2O. \quad (4)$$

These textures indicate a second granulite facies event, Stage 3 (Kamber and Biino, 1995), which was the peak of metamorphism.

A somewhat different metamorphic history for the NMZ s.s. was reported by Mkweli (1997) for the southwest part of the domain. Here, the first assemblage in metabasites is a coarse-grained intergrowth of Opx and Pl, which is interpreted to have grown after garnet during decompression heating. This is followed by an assemblage that is similar to Stage 1, which is overgrown by a symplectite of Opx, Cpx, Pl, Mag that corresponds closely to the description of Stage 3 in Kamber and Biino (1995).

"Dehydration rims" have been described around meter-sized mafic xenoliths within the charnoenderbite suite rocks (e.g., Rollinson and Blenkinsop, 1995). These 1-cm-wide finer grained rims consist of Opx, Cpx, Pl, Bt, and Qtz. The presence of Bt distinguishes them from the Stage 3 reactions given above, and they are attributed to "partial host-xenolith equilibration in which a fluid was involved" by Kamber and Biino (1995), which may have been syn- to late- or post-magmatic.

Stage 3 in the charnoenderbite suite is manifest as various degrees of melting (Kamber and Biino, 1995), creating the migmatites described previously. Metamorphic conditions in Stage 3 are constrained to >800 °C by reactions involving biotite breakdown of anhydrous pelitic assemblages, such as:

$$Bt + Sil + Qtz \rightarrow Kfs + Grt + melt, \quad (5)$$

$$Bt + Sil + Qtz \rightarrow Kfs + Crd + melt. \quad (6)$$

TABLE 2. THERMOBAROMETRIC ESTIMATES FOR THE NORTHERN MARGINAL ZONE (NMZ)

Zone	Timing	Rock type	Pmin (MPa)	Pmax (MPa)	Tmin °C	Tmax °C	Geobarometer	Geothermometer	P-T path	Reference
NLTZ	Stage 3	Metabasite	570	600	720	760	Al in Am	Pl-Am	N.D.	Mkweli (1997)
NMZss West	Stage 3	Metabasite	480	900	720	860	Opx-Cpx-Am-P-Qtz	Opx-Cpx-Am-P-Qtz	Clockwise with heating	Mkweli (1997)
NMZss East	Peak	Felsic gneiss	740	940	800	900	Gt-Opx-Plag-Qtz	Opx-Gt	Counterclockwise	Rollinson (1989)
NMZss West	Peak	Felsic gneiss	400	600	750	850	Gt-Opx-Plag-Qtz	Opx-Gt	Counterclockwise	Rollinson (1989)
NMZss	Stage 3	Metapelites		400	800		Various	Various	Counterclockwise	Kamber et al. (1995b)
NMZss	Post Stage 3	Various		800	800		Various	Various	Counterclockwise	Kamber et al. (1995b)
NMZ	Peak	Pelitic gneiss	750	790	740	790	Various	Gt-Opx-Plag-Qtz	Decreasing P and T retrograde path	Tsunogae et al. (1992)
NMZ	Peak	Charnockite	600	750	640	750	Various	Gt-Opx-Plag-Qtz	Decreasing P and T retrograde path	Tsunogae et al. (1992)
TZ (North)	2 Ga	Amphibolite	560	680	650	680	Grt-Hbl-Pl-Qtz	Grt-Hbl	N.D.	Kamber et al. (1996)
TZ (South)	2 Ga	Felsic gneiss			740	860	Grt-Cpx, Grt-Bi	Grt-Bt, Grt-Cpx	N.D.	Kamber et al. (1996)
TSZ	2 Ga	Felsic gneiss	680	960	750	900	Grt-Opx-Pl-Qtz, Cpx-Hbl-Pl-Qtz	Grt-Bt, Grt-Cpx	N.D.	Kamber et al. (1996)
TSZ	2 Ga	Mafic gneiss	1100	1330	790	890	Grt-CPx, Grt-Hbl, Grt-Pl	Grt-Hbl-Pl-Qtz, Grt-Cpx	Decreasing P and T retrograde path	Tsunogae et al. (1992)
TSZ	2 Ga	Gneiss	500	900	700	850	Various	Various	Isothermal decompression	Kamber et al. (1995a)

*N.D.—not determined; NLTZ—North Limpopo Thrust Zone; NMZss—Northern Marginal Zone sensu stricto; TZ—Transition Zone; TSZ—Triangle Shear Zone.

Peak metamorphic conditions of 740–940 MPa and 800–900 °C were measured in the eastern part of the NMZ s.s., and conditions of 400–600 MPa and 750–850 °C for the western part (Table 2; Rollinson, 1989). A counterclockwise PT loop was established from changes in Grt compositions (Rollinson, 1989). Similar temperatures were established by Kamber et al. (1995b) and Kamber and Biino (1995). The latter study suggested that pressures rose during Stage 3 from 400 MPa (suggested by the coexistence of Spl and Crd) to a maximum near 800 MPa, constrained by the lack of garnet in mafic granulites (Fig. 5A). These authors were unable to confirm the suggestion by Rollinson (1989) that peak conditions were lower in the west, and proposed instead greater levels of post-peak metamorphic exhumation southward toward the Transition Zone and Triangle Shear Zone. Mkweli (1997) obtained similar peak P-T conditions to Kamber et al. (1995b) and Kamber and Biino (1995) from the southwest part of the NMZ, which also casts doubt on a systematic increase in grade from east to west. Mkweli's results did not show a systematic cross-strike variation in grade. Mkweli interpreted the P-T path as generally clockwise, with intermittent heating events.

A different P-T path was proposed by Tsunogae et al. (1992). They suggested peak metamorphic conditions of 750–790 MPa and 740–790 °C, and linked them to lower grade conditions along a path of decreasing P and T. Their approach was criticized by Kamber et al. (1996) and Kamber and Biino (1995), because the P-T path combined granulite facies rock from Stage 3 (Archean) metamorphism with amphibolite facies assemblages that may be Proterozoic in age (see below).

Stage 4 is a retrogression recorded by Hbl replacing Stage 3 Px, and by formation of Bt (Kamber and Biino, 1995; Mkweli, 1997). Such retrogression is confined to meter–decimeter–wide shear zones where H_2O fluids were available. In the Razi granites, Stage 4 is marked by a greenschist facies assemblage of Ep, Am, Ttn, Ab, and Qtz within shear zones parallel to the main fabric. There appears to be an increase in Stage 4 metamorphic conditions toward the south: Bt changes from green-brown to reddish, and Am changes from hastingsitic Hbl to true Hbl (Kamber and Biino, 1995).

The Transition Zone

A significant proportion of rocks in the Transition Zone consists of Bt gneiss, lacking the Opx characteristic of the charnoenderbite suite in the NMZ s.s. Mafic rocks also lack a Px phase and are generally amphibolites. P-T conditions range from 560 to 680 MPa, 650 °C at the boundary with the NMZ s.s., to ~800 MPa and 740–860 °C at the boundary with the Triangle Shear Zone (Table 2), corresponding to the change from Hbl to two pyroxenes in mafic rocks from north to south across the Transition Zone.

The Triangle Shear Zone

Peak conditions of 1100–1330 MPa and 790–890 °C were determined by Tsunogae et al. (1992) from a mafic gneiss using a

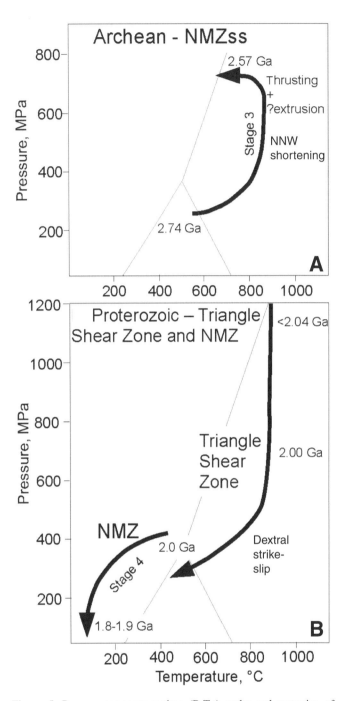

Figure 5. Pressure-temperature-time (P-T-t) paths and tectonics of Archean and Proterozoic events in the Northern Marginal Zone. (A) Archean P-T-t Stage 3 path of the NMZ s.s. from Kamber and Biino (1995). Stages 1 and 2 occurred some time before the start of Stage 3. (B) Proterozoic P-T-t paths of the NMZ (from Kamber and Biino, 1995) and the Triangle Shear Zone (from Kamber et al., 1995a; Tsunogae et al., 1992). The Proterozoic part of the NMZ path is Stage 4 of Kamber and Biino (1995).

Grt-Cpx-Pl-Qtz assemblage (the location of this sample is reported as Central Zone but corresponds to the Triangle Shear Zone of this chapter). Near-isothermal decompression was inferred. Similar temperatures (700–850 °C) were determined by Kamber et al. (1995a) for mylonitization, using a variety of mineral equilibria in felsic and mafic granulite and meta-melanosomes (Table 2), but the pressures were significantly lower (500–900 MPa) and similar to other determinations from Tsunogae et al. (1992). Pressure variations were interpreted as recording variations in exhumation during isothermal uplift, and a clockwise P-T path was inferred from textures such as coronas of Crd around Grt, and Pl around Grt (Fig. 5B; Kamber et al., 1995a).

STRUCTURE

The North Limpopo Thrust Zone

A zone of protomylonite and mylonite up to several kilometers wide marks the boundary in many places between greenschist-amphibolite-grade gneiss and greenstone of the Zimbabwe craton and the largely charnoenderbitic granulite of the NMZ s.s. (Blenkinsop et al., 1995; Mkweli, 1997). The mylonites dip moderately to the east-southeast and generally have a strong, almost perfectly downdip mineral stretching lineation, generally of Qtz (Fedo et al., 1995; Mkweli et al., 1995; Rollinson and Blenkinsop, 1995). Abundant shear sense indicators (σ and δ porphyroclasts) show a reverse shear sense. The typical mineralogy of the mylonites is Qtz, Pl, Kfs, Bt ± Hbl, Ep, Ilm, Chl, Ttn, Ap, suggesting deformation in amphibolite to greenschist facies conditions. In places the contact between the craton and the NMZ s.s. is intruded by variably foliated granite of the Razi suite, and the zone of mylonites is obscured.

The reverse shear kinematics of the contact between the NMZ and the craton was recognized by Worst (1962), Robertson (1973a), James (1975), and Coward et al. (1976), but these workers suggested that the thrust did not exist along the SW half of the craton-NMZ contact in Zimbabwe. Mkweli et al. (1995) demonstrated that the thrust zone does extend along this part of the contact. The greater abundance of Razi granite intrusions in the southwest may be one reason that earlier workers did not recognize the existence of the thrust zone there. Mkweli (1997) showed that foliations in the thrust zone dip less steeply than in the craton or than the adjacent NMZ s.s., as expected in a reverse-sense shear zone. This variation was used to estimate the displacement on the thrust by strain integration, using the shapes of deformed clasts as an additional constraint. Considering the possibility of a pure shear component, and allowing for active as well as passive behavior of the foliation, all estimates of displacement are <2 km.

The NMZ sensu stricto

A pervasive fabric is found throughout the NMZ s.s., consisting of a millimeter- to centimeter-scale gneissic banding trending generally east-northeast, parallel to the strike of the NMZ, and dipping steeply south-southeast, or subvertical (Fig. 2). A parallel foliation is defined by grain shapes of Qtz, Pl, Kfs, and Hbl in many places. These minerals may also define a weak downdip mineral lineation.

Some of the most spectacular structures of the NMZ s.s. are tight to isoclinal, or elasticus, folds with wavelengths and amplitudes of kilometers, defined by folded layers of mafic granulite and banded iron formation (Fig. 2). The axial traces of the folds are parallel to the dominant east-northeast trend of the gneissic banding, which is, however, folded itself. The hinges of the folds plunge at variable angles to the east-northeast or west-southwest. Elliptical bodies of charnockite and enderbite with dimensions of several kilometers also form prominent structures in the NMZ s.s. (Fig. 6). The long axes of the ellipses are parallel to the gneissic banding. These domal features are inferred to be diapiric in origin (Rollinson and Blenkinsop, 1995). Triangular shaped junctions of mafic granulites around the domes are similar in plan view to cleavage triple points, and some tabular bodies of mafic granulite may correspond to areas of flattening between domes (Fig. 6; cf. Brun et al., 1981).

Several different types of shear zones have been described within the NMZ s.s. (Kamber et al., 1996; Blenkinsop et al., 2004). These can be divided into four categories on the basis of inferred temperatures of deformation, relative timing, and kinematics, as follows.

High-Grade Shear Zones

A good example of this category is the Mtilikwe Shear Zone, a zone of protomylonite and mylonite at least

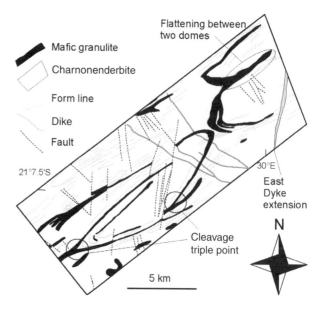

Figure 6. Geology of part of the NMZ, indicating features that could be interpreted as diapiric: domes, cleavage triple points (triangular junctions of mafic granulites), and tabular bodies of mafic granulite between two domes (from Mkweli, 1997).

25 km long and up to 500 m wide, 15 km south of the North Limpopo Thrust Zone (Blenkinsop and Kisters, 2005). The mylonitic foliation dips steeply south-southeast and carries a prominent downdip mineral lineation comprising mainly Qtz and Qtz-Fsp rods. Enderbitic protoliths of the mylonite comprise Qtz, Pl, Bt, Opx, and Cpx, whereas charnockitic protoliths consist of Qtz, Mc, Or, Pl, Bt, Opx, and Cpx, and probable metasedimentary rocks have the same assemblage but with Grt and Sil instead of Px. Zr, Ap, Ilm, and Rt are common accessory minerals. All major phases except Grt are dynamically recrystallized, forming a fabric of Qtz and Fsp ribbon grains wrapping around porphyroclasts of Grt and Px. There is little evidence for retrogression.

The Mtilikwe Shear Zone is remarkable for a variety of clear shear-sense indicators that reveal a normal sense of movement. σ and δ porphyroclasts and SC and SC′ fabrics are common. Extension microfractures in Grt porphyroclasts filled by Qtz, Fsp, and Bt of the main mineral assemblage have a consistent orientation that also indicates normal movement. A Pb-Pb step leach age of 2601 ± 5 Ma for a synkinematic garnet dates the age of deformation in this shear zone (Blenkinsop and Frei, 1996), compatible with a late Stage 3 timing, as suggested by the mineralogy.

Medium-Grade Shear Zones

Meter–decimeter–wide shear zones parallel to the east-northeast trend of the NMZ were described by Kamber and Biino (1995), Kamber et al. (1996), and Blenkinsop et al. (2004). These shear zones form an anastomosing network with a strong S>L fabric defined by Qtz, Bt, and Fsp, dipping gently to moderately south-southeast with a downdip mineral lineation. σ and δ porphyroclasts generally indicate reverse shear. Pl shows deformation twinning, Mc has well-developed undulatory extinction, and oriented myrmekite intergrowths occur at the ends of Pl porphyroclasts. Grt has ubiquitous microfractures and Qtz pressure shadows. These microstructural features indicate deformation at the upper end of the 400–500 °C range, and suggest formation as Stage 3 waned.

Strike-Slip Shear Zones

The gneissic banding in the NMZ s.s. and in the North Limpopo Thrust Zone is cut and deformed by a system of discrete shear zones, typically several meters long and a few centimeters wide (Blenkinsop and Mkweli, 1992). These shear zones have consistent orientations and kinematics over the whole of the NMZ: They are vertical and trend either northwest or north-northeast. They contain Qtz-Pl-Kfs mylonite with subhorizontal lineations, and locally they are intruded by granitic pegmatite that is also deformed. The gneissic foliation of the NMZ is deformed by these shear zones. The northwest- and north-northeast–trending shear zones are invariably dextral and sinistral, respectively, as shown by the deflection of gneissic banding, and no crosscutting relationships are seen between the two sets. These features are all consistent with formation of the shear zones as conjugate sets with a north-northwest–shortening direction, contemporary with the intrusion of the granitic pegmatites. The contemporaneous intrusive activity suggests that these shear zones were part of Stage 3. Fractures in similar orientations to these shear zones are a conspicuous feature in the Chilimanzi granites of the southern margin of the Zimbabwe Craton (Fig. 2; Robertson, 1973b).

Low-Grade Shear Zones

Low-grade shear zones are 10 cm–2 m wide with a maximum strike length of 350 m, separated by 3–5 km both along and across strike. They have been described from the area around the Renco gold mine (Blenkinsop et al., 2004). They are parallel to the trend of the NMZ, and some occur along parts of the medium-grade type of shear zone, although they are much shorter and narrower. They are characterized by cataclastic textures with centimeter-sized fragments separated by narrow fractures. Qtz is fractured, Pl is altered to Chl and Ttn, and Mc to Ser, Ep, and Chl. Am and Px are entirely replaced by Chl and Ttn. Lack of lattice deformation in feldspar suggests temperatures of deformation <400 °C. Features suggestive of pseudotachylite have been described for these shear zones. The shears were interpreted as Proterozoic in age by Blenkinsop et al. (2004), and they are similar to shear zones described by Kamber et al. (1996) that correspond to Stage 4.

An important problem posed by these shear zones is their kinematics. Kamber et al. (1996) were unable to find shear-sense indicators within these shear zones but inferred a thrust sense from feldspar clasts in the surrounding gneisses, and from asymmetric folds. Likewise, Blenkinsop et al. (2004) could not find reliable shear-sense indicators. However, Kolb et al. (2003) reported evidence for normal-sense movement on shear zones at the Renco mine, which they describe as occurring at or below 300 °C on shear zones dipping gently southeast.

The Transition Zone

The transition from the NMZ s.s. to the Triangle Shear Zone is marked by the appearance of moderately southeast-dipping foliation with lineations plunging gently to the southwest. This foliation occurs in localized zones that increase in abundance toward the south (Kamber et al., 1995b). Fabrics in the Transition Zone therefore consist of alternating zones where the foliation preserves downdip lineations typical of the NMZ s.s., and zones where the lineations are subparallel to strike. There is an overall decrease in foliation dip to the southeast within the Transition Zone, and there are also asymmetric folds with subhorizontal hinges on meter to kilometer scales, which consistently verge to the northwest. Fabrics are generally S>L. No indications of non-coaxial deformation are seen on a small scale. Fabrics within the Transition Zone have been interpreted as indicating a zone of northwest shortening and subvertical extension by Kamber et al. (1995b), associated with the 2.0 Ga orogeny in the Central Zone of the Limpopo Belt. This interpretation is supported by

Hbl Ar-Ar ages from both the downdip and strike subparallel orientations (Table 3).

The Triangle Shear Zone

In the Triangle Shear Zone the southeast dip of the foliations observed in the Transition Zone is pervasive (James, 1975). As defined by Rollinson and Blenkinsop (1995), the zone has a strike length of at least 140 km and a width of 30–50 km (Figs. 1, 2), characterized by intense mylonitic L-S fabrics (Kamber et al., 1995b) with subhorizontal Qtz and Fsp mineral lineations. Grain sizes are half those of the Transition Zone. The majority of mylonite has a Qtz, Ksp, Pl, Bt, Opx, Hbl ± Grt mineralogy and is clearly derived from the charnoenderbite suite. Rocks consisting of Bt, Grt, Sil ± Crd, Spl are interpreted as Al-rich melanosome in migmatites, with leucosome of Qtz, Ksp, Pl, Grt, and Fe-Ti oxides. Metabasic rocks consist of Cpx, Opx, Pl, Grt, Hbl ± Qtz. Quartzite, calc-silicate and metapelite have also been reported from the Triangle Shear Zone (Ridley, 1992; Rollinson and Blenkinsop, 1995). Surfaces perpendicular to the foliation and parallel to the lineation show excellent and consistent shear-sense indicators in the form of σ and δ clasts and rolling structures, always indicating dextral shear.

The Triangle Shear Zone changes in strike from east-northeast in the eastern part of the NMZ to northeast in the central part, where it is obscured by Karoo cover. It does not coincide with the Tuli-Sabi Shear Zone shown in McCourt and Vearncombe (1987, 1992), nor does the Magogaphate Shear Zone in Botswana (e.g., McCourt et al., 2004), or any other mapped shear zone, appear to be continuous with the Triangle Shear Zone. The nature of the boundary between the NMZ and the Central Zone in the west of the NMZ is therefore still problematic.

GEOPHYSICS

Resistivity, gravity, aeromagnetic, seismic-reflection–refraction, receiver-function, and shear-wave-splitting studies that include the NMZ have been carried out over the Limpopo Belt, as summarized below. The results of geophysical studies have been important in formulating tectonic models for the Limpopo Belt, especially in view of the large portion that is covered. Unfortunately the geophysical database for the whole belt is still rather sparse.

One of the first large-scale geophysical studies of the Limpopo Belt was a deep resistivity profile carried out by Van Zijl (1978). Results were interpreted to indicate a flat Moho at a depth of ~40 km under both cratons and the belt, at the base of a hydrated lower crust.

Coward and Fairhead (1980) presented a gravity map of the Limpopo Belt that had few stations in the NMZ. Relatively positive Bouguer values under the Southern Marginal Zone were interpreted as a fold above a ramp in a north-verging, crustal-scale thrust. North vergence was also interpreted in the Central Zone and the NMZ, where granulite was shown as thrusting over the Zimbabwe Craton on a contact dipping at 10° south. The concept of a north-verging collisional structure had considerable influence on subsequent concepts for the tectonic evolution of the Limpopo Belt (e.g., Roering et al., 1992; Treloar et al., 1992).

The earliest geophysical study focused on the NMZ specifically was a refraction seismic survey (Stuart and Zengeni, 1987), which suggested Moho depths of 40 and 34 km under the Zimbabwe Craton and the NMZ, respectively, and that the NMZ crust thins from 34 to 29 km under the Triangle Shear Zone. The Craton-NMZ boundary was considered to be a steeply dipping feature, in contrast to the moderate dips of structures seen at the surface, and fast traveltimes were interpreted to indicate some high-velocity material along the boundary, possibly the Buhwa Greenstone Belt. The idea of a steep boundary between the NMZ and the craton also influenced tectonic models (e.g., Rollinson, 1989; Ridley, 1992).

The detailed work of Gwavava et al. (1992, 1996) on gravity of the Limpopo Belt challenged both of these earlier studies. A much larger database of gravity stations was available, which give a negative Bouguer anomaly of the same order of magnitude under the craton and the NMZ. These data imply that there is no change in the Moho depth of 34 km across the boundary between the craton and the NMZ. Admittance and coherence studies by Gwavava et al. (1996) using the same gravity data recognized a zone of negative isostatic anomalies in the NMZ, which may have resulted from uncompensated recent erosion.

Receiver-function studies from the Kaapvaal Craton seismic experiment (James, 2003) suggested different results from the gravity and previous seismic studies for the crustal structure of the Limpopo Belt. Unfortunately only a single station in the experiment was on the NMZ itself. The results indicate that the Zimbabwe Craton has crustal thicknesses of 35–40 km (Nguuri et al., 2001; Niu and James, 2002; Stankiewicz et al., 2002; Nair et al., 2006) and that the NMZ has a crustal thickness of 37 km, the same as the adjacent Zimbabwe Craton, within error. These results are supported by shear-wave-velocity profiles obtained by joint inversion of receiver functions and Rayleigh wave group velocities (Kgaswane et al., 2010).

Receiver functions can also be used to find the Poisson's ratio of the crust, which is typical of felsic-intermediate compositions for the cratons and intermediate-mafic compositions for the Limpopo Belt. The shear-wave velocities from joint inversion suggest that most of the lower 10 km of crust under the NMZ is mafic in composition (Kgaswane et al., 2010). This contrasts with the western part of the Tokwe terrane on the Zimbabwe Craton, where the lower crust has intermediate compositions. Receiver functions can also be used to quantify the sharpness of the Moho (Nair et al., 2006). This reveals that the Moho is less well defined under the NMZ than the craton.

The Kaapvaal Craton seismic experiment also gave some insights into mantle structure below southern Africa. The cratons and the Limpopo Belt are underlain by high-velocity zones with thicknesses of 250–300 km, visible in tomographic images (James et al., 2001; James and Fouch, 2002; Fouch et al., 2004),

TABLE 3. SUMMARY OF GEOCHRONOLOGY FOR THE NORTHERN MARGINAL ZONE

U-Pb	Error	Pb-Pb	Error	Sm-Nd	Rb-Sr	Error	Rock type	Zone	Reference(s)
Magmatic ages									
2517 Ma	55	–	–	–	–	–	Razi granite	NMZss	Frei et al. (1999)
–	–	–	–	–	2553	114	Pegmatite	NMZss	Kempen et al. (1997)
2571	0.3	–	–	–	–	–	Razi granite	NMZss	Blenkinsop et al. (2004)
2571	5	–	–	–	–	–	Enderbite	NMZss	Blenkinsop and Frei (1997)
–	–	–	–	–	2574	15	Razi granite	–	Hickman (1978)
2576	31	–	–	–	–	–	Razi granite	NMZss	Frei et al. (1999)
2580	0.3	–	–	–	–	–	Gneiss	NMZss	Blenkinsop et al. (2004)
–	–	–	–	–	2583	52	Granitoid	NMZss	Kamber et al. (1992)
2589	11	–	–	–	–	–	Razi granite	NMZss	Frei et al. (1999)
2590	7	–	–	–	–	–	Razi granite	NMZss	Frei et al. (1999)
2591	4	–	–	–	–	–	Razi charnockite	TZ	Kamber et al. (1996)
2594	0.3	–	–	–	–	–	Razi granite	NMZss	Blenkinsop et al. (2004)
2595	0.3	–	–	–	–	–	Razi granite	NMZss	Blenkinsop et al. (2004)
2603	64	–	–	2.96–3.06 Ga	–	–	Charnoenderbite	NMZss	Berger et al. (1995)
2604	22	–	–	–	–	–	Razi granite	NMZss	Frei et al. (1999)
2619	2	–	–	–	–	–	Gneiss	NMZss	Blenkinsop et al. (2004)
2622	0.4	–	–	–	–	–	Enderbite	NMZss	Blenkinsop et al. (2004)
2627	7	–	–	–	–	–	Razi granite	–	Mkweli et al. (1995)
2637	19	–	–	3.05	–	–	Charnoenderbite	NMZss	Berger et al. (1995)
2653	0.3	–	–	–	–	–	Gneiss	NMZss	Blenkinsop et al. (2004)
2669	67	–	–	–	–	–	Charnockite	–	Mkweli et al. (1995); Berger et al. (1995)
–	–	2690	55	2.87–3.09	2644	39	Charnoenderbite Suite	NMZss	Berger and Rollinson (1997)
2710	38	2718	61	3.08–3.13	2669	230	Charnoenderbite	NMZss	Berger et al. (1995); Berger and Rollinson (1997)
2739	0.3	–	–	–	–	–	Enderbite	NMZss	Blenkinsop et al. (2004)
2768	112	–	–	–	–	–	Enderbite	NMZss	Berger et al. (1995)
–	–	–	–	–	2868	60	Granulite	NMZss	Hickman (1978)
–	–	–	–	–	2870	–	Unsheared granulites	NMZss	Hickman (1984)
–	–	–	–	–	2880	47	Various	NMZss	Hickman (1976)

(Continued)

258

TABLE 3. SUMMARY OF GEOCHRONOLOGY FOR THE NORTHERN MARGINAL ZONE (Continued)

Pb-Pb	Error	Sm-Nd	Error	Rb-Sr	Error	Ar-Ar	Error	Rock Type	Zone	Minerals	Reference(s)
Metamorphic ages											
—	—	—	—	1791	—	—	—	Gneiss	NMZss	Bt	Van Breemen and Dodson (1972)
1958 Ma	13	—	—	—	—	1956	97	Metabasite	NMZss	Bt, Amph	Kamber et al. (1996)
—	—	—	—	1820	8	—	—	Mylonite	NMZss	Ttn	Kamber et al. (1996)
—	—	—	—	—	—	1969	10	Qtz diorite gneiss	NMZss	Hbk, Bt	Kamber et al. (1996)
1969	12	—	—	—	—	—	—	Amphibolite	TZ	Ttn	Frei and Kamber (1995)
1971	17	—	—	1820	8	—	—	Mylonite	NMZss	Ttn	Kamber et al. (1996)
—	—	—	—	—	—	1850	9	Enderbite	NMZss	Bt	Kamber et al. (1995b)
—	—	—	—	—	—	1866	9	Granite	NMZss	Bt	Kamber et al. (1995b)
—	—	—	—	1875	6	—	—	Charnokite	NMZss	Kfs, Bt	Blenkinsop and Frei (1996)
—	—	—	—	—	—	1933	25	Amphibolite	TSZ	Hbl	Kamber et al. (1995b)
—	—	—	—	—	—	1955	8	Amphibolite	TSZ	Hbl	Kamber et al. (1995b)
—	—	—	—	—	—	1955	3	Amphibolite	TSZ	Hbl	Kamber et al. (1995b)
—	—	—	—	1955	6	—	—	Enderbite	NMZss	Bt,Pl	Blenkinsop and Frei (1996)
—	—	1974	14	—	—	—	—	Gneiss	TSZ	Grt	Van Breemen and Hawkesworth (1980)
—	—	—	—	—	—	1984	12	Gneiss	TSZ	Hbl	Kamber et al. (1995a)
—	—	1984	22	—	—	—	—	Gneiss	TSZ	Grt, Pl, Ap. II	Kamber et al. (1995a)
—	—	1988	14	—	—	—	—	Gneiss	TSZ	Grt	Van Breemen and Hawkesworth (1980)
—	—	—	—	—	—	1999	14	Amphibolite	TSZ	Hbl	Kamber et al. (1995b)
—	—	—	—	—	—	2001	11	Amphibolite	TSZ	Hbl	Kamber et al. (1995b)
—	—	—	—	—	—	2005	25	Gneiss	TSZ	Hbl	Kamber et al. (1995a)
2031	+86-91	—	—	—	—	—	—	Gneiss	TSZ	Grt, Pl	Kamber et al. (1995a)
—	—	—	—	—	—	2040	15	Gneiss	TSZ	Hbl	Kamber et al. (1995a)
2042	+80-85	—	—	—	—	—	—	Gneiss	TSZ	Grt	Kamber et al. (1995a)
2084	+80-84	—	—	—	—	—	—	Gneiss	TSZ	Grt	Kamber et al. (1995a)
—	—	—	—	—	—	2155	21	Amphibolite	TSZ	Hbl	Kamber et al. (1995a)
2216	±10	—	—	—	—	—	—	Gneiss	TSZ	Grt, Kfs	Kamber et al. (1995a)
2218	+70-74	—	—	—	—	—	—	Gneiss	TSZ	Grt, Kfs	Kamber et al. (1995a)
2248	21	—	—	—	—	—	—	Gneiss	TSZ	Grt, Pl	Kamber et al. (1995a)
2399	19	—	—	—	—	—	—	Pyroxenite	NMZss	Grt	Blenkinsop and Frei (1996)
2532	35	—	—	—	—	—	—	Skarn	NMZss	Cpx	Frei and Kamber (1995)
2544	39	—	—	—	—	—	—	Enderbite	NMZss	Px	Blenkinsop and Frei (1996)
—	—	—	—	2590	—	—	—	Various sheared rocks		—	Hickman (1976)
2601	5	—	—	—	—	—	—	Protomylonite NLTZ		—	Blenkinsop and Frei (1996)
2644	+98-106	—	—	—	—	—	—	Gneiss	TSZ	Grt, Pl	Kamber et al. (1995a)

U-Pb	Error	Pb-Pb	Error	Sm-Nd	Error	Rb-Sr	Error	Rock Type	Zone	Reference(s)
Zimbabwe Craton magmatic ages										
2540	38	—	—	—	—	—	—	Chilimanzi Granite	Masvingo	Frei et al. (1999)
2575	0.7	—	—	—	—	—	—	Great Dyke	—	Oberthuer et al. (2002)
2601	14	—	—	—	—	—	—	Chilimanzi Granite	—	Jelsma et al. (1996)
2634	17	—	—	—	—	—	—	Chilimanzi Granite	Masvingo	Horstwood (1998)
2673	10	—	—	—	—	—	—	Sesombi Tonalite	Midlands Belt	Dougherty Page (1994)
2852	64	—	—	—	—	—	—	Chingezi Suite	Shurugwi Belt	Nägler et al. (1997)
2950	100	—	—	—	—	—	—	Mashaba Tonalite	Tokwe segment	Dougherty Page (1994)
3554	45	—	—	—	—	—	—	Tonalitic gneiss	Tokwe segment	Kröner and Munyanyiwa in Blenkinsop and Wilson (1997)
3565	21	—	—	—	—	—	—	Sebakwe River gneiss	Midlands Belt	Horstwood (1998)
—	—	3088	+44/-46	3.36–3.55	—	3495	120	Shabani Gneiss	—	Taylor et al. (1991), Moorbath et al. (1977)

Note: See Table 2 for full names of tectonic zones.

which correspond to the "tectosphere" of Jordan (1975). Despite the significant differences in the crust of the cratons and the Limpopo Belt suggested by the experiment, the mantle seems to be similar under both: This conclusion is also supported by the joint inversion work (Kgaswane et al., 2010). McKenzie and Priestley (2008) show that the thickness of the lithosphere underneath the cratons is similar to that under the Limpopo Belt. James and Fouch (2002) interpreted the formation of the tectosphere as a result of depletion in subduction-related melting events during the Archean to form high-velocity mantle roots to depths of at least 200 km.

Shear-wave-splitting measurements reveal one difference between mantle under the Zimbabwe Craton and under the NMZ: The azimuths of fast polarization directions on the craton in Zimbabwe are approximately north-northeast, subparallel to the Great Dyke, whereas the NMZ station has a NW azimuth (336°), quite different from the craton, and from the northeast to east-northeast direction of the rest of the Limpopo Belt (Silver et al., 2001). The origin of the anisotropy is considered to lie in the mantle at depths of no greater than 50–100 km, and is clearly different from the direction predicted by modern movements of the African plate. The mantle anisotropy is interpreted to reflect Archean deformation patterns, both on the craton, where an earlier anisotropy was exploited by intrusion of the Great Dyke, and in the Limpopo Belt (Silver et al., 2004). It is difficult to evaluate the significance of the single NMZ shear-wave-splitting result within the NMZ: Such a northwest direction corresponds to the ca. 2.0 Ga Bubi swarm, but as pointed out by Silver et al. (2004), it is unlikely that dikes themselves can produce the necessary shear-wave anisotropy.

In summary, most geophysical evidence suggests that the NMZ has a similar crustal thickness to the Zimbabwe Craton, 35–37 km. However, the lower part of the NMZ crust seems to have a more mafic composition and a less-well-defined Moho than part of the craton. The single shear-wave-splitting result gives a different direction in the NMZ from the craton, but the mantle beneath the cratons, the NMZ, and the rest of the Limpopo Belt is otherwise rather similar and constitutes a tectosphere.

GEOCHRONOLOGY

The study of rocks from the NMZ has played a significant role on the world stage in the development of geochronological techniques. The first ever Sm-Nd ages were obtained from rocks in the Triangle Shear Zone (Van Breemen and Hawkesworth, 1980), and rocks from the NMZ s.s. were used by Frei and Kamber (1995) to establish the Pb-Pb step leach method. Significant advances in understanding the geochronology of granulite facies rocks were taken by Kamber et al. (1996, 1998), who demonstrated from studies in NMZ rocks that partial resetting below commonly accepted closure temperatures may occur in Grt, Hbl, and Zr. The influence of ionic porosity on Ar-Ar dates was investigated by Kamber et al. (1995b). Despite these pioneering studies, there have so far not been any SHRIMP zircon age determinations from the NMZ. The vast majority of U-Pb studies have been via the TIMS method, with some Kober ages (Table 3 and references therein).

Charnoenderbite suite and Razi suite granites have highly consistent T_{DM} ages of ca. 3.0 Ga (Berger et al., 1995). The oldest ages that are interpreted as magmatic from the NMZ are Rb/Sr whole rock isochrons from granulite gneiss at ca. 2870 Ma (Hickman, 1978, 1984; Table 3). These determinations were the first geochronological studies carried out in the belt, but they remain unique. The closest age by any other method is a U/Pb Zircon date 100 m.y. younger (Table 3).

Ages interpreted to be magmatic (mostly U-Pb zircon) are in the range 2.8–2.5 Ga, including crystallization of the charnoenderbite suite and the intrusion of the related Razi suite granites (Fig. 7; Table 3). Taking the dates at face value, the charnoenderbite suite formed between 2768 and 2571 Ma, and the Razi suite granites intruded between 2637 and 2517 Ma (Table 3 and references therein). The ages at the extremes of this range have large errors: A better constrained estimate for the intrusion of the charnoenderbite suite is 2739–2571 Ma, and for the Razi suite granites, 2627–2571 Ma. There is a clear overlap in ages between the two groups of rocks, with the Razi suite granites intruding toward the end of the charnoenderbite suite time interval (Fig. 7), consistent with the field observations that the Razi suite granites show little or no fabric, or magmatic fabrics, and that some Razi suite granites apparently intrude the charnoenderbite suite rocks.

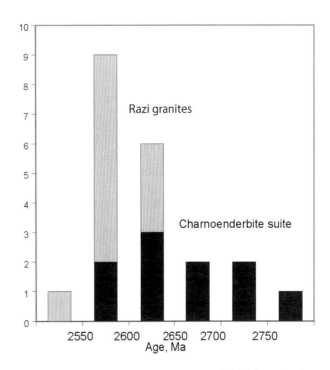

Figure 7. Histograms of U-Pb zircon ages, divided into the charnoenderbite suite and the Razi granites. Data and sources from Table 3. There is a clear overlap of the charnoenderbite and Razi suites, with the latter intruding toward the end of the former. Dates at the extreme ends of the spectrum have very large errors.

The Razi suite granites appear to have intruded diachronously, younging from west to east (Frei et al., 1999). The only magmatic rocks of early Proterozoic age in the NMZ may be the Bubi swarm dikes (Wilson et al., 1987).

Dates from the NMZ that are interpreted to reflect metamorphic mineral growth fall into two major groups: late Archean and early Proterozoic. The syntectonic charnockite intrusion age of 2591 ± 4 Ma obtained by Kamber et al. (1996) in the Transition Zone was regarded as timing peak Archean (Stage 3) metamorphism and is similar to a garnet Pb-Pb age in a protomylonite obtained by Blenkinsop and Frei (1996), and within error of a Pb-Pb age from the Triangle Shear Zone (Kamber et al., 1995a). Slightly younger, late Archean ages are given by minerals from samples in the NMZ s.s. (Frei and Kamber, 1995; Blenkinsop and Frei, 1996).

The importance of early Proterozoic ages in the NMZ was recognized by the pioneering Sm/Nd study of Van Breemen and Hawkesworth (1980) but was largely overlooked in subsequent studies (Kamber et al., 1995a). The combined data from Pb-Pb, Sm-Nd, Rb-Sr, and Ar-Ar studies now indicate a plethora of early Proterozoic metamorphic ages, from 2084 to 1956 Ma (Table 3). The older part of this spectrum comes exclusively from the Triangle Shear Zone and records a granulite facies event with dextral strike-slip movement (Kamber et al., 1995a). Ar-Ar and Pb-Pb ages in the NMZ s.s. are interpreted as recording a short period of exhumation at 1.97 Ga (Kamber et al., 1996).

Four Pb-Pb dates and one Ar-Ar date fall between the Archean and Proterozoic groups (Table 3). These dates have been interpreted as partial resetting of minerals in the Proterozoic event by Kamber et al. (1996).

DISCUSSION

Relationship of the NMZ to the Zimbabwe Craton and Archean Tectonics

There are several notable similarities between the geology of the NMZ and the Zimbabwe Craton. Both have a majority of felsic intrusive igneous rocks with a minority of mafic rocks (although there is a greater proportion of mafic rocks on the craton at the surface), and the range of tonalitic-granodioritic pluton compositions is similar. The occurrence of thin mafic units associated with metasedimentary rocks around the felsic plutons in the NMZ has its counterpart in the arrangement of granites and greenstones in the Zimbabwe Craton, and the large intrusive bodies in the NMZ are echoed by the "gregarious batholiths" of the Zimbabwe Craton (MacGregor, 1951). Diapirism has been invoked as an important process in both the craton (e.g., Jelsma et al., 1993) and the NMZ (e.g., Rollinson and Blenkinsop, 1995). Geochemical affinities between mafic and felsic rocks in the NMZ and craton have been emphasized by Rollinson and Lowry (1992), Berger et al. (1995), and Berger and Rollinson (1997).

Moreover, there are similarities in the geochronology of the NMZ with the Zimbabwe Craton, which was comprehensively compiled by Jelsma and Dirks (2002). The consistent ca. 3.0 Ga Sm-Nd age of charnoenderbite suite rocks in the NMZ (e.g., Berger et al., 1995) is similar to Sm-Nd ages of the Chingezi Tonalite (Taylor et al., 1991). The late Archean ages of the charnoenderbites overlap with zircon dates of the Sesombi suite (Dougherty-Page, 1994), and the Razi granite ages correspond closely to the range of dates for the Chilimanzi granites (e.g., Jelsma et al., 1996; Horstwood, 1998; Frei et al., 1999). It is notable that the age of the Great Dyke (2575 ± 0.7 Ma; Oberthür et al., 2002) shows that it predates some Chilimanzi and Razi suite granites, suggesting that the Great Dyke intruded before final cratonization.

One conspicuous difference between the craton and the NMZ is the lack of reliable crystallization ages greater than 2768 Ma, or 2739 Ma using better constraints (Table 3). Older Rb-Sr dates from Hickman (1978) are similar to U-Pb zircon dates of the Chingezi suite (Table 3; e.g., Nägler et al., 1997). Although the Rb-Sr dates could relate to cratonic events, such dates have never been demonstrated from U-Pb work in the NMZ, which instead suggests a major crust-forming period from 2.74 to 2.57 Ga. No counterparts of the ca. 2.95–3 Ga or ca. 3.5 Ga ages from the craton have been found, despite isotopic evidence for older crust being involved in charnoenderbite suite genesis (Berger and Rollinson, 1997). The extent to which the difference between NMZ and craton geochronology is a sampling issue may be addressed by comparing the 53 crystallization ages reported from the craton (Jelsma and Dirks, 2002) with 28 magmatic ages from the NMZ (Table 3).

Another major difference between the craton and the NMZ is the presence of a widespread penetrative fabric in the NMZ, which reflects north-northwest shortening in the late Archean. Regional shortening in a similar direction occurred on the craton (Blenkinsop et al., 1997; Wilson et al., 1995) but had a much weaker expression. Understanding this difference in late Archean mechanics of the craton and the NMZ is a significant challenge. An explanation could be sought in the mantle. One difference between the subcratonic mantle and that beneath the NMZ is the shear-wave-splitting direction (Silver et al., 2001), although the NMZ orientation is based on a single measurement. In other respects there appear to be no major variations in geophysical properties of the subcontinental lithospheric mantle between the NMZ and the craton today (James and Fouch, 2002). This observation agrees with the inference from Re-Os systematics that the subcontinental lithospheric mantle under the craton and NMZ has a similar history, involving episodes of growth from 3.8 Ga through the Archean (Nägler et al., 1997). Hence, mantle explanations for the difference between the craton and the NMZ may be inadequate.

The oldest part of the Zimbabwe craton (the Tokwe segment, 3.6 Ga) abuts the NMZ, and the presence of xenocrystic zircons predating all of the granite-greenstone cycles on the craton suggests that early crust was widespread there (Wilson et al., 1995). Possibly a greater abundance of old crust strengthened the craton relative to the NMZ in the late Archean. Alternatively or additionally, the higher heat generation of the rocks in the NMZ may have caused greater thermal weakening of its crust. A crustal explanation of the difference between the late Archean events in the craton compared with the NMZ is likely in view of the inference that the lower 10 km of the NMZ is more mafic than the craton (Kgaswane et al., 2010). This material could partly comprise supracrustal rocks such as those on the Zimbabwe Craton, which have been entrained downward by more complete convective overturn than on the craton (see below).

A striking feature of the NMZ s.s. is the consistency of the moderate to steep dips of the main foliation. There are no flat-lying foliations or recumbent folds that might indicate gravitational collapse, a feature that is increasingly seen as typical of orogenesis (e.g., Rey et al., 2001). Instead, the combination of normal-sense movement on shear zones such as the Mtilikwe Shear Zone, thrusting on the North Limpopo Thrust Zone, and horizontal shortening with steeply plunging extension, suggests subvertical extrusion in the northern part of the NMZ (Blenkinsop and Kisters, 2005). This is consistent with a more rigid Zimbabwe Craton. The lack of gravitational collapse also implies a lack of overthickening by thrusting, which may be characteristic of Archean orogeny (e.g., Choukroune et al., 1995). The distinctive features of the late Archean orogeny in the NMZ partly reflect the high heat flow of the late Archean and thus offer some insights into secular changes in orogenic style (cf. Marshak, 1999).

The shortening that generated the penetrative fabrics in the NMZ and was responsible for thrusting the NMZ over the Zimbabwe Craton has been attributed to convergence between the Zimbabwe and Kaapvaal Cratons (e.g., McCourt and Vearncombe, 1987; van Reenen et al., 1987; de Wit et al., 1992; Treloar et al., 1992), although details have been debated (e.g., Treloar and Blenkinsop, 1995). Hallmarks of a Himalayan-type continental collision are absent from the Limpopo Belt: A suture cannot be identified, and the polarity of subduction cannot even be clearly defined. This may be due to overprinting by Proterozoic events. Thus there are still fundamental questions about the geodynamics of late Archean events in the Limpopo Belt, which include the possibility of intracontinental orogeny.

Formation of Granulite

Most of the rocks in the NMZ are magmatic granulites of the charnoenderbite suite, associated with low- to medium-pressure granulite facies metamorphism. Three alternative models are discussed for formation of the granulites: an arc-backarc origin, the MARCY model of Kramers (1988), and radiogenic heating leading to diapirism.

Higher heat flow in backarc regions has been suggested as an explanation for the mechanical difference between "mobile belts" and cratons by Hyndman et al. (2005), and backarcs are identified as a locus of granulite facies metamorphism by Collins (2002). Berger and Rollinson (1997) offered an Andean setting for comparison with the NMZ on the basis of evidence for crust-mantle interaction during crustal growth. However, the possibility that the NMZ was a backarc attached to a stable craton during Stage 3 granulite formation is contradicted by the late Archean tectonics of the Zimbabwe Craton, which did not become fully cratonized until the end of the Archean (Wilson, 1990). Arc and backarc settings are indeed sometimes invoked for the greenstone sequences of the Zimbabwe Craton itself (e.g., Jelsma and Dirks, 2002). An intraoceanic arc setting for the NMZ is denied by Nd compositions of charnoenderbite suite rocks, which are between +1 and −2 ε_{ND}, compared with predicted values of +3 for 2.7 Ga intraoceanic island arcs (Berger et al., 1995).

The differences in metamorphic pressure estimates between the craton and the NMZ are taken as reflecting an original 15 km increase in crustal thickness in the NMZ by Ridley (1992), which should have contributed to an increased geothermal gradient in the NMZ. Lack of evidence for such a geotherm suggested to Ridley (1992) that the crust was thickened in the NMZ with a mantle heat-flow value lower than today's value. Such a mantle heat sink could occur above a mantle downwelling zone, which is also part of the MARCY (MAntle ReCYcling) model of Kramers (1988) and Kramers and Ridley (1989) for the formation of low- to medium-pressure granulite in the NMZ by crystallization from the base of a shallow, partially molten zone. TTG compositions such as those of the charnoenderbite suite can form by crystallization of a magma reservoir being continuously recharged by tholeitic magma from the mantle. Hydration crystallization reactions such as described above are predicted if the magma reservoir is hydrated, which might occur by incorporation of hydrothermally altered crust into the magma reservoir. This model can also generate the light REE enriched geochemistry of the charnoenderbite suite, and predicts that water activity would be buffered at low values.

The low value of the mantle heat flux required in the MARCY model is incompatible with the thermal modeling of Kramers et al. (2001), nor is the former model commensurate with the Nd isotopic evidence that requires mixing between mantle melts and preexisting continental crust (Berger and Rollinson, 1997). As described above, there is no record of an anomalous zone in the mantle below the NMZ today (James et al., 2001; James and Fouch, 2002) that might record a previous subduction event, such as is seen in the Superior province (e.g., White et al., 2003).

A different scenario for granulite facies metamorphism in the NMZ is suggested by Kramers et al. (2001), who demonstrate that lower crustal melting and advective heat flux could occur, given the high heat production of the NMZ rocks. This alternative for a source of heat for metamorphism is favored over crustal thickening because of the geochemical evidence of Berger and Rollinson (1997) that a mantle component was involved in charnoenderbite suite magmatism. The geochemical evidence for mantle input could indicate that mantle melts were an additional heat source. Widespread evidence for diapirisim in the NMZ s.s. accords well with this model (e.g., Fig. 6), which can also explain the variability in P-T conditions and paths reported for the NMZ s.s., because different parts of a diapiric system can undergo different P-T conditions and trajectories (e.g., Collins and Van Kranendonk, 1999). Mkweli (1997) considered that convection was an appropriate mechanism to explain not only variable P-T conditions in the NMZ s.s., but also the apparent small displacement on the North Limpopo Thrust Zone. In a convective model (either magmatic or solid state), large temperature differences can occur across boundaries that would not necessarily record large tectonic translations.

Diapirism involves at least two positive feedback loops, decompression melting and downward entrainment of fertile source rocks (Weinberg, 1997). These processes can generate multiple pulses of heat advection, allowing the system to maintain plutonism for long periods, and they may go some way toward explaining the ~200 m.y. duration of the charnoenderbite suite and Razi granite magmatism. Gerya et al. (2004) showed that hot continental crust may become inherently gravitationally unstable.

Diaprisim is one component of the convective overturn model of the crust proposed by Collins and Van Kranendonk (1998), in which diapiric rise of material is complemented by sinking of higher density greenstone material. Partial or complete convective overturn is likely to have played a role in the NMZ in order to explain the thorough mixing of the charnoenderbite suite protoliths, which occurred throughout their formation (Berger and Rollinson, 1997). The mafic lower crust of the NMZ (Kgaswane et al., 2010) is consistent with this model: It may represent both the accumulation of denser material during convective overturn and mantle intrusions. The latter may be the reason for the less-well-defined Moho under the NMZ than the craton (Nair et al., 2006).

The formation of granulites in the NMZ from heat advection in diapirs as part of a convective system, driven by internal heating, possibly with additional heat from mantle melts, is an attractive hypothesis that is consistent with diverse sets of observations. It differs from other mechanisms for granulite facies metamorphism in not requiring anomalous thicknesses of crust, for which there is no direct evidence, and which is consistent with the lack of gravitational collapse features.

Archean versus Proterozoic Tectonics in the NMZ

Although the tectonics of the NMZ were considered entirely Archean in several earlier studies (e.g., Coward et al., 1973; McCourt and Vearncombe, 1987; Coward and Treloar, 1990; Rollinson, 1993), geochronological results clearly indicate the importance of the 2.0 Ga event in the Triangle Shear Zone (e.g., Van Breemen and Dodson, 1972; Kamber et al., 1995a), which features in more recent tectonic models (e.g., Treloar and Blenkinsop, 1995; Kamber et al., 1995b, 1996). A lateral component of movement in the Proterozoic was recognized by Roering et al. (1992) and van Reenen et al. (1987), and subsequent studies reflected the growing appreciation of the significance of the Proterozoic event in the Central Zone (e.g., Barton et al., 1994; Holzer et al., 1998, 1999). What remains controversial for the NMZ, however, is the extent to which it has been reworked in the Proterozoic. The problem is considerably compounded because tectonics of the Proterozoic event in the northern part of the NMZ were apparently identical to the late Archean event: Both involve purely reverse movement on southeast-dipping shears (e.g., Blenkinsop et al., 2004).

Mineral ages of ca. 1.97 Ga in fabric-defining Hbl (Kamber et al., 1995b) are reasonable evidence that the Proterozoic event had an important impact on the Transition Zone. Other Proterozoic mineral ages in the NMZ s.s. are mainly from Bt and Ttn phases with lower closure temperatures, confined to meter- to decimeter-wide shear zones (Kamber et al., 1996). If these ages represent a deformation event, its significance depends on the density and distribution of these shear zones, and the displacements along them. Such shear zones are described by Kamber et al. (1996) for the Chiredzi River, but it is difficult to evaluate their tectonic significance from cross-strike profiles alone, which do not reveal their strike-parallel dimensions or spacing. The low-temperature shear zones described above have strike lengths limited to a few hundred meters and are separated by several kilometers (Blenkinsop et al., 2004). Since linking structures have not been observed between these shear zones, it is not likely that large regional strains have been accommodated on them.

An additional constraint on the tectonics of the NMZ is placed by the geometry of the southern extension of the Great Dyke and its satellites (Figs. 1, 2). Kamber et al. (1996) suggested that the dike was displaced parallel to its intersection with shear zones trending ~100° (which they had measured from the eastern part of the NMZ s.s.), and therefore did not have a mapped separation. They further argued that sinistral separations of the northwest-trending "East Dyke" were compatible with thrusting, and used these separations to derive an estimate of 10 km of vertical displacement at ca. 2.0 Ga. However, no such thrust structures have been observed in the field cutting the Great Dyke extension, and it is also likely that the northwest-trending dike is part of the Bubi–Sebanga Poort set of dikes (e.g., Pitfield, 1996). Mineral ages, remagnetization of the Great Dyke (Mushayandebvu et al., 1994), retrogression, and isolated low-temperature shear zones are all compatible with only limited deformation in the NMZ during the 2.0 Ga event. There is also the possibility that extensional structures (e.g., Kolb et al., 2003) occurred at this time.

The debate about Archean versus Proterozoic tectonics in the NMZ s.s. is reflected in the controversy over the ages of hydrothermal mineralization at the Renco gold mine, which is ~10 km south of the North Limpopo Thrust Zone. Proterozoic greenschist facies mineralization was proposed by Blenkinsop and Frei (1996, 1997), but Kisters et al. (1997, 1998, 2000) and Kolb et al. (2000, 2003) argued that mineralization was related to late Archean, amphibolite facies conditions, shortly after peak metamorphism. Blenkinsop et al. (2004) proposed that the controversy could be resolved by recognizing that there were indeed two phases of deformation at the mine, in the late Archean and early Proterozoic, respectively, but that if mineralization occurred at high temperatures, as suggested by the thermometry of Kisters et al. (1997), then it was likely to have been late Archean.

Exhumation of the NMZ

The amount of displacement in the Proterozoic is critical to understanding exhumation mechanisms in the NMZ and other high-grade gneiss terranes. Modeling suggests that granulites that form in the lower plate of an orogen, or in a single plate, are not exhumed by isostasy in a single event but require a second orogeny (Ellis, 1992). This two-phase exhumation history has been documented (e.g., Harley, 1989), and Kamber et al. (1996) suggested that the 2.0 Ga event (Stage 4) in the NMZ led to the final exhumation of the NMZ granulites. Integral to this argument is the suggestion that the Archean crust was never >45 km thick; otherwise the granulites would have been exhumed isostatically.

Several arguments against a two-phase exhumation history for the NMZ have been made. Continuity of deformation from granulite through to amphibolite facies deformation along the North Limpopo Thrust Zone is suggested by Ridley (1992), and through to greenschist facies by Mkweli et al. (1995). The lack of deformation of the Great Dyke was taken as evidence against two-stage exhumation by Blenkinsop et al. (2004), who also cited the preservation of prograde textures in granulites, which should not exist if the rocks have cooled at depth (Ellis, 1992). There are no empirical constraints on the thickness of the Archean crust to allow resolution of this issue by recourse to isostatic arguments. A comparison with the Southern Marginal Zone of the Limpopo Belt is interesting: there, thrusting of the Southern Marginal Zone over the Kaapvaal Craton occurred exclusively in the Archean over the period 2690–2600 Ma (Kreissig et al., 2001). There has been no suggestion of a Proterozoic event to exhume the Southern Marginal Zone granulites.

The history of exhumation in the NMZ can be completed with new data pertaining to Cretaceous to recent events from apatite fission-track analysis. These indicate a cooling (exhumation) event from 130 to 90 Ma, corresponding to removal of 1.3–2 km material, and a second event in post-Cretaceous time, in which 0.8–1 km of exhumation occurred (Belton, 2006). A third event may also have occurred in the Paleogene. The first event has been linked to the opening of the South Atlantic (Belton, 2006), and all three may correspond to the formation of epeirogenic uplift axes (Partridge and Maud, 1987; Moore et al., 2009). These Phanerozoic exhumation episodes therefore may have contributed 3 km of differential exhumation between the NMZ and the craton, although this does not seem compatible with the abundance of crater-facies blocks in the River Ranch kimberlite in the Central Zone (Muusha and Kopylova, 1998).

Outstanding Challenges in Northern Marginal Zone Geological Research

Few studies have examined along-strike variations in the character of the NMZ. The lack of clarity about the nature of the NMZ to the west is a major impediment to relating the NMZ to the rest of the Limpopo Belt. The boundaries of the Central Zone have been defined on the basis of the Tuli-Sabi or Triangle Shear Zone, but there is no continuity between shear zones mapped in Botswana and the Triangle Shear Zone, leaving a major gap in our understanding of the surface geometry of the Limpopo Belt. Unfortunately the area is poorly exposed or covered by Karoo deposits. The suggestion, on the basis of gravity results by Ranganai et al. (2002), that the Limpopo Belt continues as the Shashe Belt to the west in Botswana, with no Proterozoic overprint, has major implications for the issue of the westward continuation of the NMZ.

The extent of Proterozoic reworking in the NMZ s.s. is still not resolved and requires systematic detailed mapping of well-exposed areas where the length, continuity, separation, and possibly displacements across low-grade structures can be evaluated. This is a target for future research, which would preferably encompass areas at contrasting ends of the NMZ to allow evaluation of strike variations.

Understanding the relation between the Zimbabwe Craton and the NMZ may offer the most acute insights into Archean tectonics. A systematic search for a pre–2.7 Ga history in the NMZ, which is hinted at by various isotopic evidence (Berger and Rollinson, 1997; Nägler et al., 1997), would be a starting point, possibly involving SHRIMP dating. This would confirm how far back into the Archean the craton and the NMZ may have a common history. An important additional step is to integrate the NMZ and Zimbabwe Craton tectonic histories to those of the rest of the Limpopo Belt and the Kaapvaal Craton (e.g., Kreissig et al., 2001), which has not been attempted here.

Several problems remain with geophysical investigations of the NMZ. The recent results are tantalizing because they rest on a single station in the NMZ. A more robust assessment of the lower crustal composition and shear-wave-splitting fast direction in the NMZ would be most useful. A better knowledge of the deep crustal structure of the North Limpopo Thrust Zone and the Triangle Shear Zone could have implications for understanding the whole Limpopo Belt; a deep seismic reflection profile would be invaluable for global comparisons.

In order to test and refine the roles of diapirism and convection in the NMZ, detailed field studies to document the diagnostic criterion for diapirism (e.g., England, 1990; Jelsma et al., 1993; Paterson and Vernon, 1995) are needed. Relating the metamorphic textures to material paths in and around a diapir would be particularly informative, especially given the widespread observations of two stages of granulite facies metamorphism in the NMZ.

CONCLUSIONS

The charnonenderbite suite of rocks in the Northern Marginal Zone of the Limpopo Belt are granulites that crystallized directly from magma between 2.74 and 2.57 Ga. The exceptionally high-heat-producing composition of NMZ rocks led to diapirism, which produced a variety of P-T conditions and paths. Additional heat input may have been supplied from mantle intrusions. The NMZ shows considerable similarities with the

Zimbabwe Craton in terms of rock types, timing of late Archean igneous events, and the diapiric record, yet by contrast it records pervasive late Archean shortening in a uniform north-northwest direction, which occurred in association with the exhumation of granulites. The Nd isotopic record suggests that the magmatic rocks were formed by mixing of preexisting crust and mantle, but the oldest well-constrained intrusive date so far obtained from the NMZ is 2.74 Ga. The difference between the craton and the NMZ in the late Archean may therefore reflect the greater strength of the craton because it had a larger component of early crust, and indeed obvious geophysical differences between the NMZ and the craton are confined to the crust: The mantle underneath both the NMZ and the craton is a tectosphere. A lack of gravitational collapse features, implying a lack of overthickening, may have been a distinctive feature of the type of Archean orogenesis represented by the NMZ. The NMZ was affected in localized zones by amphibolite to greenschist facies retrogression and transpressive deformation at 2.0 Ga, but intense effects of this event took place only in the south part of the NMZ.

ACKNOWLEDGMENTS

Numerous colleagues have made important contributions to NMZ geology, not always completely reflected in the reference list. The following especially should be acknowledged: Michael Berger, Victor Chiwara, Paul Dirks, Roebi Frei, Oswald Gwavava, Laurence Holzer, Hielke Jelsma, Shepherd Kadzviti, Balz Kamber, Alex Kisters, Joachem Kolb, Jan Kramers, Martin Mushayandebvu, Siyanda Mwkeli, Martin Prendergast, Hugh Rollinson, Andreas Schmidt Mumm, Peter Treloar, and Jim Wilson. Hugh Rollinson and Paul Dirks provided incisive reviews. Bill Collins, Michael Rubenach, and editor Steve McCourt also provided helpful comments. Julian Lawn, Katrin Walz, and Adella Edwards are thanked for their assistance with the illustrations.

REFERENCES CITED

Barton, J.M., Jr., Holzer, L., Karnber, B., Doig, R., Kramers, J.D., and Nyfeler, D., 1994, Discrete metamorphic events in the Limpopo belt, southern Africa: Implications for the application of P-T paths in complex metamorphic terrains: Geology, v. 22, p. 1035–1038, doi:10.1130/0091-7613 (1994)022<1035:DMEITL>2.3.CO;2.

Belton, D.X., 2006, The low temperature chronology of cratonic terrains [Ph.D. thesis]: Melbourne, University of Melbourne, 403 p.

Berger, M., and Rollinson, H., 1997, Isotopic and geochemical evidence for crust-mantle interaction during late Archaean crustal growth: Geochimica et Cosmochimica Acta, v. 61, p. 4809–4829, doi:10.1016/S0016-7037(97)00271-8.

Berger, M., Kramers, J.D., and Nägler, T.F., 1995, Geochemistry and geochronology of charnoenderbites in the northern marginal zone of the Limpopo Belt, Southern Africa, and genetic models: Schweizerische Mineralogische und Petrographische Mitteilungen, v. 75, p. 17–42.

Blenkinsop, T.G., and Frei, R., 1996, Archean and Proterozoic mineralization and tectonics at the Renco Mine (northern marginal zone, Limpopo Belt, Zimbabwe): Economic Geology and the Bulletin of the Society of Economic Geologists, v. 91, p. 1225–1238.

Blenkinsop, T.G., and Frei, R., 1997, Archaean and Proterozoic mineralization and Tectonics at the Renco Mine (NMZ, Limpopo Belt), Zimbabwe: Economic Geology and the Bulletin of the Society of Economic Geologists, v. 92, p. 746–747.

Blenkinsop, T.G., and Kisters, A.F.M., 2005, Steep extrusion of late Archaean granulites in the Northern Marginal Zone, Zimbabwe; evidence for secular change in orogenic style, in Gapais, D., Brun, J.P., and Cobbold, P.R., eds., Deformation Mechanisms, Rheology and Tectonics: From Minerals to the Lithosphere: Geological Society [London] Special Publication 243, p. 193–204, doi:10.1144/GSL.SP.2005.243.01.14.

Blenkinsop, T.G., and Mkweli, S., 1992, The relationship between the Zimbabwean craton and the North marginal zone of the Limpopo Belt: Proceedings of the Second Symposium of Science and Technology, Research Council of Zimbabwe, v. 3C, p. 236–248.

Blenkinsop, T.G., and Wilson, J.F., 1997, The Limpopo Belt of Zimbabwe: Field Trip Guide: Geological Society of Zimbabwe, 22 p.

Blenkinsop, T.G., Mkweli, S., Rollinson, H.R., Fedo, C.M., Paya, B.K., Kamber, B., Kramers, J., and Berger, M., 1995, The North Limpopo Thrust Zone: The Northern Boundary of the Limpopo belt in Zimbabwe and Botswana, in Barton, J.R., and Copperthwaite, Y.E., eds., Centennial Geocongress, Volume II: Johannesburg, Geological Society of South Africa, p. 174–177.

Blenkinsop, T.G., Martin, A., Jelsma, H., and Vinyu, M., 1997, The Zimbabwe Craton: Summary of Greenstone Belt geology, in de Wit, M.J., and Ashwal, L.D., eds., Greenstone Belts: Oxford, UK, Oxford Monographs on Geology and Geophysics, Oxford University Press, p. 567–580.

Blenkinsop, T.G., Kroner, A., and Chiwara, V., 2004, Single stage, late Archaean exhumation of granulites in the Northern Marginal Zone, Limpopo Belt, Zimbabwe, and relevance to gold mineralization at Renco mine: South African Journal of Geology, v. 107, p. 377–396, doi:10.2113/107.3.377.

Brandl, G., 1995, Geology of the Limpopo Belt and Environs: Contribution to "A Field Workshop on Granulites and Deep Crustal Tectonics, 1990": Johannesburg, Council for Geoscience, map on 2 sheets, scale 1:500,000.

Brun, J.P., Gapais, D., and Le Theoff, B., 1981, The mantled gneiss domes of Kuopio (Finland): Interfering diapirs: Tectonophysics, v. 74, p. 283–304, doi:10.1016/0040-1951(81)90194-3.

Choukroune, P., Bouhallier, H., and Arndt, N.T., 1995, Soft lithosphere during periods of Archean crustal growth or crustal reworking, in Coward, M.P., and Ries, A.C., eds., Early Precambrian Processes: Geological Society [London] Special Publication 95, p. 67–86, doi:10.1144/GSL.SP.1995.095.01.05.

Collins, W.J., 2002, Hot orogens, tectonic switching and creation of continental crust: Geology, v. 30, p. 535–538, doi:10.1130/0091-7613 (2002)030<0535: HOTSAC>2.0.CO;2.

Collins, W.J., and Van Kranendonk, M.J., 1998, Partial convective overturn of Archaean crust in the east Pilbara Craton, Western Australia: Driving mechanisms and tectonic implications: Journal of Structural Geology, v. 20, p. 1405–1424, doi:10.1016/S0191-8141(98)00073-X.

Collins, W.J., and Van Kranendonk, M.J., 1999, Model for the development of kyanite during partial convective overturn of Archean granite-greenstone terranes: The Pilbara craton, Australia: Journal of Metamorphic Geology, v. 17, p. 145–156, doi:10.1046/j.1525-1314.1999.00187.x.

Coward, M.P., and Fairhead, J.D., 1980, Gravity and structural evidence for the deep structure of the Limpopo Belt, southern Africa: Tectonophysics, v. 68, p. 31–43, doi:10.1016/0040-1951(80)90007-4.

Coward, M.P., and Treloar, P.J., 1990, Tibetan models for Archean tectonics in southern Africa: The Limpopo Belt: A field workshop on granulites and deep crustal tectonics: Johannesburg, Department of Geology, Rand Afrikaans University, p. 35–38.

Coward, M.P., Graham, R.H., James, P.R., and Wakefield, J., 1973, A structural interpretation of the northern margin of the Limpopo orogenic belt, southern Africa: Philosophical Transactions of the Royal Society of London, ser. A, v. 273, p. 487–491.

Coward, M.P., James, P.R., and Wright, L., 1976, Northern margin of the Limpopo mobile belt, southern Africa: Geological Society of America Bulletin, v. 87, p. 601–611, doi:10.1130/0016-7606(1976)87<601:NMOTLM>2.0.CO;2.

Cox, K.G., Johnson, R.L., Monkman, L.J., Stillman, C.J., Vail, J.R., and Wood, D.N., 1965, The geology of the Nuanetsi igneous province: Philosophical Transactions of the Royal Society of London, ser. A, v. 257, p. 71–218.

de Wit, M.J., Roering, C., Hart, R.J., Armstrong, R.A., De Ronde, C.E.J., Green, R.W.E., Tredoux, M., Peberdy, E., and Hart, R.A., 1992, Formation of an Archean continent: Nature, v. 357, p. 553–562.

Dougherty-Page, J.S., 1994, The evolution of the Archean continental crust of northern Zimbabwe [Ph.D. thesis]: Milton Keynes, UK, Open University, 215 p.

Duncan, A.R., Erlane, A.J., and Marsh, J.S., 1984, Regional Geochemistry of the Karoo Igneous Province, in Erlank, A.J., ed., Petrogenesis of the Volcanic Rocks of the Karoo Igneous Province: Geological Survey of South Africa Special Publication 13, p. 355–368.

Ellis, D.J., 1992, Precambrian tectonics and the physicochemical evolution of the continental crust. II. Lithosphere delamination and ensialic orogeny: Precambrian Research, v. 55, p. 507–524, doi:10.1016/0301-9268(92)90042-M.

England, R.W., 1990, The identification of granitic diapirs: Journal of the Geological Society [London], v. 147, p. 931–933, doi:10.1144/gsjgs.147.6.0931.

Fedo, C.M., Eriksson, K.A., and Blenkinsop, T.G., 1995, Geologic history of the Archean Buhwa greenstone belt and surrounding granite-gneiss terrane, Zimbabwe, with implications for the evolution of the Limpopo Belt: Canadian Journal of Earth Sciences, v. 32, p. 1977–1990.

Fouch, M.J., James, D.E., Vandecar, J.C., and Van Der Lee, S., and the Kaapvaal Seismic Group, 2004, Mantle seismic structure beneath the Kaapvaal and Zimbabwe Cratons: South African Journal of Geology, v. 107, p. 33–44, doi:10.2113/107.1-2.33.

Frei, R., and Kamber, B.S., 1995, Single mineral Pb-Pb dating: Earth and Planetary Science Letters, v. 129, p. 261–268, doi:10.1016/0012-821X(94)00248-W.

Frei, R., Schoenberg, R., and Blenkinsop, T.G., 1999, Geochronology of the late Archaean Razi and Chilimanzi suites of granites in Zimbabwe; implications for the late Archaean tectonics of the Limpopo Belt and Zimbabwe Craton: South African Journal of Geology, v. 102, p. 55–63.

Gerya, T.V., Perchuk, L.L., Maresch, W.V., and Willner, A.P., 2004, Inherent gravitational instability of hot continental crust; implications for doming and diapirism in granulite facies terrains, in Whitney, D.L., Teyssier, C., and Siddoway, C.S., eds., Gneiss Domes in Orogeny: Geological Society of America Special Paper 380, p. 97–115.

Gwavava, O., Swain, C.J., Podmore, F., and Fairhead, J.D., 1992, Evidence of crustal thinning beneath the Limpopo Belt and Lebombo monocline of Southern Africa based on regional gravity studies and implications for the reconstruction of Gondwana: Tectonophysics, v. 212, p. 1–20, doi:10.1016/0040-1951(92)90136-T.

Gwavava, O., Swain, C.J., and Podmore, F., 1996, Mechanisms of isostatic compensation of the Zimbabwe and Kaapvaal cratons, the Limpopo Belt and the Mozambique basin: Geophysical Journal International, v. 127, p. 635–650, doi:10.1111/j.1365-246X.1996.tb04044.x.

Hamilton, W.B., 1998, Archean magmatism and deformation were not the products of plate tectonics: Precambrian Research, v. 91, p. 143–179, doi:10.1016/S0301-9268(98)00042-4.

Hamilton, W.B., 2003, An alternative Earth: GSA Today, v. 13, no. 11, p. 4–12, doi:10.1130/1052-5173(2003), 013<0004:AAE>2.0.CO;2.

Harley, S.D., 1989, The origins of granulites: A metamorphic perspective: Geological Magazine, v. 126, p. 215–247, doi:10.1017/S0016756800022330.

Hickman, M.H., 1976, Geochronological investigations in the Limpopo Belt and part of the Rhodesian Craton [Ph.D. thesis]: Leeds, UK, University of Leeds, 188 p.

Hickman, M.H., 1978, Isotopic evidence for crystal reworking in the Rhodesian Archean craton, southern Africa: Geology, v. 6, p. 214–216, doi:10.1130/0091-7613(1978)6<214:IEFCRI>2.0.CO;2.

Hickman, M.H., 1984, Rb-Sr chemical and isotopic response of gneisses in late Archean shear zones of the Limpopo Mobile Belt, Southern Africa: Precambrian Research, v. 24, p. 123–130, doi:10.1016/0301-9268(84)90054-8.

Holzer, L., Frei, R., Barton, J.M., and Kramers, J.D., 1998, Unravelling the record of successive high grade events in the Central Zone of the Limpopo Belt using Pb single phase dating of metamorphic minerals: Precambrian Research, v. 87, p. 87–115, doi:10.1016/S0301-9268(97)00058-2.

Holzer, L., Barton, J.M., Paya, B.K., and Kramers, J.D., 1999, Tectonothermal history of the western part of the Limpopo Belt; tectonic models and new perspectives: Journal of African Earth Sciences, v. 28, p. 383–402, doi:10.1016/S0899-5362(99)00011-1.

Horstwood, M.S.A., 1998, Stratigraphy, geochemistry and zircon geochronology of the Midlands greenstone belt, Zimbabwe [Ph.D. thesis]: Southampton, UK, University of Southampton, 215 p.

Hyndman, R.D., Currie, C.A., and Mazzoti, S.P., 2005, Subduction zone backarcs, mobile belts and orogenic heat: GSA Today, v. 15, no. 2, p. 4–10, doi:10.1130/1052-5173(2005)015<4:SZBMBA>2.0.CO;2.

James, D.E., 2003, Imaging crust and upper mantle beneath southern Africa: The southern Africa broadband seismic experiment: The Leading Edge (Society of Exploration Geophysicists), v. 22, p. 238–249, doi:10.1190/1.1564529.

James, D.E., and Fouch, M.J., 2002, Formation and evolution of Archean cratons: Insights from southern Africa, in Fowler, C.M.R., Ebinger, C.J., and Hawkesworth, C.J., eds., The Early Earth: Physical, Chemical and Biological Development: Geological Society [London] Special Publication 199, p. 1–26.

James, D.E., Fouch, M.J., Vandecar, J.C., and Van Der Lee, S., and the Kaapvaal Seismic Group, 2001, Tectonspheric structure beneath southern Africa: Geophysical Research Letters, v. 28, p. 2485–2488, doi:10.1029/2000GL012578.

James, P.R., 1975, A deformation study across the northern margin of the Limpopo mobile belt, Rhodesia [Ph.D. thesis]: Leeds, UK, University of Leeds, 303 p.

Jelsma, H.A., and Dirks, P.H.G.M., 2002, Neoarchaean tectonic evolution of the Zimbabwe Craton, in Fowler, C.M.R., Ebinger, C.J., and Hawkesworth, C.J., eds., The Early Earth; Physical, Chemical and Biological Development: Geological Society [London] Special Publication 199, p. 183–211, doi:10.1144/GSL.SP.2002.199.01.10.

Jelsma, H.A., Van Der Beek, P.A., and Vinyu, M.L., 1993, Tectonic evolution of the Bindura-Shamva greenstone belt (northern Zimbabwe): Progressive deformation around diapiric batholiths: Journal of Structural Geology, v. 15, p. 163–176, doi:10.1016/0191-8141(93)90093-P.

Jelsma, H.A., Vinyu, M.L., Valbracht, P.J., Davies, G.R., Wijbrans, J.R., and Verdumen, E.A.T., 1996, Constraints on Archean crustal evolution of the Zimbabwe craton: A U-Pb zircon, Sn-Nd and Pb-Pb whole rock isotopes study: Contributions to Mineralogy and Petrology, v. 124, p. 55–70, doi:10.1007/s004100050173.

Jordan, T.H., 1975, The continental tectosphere: Reviews of Geophysics and Space Physics, v. 13, p. 1–12, doi:10.1029/RG013i003p00001.

Kamber, B.S., and Biino, G.G., 1995, The evolution of high T–low P granulites in the Northern Marginal Zone sensu stricto, Limpopo Belt, Zimbabwe—The case for petrography: Schweizerische Mineralogische und Petrographische Mitteilungen, v. 75, p. 427–454.

Kamber, B., Blenkinsop, T., Rollinson, H., Kramers, J., and Berger, M., 1992, Dating of an important tectono-magmatic event in the Northern Marginal Zone of the Limpopo Belt, Zimbabwe: First results, in Blenkinsop, T., and Rollinson, H., eds., North Limpopo Field Workshop Field Guide and Abstracts Volume: Harare, University of Zimbabwe, 39 p.

Kamber, B.S., Kramers, J.D., Napier, R., Cliff, R.A., and Rollinson, H.R., 1995a, The Triangle Shear Zone, Zimbabwe, revisited: New data document an important event at 2.0 Ga in the Limpopo Belt: Precambrian Research, v. 70, p. 191–213, doi:10.1016/0301-9268(94)00039-T.

Kamber, B.S., Blenkinsop, T.G., Villa, I.M., and Dahl, P.S., 1995b, Proterozoic transpressive deformation in Northern Marginal Zone, Limpopo Belt, Zimbabwe: Journal of Geology, v. 103, p. 493–508, doi:10.1086/629772.

Kamber, B.S., Biino, G.G., Wijbrans, J.R., Davies, G.R., and Villa, I.M., 1996, Archaean granulites of the Limpopo Belt, Zimbabwe; one slow exhumation or two rapid events?: Tectonics, v. 15, p. 1414–1430, doi:10.1029/96TC00850.

Kamber, B.S., Frei, R., and Gibb, A.J., 1998, Pitfalls and new approaches in granulite chronometry; an example from the Limpopo Belt, Zimbabwe: Precambrian Research, v. 91, p. 269–285.

Kempen, T., Kisters, A.F.M., Glodny, J., Meyer, F.M., and Kramm, U., 1997, Lode Gold Mineralization under High Grade Metamorphic Conditions: The Renco Gold Mine, Southern Zimbabwe: Terra Nova, v. 9, 549 p.

Kgaswane, E.M., Nyblade, A.A., Julia, J., Dirks, P.H.G.M., Durrheim, R.J., and Pasyanis, M.E., 2010, Shear wave velocity structure of the lower crust in southern Africa: Evidence for compositional heterogeneity within Archean and Proterozoic terrains: Journal of Geophysical Research, v. 114, B12304, doi:10.1029/2008JB006217.

Kisters, A.F.M., Kolb, J., and Meyer, F.M., 1997, Archean and Proterozoic mineralization and tectonics at the Renco Mine (northern marginal zone, Limpopo Belt, Zimbabwe): Economic Geology and the Bulletin of the Society of Economic Geologists, v. 92, p. 745–747.

Kisters, A.F.M., Kolb, J., and Meyer, F.M., 1998, Gold mineralization in high-grade metamorphic shear zones of the Renco Mine, southern Zimbabwe: Economic Geology and the Bulletin of the Society of Economic Geologists, v. 93, p. 587–601.

Kisters, A.F.M., Kolb, J., Meyer, F.M., and Hoernes, S., 2000, Hydrologic segmentation of high-temperature shear zones; structural, geochemical and isotopic evidence from auriferous mylonites of the Renco Mine, Zimbabwe: Journal of Structural Geology, v. 22, p. 811–829, doi:10.1016/S0191-8141(00)00006-7.

Kisters, A.F.M., Stevens, G., Dziggei, A., and Armstrong, R., 2003, Extensional detachment faulting and core-complex formation in the southern Barberton granite-greenstone terrain, South Africa; evidence for a 3.2 Ga orogenic collapse: Precambrian Research, v. 127, p. 355–378, doi:10.1016/j.precamres.2003.08.002.

Kolb, J., Kisters, A.F.M., Hoernes, S., and Meyer, F.M., 2000, The origin of fluids and nature of fluid-rock interaction in mid-crustal auriferous mylonites of the Renco Mine, southern Zimbabwe: Mineralium Deposita, v. 35, p. 109–125, doi:10.1007/s001260050010.

Kolb, J., Kisters, A.F.M., Meyer, F.M., and Siemes, H., 2003, Polyphase deformation of mylonites from the Renco gold mine (Zimbabwe); identified by crystallographic preferred orientation of quartz: Journal of Structural Geology, v. 25, p. 253–262, doi:10.1016/S0191-8141(02)00031-7.

Kramers, J.D., 1988, An open-system fractional crystallization model for very early continental crust formation: Precambrian Research, v. 38, p. 281–295, doi:10.1016/0301-9268(88)90028-9.

Kramers, J.D., and Ridley, J.R., 1989, Can Archean granulites be direct crystallization products from a sialic magma layer?: Geology, v. 17, p. 442–445, doi:10.1130/0091-7613(1989)017<0442:CAGBDC>2.3.CO;2.

Kramers, J.D., Kreissig, K., and Jones, M.Q.W., 2001, Crustal heat production and style of metamorphism: A comparison between two Archean high grade provinces in the Limpopo Belt, southern Africa: Precambrian Research, v. 112, p. 149–163, doi:10.1016/S0301-9268(01)00173-5.

Kreissig, K., Thomas, F.N., Kramers, J.D., van Reenen, D.D., and Smit, C.A., 2000, An isotopic and geochemical study of the northern Kaapvaal Craton and the Southern Marginal Zone of the Limpopo Belt: Are they juxtaposed terranes?: Lithos, v. 50, p. 1–25, doi:10.1016/S0024-4937(99)00037-7.

Kreissig, K., Holzer, L., Frei, R., Villa, I.M., Kramers, J.D., Kroener, A., Smit, C.A., and van Reenen, D.D., 2001, Geochronology of the Hout River shear zone and the metamorphism in the Southern marginal zone of the Limpopo Belt, Southern Africa: Precambrian Research, v. 109, p. 145–173, doi:10.1016/S0301-9268(01)00147-4.

Kroener, A., Jaeckel, P., Hofmann, A., Nemchin, A.A., and Brandl, G., 1998, Field relationships and age of supracrustal Beit Bridge Complex and associated granitoid gneisses in the Central Zone of the Limpopo Belt, South Africa: South African Journal of Geology, v. 101, p. 201–213.

Luais, B., and Hawkesworth, C.J., 1994, The generation of continental crust: An integrated study of crust-forming processes in the Archaean of Zimbabwe: Journal of Petrology, v. 35, p. 43–93.

MacGregor, A.M., 1951, Some milestones in the Precambrian of southern Rhodesia: Transactions, Geological Society of South Africa, v. 54, p. 27–71.

MacGregor, A.M., 1953, Precambrian formations of tropical Southern Africa, in Proceedings of the 19th International Geological Congress, vol. 1, Algiers, 1952, p. 39–52.

Marshak, S., 1999, Deformation style way back then: Thoughts on the contrasts between Archean/Paleoproterozoic and contemporary orogens: Journal of Structural Geology, v. 21, p. 1175–1182, doi:10.1016/S0191-8141(99)00057-7.

Mason, R., 1973, The Limpopo mobile belt; southern Africa: Philosophical Transactions of the Royal Society of London, ser. A, v. 273, p. 463–485.

McCourt, S., and Vearncombe, J.R., 1987, Shear zones bounding the central zone of the Limpopo Mobile Belt, southern Africa: Journal of Structural Geology, v. 9, p. 127–137, doi:10.1016/0191-8141(87)90021-6.

McCourt, S., and Vearncombe, J.R., 1992, Shear zones of the Limpopo Belt and adjacent granitoid-greenstone terranes; implications for late Archaean collision tectonics in Southern Africa: Precambrian Research, v. 55, p. 553–570, doi:10.1016/0301-9268(92)90045-P.

McCourt, S., Kampunzu, A.B., Bagai, Z., and Armstrong, R.A., 2004, The crustal architecture of Archaean terranes in Northeastern Botswana: South African Journal of Geology, v. 107, p. 147–158, doi:10.2113/107.1-2.147.

McKenzie, D., and Priestley, K., 2008, The influence of lithospheric thickness variations on continental evolution: Lithos, v. 102, p. 1–11, doi:10.1016/j.lithos.2007.05.005.

Mkweli, S., 1997, The Zimbabwe craton—Northern Marginal Zone boundary and the exhumation process of lower crustal rocks [Ph.D. thesis]: Harare, University of Zimbabwe, 171 p.

Mkweli, S., Kamber, B., and Berger, M., 1995, Westward continuation of the craton—Limpopo Belt tectonic break in Zimbabwe and new age constraints on the timing of the thrusting: Journal of the Geological Society [London], v. 152, p. 77–83, doi:10.1144/gsjgs.152.1.0077.

Moorbath S., Wilson J.F., Goodwin R., and Humm M., 1977, Further Rb-Sr age and isotope data on early and late Archean rocks from the Rhodesian Craton: Precambrian Research, v. 5, p. 229–239.

Moore, A., Blenkinsop, T.G., and Cotterill, F., 2009, South African topography and erosion history: Plumes or plate tectonics?: Terra Nova, v. 21, p. 310–315, doi:10.1111/j.1365-3121.2009.00887.x.

Mushayandebvu, M.F., Jones, D.L., and Briden, J.C., 1994, A paleomagnetic study of the Umvimeela Dyke, Zimbabwe: Evidence for a Mesoproterozoic overprint: Precambrian Research, v. 69, p. 269–280, doi:10.1016/0301-9268(94)90091-4.

Muusha, M., and Kopylova, M., 1998, River Ranch Diamond Mine, in Large Mines Field Excursion Guide: Cape Town, International Kimberlite Conference, 7th, p. 33–37.

Mwatahwa, C., 1992, The geology of the area around Rhonda and the United Mines (Mberengwa) [B.S. honor's thesis]: Harare, University of Zimbabwe, 44 p.

Nägler, T.F., Kramers, J.D., Kamber, B.S., Frei, R., and Prendergast, M.D.A., 1997, Growth of the subcontinental lithospheric mantle beneath Zimbabwe started at or before 3.8 Ga: Re-Os study on chromites: Geology, v. 25, p. 983–986, doi:10.1130/0091-7613(1997)025<0983:GOSLMB>2.3.CO;2.

Nair, S.K., Gao, S.S., Liu, K.H., and Silver, P.G., 2006, Southern African crustal evolution and composition: Constraints from receiver function studies: Journal of Geophysical Research, v. 111, B02304, doi:10.1029/2005JB003802.

Nguuri, T.K., Gore, J., James, D.E., Webb, S.J., Wright, C., Zengeni, T.G., Gwavava, O., Snoke, J.A., and the Kaapvaal Seismic Group, 2001, Crustal structure beneath southern Africa and its implications for the formation and evolution of the Kaapvaal and Zimbabwe cratons: Geophysical Research Letters, v. 28, p. 2501–2504, doi:10.1029/2000GL012587.

Niu, F., and James, D.E., 2002, Fine structure of the lowermost crust beneath the Kaapvaal craton and its implications for crustal formation and evolution: Earth and Planetary Science Letters, v. 200, p. 121–130, doi:10.1016/S0012-821X(02)00584-8.

Oberthür, T., Davis, D.W., Blenkinsop, T.G., and Höhndorf, A., 2002, Precise U–Pb mineral ages, Rb–Sr and Sm–Nd systematics for the Great Dyke, Zimbabwe—Constraints on late Archean events in the Zimbabwe craton and Limpopo belt: Precambrian Research, v. 113, p. 293–305, doi:10.1016/S0301-9268(01)00215-7.

Odell, J., 1975, Explanation of the Geological Map of the Country around Bangala Dam: Rhodesian Geological Survey Short Report, v. 42, 46 p.

Partridge, T.C., and Maud, R.R., 1987, Geomorphic evolution of southern Africa since the Mesozoic: South African Journal of Geology, v. 90, p. 165–184.

Paterson, S.R., and Vernon, R.H., 1995, Bursting the bubble of ballooning plutons: A return to nested diapirs emplaced by multiple processes: Geological Society of America Bulletin, v. 107, p. 1356–1380, doi:10.1130/0016-7606(1995)107<1356:BTBOBP>2.3.CO;2.

Pitfield, P.E.J., 1996, Tectonic Map of Zimbabwe: Geological Survey of Zimbabwe, scale 1:1,000,000, 1 sheet.

Ranganai, R.T., Kampunzu, A.B., Atekwana, E.A., Paya, B.K., King, J.G., Koosimile, D.I., and Stettler, E.H., 2002, Gravity evidence for a larger Limpopo Belt in Southern Africa and geodynamic implications: Geophysical Journal International, v. 149, p. F9–F14, doi:10.1046/j.1365-246X.2002.01703.x.

Reeves, C., 1978, A failed Gondwana spreading axis in southern Africa: Nature, v. 273, p. 222–223, doi:10.1038/273222a0.

Rey, P., Vanderhaeghe, O., and Teyssler, C., 2001, Gravitational collapse of the continental crust: Definition, regimes and modes: Tectonophysics, v. 342, p. 435–449.

Ridley, J.R., 1992, On the origins and tectonic significance of the charnockite suite of the Archaean Limpopo Belt, northern marginal zone, Zimbabwe: Precambrian Research, v. 55, p. 407–427, doi:10.1016/0301-9268(92)90037-O.

Robertson, I.D.M., 1973a, The Geology of the Country around Mount Towla, Gwanda District: Bulletin of the Rhodesian Geological Survey, v. 68, 166 p.

Robertson, I.D.M., 1973b, Potash granites of the southern edge of the Rhodesian craton and the northern granulite zone of the Limpopo Mobile Belt, in Lister, L.A., ed., Symposium on Granites, Gneisses and Related Rocks: Geological Society of South Africa Special Publication 3, p. 265–276.

Robertson, I.D.M., 1974, Explanation of the Geological Map of the Country South of Chibi: Geological Survey of Rhodesia Short Report, v. 41, 40 p.

Roering, C., van Reenen, D.D., Smit, C.A., Barton, J.M., De Beer, J.H., de Wit, M.J., Stettler, E.H., Van Schalkwyk, J.F., Stevens, G., and Pretorius, S., 1992, Tectonic model for the evolution of the Limpopo Belt: Precambrian Research, v. 55, p. 539–552, doi:10.1016/0301-9268(92)90044-O.

Rollinson, H.R., 1989, Garnet-orthopyroxene thermobarometry of granulites from the northern marginal zone of the Limpopo belt, Zimbabwe, in Daly, J.S., Cliff, R.A., and Yardley, B.W.D., eds., Evolution of Metamorphic Belts: Geological Society [London] Special Publication 43, p. 331–335.

Rollinson, H.R., 1993, A terrane interpretation of the Archaean Limpopo Belt: Geological Magazine, v. 130, p. 755–765, doi:10.1017/S001675680002313X.

Rollinson, H.R., 1995a, The relationship between chromite chemistry and the tectonic settings of Archean ultramafic rocks, in Blenkinsop, T.G., and Tromp, P., eds., Sub-Saharan Economic Geology: Amsterdam, Balkema, p. 7–23.

Rollinson, H.R., 1995b, Composition and tectonic setting of chromite deposits through time—A discussion: Economic Geology and the Bulletin of the Society of Economic Geologists, v. 90, p. 2091–2092.

Rollinson, H.R., 1997, The Archean komatiite-related Inyala Chromitite, southern Zimbabwe: Economic Geology and the Bulletin of the Society of Economic Geologists, v. 92, p. 98–107.

Rollinson, H., and Blenkinsop, T., 1995, The magmatic, metamorphic and tectonic evolution of the Northern Marginal Zone of the Limpopo Belt in Zimbabwe: Journal of the Geological Society [London], v. 152, p. 65–75, doi:10.1144/gsjgs.152.1.0065.

Rollinson, H.R., and Lowry, D., 1992, Early basic magmatism in the evolution of the northern marginal zone of the Archean Limpopo Belt: Precambrian Research, v. 55, p. 33–45, doi:10.1016/0301-9268(92)90012-D.

Siivola, J., and Schmid, R., 2007, A systematic nomenclature for metamorphic rocks: 12. List of mineral abbreviations: Recommendations by the IUGS Subcommission on the Systematics of Metamorphic Rocks; available at http://www.bgs.ac.uk/scmr/home.html (accessed 1 February 2007).

Silver, P.G., Gao, S.S., Liu, K.H., and the Kaapvaal Seismic Group, 2001, Mantle deformation beneath southern Africa: Geophysical Research Letters, v. 28, p. 2493–2496, doi:10.1029/2000GL012696.

Silver, P.G., Fouch, M.J., Gao, S.S., Schmitz, M., and the Kaapvaal Seismic Group, 2004, Seismic anisotropy, mantle fabric and the magmatic evolution of Precambrian southern Africa: South African Journal of Geology, v. 107, p. 45–58, doi:10.2113/107.1-2.45.

Stankiewicz, J., Chevrot, S., Van Der Hilst, R.D., and de Wit, M.J., 2002, Crustal thickness, discontinuity depth, and upper mantle structure beneath southern Africa: Constraints from body wave conversions: Physics of the Earth and Planetary Interiors, v. 130, p. 235–251, doi:10.1016/S0031-9201(02)00012-2.

Stuart, G.W., and Zengeni, T., 1987, Seismic crustal structure of the Limpopo mobile belt, Zimbabwe: Tectonophysics, v. 144, p. 323–335, doi:10.1016/0040-1951(87)90300-3.

Taylor, P.N., Kramers, J.D., Moorbath, S., Wilson, J.F., Orpen, J.L., and Martin, A., 1991, Pb-Pb, Sm-Nd and Rb-Sr geochronology in the Archean craton of Zimbabwe: Chemical Geology, v. 87, p. 175–196.

Treloar, P.J., and Blenkinsop, T.G., 1995, Archaean deformation patterns in Zimbabwe; true indicators of Tibetan-style crustal extrusion or not? in Coward, M.P., and Ries, A.C., eds., Early Precambrian Processes: Geological Society [London] Special Publication 95, p. 87–107, doi:10.1144/GSL.SP.1995.095.01.06.

Treloar, P.J., Coward, M.P., and Harris, N.B.W., 1992, Himalayan-Tibetan analogies for the evolution of the Zimbabwe craton and Limpopo Belt: Precambrian Research, v. 55, p. 571–587, doi:10.1016/0301-9268(92)90046-Q.

Tsunogae, T., Miyano, T., and Ridley, J.R., 1992, Metamorphic P-T profiles from the Zimbabwe Craton to the Limpopo Belt, Zimbabwe: Precambrian Research, v. 55, p. 259–277, doi:10.1016/0301-9268(92)90027-L.

Van Breemen, O., and Dodson, M.H., 1972, Metamorphic chronology of the Limpopo Belt, Southern Africa: Geological Society of America Bulletin, v. 83, p. 2005–2018, doi:10.1130/0016-7606(1972)83[2005:MCOTLB]2.0.CO;2.

Van Breemen, O., and Hawkesworth, C.J., 1980, Sm-Nd isotopic study of garnets and their metamorphic host rocks: Transactions of the Royal Society of Edinburgh, v. 71, p. 97–102.

van Reenen, D.D., Barton, J.M., Jr., Roering, C., Smith, C.A., and Van Schalkwyk, J.F., 1987, Deep crustal response to continental collision: The Limpopo belt of southern Africa: Geology, v. 15, p. 11–14, doi:10.1130/0091-7613(1987)15<11:DCRTCC>2.0.CO;2.

Van Zijl, J.S.V., 1978, The relationship between deep electrical resistivity and structure and tectonic provinces in southern Africa. Part I. Results obtained by Schlumberger soundings: Transactions of the Geological Survey of South Africa, v. 81, p. 129–142.

Weinberg, R.F., 1997, Diapir-driven crustal convection: Decompression melting, renewal of magma source and the origin of nested plutons: Tectonophysics, v. 271, p. 217–229, doi:10.1016/S0040-1951(96)00269-7.

White, D.J., Musacchio, G., Helmstaedt, H.H., Harrap, R.M., Thurston, P.C., van der Velden, A., and Hall, K., 2003, Images of a lower-crustal oceanic slab: Direct evidence for tectonic accretion in the Archean western Superior province: Geology, v. 31, p. 997–1000, doi:10.1130/G20014.1.

Wilson, J.F., 1990, A craton and its cracks: Some of the behaviour of the Zimbabwe block from the Late Archaean to the Mesozoic in response to horizontal movements, and the significance of some of its mafic dyke fracture patterns: Journal of African Earth Sciences, v. 10, p. 483–501, doi:10.1016/0899-5362(90)90101-J.

Wilson, J.F., Jones, D.L., and Kramers, J.D., 1987, Mafic Dyke Swarms of Zimbabwe, in Halls, H.C., and Fahrig, W.F., eds., Mafic Dyke Swarms: Geological Society of Canada Special Paper 34, p. 433–444.

Wilson, J.F., Nesbitt, R.W., and Fanning, C.M., 1995, Zircon geochronology of Archaean felsic sequences in the Zimbabwe craton: A revision of greenstone stratigraphy and a model for crustal growth, in Coward, M.P., and Ries, A.C., eds., Early Precambrian Processes: Geological Society [London] Special Publication 95, p. 109–126, doi:10.1144/GSL.SP.1995.095.01.07.

Worst, B.G., 1962, The Geology of the Buhwa Iron Ore Deposits and Adjoining Country: Belingwe District: Bulletin of the Geological Survey of South Rhodesia, v. 53, 114 p.

MANUSCRIPT ACCEPTED BY THE SOCIETY 24 MAY 2010

Heterogeneous strain and polymetamorphism in high-grade terranes: Insight into crustal processes from the Athabasca Granulite Terrane, western Canada, and the Limpopo Complex, southern Africa

K.H. Mahan
Department of Geological Sciences, University of Colorado, Boulder, Colorado 80302, USA

C.A. Smit
Department of Geology, University of Johannesburg, P.O. Box 524, Auckland Park, 2006, Johannesburg, South Africa

M.L. Williams
Department of Geosciences, University of Massachusetts, Amherst, Massachusetts 01003, USA

G. Dumond
Department of Geosciences, University of Arkansas, 18 Ozark Hall, Fayetteville, Arkansas 72701, USA

D.D. van Reenen
Department of Geology, University of Johannesburg, P.O. Box 524, Auckland Park, 2006, South Africa

ABSTRACT

Heterogeneous strain commonly serves as an important natural instrument for unraveling complex tectonic histories in polyphase metamorphic terranes. We present key examples of multi-scale heterogeneous deformation from two classic deep-crustal granulite terranes, the Athabasca Granulite Terrane in western Canada and the Limpopo Complex in southern Africa. These examples are chosen to illustrate how localized strain and attendant metamorphism played a key role in the development and preservation of important records of deep-crustal processes. In addition, several common characteristics of these terranes are identified through this analysis and include heterogeneous deep-crustal flow, regional-scale tectonic heterogeneity, and multistage exhumation with high-resolution records developed in locally hydrated shear zones. Better recognition of the fundamental spatial and temporal heterogeneity in these and other similar polymetamorphic terranes may help to reconcile apparently conflicting interpretations and tectonic models.

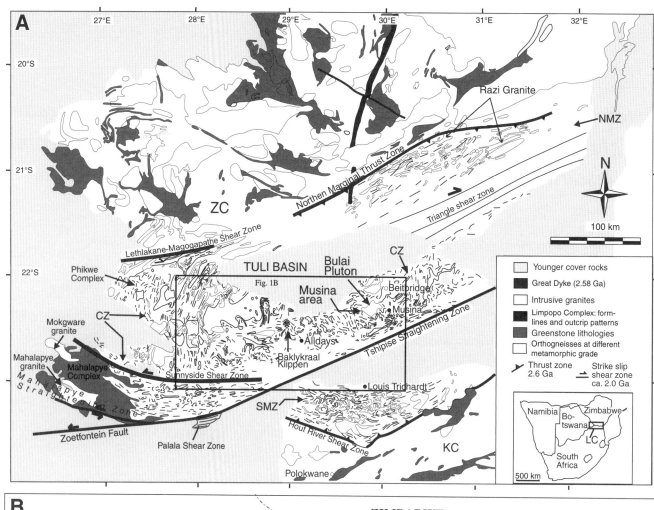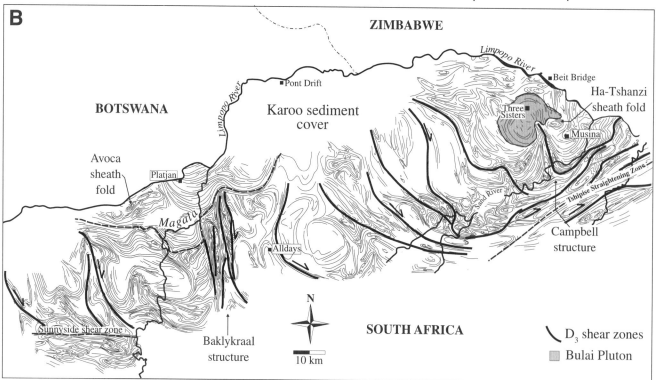

INTRODUCTION

Heterogeneous strain is a hallmark of natural deformation systems. On the scale of orogens, strain localization and reactivation may occur because of variations in lithospheric strength, gravitational potential energy, and/or mantle perturbations (e.g., Houseman and Molnar, 2001; Krabbendam, 2001). At smaller scales, the feedbacks between magmatic, metamorphic, and deformation processes become increasingly manifest (e.g., melt-enhanced flow, reaction-weakening, strain-induced reaction; White and Knipe, 1978; Hollister and Crawford, 1986; Stunitz and Tullis, 2001; Vernon, 2004). These processes commonly result in distributed high- and low-strain domains that may contain apparently contradictory records of tectonic history, particularly in high-grade polymetamorphic terranes. However, the recognition of domains with temporally distinct strain histories allows such terranes to be analyzed and interpreted to a degree not often possible in any one homogeneous domain.

In this chapter, we review examples of multi-scale interaction between heterogeneous deformation and metamorphism in two classic granulite terranes, the Limpopo Complex of southern Africa and the Athabasca Granulite Terrane in the western Canadian Shield. Our studies of these two terranes have documented polyphase pressure-temperature-time-deformation (P-T-t-D) histories at the kilometer-, outcrop-, sample-, and microstructural scales. Here we use the results of these studies to highlight the critical role of heterogeneous strain in providing insight into the tectonic evolution of these terranes, and in recording a variety of deep-crustal processes that are likely active in other orogens as well. Furthermore, a limited number of extensive and well-exposed granulite terranes have been recognized, and an important and persistent question is to what extent are these records typical of Earth processes. In addition to illustrating the utility of studying heterogeneous deformation in general, another related goal in this contribution is to highlight key characteristics and processes that are common in both terranes, and that might be important in other high-grade terranes or in the modern deep crust.

Figure 1. (A) Generalized geologic map of the Limpopo Complex and surrounding cratons (after van Reenen et al., 1990). The Central Zone (CZ) is separated from the Southern Marginal Zone (SMZ) and the Northern Marginal Zone (NMZ) by major inward-dipping, ca. 2.0 Ga strike-slip shear zones. The Limpopo Complex is separated from the adjacent Kaapvaal (KC) and Zimbabwe (ZC) Cratons by the Hout River shear zone in the south and the North Marginal Thrust in the north. LC—Limpopo Complex (in inset map). (B) Photogeological interpretation of Central Zone structure (area outlined in A) with major Neoarchean D_2 fold patterns, bounded in the south by the NE-trending D_2 Tshipise Straightening Zone. Discrete NE- and N-S-trending Paleoproterozoic D_3 shear zones, respectively, overprint D_2 structures in the Tshipise Straightening Zone and in the CZ (after Boshoff, 2008; Smit et al., this volume).

GEOLOGIC BACKGROUND AND GENERALIZED P-T-t-D PATHS

The Limpopo Complex (Fig. 1) and the Athabasca Granulite Terrane (Fig. 2) represent two equally enigmatic large-scale tectonic complexes on their respective continents. Both terranes represent the boundary zones between Archean cratonic provinces: the Limpopo Complex separates the Zimbabwe and Kaapvaal Cratons, and the Athabasca Granulite Terrane coincides with the central segment of the Snowbird Tectonic Zone, which separates the Rae and Hearne sub-provinces of the Churchill Province (craton). Domains at the tens of kilometer-scale, with distinct P-T-t-D histories, comprise both terranes, and all are characterized by early regional fabrics cut by younger kilometer-scale high-strain zones. Although the P-T-t-D paths for different domains vary in terms of (1) completeness of the record, (2) magnitudes of P and T undergone at a particular time (t), and (3) degree of strain (D), the main characteristics of each terrane can be represented with the generalized P-T-t-D paths illustrated in Figure 3 (paths shown are for selected domains). Both terranes contain domains that underwent deep-crustal granulite-facies events in the Neoarchean and again in the Paleoproterozoic. These events involved prograde heating, which in some cases occurred nearly isobarically (segments labeled 1 in Fig. 3) and in others may have initially involved significant prograde burial, although a record of the burial is commonly not well preserved in granulite-facies terranes. The paths shown for the Limpopo Complex are derived from metasedimentary rocks, and thus early burial can be inferred (Fig. 3A). Paths shown for the Athabasca terrane are derived from intrusive rocks, most of which were emplaced at depth, implying that their P-T-t-D histories began at high pressure (Fig. 3B) (e.g., Williams et al., 2000; Mahan et al., 2008). Major exhumation phases are generally characterized by decompression + cooling (segments labeled 2 in Fig. 3), and some of these domains underwent extended residence at middle- to deep-crustal levels (near isobaric cooling segments labeled 3) (e.g., Harley, 1989; Williams and Hanmer, 2005; Williams et al., 2009).

Limpopo Complex, Southern Africa

The Limpopo Complex crops out over a 700 × 300 km area and is an oblique, deep-crustal section through a Neoarchean granulite terrane (Fig. 1A) (Roering et al., 1992; Kramers et al., 2006). It is subdivided into three major zones. The Southern and Northern Marginal Zones are interpreted to constitute the high-grade metamorphic equivalents of the adjacent granite-greenstone cratons (van Reenen et al., 1990, this volume; Kreissig et al., 2000). Based on this interpretation, D_1 represents the inferred pre-Limpopo deformation related to the formation of cratonic greenstone belts and for which little evidence is preserved in the complex itself. The structure of the marginal zones is generally symmetrical, with 2.7–2.6 Ga outwardly verging contractional dip-slip shear zones (D_2) dividing the zones into subdomains. The two most prominent of these D_2 thrust systems

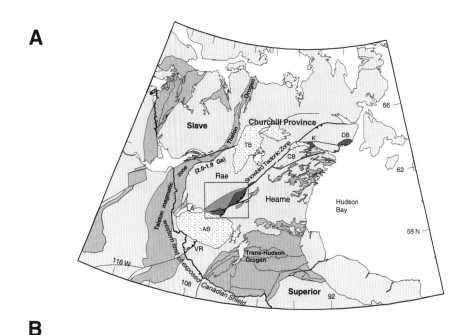

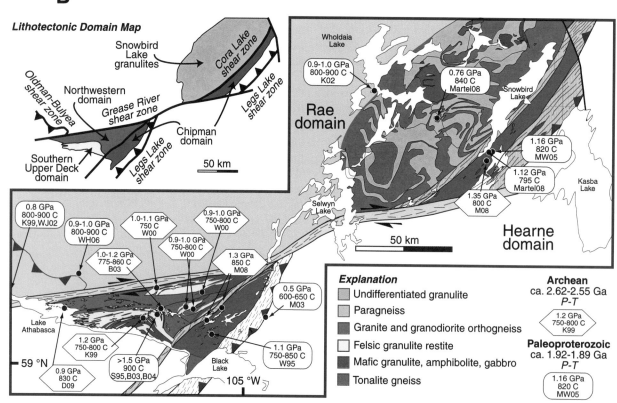

Figure 2. (A) Simplified geologic and tectonic map of the western Churchill Province. AB—Athabasca basin; CB—Chesterfield block; DB—Daly Bay complex; K—Kramanituar complex; TB—Thelon basin; VR—Virgin River shear zone. (B) Map of Athabasca granulite terrane. Compiled from Hanmer (1994), Mahan and Williams (2005), and Martel et al. (2008). Thermobarometric data are keyed to the following references: W95—Williams et al. (1995); S95—Snoeyenbos et al. (1995); K99—Kopf (1999); W00—Williams et al. (2000); K02—Krikorian (2002); WJ02—Williams and Jercinovic (2002); B03, B04—Baldwin et al. (2003, 2004); D09—Dumond et al. (2010); M03—Mahan et al. (2003); MW05—Mahan and Williams (2005); WH06—Williams and Hanmer (2005); Martel08—Martel et al. (2008); M08—Mahan et al. (2008).

are the south-vergent Hout River shear zone in the Southern Marginal Zone and the north-vergent Northern Marginal Thrust in the Northern Marginal Zone (Fig. 1A). These two systems represent the major terrane-bounding structures that separate the granulite terrane from the adjacent cratons (McCourt and Vearncombe, 1992; Roering et al., 1992; Ridley, 1992; Blenkinsop et al., 2004; Blenkinsop and Kisters, 2005; Kramers et al., 2006; Blenkinsop, this volume; van Reenen et al., this volume).

The Central Zone contains Mesoarchean supracrustal rocks intruded by Meso- to Neoarchean (3.3–2.6 Ga) granitoids (Kramers et al., 2006), many of which are not characteristic of the adjacent cratons. Major Neoarchean structures include prominent folds in the zone's interior (Fig. 1B) and the >20-km-wide NE-striking and steeply SE-dipping Tshipise Straightening Zone, which bounds the Central Zone to the south (Fig. 1A). Dominant fold types include large N-trending early D_2 recumbent structures commonly called *cross-folds* after Söhnge (1946) as well as late D_2 sheath folds with diameters up to >4 km (Roering et al., 1992; Kramers et al., 2006). In this chapter, major deformation events (e.g., D_2) are correlated throughout the Limpopo Complex on the basis of similar age constraints. The Central Zone was also affected by major Paleoproterozoic deformation (D_3). These structures include a discrete system of ca. 2.02 Ga NE- and N-striking D_3 shear zones (Fig. 1B), the overprinting effects of which are partly responsible for the unique geometry of the cross-folds (Boshoff et al., 2006; van Reenen et al., 2008). Two ca. 2.0 Ga (late D_3) inward-dipping, upper-crustal strike-slip mylonitic shear zones separate the Central Zone from the Southern and Northern Marginal Zones (Fig. 1A).

In general, the marginal zones are single-cycle monometamorphic Archean domains, whereas the Central Zone is characterized by polymetamorphic granulites. Major deformation (D_2) and metamorphism (M_2) in the marginal zones occurred in the late Archean between 2.69 and 2.60 Ga (e.g., Perchuk et al., 2000; Smit et al., 2001; Blenkinsop and Kisters, 2005; Blenkinsop, this volume; van Reenen et al., this volume). In the Central Zone, including the structures that separate it from the marginal zones, high-grade tectono-metamorphism first occurred in the Neoarchean at 2.68–2.61 Ga (M_2) and again during the Paleoproterozoic at 2.02–2.00 Ga (M_3) (e.g., Kamber et al., 1995; Holzer

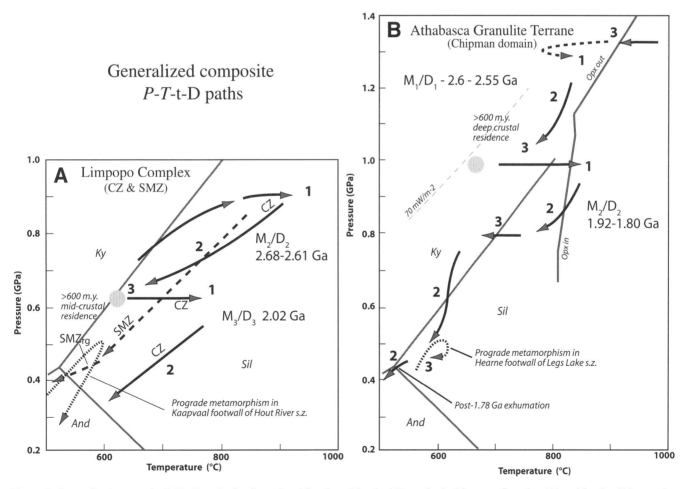

Figure 3. Generalized composite P-T-t-D paths for Central and Southern Marginal Zones in the Limpopo Complex (A) and for the Chipman domain in the Athabasca Granulite Terrane (B). And—andalusite; Ky—kyanite; CZ—Central Zone; Sil—sillimanite; SMZrg—Southern Marginal Zone retrogressed.

et al., 1998; Zeh et al., 2005; Boshoff et al., 2006; Buick et al., 2007; Mouri et al., 2009; Perchuk et al., 2008; van Reenen et al., 2008; Smit et al., this volume).

Athabasca Granulite Terrane, Western Canadian Shield

The Athabasca Granulite Terrane consists of at least 20,000 km^2 of contiguous high-pressure granulite exposed north of the 1.7 Ga intracratonic Athabasca basin in the western Canadian Shield (Fig. 2A) (Mahan and Williams, 2005; Williams and Hanmer, 2005; Dumond et al., 2008). The granulites represent the deepest exposed portion of the Rae Province where it is juxtaposed with the Hearne Province across the central segment of the Snowbird Tectonic Zone (e.g., Hoffman, 1988). The Legs Lake shear zone is the major southeastern terrane-bounding structure, which accommodated 1.85 Ga east-vergent uplift and juxtaposition of high-P rocks in the hanging wall with middle crustal rocks of the Hearne domain (Mahan et al., 2003, 2006a). Several major tectonic features were offset at 1.80 Ga by ~110 km of right-lateral displacement across the Grease River shear zone (Fig. 2B; Mahan and Williams, 2005). Details of the 1.9–1.78 Ga multi-stage regional exhumation history and late-stage strike-slip disruption of the terrane during the Trans-Hudson orogeny are presented by Mahan and Williams (2005) and Flowers et al. (2006b).

The most extensively studied portion of the Athabasca Granulite Terrane, the East Athabasca mylonite triangle (Hanmer, 1994, 1997), is divided into three lithotectonic subdomains: Southern, Northwestern, and Chipman (Fig. 2B). All three subdomains contain Meso- to Neoarchean igneous protoliths (3.2–2.6 Ga) that were strongly reworked by at least two major periods of high-P tectono-metamorphism: one in the Neoarchean at ca. 2.6–2.5 Ga, and a second in the Paleoproterozoic at ca. 1.9 Ga (Hanmer et al., 1994, 1995; Snoeyenbos et al., 1995; Williams et al., 1995, 2000; Kopf, 1999; Baldwin et al., 2003, 2004, 2006, 2007; Flowers et al., 2006a, 2006b, 2008; Mahan et al., 2006b, 2008; Dumond et al., 2008, 2010). Williams and Hanmer (2005) provide a comprehensive summary of this terrane and also emphasize many aspects of the heterogeneity.

SCALES OF HETEROGENEOUS STRAIN AND POLYMETAMORPHIC RECORDS

Heterogeneity is one of the most defining characteristics of the Limpopo Complex and the Athabasca Granulite Terrane. In the context of polyphase tectonism, outcrop to kilometer-scale examples are perhaps the most readily observable in the field and on reconnaissance geologic maps (Figs. 1, 2) (i.e., large domains with early relatively low-strain, fold-dominated fabrics bounded by shear zones with relatively younger high-strain fabrics). However, heterogeneous lithologies, mineral assemblages, and structures occur at all scales of observation. This commonly results in rocks that reflect very different parts of the tectonic history occurring in close proximity to one another. The extent and significance of this heterogeneity at smaller (centimeter to micrometer) and larger scales (e.g., 100 km) are not as readily apparent without more detailed integration of structural mapping, petrology, and geochronology.

In evaluating previous work on these two terranes (ours as well as others), we emphasize here the important role that heterogeneous deformation has played in the development and preservation of important components of the geological record. Below, we describe selected kilometer- and hand sample–scale examples of this heterogeneity in each terrane. Although clearly not a comprehensive list, these examples provide insight into important crustal processes or specific tectonic events. Regional-scale heterogeneity, which is also evident in both terranes and of particular significance to tectonic models, is addressed in the discussion.

Limpopo Complex

Kilometer-Scale Heterogeneity

Hout River shear system in Southern Marginal Zone. The southernmost late D$_2$ structure in the Southern Marginal Zone is the terrane-bounding Hout River shear zone (Smit et al., 1992). Several major features of this shear system record important components of the Archean evolution of the Southern Marginal Zone. First, exhumation of the granulite terrane was controlled by an early stage of SW-verging steep contractional deformation followed by lateral spreading of the granulites southward over the Kaapvaal greenstone terrane (Smit et al., 1992; Smit and van Reenen, 1997). Late-stage spreading was accommodated by a set of near-horizontal, out-of-sequence thrusts (Fig. 4). Second, the juxtaposition is reflected by a retrograde orthoamphibole isograd in the hanging-wall section of the Hout River shear zone and by tectonic klippen of retrogressed granulites that occur on the craton ~15 km south of the main exposure of the shear zone (Figs. 1A, 4). The retrograde isograd, marked by the replacement of orthopyroxene and cordierite by anthophyllite and Ged + Ky-Sil, respectively, in metapelitic gneisses, subdivides the zone into a northern, well-preserved granulite domain and a southern, retrogressed domain (van Reenen, 1986; mineral abbreviations after Whitney and Evans, 2010). Third, the footwall of the Hout River shear zone is characterized by metamorphosed and tectonized cratonic greenstone rocks that record a distinct prograde component of the process of terrane juxtaposition (van Reenen et al., this volume).

Exhumation from the deepest crustal level (~0.8 GPa) to the middle crust (~0.5 GPa) is recorded in metapelites throughout the granulite terrane (Smit et al., 2001). This record is preserved by static reaction textures in rocks unaffected by late D$_2$ shearing in the northern domain (Fig. 5A) and by synkinematic textures in those affected by D$_2$ in the retrogressed granulite domain (Fig. 5B). In the footwall, Grt-Ms-St-Ky schist records a narrow, clockwise P-T loop (Fig. 3A) interpreted to reflect burial and heating of rocks to a depth of ~17 km and near 600 °C (Perchuk et al., 2000; van Reenen et al., this volume). The synchronicity of 2.69–2.67 Ga thrusting in the hanging wall and peak metamorphism in the underthrust greenstone-derived schists is indicated

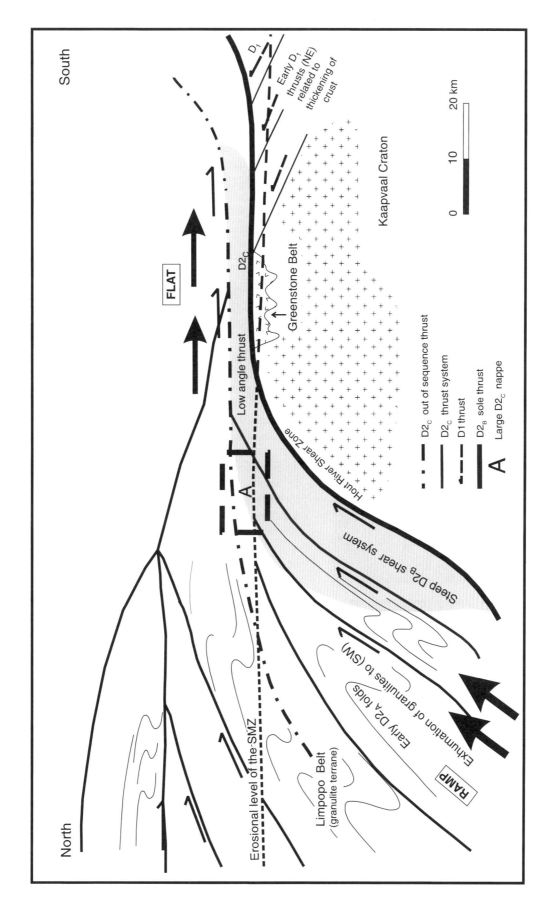

Figure 4. Schematic cross section of Southern Marginal Zone (see Fig. 1A), showing emplacement and spreading of granulites over lower grade rocks of the Kaapvaal Craton. Southern domain of hydrated granulites shown in gray (van Reenen et al., this volume).

by a variety of U-Pb and Pb-Pb dates from monazite and zircon from the granulite terrane and from synkinematic peak silicates (Grt, St, Ky) in the greenstone schist (Barton et al., 1992; Kreissig et al., 2001). Rocks in the retrogressed granulite domain record continued reaction along a more shallowly sloping decompression + cooling P-T segment (Fig. 3A). These late-stage reactions, which occurred synchronously with out-of-sequence thrusting at 2.63–2.60 Ga (Kreissig et al., 2001), are interpreted to represent accelerated cooling of the granulites by lateral spreading onto the cooler granite-greenstone craton (Perchuk et al., 2000; van Reenen et al., this volume).

The contrast in metamorphic records from lithologically similar tectonites preserved in D_2 domains versus those affected by later deformation demonstrates the critical role played by heterogeneous strain in the Central Zone. Granoblastic metapelites in early D_2 cross-folds record two distinct stages of metamorphic evolution (Boshoff et al., 2006; Perchuk et al., 2008; van Reenen et al., 2008; Smit et al., this volume). The primary record is one of a late D_2 retrograde stage, defined by the breakdown of high-pressure porphyroblastic Grt (with oriented inclusions defining a relict early S_2 fabric) to symplectitic Opx + Crd. This breakdown reflects decompression and cooling from deep-crustal conditions of >0.9 GPa, 900 °C to 0.5 GPa and <600 °C (Perchuk et al., 2008). The end of this stage is marked by emplacement of the relatively undeformed 2.61 Ga Bulai Pluton (Fig. 1B) (Millonig et al., 2008). A Paleoproterozoic stage of near-isobaric reheating at 0.5 GPa is inferred from late static M_3 reaction textures and chemical zoning in major phases (Perchuk et al., 2008).

In contrast, metapelites in kilometer-scale D_3 shear zones are completely reworked and only record a Paleoproterozoic middle-crustal metamorphic history. For example, the highest preserved pressures from D_3 gneisses are 0.55 GPa (Perchuk et al., 2008; van Reenen et al., 2008). These rocks contain synkinematic Grt-Crd-Sil-Bt-Kfs assemblages that record near-isobaric heating to >750 °C followed by decompression and cooling to 0.3 GPa and <600 °C. Monazite (U-Pb SHRIMP) and garnet (Pb-Pb step-leaching) dates from syntectonic leucosome in the D_3 shear zones record 2.02 Ga tectono-metamorphism (Boshoff et al., 2006; van Reenen et al., 2008; van Reenen and Boshoff, 2008).

In summary, the development of composite histories in the single Archean cycle of the Southern Marginal Zone and in the polymetamorphic Central Zone granulites is attributed to heterogeneously distributed deformation among major kilometer-scale structural domains. In the Southern Marginal Zone, rocks from the three major domains record distinctly different but related components of a shared tectonic history between the granulite terrane and the adjacent Kaapvaal Craton. Continued reaction in the retrogressed granulite domain is attributed to localized late D_2 deformation and infiltration of externally derived fluid from the footwall (van Reenen, 1986; van Reenen and Hollister, 1988; Baker et al., 1992; Hoernes et al., 1995; van Reenen et al., this volume). Similarly, whereas D_2 domains in the Central Zone record details of the earlier Archean high-pressure history, strain localization in D_3 shear zones promoted complete middle-crustal re-equilibration and continued reaction during exhumation to upper-crustal levels (Smit et al., this volume).

Sample-Scale Heterogeneity

Heterogeneity of strain, metamorphic assemblages, and relative age at sample or thin-section scale can be particularly revealing, and commonly forms the basis for some of the most convincing evidence for distinctly different tectonic events. For example, metapelite on the rim of a late D_2 sheath fold in the eastern Central Zone (Fig. 1B) preserves thin-section-scale evidence for the complete multi-cycle (early and late D_2 and D_3) history of this terrane (Perchuk et al., 2008; van Reenen et al., 2008). Relict early S_2 domains are characterized by porphyroblasts of Grt + Crd with oriented inclusions of early Bt, Sil, and Qtz. These domains are overprinted by a well-developed late S_2 shear fabric defined by augen of new synkinematic Grt_2 with asymmetric tails occupied by new Crd, Sil, and Qtz. This late S_2 fabric, which is associated with formation of the sheath fold, is further cut by millimeter-scale S_3 shear zones containing a finer grained third generation of Grt_3 + Sil_3 + Crd_3 + Bt_3. These microscale shear

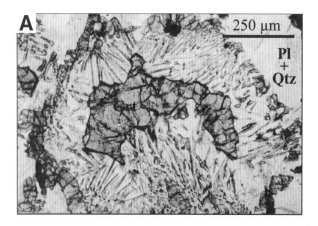

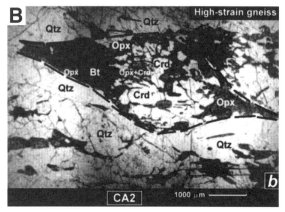

Figure 5. Photomicrographs from Southern Marginal Zone metapelites. (A) Symplectitic reaction texture, reflecting low strain (early D_2) conditions in northern domain. (B) Syntectonic reaction textures characterize late D_2 shear zones. Bt—biotite; Crd—cordierite; Opx—orthopyroxene; Pl—plagioclase; Qtz—quartz.

zones, which locally cut through Grt_2 porphyroblasts (Fig. 6), are parallel to and kinematically compatible with meter-scale and larger D_3 shear zones (Smit et al., this volume).

A composite P-T path, similar to that described from separate D_2 and D_3 domainal metapelites, is constructed for this sample. Initial decompression + cooling from >0.8 GPa, 850 °C to 0.5 GPa, 500 °C is bracketed by the first and second generation assemblages, respectively. In contrast, the compositions and textures of the third generation micro-shear zone assemblage indicate near-isobaric heating to ~650 °C at 0.55 GPa (Perchuk et al., 2008). The high-pressure, early D_2 stage predates 2.65 Ga, which is the crystallization age of granitoids that locally define the rims of the sheath folds (van Reenen et al., 2008; van Reenen and Boshoff, 2008). Late D_2 sheath fold formation may have occurred as late as 2.62 Ga, the age of L-tectonite granitoids locally coring the sheath folds (Boshoff, 2004; van Reenen et al., 2008; Smit et al., this volume). The age of D_3 shearing that overprints the sheath-fold margins is best constrained by 2.0 Ga Pb-Pb dates from titanite and clinopyroxene in sheared calc-silicate (Holzer et al., 1998).

Records of at least two distinct high-grade events in the Central Zone thus developed on a variety of scales owing in large part to the role of heterogeneous strain. In each case, older and younger generations of structures are characterized by distinct fabric elements and mineral assemblages from which P-T path segments can be determined and dated. In some situations, this heterogeneity allowed composite P-T-t-D paths to be constructed from single thin sections, providing increased confidence in the diachroneity of events. Examples of this type form the basis for the construction of composite P-T-t-D paths that can be used to constrain models for the evolution of the Central Zone (Smit et al., this volume), and similar approaches are likely to be useful in other complex high-grade terranes. The similarity in final M_2 and initial M_3 equilibration pressures at 0.6–0.5 GPa furthermore implies that much of the Central Zone resided in the middle crust for the intervening 600 m.y. period. Thus, recognizing heterogeneous deformation provides a means for gaining additional insight into Central Zone evolution and the context for these separate tectonic cycles.

Athabasca Granulite Terrane

Kilometer-Scale Heterogeneity

Two main generations of deformation have been identified throughout many parts of the Athabasca Granulite Terrane, although multiple periods of reactivation have been documented along the high-strain domain boundaries. Early structures include sub-horizontal- to NW-striking gneissic fabrics (S_1) and recumbent isoclinal folds of compositional layering (e.g., Slimmon, 1989; Kopf, 1999; Card, 2002; Mahan and Williams, 2005; Martel et al., 2008; Dumond et al., 2010). Later structures include open to isoclinal folds of S_1 and transposition of older folds and fabrics into sub-vertical, NE-striking mylonitic foliations (Hanmer et al., 1995; Hanmer, 1997; Mahan et al., 2003; Mahan and Williams, 2005; Martel et al., 2008; Dumond et al., 2010). Preserved S_1 fabric domains and the youngest discrete shear zones, such as the Legs Lake and Grease River structures, are primary examples of kilometer-scale heterogeneous strain and provide insight into otherwise missing pieces of the overall history of the region, either because they were completely overprinted by later events or because the later fabrics never developed.

S1 domains in East Athabasca mylonite triangle. A notable feature of many parts of the Athabasca Granulite Terrane is a ubiquitous, sub-horizontal to gently dipping, early granulite-grade fabric (S_1), variably overprinted by steep fabrics and localized high-strain zones (S_2) (Dumond, 2008; Mahan et al., 2008; Dumond et al., 2010). The early horizontal fabric is dominant in a particularly notable ~20 km² window along the northeast shore of Lake Athabasca in the Northwestern subdomain (Fig. 7). The window is bounded by kilometer-scale D_2 dextral high-strain zones. Internally, the D_2 overprint is minor, thus permitting direct examination of the geometry and tectonic significance of D_1 structures (Fig. 7).

The Northwestern subdomain is locally underlain by 2.63–2.60 Ga Opx-bearing granodiorite (Hanmer et al., 1994), which was emplaced, metamorphosed, and deformed at 0.9–1.1 GPa, 700–800 °C (Williams et al., 2000; Baldwin et al., 2003; Dumond et al., 2010). The S_1 foliation contains striping lineations (L_1) defined by discontinuous ribbons of recrystallized Kfs + Pl + Qtz + Amph ± Opx. Gneissic fabric development during synkinematic conversion of the granitoid into Grt-Cpx granulite was accompanied by melt-enhanced flow within interleaved meter-scale sheets of migmatitic garnet-rich felsic granulite (Williams et al., 2000; Dumond et al., 2010). Kinematic observations and metamorphic assemblages, supported by detailed monazite geochronology, indicate Neoarchean (ca. 2.60–2.55 Ga)

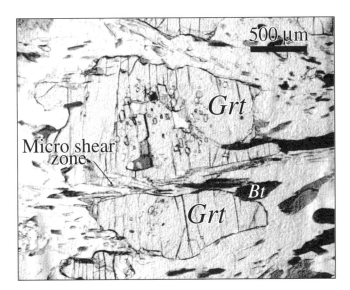

Figure 6. D_3 micro-shear zone cutting Grt_2 in late D_2 metapelite from Ha-Tshanzi sheath fold. From van Reenen and Boshoff (2008). Bt—biotite; Grt—garnet.

Figure 7. Outcrop photographs showing S_1 and local S_2 fabric relations in the S_1 window of Dumond et al. (2010). Northwestern subdomain, Athabasca Granulite Terrane.

granulite-grade (~800 °C) top-to-the-ESE, general shear strain at ~1.0 GPa (Fig. 8). The preservation of extremely fine recrystallized grain size in the granitoid mylonite (3–20 μm in K-feldspar, and 10–80 μm in plagioclase), in addition to petrologic modeling, suggests relatively H_2O-poor and/or CO_2-rich conditions in the S_1 tectonite. These conditions may help to explain subsequent strain partitioning and preservation of this low D_2 strain window. The early D_1 fabric is a directly observable record of sub-horizontal ductile flow at 30–40 km paleodepths, and thus it is hypothesized to represent a field-based analogue for lower crustal flow during collisional orogenesis or large magnitude extension (Dumond, 2008; Dumond et al., 2010).

Legs Lake and Grease River shear zones. Two major 1.9–1.8 Ga shear zones significantly influenced the present-day distribution of high-P rocks in the Athabasca Granulite Terrane (Mahan et al., 2003; Mahan and Williams, 2005). First, the 1.85 Ga Legs Lake shear zone marks the structural and metamorphic boundary between the deep-crustal (1.0 + GPa) Athabasca Granulite Terrane and the shallow-crustal (<0.5 GPa) Hearne domain (analogous to the Hout River shear zone in the Southern Marginal Zone of the Limpopo Complex). The shear zone is a 5–8-km-wide zone of intense amphibolite-facies E-vergent shearing (Mahan et al., 2003) and is part of a ~500-km-long contractional system (Mahan and Williams, 2005). Prograde metamorphism in the footwall peaked at <0.5 GPa and 600 °C and coincided with shear zone deformation (Mahan et al., 2003). The Grease River shear zone is a ~400-km-long ENE-striking, dextral strike-slip structure (Slimmon, 1989; Card, 2001; Mahan and Williams, 2005). It is a relatively long-lived structure, with an early 1.9 Ga granulite-facies strike-slip history, followed by a younger 5–7-km-wide amphibolite to greenschist-facies overprint (Dumond et al., 2008). The latest phase of strike-slip deformation within this shear zone overprints and offsets the northeastern extension of the Legs Lake shear zone (Fig. 2; Mahan and Williams, 2005).

Important details of the exhumation history of the entire Athabasca Granulite Terrane are restricted to these discrete shear zones where deformation, fluid-flow, and retrograde metamorphism were focused (Mahan et al., 2006a; Dumond et al., 2008). Hanging-wall decompression in the Legs Lake shear zone is characterized by breakdown of the peak assemblage Grt + Sil + Kfs + Pl + Qtz into the assemblage Grt + Crd + Bt ± Sil + Pl + Qtz. Similar felsic granulite occurs throughout the high-grade terrane, where the anhydrous peak assemblage is generally well preserved, but retrograde cordierite is restricted to the immediate hanging wall. Decompression of the hanging wall from peak conditions of 1.1 GPa, ~800 °C, involved several distinct stages (Mahan et al., 2006a), and the last re-equilibration occurred at 0.5–0.4 GPa, 550–650 °C (Mahan et al., 2003). Several monazite generations have been identified in felsic granulites throughout the terrane and are correlated with distinct phases of high-P metamorphism or magmatism (Baldwin et al., 2006; Mahan et al., 2006a; Martel et al., 2008; Dumond et al., 2010). However, distinctly younger 1.85 Ga and 1.80 Ga populations are present primarily within retrograde tectonites in the shear zones (Mahan et al., 2006a; Dumond et al., 2008). These late monazite generations are correlated with specific Grt-consuming metamorphic reactions (Fig. 9) (Mahan et al., 2006a) or with distinct deformation phases (Dumond et al., 2008). Thus, late-stage strain localization in the shear zones promoted deformation-enhanced retrograde reactions, and monazite provides a key tool for extracting the detailed chronological record of this process.

Sample-Scale Heterogeneity

Numerous examples of heterogeneous strain, coupled with metamorphic reaction at the hand-sample to microscopic scale, occur in the Athabasca Granulite Terrane. These include charnockitic granodiorite that records the development of millimeter-scale gneissic layering via deformation-enhanced metamorphic reaction (Williams et al., 2000; Dumond et al., 2010), local

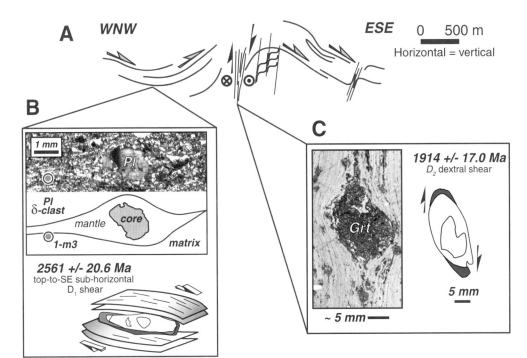

Figure 8. Summary diagram of S_1 and local S_2 fabric relations and timing in S_1 window of Dumond et al. (2010). Northwestern subdomain, Athabasca Granulite Terrane.

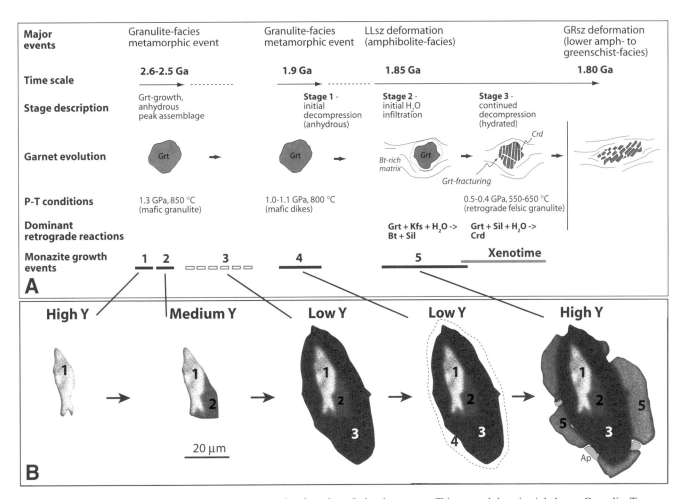

Figure 9. Felsic granulite evolution and monazite generation from Legs Lake shear zone, Chipman subdomain, Athabasca Granulite Terrane. After Mahan et al. (2006b). Bt—biotite; Crd—cordierite; GRsz—Grease River shear zone; Grt—garnet; Kfs—K-feldspar; LLsz—Legs Lake shear zone; Sil—sillimanite.

preservation of relict Archean garnet and monazite in felsic granulites that are otherwise completely overprinted by Proterozoic fabrics and assemblages (Mahan et al., 2006a), and mafic lithologies that preserve a distinct record of multiple high-P granulite-facies events owing to an intermediate stage of heterogeneous deformation and retrogression (Mahan et al., 2008; Flowers et al., 2008). The latter serves as a particularly useful case study because it provides the most conclusive evidence to date that the Athabasca Granulite Terrane underwent both Archean and Proterozoic high-P metamorphism.

Garnet-bearing, two-pyroxene mafic granulite occurs in the western Chipman domain as meter- to tens of meters–wide lenses hosted by 3.2 Ga tonalite. These mafic rocks are characterized by a three-stage history that began with development of a Grt + Cpx + Qtz ± Pl (M_1) assemblage from an Opx-bearing (igneous?) protolith at 1.3 GPa and 850–900 °C (Fig. 10) (Mahan et al., 2008). Stage 2 is characterized by heterogeneous deformation (D_2) and synkinematic partial retrogression of the peak assemblage to an amphibolite (M_2). The relict M_1 assemblage is preserved at the centimeter scale in low-strain boudins wrapped by S_2 or in F_2 fold hinges where the earlier fabric is transected by an S_2 axial planar foliation defined by M_2 amphibole. Stage 3 represents a return to granulite-facies conditions marked by the breakdown of amphibole to the M_3 assemblage Cpx + Opx + Ilm + Pl at 1.0 GPa, 800–900 °C. M_1 and M_3 are correlated with 2.55 and 1.9 Ga generations of metamorphic zircon, respectively, from the same granulites (Flowers et al., 2008).

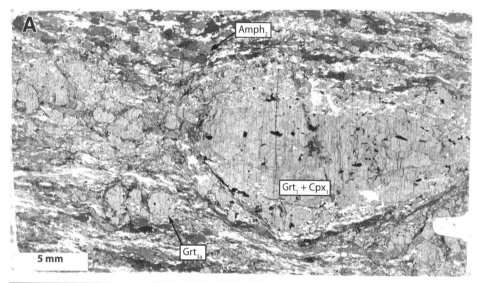

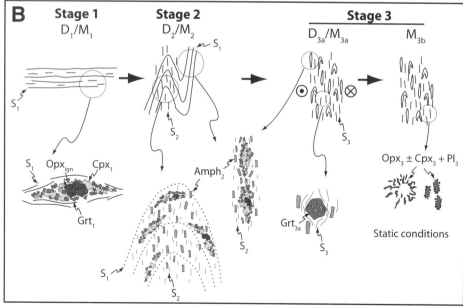

Figure 10. Mafic granulite evolution and relationship to fabric generations in Chipman subdomain, Athabasca Granulite Terrane. After Mahan et al. (2008). Amph—amphibolite; Cpx—clinopyroxene; Pl—plagioclase.

The second-generation granulite assemblage (M_3) was localized in the areas of M_2 amphibole and away from relict M_1 domains. Thus, heterogeneous strain is largely responsible for preserving the earlier M_1 assemblage, whereas localized hydration and metamorphic reaction during stage 2 predisposed the matrix to record the second granulite-facies event in the Proterozoic (Fig. 10). A similar polymetamorphic history is likely in other samples from this and other domains within the terrane, but the record is commonly less clear. The lack of distinctly decipherable textures and assemblages may be partly due to the absence of localized deformation and/or intermediate hydration stages.

DISCUSSION

Our collective studies in western Canada and southern Africa demonstrate that careful construction of P-T-t-D paths or path segments for both "high-strain" and "low-strain" domains is an effective strategy for distinguishing single-phase versus polyphase tectonism. Such a strategy necessarily begins with detailed field-based structural and microstructural observations (D) that are then texturally integrated with petrological (P-T) and geochronological (t) analyses. One of the most important aspects of this approach is the recognition and utilization of the heterogeneous strain record.

Records of Deep-Crustal Processes

Polyphase Granulite Metamorphism and Deep-Crustal Residence

Recognizing and interpreting the significance of polyphase metamorphism and deformation events are major challenges in studying Precambrian and Phanerozoic high-grade terranes (e.g., New England Appalachians: Moecher and Wintsch, 1994; Robinson et al., 1998; Grenville Orogen: Corrigan et al., 1994; deep arc terrane in Fiordland, New Zealand: Klepeis et al., 2004; Caledonides: Raimbourg et al., 2005). Long-standing debates about the number, age, and importance of high-grade tectonic events have been at the forefront of research in the two terranes highlighted in this paper (e.g., Barton and van Reenen, 1992; Barton et al., 2006; Kramers et al., 2006; Zeh and Klemd, 2008; Hanmer et al., 1995; Baldwin et al., 2003; Mahan and Williams, 2005; Mahan et al., 2008; Berman et al., 2007). Resolving these issues is critical for successful tectonic models and for understanding deep crustal processes in general.

One significant general implication of polyphase high-grade tectonism is the implied residence of such terranes at a specific crustal level for some extended time period (Williams et al., 2009). Rocks now exposed in the Athabasca Granulite Terrane initially underwent high-pressure metamorphism at >40 km depths in the Neoarchean but appear to have stabilized at ~35 km depths during the same cycle. These same rocks again underwent granulite-facies tectonism at similar pressure (1.0 GPa) at 1.9 Ga before ultimate exhumation to the Earth's surface by 1.7 Ga. Thus, parts of the terrane resided at lower crustal depths for >600 m.y., and the available data suggest that the Central Zone of the Limpopo Complex resided at mid-crustal depths for a similar time. Such "isobaric" terranes are of special importance because their residence at depth under normal geothermal gradients implies that parts of their evolution may record the processes most representative of stable continental crust (e.g., Williams and Hanmer, 2005; Williams et al., 2009). These records provide opportunities to examine the behavior of deep continental crust from Archean assembly to Paleoproterozoic reworking and ultimate exhumation (van Reenen et al., 1987; Williams and Hanmer, 2005; Flowers et al., 2008).

Lower Crustal Flow

In the Athabasca Granulite Terrane, kilometer-scale "low-strain" domains at 1.9 Ga commonly represent windows into "high-strain" processes active during the Archean. These windows preserve sub-horizontal shear fabrics that developed at high-P granulite-facies conditions, and may represent field-based analogues for lower crustal flow (Dumond et al., 2010). This type of high-temperature ductile flow, possibly in the presence of melt (Hollister and Crawford, 1986; Beaumont et al., 2004), has been hypothesized as an important mechanism for lateral redistribution of mass at deep-crustal levels in a variety of orogenic settings (e.g., North American Basin and Range—Wernicke, 1990; Kruse et al., 1991; Wernicke et al., 2008; Himalaya-Tibetan system—Royden et al., 1997; Beaumont et al., 2004; Andes—Gerbault et al., 2005). Field-based examples provide a means for evaluating these hypotheses and models. For example, two important aspects of lower crustal flow that may be addressed by studying these domains in the Athabasca Terrane are briefly highlighted here. The first relates to the processes responsible for generating lower crust that is weak enough to flow, and the second relates to the nature of lateral heterogeneities that could form rheological boundaries to domains of active flow.

Neoarchean ductile flow in the Northwestern subdomain of the Athabasca Granulite Terrane occurred in crust weakened both by advective heat associated with magma emplacement and the local presence of melt. Syn-deformational metamorphic and magmatic segregation and a subsequent period of near-isobaric cooling all served to strengthen the crust, analogous to a strain-hardening process, which significantly influenced the style of later Proterozoic deformation events (Dumond et al., 2010). Pervasive low-viscosity flow and sub-horizontal fabric development at 2.60–2.55 Ga contrast with the highly partitioned shearing of vertical shear zones that occurred at ca. 1.9 Ga even though temperatures were high during both events. Similar indications of an evolving deformation style for deep crust are present in the Limpopo Complex where early penetrative fabrics reflecting relatively homogeneous flow, commonly in the presence of melt, gave way to more heterogeneous mass transport during kilometer-scale sheath folding and finally more discrete shear zones in the Proterozoic. A similar evolution was also described in the Mesozoic arc crust of Fiordland, New Zealand (e.g., Klepeis et al., 2004, 2007).

Although commonly modeled in two dimensions, lateral variations in modern lower crustal flow regimes have been recognized or inferred from topography (e.g., Clark et al., 2005), seismology (Flesch et al., 2005), and geodesy (Davis et al., 2006). Although most of the steep structures that interrupt the S_1 fabric in the Athabasca Granulite Terrane are Proterozoic, the Cora Lake shear zone forms one of the boundaries to the S_1 "window" in the Northwestern domain and has an Archean strike-slip origin (Mahan et al., 2008). We hypothesize that this shear zone may have acted as a local boundary to sub-horizontal flow, likely due to rheological contrasts derived from lithological heterogeneity, thus representing a potential analogue for strain partitioning in modern lower crustal flow regimes (e.g., northern Tibetan Plateau, Dayem et al., 2009). Distinguishing older and younger deformation is critical for recognizing this type of heterogeneity in deep-crustal flow regimes.

Exhumation Processes

The exhumation history is an important characteristic of polymetamorphic high-grade terranes. The degree of preservation of a deep-crustal record and the extent of understanding the origin and tectonic setting of these terranes are both commonly dependent on knowledge of the mechanisms and time scales of exhumation (e.g., Handy, 1990). Major components of the exhumation commonly involve localized strain, and the attendant metamorphic re-equilibration processes that occur in major terrane-bounding fault zones (e.g., Fountain and Salisbury 1981; Handy, 1990). Whereas extensional faulting and erosion are the characteristic agents of exhumation at the Earth's surface, thrusting has long been recognized as an important driving mechanism for transporting deep-crustal material upward with respect to adjacent lower grade terranes. The Hout River and Legs Lake shear zones provide type examples here, and well-known examples from other terranes include bounding structures of the Kapuskasing Uplift (Percival and West, 1994), Ivrea Zone (Handy and Zingg, 1991), and the Fiordland granulite terrane (Klepeis et al., 1999).

Related Structures and Strain Regimes during Exhumation

In both the Athabasca Granulite Terrane and the Limpopo Complex, deformation during exhumation was commonly focused into discrete kilometer-scale structures such as the Proterozoic Legs Lake and Grease River shear zones in the former and the Archean Hout River shear system or various Proterozoic shear zones in the latter. The distribution, geometry, and kinematics of these coeval structures provide insight into the large-scale deformation regime and input for regional exhumation models.

Regional exhumation of the Athabasca Granulite Terrane from ~35 km depths occurred during at least three distinct Proterozoic phases over a ~150 m.y. period (Fig. 3B) (Mahan et al., 2006a; Flowers et al., 2006b). Early high-temperature decompression to 0.8 GPa (Mahan et al., 2008) probably occurred in an extensional or transtensional regime, as suggested by the emplacement of an extensive mafic dike swarm at 1.9 Ga (Flowers et al., 2006a). The second phase involved decompression to 0.5 GPa during transpression at 1.85 Ga driven by convergence in the Trans-Hudson orogen. The final phase of exhumation after 1.80 Ga involved major strike-slip reactivation of shear zones and lateral disruption of the granulite terrane followed by regional extensional faulting (Mahan et al., 2003; Mahan and Williams, 2005; Dumond et al., 2008).

Although the timing and nature of exhumation of deep-crustal granulites in the Limpopo Complex varied significantly between the Marginal and Central zones, multistage exhumation appears characteristic of all areas. The two marginal zones are largely single-cycle metamorphic terranes that were exhumed to upper crustal levels in the Neoarchaean (2.69–2.60 Ga). The strain regime during early exhumation at 2.69–2.67 Ga was primarily contractional, as indicated by the geometry and kinematics of major late D_2 structures (e.g., Hout River shear system: Smit and van Reenen, 1997; Smit et al., 2001; van Reenen et al., this volume; and Northern Marginal Thrust Zone: Ridley, 1992; Blenkinsop and Kisters, 2005). In the Southern Marginal Zone, late-stage exhumation to upper-crustal levels at 2.63–2.60 Ga was accommodated by lateral spreading of hanging-wall granulite onto the adjacent lower grade greenstone terranes (van Reenen et al., 1995, 2008, this volume). Proterozoic exhumational events (2.0–1.97 Ga) have also been proposed in the Northern Marginal Zone (Kamber et al., 1995, 1996). This exhumation was relatively minor and of low temperature in the main northern part of the marginal zone, associated with greenschist-facies thrusting. However, the southern half of the zone underwent a major tectonic and metamorphic overprint at 2.0 Ga, involving significant high-temperature isothermal decompression (from 0.9 GPa to 0.3–0.5 GPa) in association with development of the strike-slip Triangle shear zone (Kamber et al., 1995).

Granulites of the Central Zone underwent two major phases of exhumation, one in the Neoarchean that was responsible for emplacement of high-P rocks at mid-crustal depths of ~0.5 GPa, and a second phase at 2.0 Ga following a major period of mid-crustal high-T reworking. Neoarchaean deformation associated with exhumation in the Central Zone mainly involved kilometer-scale sheath folding and top-to-the-NE movement of high-grade material within the interval of 2.65–2.61 Ga (van Reenen et al., 2008; Smit et al., this volume). Deformation during Paleoproterozoic exhumation was expressed as a distributed transpressive system of discrete ca. 2.02 Ga shear zones within the Central Zone and late-stage 2.0 Ga strike-slip structures that bound the Central Zone (e.g., Palala shear zone; McCourt and Vearncombe, 1987, 1992; Schaller et al., 1999).

Hanging-Wall Retrograde and Footwall Prograde Processes

Shear zones provide settings for localized coupling of deformation and fluid flow, which can lead to enhanced metamorphic reaction (e.g., Etheridge et al., 1983) and production of high-resolution records for parts of the P-T-t-D history. The Hout River and Legs Lake shear zones are both examples of crustal-scale thrust systems in which synkinematic hydrous fluid infiltration was restricted to the shear zone and the immediate vicinity of the hanging-wall granulites (van Reenen, 1986; Mahan et al., 2006b). In the Southern Marginal Zone, this process produced a well-developed retrograde orthoamphibole isograd (van Reenen, 1986; van Reenen et al., this volume). In the Legs Lake shear zone, synkinematic retrograde reaction in felsic granulites is marked by Crd-Bt assemblages and dated by a new compositionally distinct population of monazite (Mahan et al., 2006b). A similarly distinct monazite record was produced in deformed granite dikes in the Grease River shear zone owing to dynamic recrystallization of feldspar and biotite (Dumond et al., 2008).

In both thrust systems the retrogressed granulites record the passage of deeper crustal levels that underwent decompression as they were emplaced over shallower crust. The shear zones acted as channels for fluid that may have been produced by dehydration of metasediments in the footwall domains. Rocks in the most proximal portions of these underthrust granite-greenstone domains (i.e., Kaapvaal Craton in the case of the Hout River system and the Hearne domain in the Legs Lake system) record a burial response via clockwise mid-crustal P-T loops (Perchuk et al., 2000; Mahan et al., 2003; van Reenen et al., this volume). In summary, important details regarding the exhumation of these terranes were apparently recorded only within high strain and hydrated zones that were active during the exhumation process.

Regional Heterogeneity and Implications for Tectonic Models

A pressing topic of current research in both the Athabasca and Limpopo Granulite Terranes concerns the significance of very large-scale heterogeneous tectono-metamorphism. This is most strikingly expressed in the form of domains from ~10 to ~100 km wide that appear to have distinctly different tectonic histories. The dichotomy presented by the contrasting histories of these domains is one of the main challenges for constraining the origin and evolution of these terranes.

The first-order tectonic heterogeneity in the Athabasca Granulite Terrane is expressed in the extensive record of 2.6 Ga high-grade tectonism in the eastern part of the terrane and its relatively limited existence in rocks northwest of the Grease River shear zone. Although the common characteristic of all domains is 1.9 Ga high-grade metamorphism, it is clear that thick continental crust already existed in the region by 2.55 Ga (Snoeyenbos et al., 1995; Williams et al., 2000; Mahan et al., 2008; Flowers et al., 2008; Dumond et al., 2010).

We suspect that at least part of the terrane records a deep crustal response to the final phases of a ca. 2.6 Ga collisional or accretionary orogenic event (Dumond et al., 2010). Regardless of the specific tectonic setting, the presence of a relatively narrow belt of 2.6–2.5 Ga granulite-facies rocks, extending the length of the Athabasca Granulite Terrane and for several hundred kilometers farther north (Stern and Berman, 2000; Sanborn-Barrie et al., 2001; Hanmer et al., 2006; Berman et al., 2007), suggests a Neoarchean origin for the terrane. Alternative interpretations include a single major continental collisional event at 1.9 Ga (Baldwin et al., 2003, 2004, 2007; Berman et al., 2007; Martel et al., 2008), but these models do not explain the record of Archean high-P tectonism in the same rocks.

Regional-scale heterogeneous tectonism (100 km scale) in the Limpopo Complex is most clearly expressed in the contrasting nature of the single-cycle Neoarchean marginal zones versus the polymetamorphic nature and extensive Proterozoic reactivation of the Central Zone. Most models to date have argued for either a largely Archean "Limpopo orogeny" with limited upper crustal Proterozoic reactivation (e.g., van Reenen et al., 1987, 1992; Barton and van Reenen, 1992; de Wit et al., 1992; McCourt and Vearncombe, 1987, 1992; Treloar et al., 1992) or a major Proterozoic orogeny with a minor Archean record (e.g., Holzer et al., 1998; Schaller et al., 1999; Zeh et al., 2004; Buick et al., 2007; Kramers et al., 2006). Another alternative is a model whereby terranes that evolved independently were assembled during a very long period (2.7–2.0 Ga) of subduction and accretion (Barton et al., 2006; termed "Turkic-type orogeny" by Sengör and Natal'in, 1996).

We suggest that the large-scale spatial heterogeneity, the similarity in ages of deformation among major zones, and the recognition of multiple granulite-facies events in the Central Zone do not favor single-cycle models for the Limpopo Complex as a whole, nor do they require the complexity of a "Turkic-type" model. The marginal zones preserve a more complete Neoarchean history and a record of early contractional orogenic processes. However, the Central Zone is now recognized to have a similar Neoarchean deep-crustal, granulite-facies deformation record. The similarities between the Archean characteristics of the Southern Marginal Zone and the Central Zone support a pop-up type of tectonic model that was related either to crustal thickening from continent-continent collision (e.g., Roering et al., 1992) or to gravitational redistribution related to a mantle plume (Gerya et al., 2000). In summary, successful tectonic models for either of the terranes discussed here will need to recognize spatial and temporal tectonic heterogeneity as a first-order characteristic.

CONCLUDING STATEMENTS

Both terranes discussed here are samples of the deep crust, and the rocks and structures that are preserved offer records of important deep-crustal events and of the exhumation processes that are responsible for their surface exposure today. We conclude that heterogeneous strain played a significant role in developing the record of these processes. Heterogeneous behavior of deformation is controlled by a variety of factors, some intrinsic and others more extrinsic. In these terranes the former include lithologic variations and the geometry of preexisting structures, whereas the latter include localization of early or late melt (i.e., melt weakening, Hollister and Crawford, 1986) and infiltration of externally derived hydrous fluids. Such heterogeneity is almost certainly ubiquitous in all orogens, but whereas recognized in many orogens (e.g., Appalachians, Grenville orogen, Fiordland deep arc terrane, and Caledonides, to name only a few), the extent and full implications of the heterogeneity have not been incorporated into many models.

Some common characteristics between the Limpopo and Athabasca granulite terranes include multistage exhumation, pervasive mass transport at deep-crustal levels, localized high-resolution records developed in hydrated zones during exhumation, and regional-scale tectonic heterogeneity. Multistage exhumation is one of the most prominent of these characteristics and has been recognized for some other major granulite terranes (e.g., Kapuskasing Uplift, Musgrave Block). Such a history essentially requires polyphase deformation and metamorphism, even if only within one related tectonic cycle. Recognizing that heterogeneous strain and metamorphic reaction can and did occur at a variety of scales may help to reconcile apparently conflicting interpretations and tectonic models for these and other similar polymetamorphic high-grade terranes.

ACKNOWLEDGMENTS

Work in the Athabasca Granulite Terrane was funded in part by Geological Society of America grants to KHM and GD, and U.S. National Science Foundation grants EAR 0001152 and 0609935 to MLW. Research in the SMZ since 1995 was carried out as a part of the RF-RSA collaboration supported by NRF grants to DDvR (GUN 2053192), and grants 08-05-00351 and 1949.2008 to LLP, respectively, from the RFBR and the RF President Program entitled "Leading Research Schools of Russia." DDvR acknowledges the consistent support of the Faculty of Science, University of Johannesburg (previously Rand Afrikaans University). Special thanks for numerous discussions on Canadian Shield topics is extended to S.A. Bowring, R.M. Flowers, P. Goncalves, M.J. Jercinovic, C. Kopf, and D. Snoeyenbos. L. Perchuk and C. Roering are especially thanked for valuable discussion regarding the Limpopo Complex.

REFERENCES CITED

Baker, J., van Reenen, D.D., van Schalkwyk, J.F., and Newton, R.C., 1992, Constraints on the composition of fluids involved in retrograde anthophyllite formation in the Limpopo Belt, South Africa: Precambrian Research, v. 55, p. 327–336, doi:10.1016/0301-9268(92)90032-J.

Baldwin, J.A., Bowring, S.A., and Williams, M.L., 2003, Petrologic and geochronologic constraints on high-pressure, high-temperature metamorphism in the Snowbird tectonic zone, Canada: Journal of Metamorphic Geology, v. 21, p. 1–19, doi:10.1046/j.1525-1314.2003.00413.x.

Baldwin, J.A., Bowring, S.A., Williams, M.L., and Williams, I.S., 2004, Eclogites of the Snowbird tectonic zone: Petrological and U-Pb geochronological evidence for Paleoproterozoic high-pressure metamorphism in the western Canadian Shield: Contributions to Mineralogy and Petrology, v. 147, p. 528–548, doi:10.1007/s00410-004-0572-4.

Baldwin, J.A., Bowring, S.A., Williams, M.L., and Mahan, K.H., 2006, Geochronological constraints on the crustal evolution of felsic high-pressure granulites, Snowbird tectonic zone, Canada: Lithos, v. 88, p. 173–200, doi:10.1016/j.lithos.2005.08.009.

Baldwin, J.A., Powell, R., Williams, M.L., and Goncalves, P., 2007, Formation of eclogite, and reaction during exhumation to mid-crustal levels, Snowbird tectonic zone, western Canadian Shield: Journal of Metamorphic Geology, v. 25, p. 953–974, doi:10.1111/j.1525-1314.2007.00737.x.

Barton, J.M., Jr., and van Reenen, D.D., 1992, When was the Limpopo Orogeny?: Precambrian Research, v. 55, p. 7–16, doi:10.1016/0301-9268(92)90010-L.

Barton, J.M., Jr., Doig, R., Smith, C.B., Bohlender, F., and van Reenen, D.D., 1992, Isotopic and REE characteristics of the intrusive charnoenderbite and enderbite geographically associated with Matok Complex: Precambrian Research, v. 55, p. 451–467, doi:10.1016/0301-9268(92)90039-Q.

Barton, J.M., Jr., Klemd, R., and Zeh, A., 2006, The Limpopo Belt: A result of Archean to Proterozoic Turkic-type orogenesis?, in Reimold, W.U., and Gibson, R.L., eds., Processes on the Early Earth: Geological Society of America Special Paper 405, p. 315–331.

Beaumont, C., Jamieson, R.A., Nguyen, M.H., and Medvedev, S., 2004, Crustal channel flows: 1. Numerical models with applications to the tectonics of the Himalayan-Tibetan orogen: Journal of Geophysical Research, Solid Earth, v. 109, B06406.

Berman, R.G., Davis, W.J., and Pehrsson, S., 2007, The collisional Snowbird tectonic zone resurrected: Growth of Laurentia during the 1.9 Ga accretionary phase of the Trans-Hudson orogen: Geology, v. 35, p. 911–914, doi:10.1130/G23771A.1.

Blenkinsop, T.G., 2011, this volume, Archean magmatic granulites, diapirism, and Proterozoic reworking in the Northern Marginal Zone of the Limpopo Belt, in van Reenen, D.D., Kramers, J.D., McCourt, S., and Perchuk, L.L., eds., Origin and Evolution of Precambrian High-Grade Gneiss Terranes, with Special Emphasis on the Limpopo Complex of Southern Africa: Geological Society of America Memoir 207, doi: 10.1130/2011.1207(13).

Blenkinsop, T.G., and Kisters, A.F.M., 2005, Steep extrusion of late Archaean granulites in the Northern Marginal Zone, Zimbabwe: Evidence for secular change in orogenic style, in Gapais, D., Brun, J.P., and Cobbold, P.R., eds., Deformation Mechanisms, Rheology and Tectonics: From Minerals to the Lithosphere: Geological Society [London] Special Publication 243, p. 193–204.

Blenkinsop, T.G., Kroner, A., and Chiwara, V., 2004, Single stage, late Archean exhumation of granulites in the Northern Marginal Zone, Limpopo Belt, Zimbabwe, and relevance to gold mineralization at Renco Mine: South African Journal of Geology, v. 107, p. 377–396, doi:10.2113/107.3.377.

Boshoff, R., 2004, Formation of major fold types during distinct geological events in the Central Zone of the Limpopo Belt, South Africa: New structural, metamorphic and geochronologic data [M.S. thesis]: Johannesburg, Rand Afrikaans University, 121 p.

Boshoff, R., 2008, The Neoarchean to Palaeoproterozoic evolution of the polymetamorphic Central Zone of the Limpopo High-Grade Terrain in South Africa [Ph.D. dissertation]: Johannesburg, University of Johannesburg, 216 p.

Boshoff, R., van Reenen, D.D., Smit, C.A., Perchuk, L.L., Kramers, J.D., and Armstrong, R., 2006, Geologic history of the Central Zone of the Limpopo Complex: The West Alldays area: Journal of Geology, v. 114, p. 699–716, doi:10.1086/507615.

Buick, I.S., Hermann, J., Maas, R., and Gibson, R.L., 2007, The timing of subsolidus hydrothermal alteration in the Central Zone, Limpopo Belt (South Africa): Constraints from titanite U-Pb geochronology and REE partitioning: Lithos, v. 98, p. 97–117, doi:10.1016/j.lithos.2007.02.002.

Card, C.D., 2001, Basement rocks to the western Athabasca basin in Saskatchewan: Saskatchewan Geological Survey Summary of Investigations, v. 2, p. 321–333.

Card, C.D., 2002, New Investigations of Basement to the Western Athabasca Basin: Saskatchewan Geological Survey Summary of Investigations 2002, v. 2, 17 p.

Clark, M.K., Bush, J.W.M., and Royden, L.H., 2005, Dynamic topography produced by lower crustal flow against rheological strength heterogeneities bordering the Tibetan Plateau: Geophysical Journal International, v. 162, p. 575–590, doi:10.1111/j.1365-246X.2005.02580.x.

Corrigan, D., Culshaw, N.G., and Mortensen, J.K., 1994, Pre-Grenvillian evolution and Grenvillian overprinting of the Parautochthonous Belt in Key Harbour, Ontario: U-Pb and field constraints: Canadian Journal of Earth Sciences, v. 31, p. 583–596.

Davis, J.L., Wernicke, B.P., Bisnath, S., Niemi, N.A., and Elosegui, P., 2006, Subcontinental-scale crustal velocity changes along the Pacific–North America plate boundary: Nature, v. 441, p. 1131–1134, doi:10.1038/nature04781.

Dayem, K.E., Houseman, G.A., and Molnar, P., 2009, Localization of shear along a lithospheric strength discontinuity: Application of a continuous deformation model to the boundary between Tibet and the Tarim Basin: Tectonics, v. 28, TC3002, doi:10.1029/2008TC002264.

de Wit, M.J., Roering, C., Hart, R.J., Armstrong, R.A., De Ronde, C.E.J., and Green, R.W.E., 1992, Formation of an Archean continent: Nature, v. 357, p. 553–562, doi:10.1038/357553a0.

Dumond, G., 2008, Tectonic and petrogenetic evolution of sub-horizontal granulite-grade fabric in lower continental crust: Constraints on lower crustal flow from the western Canadian Shield [Ph.D. thesis]: Amherst, University of Massachusetts–Amherst, 184 p.

Dumond, G., McClean, N., Williams, M.L., and Jercinovic, M.J., 2008, High-resolution dating of granite petrogenesis and deformation in a lower crustal shear zone: Athabasca granulite terrane, western Canadian Shield: Chemical Geology, v. 254, p. 175–196, doi:10.1016/j.chemgeo.2008.04.014.

Dumond, G., Goncalves, P., Williams, M.L., and Jercinovic, M.J., 2010, Subhorizontal fabric in exhumed continental lower crust and implications for lower crustal flow: Athabasca Granulite Terrane, western Canadian Shield: Tectonics, v. 29, TC2006, doi:10.1029/2009TC002514.

Etheridge, M.A., Wall, V.J., and Vernon, R.H., 1983, The role of the fluid phase during regional metamorphism and deformation: Journal of Metamorphic Geology, v. 1, p. 205–226, doi:10.1111/j.1525-1314.1983.tb00272.x.

Flesch, L.M., Holt, W.E., Silver, P.G., Stephenson, M., Wang, C.Y., and Chan, W.W., 2005, Constraining the extent of crust-mantle coupling in central Asia using GPS, geologic, and shear wave splitting data: Earth and Planetary Science Letters, v. 238, p. 248–268, doi:10.1016/j.epsl.2005.06.023.

Flowers, R.M., Bowring, S.A., and Williams, M.L., 2006a, Timescales and significance of high-pressure, high-temperature metamorphism and mafic dike anatexis, Snowbird tectonic zone, Canada: Contributions to Mineralogy and Petrology, v. 151, p. 558–581, doi:10.1007/s00410-006-0066-7.

Flowers, R.M., Mahan, K.H., Bowring, S.A., Williams, M.L., Pringle, M.S., and Hodges, K.V., 2006b, Multistage exhumation and juxtaposition of lower continental crust in the western Canadian Shield: Linking high-resolution U-Pb and ^{40}Ar/^{39}Ar thermochronometry with pressure-temperature-deformation paths: Tectonics, v. 25, TC4003, doi:10.1029/2005TC001912.

Flowers, R.M., Bowring, S.A., Mahan, K.H., and Williams, M.L., 2008, Stabilization and reactivation of cratonic lithosphere from the lower crustal record in the western Canadian shield: Contributions to Mineralogy and Petrology, p. 529–549, doi:10.1007/s00410-008-0301-5.

Fountain, D.M., and Salisbury, M.H., 1981, Exposed cross-sections through the continental crust: Implications for crustal structure, petrology, and evolution: Earth and Planetary Science Letters, v. 56, p. 263–277, doi:10.1016/0012-821X(81)90133-3.

Gerbault, M., Martinod, J., and Herail, G., 2005, Possible orogeny-parallel lower crustal flow and thickening in the Central Andes: Tectonophysics, v. 399, p. 59–72, doi:10.1016/j.tecto.2004.12.015.

Gerya, T.V., Perchuk, L.L., van Reenen, D.D., and Smit, C.A., 2000, Two-dimensional numerical modeling of pressure-temperature-time paths for the exhumation of some granulite facies terrains in the Precambrian: Journal of Geodynamics, v. 30, p. 17–35, doi:10.1016/S0264-3707(99)00025-3.

Handy, M.R., 1990, The exhumation of cross-sections of the continental crust: Structure, kinematics and rheology, in Salisbury, M.H., and Fountain, D.M., eds., Exposed Cross-Sections of the Continental Crust: Dordrecht, Kluwer Academic Publishers, p. 485–507.

Handy, M.R., and Zingg, A., 1991, The tectonic and rheological evolution of an attenuated cross-section of the continental-crust—Ivrea Crustal Section, Southern Alps, northwestern Italy and southern Switzerland: Geological Society of America Bulletin, v. 103, p. 236–253, doi:10.1130/0016-7606(1991)103<0236:TTAREO>2.3.CO;2.

Hanmer, S., 1994, Geology, East Athabasca mylonite triangle, Saskatchewan: Geological Survey of Canada Map 1859A, scale 1:100,000, 1 sheet.

Hanmer, S., 1997, Geology of the Striding-Athabasca mylonite zone, northern Saskatchewan and southeastern District of Mackenzie, Northwest Territories: Geological Survey of Canada Bulletin 501, 92 p.

Hanmer, S., Parrish, R., Williams, M.L., and Kopf, C., 1994, Striding-Athabasca mylonite zone: Complex Archean deep-crustal deformation in the East Athabasca mylonite triangle, N. Saskatchewan: Canadian Journal of Earth Sciences, v. 31, p. 1287–1300.

Hanmer, S., Williams, M.L., and Kopf, C., 1995, Striding-Athabasca mylonite zone: Implications for Archean and Early Proterozoic tectonics of the western Canadian Shield: Canadian Journal of Earth Sciences, v. 32, p. 178–196.

Hanmer, S., Tella, S., Ryan, J.J., Sandeman, H., and Berman, R., 2006, Late Neoarchean thick-skinned thrusting and Paleoproterozoic reworking in the MacQuoid supracrustal belt and Cross Bay plutonic complex, western Churchill Province, Nunavut, Canada: Precambrian Research, v. 144, p. 126–139, doi:10.1016/j.precamres.2005.10.005.

Harley, S.L., 1989, The origins of granulites: A metamorphic perspective: Geological Magazine, v. 126, p. 215–247, doi:10.1017/S0016756800022330.

Hoernes, S., Lichenstein, U., van Reenen, D.D., and Mokgatiha, K., 1995, Whole-rock/mineral O-isotope fractionations as a tool to model fluid-rock interaction in deep seated shear zones of the southern marginal zone of the Limpopo Belt, South Africa: South African Journal of Geology, v. 98, p. 488–497.

Hoffman, P.F., 1988, United plates of America, the birth of a craton: Early Proterozoic assembly and growth of Laurentia: Annual Review of Earth and Planetary Sciences, v. 16, p. 543–603, doi:10.1146/annurev.ea.16.050188.002551.

Hollister, L.S., and Crawford, M.L., 1986, Melt-enhanced deformation—A major tectonic process: Geology, v. 14, p. 558–561, doi:10.1130/0091-7613(1986)14<558:MDAMTP>2.0.CO;2.

Holzer, L., Frei, R., Barton, J.M., Jr., and Kramers, J.D., 1998, Unraveling the record of successive high grade events in the Central zone of the Limpopo belt using Pb single phase dating of metamorphic minerals: Precambrian Research, v. 87, p. 87–115, doi:10.1016/S0301-9268(97)00058-2.

Houseman, G., and Molnar, P., 2001, Mechanisms of lithospheric rejuvenation associated with continental orogeny, in Miller, J.A., Holdsworth, R.E., Buick, I.S., and Hand, M., eds., Continental Reactivation and Reworking: Geological Society [London] Special Publication 184, p. 13–38.

Kamber, B.S., Blenkinsop, T.G., Villa, I.M., and Dahl, P.S., 1995, Proterozoic transpressive deformation in the Northern Marginal Zone, Limpopo Belt, Zimbabwe: Journal of Geology, v. 103, p. 493–508, doi:10.1086/629772.

Kamber, B.S., Biino, G.G., Wijbrans, J.R., Davies, G.R., and Villa, I.M., 1996, Archaean granulites of the Limpopo Belt, Zimbabwe: One slow exhumation or two rapid events?: Tectonics, v. 15, p. 1414–1430, doi:10.1029/96TC00850.

Klepeis, K.A., Daczko, N.R., and Clarke, G.L., 1999, Kinematic vorticity and tectonic significance of superposed mylonites in a major lower crustal shear zone, northern Fiordland, New Zealand: Journal of Structural Geology, v. 21, p. 1385–1405, doi:10.1016/S0191-8141(99)00091-7.

Klepeis, K.A., Clarke, G.L., Gehrels, G., and Vervoort, J., 2004, Processes controlling vertical coupling and decoupling between the upper and lower crust of orogens: Results from Fiordland, New Zealand: Journal of Structural Geology, v. 26, p. 765–791, doi:10.1016/j.jsg.2003.08.012.

Klepeis, K., King, D., De Paoli, M., Clarke, G.L., and Gehrels, G., 2007, Interaction of strong lower and weak middle crust during lithospheric extension in western New Zealand: Tectonics, v. 26, TC4017, doi:10.1029/2006TC002003.

Kopf, C., 1999, Deformation, metamorphism, and magmatism in the East Athabasca mylonite triangle, northern Saskatchewan: Implications for the Archean and Early Proterozoic crustal structure of the Canadian Shield [Ph.D. thesis]: Amherst, University of Massachusetts–Amherst, 139 p.

Krabbendam, M., 2001, When the Wilson Cycle breaks down: How orogens can produce strong lithosphere and inhibit their future reworking, in Miller, J.A., Holdsworth, R.E., Buick, I.S., and Hand, M., eds., Continental Reactivation and Reworking: Geological Society [London] Special Publication 184, p. 57–75.

Kramers, J.D., McCourt, S., and van Reenen, D.D., 2006, The Limpopo Belt, in Johnson, M.R., Anhaeusser, C.R., and Thomas, R.J., eds., The Geology of South Africa: Pretoria, South African Geological Society and Council for Geoscience, p. 209–236.

Kreissig, K., Thomas, F.N., Kramers, J.D., van Reenen, D.D., and Smit, C.A., 2000, An isotopic and geochemical study of the northern Kaapvaal Craton and the Southern Marginal Zone of the Limpopo Belt: Are they juxtaposed terranes?: Lithos, v. 50, p. 1–25, doi:10.1016/S0024-4937(99)00037-7.

Kreissig, K., Holzer, L., Frei, R., Kramers, J.D., Kroner, A., Smit, C.A., and van Reenen, D.D., 2001, Geochronology of the Hout River shear zone and the metamorphism in the Southern Marginal zone of the Limpopo Belt, southern Africa: South African Journal of Geology, v. 101, p. 201–213.

Krikorian, L., 2002, Geology of the Wholdaia Lake segment of the Snowbird Tectonic Zone, Northwest Territories (Nunavut); a view of the deep crust during assembly and stabilization of the Laurentian craton [M.S. thesis]: Amherst, University of Massachusetts–Amherst, 90 p.

Kruse, S., McNutt, M., Phipps-Morgan, J., and Royden, L., 1991, Lithospheric extension near Lake Mead, Nevada: A model for ductile flow in the lower crust: Journal of Geophysical Research, v. 96, p. 4435–4456, doi:10.1029/90JB02621.

Mahan, K.H., and Williams, M.L., 2005, Reconstruction of a large deep crustal exposure: Implications for the Snowbird Tectonic Zone and early growth of Laurentia: Geology, v. 33, p. 385–388, doi:10.1130/G21273.1.

Mahan, K.H., Williams, M.L., and Baldwin, J.A., 2003, Contractional uplift of deep crustal rocks along the Legs Lake shear zone, western Churchill province, Canadian Shield: Canadian Journal of Earth Sciences, v. 40, p. 1085–1110, doi:10.1139/e03-039.

Mahan, K., Williams, M., Flowers, R., Jercinovic, M., Baldwin, J., and Bowring, S., 2006a, Geochronological constraints on the Legs Lake shear zone with implications for regional exhumation of lower continental crust, western Churchill Province, Canadian Shield: Contributions to Mineralogy and Petrology, v. 152, p. 1–20, doi:10.1007/s00410-006-0106-3.

Mahan, K.H., Goncalves, P., Williams, M.L., and Jercinovic, M.J., 2006b, Dating metamorphic reactions and fluid flow: Application to exhumation of high-P granulites in a crustal-scale shear zone, western Canadian Shield: Journal of Metamorphic Geology, v. 24, p. 193–217, doi:10.1111/j.1525-1314.2006.00633.x.

Mahan, K.H., Goncalves, P., Flowers, R.M., Williams, M.L., and Hoffman-Setka, D., 2008, The role of heterogeneous strain in the development and preservation of a polymetamorphic record in high-P granulites, western Canadian Shield: Journal of Metamorphic Geology, doi:10.1111/j.1525-1314.2008.00783.x.

Martel, E., van Breemen, O., Berman, R.G., and Perhsson, S., 2008, Geochronology and tectonometamorphic history of the Snowbird Lake area, Northwest Territories, Canada: New insights into the architecture and significance of the Snowbird tectonic zone: Precambrian Research, v. 161, p. 201–230, doi:10.1016/j.precamres.2007.07.007.

McCourt, S., and Vearncombe, J.R., 1987, Shear zones bounding the central zone of the Limpopo mobile belt, southern Africa: Journal of Structural Geology, v. 9, p. 127–137, doi:10.1016/0191-8141(87)90021-6.

McCourt, S., and Vearncombe, J.R., 1992, Shear zones of the Limpopo Belt and adjacent granitoid-greenstone terranes: Implications for late Archean collisional tectonics in southern Africa: Precambrian Research, v. 55, p. 553–570, doi:10.1016/0301-9268(92)90045-P.

Millonig, L., Zeh, A., Gerdes, A., and Klemd, R., 2008, Neoarchean high-grade metamorphism in the Central Zone of the Limpopo Belt (South Africa): Combined petrological and geochronological evidence from the Bulai pluton: Lithos, v. 103, p. 333–351, doi:10.1016/j.lithos.2007.10.001.

Moecher, D.P., and Wintsch, R.P., 1994, Deformation induced reconstitution and local resetting of mineral equilibria in polymetamorphic gneisses: Tectonic and metamorphic implications: Journal of Metamorphic Geology, v. 12, p. 523–538, doi:10.1111/j.1525-1314.1994.tb00040.x.

Mouri, H., Whitehouse, M.J., Brandl, G., and Rajesh, H.M., 2009, A magmatic age and four successive metamorphic events recorded in zircons from a single meta-anorthosite sample in the Central Zone of the Limpopo Belt, South Africa: Journal of the Geological Society [London], v. 166, p. 827–830, doi:10.1144/0016-76492008-148.

Perchuk, L.L., Gerya, T.V., van Reenen, D.D., Krotov, A.V., Safonov, O.G., Smit, C.A., and Shur, M.Y., 2000, Comparative petrology and metamorphic evolution of the Limpopo (South Africa) and Lapland (Fennoscandia) high-grade terrains: Mineralogy and Petrology, v. 69, p. 69–107, doi:10.1007/s007100050019.

Perchuk, L.L., van Reenen, D.D., Varlamov, D.A., van Kal, S.M., Tabatabaeimanesh, and Boshoff, R., 2008, P-T record of two high-grade metamorphic events in the Central Zone of the Limpopo Complex, South Africa: Lithos, v. 103, p. 70–105, doi:10.1016/j.lithos.2007.09.011.

Percival, J.A., and West, G.F., 1994, The Kapuskasing uplift: A geological and geophysical synthesis: Canadian Journal of Earth Sciences, v. 31, p. 1256–1286, doi:10.1139/e94-110.

Raimbourg, H., Jolivet, L., Labrousse, L., Leroy, Y., and Avigad, D., 2005, Kinematics of syneclogite deformation in the Bergen Arcs, Norway: Implications for exhumation mechanisms, in Gapais, D., Brun, J.P., and Cobbold, P.R., eds., Deformation Mechanisms, Rheology and Tectonics: From Minerals to the Lithosphere: Geological Society [London] Special Publication 243, p. 175–192.

Ridley, J., 1992, On the origin and tectonic significance of the charnockite suite of the Archean Limpopo belt, Northern Marginal Zone, Zimbabwe: Precambrian Research, v. 55, p. 407–427, doi:10.1016/0301-9268(92)90037-O.

Robinson, P.R., Tucker, R.D., Bradley, D., Berry, H.N., IV, and Osberg, P.H., 1998, Paleozoic orogens in New England, USA: Geological Society of Sweden, v. 120, p. 119–148.

Roering, C., van Reenen, D.D., Smit, C.A., Barton, J.M., de Beer, J.H., de Wit, M.J., Stettler, E.H., van Schalkwyk, J.F., Stevens, G., and Pretorius, S., 1992, Tectonic model for the evolution of the Limpopo Belt: Precambrian Research, v. 55, p. 539–552.

Royden, L.H., Burchfiel, B.C., King, R.W., Wang, E., Chen, Z.L., Shen, F., and Liu, Y.P., 1997, Surface deformation and lower crustal flow in eastern Tibet: Science, v. 276, p. 788–790, doi:10.1126/science.276.5313.788.

Sanborn-Barrie, M., Carr, S.D., and Theriault, R., 2001, Geochronological constraints on metamorphism, magmatism and exhumation of deep-crustal rocks of the Kramanituar Complex, with implications for the Paleoproterozoic evolution of the Archean western Churchill province, Canada: Contributions to Mineralogy and Petrology, v. 141, p. 592–612.

Schaller, M., Steiner, O., Studer, I., Holzer, L., Herwegh, M., and Kramers, J.D., 1999, Exhumation of Limpopo Central Zone granulites and dextral continent-scale transcurrent movement at 2.0 Ga along the Palala shear zone, Northern Province, South Africa: Precambrian Research, v. 96, p. 263–288, doi:10.1016/S0301-9268(99)00015-7.

Sengör, A.M.C., and Natal'in, B.A., 1996, Turkic-type orogeny and its role in the making of the continental crust: Annual Review of Earth and Planetary Sciences, v. 24, p. 263–337, doi:10.1146/annurev.earth.24.1.263.

Slimmon, W.L., 1989, Compilation Bedrock Geology Map Series, Fond-du-Lac NTS Area 74O: Saskatchewan Energy and Mines, scale 1:250,000, 1 sheet.

Smit, C.A., and van Reenen, D.D., 1997, Deep crustal shear zones, high grade tectonites, and associated alteration in the Limpopo belt, South Africa: Implications for deep crustal processes: Journal of Geology, v. 105, p. 37–57, doi:10.1086/606146.

Smit, C.A., Roering, C., and van Reenen, D.D., 1992, The structural framework of the southern margin of the Limpopo Belt, South Africa: Precambrian Research, v. 55, p. 51–67, doi:10.1016/0301-9268(92)90014-F.

Smit, C.A., van Reenen, D.D., Gerya, T.V., and Perchuk, L.L., 2001, P-T conditions of decompression of the Limpopo high-grade terrain: Record from shear zones: Journal of Metamorphic Geology, v. 19, p. 249–268, doi:10.1046/j.0263-4929.2000.00310.x.

Smit, C.A., van Reenen, D.D., Roering, C., Boshoff, R., and Perchuk, L.L., 2011, this volume, Neoarchean to Paleoproterozoic evolution of the polymetamorphic Central Zone of the Limpopo Complex, in van Reenen, D.D., Kramers, J.D., McCourt, S., and Perchuk, L.L., eds., Origin and Evolution of Precambrian High-Grade Gneiss Terranes, with Special Emphasis on the Limpopo Complex of Southern Africa: Geological Society of America Memoir 207, doi:10.1130/2011.1207(12).

Snoeyenbos, D.R., Williams, M.L., and Hanmer, S., 1995, An Archean eclogite facies terrane in the western Canadian Shield: European Journal of Mineralogy, v. 7, p. 1251–1272.

Söhnge, P.G., 1946, The Geology of the Messina Copper Mines and Surrounding Country: Memoirs of the Geological Survey of South Africa, v. 40, 280 p.

Stern, R.A., and Berman, R.G., 2000, Monazite U-Pb and Th-Pb geochronology by ion microprobe, with an application to in situ dating of an Archean metasedimentary rock: Chemical Geology, v. 172, p. 113–130, doi:10.1016/S0009-2541(00)00239-4.

Stunitz, H., and Tullis, J., 2001, Weakening and strain localization produced by syn-deformational reaction of plagioclase: International Journal of Earth Sciences, v. 90, p. 136–148, doi:10.1007/s005310000148.

Treloar, P.J., Coward, M.P., and Harris, N.B.W., 1992, Himalayan-Tibetan analogies for the evolution of the Zimbabwe craton and Limpopo belt: Precambrian Research, v. 55, p. 571–587, doi:10.1016/0301-9268(92)90046-Q.

van Reenen, D.D., 1986, Hydration of cordierite and hypersthene and a description of the retrograde orthoamphibole isograd in the Limpopo belt, South Africa: American Mineralogist, v. 71, p. 900–950.

van Reenen, D.D., and Boshoff, R., 2008, Limpopo International Field Workshop 2008, Field Trip Guidebook: University of Johannesburg, Department of Geology, 130 p.

van Reenen, D.D., and Hollister, L.S., 1988, Fluid inclusions in hydrated granulite-facies rocks, southern marginal zone of the Limpopo Belt, South Africa: Geochimica et Cosmochimica Acta, v. 52, p. 1057–1064, doi:10.1016/0016-7037(88)90260-8.

van Reenen, D.D., Barton, J.M., Jr., Roering, C., Smit, C.A., and van Schalkwyk, J.F., 1987, Deep crustal response to continental collision: The Limpopo belt of South Africa: Geology, v. 15, p. 11–14, doi:10.1130/0091-7613(1987)15<11:DCRTCC>2.0.CO;2.

van Reenen, D.D., Roering, C., Brandl, G., Smit, C.A., and Barton, J.M., Jr., 1990, The granulite facies rocks of the Limpopo belt, southern Africa, in Vielzeuf, D., and Vidal, Ph., eds., Granulites and Crustal Evolution: NATO ASI, Ser. C: Dordrecht, Kluwer, v. 311, p. 257–289.

van Reenen, D.D., Roering, C., Ashwal, L.D., and de Wit, M.J., 1992, The Archean Limpopo granulite belt: Tectonics and deep crustal processes: Precambrian Research, v. 55, p. 587.

van Reenen, D.D., McCourt, S., and Smit, C.A., 1995, Are the Southern and Northern Marginal Zones of the Limpopo Belt related to a single continental collisional event?: South African Journal of Geology, v. 98, p. 489–504.

van Reenen, D.D., Boshoff, R., Smit, C.A., Perchuk, L.L., Kramers, J.D., McCourt, S.M., and Armstrong, R.A., 2008, Geochronological problems in the Limpopo Complex, South Africa: Gondwana Research, v. 14, p. 644–662, doi:10.1016/j.gr.2008.01.013.

van Reenen, D.D., Smit, C.A., Perchuk, L.L., Roering, C., and Boshoff, R., 2011, this volume, Thrust exhumation of the Neoarchean ultrahigh-temperature Southern Marginal Zone, Limpopo Complex: Convergence of decompression-cooling paths in the hanging wall and prograde P-T paths in the footwall, in van Reenen, D.D., Kramers, J.D., McCourt, S., and Perchuk, L.L., eds., Origin and Evolution of Precambrian High-Grade Gneiss Terranes, with Special Emphasis on the Limpopo Complex of Southern Africa: Geological Society of America Memoir 207, doi:10.1130/2011.1207(11).

Vernon, R.H., 2004, A Practical Guide to Rock Microstructure: New York, Cambridge University Press, 594 p.

Wernicke, B.P., 1990, The fluid crustal layer and its implications for continental dynamics, in Salisbury, M.H., and Fountain, D.M., eds., Exposed Cross-Sections of the Continental Crust: London, Kluwer Academic Publishers, p. 509–544.

Wernicke, B., Davis, J.L., Niemi, N.A., Luffi, P., and Bisnath, S., 2008, Active megadetachment beneath the western United States: Journal of Geophysical Research, v. 113, B11409, doi:10.1029/2007JB005375.

White, S.H., and Knipe, R.J., 1978, Transformation- and reaction-enhanced ductility in rocks: Journal of the Geological Society [London], v. 135, p. 513–516, doi:10.1144/gsjgs.135.5.0513.

Whitney, D.L., and Evans, B.W., 2010, Abbreviations for names of rock-forming minerals: American Mineralogist, v. 95, p. 185–187.

Williams, M.L., and Hanmer, S., 2005, Structural and metamorphic processes in the lower crust: Evidence from the East Athabasca mylonite triangle, Canada, a deep-crustal isobarically cooled terrane, in Brown, M., and Rushmer, T., eds., Evolution and Differentiation of the Continental Crust: Cambridge, UK, Cambridge University Press, p. 232–268.

Williams, M.L., and Jercinovic, M.J., 2002, Microprobe monazite geochronology: Putting absolute time into microstructural analysis: Journal of Structural Geology, v. 24, p. 1013–1028, doi:10.1016/S0191-8141(01)00088-8.

Williams, M.L., Hanmer, S., Kopf, C., and Darrach, M., 1995, Syntectonic generation and segregation of tonalitic melts from amphibolite dikes in the lower crust, Striding-Athabasca mylonite zone, northern Saskatchewan: Journal of Geophysical Research, v. 100, p. 15,717–15,734, doi:10.1029/95JB00760.

Williams, M.L., Melis, E.A., Kopf, C., and Hanmer, S., 2000, Microstructural tectonometamorphic processes and the development of gneissic layering: A mechanism for metamorphic segregation: Journal of Metamorphic Geology, v. 18, p. 41–57, doi:10.1046/j.1525-1314.2000.00235.x.

Williams, M.L., Karlstrom, K.E., Dumond, G., and Mahan, K.H., 2009, Perspectives on the architecture of continental crust from integrated field studies of exposed isobaric sections, in Miller, R.B., and Snoke, A.W., eds., Crustal Cross Sections from the Western North American Cordillera and Elsewhere: Implications for Tectonic and Petrologic Processes: Geological Society of America Special Paper 465, p. 219–241, doi:10.1130/2009.2456(08).

Zeh, A., and Klemd, R., 2008, Comments on "P-T record of two high-grade metamorphic events in the Central Zone of the Limpopo Complex, South Africa" by L.L. Perchuk, D.D. van Reenen, D.A. Varlamov, S.M. van Kal, Tabatabaeimanesh, and R. Boshoff: Lithos, v. 106, p. 399–402.

Zeh, A., Klemd, R., Buhlmann, S., and Barton, J.M., Jr., 2004, Pro- and retrograde P-T evolution of granulites of the Beit Bridge Complex (Limpopo Belt, South Africa): Constraints from quantitative phase diagrams and geotectonic implications: Journal of Metamorphic Geology, v. 22, p. 79–95, doi:10.1111/j.1525-1314.2004.00501.x.

Zeh, A., Klemd, R., and Barton, J.M., Jr., 2005, Petrological evolution in the roof of the high-grade metamorphic Central Zone of the Limpopo Belt, South Africa: Geological Magazine, v. 142, p. 229–240, doi:10.1017/S001675680500052X.

MANUSCRIPT ACCEPTED BY THE SOCIETY 24 MAY 2010

Formation and evolution of Precambrian granulite terranes: A gravitational redistribution model

Leonid L. Perchuk*
Department of Petrology, Geological Faculty, Moscow State University, Vorobievy Gory, Moscow, 119899, Russia;
Institute of Experimental Mineralogy, Russian Academy of Sciences, Chernogolovka, Moscow District, 142432, Russia; and
Department of Geology, Rand Afrikaans University, Auckland Park, South Africa

Taras V. Gerya
Department of Earth Sciences, Swiss Federal Institute of Technology (ETH-Zurich), CH-8092 Zurich, Switzerland, and
Adjunct Professor of Geological Faculty, Moscow State University, Vorobievy Gory, Moscow, 119899, Russia

ABSTRACT

This paper proposes a revision of the gravitational redistribution model suggested by Leonid Perchuk for the formation, evolution, and exhumation of Precambrian high-grade terranes (HGTs) located between granite-greenstone cratons. Such HGTs are separated from greenstone belts by crustal-scale shear zones up to 10 km wide and several hundred kilometers long. Pelite samples far (>~50 km) from the bounding shear zones show coronitic and symplectitic textures that reflect a decompression-cooling (DC) pressure-temperature (P-T) path. On the other hand, samples from within ~50 km of the bounding shear zones are characterized by textures that reflect an isobaric or near-isobaric cooling (IC) path. Local mineral equilibria in the schists from the shear zones record hairpin-shaped clockwise P-T loops. The results of a numerical test of the gravitational redistribution model show the following plausible scenario: The diapiric rise of low-density, hot granulite upward in the crust causes the relatively high-density, predominantly mafic upper crust in the adjacent greenstone belt, consisting of metabasalt and komatiite, to move downward (subducted), cooling the base of the granulites along the intervening syntectonic shear zone. This causes (1) the formation of local convection cells that control the movement of some of the ascending granulite blocks near the contact with the cratonic rocks, and (2) near-isobaric cooling (IC) of the granulite blocks in the vicinity of the boundary with the colder wall rocks. Cooling of granulite blocks farther away from the contact is not arrested, and they ascend to the Earth's surface, recording DC P-T paths. In general, the results of numerical modeling provide support for the buoyant exhumation mechanism of granulites owing to gravitational redistribution within the metastable relatively hot and soft early Precambrian crust that was subjected to high-temperature (HT) and ultrahigh-temperature (UHT) metamorphism.

---

*posthumous

Perchuk, L.L., and Gerya, T.V., 2011, Formation and evolution of Precambrian granulite terranes: A gravitational redistribution model, *in* van Reenen, D.D., Kramers, J.D., McCourt, S., and Perchuk, L.L., eds., Origin and Evolution of Precambrian High-Grade Gneiss Terranes, with Special Emphasis on the Limpopo Complex of Southern Africa: Geological Society of America Memoir 207, p. 289–310, doi:10.1130/2011.1207(15). For permission to copy, contact editing@geosociety.org. © 2011 The Geological Society of America. All rights reserved.

INTRODUCTION

Many Precambrian HGTs (high grade terranes) comprise rocks that formed in the lower crust and were subsequently displaced to the Earth's surface; hence they might be expected to have recorded evidence of the prograde, peak (separate pressure and temperature peaks may exist along a P-T path), and retrograde conditions in their mineral assemblages. However, a record of the prograde and peak stages is rarely preserved, whereas the retrograde stage related to the exhumation of the HGT is commonly well recorded by both the geological structures and mineral assemblages. Apart from rare exceptions (e.g., Perchuk et al., 1985; Zeh et al., 2004), the absence of petrological evidence for the prograde stage of high-grade metamorphism is systematic and is considered to be due to the increase in temperature and therefore in entropy, providing the energy for the structural and textural homogenization at peak temperature conditions (e.g., Perchuk, 1986, 1989; Vernon et al., 2008). In contrast, the declining temperature characteristic of the retrograde stage of metamorphism provides a more suitable thermodynamic environment for the preservation of structural and textural records and mineral compositions. Correct interpretation of such retrograde records is the only way to establish and quantitatively evaluate a geodynamic model of the exhumation history of an HGT (Perchuk and van Reenen, 2008; Perchuk, this volume; Mahan et al., this volume, and references therein).

In the last decade the issue of ultrahigh-temperature (UHT) metamorphism has occupied the pages of some geological journals (e.g., Harley, 1998; Santosh et al., 2004; Sajeev and Santosh, 2006; Kelsey, 2008; Brown and White, 2008; Santosh and Omori, 2008a, 2008b, and references) as a result of the discovery of very high temperature Neoarchean metamorphic rocks (>1000 °C) that were found in Canada (e.g., Arima and Barnett, 1984), Antarctica (e.g., Harley, 1985; Harley and Motoyoshi, 2000), South Africa (e.g., Tsunogae et al., 2004), India (Sajeev et al., 2004), and Russia (Fonarev et al., 2006). Three major models have been proposed to explain the UHT metamorphism in the Neoarchaean: (1) episodic assembly and disruption of supercontinents, (2) plume activity, and (3) UHT conditions in the Neoarchean backarc zones in the course of subduction (Brown, 2006, 2008).

Among the diverse hypotheses for the formation, evolution, and exhumation of Precambrian HGTs (see a review in Thompson, 1990) the most popular are the collision-subduction models (e.g., Ellis, 1980, 1987; England and Thompson, 1984; Perchuk et al., 1985; Sandiford and Powell, 1986; Bohlen, 1987; De Wit et al., 1992; Treloar et al., 1992; Samsonov et al., 2005; Brown, 2006; Lopez et al., 2006). With few exceptions (e.g., Perchuk, 1989, 1991; van Reenen et al., 1987, 1990; Perchuk et al., 1996, 2000b; Dirks, 1995; Percival et al., 1997; Bennett et al., 2005) these models were discussed without detailed consideration of the evolution of adjacent greenstone belts against which the HGTs were frequently juxtaposed. For the particularly well-studied Limpopo HGT, Roering et al. (1992a) proposed the so-called "pop-up" model, in which the HGT exhumation was inferred to be the deep crustal response to continental collision. Although this model was able to explain some features of the Limpopo Complex, aspects of the relationship between the HGT and the cratonic wall rocks remained open.

More generally, plate tectonic models are applied to greenstone belts on the basis of petrochemical and/or geochemical data. However, occurrences of adakite or sanukitoid bodies in greenstone belts do not provide an unequivocal basis for the conclusion of a typical subduction origin (Condie, 2005; Condie and Pease, 2008, and references therein) because the magmatic rocks may have been derived from lower crust composed of garnet-bearing metagabbroids (Hamilton, 1998). Moreover, there has so far been only one report of *true Neoarchean eclogite*, which is from the Belomorian mobile belt, along the northern portion of the Karelian Craton in Russia (Volodichev et al., 2004). In this case, the Neoarchean age was proven for coexisting zircon and garnet, based on the U-Pb method and rare earth element (REE) partitioning (Rubatto, 2002; Rubatto and Hermann, 2007).

Indeed, there are several important observations that are not in accord with collisional models and that concern a number of HGTs (including UHT examples) worldwide. (The first three examples below constitute negative evidence and point toward the absence of first-order features, which should otherwise be broadly present.)

Sediments resulting from erosion of orogenic belts produced by putative Precambrian collisions have not been documented around the HGTs. For example, no sediments from the ca. 1.9 Ga Lapland HGT (Fig. 1A) have been found in either the Karelian or the Inari craton of Kola-Fennoscandia (e.g., Kozlov et al., 1990; Perchuk et al., 2000a, and references therein). Similarly, no sedimentary material from the ca. 2.6 Ga Limpopo HGT has been found within the Precambrian successions of the adjacent Kaapvaal and Zimbabwe Cratons of southern Africa (Fig. 1B), although a large amount of sediments, ~5 × 10^6 km^3, should have accumulated within the restricted period of ca. 2.69–2.65 Ga (van Reenen and Smit, 1996; Kröner et al., 1999; Kreissig et al., 2000, 2001; Dorland et al., 2004).

In addition, granulite clasts derived from the Lapland HGT have not been observed in rare early Precambrian conglomerates of the Kola Peninsula (e.g., Glebovitskii et al., 1996). Moreover, no Archean zircons have been found so far within the Mesoproterozoic Pretoria Sedimentary Formation, whereas the ca. 2 Ga detrital zircon is very common in these rocks (Dorland et al., 2004). This suggests that, in contrast to the Paleoproterozoic continental crust, no significant mountains existed or eroded above Archean HGTs.

Published stratigraphies of Archean cratons (De Wit and Ashwal, 1997, and references therein; Perchuk, 1989, 1991; Hart et al., 1990; Gerya et al., 1997) show that they are mainly composed of orthogneisses ("gray gneisses") and greenstone belt materials. The latter are mainly granite-gneiss with subordinate banded iron formation (BIF), metapelite, metabasite, metakomatiite, and marble. In contrast, xenoliths from younger kimberlite pipes that intruded granulite-facies rocks are mainly composed of garnet-bearing metabasites and peridotites (e.g., Griffin and O'Reilly, 1987a, 1987b). This may indicate that the lower crust beneath HGTs is dominated by (ultra)mafic rocks (Specius, 1998). For example, the early Paleozoic (534 Ma) (Allsopp et al., 1995) Venetia diamondiferous kimberlites that intrude the Central Zone of the Limpopo granulite complex contain ~50% mafic and ultramafic lower crustal xenoliths (Pretorius and Barton, 1997; Barton and Gerya, 2003). Similar compositional characteristics of the lower crust are typical for xenoliths from the Paleozoic Elovy diatreme (e.g., Kempton et al., 1995) that intruded the boundary between the Belomorian Complex and the Tanaelv Belt (Fig. 1A). In addition, Perchuk et al. (1999) described inclusions of relict epidote and amphibole preserved in clinopyroxene of high-grade metabasite in the giant Tanaelv shear zone (see Fig. 1A) that are similar to the widespread assemblages of low-grade metamorphic rocks from the Karelian greenstone belt. The garnet-rich metabasites from the Tanaelv Shear Zone record a P-T of ~12 kbar, 750–800 °C (Fonarev et al., 1994). Thus, the mafic greenstone material was metamorphosed at depths of ~35 km and subsequently was exhumed from the lower crust.

The granulites are typically thrusted onto the cratonic rocks, displaying intrusive-like or harpolith geometry in cross section (a *harpolith* is a large sickle-shaped igneous intrusion injected into previously deformed rocks; Tomkeieff, 1983) (Figs. 1C, 1D). This geometry has been constrained from geophysical data, including seismic reflection studies across Limpopo and Lapland HGTs (De Beer and Stettler, 1992; Durrheim et al., 1992; Pozhilenko et al., 1997). This also suggests that marginal parts of granulite-facies terranes may have been significantly transported laterally during exhumation so that they no longer overlie the lower-middle crust from which they were presumably derived.

Crustal-scale shear zones a few kilometers wide and hundreds of kilometers long separate the HGTs from the adjacent cratonic rocks; metamorphic temperature zonation in the cratonic rocks is well documented across these shear zones (e.g., van Reenen and Smit, 1996; Perchuk and Krotov, 1998; Perchuk et al., 1996, 2000b).

The Kaapvaal and Zimbabwe Cratons, between which the Limpopo HGT is located, both exhibit similar lithologies and rock chemistries to those of the granulites, suggesting that the latter are high-grade equivalents of the adjacent greenstone belt

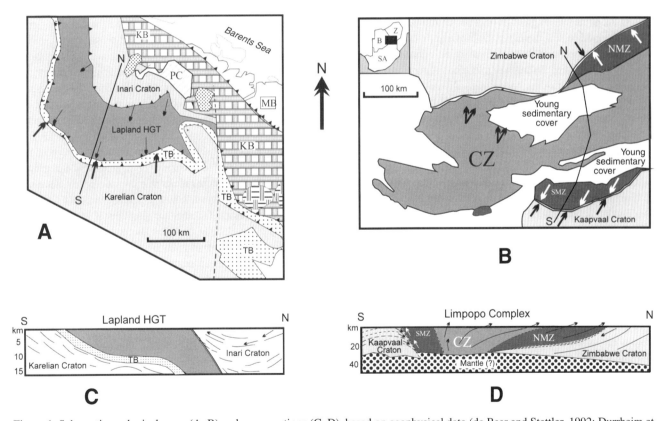

Figure 1. Schematic geological maps (A, B) and cross sections (C, D), based on geophysical data (de Beer and Stettler, 1992; Durrheim et al., 1992; Nguuri et al., 2001; Pozhilenko et al., 1997) for the Lapland and the Limpopo HGTs situated between cratons (taken from Perchuk et al., 2000a). See discussion in text. (A) Lapland HGT: PC—Pechenga Green Stone Block; KB—Kola granulite Block (ca. 2.7 Ga); TB—Tanaelv Shear Zone (1.9 Ga). (B) Limpopo HGT: NMZ—Northern Marginal Zone; CZ—Central Zone; SMZ—Southern Marginal Zone. Inset map: SA—South Africa; B—Botswana; Z—Zimbabwe; black rectangle—Limpopo HGT. Note that the crust beneath the Limpopo HGT is thinner than beneath adjacent cratons, suggesting elevation of the mantle (de Beer and Stettler, 1992). Results of seismic tomography, however, allow an interpretation as garnet-bearing metabasites and metakomatiites (Nguuri et al., 2001; Barton and Gerya, 2003).

rocks (e.g., Petrova and Levitskii, 1986; Petrova et al., 2001; Perchuk, 1989, 1991; van Reenen et al., 1990; Kreissig et al., 2000, 2001). Similar relationships were described for the Lapland HGT, situated between the Karelian and Inari Cratons (e.g., Perchuk et al., 1999, 2000a), and the Sharyzhalgay HGT at the SW shore of Lake Baikal (Petrova and Levitskii, 1986; Perchuk, 1989).

Metamorphic ages derived for the granulites are systematically younger than the greenstone belts of the adjacent cratons. Perchuk (1989) was the first to suggest this relationship as a geochronological-geodynamic rule. Subsequently Kozhevnikov and Svetov (2001) compared metamorphic ages for >50 HGTs and adjacent greenstone belts and did not find any exception to this rule.

These observations provided the basis for an alternative, gravitational redistribution model for Archean HGTs (Perchuk, 1989, 1991; Perchuk et al., 2001). In this model, the Earth's crust is considered as a multilayered system composed of layers of different density that is metastable in the gravity field under relatively low-temperature conditions because of its overall high effective viscosity (e.g., Ramberg, 1981). This viscosity, however, is exponentially lowered with increasing temperature, so that under high-grade conditions gravitational redistribution is triggered in the inherently unstable crustal configuration (e.g., Gerya et al., 2004). The driving forces of the proposed model are thus based on the principle of gravitational instability of density- and viscosity-layered materials within the Earth's crust. Gravitational redistribution is presumably triggered by an enhanced mantle-derived, fluid-heat flow; in contrast to the common term *heat flow* as a conductive heat transfer, the term *fluid-heat flow* defines both heat conduction and heat advection assisted by mantle-derived fluids and melts (Perchuk, 1976, 1989, 1991), which possibly can be related to mantle plume activity.

The gravitational redistribution model has, in particular, been applied to the origin of the Lake Baikal HGT (Perchuk, 1989); the Limpopo, Lapland, and Yenisey HGTs; and the adjacent greenstone belts (Bibikova et al., 1993; Perchuk et al., 1996, 1999, 2000a, 2000b, 2001; Perchuk and Krotov, 1998; Smit et al., 2001; Kreissig et al., 2000, 2001; Gerya and Maresch, 2004), and presumably it also applies to the Aldan HGT (Perchuk et al., 1985; Smelov, 1995) in eastern Siberia, and the Ouzzal HGT in the western Hoggar, Algeria (Ouzegane et al., 2003). The present chapter summarizes a large volume of natural data and compares them systematically to test the *gravitational redistribution model*, focusing on relationships between and within rocks from HGTs and adjacent cratons with particular focus in the context of the seven crucial points listed above. The most detailed testing was done using the abundant natural data from several well-studied granulite complexes between cratons, such as the Limpopo HGT complex of South Africa (Fig. 1B), the Yenisey Range HGT (Perchuk et al., 1989; Smit et al., 2000; Gerya and Maresch, 2004), the Lapland granulite complex of the Kola Peninsula and Fennoscandia (Fig. 1A), and the Aldan HGT (Perchuk et al., 1985; Smelov, 1995) in eastern Siberia. Apart from employing traditional geological, petrological, geochemical, and geophysical evidence, we have also used results of two-dimensional numerical modeling to more rigorously determine the critical geodynamic constraints (Perchuk et al., 1992; Gerya et al., 2000, 2001, 2002, 2004).

It should also be pointed out that crustal thickening and gravitational redistribution processes are not mutually exclusive and can complement each other (e.g., Gerya et al., 2004). Therefore, the main focus of this chapter is not to disprove collisional models proposed for the origin of granulites but rather to bring to attention gravitational mechanisms of high-grade rock movement and exhumation that were possibly dominant in the hot Precambrian crust (Perchuk, 1989, 1991).

MODEL TESTING BY NATURAL OBSERVATIONS

Structural and Lithological Tests

The critical question to be addressed here is: Do structural and seismic data reflect the tectonic history of an HGT as the result of gravitational redistribution? If yes, two major aspects should be addressed for testing geological aspects of the model: (A) the style of deformation within granulite complexes, and (B) the relationships of the complexes to adjacent granite-greenstone belts. These are discussed separately below.

MacGregor (1951) was the first to suggest gravitational redistribution as a major driving force in the formation of granite-greenstone belts. His proposal was based on structural studies of relationships between tonalite intrusions and their host metavolcanic sequences of felsic to ultramafic composition. He suggested that gravity-driven movement of this sequence was triggered by heating and partial melting of the lower crust, resulting in the formation of granite domes among greenstone rocks. On geological maps these domes exhibit oval to circular shapes (Figs. 2A, 2B) that were interpreted to reflect upward, almost vertical movement of granitic material into the overlying carapace of amphibolite- and greenschist-facies rocks that are inferred to have moved simultaneously downward.

Similar structural features are also typical for the HGTs that occur among greenstone belts. If the exhumation of an HGT resulted from gravitational redistribution of material, the sheath folds should be common regional structures that resulted from oppositely directed dome-, diapir-, and plume-like material movements driven by gravity. The sheath folds should not necessarily be vertically oriented, since horizontal movements are dominant at culminate stages in large-scale, laterally spreading, buoyancy-driven flows (e.g., Ramberg, 1981). Complementary (i.e., oppositely directed) structures are represented by narrow, intensely deformed gneissic rims that commonly surround the sheath folds (Figs. 2C, 2D). The Limpopo HGT is perhaps the best example, providing a continuous cross section from the Kaapvaal Craton in the south to the Zimbabwe Craton in the north. This cross section allows direct observation of the change in deformational style from that typical of greenstone terranes to a macro-mélange of highly attenuated granulite-facies rocks that compose the Limpopo HGT. In general, the tectonic style in granite-greenstone belts is characterized by granite domes surrounded by strongly deformed amphibolite- to greenschist-facies rocks. In contrast, the deformational style of the HGT is characterized by complex fold patterns that include mega–sheath folds formed at high grade, the shapes of which are similar to structural domes (Figs. 2C, 2D). For example, the high-grade Central Zone of the Limpopo Belt, despite its polymetamorphic history (Perchuk et al., 2006), clearly preserves the oval to cylindrical style of folding of heterogeneous multilayered crust at high-temperature granulite-facies conditions. Increasing temperature is inferred to have led to a decrease in viscosities of the hot rocks of different densities and, as the result, to the dominance of buoyancy forces. Therefore we should expect to observe this type of structure in many Archean HGTs. According to this interpretation, the cylindrical type of the sheath folds must predominate in the structural pattern. This was recently shown to be the case for the Central Zone of the Limpopo HGT (Perchuk et al., 2008). Another discussion of the structural aspects related to the gravitational redistribution model is given by Smit et al. (this volume).

In many areas around the world, the Precambrian granulite complexes have tectonic contacts with lower grade wall rocks (Kozhevnikov and Svetov, 2001). Detailed geological mapping in the contact zones between HGTs and greenstone belts (including orientations of the principal structural elements such as linear and planar fabrics, as well as asymmetric structures that permitted kinematic analysis) of contact zones between HGT and GSB (granite-greenstone belts), resulted in concluding that the zones commonly have a typical thrust origin (Kozlov et al., 1990; Roering et al., 1992b; Marker, 1991; Pozhilenko et al., 1997; van Reenen and Smit, 1996; Perchuk and Krotov, 1998; Perchuk et al., 1996, 1999, 2000b; Kreissig et al., 2000, 2001; Smit et al., 2000). Systematic structural analyses of these boundaries have resulted in the conclusion that the granulite complexes overrode the granite-greenstone regions along the steep to moderately dipping crustal-scale shear zones that separated the greenstone belt from the HGT. The maps and sections in Figure 1 illustrate the Tanaelv Shear Zone separating the Lapland HGT from the

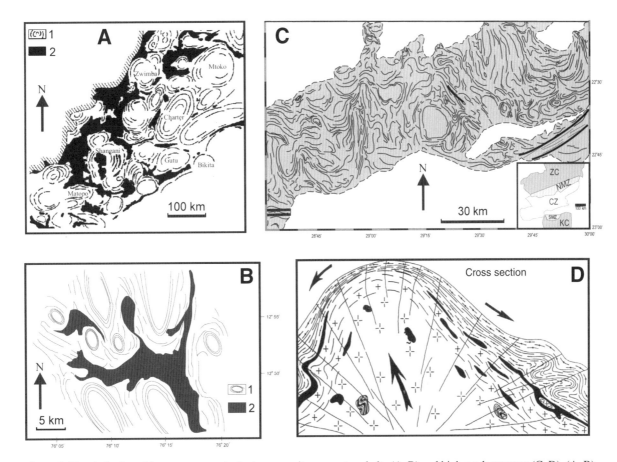

Figure 2. Morphologies of dome structures in Archean granite-greenstone belts (A, B) and high-grade terranes (C, D). (A, B) Granite domes (1—white) within greenstone belts (2—black) of the Zimbabwe Craton, southern Africa (A), and the Western Dharwar Craton, India (B) (MacGregor, 1951). (C) Sheath folds in the Central Zone of the Limpopo HGT, southern Africa (van Reenen et al., 2004). (D) Vertical cross section of a small granite cupola within the Sharyzhalgay granulite complex, SW Lake Baikal (Perchuk et al., 1992). First-order similarities between structural patterns of greenstone belts (A, B) and high-grade terranes (C, D) suggest that gravity-driven tectonics originally proposed for greenstone belts (A, B; e.g., MacGregor, 1951; Ramberg, 1981) may also be relevant for granulite-facies terranes (C, D; Perchuk, 1989, 1991).

Karelian greenstone belt (Fig. 1A), and the Hout River Shear Zone separating the Limpopo HGT from the Kaapvaal granite-greenstone terrane (Fig. 1B). In all cases, the contacting rocks on either side of the shear zone underwent synchronous but oppositely directed movement. Moreover the displacement was synchronous with the formation of sheath folds in gneisses and migmatites in the mid-crust.

Thus, we conclude that structural data from several HGTs and adjacent greenstone belts strongly support synchronous, but oppositely directed, movement of granulites (upward) on the one hand, and host greenstones (downward) on the other.

Geochronological Test

For years, high-grade rocks were considered as the basement of the continental crust and to be older and to have been metamorphosed before the overlying greenstone belts (e.g., De Wit et al., 1992). The gravitational redistribution model, however, suggests the opposite relationships—i.e., that the low-grade metamorphism in the greenstone belt should be older than the high-grade metamorphism in the HGT. The critical question is, therefore: Do the geochronological data document a coherent geochronological history of major geological events that reflect gravitational emplacement of the HGT? A large volume of geochronological data exists for many HGTs and hosting GSBs. Kozhevnikov and Svetov (2001) summarized metamorphic age data for 59 HGT + GSB pairs and clearly showed that in *all* of the pairs the granulites are younger than the contacting greenstones. Table 1 exemplifies this rule with data on the Limpopo and Lapland HGTs. In contrast, the rocks from shear zones separating the GSB from the HGT have similar or identical metamorphic ages with the HGT. For example, the Lapland HGT (1.9 Ga) is much younger than both the Karelian and the Inari adjacent GSBs (more than 2.7 Ga) and shows the same age as mica schists from the Tanaelv Shear Zone (Table 1). The Limpopo HGT (2.69 Ga) is younger than the GSB from both the adjacent Zimbabwe and Kaapvaal Cratons (more than 2.75 Ga), but little difference in age exists for the formation of the Hout River Shear Zone (2.67 Ga), which separates the South Marginal Zone (SMZ) of the Limpopo HGT from the Kaapvaal Craton (Table 1). Similar isotopic age data show rocks from the Northern Marginal Zone (NMZ) (2.62–2.71 Ga) and the adjacent Zimbabwe Craton (more than 2.7 Ga) (Table 1). Thus, geochronological data support an idea that HGTs have been emplaced within GSBs along shear zones much later than the formation of the cratons.

Textural Test for Exhumation

Hundreds of retrograde reaction textures and corresponding P-T paths have been published for the Precambrian HGTs, but there are only a few examples of prograde textures documented in granulites. These include data from the Aldan Shield (Perchuk et al., 1985), Palni Hill Ranges of southern India (Raith et al., 1997), and the Central Zone of the Limpopo Complex (Zeh et al., 2004; Perchuk et al., 2008). Because the exponential increase in diffusion rates above 700–750 °C leads to mineralogical homogenization and microstructural recrystallization, the majority of reaction textures preserved in granulites formed after the peak metamorphic T (e.g., Perchuk et al., 1985; Perchuk, 1989; Harley, 1989).

Metapelites in many HGTs commonly show evidence (Fig. 3) for two mineral reactions (e.g., Harley, 1989):

$$Grt + Qtz \rightarrow Opx + Crd, \quad (1)$$

i.e., $2(Mg, Fe)_3Al_2Si_3O_{12} + 3SiO_2 =$
$2(Mg, Fe)_2Si_2O_6 + (Mg, Fe)_2Al_4Si_5O_{12},$

and

$$Grt + Qtz + Sil \rightarrow Crd, \quad (2)$$

i.e., $2(Mg, Fe)_3Al_2Si_3O_{12} + 5SiO_2 +$
$4Al_2SiO_5 = 3(Mg, Fe)_2Al_4Si_5O_{12}.$

Depending on change of P-T parameters, these reactions can be displaced to either side, producing well-developed and distinctive textures (e.g., Harley, 1989; Perchuk et al., 1985,

TABLE 1. U-Pb METAMORPHIC AGES OF ROCKS FROM SHEAR ZONES SEPARATING HIGH-GRADE TERRANES FROM GREENSTONE BELTS

Rock	Zone/Group	Age (Ga)	References
Lapland area (Russia, Finland)			
Ky-bearing mica schist after GSB	Shear zone, western part of the Tanaelv Belt (Finland)	1.9	Bernard-Griffiths et al. (1984)
Ky-bearing mica schist after GSB	Korva Tundra Group (shear zone), the Tanaelv Belt (Russia)	1.91	Volodichev (1990)
Grt amphibolite (sample Lap34)	Kandalaksha Group (shear zone), the Tanaelv Belt (Russia)	1.911	Perchuk et al. (2006)
Opx-Bt plagiogneiss (Lap-9)	Pados, southern part of the Lapland HGT (Russia)	1.91	
Granulitic gneiss	Tupaya Guba, southern part of the Lapland HGT	1.916	Bibikova et al. (1993)
Gneisses-tonalites	Karelian Craton (Belomorides)	2.9–2.7	Kozlov et al. (1990)
Limpopo area (southern Africa)			
Ky-bearing mica schist after KVC	Hout River Shear Zone, at the contact with granulite body	2.689	Kreissig et al. (2000, 2001)
Granulitic metapelite	Southern Marginal Zone (SMZ) of the Limpopo granulite belt	2.691	
Charno-enderbite	Matok Pluton (SMZ)	2.671	Barton et al. (1992)
Metavolcanics	Kaapvaal Craton	2.75–3.54	de Wit et al. (1992)

Note: HGT—high-grade terrane; GSB—greenstone belt; KVC—Kaapvaal Craton; Ky—kyanite; Grt—garnet; Opx—orthopyroxene; Bt—biotite.

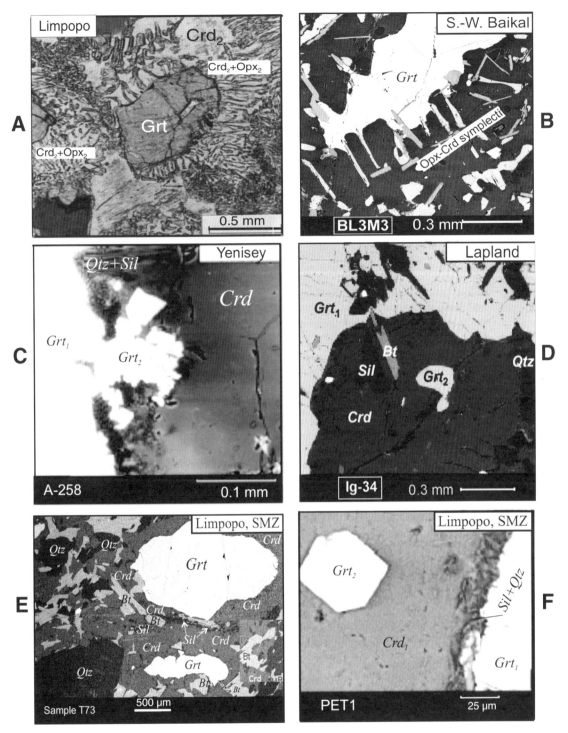

Figure 3. Typical reaction textures $Grt + Qtz \rightarrow Opx + Crd$ (reaction 1), $Grt + Sil + Qtz \rightarrow Crd$ (reaction 2), and $Crd \rightarrow Grt + Sil + Qtz$ (reaction 3) occurring in high-grade metapelites are recorded in Precambrian HGTs worldwide. Grt_1 (Grt) and Crd_1 (Crd) are stable at the metamorphic peak, and Grt_2 formed from the breakdown of Crd_1 during isobaric cooling and compression cooling. Crd_2 recrystallized from Crd_1 during the growth of Grt_2. (A) Texture resulting from reaction 1, composed of three zones: (i) Grt core $\rightarrow$ (ii) Crd-Opx symplectitic zone $\rightarrow$ (iii) Opx rim, CZ, Limpopo HGT, southern Africa (compiled from Perchuk et al., 2006). (B) The same as (A): Sharyzhalgay HGT, SW Lake Baikal, eastern Siberia (from Perchuk, 1989). (C) Reaction texture 3 in metapelite from the Atamanovskaia Group, the Kanskiy HGT, Yenisey River, eastern Siberia. (D) Reaction texture 3 in metapelite from northern portion of the Lapland HGT, Kola Peninsula, Russia (from Perchuk et al., 1999). (E) Texture resulting from reaction 2; i.e., rim of Crd around Grt containing $Sil + Qtz$ in sample from the 2 Ga Baklykraal regional shear structure; CZ, Limpopo HGT, southern Africa (from van Reenen et al., 2004). (F) Texture resulting from reaction 3 in metapelite from the Petronela Shear Zone, SMZ, Limpopo HGT (Smit et al., 2001); new euhedral Grt_2 and $Sil + Qtz$ intergrowth appeared within Crd matrix at the contact with Grt_1: $X_{Mg}^{Grt1} < X_{Mg}^{Grt2}$.

1989, 1996; Raith et al., 1997). Decompression cooling (DC) of an HGT appears the most common style of exhumation type in Precambrian granulite-facies terranes. Metapelites from the SMZ of the Limpopo HGT show textures that reflect *both* reactions (1) and (2). Orthopyroxene coronas and *Opx* + *Crd* symplectites are well developed between garnet and quartz (reaction 1, Figs. 3A, 3B). At a relatively high Ca content in garnet, a plagioclase zone can be formed between the *Opx* corona and the *Crd*-*Opx* symplectite zone. The commonly observed growth of cordierite at the contact of garnet with sillimanite *(Sil)* and quartz *(Qtz)* is also documented in the metapelites (reaction 2, Fig. 3E).

On the other hand, the reaction:

$$Crd \rightarrow Grt + Qtz + Sil \qquad (3)$$

produces textures that are composed of a very fine grained (<100 μm) skeletal intergrowth of garnet, quartz, and sillimanite on the rims of cordierite (Fig. 3C). This texture was first observed in metapelites of the Yenisey Range in eastern Siberia (Fig. 3C; Perchuk et al., 1989) and then in similar rocks of the northern part of the Lapland HGT (Fig. 3D; Perchuk et al., 1999, 2000a) and in the SMZ of the Limpopo Complex (Fig. 3F; Perchuk et al., 1996). An important observation is that samples with the reaction texture (3) were only collected close (<~50 km) to the boundaries of the HGTs with greenstone-belt wall rocks, whereas samples with reaction textures (1) and (2) commonly occur farther away (>~50 km) from the boundaries.

P-T History

P-T parameters were uniformly calculated for both the high-grade rocks and the rocks from shear zones separating HGTs from the cratons using methods and approaches discussed in detail elsewhere in this volume (Perchuk, this volume).

Granulite Complexes

In the rocks from the HGTs studied we found two groups of P-T paths (Fig. 4): decompression cooling (DC) and near-isobaric cooling (IC) paths that are strongly correlated with the two groups of reaction textures noted above, and which are typical for many *different* granulite facies complexes (Harley, 1989). However, both kinds of P-T path occur in a single HGT depending upon their location within the terrane (e.g., Perchuk et al., 1996, 1999, 2000a). For example, in the Limpopo HGT the samples preserving a record of the DC PT path (Fig. 4C) occur far (>~50 km) from the boundary with the craton, whereas samples located close to the boundary (<~50 km) also exhibit IC paths (Fig. 4D). This can be explained if some crustal blocks were exhumed sufficiently slowly so they could equilibrate with the adjacent crust during their emplacement (DC path), whereas the emplacement of others was arrested at crustal levels of ~13–15 km (IC path). We infer that the large temperature gradient between the hot granulite and the cooler wall rocks promoted heat flow from the Limpopo HGT toward the greenschist-facies footwall of the Kaapvaal Craton.

This resulted in isobaric cooling of the granulite (IC portion of the P-T path). Hydration reactions (involving the formation of cummingtonite, anthophyllite, gedrite, secondary biotite, water-bearing cordierite, etc.) that characterize many samples from the southern part of the SMZ (van Reenen, 1986) reflect the involvement of water-rich fluids in the exhumation process (Perchuk et al., 1996). This suggestion was later supported by numerical modeling (Gerya et al., 2000) and additional petrological observations in other HGTs (e.g., Perchuk et al., 1999, 2000a; Smit et al., 2000; Gerya and Maresch, 2004).

Rarely, granulites near contacts with cool wall rock may preserve evidence for a compression-cooling (CC) segment of the P-T path in place of an IC segment (Fig. 4B). Figures 4E and 4F, determined for mineral assemblages from the Kanskiy Complex, Yenisey Range, eastern Siberia (Gerya and Maresch, 2004), are inferred to reflect differential vertical movements within parts of the granulite body during its emplacement in the mid-crust. Both the IC and CC portions of P-T paths therefore characterize complications in the flow-cooling patterns of marginal zones of the HGT caused by their thermomechanical interactions with colder and stronger cratonic rocks.

Shear-Zone Wall Rocks

The P-T loop for the footwall rocks in the Tanaelv Shear Zone underlying the Lapland HGT (Fig. 1A) was first derived by Perchuk and Krotov (1998) on the basis of the local equilibrium *Bt* + *Ky* + *Qtz* = *Ms* + *Grt* within the studied samples. Figure 5 exemplifies all stages of the evolution of these garnet-bearing schists from the chlorite-staurolite and kyanite-staurolite zones in the shear zone. Figure 5 documents both the prograde and retrograde stages of the evolution of the shear zone rocks in the footwall in terms of *Grt* morphology, chemical zoning, and P-T path. The rotated garnets with quartz-rich inclusion trails (Fig. 5A) are associated with the prograde stage of metamorphism, whereas snowball (Fig. 5B) and inclusion-free (Fig. 5C) garnet reflects peak and subsequent retrograde metamorphism. Similar textural relationships occur in the Hout River Shear Zone, separating the Limpopo HGT from the adjacent Kaapvaal Craton Perchuk et al., 2000b). Figures 5D–5F demonstrates movement of the shearing rocks downward to the mid-crust and subsequent return toward the surface along with the simultaneous exhumation of granulite (see Table 1 for isotopic data indicating the contemporaneity of decompressional metamorphism in the HGT and prograde metamorphism in the subjacent footwall shear zone).

P-T paths in Figures 5D–5F were calculated using the *Bt-Ky-Qtz-Grt-Ms* geothermobarometer (Perchuk, 1973, 1977), which was recalibrated using new thermodynamic data (Perchuk and Krotov, 1998; Perchuk et al., 2000b). These P-T paths define very tight, hairpin-shaped loops, indicating that the cool subducting plate, composed of greenstone material, and the hot ascending granulite diapir moved along virtually the same P-T gradient. This result is, in a way, similar to the constraint imposed upon a subducting oceanic plate in which both the prograde (downward)

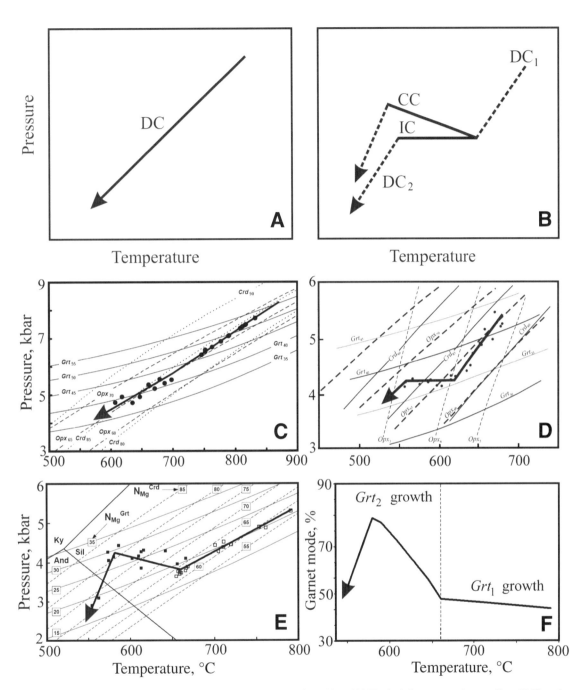

Figure 4. Typical styles of P-T paths for monometamorphic HGTs. (A) Typical decompression-cooling (DC) path. (B) Isobaric (IC) and compression-cooling (CC) portions of the DC P-T paths (see Perchuk, this volume). (C) DC P-T path for metapelite DR45 from the SMZ of the Limpopo HGT. (D) DC_1-IC-DC_2 segments of P-T path for metapelite LW7 collected from the most marginal part of the SMZ of the Limpopo Complex. Both C and D after Perchuk et al. (1996). (E) P-T path derived for granulite A-275 is from the Kanskiy HGT (Yenisey Range, eastern Siberia); open rectangles indicate P-T data for the early generation of garnet (Grt_1), and solid rectangles indicate those for the later generation (Grt_2). (F) Calculated equilibrium garnet modes [atomic ratio $100(Mg + Fe)_{in\,garnet}/(Mg + Fe)_{in\,rock}$] along the P-T path in E. Diagrams E and F after Gerya and Maresch (2004). Relative uncertainties of P-T paths (±0.5 kbar and ±25 °C) correspond to the scattering of individual P-T points in diagrams C, D, and E. See text for discussion.

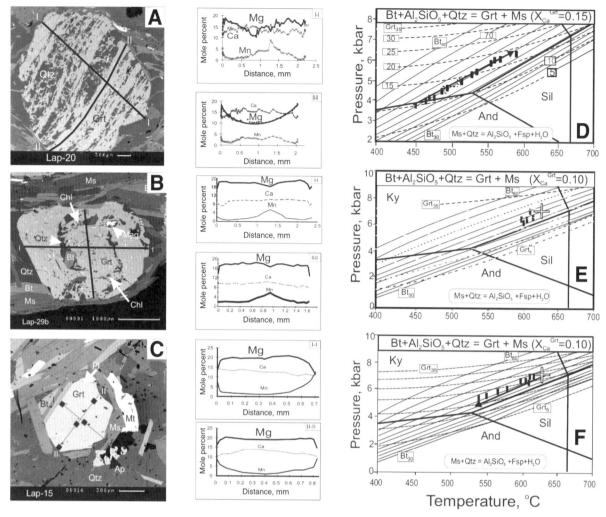

Figure 5. Textural, chemical, and P-T evolution of the mineral assemblage $Bt + Ky + Qtz + Grt + Ms$ in mica schists from the Tanaelv Shear Zone, separating the Lapland HGT from the Karelian Greenstone Belt (modified from Perchuk and Krotov, 1998). (A–C) Backscattered electron images of garnet from mica schist, illustrating morphology of the Grt porphyroblasts and their textural characteristics in the prograde (A), peak (B), and retrograde (C) stages of metamorphic evolution. *Rotated* garnet porphyroblasts from the chlorite-staurolite zone characterize the prograde stage. *Isometric* inclusion-rich snowball Grt porphyroblasts occur in the biotite-kyanite zone and indicate both partly prograde and peak stages of metamorphism. *Euhedral* inclusions-free Grt porphyroblasts from the chlorite-staurolite zone indicate peak to retrograde stages of metamorphism. Lines indicated as I–I, II–II, etc., correspond to the profiles along which detailed microprobe analyses were done (see second column in the figure). Each path in the (D–F) diagrams relates to textural, chemical, and morphological evolution of the Grt porphyroblasts shown in A–C, from prograde (A), through peak (B), to retrograde (C) P-T conditions. Crosses on the P-T diagrams indicate precision of P-T estimates (±0.5 kbar and ±25 °C). See text for discussion.

and the retrograde (upward) histories are similar but reversed (e.g., Cloos, 1982).

P-T diagrams in Figure 6 demonstrate that the peak metamorphic conditions for the footwall rocks coincide with the P-T minimum recorded by the granulite assemblages. Considering the similar ages of metamorphism in both cases (Table 1), we infer that this convergence of P-T conditions implies that the P-T loops for the footwall rocks resulted from the emplacement of the granulite body into the upper crust.

Geophysical Test

The gravitational redistribution (or diapiric) model assumes that the gravitational redistribution of material occurred within the normal-thickness continental crust, produced intrusion-shaped geometries caused by the reduced viscosity of the hot granulite bodies, and was triggered by the activity of large mantle plumes (Perchuk, 1989; Perchuk et al., 2001). This suggestion is exemplified by seismic data for both the Limpopo (e.g.,

De Beer and Stettler, 1992; Durrheim et al., 1992; Nguuri et al., 2001) and Lapland (Pozhilenko et al., 1997) granulite complexes (Fig. 1). The interpretation of the seismic profiles, coupled with structural and geochronological data, suggests that the Limpopo HGT resembles the shape of a laccolith or harpolith (Tomkeieff, 1983) emplaced between the Kaapvaal and Zimbabwe Cratons (Fig. 1D). The geometry of the Lapland granulite complex, between the Karelian and Inari greenstone belts, is a mirror image of the Limpopo HGT. On the southern and western margins the Tanaelv listric shear zone dips northward under the Lapland HGT at an angle ranging from 60° to 12°. However, the northern boundary of the Lapland HGT with the Inari greenstone belt is still an unresolved problem: According to the structural and seismic data from Finland (Marker, 1991; Mints et al., 1996; Pozhilenko et al., 1997) the northern boundary of the Lapland HGT dips steeply NE, whereas the Russian seismic data suggest that the contact is vertical to steeply S-dipping. A cross section through the Lapland HGT also suggests a harpolith shape (Fig. 1C). Thus, both the Limpopo and Lapland HGTs form bodies which resemble crustal-scale harpoliths that plausibly reflect intrusive-like mechanisms of the emplacement of the intervening HGTs.

In reality, the mantle-derived fluid-heat flow that triggers gravitational redistribution in our model may also associate with either horizontal or vertical movement of crustal material by other processes such as thrust thickening, extensional thinning, erosion, magmatic underplating, or crustal delamination. For example, if the formation of the Limpopo HGT resulted from the collision (horizontal movement) of the Kaapvaal and Zimbabwe Cratons (Treloar et al., 1992), the locations of the plate boundaries are expected to be imaged by deep-seated seismic profiles across the "craton-orogen-craton" system. However, the detailed seismic tomographies of this portion of the African continent show the mantle lithosphere to be a single unit, and there is no seismic discontinuity at the Limpopo HGT (James et al., 2001). Consequently, if no subsequent overprinting affected this lithosphere, it was acting as a single tectonic unit during the exhumation of the Limpopo HGT.

MODEL TESTING BY NUMERICAL EXPERIMENTS

Gravitational mechanism of granulite exhumation was extensively studied numerically (Perchuk et al., 1992, 1999; Gerya et al., 2000, 2004), which is reviewed below. The first numerical testing of a gravitational redistribution model for granulites was performed by Perchuk et al. (1992). Using a mechanical numerical model and a gravitationally metastable model set-up with a rhythmic multilayered crust (e.g., Perchuk 1989) comprising rocks of different density and viscosity, and assuming that the gravitational redistribution process was triggered by a mantle-derived fluid-heat flow (e.g., England and Thompson, 1984; Perchuk, 1976, 1991; Hoernes et al., 1995; Pili et al., 1997), Perchuk et al. demonstrated that the growth dynamics of multiwavelength gravity structures (polydiapirs) (Weinberg and Shmelling, 1992) is accelerated by a chain-reaction mechanism (Perchuk, 1991). The modeling showed that the formation of some granulite complexes as the result of the continuous evolution of an initial metastable multilayered craton structure triggered by fluid-heat flow is indeed feasible. Similar results can also be obtained with more realistic temperature- and stress-dependent rheologies (Fig. 7) with a gravitational redistribution process being triggered in the lower crust where effective viscosity for all rock types is the lowest because of the high temperature. The effective viscosity of the upper crust (cf. Figs. 7A, 7B) mainly regulates the potential for the penetration of large, strongly internally deformed granulite bodies into the

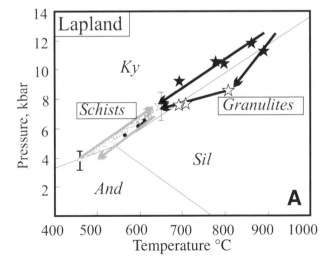
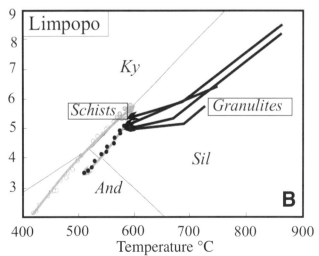

Figure 6. Integrated P-T paths, reflecting simultaneous tectonic histories of the high-grade terranes and schists from the footwall shear zones separating granulite bodies from cratonic wall rocks. (A) Lapland; (B) Limpopo; modified from Perchuk et al., 2000b. Note that the burial and exhumation branches of paths for mica schists are nearly identical. Also note that P-T scales in A and B are different.

shallower levels of the crust. Detailed petrological data for granulite complexes and adjacent greenstone belts (Perchuk et al., 1996, 1999, 2000b; Perchuk and Krotov, 1998, and summaries in this chapter) created a basis for thermomechanical numerical modeling (including simulation of metamorphic P-T paths) for the generation, exhumation, and emplacement of granulite complexes within greenstone belts (Gerya et al., 2000, 2004).

Based on the typical stratigraphy for most well-known greenstone belts (e.g., De Wit et al., 1992; De Wit and Ashwall, 1997; Perchuk et al., 2000a), a rhythmic layered succession was used (Gerya et al., 2000) in the initial design (Fig. 8A, for 0 m.y.). This figure shows a general scenario for the gravitational redistribution of rocks in the Earth's crust (see Gerya et al., 2000, 2004, for the choice of numerical values and sensitivity analysis). Both the

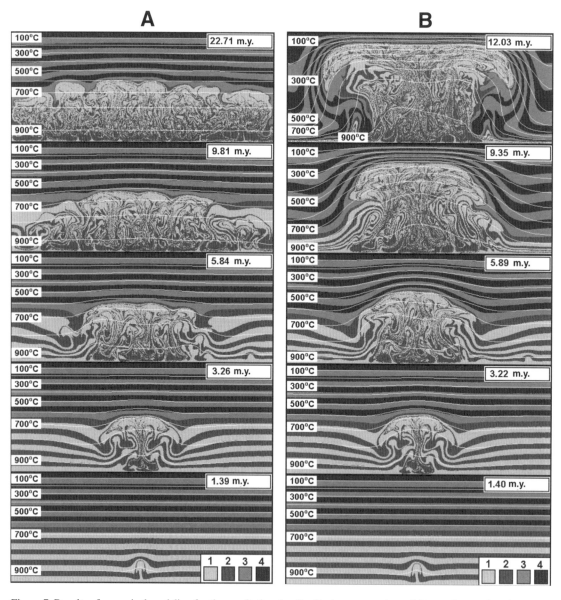

Figure 7. Results of numerical modeling for the gravitational redistribution process in multilayered crust, forming either a highly deformed granulite layer in the lower-middle crust (A) or a crustal-scale granulite diapir penetrating into the upper crust (B). Model size is 100 × 40 km. Color code corresponds to different lithologies of upper (1, 2) and lower (3, 4) crust with the following properties (see Gerya et al., 2000, 2004, for the choice of numerical values and sensitivity analysis): 1 and 3, density = 2820 kg/m^3, thermal conductivity = 2 W/m/K, radiogenic heat production = 9×10^{-10} W/kg, heat capacity = 1100 J/kg, flow law = wet granite (Ranalli, 1995); 2 and 4, density = 2870 kg/m^3, thermal conductivity = 2 W/m/K, radiogenic heat production = 3×10^{-10} W/kg, heat capacity = 1100 J/kg, flow law = diorite (Ranalli, 1995). Maximum model viscosity contrast between the upper and lower crust is 10^4 for A and 10^3 for B. Instability is initiated by small thermal perturbations (placed at the bottom of the model). No lateral forcing is applied in the models.

presence of three major initial layer sequences (rhythms) in the modeled stratigraphic succession and the difference in densities between layers of each rhythm provide acceleration of the process of gravitational redistribution (Perchuk et al., 1992). The interaction between the rhythms and individual layers allows a rapid large-scale flow over the entire sequence (Fig. 8A, 1.7–8.9 m.y.; see also Fig. 7B). The model development shows thrusting of the hot, flattened granulitic diapir onto the colder, upper crustal cratonic rocks. Figure 8B demonstrates oppositely directed vertical movement of mafic and felsic material from different crustal levels: granulites moving up to the surface, and cold metabasites and metakomatiites of the upper part of the craton moving down under hot granulites along the shear zone separating them.

The numerical modeling of gravitational redistribution allows the calculation of P-T-time paths for both the granulites and the cratonic rocks (Figs. 9, 10), which then can be compared with natural data. Numerical experiments (Gerya et al., 2000) showed that the geometry of the modeled P-T paths (Fig. 9) compares closely with petrological P-T paths (Figs. 4, 6). Figure 8B shows oppositely directed movement of mafic and felsic material from different crustal levels: While granulites move up toward the surface, metabasaltic and metakomatiitic cratonic rocks move down and cool the granulites along the shear zone separating the two terranes. This causes a local convective cell that changes the movement of some uprising granulite fragments: The square marker in Figure 8B (1.7–2.5 m.y.) moves from 700 °C to 600 °C toward the downward moving, cooler greenstone plate. In the case of the Limpopo HGT the upward movement of the two Marginal Zones along the contact with the adjacent cratons was accompanied by a narrow zone of amphibolite-facies retrogression of the uprising granulites (see Figs. 4, 6; van Reenen, 1986; van Reenen et al., this volume). This retrogression was caused

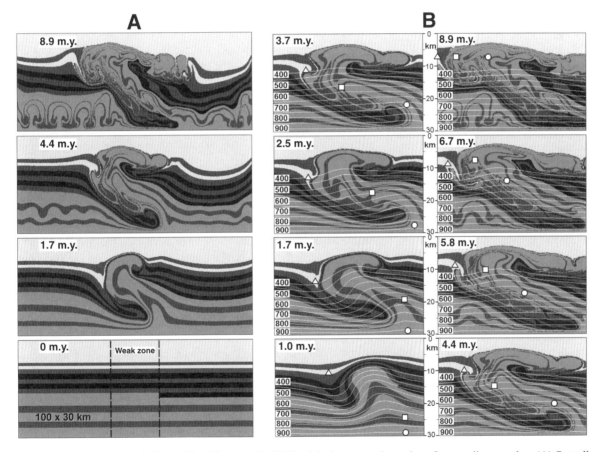

Figure 8. Results of numerical modeling (Gerya et al., 2000) of the buoyant exhumation of a granulite complex: (A) Overall model development; (B) enlargement of the granulite diapir. Model design in A: Size, 100 × 30 km; grid resolution, 100 × 30 nodes, 500 × 150 markers; weak zone mimics preexisting tectonic boundary between two cratons; instability is initiated owing to different numbers of ultramafic layers in the cratons; no lateral forcing is applied in the model. Rock types and properties (see Gerya et al., 2000, for the choice of numerical values and sensitivity analysis): sediments (white, ρ = 2700 kg/m^3, η = 10^{19} Pa s, where ρ is density and η is viscosity), felsic granulites (light gray, ρ = 2800 kg/m^3, η = 10^{19} Pa s), metabasites (dark gray, ρ = 3000 kg/m^3, η = 10^{19} Pa s at T >600 °C in granulite sequence, η = 10^{21} Pa s at T <600 °C in greenstone sequence), metakomatiites (black, ρ = 3300 kg/m^3, η = 10^{21} Pa s), weak tectonic zone (dashed, η = 10^{19} Pa s for all rock types). Heat conductivity of rocks, 4 W/m/K; isobaric heat capacity of rocks, 1100 J/kg/K. Symbols (triangle, circle, and square) in B show movement of representative rock units with P-T-time paths displayed in Figure 9.

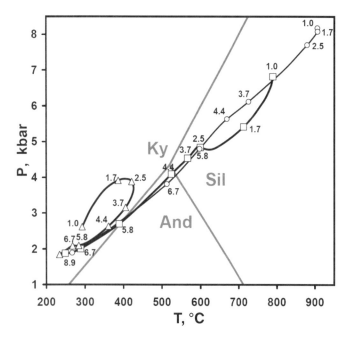

Figure 9. Results of numerical modeling of P-T-time paths for representative cratonic rock units (triangles) and two markers in the granulite fragments (circles and squares) shown in Figure 8 (compare with Figs. 4–6). See Gerya et al. (2000) and Gerya and Maresch (2004) for comparison of natural and numerical P-T paths.

by H_2O-rich fluids that resulted from dehydration reactions in the downward movement of the greenstones (see movement of the triangle marker in Fig. 8B) in the greenstone lithologies plate. These movements result in a near-isobaric portion of the P-T path for the upper granulite fragment (see steplike lines in Fig. 4B and kinked P-T path with squares in Fig. 9). On the other hand, a high-temperature, deep-seated fragment moves from the 900 °C isotherm directly to a near-surface level (see circle marker in Fig. 8B), producing a decompression-cooling P-T path (see straight line in Fig. 4A and linear P-T path with circles in Fig. 9).

An important detail is illustrated in Figure 8A (8.9 m.y.): The low-grade greenstone block of the craton on the right side of the model is bent upward by the flow of exhuming granulite, and its leftmost part is exhumed close to the middle of the flattened granulite complex. If this feature is realistic, one should be able to observe comparable features among granulites (i.e., higher grade greenstone rocks that were metamorphosed under amphibolite to epidote-amphibolite facies). A possible example is greenstone rocks that occur in the Central Zone of the Limpopo HGT, part of the so-called Venetia Klippen Complex, for which a well-constrained P-T loop with maximum P-T estimates of 6 kbar and 650 °C was inferred (Zeh et al., 2004).

According to the P-T paths in Figures 5 and 6, the model in Figure 8 must reflect the circulation of some of the cratonic wall rocks within shear zones separating cratons from granulite complexes. Indeed, triangle markers in Figure 8B trace the movement of the rocks, repeating a clockwise P-T loop for the cratonic rocks shown in Figures 5 and 6. Overall the variability of P-T paths in the numerical model (Fig. 10) is quite large and coincides well with variations in P-T trajectories discovered within and outside of each studied granulite complex and associated greenstone belt (e.g., Perchuk et al., 1989, 1996; Perchuk and Krotov, 1998; Gerya et al., 2004; see also Perchuk, this volume).

Seismic profiling, electric studies, and gravity data show that the continental crust beneath some granulite terranes is thinner than that under the adjacent cratons (e.g., Fig. 11A). The correct position of the Moho boundary beneath such crustal-scale complexes, however, is unknown (e.g., Griffin and O'Reilly, 1987a, 1987b) presumably because their present lower parts are composed of garnet-bearing mafic and ultramafic rocks (Pretorius and Barton, 1995; Specius, 1998) whose rheologies and densities are similar to those of the upper mantle. Figure 11A illustrates a N-S cross section through the Limpopo HGT compared with the results of numerical modeling of gravitational redistribution of the Archean greenstone belts. Figure 11B shows a NE-SW cross section through the Lapland HGT compared with the results of numerical modeling (Perchuk et al., 1999) of gravitational redistribution of rocks within the Archean Karelian-Inari Cratons. We contend that the morphologies of these two examples, for which good-quality seismic data are available (De Beer and Stettler, 1992; Pozhilenko et al., 1997; James et al., 2001; Nguuri et al., 2001), are in accord with the gravitational redistribution mechanism of HGT formation.

Thus the results of numerical modeling suggest that granulite-facies terranes can be formed (and then exhumed) within essentially normal-thickness (35–40 km), hot continental crust affected by the activity of mantle-derived fluid-heat flow. Such a process results in the formation of granulitic bodies with intrusion-like shapes (Figs. 8, 11) through a mechanism of gravitational redistribution of material in the crust during a time period of ~10 m.y. The results of our modeling are also in accordance with the conclusion of England and Thompson (1984) and Henry et al. (1997) that *thick* (60–70 km) continental crust in many cases is not necessarily the most appropriate setting for the formation of granulite-facies terranes. Thickening of such crust initially results in a relatively cold geotherm, which is more suitable for the origin of eclogite and HP granulite along a prograde PT path, followed by crustal thinning (e.g., owing to strong mantle plume activity) that in turn leads to the formation of garnet amphibolite (Perchuk, 1977, 1989). In some cases (e.g., Tibet) such eclogites can subsequently undergo granulite-facies conditions (e.g., Henry et al., 1997).

DISCUSSION

Granulite-facies terranes occur in a variety of geological settings of different age. However, the majority of them were formed in the Precambrian, when the Earth's crust was hotter than today. Thompson (1990) reviewed several tectonic models (collision followed by erosional or extensional crustal thinning, multiple episodes of crustal thickening, magmatic underplating, and crustal delamination) and concluded that there are limited possibilities, if they exist at all, to form and exhume lower crustal granulites into the upper Earth's crust during a single tectonic cycle, and most granulite-facies terranes were exposed tectonically. Harley (1989), on the basis of a study of 90 P-T paths from different HGTs, concluded that no single universal tectonic model could be proposed for the genesis of granulites. Nevertheless, despite the fact that eclogites as markers of subduction-collision mechanisms occur extremely rarely in Precambrian continental crust in general, and for the Archean, in particular, many petrologists still believe the plate tectonic models. This may be the result of systematic comparison of the Precambrian granulites with significantly younger granulites from Phanerozoic fold belts, but the rheology of these rocks was very different.

Recently Sizova et al. (2010), based on a numerical modeling study, demonstrated that there should be major transitions in geodynamic styles back in Earth history related to the cooling of sub-lithospheric mantle (see also, e.g., Hamilton, 1998; Brown, 2006). These authors identified a first-order transition from a "no-subduction" tectonic regime through a "pre-subduction" tectonic regime to the modern style of subduction. The first transition is gradual, and occurs at upper mantle temperatures between 250 and 200 K above present-day values, whereas the second transition is more abrupt and occurs at 175–160 K above present-day values. The link between geological observations and model

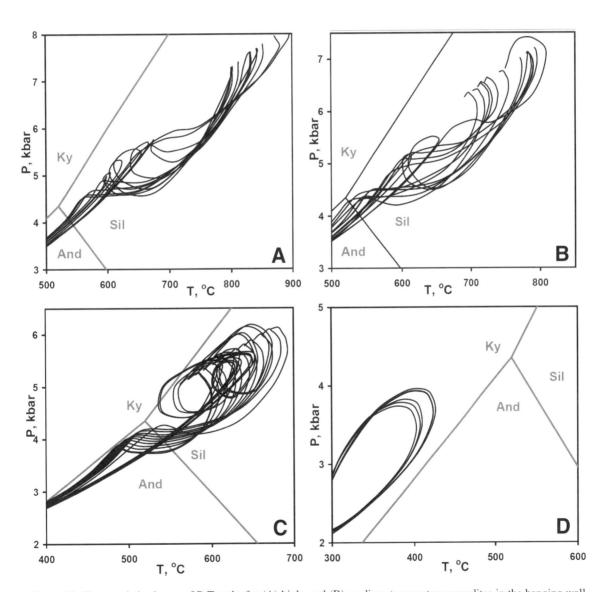

Figure 10. Characteristic shapes of P-T paths for (A) high- and (B) medium-temperature granulites in the hanging wall of the shear zone for the model shown in Figure 8. Characteristic P-T paths for (C) mid-crustal and (D) upper crustal cratonic footwall rocks for the model shown in Figure 8 (compare with Figs. 4–6). See Gerya et al. (2000) and Gerya and Maresch (2004) for comparison of natural and numerical P-T paths.

results suggests that the transition to the modern plate tectonics regime might have occurred during Mesoarchean–Neoarchean time (ca. 3.2–2.5 Ga). For the "pre-subduction" tectonic regime (upper-mantle temperature 175–250 K above present) the plates are weakened by intense percolation of melts derived from the underlying hot melt–bearing sub-lithospheric mantle. In such cases, convergence does not produce self-sustaining, one-sided subduction, but rather results in shallow underthrusting of the oceanic plate under the continental plate. Further increase in the upper-mantle temperature (>250 K above present) causes a transition to a "no subduction" regime in which horizontal movements of small deformable plate fragments are accommodated by internal strain, and even shallow underthrusts do not form under convergence. Taking this into account, one can expect that the geodynamic origin of many Precambrian granulites may have differed markedly from their modern analogues.

Inherent gravitational instability of hot continental crust was confirmed on the basis of modeling of in situ rock properties using a Gibbs free-energy minimization approach (Gerya et al., 2001, 2002, 2004). This modeling showed that regional metamorphism of granulite-facies crust may critically enhance the decrease of crustal density with depth. This leads to a gravitational instability of hot continental crust, resulting in regional doming and diapirism. Two types of crustal models were studied: (1) lithologically homogeneous crust, and (2) lithologically heterogeneous, multilayered crust. Gravitational instability of relatively homogeneous continental crust is related to a vertical density contrast developed during prograde changes in mineral

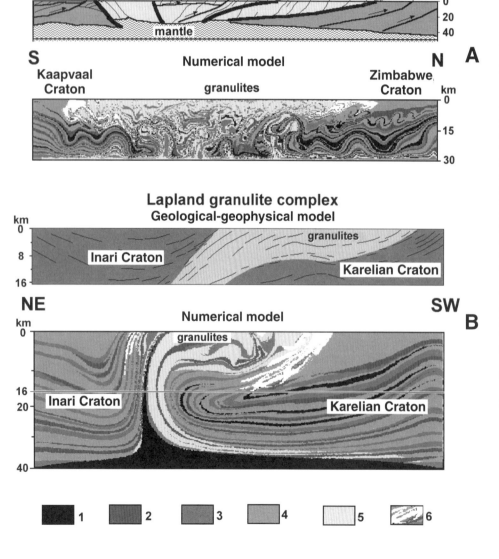

Figure 11. Geological-geophysical cross sections of (A) the Limpopo HGT (Roering et al., 1992a), and (B) the Lapland granulite complex (Pozhilenko et al., 1997) compared with results of numerical experiments (Perchuk et al., 1999) conducted with the gravitational redistribution models. In the presented models, granulite-facies rocks were originally in the lower portion of a metastable layered 30–40-km-thick crust. No lateral forcing is applied in the models. Color code for numerical models and rock properties (see Perchuk et al., 1999; Gerya et al., 2000, for the choice of numerical values and sensitivity analysis): 1 = ultramafic rocks ($\rho = 3300$ kg/m^3, $\eta = 10^{21}$ Pa s); 2 = metabasic rocks ($\rho = 3000$ kg/m^3, $\eta = 10^{21}$ Pa s); 3 = andesitic rocks ($\rho = 2800$ kg/m^3, $\eta = 10^{21}$ Pa s); 4 = felsic and sedimentary rocks ($\rho = 2700$ kg/m^3, $\eta = 10^{19}$ Pa s); 5 = felsic, partially molten granulites ($\rho = 2650$ kg/m^3, $\eta = 10^{19}$ Pa s); 6 = weak tectonic zones in rocks of various lithologies ($\eta = 10^{19}$ Pa s).

assemblages and the thermal expansion of minerals with increasing temperature. Gravitational instability of lithologically heterogeneous multilayered crust is related to an initial density contrast of dissimilar intercalated layers enhanced by high-temperature phase transformations. In addition, the thermal regime of heterogeneous crust strongly depends on the pattern of vertical interlayering: A strong positive correlation between temperature and the estimated degree of lithological gravitational instability is indicated. It has also been shown (Gerya et al., 2004) that exponential lowering of viscosity with increasing temperature, in conjunction with prograde changes in metamorphic mineral assemblages during thermal relaxation after collisional thickening of the crust, provides a positive feedback mechanism leading to regional doming and diapirism that contribute to the exhumation of high-grade metamorphic rocks.

It is important to mention that collisional (i.e., driven by external forces) and gravitational (i.e., driven by internal forces) mechanisms of rock deformation are not mutually exclusive. Gravitationally unstable crust is expected to result, above all, from collisional events involving initially stable sections of continental crust where regional thrusting, multiple stacking, and regional folding occur (e.g., the double-stacked crust of England and Thompson, 1984; Le Pichon et al., 1997). This suggests a strong causal and temporal link between external collisional and internal gravitational mechanisms of rock deformation in high-grade metamorphic regions. Collisional mechanisms should operate during the early prograde stages of a tectono-metamorphic cycle, causing thickening of the crust and a corresponding increase in radiogenic thermal supply, whereas gravitational mechanisms should dominate during the later thermal peak and retrograde stages, providing an important factor for regional doming and diapirism that contributes to the exhumation of high-grade rocks (Figs. 7, 8). This also extends validity of gravitational redistribution models for relatively hot young orogens where gravitational tectonics is activated (e.g., Ramberg, 1981) owing to thickening and heating of initially cold continental crust (Burg and Gerya, 2005; Gerya et al., 2004, 2008; Faccenda et al., 2008).

For this discussion it should be recalled that Ramberg (1981) extended the diapiric model of MacGregor (1951) originally proposed for granite-greenstone belts to explain the structural pattern of many orogenic belts. More recently this model has been invoked to explain the formation of granite domes within greenstone belts (Hunter and Stowe, 1997). In essence, an increase in temperature leads to a decrease in both density and viscosity of the rocks in the lower-middle crust, thus triggering their movement within the gravity field. This softening of crustal rocks may generate their upward movement if the density and viscosity contrasts are appropriate (e.g., Fig. 7). This simple idea is also the background of the gravitational redistribution model suggested for granulites (Perchuk, 1989, 1991; Perchuk et al., 1992). Crustal (radioactive decay) and mantle heat flow (England and Thompson, 1984; Sandiford and Powell, 1986; Harley, 1989; Thompson, 1990; Pili et al., 1997; Gerya et al., 2002, 2004) are both considered to be the driving forces for the crustal heating process. The gravitational redistribution model suggests a strong mantle-derived fluid-heat flow as the trigger of high-grade metamorphism and both horizontal and vertical movements of rocks within convecting continental crust. We can easily determine the high temperature of the metamorphic process from mineral assemblages, but the contribution of the fluid component of the flow can only be proven on the basis of petrological and isotopic data. For example, Pili et al. (1997) demonstrated the existence of CO_2 flow by stable isotopic data obtained for the rocks from regional-scale shear zones in the Madagascar granulites of similar age.

Because most of the Precambrian HGTs lie between granite-greenstone belts, we have to consider their genetic connection in terms of evolution of the upper mantle as the source of heat and fluids. It is clear that the heat cannot be introduced in the giant metamorphic system by conductivity only. Therefore, crustal radiogenic decay (e.g., Gerya et al., 2001, 2002; Sizova et al., 2010) and mantle-derived fluid flows are necessary physical mechanisms to supply a deep-seated metamorphic system with heat and volatiles, thereby supporting high-temperature conditions. In relation to this, partially molten sub-lithospheric mantle plumes are expected to have played crucial roles in producing magmatism and metamorphism in the Paleoarchean, and even Hadean, crust (3.8–4.4 Ga), which follows from discovery of the 4.4 Ga detrital zircon derived from granite in the Yilgarn Craton of Western Australia (Wilde et al., 2001). These plumes produced large amounts of mafic and ultramafic magmas that could have initiated growth of the continental crust on the one hand (e.g., Bogatikov and Sharkov, 2008), and that partially melted the crust, producing tonalite-trondhjemite-granodiorite cupolas on the other (MacGregor, 1951; Ramberg, 1981; Hunter and Stowe, 1997). Large amounts of sub-lithospheric melts produced by these plumes are expected also to have notably weakened the Precambrian lithosphere, causing distinct tectonic styles (such as pre-subduction and no-subduction regimes) with pronounced horizontal movements (Sizova et al., 2010). Plume tectonics should induce horizontal flows in the rheologically weakened overlying lithosphere and create areas of assembled, deforming cratonic crust between mantle upwellings (Fig. 12). Such crustal assembly areas (where HGTs were subsequently generated) are precursors of modern collision zones, but in contrast to them they are not characterized by greatly thickened crust and high topography, and they are associated with gentle two-sided mantle downwellings rather than with asymmetric subduction and collision of rigid lithospheric plates (e.g., Sizova et al., 2010). Figure 12 illustrates this idea, showing both processes beneath two cratons and the HGT located between them. Ascent of the large plumes is accompanied by their partial and then total melting, which provides growth of the crust by mafic and ultramafic magma additions. At earlier stages the plumes provide enough heat for melting of ancient crust, resulting in felsic magma production. All these provide a magmatic basis for the formation of granite-greenstone terranes. At the later stages these terranes could be assembled in inter-plume areas

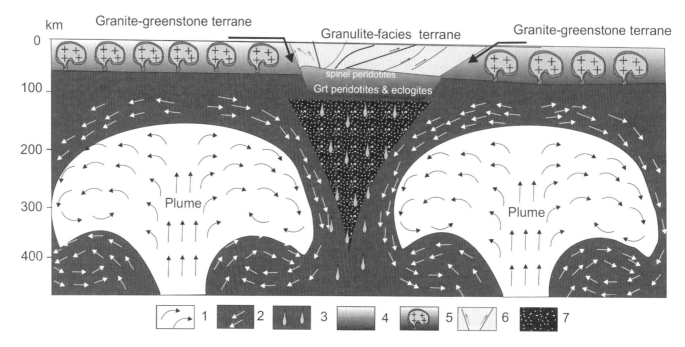

Figure 12. Conceptual scheme for the formation of the Precambrian continental crust resulting from plume tectonics: (1) arrows reflecting movement directions of plume material; (2) upper mantle rocks and direction of their movements in response to the plume activity; (3) crustal eclogites descending into the deep mantle; (4) continental crust composed mainly of volcanic rocks in layered successions, as discussed in the text (see also Perchuk, 1989); (5) ascending TTG (tonalite-trondhjemite-granodiorite) cupola within the continental crust, resulting from the hot plume activities; (6) HGT with shear zones (straight gneisses, Smit and van Reenen, 1997) recording its exhumation; (7) upward-directed heat-fluid flow that triggers HG metamorphism and exhumation of granulites within the continental crust from mechanisms of gravitational redistribution (fluids are presumably released owing to eclogitization reactions and crystallization of residual mantle melts in the downwelling zone). Dimension and shape of the granulite complex approximately correspond to those of the Limpopo HGT (Fig. 11A).

and involved in the gravitational redistribution (Fig. 12) triggered by remnant fluid-heat flows separated from descending and crystallizing mantle plume material (Fig. 12). The conceptual scenario of plume tectonics delineated in Figure 12 obviously needs further assessment and testing on the basis of both field observation and numerical modeling, including detailed studies of mutual structural and metamorphic aspects of granulite terranes and adjacent greenstone belts (e.g., Roering et al., 1992b; Smit et al., 2000; Bennett et al., 2005; van Reenen et al., this volume).

CONCLUSIONS

Precambrian HGTs can be formed and exhumed during a single cycle within a hot continental crust that can be relatively thin (35–40 km thick). The driving forces of the process are (1) a potential gravitational instability of an initial metastable stratigraphic/tectonic succession, and (2) gravitational redistribution of the crustal rocks owing to a mantle-derived fluid-heat flow (a mantle plume). This conclusion is based on the following points:

1. Many HGTs have a laccolith or harpolith shape in vertical cross section. Hence, such complexes could represent large diapirs intruded into the upper crust.
2. The HGTs exhibit DC and DC-IC retrograde paths compatible with a single episode exhumation. DC paths are >~50 km from the footwall of the HGT, whereas IC-DC paths are <~50 km from the footwall.
3. The greenstone belts in the footwall adjacent to the granulite complexes show *strongly non-isobaric* metamorphic zoning that is time- and mechanism-related to the emplacement of the granulite diapir. The peak metamorphic conditions of the prograde path for the footwall cratonic rocks coincide with the P-T minimum recorded by the granulite assemblages. This convergence of P-T conditions implies that the P-T loops for the footwall rocks resulted from the emplacement of the granulite diapir into the upper crust.
4. Collisional and gravitational mechanisms of rock deformation are not mutually exclusive: Collisional mechanisms may operate during the early prograde stages of a tectono-metamorphic cycle, whereas gravitational mechanisms should dominate during the later thermal peak and retrograde stages.
5. The gravitational redistribution model is well supported by multidisciplinary tests in many HGTs, including those in the Limpopo and Lapland HGTs.

ACKNOWLEDGMENTS

This chapter has benefited greatly from discussions with many colleagues around the world. Since 1995 this work has been carried out as part of the RF-RSA collaboration supported by the NRF (grant 68288), Gencor and JCI grants to Dirk van Reenen, and RFBR grants 08-05-00351 to Leonid Perchuk and 09-05-00991 to Taras Gerya. Particularly important support has come from the University of Johannesburg (to Dirk van Reenen and Leonid Perchuk) and the Program of the RF President entitled "Leading Research Schools of Russia" (grant 1949.2008 to Leonid Perchuk). We are truly thankful to Toby Rivers for his detailed comments, important suggestions, and English corrections. Constructive reviews from Weronika Gorczyk and Rebecca Jamieson are also greatly appreciated.

REFERENCES CITED

Allsopp, H.L., Seggle, A.G., Skinner, E.M.W., and Colgan, E.A., 1995, The emplacement age and geochemical character of the Venetia kimberlite bodies, Limpopo belt, Northern Transvaal: South African Journal of Geology, v. 98, p. 239–244.

Arima, M., and Barnett, R.L., 1984, Sapphirine bearing granulites from the Sipiwesk Lake area of the late Archean Pikwitonei granulite terrain, Manitoba, Canada: Contributions to Mineralogy and Petrology, v. 88, p. 102–112, doi:10.1007/BF00371415.

Barton, J.M., Jr., and Gerya, T.V., 2003, Mylonitization and isochemical reaction of garnet: Evidence for rapid deformation and entrainment of mantle garnet-harzburgite by kimberlite magma, K1 pipe, Venetia Mine, South Africa: South African Journal of Geology, v. 106, p. 231–242, doi:10.2113/106.2-3.231.

Barton, J.M., Jr., Doig, R., Smith, C.B., and Bohlender, F., 1992, Isotopic and REE characteristics of the intrusive charnoenderbite and enderbite geographically associated with Matok Complex: Precambrian Research, v. 55, p. 451–467.

Bennett, V., Jackson, V.A., Rivers, T., Relf, C., Horan, P., and Tubrett, M., 2005, Geology and U-Pb geochronology of the Neoarchean Snare River terrane: Tracking evolving tectonic regimes and crustal growth mechanisms: Canadian Journal of Earth Sciences, v. 42, p. 895–934, doi:10.1139/e04-065.

Bernard-Griffiths, J., Peucat, J.J., Postaire, B., Vidal, Ph., Convert, J., and Moreau, B., 1984, Isotopic data (U-Pb, Rb-Sr, Pb-Pb and Sm-Nd) on mafic granulites from Finnish Lapland: Precambrian Research, v. 23, p. 225–348.

Bibikova, E.V., Melnikov, V.F., and Avakyan, K.Kh., 1993, Lapland granulites: Petrology, geochemistry, and isotopic age: Petrology, v. 1, p. 215–234.

Bogatikov, O.A., and Sharkov, E.B., 2008, Irreversible evolution of tectono-magmatic processes at the Earth and Moon: Petrological data: Petrology, v. 16, p. 629–651, doi:10.1134/S0869591108070011.

Bohlen, S.R., 1987, Pressure-temperature-time paths and a tectonic model for the evolution of granulites: Journal of Geology, v. 95, p. 617–632, doi:10.1086/629159.

Brown, M., 2006, Duality of thermal regimes is the distinctive characteristic of plate tectonics since the Neoarchean: Geology, v. 34, p. 961–964, doi:10.1130/G22853A.1.

Brown, M., 2008, Characteristic thermal regimes of plate tectonics and their metamorphic imprint throughout Earth history: When did Earth first adopt a plate tectonics mode of behavior?, in Condie, K.C., and Pease, V., eds., When Did Plate Tectonics Begin on Planet Earth?: Geological Society of America Special Paper 440, p. 97–121.

Brown, M., and White, R.W., 2008, Processes in granulite metamorphism: Journal of Metamorphic Geology, v. 26, p. 121–124, doi:10.1111/j.1525-1314.2007.00760.x.

Burg, J.-P., and Gerya, T.V., 2005, Viscous heating and thermal doming in orogenic metamorphism: Numerical modeling and geological implications: Journal of Metamorphic Geology, v. 23, p. 75–95, doi:10.1111/j.1525-1314.2005.00563.x.

Cloos, M., 1982, Flow melanges: Numerical modeling and geologic constraints on their origin in the Franciscan subduction complex, California: Geological Society of America Bulletin, v. 93, p. 330–345, doi:10.1130/0016-7606(1982)93<330:FMNMAG>2.0.CO;2.

Condie, K.C., 2005, TTGs and adakites: Are they both slab melts?: Lithos, v. 80, p. 33–44, doi:10.1016/j.lithos.2003.11.001.

Condie, K.C., and Pease, V., eds., 2008, When Did Plate Tectonics Begin on Planet Earth?: Geological Society of America Special Paper 440, 294 p.

De Beer, J.H., and Stettler, E.H., 1992, The deep structure of the Limpopo belt from geophysical studies: Journal of Precambrian Research, v. 55, p. 173–186, doi:10.1016/0301-9268(92)90022-G.

De Wit, M.J., and Ashwal, L.D., eds., 1997, Greenstone Belts: Oxford, UK, Clarendon Press, 809 p.

De Wit, M.J., Roering, C., Hart, R.J., Armstrong, R.A., de Ronde, C.J., Green, R.W.E., Tredoux, M., Pederdy, E., and Hart, R.A., 1992, Formation of an Archean continent: Nature, v. 357, p. 553–562, doi:10.1038/357553a0.

Dirks, P., 1995, Crustal convection: Evidence from granulite terrains: Geological Society of South Africa, Centennial Geocongress, Extended Abstracts, v. 2, p. 612–615.

Dorland, H.C., Beukes, N.J., Gutzmer, J., Evans, D.A.D., and Armstrong, R.A., 2004, Trends in detrital zircon provenance from Neoarchean-Paleoproterozoic sedimentary successions of the Kaapvaal Craton: Johannesburg, University of the Witwatersrand, Geoscience Africa Abstracts Volume, p. 176–177.

Durrheim, R.J., Barker, W.H., and Green, R.W.E., 1992, Seismic studies in the Limpopo belt: Precambrian Research, v. 55, p. 187–200, doi:10.1016/0301-9268(92)90023-H.

Ellis, D.J., 1980, Osumilite-sapphirine-quartz granulites from Enderby Land, Antarctica: P-T conditions of metamorphism, implication for garnet–cordierite equilibria and the evolution of the deep crust: Contributions to Mineralogy and Petrology, v. 74, p. 201–210, doi:10.1007/BF01132005.

Ellis, D.J., 1987, Origin and evolution of granulites in normal and thickened crust: Geology, v. 15, p. 167–170, doi:10.1130/0091-7613(1987)15<167:OAEOGI>2.0.CO;2.

England, P.C., and Thompson, A.B., 1984, Pressure-temperature-time paths of regional metamorphism: I. Heat transfer during the evolution of regions of thickened continental crust: Journal of Petrology, v. 25, p. 894–928.

Faccenda, M., Gerya, T.V., and Chakraborty, S., 2008, Styles of post-subduction collisional orogeny: Influence of convergence velocity, crustal rheology and radiogenic heat production: Lithos, v. 103, p. 257–287, doi:10.1016/j.lithos.2007.09.009.

Fonarev, V.I., Grafchikov, A.A., and Kopnilov, A.N., 1994, Experimental studies of mineral solid solutions and geothermobarometry, in Zharikov, V.A., and Fedkin, V.V., eds., Experimental Problems in Geology: Moscow, Nauka Press, p. 323–355.

Fonarev, V.I., Pilugin, S.M., Savko, K.A., and Novikova, M.A., 2006, Exsolution textures of ortho- and clinopyroxene in high grade BIF of the Voronezh Crystalline Massif: Evidence of ultrahigh-temperature metamorphism: Journal of Metamorphic Geology, v. 24, p. 135–151, doi:10.1111/j.1525-1314.2006.00630.x.

Gerya, T.V., and Maresch, W.V., 2004, Metapelites of the Kanskiy granulite complex (Eastern Siberia): Kinked P-T paths and geodynamic model: Journal of Petrology, v. 45, p. 1393–1412, doi:10.1093/petrology/egh017.

Gerya, T.V., Perchuk, L.L., Triboulet, C., Audren, C., and Ses'ko, A.J., 1997, Petrology of the Tumanshet metamorphic complex, eastern Sayan, Siberia: Petrology, v. 5, p. 503–533.

Gerya, T.V., Perchuk, L.L., van Reenen, D.D., and Smit, C.A., 2000, Two-dimensional numerical modeling of pressure-temperature-time paths for the exhumation of some granulite facies terrains in the Precambrian: Journal of Geodynamics, v. 30, p. 17–35, doi:10.1016/S0264-3707(99)00025-3.

Gerya, T.V., Maresch, W.V., Willner, A.P., van Reenen, D.D., and Smit, C.A., 2001, Inherent gravitational instability of thickened continental crust with regionally developed low- to medium-pressure granulite facies metamorphism: Earth and Planetary Science Letters, v. 190, p. 221–235, doi:10.1016/S0012-821X(01)00394-6.

Gerya, T.V., Perchuk, L.L., Maresch, W.V., Willner, A.P., van Reenen, D.D., and Smit, C.A., 2002, Thermal regime and gravitational instability of multi-layered continental crust: Implications for the buoyant exhumation of high-grade metamorphic rocks: European Journal of Mineralogy, v. 14, p. 687–699, doi:10.1127/0935-1221/2002/0014-0687.

Gerya, T.V., Perchuk, L.L., Maresch, W.V., and Willner, A.P., 2004, Inherent gravitational instability of hot continental crust: Implication for doming and diapirism in granulite facies terrains, in Whitney, D., Teyssier, C., and Siddoway, C.S., eds., Gneiss Domes in Orogeny: Geological Society of America Special Paper 380, p. 97–115.

Gerya, T.V., Perchuk, L.L., and Burg, J.-P., 2008, Transient hot channels: Perpetrating and regurgitating ultrahigh-pressure, high temperature crust-mantle associations in collision belts: Lithos, v. 103, p. 236–256, doi:10.1016/j.lithos.2007.09.017.

Glebovitskii, V.A., Miller, Yu.V., and Drugova, G.M., 1996, Structure and metamorphism of the Belomorian-Lapland collision zone: Geotectonica, no. 1, p. 63–75.

Griffin, W.L., and O'Reilly, S.Y., 1987a, Is the continental Moho the crust-mantle boundary?: Geology, v. 15, p. 241–244, doi:10.1130/0091-7613 (1987)15<241:ITCMTC>2.0.CO;2.

Griffin, W.L., and O'Reilly, S.Y., 1987b, The lower crust in eastern Australia: Xenolith evidence, in Dawson, J.B., Carswell, D.A., Wall, J., and Wedepohl, K.H., eds., The Nature of Lower Continental Crust: Geological Society [London] Special Publication 24, p. 241–244.

Hamilton, W.B., 1998, Archean magmatism and deformation were not products of plate tectonics: Precambrian Research, v. 91, p. 143–179, doi:10.1016/S0301-9268(98)00042-4.

Harley, S.L., 1985, Garnet ortho-pyroxene bearing granulites from Enderby Land, Antarctica—Metamorphic pressure temperature time evolution of the Archean Napier Complex: Journal of Petrology, v. 26, p. 819–856.

Harley, S.L., 1989, The origin of granulites: A metamorphic perspective: Geological Magazine, v. 126, p. 215–231, doi:10.1017/S0016756800022330.

Harley, S.L., 1998, On the occurrence and characterization of ultrahigh-temperature crustal metamorphism, in Treloar, P.J., and O'Brien, P.J., eds., What Drives Metamorphism and Metamorphic Reactions?: Geological Society [London] Special Publication 138, p. 81–107.

Harley, S.L., and Motoyoshi, Y., 2000, Al zoning in orthopyroxene in a sapphirine quartzite: Evidence for >1120°C UHT metamorphism in the Napier complex, Antarctica, and implications for the entropy of sapphirine: Contributions to Mineralogy and Petrology, v. 138, p. 293–307, doi:10.1007/s004100050564.

Hart, R.J., Andreoli, M.A.G., Smith, C.B., Tredoux, M., and de Wit, M.J., 1990, Geochemistry across an exposed section of Archaean crust at Vredefort, South Africa, with implications for mid-crustal discontinuities: Chemical Geology, v. 82, p. 21–50, doi:10.1016/0009-2541(90)90072-F.

Henry, P., Le Pichon, X., and Goffe, B., 1997, Kinematics, thermal and petrological model of the Himalayas: Constraints related to metamorphism within the underthrust Indian crust and topographic elevation: Tectonophysics, v. 273, p. 31–56, doi:10.1016/S0040-1951(96)00287-9.

Hoernes, S., Lichtenstein, U., van Reenen, D.D., and Mokgatlha, K., 1995, Whole-rock/mineral O-isotope fractionations as a tool to model fluid-rock interactions in deep shear zones of the Southern Marginal Zone of the Limpopo Belt, South Africa: South African Journal of Geology, v. 98, p. 488–497.

Hunter, D.R., and Stowe, C.W., 1997, A historical overview of the origin, composition, and setting of Archaean greenstone belts (pre-1980), in de Wit, M.J., and Ashwal, L.D., eds., Greenstone Belts: Oxford, UK, Clarendon Press, p. 5–30.

James, D.E., Fouch, M.J., VanDecar, J.C., van der Lee, S., and Kaapvaal Seismic Group, 2001, Tectonospheric structure beneath southern Africa: Geophysical Research Letters, v. 28, p. 2485–2488.

Kelsey, D.E., 2008, On ultra-high temperature crustal metamorphism: Gondwana Research, v. 13, p. 1–29, doi:10.1016/j.gr.2007.06.001.

Kempton, P.D., Downes, H., Sharkov, E.V., Vetrin, V.R., Ionov, D.A., Carswell, D.A., and Beard, A., 1995, Petrology and geochemistry of xenoliths from the northern Baltic shield: Evidence for partial melting and metasomatism in the lower crust beneath an Archean terrane: Lithos, v. 36, p. 157–184, doi:10.1016/0024-4937(95)00016-X.

Kozhevnikov, V.N., and Svetov, C.A., 2001, Mantle and crustal thermal anomalies in the Archean and Early Proterozoic: Regional analysis, global correlation and metallogenic consequences, in Golubev, A.I., ed., Geology and Economic Resources of Karelia: Petrosavodsk, KSC, Russian Academy of Sciences, Issue 4, p. 3–17.

Kozlov, N.E., Ivanov, A.A., and Nerovich, L.I., 1990, The Lapland Granulite Belt (the Primary Origin and Evolution): Kola Scientific Center, RAS Press, Apatity, 231 p.

Kreissig, K., Nögler, T.F., Kramers, J.D., van Reenen, D.D., and Smit, A.S., 2000, An isotopic and geochemical study of the northern Kaapvaal craton and the Southern Marginal Zone of the Limpopo Belt: Are they juxtaposed terranes?: Lithos, v. 50, p. 1–25, doi:10.1016/S0024-4937(99)00037-7.

Kreissig, K., Holzer, L., Frei, R., Villa, I.M., Kramers, J.D., Kröner, A., Smit, A.S., and van Reenen, D.D., 2001, Geochronology of the Hout River shear zone and metamorphism in the Southern Marginal Zone of the Limpopo Belt, South Africa: Precambrian Research, v. 109, p. 145–173, doi:10.1016/S0301-9268(01)00147-4.

Kröner, A., Jaeckel, P., Brandl, G., Nemchin, A.A., and Pidgeon, R.T., 1999, Single zircon ages for granitoid gneisses in the Central Zone of the Limpopo Belt, southern Africa, and geodynamic significance: Precambrian Research, v. 93, p. 299–337, doi:10.1016/S0301-9268(98)00102-8.

Le Pichon, X., Henry, P., and Goffe, B., 1997, Uplift of Tibet: From eclogites to granulites—Implications for the Andean Plateau and the Variscan belt: Tectonophysics, v. 273, p. 57–76, doi:10.1016/S0040-1951(96)00288-0.

Lopez, S., Fernandez, C., and Castro, A., 2006, Evolution of the Archaean continental crust: Insights from the experimental study of Archaean granitoids: Current Science, v. 91, p. 607–621.

MacGregor, A.M., 1951, Some milestones in the Precambrian of Southern Rhodesia: Transactions—Geological Society of South Africa, v. 54, p. 27–71.

Marker, M., 1991, Metamorphism, deformations and structure of the crust, in Tuisku, P., and Laajoki, K., eds., Excursion Guide to Lapland: University of Oulu, Finland, Department of Geology, Res Terrae, Ser. A, no. 6, p. 38–66.

Mahan, K.H., Smit, C.A., Williams, M.L., Dumond, G., and van Reenen, D.D., 2011, this volume, Heterogeneous strain and polymetamorphism in high-grade terranes: Insight into crustal processes from the Athabasca Granulite Terrane, western Canada, and the Limpopo Complex, in van Reenen, D.D., Kramers, J.D., McCourt, S., and Perchuk, L.L., eds., Origin and Evolution of Precambrian High-Grade Gneiss Terranes, with Special Emphasis on the Limpopo Complex of Southern Africa: Geological Society of America Memoir 207, doi:10.1130/2011.1207(14).

Mints, M.V., Glaznev, V.N., and Konilov, A.N., Kunina, A.N., Nikitichev, N.M., Raevsky, A.P., Sedich, A.B., Yu, N., Stupak, V.M., and Fonarev, V.I., 1996, The Early Precambrian of the Northeastern Baltic Shield: Paleodynamics, Crustal Structure, and Evolution: Moscow, Scientific World Press (in Russian with English extended abstracts of introduction, conclusion, and each chapter).

Nguuri, T., Gore, J., James, D.E., Webb, S.J., Wright, C., Zengeni, T.G., Gwavava, O., Snoke, J.A., and the Kaapvaal Seismic Group, 2001, Crustal structure beneath southern Africa and its application for the formation and evolution of the Kaapvaal and Zimbabwe cratons: Geophysical Research Letters, v. 28, p. 2501–2504, doi:10.1029/2000GL012587.

Ouzegane, Kh., Kienast, J.-R., Bendaoud, A., and Drareni, A., 2003, A review of Archaean and Paleoproterozoic evolution of the Ouzzal granulitic terrain (Western Hoggar, Algeria): Journal of African Earth Sciences, v. 37, p. 207–227, doi:10.1016/j.jafrearsci.2003.05.002.

Perchuk, L.L., 1973, Thermodynamic Regime of Deep-Seated Petrogenesis: Moscow, Nauka Press, 300 p.

Perchuk, L.L., 1976, Gas-mineral equilibria and possible geochemical model of the Earth's Interior: Physics of the Earth and Planetary Interiors, v. 13, p. 232–239, doi:10.1016/0031-9201(76)90097-2.

Perchuk, L.L., 1977, Thermodynamic control of metamorphic processes, in Saxena, S.K., and Bhattacharj, S., eds., Energetics of Geological Processes: New York, Springer, p. 285–352.

Perchuk, L.L., 1986, The course of metamorphism: International Geology Review, v. 28, p. 1377–1400, doi:10.1080/00206818609466374.

Perchuk, L.L., 1989, P-T-fluid regimes of metamorphism and related magmatism with specific reference to the Baikal Lake granulites, in Daly, S., Yardley, D.W.D., and Cliff, B.O., eds., Evolution of Metamorphic Belts: Geological Society [London] Special Publication 43, p. 275–291.

Perchuk, L.L., 1991, Studies in magmatism, metamorphism, and geodynamics: International Geology Review, v. 33, p. 311–374, doi:10.1080/00206819109465695.

Perchuk, L.L., 2011, this volume, Local mineral equilibria and P-T paths: Fundamental principles and their application to high-grade metamorphic terranes, in van Reenen, D.D., Kramers, J.D., McCourt, S., and Perchuk, L.L., eds., Origin and Evolution of Precambrian High-Grade Gneiss Terranes, with Special Emphasis on the Limpopo Complex of Southern Africa: Geological Society of America Memoir 207, doi:10.1130/2011.1207(05).

Perchuk, L.L., and Krotov, A., 1998, Petrology of the mica schists of the Tanaelv belt in the southern tectonic framing of the Lapland granulite complex: Petrology, v. 6, p. 149–179.

Perchuk, L.L., and van Reenen, D.D., 2008, Reply to comments on "P–T record of two high-grade metamorphic events in the Central Zone of the Limpopo Complex, South Africa" by Armin Zeh and Reiner Klemd: Lithos, v. 106, p. 403–410, doi:10.1016/j.lithos.2008.07.011.

Perchuk, L.L., Aranovich, L.Ya., Podlesskii, K.K., Lavrent'eva, I.V., Gerasimov, V.Yu., Fed'kin, V.V., Kitsul, V.N., and Karsakov, L.P., 1985, Precambrian granulites of the Aldan shield, eastern Siberia, USSR: Journal of Metamorphic Geology, v. 3, p. 265–310, doi:10.1111/j.1525-1314.1985.tb00321.x.

Perchuk, L.L., Gerya, T.V., and Nozhkin, A.D., 1989, Petrology and retrogression in granulites of the Kanskiy Formation, Yenisey Range, Eastern Siberia: Journal of Metamorphic Geology, v. 7, p. 599–617, doi:10.1111/j.1525-1314.1989.tb00621.x.

Perchuk, L.L., Podladchikov, Y.Yu., and Polaykov, A.N., 1992, Geodynamic modeling of some metamorphic processes: Journal of Metamorphic Geology, v. 10, p. 311–319, doi:10.1111/j.1525-1314.1992.tb00086.x.

Perchuk, L.L., Gerya, T.V., van Reenen, D.D., Safonov, O.G., and Smit, C.A., 1996, The Limpopo metamorphic complex, South Africa: 2. Decompression/cooling regimes of granulites and adjacent rocks of the Kaapvaal craton: Petrology, v. 4, p. 571–599.

Perchuk, L.L., Krotov, A.V., and Gerya, T.V., 1999, Petrology of amphibolites of the Tanaelv Belt and granulites of the Lapland complex: Petrology, v. 7, p. 539–563.

Perchuk, L.L., Gerya, T.V., van Reenen, D.D., Smit, C.A., Krotov, A.V., Safonov, O.G., and Shur, M.Yu., 2000a, Comparable petrology and metamorphic evolution of the Limpopo (South Africa) and Lapland (Fennoscandia) high-grade terrains: Mineralogy and Petrology, v. 69, p. 69–107, doi:10.1007/s007100050019.

Perchuk, L.L., Gerya, T.V., van Reenen, D.D., Smit, C.A., and Krotov, A.V., 2000b, P-T paths and tectonic evolution of shear zones separating high-grade terrains from cratons: Examples from Kola Peninsula (Russia) and Limpopo Region (South Africa): Mineralogy and Petrology, v. 69, p. 109–142, doi:10.1007/s007100050020.

Perchuk, L.L., Gerya, T.V., van Reenen, D.D., and Smit, C.A., 2001, Formation and dynamics of granulite complexes within cratons: Gondwana Research, v. 4, p. 729–732, doi:10.1016/S1342-937X(05)70524-4.

Perchuk, L.L., Gerya, T.V., van Reenen, D.D., and Smit, C.A., 2006, P-T paths and problems of high-temperature polymetamorphism: Petrology, v. 14, p. 117–153, doi:10.1134/S0869591106020019.

Perchuk, L.L., van Reenen, D.D., Varlamov, D.A., van Kal, S.M., Tabatabaeimanesh, S.M., and Boshoff, R., 2008, P-T record of two high-grade metamorphic events in the Central Zone of the Limpopo Complex, South Africa: Lithos, v. 103, p. 70–105.

Percival, J.A., Roering, C., van Reenen, D.D., and Smit, C.A., 1997, Tectonic evolution of associated greenstone belts and high-grade terrains, *in* de Wit, M.J., and Ashwal, L.D., eds., Greenstone Belts: Oxford, UK, Oxford University Press, p. 398–421.

Petrova, Z.I., and Levitskii, V.I., 1986, Mafic crystalline schists in granulite-gneissic complexes of the Siberian Platform and their primary nature, *in* Tauson, L.V., ed., Geochemistry of Volcanics from Different Geodynamic Settings: Novosibirsk, Nauka Press, p. 17–34.

Petrova, Z.I., Reznitskii, L.Z., and Makrygina, V.A., 2001, Geochemical parameters of metaterrigenous rocks from the Sliudianka Group as indicator of source and conditions of the formation of protholith (S.W. Pribaikalie): Geokchimia, v. 4, p. 1–12.

Pili, E., Shepard, S.M.F., Lardeaux, J.-M., Martelat, J.-E., and Nicollet, C., 1997, Fluid flow vs. scale of shear zones in the lower continental crust and the granulite paradox: Geology, v. 25, p. 15–18, doi:10.1130/0091-7613(1997)025<0015:FFVSOS>2.3.CO;2.

Pozhilenko, V.I., Smolkin, V.F., and Sharov, N.V., 1997, Seismic-geological models for the Earth's crust in the Lapland-Pechenga region, Russia, *in* Sharov, N.V., ed., A Seismic Model of the Lithosphere of Northern Europe: Lapland-Pechenga Region: Apatity, Russian Academy of Sciences, Kola Scientific Center, p. 181–208.

Pretorius, W., and Barton, J.M., Jr., 1995, Lithospheric structure and geothermal gradient at 530 Ma beneath a portion of the Central Zone of the Limpopo Belt, as deduced from crustal and upper mantle xenoliths in Venetia kimberlite pipes: Geological Society of South Africa, Centennial Geocongress, Extended Abstracts, v. 2, p. 335–338.

Pretorius, W., and Barton, J.M., Jr., 1997, The lower unconformity-bounded sequence of the Soutpansberg group and its correlatives—Remnants of a Proterozoic large igneous province: South African Journal of Geology, v. 100, p. 335–339.

Raith, M., Karmakar, S., and Brown, M., 1997, Ultra-high-temperature metamorphism and multistage decompressional evolution of sapphirine granulites from the Palni Hill Ranges, southern India: Journal of Metamorphic Geology, v. 15, p. 379–399, doi:10.1111/j.1525-1314.1997.00027.x.

Ramberg, H., 1981, Gravity, Deformation and the Earth's Crust: London, Academic Press, 452 p.

Ranalli, G., 1995, Rheology of the Earth (2nd edition): London, Chapman and Hall, 413 p.

Roering, C., van Reenen, D.D., Smit, C.A., Barton, J.M., de Beer, J.H., de Wit, M.J., Stettler, E.H., van Schalkwyk, J.F., Stevens, G., and Pretorius, S., 1992a, Tectonic model for the evolution of the Limpopo Belt: Precambrian Research, v. 55, p. 539–552.

Roering, C., van Reenen, D.D., de Wit, M.J., Smit, C.A., de Beer, J.H., and Van Schalkwyk, J.F., 1992b, Structural geological and metamorphic significance of the Kaapvaal Craton–Limpopo Belt contact: Precambrian Research, v. 55, p. 69–80, doi:10.1016/0301-9268(92)90015-G.

Rubatto, D., 2002, Zircon trace element geochemistry: Partitioning with garnet and the link between U–Pb ages and metamorphism: Chemical Geology, v. 184, p. 123–138, doi:10.1016/S0009-2541(01)00355-2.

Rubatto, D., and Hermann, J., 2007, Experimental zircon/melt and zircon/garnet trace element partitioning and implications for the geochronology of crustal rocks: Chemical Geology, v. 241, p. 38–61, doi:10.1016/j.chemgeo.2007.01.027.

Sajeev, K., and Santosh, M., 2006, Extreme crustal metamorphism and related crust-mantle processes: Lithos, v. 92, p. v–ix.

Sajeev, K., Osanai, Y., and Santosh, M., 2004, Ultrahigh-temperature metamorphism followed by two-stage decompression of garnet-orthopyroxene-sillimanite granulites from Ganguvarpatti, Madurai block, southern India: Contributions to Mineralogy and Petrology, v. 148, p. 29–46, doi:10.1007/s00410-004-0592-0.

Samsonov, A.V., Bogina, M.M., Bibikova, E.V., Petrova, A.Y., and Shchipansky, A.A., 2005, The relationship between adakitic, calc-alkaline volcanic rocks and TTGs: Implications for the tectonic setting of the Karelian greenstone belts, Baltic Shield: Lithos, v. 79, p. 83–106, doi:10.1016/j.lithos.2004.04.051.

Sandiford, M., and Powell, R., 1986, Deep crustal metamorphism during continental extension: Ancient and modern examples: Earth and Planetary Science Letters, v. 79, p. 151–158, doi:10.1016/0012-821X(86)90048-8.

Santosh, M., and Omori, S., 2008a, CO_2 flushing: A plate tectonic perspective: Gondwana Research, v. 13, p. 86–102, doi:10.1016/j.gr.2007.07.003.

Santosh, M., and Omori, S., 2008b, CO_2 windows from mantle to atmosphere: Models on ultrahigh-temperature metamorphism and speculations on the link with melting of snowball Earth: Gondwana Research, v. 14, doi:10.1016/j.gr.2007.11.001.

Santosh, M., Osanai, Y., and Tsunogae, T., 2004, Ultrahigh temperature metamorphism and deep crustal processes: Journal of Mineralogical and Petrological Sciences, v. 99, p. 137–139, doi:10.2465/jmps.99.137.

Sizova, E., Gerya, T., Brown, M., and Perchuk, L.L., 2010, Subduction styles in the Precambrian: Insight from numerical experiments: Lithos, v. 116, p. 209–229, doi:10.1016/j.lithos.2009.05.028.

Smelov, A.P., 1995, The influence of deformations on metamorphism of the Tungurcha greenstone belt, Olekma granite-greenstone terrain, the Aldan Shield, Siberia: Johannesburg, Centennial Geocongress, Extended Abstracts, v. 2, p. 684–687.

Smit, C.A., and van Reenen, D.D., 1997, Deep crustal shear zones high-grade tectonites and associated alteration in the Limpopo belt, South Africa: Implication for deep crustal processes: Journal of Geology, v. 105, p. 37–57, doi:10.1086/606146.

Smit, C.A., van Reenen, D.D., Gerya, T.V., Varlamov, D.A., and Fed'kin, A.V., 2000, Structural-metamorphic evolution of the Southern Yenisey Range of Eastern Siberia: Implications for the emplacement of the Kanskiy granulite complex: Mineralogy and Petrology, v. 69, p. 35–67, doi:10.1007/s007100050018.

Smit, C.A., van Reenen, D.D., Gerya, T.V., and Perchuk, L.L., 2001, P-T conditions of decompression of the Limpopo high grade terrain: Record from shear zones: Journal of Metamorphic Geology, v. 19, p. 249–268, doi:10.1046/j.0263-4929.2000.00310.x.

Smit, C.A., van Reenen, D.D., Roering, C., Boshoff, R., and Perchuk, L.L., 2011, this volume, Neoarchean to Paleoproterozoic evolution of the polymetamorphic Central Zone of the Limpopo Complex, *in* van Reenen, D.D., Kramers, J.D., McCourt, S., and Perchuk, L.L., eds., Origin and Evolution of Precambrian High-Grade Gneiss Terranes, with Special

Emphasis on the Limpopo Complex of Southern Africa: Geological Society of America Memoir 207, doi:10.1130/2011.1207(12).

Specius, Z.V., 1998, Nature of crust and crust/mantle boundary beneath the Siberian craton: Evidence from kimberlite xenoliths, *in* Evolution of the Deep Crust in the Central and Eastern Alps (International Workshop and Field Excursion) Abstracts: Padova, Italy, p. 18–19.

Thompson, A.B., 1990, Heat, fluids, and melting in the granulite facies, *in* Vielzeuf, D., and Vidal, Ph., eds., Granulites and Crustal Evolution (NATO ASI Ser. C): Dordrecht, Kluwer, v. 311, p. 37–58.

Tomkeieff, S.I., 1983, Dictionary of Petrology: Chichester, New York, Wiley and Sons, Interscience Publication, 680 p.

Treloar, P.J., Coward, M.J., and Harris, B.W., 1992, Himalayan-Tibetian analogies for the evolution of the Zimbabwe craton and Limpopo belt: Precambrian Research, v. 55, p. 571–587, doi:10.1016/0301-9268(92)90046-Q.

Tsunogae, T., Miyano, T., van Reenen, D.D., and Smit, C.A., 2004, Ultrahigh-temperature metamorphism of the Southern Marginal Zone of the Archean Limpopo Belt, South Africa: Journal of Mineralogical and Petrological Sciences, v. 99, p. 213–224, doi:10.2465/jmps.99.213.

van Reenen, D.D., 1986, Hydration of cordierite and hypersthene and description of the retrograde orthoamphibole isograd in the Limpopo Belt, South Africa: American Mineralogist, v. 71, p. 900–915.

van Reenen, D.D., and Smit, C.A., 1996, The Limpopo metamorphic complex, South Africa: 1. Geological setting and relationships between the granulite complex and the Kaapvaal and Zimbabwe cratons: Petrology, v. 4, p. 562–570.

van Reenen, D.D., Barton, J.M., Jr., Roering, C., Smit, C.A., and van Schalkwyk, J.F., 1987, Deep-crustal response to continental collision: The Limpopo Belt of southern Africa: Geology, v. 15, p. 11–14, doi:10.1130/0091-7613(1987)15<11:DCRTCC>2.0.CO;2.

van Reenen, D.D., Roering, C., Brandl, G., Smit, C.A., and Barton, J.M., Jr., 1990, The granulite facies rocks of the Limpopo belt, southern Africa, *in* Vielzeuf, D., and Vidal, Ph., eds., Granulites and Crustal Evolution (NATO ASI Ser. C): Dordrecht, Kluwer, v. 311, p. 257–289.

van Reenen, D.D., Perchuk, L.L., Smit, C.A., Varlamov, D.A., Boshoff, R., Huizenga, J.M., and Gerya, T.V., 2004, Structural and P-T evolution of a major cross fold in the Central Zone of the Limpopo high-grade terrain, South Africa: Journal of Petrology, v. 45, p. 1413–1439, doi:10.1093/petrology/egh028.

van Reenen, D.D., Smit, C.A., Perchuk, L.L., Roering, C., and Boshoff, R., 2011, this volume, Thrust exhumation of the Neoarchean ultrahigh-temperature Southern Marginal Zone, Limpopo Complex: Convergence of decompression-cooling paths in the hanging wall and prograde P-T paths in the footwall, *in* van Reenen, D.D., Kramers, J.D., McCourt, S., and Perchuk, L.L., eds., Origin and Evolution of Precambrian High-Grade Gneiss Terranes, with Special Emphasis on the Limpopo Complex of Southern Africa: Geological Society of America Memoir 207, doi:10.1130/2011.1207(11).

Vernon, R.H., White, R.W., and Clarke, G.L., 2008, False metamorphic events inferred from misinterpretation of microstructural evidence and P–T data: Journal of Metamorphic Geology, v. 26, p. 437–449, doi:10.1111/j.1525-1314.2008.00762.x.

Volodichev, O.I., 1990, Belomorian Complex of Karelia (Geology and Petrology): Leningrad, Nauka Press, 243 p. (in Russian).

Volodichev, O.I., Slabunov, A.I., Bibikova, E.V., Konilov, A.N., and Kusenko, T.I., 2004, Archaean eclogites from the Belomorian mobile belt, the Baltic shield: Petrology, v. 12, p. 540–560.

Weinberg, R.B., and Shmelling, H., 1992, Polydiapirs: Multiwavelength gravity structures: Journal of Structural Geology, v. 14, p. 425–436, doi:10.1016/0191-8141(92)90103-4.

Wilde, S.A., Valley, J.W., Peck, W.H., and Graham, C.M., 2001, Evidence from detrital zircons for existence of continental crust and oceans on the Earth 4.4 Gyr ago: Nature, v. 409, p. 175–178, doi:10.1038/35051550.

Zeh, A., Klemd, R., Buhlmann, S., and Barton, J.M., 2004, Prograde-retrograde *P–T* evolution of granulites of the Beit Bridge Complex (Limpopo Belt, South Africa): Constraints from quantitative phase diagrams and geotectonic implications: Journal of Metamorphic Geology, v. 22, p. 79–95, doi:10.1111/j.1525-1314.2004.00501.x.

MANUSCRIPT ACCEPTED BY THE SOCIETY 24 MAY 2010

# Tectonic models proposed for the Limpopo Complex: Mutual compatibilities and constraints

Jan D. Kramers
Department of Geology, University of Johannesburg, P.O. Box 524, Auckland Park, 2006, Johannesburg, South Africa, and School of Geosciences, University of the Witwatersrand, Private Bag 3, Wits 2050, South Africa

Stephen McCourt
School of Geological Sciences, University of KwaZulu-Natal, Private Bag X 54001, Durban 4000, South Africa

Chris Roering
C. André Smit
Dirk D. van Reenen
Department of Geology, University of Johannesburg, P.O. Box 524, Auckland Park, 2006, Johannesburg, South Africa

ABSTRACT

Published models for the Limpopo Complex as a whole include Neoarchean (ca. 2.65 Ga) continent-continent collision, Turkic-type terrane accretion, and plume-related gravitational redistribution within the crust. Hypotheses proposed for parts of the complex are Paleoproterozoic (ca. 2.0 Ga) dextral transpression for the Central Zone, westward emplacement of the Central Zone as a giant nappe, and gravitational redistribution scenarios. In this chapter these models and hypotheses are reviewed and tested against new data from geophysics (chiefly seismics and gravity), isotope geochemistry (mainly Sm-Nd and Lu-Hf data), geochronology, and petrology. Among the whole-complex models, the plume-related gravitational redistribution model and the Turkic-type terrane accretion model do not satisfy the constraints. The Neoarchean collision model remains as a viable working hypothesis, whereby (in contrast to published versions) the Zimbabwe Craton appears to be the overriding plate, with the Northern Marginal and Central Zones of the Limpopo Complex as its (possibly Andean-type) active margin and shelf, respectively. Of the partial models, gravitational redistribution in the context of crustal thickening is compatible with Neoarchean collision and can explain features at the Complex–Kaapvaal Craton boundary. Paleoproterozoic dextral transpression in the Central Zone can be superimposed on Neoarchean collision, provided that it does not itself entail a continent collision. The Paleoproterozoic metamorphism is characterized by near-isobaric prograde paths, which (along with combined teleseismic and gravity data) suggest magmatic underplating. This could be related to the Bushveld Complex, and may have weakened the crust, leading to the focusing of regional strain into transcurrent movement in the Central Zone.

Kramers, J.D., McCourt, S., Roering, C., Smit, C.A., and van Reenen, D.D., 2011, Tectonic models proposed for the Limpopo Complex: Mutual compatibilities and constraints, *in* van Reenen, D.D., Kramers, J.D., McCourt, S., and Perchuk, L.L., eds., Origin and Evolution of Precambrian High-Grade Gneiss Terranes, with Special Emphasis on the Limpopo Complex of Southern Africa: Geological Society of America Memoir 207, p. 311–324, doi:10.1130/2011.1207(16). For permission to copy, contact editing@geosociety.org. © 2011 The Geological Society of America. All rights reserved.

INTRODUCTION

A most important aim of the study of early Precambrian metamorphic provinces is to achieve a better understanding of geotectonic processes in the early Earth, when both mantle heat flow and intracrustal heat production were significantly higher than today. In this context, a lively debate exists on the onset and the early character of plate tectonics. For instance, geochemical data and metamorphic studies suggest that subduction-related crust formation started in the early to mid-Archean (Shirey et al., 2008; Foley, 2008; Pease et al., 2008). On the other hand, a late onset of plate tectonics has been argued from the lack of ophiolites, ultrahigh-pressure terranes, eclogites, and paired metamorphic belts in most of early Earth history (Stern, 2008). In this chapter we join this debate by discussing and testing existing models (both within and outside the plate tectonics framework) for the formation of the Limpopo Complex of southern Africa.

The Limpopo Complex is a high-grade metamorphic province situated between the Zimbabwe and Kaapvaal Cratons in southern Africa (Fig. 1). It consists of three zones, which are separated from each other and from the adjoining cratons by complex shear zones, with transcurrent as well as thrust-sense movement:

1. A Northern Marginal Zone (NMZ; Blenkinsop, this volume) consisting chiefly of magmatic enderbites that intruded in diapir-fashion between 2.6 and 2.7 Ga (coeval with high-grade metamorphism in metasediments) and carrying a weak but pervasive WSW-ENE fabric. A second metamorphic event, at 2.0 Ga, increasing southward from low to medium grade, is detected in shear zones.
2. A Central Zone (CZ; Smit et al., this volume), consisting mainly of metasediments with deposition ages up to ca. 3.4 Ga and interleaved S-type granitoids also formed at 2.6–2.7 Ga, a period of high-grade metamorphism. Map-scale structures are irregular, and many are perpendicular to the general WSW-ENE trend of the complex (Fig. 1). Besides the Neoarchean high-grade metamorphism, a second high-grade metamorphism at 2.0 Ga is seen, and it appears that large scale, WSW-ENE shear zones are of this age.
3. A Southern Marginal Zone (SMZ; van Reenen et al., this volume), consisting of strongly (mainly WSW-ENE) foliated enderbitic and charnockitic gneisses and intercalated metasediments. This zone underwent metamorphism and tectonism between 2.72 and 2.65 Ga only.

An important regional difference appears in Pb isotope characteristics. The Kaapvaal Craton and the SMZ have generally average $^{207}Pb/^{204}Pb$ ratios at given $^{206}Pb/^{204}Pb$ ratios, whereas those of the Zimbabwe Craton, the NMZ, and the CZ are mainly elevated (Barton et al., 2006). This indicates that the latter three units constitute a separate class of continental crust, as they were wholly or partly derived from a mantle or crustal province with anomalously high U/Pb ratios since at least 3.5 Ga.

The tectonic interpretation of high-grade metamorphic terranes poses special problems, as stratigraphic relationships are usually unclear, much upper crust has been removed, and textures may be annealed. In the case of the Central Zone of the Limpopo Complex, the superposition of at least two high-grade metamorphic events (see Kramers and Mouri, this volume, for a review) creates additional difficulties. Models for the tectonic processes that shaped the Limpopo Complex are varied, and have evolved with geological knowledge.

Mason (1973) noted the remarkable north-south symmetry of the belt, its polymetamorphic nature, and the major shear belts bounding its Central Zone. He expressed skepticism about plate tectonics scenarios for its origin and suggested major transcurrent shearing as the dominant mechanism. Later, as evidence of medium- and even high-pressure metamorphism of supracrustal rocks accumulated, the concept of crustal thickening by thrust stacking became central to models. Coward (1976, 1984), Coward and Fairhead (1980), and Light (1982) related crustal thickening to the thrusting of the Kaapvaal Craton over the Zimbabwe Craton. On the basis of available dates on the Bulai (Fig. 1) and Singelele granitoids (present throughout the CZ), used as time markers at the time, these events were postulated to have occurred at ca. 2.6–2.7 Ga. Light (1982) proposed that this thrusting was related to a subduction zone dipping south below the Kaapvaal Craton. Watkeys (1984), on the other hand, suggested that the Central Zone collided with the Zimbabwe Craton at ca. 2.7 Ga and that the combined block was juxtaposed with the Kaapvaal Craton along the line of the Palala Shear Zone at ca. 2 Ga.

Van Reenen et al. (1987) proposed that the shear zones defining the external boundaries of the Limpopo Complex were a response to crustal thickening, rather than structures directly related to collision between the Kaapvaal and Zimbabwe Cratons. Roering et al. (1992) and Treloar et al. (1992) further elaborated collisional models, the latter proposing an analogy to Himalayan-style tectonics in which the present geometry of the Limpopo Belt was due to NNW-SSE–directed shortening in response to collision, followed by crustal extrusion (escape tectonics) to the WSW along WSW-ENE–oriented strike-slip zones, including the Palala and Triangle Shear Zones. Following this, the Limpopo Belt was proposed as an example of Archean plate tectonics (Windley, 1993).

Rollinson (1993) and Barton et al. (2006) suggested that provinces of the Limpopo Belt represent separate terranes and that they, and the Zimbabwe block, accreted against the stable Kaapvaal Craton. In this model, the 2600 Ma "late granites" common to all five terranes are regarded as the key to the timing of terrane assembly. In a related concept, McCourt and Vearncombe (1987, 1992) proposed that the Central Zone was an exotic terrane emplaced into its present position by westward thrusting, and interpreted the Triangle and Palala Shear Zones as lateral ramps related to this emplacement.

In contrast to the above models of Neoarchean tectonism, a model of a transpressive orogeny in the Paleoproterozoic was proposed by Kamber et al. (1995a, 1995b), Holzer et al. (1998,

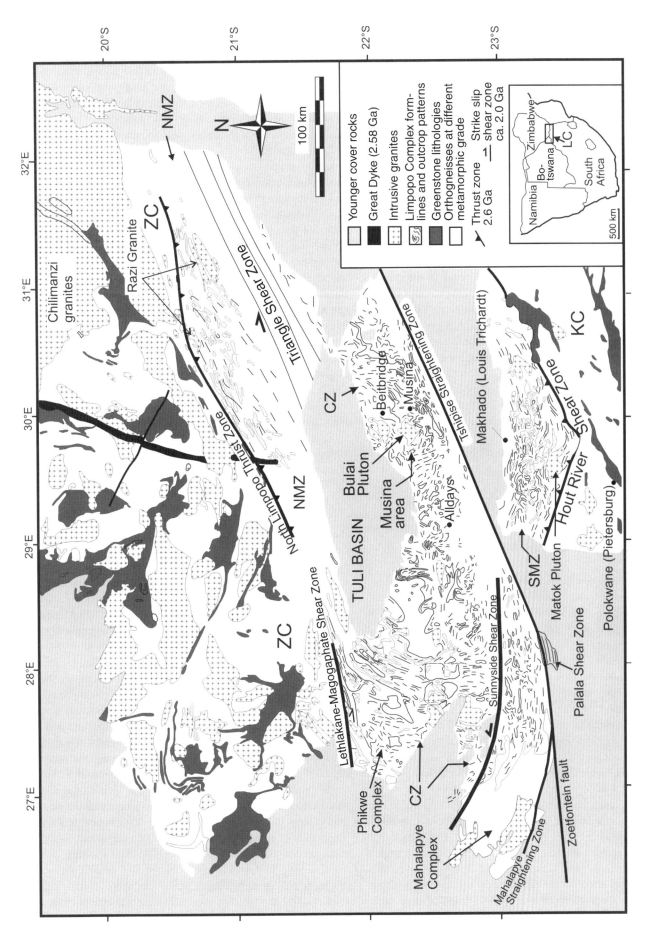

Figure 1. Map of the Limpopo Complex and adjoining areas of the Zimbabwe and Kaapvaal Cratons, showing geological units mentioned in the text. NMZ—Northern Marginal Zone; CZ—Central Zone; SMZ—Southern Marginal Zone; ZC—Zimbabwe Craton; KC—Kaapvaal Craton; LC—Limpopo Complex (inset map).

1999) and Schaller et al. (1999). This interpretation was based on consistent ages close to 2.0 Ga obtained on metamorphic minerals from strongly deformed rocks. These are mainly associated with transcurrent shear zones, but also occur more widely in the CZ, as well as locally in the southern part of the NMZ. The Paleoproterozoic transpression model does not address the evidence for Neoarchean tectono-metamorphism in all three zones of the Limpopo Complex.

Whereas in all the above models plate tectonics scenarios are invoked, a gravitational redistribution model was proposed (Perchuk, 1989, 1991; Perchuk et al., 2001) in which observed tectonism and metamorphism were explained as end results of thermal perturbation at the base of the crust alone, e.g., by a mantle plume. Vertical movements could occur in the crust as a result of the density inversion caused in many rock types by medium pressure granulite facies metamorphism (Gerya et al., 2004). Apart from a model for the entire Limpopo Complex as cited above, the concept has also been invoked to explain tectonic features in the framework of crustal thickening caused by collision tectonics (Perchuk and Gerya, this volume).

These models are briefly described below, along with the main features that they are set up to account for. Then, mutual compatibilities and incompatibilities are noted, and finally, the models are discussed and tested in the light of new data that have become available and that are mainly reviewed in contributions in this volume.

MODELS CREATED TO EXPLAIN THE ENTIRE LIMPOPO COMPLEX

The Neoarchean "Himalayan" Model

Light (1982) and Roering et al. (1992) argued that the Limpopo Complex exposes a deep crustal section through a collisional mountain belt formed by processes not dissimilar to Phanerozoic-style plate tectonics and proposed that the crustal thickening responsible for the formation of the Limpopo granulite terrane resulted from the thrusting of the Kaapvaal Craton over the Zimbabwe Craton along the south-dipping Triangle Shear Zone. They acknowledged, however, that the early history of the Limpopo Complex is very difficult to decipher and cautioned against using the kinematics of preserved structures in the belt to define the direction of plate movement during collision.

In their analysis of constraints on geotectonic models for the Limpopo Complex, Roering et al. (1992) identified a number of aspects of Limpopo geology that in their opinion were best explained by what has become known as the "pop-up" model by Limpopo researchers. Important among these constraints were widespread evidence (at the time) for a single metamorphic cycle in rocks from the Central Zone and Southern Marginal Zone of the complex and a large database of structural data indicating regional-scale ductile deformation in response to thrust-sense simple shear. Roering et al. (1992) modeled deep burial of the supracrustal rocks as a result of crustal thickening along a system of northward-verging thrusts and ductile shear zones that included the shear zones along the boundaries of the Central Zone, as well as the thrust zone separating the Northern Marginal Zone from the Zimbabwe Craton. The movement vector during this period of crustal thickening was to present day 40° (oblique to the regional strike of the complex), as indicated by the geometry of kilometer-scale circular structures now exposed at the surface in the Central Zone (see Roering et al., 1992, and Smit et al., this volume, for details).

Uplift of the granulite facies terrane was proposed to have occurred through a combination of processes, including widespread anatexis, which generated melts that increased the buoyancy of the Limpopo terrane, and thrusting of the granulites onto the adjacent cratons in response to continued compression. This thrusting occurred at the northern and southern margins of the orogen such that displacement along both the North Limpopo Thrust Zone and the Hout River Shear Zone (Fig. 1) produced a regional pop-up structure. Age constraints on deformation along the bounding shear zones to this pop-up structure constrain post-collisional uplift in the Limpopo Complex to the Neoarchean (2.62–2.72 Ga).

Subsequent to regional uplift, the shear zones along both margins of the Central Zone were reactivated, and as a result the Central Zone was displaced westward (van Reenen et al., 1987). In the model of Roering et al. (1992), therefore, the westward displacement of the Central Zone owing to displacement along the Triangle and Palala Shear Zones (Fig. 1) was not regarded as a component of the Limpopo orogeny.

Treloar et al. (1992) followed Coward (1976) in linking the tectonics of the Limpopo Complex with those of the adjacent cratons. They recognized a cycle of continental growth by accretion, followed by crustal shortening (folding and regional thrusting with associated crustal extrusion), and compared the Archean tectonics responsible for the formation of the Limpopo Complex with the tectonic evolution of Tibet since the Mesozoic. The model of Treloar et al. (1992) was the first to explicitly link the Limpopo Complex to Himalayan style tectonics.

An important element of the argument presented by Treloar et al. (1992) was the contention that the Limpopo Complex should be considered not as a narrow belt of deformation comprising the Northern and Southern Marginal Zones plus the Central Zone but as part of a much larger region of crustal scale deformation that includes much of the adjacent cratons. In a discussion of their model, Treloar et al. (1992) point out that at the regional scale the crustal section between the Witwatersrand Basin on the Kaapvaal Craton northward to the Harare Greenstone Belt on the Zimbabwe Craton is dominated by SE-dipping planar features and the map pattern by ENE- and NE-trending deformation zones. The kinematics of these structural features are consistent along the entire section and correspond to either NW-SE shortening involving thrusting, folding, and flattening or SW- to WSW-directed displacement on sub-vertical strike-slip shear zones. Treloar et al. (1992) argued that these different movement directions can be related to a single deformation event in which NW-SE compression was accommodated by both crustal thickening involving folding and NW-directed thrusting and lateral extrusion of crustal blocks along SW- to WSW-trending shear zones. They point out that this kinematic pattern, which can be constrained to the Neoarchean, is essentially analogous to Mesozoic tectonics in the Tibetan region of the India-Asia collision zone.

Treloar et al. (1992) concluded that Archean crustal growth in southern Africa was achieved through accretion processes that migrated northwestward over time from the central Kaapvaal Craton. They modeled the Limpopo orogeny not as a discrete collision between the Kaapvaal and Zimbabwe cratons at ca. 2.65 Ga as suggested by all previous workers (e.g., Light, 1982; van Reenen et al., 1987; Roering et al., 1992), but as the internal deformation of a crustal block comprising the Kaapvaal Craton, the Central Zone, and the southern part of the Zimbabwe Craton. This crustal block formed prior to 2.85 Ga and underwent internal deformation at ca. 2.65 Ga in response to NW-SE shortening linked to the accretion tectonics responsible for the granite-greenstone terranes that characterize the northern part of the Zimbabwe Craton. In contrast to Roering et al. (1992), Treloar et al. (1992) linked the lateral extrusion of the Central Zone along the Triangle and Palala Shear Zones to the ca. 2.65 Ga Limpopo event.

The Gravitational Redistribution Model

A gravitational redistribution model was first applied to the origin of the Lake Baikal high-grade metamorphic province (Perchuk, 1989). A number of features led to the proposal that such a model might account for the high-grade metamorphism in other Precambrian provinces, including the Lapland, Yenisey (Perchuk, 1991), and Limpopo high-grade belts, as described in Perchuk and Gerya (this volume).

Many Precambrian high-grade provinces are situated between granite-greenstone terranes, and are separated from these by crustal-scale shear zones up to 10 km wide and several hundred kilometers long, which have a thrust component. A fundamental observation that prompted the proposal of gravitational redistribution is that the high-grade metapelites show two types of pressure-temperature (P-T) paths, distinguished by their post-peak corona texture. At >50 km from the crustal-scale shear zones that bound the granulite terranes a decompression-cooling P-T path is indicated by the reactions Grt + Qtz + Sil → Crd and Grt + Qtz → Opx + Crd. In contrast, at <50 km distance from these shear zones, textures indicating the reverse reaction Crd → Grt + Qtz + Sil prevail, reflecting near-isobaric cooling paths (Perchuk and Gerya, this volume, and references therein). Further, mineral equilibria in schists within the shear zones themselves or in the immediately adjacent granite greenstone terrane may record hairpin-shaped, clockwise P-T loops (van Reenen et al., this volume).

The interpretation in the framework of the gravitational redistribution model is that granulites from the lower crust rose in giant diapirs (describing decompression cooling paths) and were thrust over upper crustal rocks (the granite-greenstone terranes) as these diapirs spread. Marginal parts of the diapir would undergo less uplift and lose heat to the adjacent upper crust, resulting in near-isobaric cooling. The upper crust, containing metasediments and metavolcanics of higher density than the lower crustal granulites, would sink to lower crustal levels and hence become metamorphosed. The prograde part of clockwise hairpin P-T loops found in and near the bounding shear zones could reflect burial under a spreading diapir of granulite-facies rocks, with simultaneous heating by this diapir.

In the first versions of such models (Perchuk, 1989, 1991) the crust was considered as consisting of layers of different density prior to the process. Under relatively low-temperature conditions gravitational inversions could persist metastably because of its overall high effective viscosity. Gravitational redistribution would then be triggered by an increase in temperature, greatly reducing the lower crustal viscosity (e.g., Whitney et al., 2004). As a possible cause, an enhanced mantle-derived heat input was envisaged (including heat advection by mantle-derived fluids and melts), possibly related to plume activity. In this "pure" form, the gravitational redistribution model therefore does not postulate crustal thickening by any cause, and is a non–plate tectonics scenario. The application of the model to the Limpopo Complex as a whole was discussed in this sense by Perchuk et al. (2001). Elements of the model have also been applied to parts of the complex in the context of crustal thickening (Perchuk and Gerya, this volume), as briefly described further below.

Terrane Accretion Models

Interpretations of the Limpopo Complex as an amalgamation of terranes were proposed by Rollinson (1993) and Barton et al. (2006). Rollinson (1993) proposed that the northern Kaapvaal Craton; the SMZ, CZ, and NMZ of the Limpopo Complex; and the Zimbabwe Craton all constitute separate terranes (microcontinents) that accreted at ca. 2.65 Ga. He based this proposal on two main observations: first, apparently different crustal histories prior to the Neoarchean tectonism and metamorphism, and second, the separation of all these proposed terranes by prominent thrusts and/or strike-slip shear zones. Further, he argued that a model of continent collision at ca. 2.65 Ga, as proposed by Roering et al. (1992) and Treloar et al. (1992), was contradicted by the occurrence of widespread granitic magmatism in the Zimbabwe Craton at 2.6 Ga (the Chilimanzi Granite suite), indicating that this craton did not constitute a stable continent at the time of the proposed collision.

The Turkic-type model of Barton et al. (2006) invokes a larger number of terranes, which are proposed to have accreted in the period between ca. 2.7 and 2.04 Ga. Turkic-type orogenies are defined (Şengör and Natal'in, 1996, p. 263) as "a class of collisional mountain building, in which the precollision history of one or both of the colliding continents involved the growth of very large, subcontinent-size subduction-accretion complexes, into which magmatic arc complexes commonly migrate and thus enlarge the continent to which they are attached." Şengör and Natal'in (1996) cite the Siberian Altaide Complex, the Nipponides, the Neoproterozoic East African and Arabian Shield, and the Archean Yilgarn Craton as examples. In addition to the terranes defined by Rollinson (1993), Barton et al. (2006) proposed separate terranes in the westernmost Zimbabwe Craton as well as in the CZ, where, following Aldiss (1991), the Beit Bridge and Phikwe Complexes are treated as separate terranes. The long history of amalgamation proposed by Barton et al. (2006) accommodates the Neoarchean and Paleoproterozoic periods of tectonism and metamorphism recognized in the CZ and the NMZ (see Kramers and Mouri, this volume, and references therein). Criteria cited by Barton et al. (2006) as arguments against a continent-continent collision model and in favor of a Turkic-type terrane amalgamation scenario are the long duration; the (perceived) dominance of I-type, calc-alkaline, and absence of S-type granitoid magmatism; the absence of ophiolites; the absence of syntectonic foreland sedimentary basins; a scarcity of horizontal thrust tectonics; and low apparent uplift rates of <1 mm/yr.

MODELS FOR PARTS OF THE LIMPOPO COMPLEX

The Central Zone as a Westward-Emplaced Thrust Sheet

In a discussion of the kinematic pattern of the Limpopo Complex, McCourt and Vearncombe (1992) noted that the different crustal terranes that constitute the belt are separated from each other, and from the adjacent cratons, by kilometer-scale ductile shear zones, thought to have been active during the 2.6–2.7 Ga period of high-grade metamorphism. McCourt and Vearncombe (1992) argued that although the gneissic shear zones forming the external boundaries to the Limpopo Complex have different movement directions, the sense of movement on these shear zones is consistently dip- or oblique-slip and reverse such that the kinematic pattern of the Limpopo Complex can be accommodated in an orogen-scale thrust-sense system. The Central Zone is then the structurally highest unit in this thrust system.

On the basis of published Pb-isotope data for meta-igneous rocks in the Central Zone, McCourt and Vearncombe (1992) argued it was an exotic terrane and could represent part of the overriding plate during the Neoarchean Limpopo orogeny. Fundamental to this interpretation was a set of structural data from the shear zones that defines the boundaries to the Central Zone, namely, the Triangle Shear Zone to the north and the Palala Shear Zone to the south. Both are strike-slip shear zones. The Triangle Shear Zone clearly shows a dextral shear sense, and McCourt and Vearncombe (1987, 1992) determined predominantly sinistral movement in the Palala Zone. They interpreted these shear zones as lateral ramps on a crustal-scale thrust sheet emplaced westward during the Limpopo orogeny. Observations at the western boundary of the Central Zone (the Magogaphate and Dikalata Hills shear zones) appeared to be geometrically compatible with this interpretation such that the western boundary could be modeled as a convex west frontal ramp structure (McCourt and Vearncombe, 1992).

The CZ was metamorphosed at ca. 2.65 Ga, as were the NMZ and the SMZ. This would be a highly unlikely coincidence if its emplacement had occurred after the Neoarchean; therefore the dating of bounding shear zones can provide a test for the model.

Gravitational Redistribution within Individual Zones

Forms of gravitational redistribution, or doming, have been proposed for the NMZ (Blenkinsop, this volume), the CZ (Andreoli et al., this volume), and the SMZ (Perchuk and Gerya, this volume). In the NMZ, the configuration of enderbitic bodies and interstitial supracrustal septa was interpreted as a result of widespread diapirism whereby intracrustal radioactivity might have been sufficient to provide the required lower crustal heating even with moderate crustal thickening, and crustal anatexis is envisaged (Blenkinsop, this volume; Kramers et al., 2001). For the CZ, it was suggested that steep structures, including closed structures previously interpreted as sheath folds, could have resulted from diapir-type flow (Perchuk and Gerya, this volume) whereby, as in the NMZ, the intracrustal heat production during the Neoarchean metamorphic period was probably high enough to cause lower crustal melting, given ~30%–50% crustal thickening (Ridley, 1992; Andreoli et al., this volume).

For the SMZ, and in particular its boundary with the Kaapvaal Craton, numerical modeling has been carried out in the context of tectonic crustal thickening (Perchuk and Gerya, this volume), taking into account density inversions enhanced by granulite metamorphism (Gerya et al., 2004). This modeling indicated the plausibility of diapiric rise of low-density, hot granulite in the SMZ and consequent downward movement of relatively high-density adjacent greenstone belt crust, consisting of metabasalt and komatiite. The base of the granulites was cooled along the bounding, inclined syntectonic shear zone. In the numerical models, local convection cells formed, which limited ascent along the bounding zones, resulting in near-isobaric cooling of granulite units near the contact with the cratonic rocks. Granulite units farther away from the contact ascended, recording decompression-cooling P-T paths. The numerical modeling results thus could reproduce many of the field observations of the SMZ. Further, the metamorphic dehydration of supracrustal rocks flowing underneath the buoyant granulite mass could account for the rehydration seen in the orthoamphibole zone of the SMZ.

The Paleoproterozoic Transpression Model

A model of a transpressive orogeny at 2.0 Ga was proposed by Kamber et al. (1995a, 1995b) and was further elaborated by Holzer et al. (1998, 1999) and Schaller et al. (1999). It was argued that the Central Zone underwent collision-related crustal thickening, uplift, and final exhumation during the

Paleoproterozoic between ca. 2.03 and 1.95 Ga. The collisional event was defined as transpressive because dextral transcurrent movement was seen to be dominant along both the Triangle and (in contrast to the interpretation of McCourt and Vearncombe, 1992) the Palala-Tshipise shear systems, which were interpreted as Paleoproterozoic suture zones bounding the CZ to the north and south. Thus it was proposed that during the perceived final collision, the Zimbabwe Craton moved eastward relative to the Kaapvaal Craton, a movement related to the coeval Magondi orogeny at the western edge of the Zimbabwe Craton (Holzer et al., 1999).

This model was proposed to reconcile a number of new findings. Kamber et al. (1995a) showed that the transcurrent shearing in the Triangle Shear Zone occurred under high-grade metamorphic conditions at ca. 2.0 Ga, and in the southern part of the NMZ, and both north-vergent thrusting and dextral transcurrent movement occurred coevally under amphibolite facies conditions (Kamber et al., 1995b). In the south, the Palala Shear Zone represents a fundamental break. To the north of it, in the Central Zone, both high-grade 2.0 Ga and late Archean tectono-metamorphism has been documented (Schaller et al., 1999; Chavagnac et al., 2001), whereas to the south (notwithstanding the intrusion of the Bushveld Complex) the tectono-metamorphism is purely Archean. The transition occurs over a distance of only 15 km, which documents uplift along this zone at ca. 2.0 Ga. Within the CZ, in the Musina area, in the Alldays area, and in the Phikwe Complex, mineral dates from high-grade shear zones also yielded 2.0 Ga ages (Boshoff et al., 2006; Holzer et al., 1998, 1999). High-grade parageneses now found at the surface and dated at 2.0 Ga would necessarily imply great uplift and erosion following the Paleoproterozoic event. An interesting aspect of the Proterozoic transpression model is that at its late stages, when after uplift the lower temperature of transcurrent movement would cause brittle faulting, the ca. 1.9 Ga Palapye and Soutpansberg basins could have been created in the extensional sectors of the transpressive shear-zone system.

The very large body of evidence documenting Neoarchean magmatism and high-grade metamorphism in the Limpopo Complex, on which the earlier models were based, is not negated in this model, but for the CZ it remains unclear which tectonic features might be attributed to the Neoarchean period. Further, it requires that the three zones of the Limpopo Complex as well as the two cratons would, prior to ca. 2.0 Ga, be in a very different spatial relation to each other from what they are today. Although the tectono-metamorphism in the NMZ is chiefly, and in the SMZ exclusively, of Neoarchean age, the Paleoproterozoic transpressive deformation thus complicates the building of models for a Neoarchean event in the Limpopo Complex.

MUTUAL COMPATIBILITIES OF MODELS

It is clear that the models proposed to explain all geological features of the entire Limpopo Complex are mutually exclusive. A gravitational redistribution model in which high mantle heat advection is invoked as a trigger for tectonic and metamorphic processes is incompatible with a model of crustal stacking by continent collision. Likewise, neither of these models can be accommodated within a Turkic-type terrane accretion model.

Among the partial models, that of a westward nappe-like emplacement of the CZ (McCourt and Vearncombe, 1992) and the dextral Paleoproterozoic transpressive model are mutually incompatible not merely in the model construction but even in the observations on which they are based: Schaller et al. (1999) report a predominantly dextral shear sense in the Palala Shear Zone, described as mainly sinistral by McCourt and Vearncombe (1987, 1992). Further, the shear zones in the westernmost exposed CZ, interpreted as a west-vergent frontal ramp by these authors, are described as east-vergent thrusts by Holzer et al. (1999). These discrepancies are in addition to the difference in ages of major shear-zone movement in the two models.

However, components of one model can in many cases be built into another one to explain features that remain unaccounted for by any of the models in its "pure" form. For instance, a continent-continent collision model does not preclude the presence of terranes within it (for instance, Afghanistan and Tibet have been proposed as terranes). Also, Paleoproterozoic transpression could have been superimposed on a Neoarchean continental collision zone, and such superposition is consistent with the data on the Limpopo Complex, provided that this transpression did not also involve continent-continent collision. Further, gravitational redistribution, or doming, can occur as a consequence of heating by crustal thickening, and thus form part of collisional tectonics (Gerya et al., 2004; Whitney et al., 2004; Perchuk and Gerya, this volume).

Geological models rarely account for all observations and generally evolve as more data become available. The discussion of models in the light of constraints must therefore not only include "full" models for the Limpopo complex but should also address possible combinations of parts of models in this sense.

DISCUSSION IN THE LIGHT OF AVAILABLE CONSTRAINTS

The various tectonic models were proposed to account for structural and metamorphic observations, as well as available geophysical data and geochronology. In this discussion, they are tested against constraints that have become available since their first proposal, that is, new geophysical, isotope geochemical, geochronological, and petrological data sets.

Geophysics

The geophysical data reviewed by De Beer and Stettler (1992) and Durrheim et al. (1992), which form one of the bases for the Neoarchean continent-continent collision model (Roering et al., 1992), include refraction and reflection seismics, gravity, geoelectrics, and aeromagnetics. Since 1992, teleseismic work has refined the crustal structure from the Kaapvaal to the

Zimbabwe Craton (Nguuri et al., 2001; Gore et al., 2009), has provided mantle tomographic imagery (Fouch et al., 2004), and has yielded data on the anisotropy of the subcontinental mantle (Silver et al., 2004).

The geoelectric and gravity transects of the Limpopo Complex show that the major structures defining its margins dip inward, i.e., the North Limpopo thrust zone dips to the south, and the Hout River Shear Zone to the north (De Beer and Stettler, 1992). Seismic reflection work also shows the Hout River Shear Zone dipping northward, and the absence of reflections along the Palala Shear Zone confirms that it is vertical (Durrheim et al., 1992). High-density rocks (~2950 kgm^{-3}) prevail in the lower crust throughout the transect. High-density zones of relatively low resistivity several kilometers in thickness accompany the depth continuations of the North Limpopo and Hout River zones and are interpreted by De Beer and Stettler (1992) as zones of high-grade metamorphic rocks. The apparent symmetry was an important cornerstone in defining the pop-up component of the Neoarchean continent-continent collision model.

A seismic refraction study from the CZ to the Zimbabwe Craton (Stuart and Zengeni, 1987) showed the crust in the NMZ to be about equal to that of the Zimbabwe Craton (35–40 km) but indicated a reduced crustal thickness of ~30 km in the CZ. The teleseismic studies of Nguuri et al., (2001) and Gore et al. (2009) yielded crustal thicknesses of ~37 and ~40 km for the NMZ and SMZ, respectively, likewise similar to the adjacent cratons. For the CZ, however, they found crustal thicknesses to be >45 km (up to 52 km in Gore et al., 2009), with a locally poorly defined and apparently structurally complex Moho. The discrepancy between the results for the CZ from Stuart and Zengeni (1987) on the one hand and Nguuri et al. (2001) and Gore et al. (2009) on the other can probably be resolved, as a dense lower crust, as required by the latter, works to provide isostatic compensation. Such a lower crust could affect the interpretation of refraction seismic data. A thick mafic lower crust with the appropriate seismic properties was also proposed to underlie the CZ at >10 km depth by Pretorius and Barton (2003) on the basis of xenoliths from the Venetia kimberlite.

Mantle tomography to a depth of 700 km (Fouch et al., 2004) reveals a deep lithospheric mantle keel that ranges in depth from ~250–300 km under the Kaapvaal Craton to ~225–250 km under the Zimbabwe Craton. This shows a break under the Bushveld Complex, but is continuous under the Limpopo Complex. Any structures related to a fossil subduction zone are either absent or not resolved. However, the pattern of seismic anisotropy in the mantle keel in the Limpopo Complex region and south of it (shear-wave-splitting fast polarization directions strike between 50° and 80°) is in accord with NNW-SSE collision and/or shearing parallel to the general WSW-ENE Limpopo Complex trend (Silver et al., 2004).

Apart from being in accord with a pop-up model related to collision, the gravity and geoelectric data for the NMZ and SMZ also do not contradict models of gravitational redistribution, or intracrustal diapirism, within these zones (Blenkinsop, this volume; Perchuk and Gerya, this volume). In this interpretation, the dipping high-density, high-conductivity domains in both zones could represent high-grade metamorphic supracrustal rocks. However, the observed presence of a continuous deep, lithospheric-mantle keel underneath the Limpopo Complex is not in accord with the notion of massive heat advection to the lower crust by plume activity and therefore prohibits a plume-related gravitational redistribution model for the complex as a whole in the sense of Perchuk (1989, 1991) and Perchuk et al. (2001).

A problem is presented by the apparent shallowness of the felsic upper crust in the CZ (only ~10 km; de Beer and Stettler, 1992; Nguuri et al., 2001; Gore et al., 2009; Pretorius and Barton, 2002). The proposal that the CZ upper crust represents a nappe (Pretorius and Barton, 2003) may be a solution, but it is in apparent contrast to the steep structures prevailing in that zone. The assumption that some or all the mafic lower crust under the CZ is the result of magmatic underplating, postdating the Neoarchean orogeny, also potentially resolves the problem. Nguuri et al. (2001) and Gore et al. (2009) likened the lower crustal structure of the CZ to that underneath the Bushveld Complex. This intriguing comparison has a bearing on possible tectonometamorphic interpretations for the Paleoproterozoic event, as outlined further below.

Sm-Nd and Lu-Hf Isotope Constraints

Isotope data relevant to questions of crustal growth and accretion, and geotectonics in general, are whole-rock Sm-Nd studies from the Zimbabwe Craton (Taylor et al., 1991; Zhai et al., 2006), the NMZ (Berger et al., 1995), the CZ (Barton, 1996; Harris et al., 1987; Kröner et al., 1999), the SMZ and the northern Kaapvaal Craton (Kreissig et al., 2000) and Lu-Hf studies on zircons in the Central Zone (Zeh et al., 2007, 2008, 2010; Millonig et al., 2010), as well as in the Southern Marginal Zone and the Kaapvaal and Zimbabwe Cratons (Zeh et al., 2009). These data and their implications are reviewed by Kramers and Zeh (this volume). Here we summarize the principal constraints that arise from them to tectonic models.

Both the Zimbabwe and Kaapvaal cratons display the typical pattern of continental growth at accretionary margins, with ca. 3.5 Ga core areas flanked by progressively younger provinces in which juvenile magma with mantle Nd and Hf isotope compositions is interpreted to mix with older continental crustal matter as in a Cordilleran-Andean setting. The northern Kaapvaal Craton appears to have been such an accretionary margin at 2.8–2.9 Ga, and the SMZ of the Limpopo Complex has identical Sm-Nd and Lu-Hf isotope characteristics, confirming the notion that it represents a high-grade metamorphic equivalent of the northern Kaapvaal Craton (Kreissig et al., 2000). The SMZ does not show any evidence of juvenile continental crust in the 2.6–2.7 Ga age range. The 2.69 Ga Matok Pluton appears to have been purely crustally derived (Zeh et al., 2009). In contrast, the geochemistry, Sm-Nd, and zircon Lu-Hf systematics of the westernmost edge of the Zimbabwe Craton indicate the existence of a magmatic arc active at 2.65–2.7 Ga (Bagai et al., 2002; Kampunzu et al., 2003; Zhai et al., 2006; Zeh et al., 2009), and Sm-Nd systematics of charnockites and enderbites of the NMZ suggest very strongly that the NMZ represents the easterly extension of this same active margin on the south side of the Zimbabwe Craton (Berger et al., 1995; Berger and Rollinson, 1997; Kramers and Zeh, this volume).

In the CZ, Sm-Nd data on the enderbitic portion of the 2.6 Ga Bulai Pluton indicate that this unit represents juvenile crust (Harris et al., 1987), whereas zircon Lu-Hf data on a charnockitic Bulai sample show extensive remelting of older crust at the same time (Zeh et al., 2007). Quartzofeldspathic gneisses of Singelele type appear to represent mainly anatexis at 2.6–2.7 Ga of older crust (pure S-type granitoids), whereas in the so-called gray gneisses a variable component of juvenile crust, derived from the mantle close to 2.65 Ga, is suggested by the data (Zeh et al., 2007). As the magmatism in the CZ is broadly coeval with the equally diachronous plutonism in the NMZ and with granitoid magmatism in the Francistown area (Kampunzu et al., 2003), and the CZ and NMZ share the feature of generally high U/Pb ratios and U, Th concentrations with the Zimbabwe Craton (Barton et al., 2006; Kramers et al., 2001; Andreoli et al., this volume), the hypothesis that the CZ was the shelf (Pretorius and Barton, 2003) associated with the NMZ as an Andean-type margin of the Zimbabwe Craton at 2.65–2.7 Ga merits serious consideration. The presence of detrital zircons up to almost 3.9 Ga in age in quartzites of both the Zimbabwe Craton and the eastern part of the CZ (Dodson et al., 1988; Zeh et al., 2008) is further in accord with this notion. An implication of this interpretation is that the subduction zone associated with collision dipped underneath the Zimbabwe Craton, i.e., to what is now the NNW. This is at variance with the assumption made by Coward and Fairhead (1980) and adopted by Roering et al. (1992). However, the structures resolved in the crustal geophysical surveys, summarized above, are interpreted as pop-up structures. Further, the deeper tomographic and mantle anisotropic study do not reveal features that can be related to subduction. Therefore, geophysical data do not constrain the direction of Neoarchean subduction in the Limpopo Complex.

In the Mahalapye Complex, at the western extremity of the exposed CZ, Sm-Nd and zircon Lu-Hf data on 2.0 Ga magmatic rocks clearly display a mantle-derived component mixed with older crust at that time (Chavagnac et al., 2001; Zeh et al., 2007, 2009; Millonig et al., 2010). The last authors, also on the basis of a prograde decompression-heating P-T path, suggested Bushveld-age magmatic underplating as a cause, and this is in accord with the requirement from teleseismic data (Nguuri et al., 2001; Gore et al., 2009) mentioned above.

Sm-Nd and zircon Lu-Hf characteristics, in combination with Pb isotope data, are also eminently suited to detect separate terranes. For provinces that are uniform with respect to these isotope systems, even if they appear subdivided by shear zones, there is no requirement to postulate exotic terranes within them.

It is important to discuss the Turkic-type terrane accretion model (Barton et al., 2006) in this light.

The NMZ and the Francistown area can be interpreted as parts of a Neoarchean active margin to the Zimbabwe Craton, as discussed above. Postulating the status of a separate terrane for the NMZ thus is not required by the isotope data. The SMZ is similar to the northern Kaapvaal Craton in Sm-Nd and Pb isotope systematics, as well as in general geochemistry. Irrespective of the acceptance of the gravitational redistribution scenario for this zone, there are no data contradicting the notion that it represents a high-grade metamorphic equivalent of the northern Kaapvaal Craton (e.g., Kreissig et al., 2000; Roering et al., 1992; van Reenen et al., this volume). Therefore there is no reason for the postulation of a separate SMZ terrane.

Throughout the CZ, the isotope data summarized by Kramers and Zeh (this volume) show a remarkable uniformity of principal features (age of the oldest crust as well as of anatexis). From these data, there is thus (notwithstanding the presence of transcurrent shear zones) no support for the discrimination of terranes within the CZ as suggested by Barton et al. (2006). For the CZ as a whole, an association with the NMZ and the Zimbabwe Craton prior to ca. 2.65 Ga, possibly as a leading-edge continental shelf (Pretorius and Barton, 2003), is not contradicted by any of the isotope or geochemical evidence. Thus there is no need to propose a separate terrane status for the CZ.

With regard to U, Th contents and the related Pb isotope compositions, it is however important to remember that there are two main crustal groupings in the complete assembly: a high [U, Th] group comprising the Zimbabwe Craton, NMZ, and CZ on the one hand, and a low [U, Th] group formed by the Kaapvaal Craton and SMZ (Barton et al., 2006; Kramers et al., 2001; Andreoli et al., this volume). Therefore, although we cannot distinguish a multitude of terranes, we can conclude that if two continents collided in the Neoarchean, they had distinct geochemical characteristics. These define the suture to lie between the CZ and the SMZ or Kaapvaal Craton, along the Palala Shear Zone and the Tshipise Straightening Zone.

A further aspect of the data that militates against a Turkic-type terrane-accretion process for the Limpopo Complex lies in the style of magmatism. In the Turkic-type scenario, magmatic rocks should be chiefly mantle derived, associated with rifting (see above, Şengör and Natal'in, 1996; Barton et al., 2006). The predominance of crustal anatexis over mantle-derived magmatism in the ca. 2.65 Ga event, as found from the Nd and Hf evidence reviewed by Kramers and Zeh (this volume), is in sharp contrast to this and thus effectively excludes this variant of the terrane-accretion model.

Geochronological and Petrological Constraints

A large body of geochronological data on the Limpopo Complex, reviewed by Kramers and Mouri (this volume), shows different age patterns in its three zones. In the SMZ, ages of metamorphism and syntectonic, synmetamorphic intrusions are exclusively Neoarchean. The NMZ shows Neoarchean emplacement ages for igneous charnockites and enderbites, and Paleoproterozoic ages close to 2.0 Ga for shear zones that range from low-grade metamorphism in the north to progressively higher grade toward the Triangle Shear Zone in the south. In the CZ, high-grade metamorphism in both the Neoarchean and Paleoproterozoic is demonstrated. The latter is mainly sharply clustered about 2.02 Ga, and many of the dates are obtained on metamorphic mineral assemblages. Of the Neoarchean dates, ranging from 2.55 to 2.7 Ga, some are from metamorphic assemblages in metapelites, but most are from zircons, giving the emplacement age of anatectic granitoids.

The interpretation of the Central Zone as a terrane emplaced westward during the Neoarchean Limpopo orogeny (McCourt and Vearncombe, 1992) can be tested by new data on the proposed lateral ramps. Deformation along the Triangle Shear Zone is constrained by Pb-Pb and Sm-Nd ages on garnets associated with quartzofeldspathic gneisses within the shear zone and Ar-Ar dating of hornblende defining the mineral elongation lineation on some of the ultramylonites (Kamber et al., 1995a). These data indicate deformation and high-temperature metamorphism along the Triangle Shear Zone between ca. 2.25 and 2.0 Ga. In the Palala Shear Zone, the presence of deformed Bushveld-age Palala granite and related low-grade quartz-sericite ultramylonite in the southern section of the shear zone constrains the maximum age of the youngest ductile deformation along this section to 2.04 Ga. Further, a syntectonically retrogressed charnockite in the northern part of the zone yielded a hornblende Ar/Ar date of 2.028 ± 0.006 Ga (Belluso et al., 2000), and fabric-forming hornblende in the Tshipise Straightening Zone has likewise yielded 2.0 Ga dates (Holzer, 1998; see summary in Kramers and Mouri, this volume). In the light of these data, the Neoarchean westward-emplacement model for the CZ now appears untenable.

In petrology, detailed work on small-scale local mineral equilibria (Perchuk et al., 2008; Perchuk, this volume; Smit et al., this volume) has shown evidence of two sequential P-T paths in some samples: first, a retrograde path from medium- to high-pressure, high-grade peak conditions to ~600 °C at mid-crustal level, then near-isobaric heating back to high grade at this same level, and finally decompression cooling. In contrast, work using the pseudosection approach (Zeh et al., 2004) has never yielded more than a single P-T loop per sample, even in areas where the local equilibria approach showed two. Whereas the difference in results from the two interpretive approaches has not yet been fully explained, a pseudosection P-T loop for a Paleoproterozoic (mono)metamorphic rock unit in the Mahalapye Complex, Botswana (Millonig et al., 2010), shows a decompression-heating path followed by decompression cooling, which is rather similar to the second loops yielded by local equilibrium studies of polymetamorphic samples. Millonig et al. (2010) interpret this result as an indication of magmatic underplating, which is in accord with the interpretation by Nguuri et al. (2001) and Gore et al. (2009) of the teleseismic data, mentioned above.

Given that the Limpopo Complex is viewed as an orogenic belt, a controversy has arisen about which of the two age clusters actually dates the collision event. The two exponent models, the Neoarchean collision model and the Paleoproterozoic transpression model, are described above. We propose a resolution here. High-grade metamorphism with decompression-cooling paths as well as widespread anatexis, resulting from collision-related crustal thickening, characterizes the Neoarchean event in the CZ, whereas the 2.02 Ga episode appears to be one of transcurrent movement, clearly also associated with folding, superimposed on this, and only moderate pressure (Boshoff et al., 2006; Perchuk et al., 2008; Smit et al., this volume). Multiple lines of evidence for magmatic underplating of the CZ crust (Nguuri et al., 2001; Gore et al., 2009; Millonig et al., 2010; Kramers and Zeh, this volume) have been discussed above. Such underplating would allow explanation of the sharp isochronism of the 2.02 Ga metamorphism, and the process would clearly lead to a regional weakening of the crust, focusing regional strain in localized transcurrent movement. It could also account for subsequent uplift and erosion, independently of whether there was an earlier phase of tectonic crustal thickening.

In this suggested solution of the age controversy, a Neoarchean collision thus appears to have been the main event that shaped the Limpopo Complex, and the suture between the two continents, with contrasting U, Th concentrations and Pb isotope systematics, would be located along the boundary between the CZ and the SMZ, i.e., the Palala Shear Zone and the Tshipise Straightening Zone. Significant transcurrent movement occurred, however, along this set of shear zones at 2.0 Ga (see above). This movement, and that along the Triangle Shear Zone, means that the three zones of the Limpopo Complex might have shifted considerably relative to each other in ENE and/or WSW directions. Spatial relationships between the three zones along a present-day north-south axis therefore provide a weak basis for a detailed model of the Neoarchean collision orogeny. Only a generic collision model appears to be appropriate at present. Whereas the pop-up structure proposed by Roering et al. (1992) is not contradicted by any data, the Neoarchean Himalayan analogy of Treloar et al. (1992) is flawed. First, at least some of the proposed escape structures are found to be 2.0 Ga, and second, one of the colliding continental blocks (the Zimbabwe Craton) may not have been quite rigid at the time of collision. This is shown by the widespread Chilimanzi suite of granites (Fig. 1), the product of crustal anatexis at 2.6 Ga.

A question also arises with regard to the timing and environment of high-grade metamorphism in the SMZ in the context of this collision model. The Sm-Nd and Lu-Hf data reviewed by Kramers and Zeh (this volume), as outlined above, make it highly plausible that a subduction zone dipped to the (present day) NNW underneath the Zimbabwe Craton, generating magmatism in this convergent margin prior to collision. In the context of a collision model, this is in accord with the corresponding data on the SMZ, which show clearly that this zone was *not* a magmatic convergent margin at that time. The zircon dates on the NMZ and the

Francistown area reflect magmatism during convergence at the Zimbabwe craton margin, i.e., prior to continental collision. The age range is 2.65–2.7 Ga for the Francistown area and 2.6–2.7 Ga for the NMZ. The Neoarchean high-grade metamorphism and magmatism in the CZ is predominantly related to crustal thickening associated with the collision (Kramers and Zeh, this volume; Andreoli et al., this volume), and the age range extends to ca. 2.55 Ga. This age range is expected, as subduction underneath the Zimbabwe Craton must have preceded collision. However, the tectonism, syntectonic metamorphism, and anatexis in the SMZ and along the Hout River Shear Zone (excluding amphibole Ar/Ar dates) range from 2.65 to 2.72 Ga. The 2.65 Ga dates are muscovite Rb-Sr ages on pegmatites that crosscut the SMZ and Hout River Shear Zone foliation (Barton and van Reenen, 1992), and thus are clear minimum ages for the tectonism. The metamorphism and tectonism in the SMZ therefore appear to have occurred too early to be caused by a collision that was preceded by 2.65–2.7 Ga subduction underneath the Zimbabwe Craton. Possibly this problem can be solved by considering the long time span of magmatism in the NMZ, from ca. 2.7 to ca. 2.6 Ga. If the younger magmatism in this age range was related to collisional thickening (as in the CZ), the paradox would be weakened. Also, the Paleoproterozoic transcurrent movement mentioned above can potentially loosen constraints of this nature. Even if the controversy on the two high-grade metamorphic episodes is to a large extent resolved, the later one still hinders detailed modeling of the earlier one.

CONCLUSIONS AND THE WAY FORWARD

The three models proposed to account for all observed features of the Limpopo Complex are mutually incompatible, and no single one on its own satisfies all the data. The constraints discussed here nevertheless allow the effective exclusion of two models. A subdivision into many terranes, as proposed in the Turkic-type model, is not required by geochemistry and isotope data, except for two main groupings: the Kaapvaal Craton and SMZ on the one hand, and the Zimbabwe Craton, NMZ, and CZ on the other (which could have been continents). Further, the Turkic-type accretion model was proposed to explain (among other things) abundant mantle-derived magmatism in the Limpopo Complex, and later work has shown crustal anatexis to be dominant, which is more in accord with a continental collision scenario. Regarding a plume-related gravitational distribution model, the objection here is that a giant plume is unlikely, as a subcontinental lithosphere mantle keel is continuous throughout the Limpopo Complex.

The Neoarchean continental collision model remains a valid working hypothesis, with the Zimbabwe Craton being the overriding plate, the NMZ (+ the Francistown region) its active margin, and the CZ its leading shelf edge. The suture would be located between the CZ and the SMZ, in a zone that was reactivated in the Paleoproterozoic as the Palala Shear Zone and the Tshipise Straightening Zone. The metamorphism in the SMZ appears to be somewhat too old in this context. However, the period of Neoarchean metamorphism and magmatism in the Limpopo Complex was protracted, and the spatial relationships in the Neoarchean (owing to Paleoproterozoic transcurrent movement) uncertain, so that the collision model cannot be rejected on this basis. By the same token, it will be difficult to refine.

In the context of (collision-related) crustal thickening, the gravitational redistribution or diapir model, applied to the SMZ, is able to predict many structural and petrological observations in that zone and the adjacent Kaapvaal Craton. Diapirism has also been plausibly suggested as a mechanism to explain important structural features in the CZ and SMZ, but has not been numerically modeled in these zones.

The hypothesis of Paleoproterozoic dextral transpression in the CZ is compatible with the Neoarchean collision model, provided that it does not itself entail a continent collision. The apparent age paradox (ca. 2.65 Ga versus 2.0 Ga) is resolved, as metamorphic parageneses showing the younger age are most common in or near large transcurrent shear zones. The P-T loops associated with the 2.0 Ga event show lower peak pressures than those of the ca. 2.65 Ga period, and near-isobaric prograde paths. This is in accord with the associated hypothesis of magmatic underplating of Bushveld age, which could also account for the sharp isochronism of the 2.0 Ga metamorphism. Further, it would necessarily have caused weakening of the CZ crust, possibly resulting in the focusing of regional stresses into transcurrent movement in this region.

From this brief summary, a number of points emerge that may help to define the way forward in research attempting to define tectonic models of the Limpopo Complex in particular, and polymetamorphic Precambrian high-grade terranes in general.

In petrology, there is a need to analyze the reason why pseudosection-based P-T studies can apparently not reveal multiple P-T histories in a single sample. A better consensus is to be reached by petrologists on the scope of both pseudosection and local equilibrium techniques. Combined mineral geochronological and petrological work has contributed greatly to our understanding of the Limpopo Complex and should be continued and improved. Directly related to this, detailed thermal and rheological modeling of the consequences of collision and crustal thickening in the Neoarchean (as done for the SMZ by Perchuk and Gerya, this volume) should be extended to other zones of the Limpopo Complex, as this will greatly enhance our understanding of the intracrustal processes.

Fieldwork remains the most important starting point, and the science has now reached the stage where fieldwork can in part be targeted to reassess sites that have yielded problematic or paradoxical petrological and geochronological results, or test structural and metamorphic predictions from modeling. Good models often make more predictions than what they are set up to explain.

It may be a valid criticism for a number of chapters in this volume (including the present one) that insufficient comparisons to other early Precambrian orogenic complexes are made. As the literature is vast, broadly comparative reviews should be encouraged and given the status of research projects. Further, specifically addressing identified paradoxes (such as, for instance, the timing problem for the metamorphism of the SMZ in the Neoarchean collision model) will increase our understanding of early tectonic processes.

Lastly, the work on the Neoarchean period of metamorphism and associated processes in the Limpopo Complex has revealed an interesting mix of plate tectonic processes and scenarios that defy a strictly uniformitarianist approach. In view of the ongoing debate on the nature of plate tectonics in the early history of the Earth, it appears useful to make plate tectonics–related interpretations only when the data strongly suggest them, and less useful to impose plate tectonics scenarios at all costs.

REFERENCES CITED

Aldiss, D.T., 1991, The Motlutse Complex and the Zimbabwe craton/Limpopo Belt transition in Botswana: Precambrian Research, v. 50, p. 89–109, doi:10.1016/0301-9268(91)90049-G.

Andreoli, M.A.G., Brandl, G., Coetzee, H., Kramers, J.D., and Mouri, H., 2011, this volume, Intracrustal radioactivity as an important heat source for Neoarchean metamorphism in the Central Zone of the Limpopo Complex, in van Reenen, D.D., Kramers, J.D., McCourt, S., and Perchuk, L.L., eds., Origin and Evolution of Precambrian High-Grade Gneiss Terranes, with Special Emphasis on the Limpopo Complex of Southern Africa: Geological Society of America Memoir 207, doi: 10.1130/2011.1207(09).

Bagai, Z., Armstrong, R., and Kampunzu, A.B., 2002, U-Pb single zircon geochronology of granitoids in the Vumba granite-greenstone terrain (NE Botswana): Implications for the Archaean Zimbabwe Craton: Precambrian Research, v. 118, p. 149–168, doi:10.1016/S0301-9268(02)00074-8.

Barton, J.M., Jr., 1996, The Messina Layered Intrusion, Limpopo Belt, South Africa: An example of in-situ contamination of an Archean anorthosite complex by continental crust: Precambrian Research, v. 78, p. 139–150, doi:10.1016/0301-9268(95)00074-7.

Barton, J.M., Jr., and van Reenen, D.D., 1992, When was the Limpopo Orogeny?: Precambrian Research, v. 55, p. 7–16, doi:10.1016/0301-9268(92)90010-L.

Barton, J.M., Jr., Klemd, R., and Zeh, A., 2006, The Limpopo Belt: A result of Archean to Proterozoic, Turkic-type orogenesis?, in Reimold, W.U., and Gibson, R.L., eds., Processes on the Early Earth: Geological Society of America Special Paper 405, p. 315–332.

Belluso, E., Ruffini, R., Schaller, M., and Villa, I.M., 2000, Electron-microscope and Ar isotope characterization of chemically heterogeneous amphiboles from the Palala shear zone, Limpopo Belt, South Africa: European Journal of Mineralogy, v. 12, p. 45–62.

Berger, M., and Rollinson, R., 1997, Isotopic and Geochemical evidence for crust-mantle interaction during late Archean crustal growth: Geochimica et Cosmochimica Acta, v. 61, p. 4809–4829, doi:10.1016/S0016-7037(97)00271-8.

Berger, M., Kramers, J.D., and Nägler, Th., 1995, Geochemistry and geochronology of charnoenderbites in the Northern Marginal Zone of the Limpopo Belt, Southern Africa, and genetic models: Schweiz, Mineralogische und Petrographische Mitteilungen, v. 75, p. 17–42.

Blenkinsop, T.G., 2011, this volume, Archean magmatic granulites, diapirism, and Proterozoic reworking in the Northern Marginal Zone of the Limpopo Belt, in van Reenen, D.D., Kramers, J.D., McCourt, S., and Perchuk, L.L., eds., Origin and Evolution of Precambrian High-Grade Gneiss Terranes, with Special Emphasis on the Limpopo Complex of Southern Africa: Geological Society of America Memoir 207, doi:10.1130/2011.1207(13).

Boshoff, R., van Reenen, D.D., Smit, C.A., Perchuk, L.L., Kramers, J.D., and Armstrong, R., 2006, Geologic history of the Central Zone of the Limpopo Complex: The West Alldays area: Journal of Geology, v. 114, p. 699–716, doi:10.1086/507615.

Chavagnac, V., Kramers, J.D., Naegler, Th.F., and Holzer, L., 2001, The behaviour of Nd and Pb isotopes during 2.0 Ga migmatization in paragneisses of the Central Zone of the Limpopo Belt (South Africa and Botswana): Precambrian Research, v. 112, p. 51–86, doi:10.1016/S0301-9268(01)00170-X.

Coward, M.P., 1976, Archaean deformation patterns in southern Africa: Philosophical Transactions of the Royal Society of London, ser. A, v. 283, p. 313–331.

Coward, M.P., 1984, Major shear zones in the Precambrian crust: Examples from NW Scotland and southern Africa and their significance, in Kröner, A., and Greiling, R., eds., Precambrian Tectonics Illustrated: Stuttgart, Schweizerbart'sche Verlagsbuchhandlung, p. 207–235.

Coward, M.P., and Fairhead, J.D., 1980, Gravity and structural evidence for the deep structure of the Limpopo Belt, southern Africa: Tectonophysics, v. 68, p. 31–43, doi:10.1016/0040-1951(80)90007-4.

de Beer, J.H., and Stettler, E.H., 1992, The deep structure of the Limpopo Belt from geophysical studies: Precambrian Research, v. 55, p. 173–186, doi:10.1016/0301-9268(92)90022-G.

Dodson, M.H., Compston, W., Williams, I.S., and Wilson, J.F., 1988, A search for ancient detrital zircons in Zimbabwean sediments: Journal of the Geological Society [London], v. 145, p. 977–983, doi:10.1144/gsjgs.145.6.0977.

Durrheim, R.J., Barker, W.H., and Green, R.W.E., 1992, Seismic studies in the Limpopo Belt: Precambrian Research, v. 55, p. 187–200, doi:10.1016/0301-9268(92)90023-H.

Foley, S., 2008, A trace element perspective on Archean crust formation and on the presence or absence of Archean subduction, in Condie, K.C., and Pease, V., eds., When Did Plate Tectonics Begin on Planet Earth?: Geological Society of America Special Paper 440, p. 31–50.

Fouch, M.J., James, D.E., Van Decar, J.C., van der Lee, S., and Kaapvaal Seismic Group, 2004, Mantle seismic structure beneath the Kaapvaal and Zimbabwe Cratons: South African Journal of Geology, v. 107, p. 33–44, doi:10.2113/107.1-2.33.

Gerya, T.V., Perchuk, L.L., Maresch, W.V., and Willner, A.P., 2004, Inherent gravitational instability of hot continental crust: Implications for doming and diapirism in granulite facies terrains, in Whitney, D.L., Teyssier, C., and Siddoway, C.S., eds., Gneiss Domes in Orogeny: Geological Society of America Special Paper 380, p. 97–115.

Gore, J., James, D.E., Zengeni, T.G., and Gwavava, O., 2009, Crustal structure of the Zimbabwe Craton and the Limpopo Belt of Southern Africa: New constraints from seismic data and implications for its evolution: South African Journal of Geology, v. 112, p. 213–228, doi:10.2113/gssajg.112.3-4.213.

Harris, N.B.W., Hawkesworth, C.J., van Calsteren, P., and McDermott, F., 1987, Evolution of continental crust in southern Africa: Earth and Planetary Science Letters, v. 83, p. 85–93, doi:10.1016/0012-821X(87)90053-7.

Holzer, L., 1998, The transpressive orogeny at 2 Ga in the Limpopo Belt, southern Africa [Ph.D. thesis]: Bern, Switzerland, University of Bern, 200 p.

Holzer, L., Frei, R., Barton, J.M., Jr., and Kramers, J.D., 1998, Unravelling the record of successive high-grade events in the Central Zone of the Limpopo Belt using Pb single-phase dating of metamorphic minerals: Precambrian Research, v. 87, p. 87–115, doi:10.1016/S0301-9268(97)00058-2.

Holzer, L., Barton, J.M., Jr., Paya, B.K., and Kramers, J.D., 1999, Tectonothermal history in the western part of the Limpopo Belt: Test of the tectonic models and new perspectives: Journal of African Earth Sciences, v. 28, p. 383–402, doi:10.1016/S0899-5362(99)00011-1.

Kamber, B.S., Kramers, J.D., Napier, R., Cliff, R.A., and Rollinson, H.R., 1995a, The Triangle Shear Zone, Zimbabwe, revisited: New data document an important event at 2.0 Ga in the Limpopo Belt: Precambrian Research, v. 70, p. 191–213, doi:10.1016/0301-9268(94)00039-T.

Kamber, B.S., Blenkinsop, T.G., Villa, I.M., and Dahl, P.S., 1995b, Proterozoic transpressive deformation in the Northern Marginal Zone, Limpopo Belt, Zimbabwe: Journal of Geology, v. 103, p. 493–508, doi:10.1086/629772.

Kampunzu, A.B., Tombale, A.R., Zhai, M., Majaule, T., and Modisi, M.P., 2003, Major and trace element geochemistry of plutonic rocks from Francistown, NE Botswana: Evidence for a Neoarchaean continental active margin in the Zimbabwe craton: Lithos, v. 71, p. 431–460, doi:10.1016/S0024-4937(03)00125-7.

Kramers, J.D., and Mouri, H., 2011, this volume, The geochronology of the Limpopo Complex: A controversy solved, in van Reenen, D.D., Kramers, J.D., McCourt, S., and Perchuk, L.L., eds., Origin and Evolution of Precambrian High-Grade Gneiss Terranes, with Special Emphasis on the Limpopo Complex of Southern Africa: Geological Society of America Memoir 207, doi:10.1130/2011.1207(06).

Kramers, J.D., and Zeh, A., 2011, this volume, A review of Sm-Nd and Lu-Hf isotope studies in the Limpopo Complex and adjoining cratonic areas, and their bearing on models of crustal evolution and tectonism, *in* van Reenen, D.D., Kramers, J.D., McCourt, S., and Perchuk, L.L., eds., Origin and Evolution of Precambrian High-Grade Gneiss Terranes, with Special Emphasis on the Limpopo Complex of Southern Africa: Geological Society of America Memoir 207, doi: 10.1130/2011.1207(10).

Kramers, J.D., Kreissig, K., and Jones, M.Q.W., 2001, Crustal heat production and style of metamorphism: A comparison between two Archean high grade provinces in the Limpopo Belt, Southern Africa: Precambrian Research, v. 112, p. 149–163, doi:10.1016/S0301-9268(01)00173-5.

Kreissig, K., Nägler, Th.F., Kramers, J.D., van Reenen, D.D., and Smit, C.A., 2000, An isotopic and geochemical study of the northern Kaapvaal Craton and the Southern Marginal Zone of the Limpopo Belt: Are they juxtaposed terranes? Lithos, v. 50, p. 1–25.

Kröner, A., Jaeckel, P., Brandl, G., Nemchin, A.A., and Pidgeon, R.T., 1999, Single zircon ages for granitoid gneisses in the Central Zone of the Limpopo Belt, southern Africa and geodynamic significance: Precambrian Research, v. 93, p. 299–337, doi:10.1016/S0301-9268(98)00102-8.

Light, M.P.R., 1982, The Limpopo Mobile Belt: A result of continental collision: Tectonics, v. 1, p. 325–342, doi:10.1029/TC001i004p00325.

Mason, R., 1973, The Limpopo Mobile Belt, southern Africa: Philosophical Transactions of the Royal Society of London, ser. A, Mathematical and Physical Sciences, v. 273, p. 463–485, doi:10.1098/rsta.1973.0012.

McCourt, S., and Vearncombe, J.R., 1987, Shear zones bounding the Central Zone of the Limpopo Mobile Belt, southern Africa: Journal of Structural Geology, v. 9, p. 127–137, doi:10.1016/0191-8141(87)90021-6.

McCourt, S., and Vearncombe, J.R., 1992, Shear zones of the Limpopo Belt and adjacent granitoid-greenstone terranes: Implications for late Archaean collision tectonics in southern Africa: Precambrian Research, v. 55, p. 553–570, doi:10.1016/0301-9268(92)90045-P.

Millonig, L., Zeh, A., Gerdes, A., Klemd, R., and Barton, J.M., Jr., 2010, Decompressional heating of the Mahalapye Complex (Limpopo Belt, Botswana): A response to Palaeoproterozoic magmatic underplating?: Journal of Petrology, v. 51, p. 703–729, doi:10.1093/petrology/egp097.

Nguuri, T.K., Gore, J., James, D.E., Webb, S.J., Wright, C., Zengeni, T.G., Gwavava, O., Snoke, J.A., and the Kaapvaal Seismic Group, 2001, Crustal structure beneath southern Africa and its implications for the formation of the Kaapvaal and Zimbabwe Cratons: Geophysical Research Letters, v. 28, p. 2501–2504, doi:10.1029/2000GL012587.

Pease, V., Percival, J., Smithies, H., Stevens, G., and Van Kranendonk, M., 2008, When did plate tectonics begin? Evidence from the orogenic record, *in* Condie, K.C., and Pease, V., eds., When Did Plate Tectonics Begin on Planet Earth?: Geological Society of America Special Paper 440, p. 199–228.

Perchuk, L.L., 1989, P-T-fluid regimes of metamorphism and related magmatism with specific reference to the Baikal Lake granulites, *in* Daly, S., Yardley, D.W.D., and Cliff, B.O., eds., Evolution of Metamorphic Belts: Geological Society [London] Special Publication 43, p. 275–291.

Perchuk, L.L., 1991, Studies in magmatism, metamorphism, and geodynamics: International Geology Review, v. 33, p. 311–374, doi:10.1080/00206819109465695.

Perchuk, L.L., 2011, this volume, Local mineral equilibria and P-T paths: Fundamental principles and applications to high-grade metamorphic terranes, *in* van Reenen, D.D., Kramers, J.D., McCourt, S., and Perchuk, L.L., eds., Origin and Evolution of Precambrian High-Grade Gneiss Terranes, with Special Emphasis on the Limpopo Complex of Southern Africa: Geological Society of America Memoir 207, doi:10.1130/2011.1207(05).

Perchuk, L.L., and Gerya, T.V., 2011, this volume, Formation and evolution of Precambrian granulite terranes: A gravitational redistribution model, *in* van Reenen, D.D., Kramers, J.D., McCourt, S., and Perchuk, L.L., eds., Origin and Evolution of Precambrian High-Grade Gneiss Terranes, with Special Emphasis on the Limpopo Complex of Southern Africa: Geological Society of America Memoir 207, doi:10.1130/2011.1207(15).

Perchuk, L.L., Gerya, T.V., van Reenen, D.D., and Smit, C.A., 2001, Formation and dynamics of granulite complexes within cratons: Gondwana Research, v. 4, p. 729–732, doi:10.1016/S1342-937X(05)70524-4.

Perchuk, L.L., van Reenen, D.D., Varlamov, D.A., Van Kal, S.M., Boshoff, R., and Tabatabaeimanesh, S.M., 2008, P-T record of two high-grade metamorphic events in the Central Zone of the Limpopo Complex, South Africa: Lithos, v. 103, p. 70–105, doi:10.1016/j.lithos.2007.09.011.

Pretorius, W., and Barton, J.M., Jr., 2002, Measured and calculated compressional wave velocities of crustal and upper mantle rocks in the Central Zone of the Limpopo Belt, South Africa—Implications for lithospheric structure: South African Journal of Geology, v. 105, p. 303–310.

Pretorius, W., and Barton, J.M., Jr., 2003, Petrology and geochemistry of crustal and upper mantle xenoliths from the Venetia Diamond Mine—Evidence for Archean crustal growth and subduction: South African Journal of Geology, v. 106, p. 213–230, doi:10.2113/106.2-3.213.

Ridley, J.R., 1992, On the origins and tectonic significance of the charnockite suite of the Archaean Limpopo Belt, Northern Marginal Zone, Zimbabwe: Precambrian Research, v. 55, p. 407–427, doi:10.1016/0301-9268(92)90037-O.

Roering, C., van Reenen, D.D., Smit, C.A., Barton, J.M., Jr., De Beer, J.H., De Wit, M.J., Stettler, E.H., Van Schalkwyk, J.F., Stevens, G., and Pretorius, S.J., 1992, Tectonic model for the evolution of the Limpopo Belt: Precambrian Research, v. 55, p. 539–552, doi:10.1016/0301-9268(92)90044-O.

Rollinson, H.R., 1993, A terrane interpretation of the Archaean Limpopo Belt: Geological Magazine, v. 130, p. 755–765, doi:10.1017/S001675680002313X.

Schaller, M., Steiner, O., Studer, I., Holzer, L., Herwegh, M., and Kramers, J.D., 1999, Exhumation of Limpopo Central Zone granulites and dextral continental-scale transcurrent movement at 2.0 Ga along the Palala Shear Zone, Northern Province, South Africa: Precambrian Research, v. 96, p. 263–288, doi:10.1016/S0301-9268(99)00015-7.

Şengör, A.M.C., and Natal'in, B.A., 1996, Turkic-type orogeny and its role in the making of the continental crust: Annual Review of Earth and Planetary Sciences, v. 24, p. 263–337, doi:10.1146/annurev.earth.24.1.263.

Shirey, S.B., Kamber, B.S., Whitehouse, M.J., Mueller, P.A., and Basu, A.R., 2008, A review of the isotopic and trace element evidence for mantle and crustal processes in the Hadean and Archean: Implications for the onset of plate tectonic subduction, *in* Condie, K.C., and Pease, V., eds., When Did Plate Tectonics Begin on Planet Earth?: Geological Society of America Special Paper 440, p. 1–29.

Silver, P.G., Fouch, M.J., Gao, S.S., Schmitz, M., and Kaapvaal Seismic Group, 2004, Seismic anisotropy, mantle fabric, and the magmatic evolution of Precambrian southern Africa: South African Journal of Geology, v. 107, p. 45–58, doi:10.2113/107.1-2.45.

Smit, C.A., van Reenen, D.D., Roering, C., Boshoff, R., and Perchuk, L.L., 2011, this volume, Neoarchean to Paleoproterozoic evolution of the polymetamorphic Central Zone of the Limpopo Complex, *in* van Reenen, D.D., Kramers, J.D., McCourt, S., and Perchuk, L.L., eds., Origin and Evolution of Precambrian High-Grade Gneiss Terranes, with Special Emphasis on the Limpopo Complex of Southern Africa: Geological Society of America Memoir 207, doi:10.1130/2011.1207(12).

Stern, R.J., 2008, Modern-style plate tectonics began in Neoproterozoic time: An alternative interpretation of Earth's tectonic history, *in* Condie, K.C., and Pease, V., eds., When Did Plate Tectonics Begin on Planet Earth?: Geological Society of America Special Paper 440, p. 265–280.

Stuart, G.V,. and Zengeni, T.G., 1987, Seismic crustal structure of the Limpopo Mobile Belt, Zimbabwe: Tectonophysics, v. 144, p. 323–335.

Taylor, P.N., Kramers, J.D., Moorbath, S., Wilson, J.F., Orpen, J.L., and Martin, A., 1991, Pb/Pb, Sm-Nd and Rb-Sr geochronology in the Archean Craton of Zimbabwe: Chemical Geology, v. 87, p. 175–196.

Treloar, P.J., Coward, M.P., and Harris, N.B.W., 1992, Himalayan–Tibetan analogies for the evolution of the Zimbabwe Craton and the Limpopo Belt: Precambrian Research, v. 55, p. 571–587, doi:10.1016/0301-9268(92)90046-Q.

van Reenen, D.D., Barton, J.M., Jr., Roering, C., Smit, C.A., and Van Schalkwyk, J.F., 1987, Deep crustal response to continental collision: The Limpopo Belt of southern Africa: Geology, v. 15, p. 11–14, doi:10.1130/0091-7613(1987)15<11:DCRTCC>2.0.CO;2.

van Reenen, D.D., Smit, C.A., Perchuk, L.L., Roering, C., and Boshoff, R., 2011, this volume, Thrust exhumation of the Neoarchean ultrahigh-temperature Southern Marginal Zone, Limpopo Complex: Convergence of decompression-cooling paths in the hanging wall and prograde P-T paths in the footwall, *in* van Reenen, D.D., Kramers, J.D., McCourt, S., and Perchuk, L.L., eds., Origin and Evolution of Precambrian High-Grade Gneiss Terranes, with Special Emphasis on the Limpopo Complex of Southern Africa: Geological Society of America Memoir 207, doi:10.1130/2011.1207(11).

Watkeys, M.K., 1984, The Precambrian geology of the Limpopo Belt north and west of Messina [Ph.D. thesis]: Johannesburg, University of the Witwatersrand, 349 p.

Whitney, D.L., Teyssier, C., and Vanderhaeghe, O., 2004, Gneiss domes and crustal flow, *in* Whitney, D.L., Teyssier, C., and Siddoway, C.S., eds., Gneiss Domes in Orogeny: Geological Society of America Special Paper 380, p. 15–33.

Windley, B.F., 1993, Uniformitarianism today: Plate tectonics is the key to the past: Journal of the Geological Society [London], v. 150, p. 9–19, doi:10.1144/gsjgs.150.1.0007.

Zeh, A., Klemd, R., Buhlmann, S., and Barton, J.M., Jr., 2004, Pro- and retrograde P-T evolution of granulites of the Beit Bridge Complex (Limpopo Belt, South Africa): Constraints from quantitative phase diagrams and geotectonic implications: Journal of Metamorphic Geology, v. 22, p. 79–95, doi:10.1111/j.1525-1314.2004.00501.x.

Zeh, A., Gerdes, A., Klemd, R., and Barton, J.M., Jr., 2007, Archaean to Proterozoic crustal evolution in the Central Zone of the Limpopo Belt (South Africa–Botswana): Constraints from combined U-Pb and Lu-Hf isotope analyses of zircon: Journal of Petrology, v. 48, p. 1605–1639, doi:10.1093/petrology/egm032.

Zeh, A., Gerdes, A., Klemd, R., and Barton, J.R., Jr., 2008, U–Pb and Lu–Hf isotope record of detrital zircon grains from the Limpopo Belt—Evidence for crustal recycling at the Hadean to early-Archean transition: Geochimica et Cosmochimica Acta, v. 72, p. 5304–5329, doi:10.1016/j.gca.2008.07.033.

Zeh, A., Gerdes, A., and Barton, J.M., Jr., 2009, Archean accretion and crustal evolution of the Kalahari Craton—The zircon age and Hf isotope record of granitic rocks from Barberton/Swaziland to the Francistown Arc: Journal of Petrology, v. 50, p. 933–966, doi:10.1093/petrology/egp027.

Zeh, A., Gerdes, A., Barton, J.M., Jr., and Klemd, R., 2010, U–Th–Pb and Lu–Hf systematics of zircon from TTG's, leucosomes, meta-anorthosites and quartzites of the Limpopo Belt (South Africa): Constraints for the formation, recycling and metamorphism of Palaeoarchaean crust: Precambrian Research, v. 179, p. 50–68, doi:10.1016/j.precamres.2010.02.012.

Zhai, M., Kampunzu, A.B., Modisi, M.P., and Bagai, Z., 2006, Sr and Nd isotope systematics of Francistown plutonic rocks, Botswana: Implications for Neoarchaean crustal evolution of the Zimbabwe craton: International Journal of Earth Sciences, v. 95, p. 355–369, doi:10.1007/s00531-005-0054-6.

MANUSCRIPT ACCEPTED BY THE SOCIETY 24 MAY 2010